# Chemical Oceanography

VOLUME 1

2ND EDITION

# Chemical Oceanography

*Edited by*
## J. P. RILEY
*Department of Oceanography,*
*The University of Liverpool, England*

*and*

## G. SKIRROW
*Department of Inorganic, Physical and Industrial Chemistry,*
*The University of Liverpool, England*

VOLUME I

## 2ND EDITION

1975

ACADEMIC PRESS
LONDON    NEW YORK    SAN FRANCISCO
*A Subsidiary of Harcourt Brace Jovanovich, Publishers*

A*

ACADEMIC PRESS INC. (LONDON) LTD.
24/28 Oval Road,
London NW1

*United States Edition published by*
ACADEMIC PRESS INC.
111 Fifth Avenue
New York, New York 10003

Library of Congress Catalog Card Number: 74–5679
ISBN: 0–12–588601–2

Printed in Great Britain by
PAGE BROS (NORWICH) LTD
NORWICH

FN0113

# Contributors to Volume 1

K. F. BOWDEN, *Department of Oceanography, The University of Liverpool, Liverpool, England*

PHYLLIS A. BRAUNER, *Chemistry Department, Simmons College, Boston, U.S.A.*

P. G. BREWER, *Woods Hole Oceanographic Institute, Woods Hole, Massachusetts 02543, U.S.A.*

D. R. KESTER, *Graduate School of Oceanography, University of Rhode Island, Kingston, RI 02881, U.S.A.*

F. T. MACKENZIE, *North Western University, Evanston, Illinois 60201, U.S.A.*

G. A. PARKS, *Department of Mineral Engineering, Stanford University, Stanford, California 14305, U.S.A.*

W. STUMM, *Swiss Federal Institute for Aquatic Research and Water Pollution CH-8600 Dübendorf, Zürich, Switzerland*

M. WHITFIELD, *Marine Biological Laboratory, Citadel Hill, Plymouth PL1 2PB, Devon, England*

T. R. S. WILSON, *Institute of Oceanographic Sciences, Wormley, Godalming, Surrey, England*

# Preface to the Second Edition

Rapid progress has occurred in all branches of Chemical Oceanography since the publication of the first edition of this book a decade ago. Particularly noteworthy has been the tendency to treat the subject in a much more quantitative fashion; this has become possible because of our much improved understanding of the physical chemistry of sea water systems in terms of ionic and molecular theories. For these reasons chapters dealing with sea water as an electrolyte system, with speciation and with aspects of colloid chemistry are now to be considered as essential in any up-to-date treatment of the subject. Fields of research which were little more than embryonic only ten years ago, for example sea surface chemistry, have now expanded so much that they merit separate consideration. Since the previous edition, there has arisen a general awareness of the potential threat to the sea caused by man's activities, in particular its use as a "rubbish bin" and a receptacle for toxic wastes. Although it was inevitable that there should be some over-reaction to this, there is real cause for concern. Clearly, it is desirable to have available reasoned discussions of this topic and also an examination of the role of the sea as a potential source of raw materials in view of the imminent exhaustion of many high grade ores; these subjects are treated in the second, third and fourth volumes.

Most branches of marine chemistry make use of analytical techniques; the number and range of these has increased dramatically over recent years. Consequently it has been necessary to expand greatly and restructure the sections dealing with analytical methodology. These developments are extending increasingly into the very important and rapidly developing area of organic chemistry.

Rapid advances which have taken place in geochemistry, particularly those that have stemmed from the Deep Sea Drilling Project, have made it necessary to devote a whole volume to topics in sedimentary geochemistry.

Both the range and accuracy of the physical constants available have increased since the first edition and a selection of tabulated values of these constants are to be found at the end of each of the first four volumes.

No attempt has been made to discuss Physical Oceanography except where a grasp of the physical concepts is necessary for a better understanding of the chemistry. For a treatment of the physical processes occurring in the sea the reader is referred to the numerous excellent texts now available on physical oceanography. Likewise, since the distribution of salinity in the sea is of greater relevance to the physical oceanographer and is well discussed in these texts, it will not be considered in the present volumes.

This series is not intended to serve as a practical handbook of Marine Chemistry, and if practical details are required the original references given in the text should be consulted. In passing, it should be mentioned that, although those practical aspects of sea water chemistry which are of interest to biologists are reasonably adequately covered in the "Manual of Sea Water Analysis" by Strickland and Parsons, there is an urgent need for a more general laboratory manual.

The editors are most grateful to the various authors for their helpful co-operation which has greatly facilitated the preparation of this book. They would particularly like to thank Messrs R. F. C. Mantoura and A. Dickson for their willing assistance with the arduous task of proof reading, without their aid many errors would have escaped detection. They would also like to acknowledge the courtesy of the various copyright holders, both authors and publishers, for permission to use tables, figures, and photographs. In conclusion, they wish to thank Academic Press, and in particular Mr. E. A. S. Cotton, for their efficiency and ready co-operation which has much lightened the task of preparing this book for publication.

*Liverpool*                                                    J. P. RILEY
*November*, 1974                                              G. SKIRROW

# CONTENTS

Chapter 1 *by* K. F. BOWDEN

**Oceanic and Estuarine Mixing Processes**

Chapter 2 *by* M. WHITFIELD

**Sea Water as an Electrolyte Solution**

Chapter 6 *by* T. R. S. WILSON

## Salinity and the Major Elements of Sea Water

Chapter 7 *by* PETER G. BREWER

## Minor Elements in Sea Water

Chapter 8 *by* DANA R. KESTER

## Dissolved Gases Other Than $CO_2$

Appendix *compiled by* J. P. RILEY

### Tables of Physical and Chemical Constants Relevant to Marine Chemistry

# Contents of Volume 2

# Contents of Volume 3

# Contents of Volume 4

# Symbols and units used in the text

A list of the more important symbols used in the text is given below. It is not exhaustive and inevitably there is some duplication of usage since some symbols have different accepted usages in two or more disciplines. The generally accepted symbols have been altered only when there is a possibility of ambiguity.

*Concentration.* There are several systems in common use for expressing concentration. The more important of these are the molarity scale (g molecules $l^{-1}$ of solution $=$ mol $l^{-1}$) usually designated by $c_i$, the molality scale (g molecules $kg^{-1}$ of solution $=$ mol $kg^{-1}$) designated by $m_i$, and the mole fraction scale usually denoted by $x_i$, which is of more fundamental significance in physical chemistry. In each instance the subscript $i$ indicates the solute species; when $i$ is an ion the charge is not included in the subscript unless confusion is likely to arise. Some other means of indicating the concentration are also to be found in the text, these include: g or mg $kg^{-1}$ of solution (for major components), µg or ng $l^{-1}$ or $kg^{-1}$ of solution (for trace elements and nutrients) and µg-at $l^{-1}$ of solution (for nutrients). Factors for conversion of µg to µg-at are to be found in Appendix Tables 4 and 5.

*Activity.* When an activity or activity coefficient is associated with a species the symbols $a_i$ and $\gamma_i$ are used respectively regardless of the method of expressing concentration, where the subscript $i$ has the significance indicated above. Further qualifying symbols may be added as superscripts and/or subscripts as circumstances demand. It is important to realize that the numerical value of the activity and activity coefficient depend on the standard state chosen. It should also be noted that since activity is a relative quantity it is dimensionless.

## UNITS

Where practicable, SI units (and the associated notations) have been adopted in the text, except where their usage goes contrary to established oceanographic practice.

## LENGTH

| | | |
|---|---|---|
| Å | = Ångstrom unit | $= 10^{-10}$ m |
| nm | = nanometre | $= 10^{-9}$ m |
| μm | = micrometre | $= 10^{-6}$ m |
| mm | = millimetre | $= 10^{-3}$ m |
| cm | = centimetre | $= 10^{-2}$ m |
| m | = metre | |
| km | = kilometre | $= 10^3$ m |
| mi | = nautical mile (6080 ft) | $= 1{\cdot}85$ km |

## WEIGHT

| | | |
|---|---|---|
| pg | = picogram | $= 10^{-12}$ g |
| ng | = nanogram | $= 10^{-9}$ g |
| μg | = microgram | $= 10^{-6}$ g |
| mg | = milligram | $= 10^{-3}$ g |
| g | = gram | |
| kg | = kilogram | $= 10^3$ g |
| ton | = metric ton | $= 10^6$ g |

## VOLUME

| | | |
|---|---|---|
| μl | = microlitre | $= 10^{-6}$ l |
| ml | = millilitre | $= 10^{-3}$ l |
| l | = litre | |
| dm$^3$ | = litre | |

## CONCENTRATION

| | |
|---|---|
| ppm | = parts per million ($\mu$g g$^{-1}$ or mg l$^{-1}$) |
| ppb | = parts per billion (ng g$^{-1}$ or $\mu$g l$^{-1}$) |
| μg-at l$^{-1}$ | = μg atoms l$^{-1}$ = $\mu$ g/atomic weight l$^{-1}$ |

## ELECTRICAL

| | |
|---|---|
| V | = volt |
| A | = ampere |
| Ω | = ohm |

## TIME

| | |
|---|---|
| s | = second |
| min | = minute |
| h | = hour |
| d | = day |
| yr | = year |

## ENERGY AND FORCE

| | | |
|---|---|---|
| J | = Joule | = 0·2390 cal |
| N | = Newton | = $10^5$ dynes |
| W | = Watt | |

## LIGHT FLUX

| | |
|---|---|
| klux | = kilolux |

## GENERAL SYMBOLS

| | |
|---|---|
| $A$ | Helmholtz free energy |
| $A$ | absorbance = optical density ($\log I_0/I$) |
| $a$ | Scatchard coefficient (see p. 125 ff.) |
| $a_i$ | activity of component $i$ |
| $\overset{\circ}{a}$ | ionic size (Debye–Hückel equation (see p. 93 ff.)) |
| AOU | apparent oxygen utilization (see p. 533) |
| $b$ | Scatchard coefficient |
| B.O.D. | biochemical oxygen demand |
| $C$ | electrical conductivity |
| $C$ | heat capacity |
| $c_G^*$ | molar concentration of gas, $G$, at saturation |
| $c_i$ | molar concentration of species $i$; charges of the species $i$ are omitted where these are obvious |
| CA | carbonate alkalinity (meq $l^{-1}$) |
| Cl | chlorinity (g kg$^{-1}$ = ‰) |
| $D_G$ | molecular diffusion coefficient of a gas, $G$, in the gas phase. |
| DH | $-\mathscr{A}\,\lvert z_+ z_- \rvert\, I^{\frac{1}{2}}$ (Debye–Hückel limiting law—see p. 85) |
| d.p.m. | radioactive disintegrations per minute |
| $E$ | the E.M.F. of a cell |
| $E^\circ$ | standard potential |
| $E_G$ | exchange coefficient of a gas, $G$ |

| | |
|---|---|
| $E_h$ | redox potential |
| $E_j$ | liquid junction potential |
| $E_{\frac{1}{2}}$ | half wave potential |
| $e_G$ | exit coefficient of a gas (see p. 512) |
| $F$ | Faraday equivalent of electric charge |
| $F_i$ | formal concentration of ion $i$ |
| $f_G$ | mole fraction of a gas in dry air (p. 499) |
| $f_i$ | rational activity coefficient of component $i$ (p. 51) |
| $G$ | Gibbs free energy |
| $H$ | enthalpy |
| $h$ | Planck's constant |
| $h$ | relative humidity (‰) |
| $I$ | ionic strength or attenuated light intensity |
| $I_k$ | light intensity for saturation (Vol. 2, p. 390) |
| $I_{mpc}$ | light intensity of most penetrating component |
| $I_0$ | light intensity at surface or initially |
| $I_z$ | light intensity at depth $z$ |
| $K$ | equilibrium constant |
| $K^*$ | stoichiometric equilibrium constant |
| $K_G$ | Henry's Law constant for gas, $G$ |
| $K_x$ | coefficient of eddy diffusion in the $x$-direction (similarly for $K_y$ and $K_z$) |
| $k$ | relative growth constant of algae |
| $k$ | Boltzman constant |
| $k_e$ | vertical extinction coefficient |
| $k_{ob}$ | relative growth constant of algae |
| $m_i$ | molarity of component, $i$ |
| $N_a$ | Avogadro's number |
| $n_i$ | number of moles of component $i$ in a mixture |
| $P$ | primary production ($mg\,C\,m^{-2}\,day^{-1}$) |
| $P$ | total phytoplankton population per unit area |
| $P_G$ | partial pressure of gas, $G$, in solution |
| $P_h$ | photosynthetic quotient |
| $P_i$ | polarizability of an ion $i$ |
| $P_{max}$ | maximum rate of photosynthesis |
| $P_T$ | total pressure |
| $p_G$ | partial pressure of gas, $G$, in the atmosphere |
| $p_0$ | saturated vapour pressure of pure water |
| $p_s$ | saturated vapour pressure of sea water or an aqueous solution |
| $p_t$ | total pressure of a gas |
| PZC | point of zero charge |

| PZR | point of zeta reversal |
| $\mathscr{Q}$ | partition function (see p. 81) |
| R | rate of production or loss of material |
| R | gas constant |
| R | relative photosynthesis (see Vol. 2, p. 423) |
| S | entropy |
| S | salinity (g kg$^{-1}$ = ‰) |
| T | temperature in K |
| t | temperature in °C |
| t | time |
| U | energy |
| u | component of velocity of water in $x$-direction |
| V | volume |
| v, w | components of velocity of water in $y$- and $z$-directions respectively |
| x | mole fraction |
| Y | see p. 60 |
| Z | atomic number |
| z | depth |
| z | charge on an ion |

*Greek Symbols*

| α | generalized solubility coefficient of a gas in a liquid (mol l$^{-1}$ atm$^{-1}$) |
| $\alpha_f$ | isotopic fractionation factor |
| $\alpha_{ik}$ | Harned rule coefficients (see p. 106) |
| $\beta_G$ | Bunsen coefficient of a gas, $G$ (see p. 501) |
| β | stability constant of a complex (see p. 193) |
| β | van Slyke buffer capacity |
| Γ | total adsorption density (see p. 248) |
| $\gamma_i$ | activity coefficient of species $i$ |
| Δ | change of (as in ΔG) |
| $\Delta_G$ | saturation anomaly of gas, $G$ (see p. 517) |
| $\delta^{13}C$ | permillage enrichment of $^{13}C$ relative to a given standard |
| $\delta^{14}C$ | permillage enrichment of $^{14}C$ relative to a given standard |
| $\delta(G)$ | isotope anomaly for gas, $G$ (see, e.g. p. 527) |
| $\varepsilon_M$ | electrode potential |
| $\varepsilon_j$ | junction potential |
| $\epsilon$ | molar absorptivity |
| η | viscosity |
| κ | electrical conductivity |

| | |
|---|---|
| $\lambda$ | wavelength |
| $\lambda$ | ionic conductance |
| $\lambda$ | radio-nuclide decay constant |
| $\mu$ | ionic strength (more correctly $I$) |
| $\mu$ | relative growth constant of plankton |
| $\boldsymbol{\mu}_i$ | chemical potential of component $i$ |
| $v$ | stoichiometric coefficient |
| $v$ | frequency |
| $\rho$ | density |
| $\sum CO_2$ | total carbon dioxide content of a solution $=$ $c_{H_2CO_3} + c_{CO_2} + c_{HCO_3(T)} + c_{CO_3(T)}$ |
| $\sigma$ | surface charge density of solid |
| $\sigma_G$ | degree of saturation of liquid with gas, $G$ (see p. 217) |
| $\sigma_t$ | specific gravity of sea water at $t°C$ |
| $\tau$ | residence time of an element in sea water |
| $\phi$ | osmotic coefficient |
| $\Psi_x$ | Gouy layer (psi) potential at a distance $x$ (see p. 248) |

*Superscripts*

| | |
|---|---|
| $\ominus$ | standard state |
| — | partial molar quantity (e.g. $\overline{V}$, etc.) |

Chapter 1

# Oceanic and Estuarine Mixing Processes

K. F. BOWDEN

*Department of Oceanography*
*University of Liverpool, England*

## 1.1. INTRODUCTION

There are two kinds of physical processes which influence the distribution of substances dissolved or suspended in sea water. In advective processes large scale movements of water occur carrying the dissolved or suspended matter with them, whereas in diffusive processes an exchange of properties

1

takes place without any overall transport of water. The first class of processes includes ocean currents at all depths and also the vertical movements of upwelling or sinking of water masses. Diffusive effects are produced by turbulent mixing both in the vertical and horizontal directions on a very wide range of scales. Molecular diffusion usually takes place so slowly compared with turbulent mixing that its direct influence may be neglected, except in a few specialized circumstances, such as gas exchange across the sea surface.

The oceans and seas of the world cover a total area of $360.8 \times 10^6 \, \text{km}^2$, which represents $70.8\%$ of the earth's surface. Of this area, $7.6\%$ is over the continental shelf, less than 200 m deep, $8.5\%$ has a depth between 200 m and 2,000 m and the remaining $83.9\%$ is more than 2,000 m deep. The greatest depth recorded to date is 11,034 m in the Marianas Trench. The areas, volumes and depths of the three oceans and their adjacent seas are given in Table 1.1 (from Defant, 1961).

TABLE 1.1
*Areas, volumes and depths of the oceans*

| Ocean, including adjacent seas | Area $10^6 \, \text{km}^2$ | Volume $10^6 \, \text{km}^3$ | Mean depth m | Greatest depth m |
|---|---|---|---|---|
| Atlantic | 106·2 | 353·5 | 3,331 | 8,526 |
| Indian | 74·9 | 291·9 | 3,897 | 7,450 |
| Pacific | 179·7 | 723·7 | 4,028 | 11,034 |
| All oceans | 360·8 | 1,369·1 | 3,795 | — |

## 1.2. SURFACE CURRENTS OF THE OCEANS

### 1.2.1. GENERAL CIRCULATION

The major oceanic circulations in the surface layers are driven by the prevailing wind systems and are influenced by the form of the ocean boundaries. The general patterns in all the oceans are similar. In the North Atlantic, for example (Fig. 1.1), there is a general clockwise circulation. The north-east trade winds drive the North Equatorial Current, which flows westwards from Africa towards the Windward and Leeward Islands of the West Indies, where it divides into two branches, one entering the Caribbean Sea and the Gulf of Mexico, from which it emerges through the Straits of Florida to form one source of the Gulf Stream. It is joined here by the other branch, which has turned northwards outside the Antilles arc of islands. The combined branches form the Gulf Stream, which flows parallel to the American coastline as far as Cape Hatteras and then turns eastwards across the ocean. In this part of its course, the current is being driven by the southwesterly winds of temperate latitudes, and on passing eastwards of the Grand Banks, off

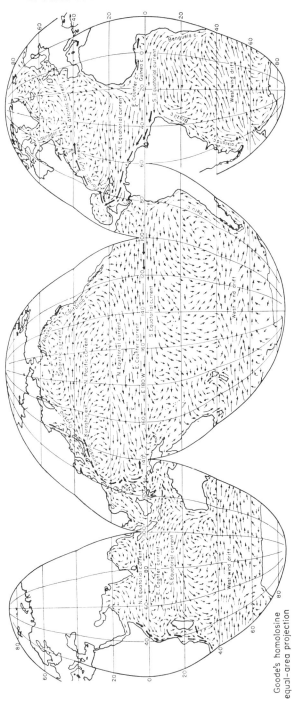

FIG. 1.1. Surface currents of the oceans in February–March (Sverdrup, 1945). By permission of George Allen & Unwin Ltd.

Newfoundland, it is known as the North Atlantic Current. The circulation is completed by the more diffuse, south-flowing Canaries Current, feeding the start of the North Equatorial Current. Another branch of the North Atlantic Current turns northwards past the British Isles and enters the Norwegian Sea.

In the South Atlantic there is an approximate mirror image of the above circulation, with the South Equatorial Current flowing westwards, driven by the south-east trades. The major part of this current turns southwards on approaching the coast of South America, and forms the Brazil Current, which leaves the neighbourhood of the coast between 30 and 40° S to join the east-flowing South Atlantic Current or West Wind Drift (which is driven by the westerly winds of these latitudes). Nearing the coast of South Africa, part of the current turns northwards, and becomes the Benguela Current which completes the anticlockwise circulation and provides the source of the South Equatorial Current.

In the region of comparatively light winds between the North and South Equatorial Currents, particularly on the east side of the ocean, the Equatorial Counter-current is found, flowing in the opposite direction, towards the east. This current is fed, at the westward end, by backward turning portions of the North and South Equatorial Currents. It will be noted that the axis of the Counter-current, at 7° N is an appreciable distance north of the equator, this asymmetry reflecting a similar lack of symmetry about the equator in the wind systems and in the distribution of barometric pressure. An important link between the circulations in the North and South Atlantic is found on the western side of the ocean, where a part of the South Equatorial Current flows northwards across the equator to join the North Equatorial Current.

The circulation in the Pacific Ocean shows similar features. The North Equatorial Current flows westwards across the whole width of the ocean. Off the Philippines part of the current turns southwards to feed the return flow in the Equatorial Counter-current, which is more strongly developed than in the Atlantic. The remaining part of the North Equatorial Current turns northwards to form the Kuroshio, a strong, comparatively narrow current which corresponds to the Gulf Stream in the Atlantic. Leaving the vicinity of Japan, the Kuroshio broadens out into the North Pacific Current, which continues eastwards across the ocean towards the coast of North America. Here, part of it turns northwards into the Gulf of Alaska, while the remaining part turns southwards and, as the California Current, completes the clockwise circulation in the North Pacific. A broadly similar circulation, in the anticlockwise direction, is formed in the South Pacific by the South Equatorial Current, the East Australia Current, the West Wind Drift and the Peru Current. In both the North and South Pacific oceans, there is a tendency for the main circulation to break up into two gyrals. Thus, in the North Pacific some of the water in the North Pacific Current turns off to the south to rejoin the North Equatorial Current in the mid part of the ocean.

In the northern part of the Indian Ocean, the circulation is largely influenced by the surrounding land masses, and there are important seasonal variations, which will be considered later. In the southern part there is an anticlockwise circulation which resembles that in the other two oceans, and which comprises the South Equatorial, Mozambique and Agulhas Currents, the West Wind Drift and the West Australia Current.

All three oceans open out at their southern ends into the Antarctic, or Southern, Ocean, in which the West Wind Drift forms a circumpolar current. Its narrowest constriction is in the Drake Passage, between Cape Horn and the South Shetland Islands. The West Wind Drift in the Southern Ocean is, therefore, largely a closed circulation, although it has offshoots into the other three oceans and is joined by certain outflowing currents from them.

It is well known that the trade winds blow from northeast and southeast, rather than from directly north or south, as a consequence of the *geostrophic* or *Coriolis effect*, arising from the earth's rotation. For the same reason, wind-driven surface currents tend to flow in a direction 45° to the right of the wind in the northern hemisphere and 45° to the left in the southern hemisphere. A striking feature of the circulation in each of the oceans is the lack of symmetry between the currents on the western and eastern sides of the ocean. While the western boundary currents, such as the Gulf Stream and Kuroshio in the northern hemisphere and the Brazil Current and Agulhas Current in the southern hemisphere, are narrow, fast-flowing currents, the return flow on the eastern sides of the oceans is more diffuse. It was first shown by Stommel (1948) that this westward intensification of the circulation was a further consequence of the geostrophic effect. Fuller theoretical treatments of the wind-driven circulation were given shortly afterwards by Munk (1950) and by Munk and Carrier (1950). These showed clearly the presence and location of the main oceanic gyres and the westward intensification. Quantitatively, however, using available data on wind stress over the oceans, they underestimated the volume transport of the western boundary currents by a factor of at least 2. More recently, a number of investigations of circulation using numerical models have been made and the discrepancies between computed and observed values of transport are much reduced when account is taken of non-linear effects (i.e. the rapid changes of velocity near the western boundary). A review of such models was given by Gill (1971).

The highest velocities of flow are found in the western boundary currents. The Florida Current reaches speeds greater than $1.5 \, \mathrm{m \, s^{-1}}$ in flowing through the Straits of Florida, where the volume of water transported varies between 20 and 40 × $10^6$ $m^3$ $s^{-1}$. Earlier observations indicated that the Gulf Stream attains its greatest volume off Cape Hatteras where it transports about 70 × $10^6$ $m^3$ $s^{-1}$. Some observations indicate that the volume continues to increase eastwards to possibly twice this value at longitude 65°W (Knauss, 1969). On the basis of observations averaged over a considerable period, the Gulf Stream appears to be about 150 km wide and to have a maximum surface velocity of about 1·5 m $s^{-1}$. More detailed measurements made by research ships in recent years, however, have shown that at a particular time its width may be as little as 50 km with velocities up to 3·0 m $s^{-1}$. Continuing eastwards from the region south of the Grand Banks, the Gulf Stream becomes much broader and correspondingly slower, so that in its continuation, the North Atlantic Current, the velocity is only about 0·15 m $s^{-1}$, decreasing to 0·05 m $s^{-1}$ as it approaches Ireland. A detailed account of the Gulf Stream has been given by Stommel (1965).

The structure of the Kuroshio in the North Pacific shows features similar to those of the Gulf Stream (Stommel and Yoshida, 1972). In most of the other major ocean currents the velocities seldom exceed 0·5 m $s^{-1}$ and are often considerably less, although because of their width and depth they transport very large volumes of water. The North Equatorial Current in the Pacific, for example, transports 45 × $10^6$ $m^3 s^{-1}$, and the Countercurrent 25 × $10^6$ $m^3$ $s^{-1}$. The greatest rate of transport is found in the Antarctic Circumpolar Current which, in the Drake Passage, is estimated to be 165 × $10^6$ $m^3$ $s^{-1}$, with the transport increasing to as much as 200 × $10^6$ $m^3$ $s^{-1}$ south of Africa (Neumann and Pierson, 1966).

1.2.2. SEASONAL VARIATIONS

In certain regions, significant seasonal variations occur in the currents. In extreme cases, a current may disappear or reverse its direction at a particular season, whereas in other instances the variation is limited to a change in the velocity or in the volume of water transported, or to a shift in its position. The most striking seasonal variations occur in the Indian Ocean, where they are associated with the monsoons. During the winter months, when the north-east monsoon winds prevail over the northern part of the ocean, the North Equatorial Current is well developed, while to the south of it, the Equatorial Countercurrent flows eastwards, with its axis at 7°S. Along the African coast, south of the Gulf of Aden, the North Equatorial Current is deflected to flow southwards. In the summer months, when a low pressure area has become established over central Asia and the south-west monsoon wind blows across the ocean, the North Equatorial current is replaced by the Monsoon Current, which flows from west to east. It appears to take about a month for the current system to become adjusted to the new distribution of winds. Along the African coast, from 10°S northwards, the flow is directed towards the north, and much water from the South Equatorial Current crosses the equator. In the Somali Current, flowing into the Arabian Gulf, the velocity may exceed $2 \cdot 0 \, \mathrm{m \, s^{-1}}$. The Equatorial Countercurrent cannot be recognized in this season. Certain changes also occur in the southern part of the Indian Ocean.

Another example of a seasonal current occurs off the west coast of North America, where the relatively cold California Current forms part of the general clockwise circulation in the North Pacific. During the months of November, December and January, a surface counter-current, the Davidson Current, develops and flows northwards between the California Current and the coast. Upwelling occurs during the rest of the year, and is associated with a series of eddies in the surface layer, although a subsurface counter current flows northwards below about 200 m.

1.2.3. IRREGULAR VARIATIONS

Apart from seasonal changes, irregular variations of currents occur, sometimes with far-reaching results. An outstanding example is El Niño, a seasonal current which, from January to March, flows off the west coast of South America, carrying warm, low salinity water to the south. It does not usually extend further than a few degrees south of the equator, but occasionally it penetrates beyond 12°S displacing the relatively cold Peru Current. The effects on marine life and fisheries are disastrous and excessive rainfall and flooding occur in the normally very dry coastal area. The most recent catastrophe of this kind occurred in the early months of 1973.

A different kind of variation arises from the instability which appears to be inherent in the positions and intensities of certain currents, as in the Gulf Stream east of Cape Hatteras. The current takes a meandering course, and the position of the axis of the current may shift by up to 60 km in a week. From time to time the meanders may form a closed loop, which then develops into an eddy and becomes detached from the main current.

Most of the data from which charts of the surface currents have been constructed have been obtained from records of ships' drift collected over a number of years. In this method, the vector distance travelled by a ship in a given time as determined by astronomical fixes or, in recent years, by radio navigational techniques including fixes from orbiting satellites, is compared with that estimated from the ship's course and speed. The vector difference between these two quantities, corrected if necessary for the effect of wind on the ship, is attributed to the current. On research vessels other methods are available for measuring currents with greater accuracy and in more detail. This greater accuracy and detail is necessary in order to determine comparatively rapid changes in the current, such as those occurring in the Gulf Stream. Descriptions of methods of measuring currents and further details of the currents themselves will be found in the books by Sverdrup *et al.* (1942), Defant (1961), Von Arx (1962) and Neumann (1968).

### 1.3. SALINITY DISTRIBUTION IN THE SURFACE LAYER

The distribution of salinity is determined by processes which take place at the sea surface and by currents and mixing. Although small but significant variations occur in some of the major constituents of sea water, these may be neglected when considering the general distribution of salinity. The processes taking place at the surface which need to be considered are, therefore, those which decrease or increase the salinity by the addition or removal of fresh water. The salinity is decreased by rainfall, or other forms of precipitation, the influx of fresh water from the land and by the melting of ice. Conversely, it is increased by evaporation and by the formation of ice, since the ice has a lower salt content than the water from which it is formed, leaving a higher concentration of salt in the remaining water.

Figure 1.2 shows the surface salinity of the oceans in northern summer. Certain seasonal changes occur but, except in coastal waters, they are not large. In the open ocean the surface salinity ranges from 32 to 37·5‰. The highest salinities are found in tropical latitudes, about 20–30°N and 15–20°S, where the rate of evaporation is high owing to the high temperatures and strong trade winds. In the equatorial zone, the salinity is lower because of the greater rainfall and lower wind speeds. Polewards of the tropical regions the salinity decreases due to the excess of rainfall over evaporation. Near the coasts of the continents the salinity is reduced owing to the influence of run-off from the land. In some areas, the effects of ocean currents are apparent. Off the north-east coast of North America, for example, the salinity near the coast is low owing to the influence of the Labrador Current and there is a sharp increase in salinity across the Gulf Stream.

Salinities higher than those in the open ocean occur in partially enclosed seas in mid-latitudes, where evaporation greatly exceeds rainfall and run-off. Thus, the Mediterranean Sea has a salinity of 37 to 39‰ and the Red Sea a salinity of 40–41‰. In coastal seas in temperate latitudes, the salinity is reduced by the influx of land water, and there may be a range of conditions from almost completely fresh water, as in the inner parts of the Baltic Sea, to the salinity of the open ocean.

In the deeper layers of the ocean, the variations in salinity are smaller than near the surface, but are highly significant in their relation to the general circulation.

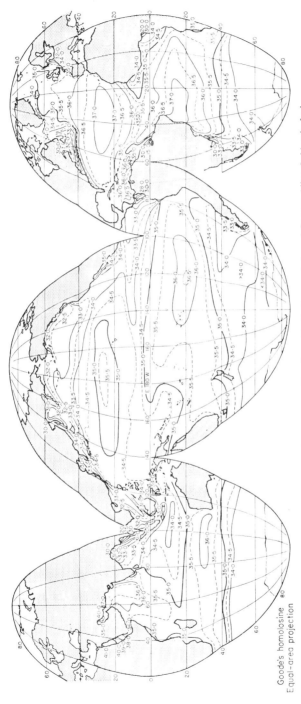

FIG. 1.2. Surface salinity of the oceans in the northern summer (Sverdrup, 1945). By permission of George Allen & Unwin Ltd.

Goode's homolosine
Equal-area projection

## 1.4. Deep Water Circulation

The wind-driven circulation extends, over much of the oceans, to a comparatively small fraction of the total depth. In general, its depth is least in equatorial regions, where the surface currents extend through a layer varying
from 20 to 200 m in depth, and in temperate latitudes it is greatest in the
western boundary currents. The Gulf Stream, for example, extends to a depth
exceeding 1,000 m and possibly to 3,000–4,000 m in its downstream part,
north-east of Cape Hatteras, but at these depths the velocity is much reduced.
The Antarctic Circumpolar Current is remarkable in that it flows in the same
direction at all depths, from the surface to the bottom at more than 4,000 m,
and has a very high transport. The currents are closely related to the distribution of density, which in turn depends on the distribution of temperature
and salinity. Observations of these two variables at different depths can be
made more easily than can direct current measurements, so that charts or
sections of temperature and salinity, or of density derived from them, are often
used to indicate the positions and velocity of the current. Other properties,
such as the content of oxygen, silicate, and phosphate, have also been used as
indicators of water movements.

Before 1955 the only practicable method of making direct measurements of
deep currents was by using current meters suspended from an anchored ship.
This method is costly in ship's time and involves the difficulty of separating
spurious components due to movements of the ship and current meters from
the true currents. The introduction in 1955 of the neutral buoyancy float
(Swallow, 1955) which drifts freely at any predetermined depth and is tracked
acoustically from a moving ship, provided a convenient method of current
measurement which has been widely used. Since 1960 increasing use has
been made of arrays of current meters suspended from moored buoys,
which can be left unattended to record automatically for periods up to several
months (Richardson et al., 1963). Now that the initial difficulties have been
largely overcome, this technique is in common use, both in deep oceanic
water and in continental shelf seas.

In many parts of the oceans it is possible to distinguish five layers: surface,
upper, intermediate, deep and bottom, each of which has a distinctive water
mass and movement of its own. The persistence of these layers is due to the
stable density gradients which greatly inhibit vertical mixing between water
masses. Figure 1.3, which shows the salinity distribution in a north–south
section of the Atlantic Ocean, provides evidence of these various layers.
The surface layer, down to 100 or 200 m, is subject to considerable seasonal
variations in temperature and salinity, owing to direct interaction with the

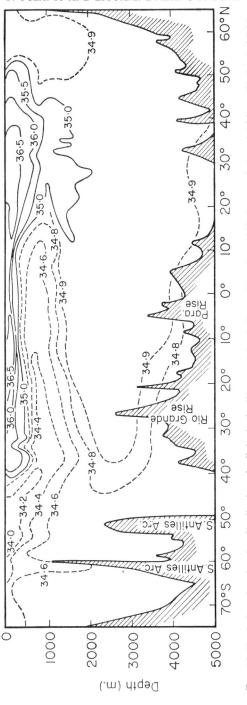

F̄ıɢ. 1.3. Vertical section showing distribution of salinity in the Western Atlantic Ocean (Von Arx, 1962). By permission of Addison-Wesley Publishing Co., Inc.

atmosphere. The upper layer extends from 100 or 200 m down to about 800 m, and is not subject to these seasonal changes, but it does take part with the surface layer in the wind-driven surface circulation, the velocity of which usually decreases with depth.

In the Atlantic, the main feature of the intermediate layer is the Antarctic intermediate water, which sinks from the surface at the Antarctic Convergence at about 50° S and then spreads northwards at about 1,000 m. It is characterized by a salinity minimum and its influence can be traced well north of the equator. Below it, at depths of 2,000 to 4,000 m, the North Atlantic deep water, formed by sinking in the Labrador and Irminger Seas, flows southwards. In the layer adjacent to the bottom, there is a northward flow of Antarctic bottom water, which is formed by sinking at the edge of the Antarctic continental shelf, especially in the Weddell Sea. Its high density is a consequence of its very low temperature and the comparatively high salinity which results from the abstraction of water during the formation of ice.

It would be misleading to think that these deep water movements are spread uniformly over the whole width of the ocean. Both calculations and, more recently, direct measurements indicate that deep currents, like those in the upper layer, tend to be concentrated in particular regions. From the density distribution and certain reasonable assumptions about the depths at which the velocity was zero, Defant (1941) constructed charts of the currents at 800 m and 2,000 m in the Atlantic. The former, reproduced in Fig. 1.4(a), shows the Antarctic intermediate water flowing northwards across the equator as a concentrated current with velocities of 6–12 cm s$^{-1}$ on the west side of the ocean. Similarly, the chart for 2,000 m (Fig. 1.4 (b)) shows the North Atlantic deep water travelling south in three distinct streams; one on the west side which follows the continental slope off North America, one on the east side of the Mid-Atlantic Ridge and another in the East Atlantic Trough. The velocities indicated are less than 2 cm s$^{-1}$ in most areas, but in a few localities exceed 10 cm s$^{-1}$. Stommel has shown that dynamical theory leads one to expect that deep currents, like surface ones, would be concentrated on the western side of the ocean basins. Swallow and Worthington (1957, 1961) showed by direct current measurements the existence of a countercurrent flowing southwards off South Carolina, slightly further offshore than the Gulf Stream, at 2,000 m with speeds of up to 17 cm s$^{-1}$.

Although the circulation of the surface and upper layers is brought about by energy received from the wind, thermodynamic processes are more important in maintaining the circulation of the deeper layers. Conditions in the ocean are less favourable than are those in the atmosphere for the maintenance of such a circulation. It can be shown, from the first and second laws of thermodynamics, that heat must be absorbed at higher pressures and higher

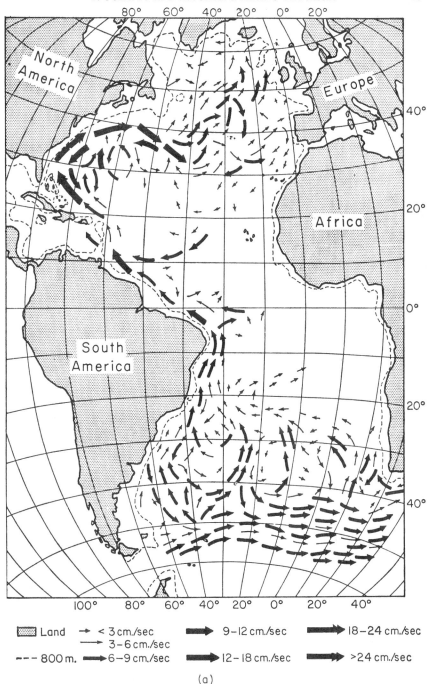

Fig. 1.4. (a) Current field at 800 m in the Atlantic Ocean. (Defant, 1961). By permission of Pergamon Press Ltd.

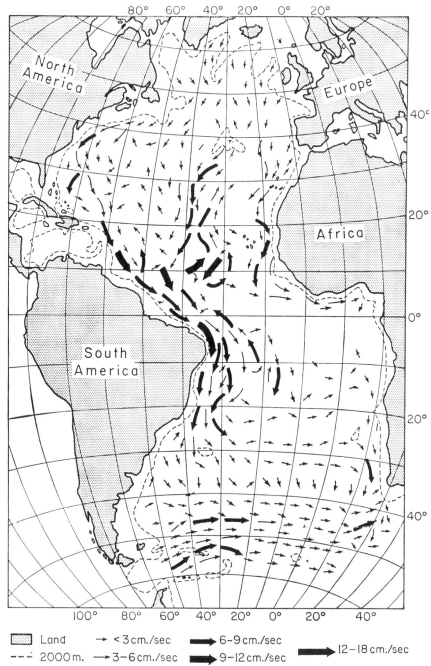

FIG. 1.4. (b) Current field at 2,000 m in the Atlantic Ocean (Defant, 1961). By permission of Pergamon Press Ltd.

temperatures than those at which it is released. In the atmosphere this is clearly the case, since heat is gained from solar radiation and by evaporation mainly near the surface of the earth and is lost by the condensation of water vapour and by back radiation to space at higher altitudes. In the oceans, on the other hand, both the main intake and the loss of heat occur in a shallow surface layer, the former in tropical and the latter in polar latitudes. When the complex circulation in the various layers is considered, however, it is found, as shown by Proudman (1953), that the thermodynamic requirements are satisfied. In the Atlantic, for example, there is a surface and mid-depth circulation formed by the Antarctic intermediate water, which sinks at 50° S and flows northwards below the warmer waters of the upper layer in the tropical region and receives heat from this layer by eddy conduction. It then becomes entrained in the North Atlantic deep water and returns to flow southwards at a depth of 2,000 to 3,000 m. At the Antarctic Convergence, it rises above the Antarctic bottom water and reaches the surface in the Sub-antarctic region, where it loses heat to the atmosphere, mainly by back radiation. Thus, for this circulation heat is taken in at higher pressure and at higher temperature, i.e. below 1,000 m in the tropics, and is given out at atmospheric pressure and lower temperature in the Sub-antarctic region. A schematic picture of the overall circulation in the Atlantic Ocean is given in Fig. 1.5.

The physical processes which change the properties of water masses take place mainly at the surface. The temperature can be increased by absorption of heat from solar radiation and reduced by evaporation and by conduction to the atmosphere. Evaporation or the formation of ice will increase the salinity of the water, and rainfall, run-off from land or the melting of ice will reduce it. If the density of the water is sufficiently increased, by cooling or evaporation for example, convection currents are set up which may extend, in some cases to the bottom, or in others to some intermediate depth. The sinking water will then spread out horizontally at a level appropriate to its density.

In other regions, often under the influence of wind, the converse process of upwelling occurs. Thus, in the Benguela Current, off the south-west coast of Africa, and in the Peru Current, off South America, the action of the south-east trade winds causes a flow of surface water away from the coast. Its place is taken by sub-surface water which rises from a depth of 200 to 300 m. This upwelling is very important from a biological point of view, since the upwelling water brings a fresh supply of nutrient substances into the euphotic zone. Estimates of the vertical velocity of upwelling are in the range 10 to 80 m per month (approximately 0·4 to $3 \times 10^{-3}$ cm s$^{-1}$). In the northern hemisphere similar upwelling regions occur off the coasts of California and of north-west Africa. There is renewed interest in areas of coastal upwelling at

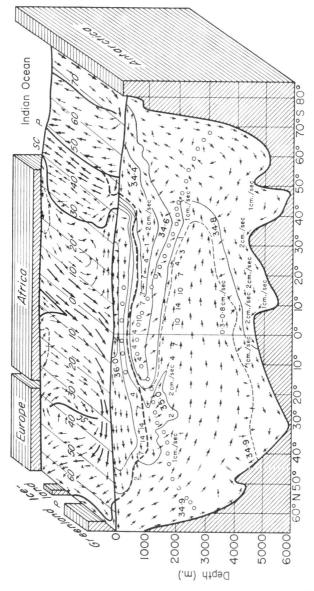

Fig. 1.5. Block diagram of the Atlantic Circulation (Defant, 1961). By permission of Pergamon Press Ltd.

present in view of the concentration of living resources associated with them. The occurrence of upwelling is subject to considerable fluctuations in space and time, and coordinated physical, chemical and biological investigations of the upwelling processes and their effects are needed (Smith, 1968; Jones, 1970).

Recent developments in knowledge of the oceanic circulation have included the discovery of the Equatorial Undercurrents, first in the Pacific and then in the Atlantic and Indian Oceans. These currents flow eastwards along the equator, underneath the west-flowing Equatorial Currents. In the Pacific, the course of the current, which has been named the Cromwell Current, has been traced for over 12,000 km from about 140° E to the Galapagos Islands. The upper boundary of the current is at about 30 m and its maximum velocity, which occurs at 100 m, reaches $1\cdot5$ m s$^{-1}$; three times as fast as the South Equatorial Current, which flows in the opposite direction above it. In the horizontal plane, its velocity is greatest at the equator but decreases to almost zero at 1–2° N and S. The volume of water transported by the current is estimated to be $40 \times 10^6$ m$^3$ s$^{-1}$. A detailed description has been given by Knauss (1960, 1966).

On the theoretical side, Stommel (1957, 1958) has built up a comprehensive, although necessarily idealised, picture of the deep water circulation. He assumed only two sources of deep water, one in the Antarctic corresponding to the Antarctic bottom water, and the other in the North Atlantic, corresponding to the North Atlantic deep water. The circulation deduced from dynamical theory is shown in Fig. 1.6. Its striking features are the strong western boundary currents and the broad diffuse flow over the greater part of the width of the oceans. Reference has been made above to the discovery in 1957 of the deep countercurrent below the Gulf Stream, in accordance with Stommel's scheme. It had long been suggested, on the basis of the temperature distribution, that the deep and bottom water of the North Pacific had its origin in the Antarctic (Wüst, 1929). In Stommel's scheme, the bottom water passes from the South to the North Pacific as a northward flowing western boundary current and much indirect evidence has been accumulated to support this point of view (Warren, 1970). In 1968 a series of current measurements made near the bottom in a narrow channel between the Samoan and Tuamotu islands in the south-west Pacific provided direct evidence of such a flow at depths greater than 4,800 m (Reid, 1969). Averaged velocities of 5 to 15 cm s$^{-1}$ were recorded in a northerly to north-easterly direction.

Fig. 1.6. The deep circulation of the world ocean, according to Stommel (1958). By permission of Pergamon Press Ltd.

## 1.5. Currents in Coastal Seas and Estuaries

### 1.5.1. tidal currents

Tidal currents are present everywhere in the oceans and occur at all depths, but in the open ocean their velocities seldom exceed 10 cm s$^{-1}$ and are often only 2 or 3 cm s$^{-1}$. In shallow waters they are stronger, frequently reaching 1·0 to 1·5 m s$^{-1}$ and in narrow straits they may reach 5 m s$^{-1}$ or more. Tidal currents are periodic, usually with a semidiurnal period of half a lunar day (12 hours 25 minutes), although an appreciable diurnal constituent is often superimposed and in some areas is predominant. In general, the current may be resolved into two components at right angles; these oscillate in different phases, so that the current vector traces out an ellipse during the tidal period. Apart from a small mass transport effect, which is likely to be appreciable only in shallow seas, the oscillatory tidal movements have an insignificant effect in transporting water masses by advection. The turbulence associated with their flow, however, has an important influence on both vertical and horizontal mixing.

### 1.5.2. local wind-driven currents

In many coastal areas, the permanent currents are weak or ill-defined and can be masked by currents produced locally by the wind. In the open sea the current tends to flow at an angle to the right of the wind (to the left in the southern hemisphere) and the wind factor, i.e. the ratio of surface current speed to wind speed, is about 0·02. In shallow water, the direction of the surface current deviates less from that of the wind and the wind factor depends on the coastal boundaries and the degree of obstruction to the flow. It is largest where there is a free passage for the water, as for example, with a southerly wind over the Irish Sea, or with a south-westerly wind over the English Channel. In the Straits of Dover the velocity of flow is increased by the throttling effect of the constriction of the channel, and with strong winds blowing along the channel a current of 1·0 to 1·5 m s$^{-1}$ may be produced. Even if the wind is blowing towards a land barrier, e.g. towards the head of a gulf or fjord, there will be a surface current in the direction of the wind but a compensating current will flow in the opposite direction below it.

### 1.5.3. DENSITY CURRENTS

Density currents are generated in areas where large horizontal gradients of density occur. In an estuary with a source of fresh water at its head, the effect is to produce an outflow of lower salinity water in the upper layer and an inflow of higher salinity water below it. The situation is often complicated by the action of tidal currents, and estuary circulation will be considered in a later section. Density currents may also develop off an open coast where there is an influx of fresh water. Near the coast there is a surface flow from the shore and a deep flow towards it but, owing to the earth's rotation, there is a tendency for both these currents to turn to the right, in the northern hemisphere. A coastal current is therefore, set up, flowing with the fresh water inflow on its right, and with an undercurrent in the opposite direction. In the southern hemisphere these directions are reversed.

Currents which are similar in origin, but larger in scale, occur in the straits connecting a land-locked sea to the open ocean*. The best known example is in the Straits of Gibraltar. Because of the intense evaporation in the Mediterranean Sea, the surface waters have a salinity and a density which are considerably higher than those in the Atlantic Ocean. There is, therefore, a surface inflow of lower salinity water from the Atlantic and a deep outflow of the higher salinity Mediterranean water. The average velocity of the surface inflow appears to be about 0·5 m s$^{-1}$, with a similar velocity in the deep outflow, but there are large variations associated with changing wind conditions (Lacombe, 1961). Similar conditions occur in the Straits of Bab el Mandeb, connecting the Red Sea to the Gulf of Aden. The reverse type of flow is found in the Kattegat, where there is a surface outflow of low salinity water from the Baltic Sea (in which precipitation and run-off greatly exceed evaporation) and an deep inflow of higher salinity water from the North Sea.

* See Chapter 15.

1.5.4. ESTUARY CIRCULATION

In an estuary, the basic factor in determining the type of circulation is the part played by tidal currents relative to that of river flow. If other influences were absent, the river water would tend to flow seawards as a layer of fresh water, separated by a fairly distinct interface from the salt water below. Tidal currents exert an important influence through turbulent mixing, which tends to break down the interface and cause a mixing of the river water and sea water within a vertical column. The interaction between river flow and tidal currents is influenced by two other factors: the physical dimensions of the estuary and the effect of the earth's rotation as represented by the Coriolis force. The following system of classification of estuaries, from a dynamical point of view, follows that of Pritchard (1955) which has come into general use (Cameron and Pritchard, 1963; Bowden, 1967; Dyer, 1973).

Where the river flow dominates the circulation almost completely, an estuary of the "salt wedge" type occurs as in Fig. 1.7(a). Below the fresh water outflow, the salt water extends into the river as a wedge, with the interface sloping slightly downwards in the upstream direction, due to the small amount of friction existing between the layers. The steep density gradient at the interface, amounting almost to a discontinuity, reduces the turbulence and mixing to a very low level. The Coriolis force has the effect of causing the interface to slope downwards to the right, in the northern hemisphere, looking towards the sea. The best known example of a salt wedge estuary is that of the Mississippi, where the wedge may extend for 70 km up river.

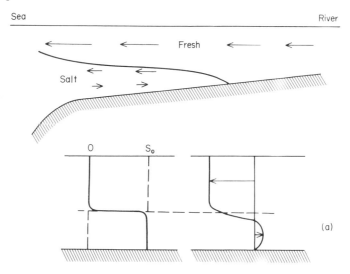

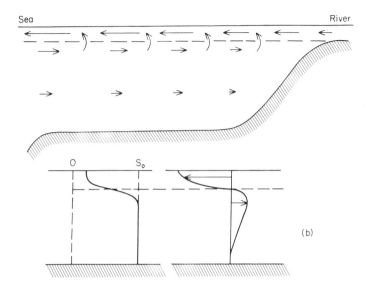

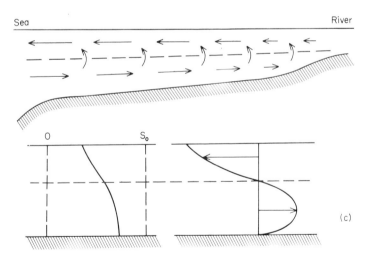

FIG. 1.7. Types of estuary circulation: (a) salt wedge, estuary, (b) two-layer flow with entrainment, (c) partially mixed estuary with entrainment and mixing. In each case the upper figure is a section along the estuary and the lower one shows typical salinity and velocity profiles (Bowden, 1967). Copyright 1967 by the American Association for the Advancement of Science.

A two-layer pattern of flow, with entrainment of higher salinity water from the lower layer into the upper, is developed when the velocity of the seaward moving river water is sufficient to cause internal waves at the interface to break. The salinity of water in the upper layer is increased by entrainment as it moves seaward, and its volume also increases. The salinity of the lower layer is almost unchanged, but there is a slow movement of water upstream to compensate for entrainment. The circulation in many fjords is essentially of this type, but it is unusual for the interface to remain so sharp that the process is purely one of entrainment. A certain amount of mixing occurs with a small proportion of the fresher water penetrating downwards. The interface becomes replaced by an intermediate layer known as the halocline in which the salinity gradient is very steep (see Fig. 1.7(b)). In deep fjords the whole two-layer pattern of flow may occupy a relatively small fraction of the total depth, with a deep stagnant layer below. In other cases a more complex circulation develops, with a three or four layer structure. In fjords with a shallow sill, the renewal of water in the inner basin may be an intermittent process, occurring seasonally or at irregular intervals under the influence of meteorological conditions. Characteristic features of Norwegian fjords have been described by Saelen (1967), and accounts have been given of inlets of the fjord type elsewhere, including British Columbia (Pickard, 1961) and Chile (Pickard, 1971). (See also Section 15.3).

In shallower estuaries, with comparatively strong tidal currents, vertical mixing extends throughout the whole depth. There are still two layers of flow, the surface of no motion which separates the seaward flowing upper layer from the upstream moving lower layer, usually occurring somewhat above mid-depth. The salinity increases continuously with depth, but often without a distinct halocline (see Fig. 1.7c). The volumes of water involved in each of the layers of flow are often many times greater than the river discharge. The estuaries of the Chesapeake Bay system, on the east coast of U.S.A., and those of the Tees, Thames and Mersey, in England, are among those of this type. In a typical case the velocity of the residual seaward flow in the upper layer is of the order of 10 cm s$^{-1}$ whereas the velocity due to the river flow alone would be less than 1 cm s$^{-1}$ and the amplitude of the tidal current may be 1 m s$^{-1}$ or greater. When the tidal currents are very strong relative to the river flow, the vertical mixing may become so intense that there is no measurable variation of salinity from surface to bottom. If the estuary is sufficiently wide, the Coriolis force gives rise to a lateral variation of salinity, with the lower salinity water on the right-hand side, looking towards the sea (in the northern hemisphere). There is a net seaward flow of fresher water on the right and a compensating flow of higher salinity water on the left.

## 1.6. Turbulent Mixing

### 1.6.1. Nature of Oceanic Turbulence

All processes which contribute to mixing and dispersion in a fluid may be described as forms of "motion". They cover a wide range of scales, which may be classified in general terms as: mean flow, turbulence and molecular motion. In the molecular case, the motion takes place on such a small scale that there can be no question of taking account of the movements of individual molecules. A statistical approach is necessary, and the effect of molecular movements on the dynamics of the mean flow is parameterized by introducing a coefficient of viscosity. Similarly, their mixing effect is represented by a coefficient of diffusion and both coefficients can be related to statistical properties of the molecular motion. For the diffusion of a given substance, the coefficient of diffusion is a physical property of the fluid, dependent on temperature, salinity and pressure, but not on position in the fluid or on the pattern of flow. Sea water of salinity $35\%_0$ and temperature $15°C$ has a kinematic viscosity of $1·2 \times 10^{-2} \, cm^2 \, s^{-1}$, a thermal diffusivity of $1·4 \times 10^{-3} \, cm^2 \, s^{-1}$ and a diffusivity for salt of $2 \times 10^{-5} \, cm^2 \, s^{-1}$.

The separation of a field of flow into mean and turbulent components, on the other hand, involves averaging over certain intervals of time or space. The mean velocities and concentrations determined in this way are then treated explicitly, as functions of position and time. The fluctuations from the mean are treated statistically, and their influences on mean flow or on mean concentration of a dissolved substance, for example, are parametrized by introducing eddy coefficients of viscosity and diffusion. The analogy between the eddy coefficients and the molecular coefficients, however, is only partial, in that the eddy coefficients depend on the scale of motion being considered and they differ in general with direction, with position in the field of flow and with the overall pattern of flow. In spite of the somewhat unsatisfactory nature of the concept of eddy diffusivity, the use of eddy coefficients of diffusion is often the only practicable way to deal with problems of mixing in the sea.

The closer the network of measurements in space and time, the more detailed are the observed patterns of water movements, salinity distribution or the concentration of a particular substance. Fairly coarse horizontal station spacings were sufficient to detect the large eddies associated with the Gulf Stream, just as the large vertical separations of water bottles in a typical hydrographic cast were adequate to distinguish the main water masses of the oceans. The use of high resolution instruments, some examples of which are given later, has shown an increasing degree of fine structure, both in the vertical and horizontal dimensions. It appears that turbulent movements can occur almost anywhere in the ocean, although in the more stable regions their occurrence may be intermittent in time and patchy in spatial distribution. Discussions of the nature of oceanic turbulence and its role in mixing processes are given in the papers by Munk (1966), Okubo (1971a) and Woods and Wiley (1972). There is usually a difference of several orders of magnitude between the vertical and horizontal scales of the turbulent movements, and similar differences in the effects which they produce. This is due partly to the large ratio of width to depth of the ocean and also to the influence of the stable density gradient, which greatly inhibits turbulent movements in the vertical without affecting appreciably the horizontal components. Because of this difference, it is often convenient to treat vertical and horizontal mixing separately.

1.6.2. VERTICAL MIXING

The turbulence responsible for vertical mixing is generated in two main ways: (a) by the stress of the wind on the sea surface, acting through wave motion and the surface drift; (b) by vertical shear in currents, arising in various ways in the interior of the sea and from the action of bottom friction. Wind acting on the surface of the sea generates waves, which cause oscillatory movements of the water particles as the wave form progresses across the sea; a further effect of the wind is to produce a steady drift of the surface water. In waves on deep water, the water particles move in approximately circular orbits in a vertical plane, the water moving forwards at the crest of a wave and in the opposite direction in a trough. At the surface, the diameter of the orbit is equal to the height of the wave, but the movements decrease rapidly with depth, and fall to $e^{-\pi}$ (approximately 0·043) of their surface values at a depth equal to half a wavelength. The orbits are not completely closed and the water particles near the surface have a small net movement in the direction of propagation of the waves. This "mass transport" velocity falls off even more rapidly with depth. A regular train of waves of comparatively small height would have very little mixing effect, the elements of water simply being distorted in shape as the waves pass. In a wind-driven sea, however, waves of many different periods are superimposed, producing an ever-changing pattern of movements in the water and so enhancing the mixing effect. If the waves are also breaking, much of their energy is being converted into turbulence, although the scale is probably fairly small and directly affects only a comparatively shallow layer of water.

The stress of the wind on the sea also produces a drift of the surface water which, in the simplified conditions assumed in Ekman's theory, is directed 45° to the right of the wind in the northern hemisphere and to the left in the southern hemisphere. The water below the surface is brought into motion by shearing stresses communicated from the water above through the action of eddy viscosity, but its movement is slower and deviates further from the wind direction. At a certain depth, called the "depth of frictional influence" and denoted by D, the drift would be in the opposite direction to that at the surface, but its velocity would be only $e^{-\pi}$ of the surface velocity. Below this depth the movement is small, and for practical purposes one may consider that the mixing effect of the wind current is confined to a layer of depth D. The value of D depends, among other factors, on the strength of the wind, and estimates made for it in the open ocean are mostly between 50 and 200 m.

When the sea is gaining heat by the absorption of solar radiation, most of the heat is absorbed in the first few metres. Molecular conductivity alone could provide only a very slow transport of heat downwards, and a steep temperature gradient would occur just below the surface. Turbulence gives rise to an "eddy conductivity" which is much larger, and the heat becomes distributed through a deeper layer. A certain temperature gradient will develop, however, and the downward decrease in temperature causes an increase in density. Under these conditions, vertical mixing involves an increase in the potential energy of a vertical column of water and this energy must be supplied by the turbulence. A stable temperature gradient has, therefore, an inhibiting effect on the turbulence. If the turbulence is being generated by the action of wind on the surface, it may happen that at a certain depth the turbulent energy is no longer intense enough to overcome the stability effect. A steep temperature gradient will then develop at this depth, and separate the warmer water above from the colder water below. In this transition layer, known as the thermocline, vertical turbulence is almost entirely suppressed and mixing of water across it can take place only very slowly.

In many areas, a thermocline is formed at a depth of 30 to 50 m in summer by this process. By diving and underwater photography in the Mediterranean, off Malta, Woods (1968) observed the occurrence of internal waves in the summer thermocline. Intermittently these waves would break at the crests, giving rise to patches of turbulence. It has been suggested that vertical mixing through the thermocline takes place almost entirely by eddy conduction and diffusion through such patches, which resemble "windows" in an otherwise laminar flow. In autumn, when the sea surface is cooled, convection currents are set up which break down the temperature gradient. In addition to seasonal thermoclines of this nature, a main or permanent thermocline occurs over widespread areas of the ocean. This is at a greater depth and is little affected by seasonal influences. The depth of the thermocline varies considerably from one region to another; its upper limit may be less than 100 m and its lower limit may be deeper than 1,000 m. A typical vertical profile of

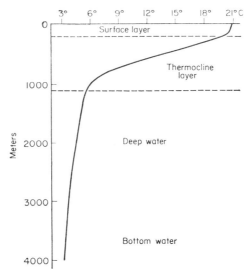

Fɪɢ. 1.8. The vertical subdivisions of the ocean (Von Arx, 1962). By permission of Addison-Wesley Publishing Co. Inc.

temperature, indicating the thermocline, is given in Fig. 1.8. The processes maintaining the large-scale features of the permanent thermocline are closely related to the general circulation of the oceans.

The increasing use of continuously-profiling instruments such as the S.T.D. (salinity-temperature-depth) probe (see Ch. 6) in recent years has revealed the widespread occurrence of "*microstructure*" in the ocean, super-imposed on the large-scale features. Very often this fine structure takes the form of a series of temperature and salinity layers, on a vertical scale of 1 m to several tens of metres and on a horizontal scale of the order of 1 to 10 km (Tait and Howe, 1968). Various theories have been proposed for the forma-tion and persistence of such layers, including the processes of salt fingering and doubly diffusive layering. Measurements with instruments of higher resolution have shown the presence of finer scale temperature, salinity and velocity fluctuations on scales down to 1 cm or less (Grant *et al.*, 1968; Gregg and Cox, 1971). It is characteristic of this small-scale turbulence that it is intermittent in space and time, rather like the patches in Woods' observations.

In shallow water, the effect of bottom friction on currents is to generate turbulence near the bottom, some of which diffuses upwards. Taking into account also the wind-generated turbulence, which is at its greatest intensity near the surface, there may be sufficient turbulence at all depths to maintain thorough vertical mixing and prevent the formation of a thermocline. This occurs, for example, in the English Channel and in those parts of the southern North Sea where the tidal currents exceed about $1 \, \mathrm{m \, s^{-1}}$.

### 1.6.3. HORIZONTAL MIXING

A very broad spectrum of horizontal motions occurs in the ocean. On the largest scale are the main wind-driven circulations of ocean-wide dimensions. Developing from these are eddies of about 100 km in diameter, such as those which occur in the Gulf Stream after it has left the continental slope and has turned eastwards across the Atlantic. Eddies of similar dimensions occur in latitudes 30 to 40°S between the South Equatorial Current and the West Wind Drift. Water movements on this scale may also be set up by moving storm areas. Such eddies break down into smaller ones, of the order of 10 km in diameter, and so on through a cascade of decreasing sizes.

When horizontal turbulence covering a wide spectrum is present, its mixing effect will depend on the scale of the process concerned. A patch of dye, for example, will become dispersed by the action of eddies of a scale smaller than its own dimensions, while those of comparable size will distort the patch and larger eddies will transport it as a whole. As the patch expands, however, increasingly larger eddies will take part in its dispersion. The effective coefficient of horizontal diffusion therefore increases with the size of the patch itself. The idea of a spectrum of turbulent eddies and a coefficient of eddy diffusion increasing with the scale of the phenomena is a feature of the theory of locally isotropic turbulence introduced by Kolmogoroff. The application of this theory to the ocean was discussed by Stommel (1949) and Defant (1954). More recent reviews of turbulence in the sea and its diffusive effects have been published by Bowden (1970) and Okubo (1971a).

## 1.7. QUANTITATIVE TREATMENT OF MIXING PROBLEMS

### 1.7.1. ADVECTION–DIFFUSION EQUATION

Let rectangular axes $OX$, $OY$, $OZ$ be taken, with $OX$ and $OY$ in a horizontal plane and $OZ$ vertically downwards. Let $u$, $v$, $w$ be the components of the velocity of the water parallel to these axes, and consider the distribution of some property which has a concentration $c$ per unit mass. By analogy with molecular diffusion, the turbulent transport of this property may be regarded as a process of eddy diffusion. Across a plane perpendicular to $OZ$, for example, the rate of transport per unit area may be represented by

$$T_z = -\rho K_z \frac{\partial c}{\partial z}$$

where $\rho$ is the density of the water. $K_z$ is defined as the coefficient of eddy diffusion in the vertical direction or, more simply, the vertical eddy diffusivity. Coefficients $K_x$ and $K_y$ may be defined similarly for turbulent transport across planes perpendicular to $OX$ and $OY$ respectively. The three coefficients $K_x$, $K_y$ and $K_z$ will, in general, differ from one another, vary with position, and depend on the type and scale of motion and on the stability. Their numerical values cover a very wide range, but are always large compared with the molecular diffusivity. In the sea, $K_z$ is usually in the range $0.1$ to $10^3 \, \text{cm}^2 \, \text{s}^{-1}$, while $K_x$ and $K_y$ fall within the range $10^4$ to $10^8 \, \text{cm}^2 \, \text{s}^{-1}$. For comparison, the coefficient of molecular diffusion in water is about $2 \times 10^{-5} \, \text{cm}^2 \, \text{s}^{-1}$.

The general equation governing the concentration, $c$, of a particular constituent at any point in the sea may be written:

$$\frac{\partial c}{\partial t} + u\frac{\partial c}{\partial x} + v\frac{\partial c}{\partial y} + w\frac{\partial c}{\partial z} = \frac{\partial}{\partial x}\left(K_x \frac{\partial c}{\partial x}\right) + \frac{\partial}{\partial y}\left(K_y \frac{\partial c}{\partial y}\right) + \frac{\partial}{\partial z}\left(K_z \frac{\partial c}{\partial z}\right) + R$$

(1.1)

where $R$ denotes the rate of production (or loss) of $c$ within a small volume surrounding the point itself. For the oxygen content, for example, $R$ will be positive if there is a net production of oxygen by photosynthesis and negative if there is a net consumption by respiration and oxidation processes. The decay of a radioactive substance would be represented by $R$ equal to $-\lambda c$, where $\lambda$ is the decay constant. For a conservative property, such as salinity, $R = 0$.

The above equation, which allows for both advection and turbulent mixing in three dimensions, is too complicated to be solved in its general form. In a particular problem, a preliminary examination will usually suggest that some of the terms are likely to be small compared with those representing the main processes and can be neglected; a few particular solutions are given below.

### 1.7.2. STEADY STATE SOLUTIONS

If the distribution is not changing with time, $\partial c/\partial t = 0$.

#### 1.7.2.1 Horizontal flow and vertical mixing
It is assumed that the current is in the OX direction while turbulent mixing takes place in the OZ direction only. For a conservative property such as salinity $S$, $R = 0$ and equation (1.1) reduces to

$$u\frac{\partial S}{\partial x} = \frac{\partial}{\partial z}\left(K_z \frac{\partial S}{\partial z}\right)$$

(1.2)

Now

$$\frac{\partial}{\partial z}\left(K_z \frac{\partial S}{\partial z}\right) = K_z \frac{\partial^2 S}{\partial z^2} + \frac{\partial K_z}{\partial z} \cdot \frac{\partial S}{\partial z} \tag{1.3}$$

If it is assumed that $K_z$ does not vary with $z$, equation (1.2) becomes

$$u \frac{\partial S}{\partial x} = K_z \frac{\partial^2 S}{\partial z^2} \tag{1.4}$$

and an analytical solution can be found, corresponding to given boundary conditions. For example

$$S = S_0 + S_1 e^{-\alpha x} \cos \frac{\pi z}{2l} \tag{1.5}$$

where

$$\alpha = \frac{\pi^2 K_z}{4l^2 u},$$

represents the distribution of the property $S$ in a current extending over a depth $2l$, flowing with velocity $u$, while mixing takes place with the water above and below it. The original value of $S$ in the current is $S_0 + S_1$, whereas the water above and below has the value $S_0$. This solution is shown in Fig. 1.9 for the numerical values $K_z = 4\,\mathrm{cm}^2\,\mathrm{s}^{-1}$, $u = 10\,\mathrm{cm}\,\mathrm{s}^{-1}$ and $l = 200\,\mathrm{m}$.

An example to which this solution applies approximately is the movement of the Antarctic Intermediate Water northwards, at a depth of about 1,000 m after sinking at the Antarctic Convergence at 50°S. This water is characterized by a salinity minimum (i.e. $S_1$ is negative), and its salinity increases slowly as it moves northwards, owing to vertical mixing with higher salinity water both above and below it.

In practice, $K_z$ is usually unknown, and an observed distribution may be used to estimate its value. One cannot assume, in general, that $\partial K_z/\partial z = 0$ but it is seen from equation (1.3) that the second term on the right hand side may also be neglected if $\partial S/\partial z = 0$. This implies that $S$ is a maximum or a minimum at the depth considered, and equation (1.4) may then be applied. Then

$$\frac{K_z}{u} = \frac{\partial S}{\partial x} \bigg/ \frac{\partial^2 S}{\partial z^2} \tag{1.6}$$

and the differential coefficients may be evaluated if the distribution of $S$ in the XOZ plane is known. Methods of applying this approach have been described by Proudman (1953). It should be noted that a knowledge of the distribution alone enables only the ratio $K_z/u$ to be computed. In order to

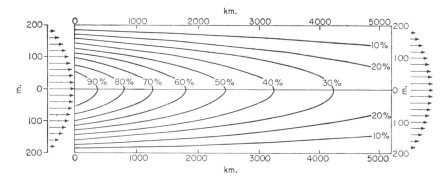

FIG. 1.9. Distribution of a property of sea water produced by horizontal flow and vertical mixing (Defant, 1961). By permission of Pergamon Press Ltd.

determine $K_z$ itself, the current $u$ must be determined independently. For the Antarctic Intermediate Water mentioned above, $K_z/u$ in the South Atlantic has been estimated to average 1 cm. Since $u$ has been estimated by other means to be about 4 cm s$^{-1}$, $K_z = 4$ cm$^2$ s$^{-1}$.

If measurements of the current at various depths are available, as well as the salinity distribution, $K_z$ may be determined as a function of $z$ by integrating equation (1.2) directly. If $\partial S/\partial z = 0$ at some depth $z_0$, then $K_z$ at depth $z$ is given by

$$K_z = \left(\frac{\partial S}{\partial z}\right)^{-1} \int_{z_0}^{z} u \frac{\partial S}{\partial x} \, dz \qquad (1.7)$$

This method was applied by Jacobsen in 1913 to the salinity distribution in the Kattegat, and yielded values of $K_z$ lying between 0·04 and 0·1 cm$^2$ s$^{-1}$ in these highly stable conditions. It has since been applied to many other areas, for both open sea and estuarine conditions.

### 1.7.2.2 Horizontal flow and lateral mixing

If the current is flowing in the OX direction, but mixing is taking place laterally, in the OY direction, the equation becomes

$$u \frac{\partial S}{\partial x} = \frac{\partial}{\partial y}\left(K_y \frac{\partial S}{\partial y}\right) \qquad (1.8)$$

This equation is applicable, for example, to a current such as the Equatorial Counter-current, flowing between two other water masses with which lateral mixing takes place. Mathematically this equation is identical with (1.2), with the $y$ co-ordinate replacing $z$, and analogous solutions may be found.

This method has given values of $K_y$ of the order $3 \times 10^7 \, \text{cm}^2 \, \text{s}^{-1}$ for the Equatorial Counter-current in the Atlantic.

Arons and Stommel (1967) set up an advection–lateral mixing model to interpret the distribution of dissolved oxygen and radiocarbon in the deep waters of the North Atlantic ocean (below 2,000 m). The pattern of flow, based on their earlier papers on the abyssal circulation consisted of a southward flowing western boundary current feeding a weak eastward flow across the whole ocean, from the equator to 56°N. To allow for this advective pattern and retain a decay term the formulation of the problem was more complex than equation (1.8) although similar in principle. The results from the model agree with the observed distribution of dissolved oxygen, which decreases from west to east, if the horizontal eddy diffusivity $K_h$ is taken as 6 to $7 \times 10^6$ $\text{cm}^2 \, \text{s}^{-1}$ and the rate of decay of dissolved oxygen as 2·0 to 2·5 $\times 10^{-3} \, \text{ml} \, \text{l}^{-1}$ $\text{year}^{-1}$. The model is also consistent with the radiocarbon data.

### 1.7.2.3 *Radioactive decay and vertical mixing*

If the concentration of a radioactive substance in the sea is $c$ and its decay constant is $\lambda$, then in equation (1.1) $R = -\lambda c$. If vertical mixing is the only other process affecting the distribution and a steady state is maintained,

$$\frac{\partial}{\partial z}\left(K_z \frac{\partial c}{\partial z}\right) - \lambda c = 0 \qquad (1.9)$$

If $K_z$ is independent of depth

$$K_z \frac{\partial^2 c}{\partial z^2} - \lambda c = 0 \qquad (1.10)$$

If the concentration is greatest at the bottom, where it has the value $c_0$ and decreases to very small values at a great distance from the bottom, the appropriate solution of (1.10) is

$$c = c_0 \, e^{-\alpha z'} \qquad (1.11)$$

where $\alpha = (\lambda/K_z)^{\frac{1}{2}}$ and $z'$ is the distance measured upwards from the bottom. The flux of the radioactive substance through the bottom would then be $(\lambda K_z)^{\frac{1}{2}} c_0$, in the appropriate units. Koczy (1956, 1958) applied equation (1.9) directly to the observed distribution of radium at several oceanic stations and derived estimates of $K_z$ as a function of depth.

### 1.7.2.4 *Vertical flow and vertical mixing.*

If one assumes that advective and diffusive processes in the horizontal direction are negligible compared with those in the vertical, then equation (1.1), taking the vertical eddy diffusion coefficient to be constant with depth and retaining the source (or decay) term, becomes:

$$w\frac{dc}{dz} = K_z\frac{d^2c}{dz^2} + R$$

This equation was used by Munk (1966) when considering the influence of vertical diffusion at depths of 1,000 to 4,000 m in the central Pacific, well away from the surface and bottom boundaries and where the horizontal currents are believed to be weak. He took into account the evidence from temperature and salinity, for which $R = 0$, dissolved oxygen with $R$ a negative constant, and carbon-14 and radium-226, for which $R = -\lambda c$ where $\lambda$ is the appropriate decay constant. Comparing the results from the model with the observed distributions, Munk estimated $w$ to be $1\cdot4 \times 10^{-5}$ cm s$^{-1}$ and $K_z$ to be $1\cdot3$ cm$^2$ s$^{-1}$.

### 1.7.3. TIME-DEPENDENT SOLUTIONS

#### 1.7.3.1. Vertical mixing only.
Under these conditions equation (1.1) reduces to

$$\frac{\partial c}{\partial t} = \frac{\partial}{\partial z}\left(K_z\frac{\partial c}{\partial z}\right) \tag{1.13}$$

If, again, $K_z$ is independent of $z$

$$\frac{\partial c}{\partial t} = K_z\frac{\partial^2 c}{\partial z^2} \tag{1.14}$$

This is essentially the heat conduction equation in one dimension and a number of solutions are possible, depending on the initial and boundary conditions.

In the particular case of a periodic oscillation of period $T$, the solution is

$$c = c_0 + c_1 e^{-\alpha z}\cos\left(\frac{2\pi t}{T} - \alpha z\right) \tag{1.15}$$

where $\alpha = (\pi K_z/T)^{\frac{1}{2}}$.

At $z = 0$, $c = c_0 + c_1 \cos(2\pi t/T)$ and with increasing $z$, the oscillation decreases in amplitude and becomes later in phase. A practical example of this solution is the penetration downwards of a seasonal variation of temperature owing to an alternating gain and loss of heat through the sea surface, or of salinity arising from seasonal changes in precipitation and evaporation. The assumption that $K_z$ is independent of depth is not very realistic but, if the seasonal variation at various depths is known for a particular area, equation (1.13) may be solved numerically to determine $K_z$.

1.7.3.2 *Spreading of a patch by horizontal diffusion.*

A particular case of interest is that of the spreading of a patch of some contaminant under the action of horizontal turbulence. It will be assumed that the turbulence is the same in all directions and that the patch has radial symmetry. If the distance of any point from the centre is denoted by $r$ and $K_x = K_y = K_r$, the diffusion equation may be written

$$\frac{\partial c}{\partial t} = \frac{1}{r} \frac{\partial}{\partial r} \left( K_r \frac{\partial c}{\partial r} \right) \tag{1.16}$$

If $K_r$ is constant, the solution of (1.16) is

$$c(r, t) = \frac{c_0}{4\pi K_r t} e^{-r^2/4K_r t} \tag{1.17}$$

where $c(r, t)$ is the concentration at a distance $r$ from the centre at time $t$.

It was suggested by Sverdrup (1946, unpublished) that $K_r$ should be taken as proportional to $r$, i.e. $K_r = Pr$ where $P$ is a constant which has the dimensions of velocity. Joseph and Sendner (1958) introduced postulates leading to the same form for $K_r$. If this is inserted in equation (1.16) the solution becomes

$$c(r, t) = \frac{c_0}{2\pi P^2 t^2} e^{-r/Pt} \tag{1.18}$$

Joseph and Sendner gave a number of examples of horizontal mixing in the sea which appeared to confirm their theory over a range of $r$ from 10–1,500 km with $P$ having the common value of $1 \pm 0.5$ cm s$^{-1}$. The theory has been used by a number of other investigators for treating problems such as the spread of effluents, including radioactive materials. Other relations for $K_r$ as a function of $r$ based on various hypotheses have been suggested and have led to solutions which differ in the rate of decrease of concentration with time at the centre of the patch and in the decrease of concentration with distance from the centre at a given time. A review of the various theories of horizontal diffusion and of their application to observed distributions was given by Okubo (1962).

Subsequently Okubo (1971b) reviewed the data from a large number of instantaneous releases in various areas, covering time scales ranging from 1 hour to 1 month and length scales from 100 m to 100 km. He found that the rate of increase of $\sigma_{rc}^2$, the variance of the radial distribution of concentration in the patch, with time $t$ could be fitted reasonably well in all cases by the relation

$$\sigma_{rc}^2 = 0.0108\, t^{2.34} \tag{1.19}$$

c

where $\sigma_{rc}^2$ is in cm$^2$ and $t$ in seconds. This relation corresponds to the apparent horizontal eddy diffusivity $K_a$ increasing with the horizontal scale according to the equation

$$K_a = 0.0103 l^{1.15} \tag{1.20}$$

where $K_a$ is in cm$^2$ s$^{-1}$ and $l$, defined by $l = 3\sigma_{rc}$, is in cm.

It is remarkable that the relationships (1.19) and (1.20) should hold as well as they do over such a wide range, but in an individual experiment the results may differ by an order of magnitude from those given by these equations. The discrepancies arise from differing physical parameters such as stability, current speed, wind speed, depth of water and so on. It is also exceptional for the dispersion to be radially symmetrical; in the presence of a current, for example, dispersion usually takes place more rapidly in the direction of the current than transverse to it. Further studies of the rate of diffusion in relation to physical features of the environment, such as those made by Kullenberg (1971) and Meerburg (1972), are needed.

The above results were derived from releases within the surface mixed layer. The few experiments which have been made in the thermocline or in the lower layer indicate that the rate of horizontal mixing is smaller by an order of magnitude.

### 1.7.4. COMBINED EFFECTS OF DIFFUSION AND SHEAR

When dye experiments are carried out in a current, the velocity of which varies with depth, it is observed that not only is the centre of the dye patch advected by the current but the patch becomes elongated in the direction of the axis of flow. The longitudinal dispersion of the dye can be described in terms of an effective diffusion coefficient $K_x$ which is much larger than that due to the turbulent fluctuations alone. This effect is essentially the same as that first described for laminar and later turbulent flow in a pipe (Taylor, 1954). It is due to the interaction of vertical shear in a horizontal current with vertical mixing and was termed the "shear effect" by Bowles et al. (1958) in their study of dispersion in a tidal current in the English Channel.

The basic processes producing the shear effect may be illustrated by a simple model due to Okubo and Carter (1966) shown in Figure 1.10. The water is assumed to be flowing in a shallow basin of uniform depth $H$, with the horizontal velocity $U$ decreasing linearly with depth. The contaminant substance is introduced as a vertical column $AB$, as in Fig. 1.10(a). For simplicity the shearing and diffusing effects are assumed to take place alternately in discrete steps. In the first step the shear, acting for a time $\tau$, distorts the column into the oblique form $AC$ (Fig. 1.10b). In the second step the shear ceases for an instant while vertical mixing immediately distributes

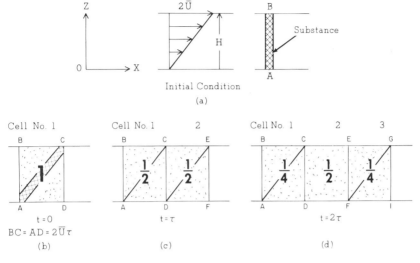

FIG. 1.10. Longitudinal dispersion of a substance by the "shear effect". (Okubo and Carter, 1966). By permission of the American Geophysical Union.

the substance uniformly throughout the cell $ABCD$, of length $2\overline{U}\tau$ where $\overline{U}$ is the depth-mean velocity. During the next interval $\tau$ from the new origin, the rectangular patch $ABCD$ is distorted to the parallelogram $ACED$ (Fig. 1.10c). The subsequent mixing disperses the substance uniformly throughout two cells, as shown. Repetition of the mixing steps thus adds a new cell of length $2\overline{U}\tau$ to the patch for each time interval $\tau$.

It can be shown that, after a time $t = n\tau$, the distribution of substance over the cells is represented by a binomial distribution with a variance $\overline{U}^2\tau t$. The time interval $\tau$ is a measure of the time required for vertical diffusion to produce substantial homogeneity over the depth H, and is related to the coefficient of vertical eddy diffusion $K_z$ by

$$\tau = H^2/\pi^2 K_z \qquad (1.21)$$

It follows that for large times the longitudinal distribution of contaminant approaches a Gaussian form with an effective longitudinal diffusivity $K_x$ given by

$$K_x = \overline{U}^2 H^2/20 K_z \qquad (1.22)$$

A more rigorous treatment shows that the numerical factor 20 in the denominator should be replaced by 30 (Saffmann, 1962).

In general, the velocity gradient need not be linear, as in Fig. 1.10, and the vertical diffusion coefficient $K_z$ may vary with depth. A derivation applicable to various velocity profiles was given by Bowden (1965), who also showed that an alternating flow, such as tidal current, could give rise to an effective longitudinal dispersion. It is characteristic of the shear effect that when the spreading is bounded in the vertical direction the effective longitudinal diffusivity $K_x$ is *inversely* proportional to the vertical diffusivity $K_z$. Thus, a stable stratification which tends to reduce $K_z$ will increase the effective $K_x$ and this effect is important in considering longitudinal dispersion in estuaries where $K_x$ may be of the order of $10^6$ cm$^2$ s$^{-1}$. As $K_z$ decreases, however, equation (1.21) shows that $\tau$ becomes larger, so that the time taken for a contaminant released at a certain depth to become mixed throughout the total depth $H$ may become considerable. The solution for the shear effect when vertical diffusion does not extend to the surface or bottom boundaries shows that the effective $K_x$ is then *directly* proportional to $K_z$ (Saffmann, 1962).

A more comprehensive treatment of the effect in an oscillatory current relative to that in a steady current was given by Okubo (1967), who considered both the bounded and unbounded cases. It is clear that horizontal variations of velocity, transverse to the axis of flow, can give rise to similar effects on the longitudinal dispersion. The contribution of the various shear effects to the total longitudinal dispersion in a tidal estuary has been considered by Holley *et al.* (1970) and Fischer (1972).

A shear–diffusion model of a more general kind was introduced by Carter and Okubo (1965) on the grounds of a hypothetical spectrum of turbulence which could be separated into two parts–eddies of scales large compared with that of the diffusing patch considered and those of much smaller scale. The large-scale eddies provided a non-uniform field of mean flow, as far as the patch was concerned, while the effect of the small-scale eddies was represented by eddy coefficients of diffusion. Solutions of the shear–diffusion equation corresponding to this model were derived by Carter and Okubo and were applied to the analysis of a number of dye release experiments in the Cape Kennedy area with satisfactory results. Further implications of the shear–diffusion model were considered in a later paper by Okubo (1968) in which he suggested that the large-scale horizontal mixing of water masses may be the result of velocity shears associated with major current systems, such as the Kuroshio and Oyashio in the North Pacific. Similar processes may also contribute, on a much smaller scale, to the formation of some features of oceanic microstructure.

## 1.8. OVERALL MIXING PROBLEMS

The thermocline has been described above as a transition layer in which the temperature gradient is large, the stability high and the coefficient of vertical eddy diffusion very small. Vertical mixing can take place more readily either below or above the thermocline than within it, so that it acts as a partial barrier between the water masses above and below. This fact led to the introduction of box type models of the ocean when considering the overall distribution of a constituent, such as carbon dioxide. In the simplest model, the world ocean is represented by two boxes, or reservoirs, the first representing the surface layers to a depth of 100 or 200 m and the second the deeper layers from this depth to the bottom, with the thermocline separating them. Complete mixing within each box is assumed to take place very rapidly, but exchange between the two boxes takes place comparatively slowly. The concentration of a given property within a box is given by a single value–the mean concentration for the box–and the exchange between two boxes is specified using exchange-rate or transfer-rate constants. The rate of consumption or decay of a non-conservative constituent within a box is represented by a suitable parameter. A box model may be regarded as a finite difference analogue of the advection–diffusion equation (1.1), which was expressed for continuously variable functions.

The two-box model of the ocean with a third box representing the atmosphere was used by Craig (1957) and Arnold and Anderson (1957) in a study of the natural distribution of radiocarbon, making some allowance also for the biosphere and humus. They were able to make estimates of the rates of exchange, and hence the average residence time of a molecule in each box, by inserting data on the observed concentrations of $^{14}C$. Atmospheric carbon dioxide, which contains a known concentration of $^{14}C$, enters the surface layer through the air-sea interface and becomes evenly distributed throughout the layer. Passage downwards into the deep layer takes place more slowly and there is a corresponding passage upwards from the deep layer. The molecules escaping upwards, however, will have been, on the average, in the deep layer for a considerable number of years and their $^{14}C$ content will have been reduced appreciably by radioactivity decay. The $^{14}C/^{12}C$ ratio is, therefore, lower in the deep water than in the atmosphere. In the surface layer the $^{14}C/^{12}C$ ratio is intermediate between the two, because a fraction of the carbon atoms have come from the lower layer and the others directly from the atmosphere. These differences are measurable, and from them the exchange rates and residence times may be calculated.

The main discrepancy between the simple two-box model and the real ocean is that in the Arctic and Antarctic regions the thermocline is absent and the deep water extends upwards to the free surface. In these areas, where deep and bottom water are being formed, convection currents extend from the surface to the bottom. This situation has been allowed for in an outcrop model (Craig, 1958) in which the deep box extends to the surface over 25 per cent of its area, representing the polar regions. An alternative representation is a three-box model, in which the polar regions are represented by a separate box. A more complex model includes separate boxes for the Arctic and Antarctic regions and for the Atlantic and Pacific + Indian Oceans. Each additional box involves at least one additional interface with an unknown exchange rate and so the more complex the model, the more detailed are the observational data required in order to make use of it. In this connection it is an advantage to be able to insert data on several chemical tracers into the same model and radium-226 has been used for this purpose. A survey of such models was given by Broecker (1963), who summarised the radiocarbon data available at that time as indicating that the average residence time of a molecule in the deep layer was at least 500 years in the Atlantic and 800 years in the Pacific. The estimated time in the surface layer was more sensitive to the assumptions made in the different models, but was likely to be in the range 10 to 100 years for dissolved solids. The time for carbon dioxide would be less by a factor of 2 or 3 because of the exchange with the atmosphere.

It is possible to include advective processes in a box model by allowing for the mass transport of water from one box to another, as well as the transfer of a property by diffusive processes. Bolin and Stommel (1961) developed a box model to study the origin and rate of circulation of the deep ocean water. They divided the ocean into four boxes containing (1) North Atlantic Deep Water, (2) Antarctic Bottom Water, (3) Pacific and Indian Intermediate Water and (4) Common Water. The name "Common Water" had been given by Montgomery (1958) to the deep water of the Indian and Pacific Oceans, which is the most voluminous water mass in the world ocean and is believed to be a mixture of Atlantic deep water and Antarctic bottom water. Bolin and Stommel formulated equations representing the conservation of the fluxes of water, salt, heat content and radiocarbon between the four boxes. Allowance was made for the geothermal heat flux through the sea bed and for the radioactive decay of $^{14}C$. They concluded from the calculations that the main source of the Common Water is the Antarctic Bottom Water which flows into box (4) at a rate of $10 \times 10^6 \, m^3 \, s^{-1}$. The North Atlantic Deep Water contributed $4 \times 10^6 \, m^3 \, s^{-1}$ and the Intermediate Water only $3 \times 10^6 \, m^3 \, s^{-1}$. The residence time in the Common Water was estimated to be 1200 years. Using another model Bolin and Stommel estimated the rate of transport of Antarctic Intermediate Water northwards at latitudes 40 to 50° S in the Atlantic ocean to be between 1·5 and $5·0 \times 10^6 \, m^3 \, s^{-1}$ and the residence time in the Intermediate Water to be between 100 and 400 years.

A comprehensive treatment of the theoretical foundation of box models was given in a paper by Keeling and Bolin (1967), who pointed out the advantage of using data for a number of chemical tracers in a set of simultaneous equations. In a second paper (Keeling and Bolin, 1968), they gave a detailed example of the method, with one atmospheric and three oceanic reservoirs and using eight chemical tracers: salinity, oxygen, inorganic phosphorus, organic and inorganic carbon, carbonate alkalinity, insoluble carbonate and radiocarbon. Horizontal and vertical advective and eddy diffusive transports were taken into account, as well as gravitational settling of insoluble chemicals in the ocean. The residence time in the deep water was deduced to be about 1,100 years, in good agreement with other estimates, and a number of deductions were made about the various constituents.

According to present ideas on the general circulation of the oceans, the renewal of the deep water by sinking from the surface layer takes place in a small number of fairly well defined areas in polar regions. To preserve continuity, there is a slow rise of deep water over the rest of the ocean. Since the main thermocline remains at the same average depth from year to year, the uprising of the deep water must take place through it. The temperature structure in the thermocline itself is maintained by a balance between the downward flux of heat by eddy conduction and an upward advective flux by the rising water. Theoretical models of the thermocline based on reasoning of this kind have been proposed by a number of authors including Robinson and Stommel (1959), Welander (1959) and Overstreet and Rattray (1969). The upward velocity required is of the order of $10^{-4}$ to $10^{-5}$ cm s$^{-1}$ which is below the threshold of direct measurement at the present time.

REFERENCES

Arnold, J. R. and Anderson, E. C. (1957). *Tellus,* **9**, 28–32.
Arons, A. B. and Stommel, H. (1967). *Deep-Sea Res.* **14**, 441–458.
Bolin, B. and Stommel, H. (1961). *Deep-Sea Res.* **8**, 95–110.
Bowden, K. F. (1965). *J. Fluid Mech.* **21**, 83–95.
Bowden, K. F. (1967). *In* "Estuaries" (G. H. Lauff, ed.) pp. 15–36. American Association for the Advancement of Science, Washington D.C.
Bowden, K. F. (1970). *Oceanogr. Mar. Biol. Ann. Rev.* **8**, 11–32.
Bowles, P., Burns, R. H., Hudswell, F. and Whipple, R. T. P. (1958). *Proc. 2nd U.N. Internat. Conf. Peaceful Uses Atomic Energy,* **18**, 376–389.
Broecker, W. (1963). *In* "The Sea", (M. N. Hill, ed.) Vol. 2, pp. 88–108. Interscience Publishers, New York, London.
Carter, H. H., and Okubo, A. (1965). Chesapeake Bay Institute, Johns Hopkins University, Report Ref. 65-2, 150 pp.
Cameron, W. M. and Pritchard, D. W. (1963). *In* "The Sea" (M. N. Hill, ed.) Vol. 2 pp. 306–324. Interscience Publishers, London, New York.
Craig, H. (1957). *Tellus,* **9**, 1–17.
Craig, H. (1958). *Proc. 2nd U.N. Internat. Conf. Peaceful Uses Atomic Energy,* **18**, 358–363.
Defant, A. (1941). *"Meteor" Rep.* **6**, (2), 5 Lief.

Defant, A. (1954). *Deut. Hydrog. Z.* **7**, 2–14.
Defant, A. (1961). "Physical Oceanography", Vol. 1. Pergamon Press, Oxford, London, New York, Paris.
Dyer, K. R. (1973). "Estuaries: a Physical Introduction". John Wiley & Sons, London, New York, Sydney, Toronto, 140 pp.
Fischer, H. B. (1972). *J. Fluid Mech.* **53**, 671–687.
Gill, A. E. (1971). *Phil. Trans. R. Soc. Lond.* A **270**, 391–413.
Grant, H. L., Moilliet, A. and Vogel, W. M. (1968). *J. Fluid Mech.* **34**, 443–448.
Gregg, M. C. and Cox, C. S. (1971). *Deep-Sea Res.* **18**, 925–934.
Holley, E. R., Harleman, D. R. F. and Fischer, H. B. (1970). *J. Hydraul. Div., Proc. ASCE,* **96**, 1691–1709.
Jones, P. G. W. (1972). *Deep-Sea Res.* **19**, 405–431.
Joseph, J. and Sendner, H. (1958). *Deut. Hydrog. Z.* **11**, 49–77.
Keeling, G. D. and Bolin, B. (1967). *Tellus,* **19**, 566–581.
Keeling, G. D. and Bolin, B. (1968). *Tellus,* **20**, 17–54.
Knauss, J. A. (1960). *Deep-Sea Res.* **6**, 265–286.
Knauss, J. A. (1966). *J. Mar. Res.* **24**, 205–240.
Knauss, J. A. (1969). *Deep-Sea Res., Suppl. to Vol. 16,* 117–123.
Koczy, F. F. (1956). *Nature, Lond.* **178**, 585–586.
Koczy, F. F. (1958). *Proc. 2nd U.N. Internat. Conf. Peaceful Uses Atomic Energy,* **18**, 336–343.
Kullenberg, G. (1971). Københavns Universitet, *Rep. Inst. Fys. Oceanogr.* **12.**
Lacombe, H. (1961). *Cahiers Océanogr.* **13**, 73–107.
Meerburg, A. J. (1972). *Netherlands J. Sea Pes.* **5**, 492–509.
Montgomery, R. B. (1958). *Deep-Sea Res.* **5**, 134–148.
Munk, W. H. (1950). *J. Met.* **7**, 79–93.
Munk, W. H. and Carrier, G. F. (1950). *Tellus,* **2**, 158–167.
Munk, W. H. (1966). *Deep-Sea Res.* **13**, 707–730.
Neuman, G. (1968). "Ocean Currents". Elsevier Publishing Company, Amsterdam, London, New York, 352 pp.
Neumann, G. and Pierson, W. J. (1966). "Principles of Physical Oceanography", Prentice-Hall Inc., Englewood Cliffs, N.J., 545 pp.
Okubo, A. (1962). *J. Oceanogr. Soc. Jap. 20th Anniv. vol.,* pp. 286–320.
Okubo, A. (1967). *Int. J. Oceanol. Limnol.* **1**, 194–204.
Okubo, A. (1968). *J. Oceanogr. Soc. Jap.* **24**, 60–69.
Okubo, A. (1971a). *In* "Impingement of Man on the Oceans" (D. W. Hood ed.) pp. 89–168. John Wiley & Sons, Inc., New York.
Okubo, A. (1971b). *Deep-Sea Res.* **18**, 789–802.
Okubo, A. and Carter, H. H. (1966). *J. Geophys. Res.* **71**, 5267–5270.
Overstreet, R. and Rattray, M. (1969). *J. Mar. Res.* **27**, 172–190.
Pickard, G. L. (1961). *J. Fish. Res. Bd. Can.* **18**, 907–999.
Pickard, G. L. (1971). *J. Fish. Res. Bd. Can.* **28**, 1077–1106.
Pritchard, D. W. (1955). *Proc. Amer. Soc. Civil Eng.* **81**, No. 717.
Proudman, J. (1953). "Dynamical Oceanography". Methuen, London; John Wiley, New York, 409 pp.
Reid, J. L. (1969). *Nature, Lond.* **221**, 848.
Richardson, W. S., Stimson, P. B. and Wilkins, C. H. (1963). *Deep-Sea Res.* **10**, 369–388.
Robinson, A. and Stommel, H. (1959). *Tellus,* **11**, 295–308.

Saelen, O. H. (1967). *In* "Estuaries" (G. H. Lauff, ed.) pp. 63–70. American Association for the Advancement of Science, Washington, D.C.

Saffmann, P. G. (1962). *Quart. J. R. Meteorol. Soc.* **88**, 382–393.

Smith, R. L. (1968). *Oceanogr. Mar. Biol. Ann. Rev.* **6**, 11–46.

Stommel, H. (1948). *Trans. Amer. Geophys. Un.* **29**, 202–206.

Stommel, H. (1949). *J. Mar. Res.* **8**, 199–225.

Stommel, H. (1957). *Deep-Sea Res.* **4**, 149–184.

Stommel, H. (1958). *Deep-Sea Res.* **5**, 80–82.

Stommel, H. (1965). "The Gulf Stream" 2nd Edn. University of California Press, Berkeley and Los Angeles; Cambridge University Press, London, 248 pp.

Stommel, H. and Yoshida, K. (eds). (1972). "Kuroshio, its Physical Aspects". University Press, Tokyo, 517 pp.

Sverdrup, H. U. (1945). "Oceanography for Meteorologists". George Allen and Unwin, London, 235 pp.

Sverdrup, H. U., Johnson, M. W. and Fleming, R. H. (1942). "The Oceans, their Physics, Chemistry and General Biology", Prentice Hall, New York, 1087 pp.

Swallow, J. C. (1955). *Deep-Sea Res.* **3**, 74–81.

Swallow, J. C. and Worthington, L. V. (1957). *Nature, Lond.* **179**, 1183–1184.

Swallow, J. C. and Worthington, L. V. (1961). *Deep-Sea Res.* **8**, 1–19.

Tait, R. I. and Howe, M. R. (1968). *Deep-Sea Res.* **15**, 275–280.

Taylor, G. I. (1954). *Proc. Roy. Soc.* A **223**, 446–468.

Von Arx, W. S. (1962). "Introduction to Physical Oceanography". Addison-Wesley, Reading (Mass.), London, 422 pp.

Warren, B. A. (1970). *In* "Scientific exploration of the South Pacific" (W. S. Wooster, ed.). National Academy of Sciences, Washington, D. C.

Welander, P. (1959). *Tellus*, **11**, 309–318.

Woods, J. D. (1968). *J. Fluid Mech.* **32**, 791–800.

Woods, J. D. and Wiley, R. L. (1972). *Deep-Sea Res.* **19**, 87–121.

Wüst, G. (1929). *Veroff. Inst. Meeresk. Univ. Berlin*, A **20**, 1–63.

# Chapter 2

# Sea Water as an Electrolyte Solution

## M. WHITFIELD

*The Laboratory, Citadel Hill, Plymouth, Devon, England*

## 2.1. INTRODUCTION

Sea water is an electrolyte solution. Its chemistry is dominated by the presence of six ions ($Na^+$, $K^+$, $Mg^{2+}$, $Ca^{2+}$, $Cl^-$, $SO_4^{2-}$) which constitute more than $99.5\%$ of the dissolved constituents. These electrolytes have a significant influence on the myriad of interlocking physical, chemical, biological and geological processes that control the chemistry of the oceans. It is precisely this influence that differentiates marine chemistry from the chemistry of other electrolyte solutions. Despite their fundamental importance the major components have played a strangely anonymous role in the development of marine chemistry since the turn of the century (Whitfield, 1972) because the forces controlling their behaviour have been inadequately understood.

Initially, sea water was considered to be a solution of neutral salt molecules which could be completely described in terms of the concentrations of the component salts. Following the classic work of Dittmar (1884), the establishment of the salinity concept (Forch et al., 1902) enabled the total concentration of the dissolved salts in sea water to be defined quite simply on the basis of a single analytical measurement (the chlorinity). For a short time this definition appeared adequate to specify the rôle of the major electrolyte component.

However, in the first quarter of this century it became apparent that solutions of salts contained not neutral salt molecules but charged ions. The significant interactions resulting from the presence of these charges have a marked effect on the solution properties so that the gross stoichiometric composition cannot provide an adequate basis for interpreting the behaviour of the solution. Over the past fifty years there has been a steadily growing quantitative understanding of the forces at work in electrolyte solutions. The ideas have developed slowly to encompass electrolyte mixtures and, in the past decade, have reached the stage where the properties of mixtures as complex as sea water may be predicted. The question can now be posed, how will these developments affect the progress of marine chemistry?

In the past, it has been the general practice to specify sea water in terms of its salinity and then to investigate the properties of particular reactions in this solvent. This pragmatic and simple approach has provided what is essentially a catalogue of reactions in the marine environment at particular salinities, temperatures and pressures. The experimental procedures have not always been consistent so that a confusing and fragmentary picture of the chemical processes frequently emerges. The carbon dioxide system is a classic example of such confusion (Hansson, 1973; Edmond, 1972). Sillén (1967) suggested a rigorous and self consistent experimental approach (the ionic medium method) that should help to remove some of these anomalies. However, since the effects of the major electrolyte components (i.e. the medium effects) are estimated empirically, sea waters with different salinities must be treated as separate solvents. In coastal or estuarine waters where the salinity concept might itself break down, new investigations are required for each electrolyte composition. Therefore, although the ionic medium method may provide a simple solution for a specific problem (e.g. the speciation of the carbon dioxide system in the open ocean, Hansson, 1973) the information it provides is descriptive rather than explanatory. By treating *each* sea water as a unique solvent, the barriers isolating marine chemistry from the broad scope of solution chemistry are perpetuated. Consequently, it is seldom possible to draw on the great wealth of data on chemical reactions in other electrolyte solutions and we are forced to undertake the painstaking job of repeating all the experiments in each new solvent.

To supplement and substantiate the empirical approach a more fundamental treatment of marine chemistry is required. In this basic approach sea water must be considered as an electrolyte solution rather than as an ionic medium and the interactions between the solution components must be examined in detail. The procedures will necessarily be more complex than those used in the ionic medium method, but they will provide the basic understanding required to explain and organize the accumulating mass of empirical observations.

The intention of this chapter is to show how far we have progressed in the development of such a general quantitative approach to marine chemistry. The aim throughout is to shed light on the chemical fine structure of sea water and to treat it as a mixture of individual ionic components rather than as a solution of a composite sea-salt. For this reason this Chapter follows a sequence in which the complexity of sea water is reached in easy stages from pure water. This procedure, which is necessary for a clear appreciation of the strengths and weaknesses of the various quantitative models employed, restricts the treatment of sea water itself to the last two sections. Although this pattern of development may be less direct than the title of the chapter implies, it seems inevitable if the aim is to promote understanding rather than to provide a catalogue of data.

## 2.2. The Activity Coefficient

### 2.2.1. DEFINITION

The consequences of interactions between the various components in an electrolyte solution can be most simply described in terms of equilibrium thermodynamics which is entirely concerned with the energetics of chemical reactions, with the initial and final states, and slides irreverently over problems of reaction mechanism and reaction rate (Lee, 1959). With such complexities removed it is possible to define quite simply the driving force behind a chemical reaction.

When dealing with the chemistry of sea water we will be mainly concerned with systems of defined pressure $(P)$ and temperature $(T)$. Under these circumstances, the fundamental thermodynamic function is the *Gibbs free energy* $(G)$. At constant temperature and pressure, systems will move spontaneously towards an equilibrium configuration where $G$ is at a minimum. Free energy changes associated with small fluctuations about the equilibrium position will be zero (i.e. $dG = 0$). The free energy change $(\Delta G)$ accompanying a chemical reaction is a composite quantity and takes account of both the enthalpy change $(\Delta H)$ and the entropy change $(\Delta S)$,

$$\Delta G = \Delta H - T \Delta S \qquad (2.1)$$

$\Delta H$ is a measure of the heat change involved in the reaction and depends on the relative strength of interactions between the various components of the system. $\Delta S$ is a measure of the degree of randomness introduced by the rearrangement of the molecules during the reaction. It is, therefore, intimately associated with the relative abilities of the reactants and products to promote organized structures within the solution at constant enthalpy. Spontaneous processes tend to proceed with an increase in entropy (i.e. to move towards a more disordered state).

A fourth parameter, fundamental to the description of chemical systems, is the volume change $(\Delta V)$ resulting from alterations in the spatial organization of the molecules. The parameters $G$, $H$, $S$, $V$ form the basis of the complete energetic description of a chemical system whose environment can be completely defined in terms of temperature $(T)$, pressure $(P)$ and composition $(n_i)$. The temperature and pressure coefficients of these parameters are listed in Table 2.1.

The thermodynamic functions introduced so far are *extensive* variables—dependent on the mass of material present. If the fundamental parameters $G, H, S$ and $V$ are differentiated with respect to solution composition (Table 2.1) partial molal quantities $(\overline{Y}_i)$ are obtained which are independent of the mass of material present. These *intensive* variables are particularly convenient for

TABLE 2.1

*Pressure, temperature and composition coefficients of extensive thermodynamic functions*

| Y | (1) $(\partial Y/\partial T)_{P,\,n_i}$ | (2) $(\partial Y/\partial P)_{T,\,n_i}$ | (3) $(\partial Y/\partial n_i)^a_{P,\,T,\,n_j}$ |
|---|---|---|---|
| $G$ | $-S$ | $V$ | $G_i$ (or $\mu_i$)[b] |
| $G/T$ | $H/T^2$ | $V/T$ | $\mu_i/T$ |
| $H$ | $C_P^c$ | $V - ET$ | $\bar{H}_i$ |
| $S$ | $-C_P/T$ | $-E^d$ | $\bar{S}_i$ |
| $V$ | $E$ | $-\kappa^e$ | $\bar{V}_i$ |

[a] Partial molal quantities $(\bar{Y}_i)$  [b] Chemical potential
[c] Specific heat at constant pressure
[d] Expansibility; coefficient of thermal expansion $= \bar{\alpha} = E/V$
[e] Compressibility; coefficient of isothermal compressibility $= \bar{\beta} = \kappa/V$
*Note.*

$$(\partial^2 G/\partial T^2)_{P,\,n_i} = -(\partial S/\partial T)_{P,\,n_i} = +C_p/T$$

$$(\partial^2 G/\partial P^2)_{T,\,n_i} = (\partial V/\partial P)_{T,\,n_i} = -\kappa$$

The mathematical properties of functions of several variables have been treated in detailed by Klotz (1964), Lewis and Randall (1961, pp. 23–27) and Guggenheim (1956, pp. 83–95).

use in the description of chemical systems (Klotz, 1964, pp. 243–249).

In an electrolyte solution the intensive thermodynamic function, $\bar{Y}_i$, is related to the corresponding extensive thermodynamic function, $Y$, by the equation

$$Y = m_w\bar{Y}_w + \sum_i v_i m_i \bar{Y}_i \qquad (2.2)$$

$m$ is the molality of the subscripted component and $v$ is the stoichiometric number of the salt under consideration. The subscripts $w$ and $i$ refer to the properties of the water and those of the ionic solutes respectively. The greatest centres around the partial molal free energy or chemical potential ($\mu_i = (\partial G/\partial n_i)_{T,\,P,\,n_j}$). In which $n_j$ implies the constancy of concentration of all components other than $i$ during the addition of $\partial n$ moles of $i$.

For ideal systems in which the solute and solvent do not interact but mix to form a simple physical *mélange*, the chemical potential of the component $i$ may be related to its concentration ($m_i$) by the equation

$$\mu_i - \mu_i^\ominus = RT \ln (m_i/m_i^\ominus) \qquad (2.3)$$

where the superscript $\ominus$ refers to properties in some as yet unspecified standard state.

In real systems, particularly in electrolyte solutions, strong interactions may exist so that the simple relationship (equation 2.3) will no longer be valid. Under these circumstances it is convenient to define a correction factor known as the *activity coefficient* ($\gamma_i$) which enables the simple form of equation (2.3) to be retained.

$$\mu_i - \mu_i^{\ominus} = RT \ln(m_i/m_i^{\ominus}) + RT \ln(\gamma_i/\gamma_i^{\ominus})$$
$$= RT \ln(a_i/a_i^{\ominus}) \qquad (2.4)$$

where $a_i$ ($= \gamma_i m_i$) is the activity of component $i$ and has the dimensions of concentration. $\gamma_i$ is dimensionless.

### 2.2.2. STANDARD STATES AND REFERENCE STATES

At a given temperature, pressure and solution composition the value of $\mu_i$ will be fixed, but the individual values of $\mu_i^{\ominus}$ and $\gamma_i$ are variable and may be defined at the convenience of the experimenter. This flexibility is of considerable help to the thermodynamicist (see for example Skirrow, 1965) but, because of the pivotal importance of $\gamma_i$, it is likely to cause confusion amongst those who use the information unless the definitions are clearly presented.

The first step in the definition of the *standard state* is to set $m_i^{\ominus} = \gamma_i^{\ominus} = a_i^{\ominus} = 1$. This simplifies the form of equation 2.4 but we must remember the existence of these terms in the denominator as they affect the dimensions of the equation. We can now write

$$\mu_i = \mu_i^{\ominus} + RT \ln m_i \gamma_i = \mu_i^{\ominus} + RT \ln a_i \qquad (2.5)$$

Three matters must be settled before a unique value can be ascribed to $\gamma_i$.

    (i) a *standard state* must be defined where the solution behaves ideally (according to equation 2.3) so that a value can be assigned to $\mu_i^{\ominus}$,

    (ii) a concentration scale must be selected to fix the value of $m_i$, and

    (iii) a physically accessible *reference state* must be located where the real solution approaches ideal behaviour. This will provide an experimental extrapolation limit for the determination of standard state properties.

Clearly, $\mu_i = \mu_i^{\ominus}$ when $a_i = 1$; therefore, the simplest way of defining the standard state is to choose the solution such that $\gamma_i = 1$ when $m_i = 1$. Since $\gamma_i$ is used to correct deviations from ideal behaviour we have defined a *hypothetical* ideal solution of unit concentration.

Three concentration scales are in common use. The molar scale ($c_i$, moles per unit volume of solution at a specified temperature and pressure) is widely used because of the analytical convenience of volumetric techniques. The molal scale ($m_i$, moles per kilogram of solvent) gives concentrations that are independent of temperature and pressure and is used in precise physicochemical calculations. The mole fraction scale ($x_i = n_i/(n_i + \sum_j n_j)$), i.e. moles of solute per mole of solution) is fundamentally the most suitable scale

for expressing deviations from ideality, but it is inconvenient to use experimentally. Although activity coefficients defined on the various scales are
nearly equal in dilute solutions (less than 0·1 M), they may deviate widely at
higher concentrations.

Let us look first at the selection of a concentration scale for water. The
obvious choice is the mole fraction scale since the hypothetical ideal state
of unit concentration ($x_w = 1$) would coincide with pure water. The molal
scale would lead to the unlikely situation in which the standard state was
equivalent to the ideal solution of one mole of water in 1000 grams of water!
However, the use of the mole fraction scale for the solute leads to the equally
indigestible concept of a phase, consisting entirely of the solute, with the
properties of an ideal solution. Consequently the activity coefficients of the
solutes in electrolyte solutions are normally defined on the molal scale
(Table 2.2).

Reference states are physically accessible states where the components of
the real system closely approach ideal behaviour as defined by equation (2.3)
(see Fig. 2.1). Since deviations from ideal behaviour are caused by interactions
between the solute and solvent and between the solutes themselves, it is

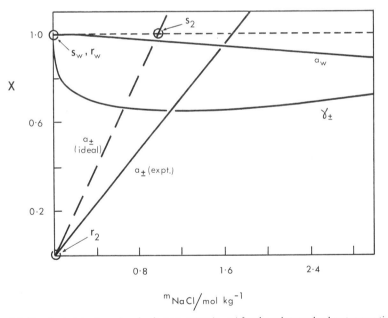

FIG. 2.1. Standard states ($s_2$, $s_w$) and reference states ($r_2$, $r_w$) for the solute and solvent respectively
in sodium chloride solutions at 298 K and 1 atm pressure. The function, $X$, plotted is identified
on the individual curves. Data taken from Robinson and Stokes (1965, pp. 476, 492).

logical to choose the infinitely dilute solution as the reference state. Since $m_i$ tends to zero in the reference state $\mu_i$ will tend to minus infinity (equation (2.5))—this explains the need for a *hypothetical* ideal state for the definition of $\mu_i^\ominus$. If we are looking at the activity coefficient of a solute in a multi-component solution then we can regard either the water or the solution made up of the other components as the solvent (Table 2.2). The activity coefficients used in this chapter are based on water as the solvent in order that sea water may be treated on the same basis as other electrolyte solutions.

Standard and reference states are normally defined at the temperature of the solution (Anderson, 1970, pp. 676–677), but there is no common convention for fixing the pressure. The choice depends on the circumstances and clear discussions of the merits of fixed (Rock, 1967) and variable (Anderson, 1970) pressure standard states have been presented. A more specialized discussion bearing on the selection of the standard pressure is given by Prausnitz (1960). Pytkowicz (1968) has described alternative definitions of fixed pressure standard states in oceanographic systems.

Since, according to Anderson (1970), 'there are no hard and fast rules, only good and bad judgement', the main point is to be aware of the difference between the two states. If we represent the standard chemical potentials of the fixed ($\mu_i^\ominus$) and variable ($\mu_i^{\ominus\prime}$) pressure standard states schematically (Fig. 2.2), then the relationship between the corresponding activity coefficients ($\gamma_i$ and $\gamma_i'$) is clear. It is important to note that neither $\mu_i$ nor $d\mu_i/dP$ are dependent on the choice of standard states.

One further point which affects the definition of $\gamma_i$ is the identification of the component $i$. When non-electrolytes are being considered there is no problem since the substance that is dissolved retains its identity in solution. However, when an electrolyte such as sodium chloride is dissolved, the solute is present not as neutral NaCl molecules but as $Na^+$ and $Cl^-$ ions. From the mechanistic point of view these ions are the dissolved components and we should write the equations in terms of single-ion activities. However, thermodynamics is not concerned with mechanisms. The solutions can *only* be prepared by the addition of the neutral salt. The ions cannot be added independently of each other, and so the only *measurable* bulk properties of the solution are those associated with the neutral solute. $\gamma_i$ is, therefore, defined in terms of the *mean-ion* activity ($a_\pm$) of sodium chloride. This may be related in the general case to the component single-ion activities by

$$a_\pm = (a_+^{\nu_+} a_-^{\nu_-})^{1/\nu} = a_2^{1/\nu} \qquad (2.6)$$

where the subscripts $+$ and $-$ refer to the cation and anion respectively, $\nu$ represented the stoichiometric number ($\nu = \nu_+ + \nu_-$) and $a_2$ is the experimentally determined solute activity. Similar equations can be used to

TABLE 2.2

*Definition of standard states and reference states in aqueous solution*

| Substance | Concentration scale | Activity coefficient | Standard[a] state | Reference[a] state |
|---|---|---|---|---|
| 1. Water | Mole fraction | $f_w$ (Rational) | Ideal pure liquid $a_w = a_w^\ominus = 1$ $f_w = f_w^\ominus = 1$ $\mu_w = \mu_w^\ominus$ | As for standard state |
| 2. Solute[b] (a) Single involatile solute | Molal | $\gamma_2$ (Practical) | Hypothetical ideal solution with $a_2 = a_2^\ominus = 1$ $\gamma_2 = \gamma_2^\ominus = 1$ $\mu_2 = \mu_2^\ominus$ when $m_2 = 1$ | Infinitely dilute solution where $\gamma_2 \to 1$ as $m_2 \to 0$ |
| (b) Single volatile solute | Expressed as partial pressure | $f_2$ (fugacity) | Hypothetical ideal gas at 1 atm. $a_2 = a_2^\ominus = 1$ $f_2 = f_2^\ominus = 1,$ $\mu_2 = \mu_2^\ominus$ when $p_2 = 1$ | Extremely low pressure where $f_2 \to 1$ as $p_2 \to 0$ |
| (c) Several solutes present | Molal | $\gamma_i$ | As for (a) | (i) *Infinite dilution scale* Infinitely dilute solution of *all* solutes i.e. $\gamma_2 \to 1$ as $(m_2 + \sum_i v_i m_i) \to 0$ (ii) *Ionic medium scale* Infinitely dilute solution of component 2 in the solvent comprising the remaining solutes. i.e. $\gamma_2 \to 1$ as $m_2 \to 0$ in a solution where the total concentration is still $\sum_i v_i m_i$ |

[a] At the same temperature and pressure as the solution
[b] Neutral electrolyte, ion or non-electrolyte
General references—Klotz (1964, pp. 346, 388–397), Lee (1969, pp. 215–217)
Lewis and Randall (1961, pp. 244–248)

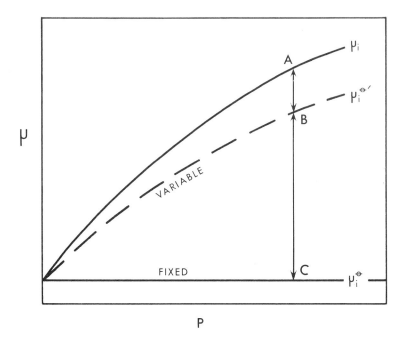

FIG. 2.2. The relationship between activities defined according to variable pressure $(a_i')$ and fixed pressure $(a_i)$ standard states. For the variable pressure standard state (span AB)

$$RT \ln a_i' = RT \ln m_i \gamma_i' = \mu_i - \mu^{\ominus\prime}$$

For the fixed pressure standard state (span AC)

$$RT \ln a_i = RT(\ln m_i \gamma_i' + \ln m_i \Gamma_i) = \mu_i - \mu_i^{\ominus}$$

The term $RT \ln m_i \Gamma_i = \mu_i^{\ominus\prime} - \mu_i^{\ominus}$ (span BC) allows for the effect of pressure on the standard chemical potential (Rock, 1967, p. 105; Lewis and Randall, 1961, pp. 249–251). In electrolyte solutions $\mu_i$ is frequently *smaller* than $\mu_i^{\ominus}$ since $\gamma_i$ is usually less than unity.

express the mean-ion activity coefficient $(\gamma_{\pm})$ and mean-ion concentration $(m_{\pm})$. The mean-ion activity coefficients of the major sea water components (Fig. 2.3.) indicate how widely these solutions deviate from ideal behaviour. The deviation becomes greater the more highly charged the solution components.

Unfortunately, the use of $\gamma_{\pm}$ values results in ambiguous chemical descriptions for electrolyte mixtures. For example, a solution containing NaCl and $CaCl_2$ can be described with rigorous thermodynamic consistency by the mean-ion activity coefficients $\gamma_{\pm(NaCl)}$ and $\gamma_{\pm(CaCl_2)}$. However if one is interested in a process that is sensitive to the activity of chloride ions a number of alternative, and incompatible definitions of $\gamma_{Cl^-}$ are possible (Section 2.5.5).

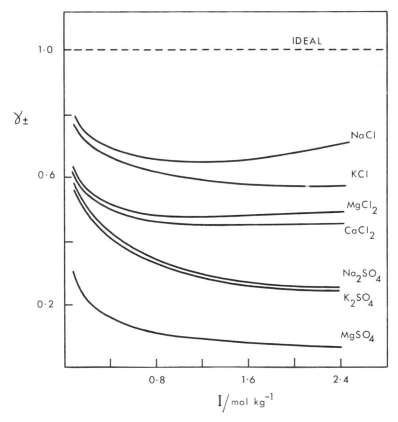

FIG. 2.3. Mean-ion activity coefficients of the major sea-water components at 298 K and 1 atm expressed on the molal scale (Robinson and Stokes, 1965). All the curves tend to $\gamma_\pm = 1$ when I tends to zero. The portions of the curves below an ionic strength of 0·1 M have been omitted for clarity.

To resolve this difficulty some assumptions must be made about the behaviour of individual ions in solution so that the ions and not the neutral salts may be treated as the solution components. The definition of single-ion activity coefficients will be considered after the structural complexity of electrolyte solutions has been discussed.

2.2.3. APPLICATION

For component $i$ at constant temperature and pressure

$$dG = (\partial G/\partial n_i)_{T, P, n_j}\, dn_i = \mu_i\, dn_i \qquad (2.7)$$

$\mu_i \, dn_i$ therefore represents the free energy change associated with the transfer of $dn_i$ moles of material. The component $i$ will move spontaneously from points of high $\mu_i$ to points of low $\mu_i$ within a heterogeneous system, and the classic laws of diffusion (Fick's first and second laws) should be defined in terms of chemical potentials rather than concentrations when electrolyte solutions are being considered (see Bockris and Reddy, 1970, pp 296–299). Therefore, the chemical potential provides the key to a whole range of transport phenomena (e.g. membrane transport, electrical conductivity, gaseous diffusion; see Haase, 1969).

The overall free energy change accompanying a chemical reaction at constant temperature and pressure is given by,

$$dG = \sum_{\substack{\text{prod} \\ i}} \mu_i \, dn_i - \sum_{\substack{\text{react} \\ i}} \mu_i \, dn_i \qquad (2.8)$$

If $dG$ is negative the reaction will move towards the products since $G$ tends to move towards a minimum value. Given the chemical potentials of the reactants and products we can, therefore, predict the ultimate course of chemical reactions—although we can say nothing about the *rate* at which these reactions will occur nor can we elucidate their mechanisms (Lee, 1959, pp. 187–202).

For systems at equilibrium the distribution of reactants and products can also be predicted. Integration of equation (2.8) at constant temperature and pressure gives, for the free energy change accompanying the combination of finite concentrations of the reactants

$$\Delta G = \sum_{\substack{\text{prod} \\ i}} \mu_i n_i - \sum_{\substack{\text{react} \\ i}} \mu_i n_i \qquad (2.9)$$

$$= 0 \quad \text{at equilibrium.}$$

Combining equations (2.4) and (2.9) we find

$$\sum_{\substack{\text{prod} \\ i}} \mu_i^{\ominus} n_i - \sum_{\substack{\text{react} \\ i}} \mu_i^{\ominus} n_i = -RT \ln \left[ \sum_{\substack{\text{prod} \\ i}} a_i \Big/ \sum_{\substack{\text{react} \\ i}} a_i \right] \qquad (2.10)$$

If the term on the left-hand side is defined as $\Delta G^{\ominus}$ (by analogy with equation 2.9) and the term in parenthesis on the right-hand side as the equilibrium constant ($K^{\ominus}$) then equation (2.10) simplifies to

$$\Delta G^{\ominus} = -RT \ln K^{\ominus} \qquad (2.11)$$

An interesting discussion of this equation has been given by Rock (1967). Using the definition of $a_i$ (equation (2.4)) the equilibrium constant may be written as

$$\ln K^{\ominus} = \ln K^* + \ln \left[ \sum_{\substack{\text{prod} \\ i}} \gamma_i \Big/ \sum_{\substack{\text{react} \\ i}} \gamma_i \right]$$

where

$$\ln K^* = \ln \left[ \sum_{i}{}_{\text{prod}} m_i / \sum_{i}{}_{\text{react}} m_i \right]$$ (2.12)

The stoichiometric equilibrium constant $(K^*)$ is the parameter that is determined when sea water is treated as an ionic medium.

A knowledge of the chemical potential therefore forms the basis of quantitative predictions about the procedure of transport processes and spontaneous reactions and about the chemical composition of a system at equilibrium. Equation (2.5) relates chemical potential to solution composition and identifies the activity coefficient $(\gamma_i)$ as the key parameter. The determination, prediction and interpretation of this parameter form the central theme of this chapter. The importance of this approach is clearly emphasized by the following quotation (Wu et al., 1965a, p. 1871).

'The principal ultimate objective of these studies of electrolyte mixtures is the development and improvement of methods for the estimation of activity coefficients of electrolytes in the presence, and also in the absence, of other electrolytes. The activity coefficients are required for the understanding of the effects of salts upon chemical equilibria. The prediction of such effects remains one of the most important problems in the physical chemistry of solutions.'

### 2.2.4. EXPERIMENTAL DETERMINATION

It is possible to determine the mean-ion activity coefficients of electrolytes *directly* by potentiometric measurements, provided electrodes are available that respond specifically to the anion and cation in question (Table 2.3). The technique was extensively used by Harned and his co-workers (Harned and Owen, 1958) in early studies of electrolyte mixtures following the very precise work of Güntelberg (1926). Conventional single-ion activity coefficients can also be estimated if some theoretical procedure is adopted for calculating the liquid junction potential at the reference electrode (Table 2.3). In the past the method was severely restricted because few suitable electrodes were available.

Recent advances in electrode technology have resulted in the development of a much wider range of ion-selective electrodes (Durst, 1969; Whitfield, 1971; Moody and Thomas, 1972). Unfortunately, few of these electrodes are ion-*specific*, and they may respond to more than one ion in a mixture. In addition, they have a rather complex internal construction and consequently a poorer reproducibility ($\sim 0.1$ mV) than the electrodes used in earlier studies ($\sim 0.01$ mV). The day to day drifts in electrode potential and the composite nature of the electrodes rob the standard potential of its precise meaning and

TABLE 2.3

*Solute activity coefficients from ion-selective electrode measurements*

A. *Mean-ion activity coefficients*

(i) Cell I

| Electrode selective to $M^+$ ions | Solution containing $M^+$ and $X^-$ ions etc. | Electrode selective to $X^-$ ions |

(ii) Cell potential

$$E = (\varepsilon_X^\ominus - \varepsilon_M^\ominus) - (vRT/nF)\ln a_\pm \qquad (1)$$

$n$ = stoichiometric number of electrons transferred in the cell reaction

(iii) Determination of standard potential, $E^\ominus (= \varepsilon_X^\ominus - \varepsilon_M^\ominus)$

(a) By extrapolation to the reference state (Ives and Janz, 1961 pp. 31–48; Klotz, 1964 pp. 399–401)

Rewrite (1) to give

$$\underbrace{E^{\ominus\prime} = E + (vRT/nF)\ln m_\pm}_{I} = \underbrace{E^\ominus - (vRT/nF)\ln \gamma_\pm}_{II}$$

$E^{\ominus\prime}$ is calculated from $I$, which contains the experimental parameters, and is plotted against $m_\pm^{\frac{1}{2}}$ (see however Harned and Owen, 1958 pp. 430–431). Extrapolation to the reference state where $m_\pm = 0$ gives $E^\ominus$.

(b) By calibration.

The cell is immersed in a reference solution of known $a_\pm$ at the required temperature and pressure and $E^\ominus$ can be calculated directly.

B. *Conventional single-ion activities*

(i) Cell II

| Electrode selective to $M^+$ ions | Solution containing $M^+$ ions etc. | Solution containing $Cl^-$ ions | Electrode selective to $Cl^-$ ions |

Liquid junction

(ii) Cell potential

$$E = (\varepsilon_M^\ominus + \varepsilon_R + \varepsilon_J) + (RT/nF)\ln a_{M^+} \qquad (2)$$

(iii) Liquid junction potential, $\varepsilon_J$

Varies with solution composition in an, as yet, unpredictable manner and makes accurate determination of the intercept potential $(E' = \varepsilon_M^\ominus + \varepsilon_R + \varepsilon_J)$ impossible.

$\varepsilon_J$ can be *minimized* by experimental technique (see e.g. Whitfield, 1971) but not eliminated. Single-ion activities measured in solution by this technique are therefore conventional in that they are based on estimated $\varepsilon_J$ values.

---

render the extrapolation procedure outlined in Table 2.3 unworkable. However, these electrodes have provided information, although not of the

TABLE 2.4

*Determination of solute activities from the activity of water*

(i)  *Gibbs–Duhem equation*
From equation 2.2 it can be shown that,

$$m_w \, d\mu_w + \sum_i v_i m_i \, d\mu_i = 0 \tag{1}$$

(see Klotz, 1964, pp. 250–252, Lewis and Randall, 1961, pp. 210–212, 552)
In terms of activities (equation 2.5)

$$m_w \, d \ln a_w + \sum_i v_i m_i \, d \ln a_i = 0 \tag{2}$$

(ii) *Define osmotic coefficient* $(\phi)$

$$n_w \ln a_w + n_2 \phi = 0 \tag{3}$$

or 
$$\phi = -(n_w/n_2) \ln a_w \tag{4}$$

where $n$ is the mole fraction of the subscriptic component convert to molal scale

$$\phi = -55 \cdot 556 \, (v m_2)^{-1} \ln a_w \tag{5}$$

for a multicomponent solution

$$\phi = -55 \cdot 556 \left( \sum_i v_i m_i \right)^{-1} \ln a_w \tag{6}$$

Gibbs–Duhem equation in terms of osmotic coefficient

$$-\sum_i v_i m_i \, d \ln \gamma_i = d\left[ (1 - \phi) \sum_i v_i m_i \right] \tag{7}$$

For non-electrolytes and for calculations where the single ions are treated as components, $v_i = 1$.

(iii) *Relationship between solute activity and osmotic coefficient*
For a solution containing a single electrolyte successive differentiation (Robinson and Stokes, 1965 p. 34) gives

$$(\phi - 1) \, dm_2/m_2 + d\phi = d \ln \gamma_2 \tag{8}$$

Integration of (8) gives

$$\ln \gamma_2 = (\phi - 1) + \int_0^{m_2} (\phi - 1) \, d \ln m_2 \tag{9}$$

since $\phi$ tends to unity as $m_2$ tends to zero (equation 3).

highest precision, where none existed before. The pitfalls and benefits of the potentiometric method for the determination of solute activities have been discussed by Butler (1969).

The most precise values for mean-ion activity coefficients have been calculated from the activity of water in the solution *via* the Gibbs–Duhem equation (Table 2.4). The introduction of the osmotic coefficient $(\phi)$ simplifies the mathematics and provides a function of water activity which differs significantly from unity at low concentrations (cf. Fig. 2.1. see also Robinson and Stokes, 1965, p. 29).

All four colligative properties of the solution—vapour pressure, freezing point, boiling point, osmotic pressure—have been used to determine water and solute activities. The freezing point method is the most precise but gives results at one temperature only. The boiling point method can be used over a range of temperature since it is very sensitive to changes in external pressure. However, the molal elevation of the boiling point (0·513 K) is only about one quarter of the molal depression of the freezing point and this, together with the pressure sensitivity, makes precise measurements difficult. The osmotic pressure technique is potentially extremely useful because large pressures are involved (tens of atmospheres). However, temperature uniformity is critical ($10^{-6}$ K throughout the system, Robinson and Stokes, 1965, p 208) and the technique is seldom used.

The direct vapour pressure method has been used, along with other techniques, to establish precise values for the activity of water in certain reference electrolyte solutions (e.g. KCl and $CaCl_2$, Lewis and Randall, 1961, pp 328–330; Robinson and Stokes, 1965, pp 213–216, 218). Once these values have been established, the water activity of other salt solutions can be measured by a process of isothermal distillation. Separate vessels containing solutions of the reference electrolyte (A) and the test electrolyte (B) are placed in a chamber which is subsequently evacuated. Water distils slowly from the solution with the highest vapour pressure to the solution of lowest vapour pressure until both solutions have the same water activity. The distillation process is normally complete within a few days and the solutions are removed and analyzed. The osmotic coefficients of the two solutions (A and B) after equilibration are related by the equation

$$\phi_B = (v_A m_A / v_B m_B) \phi_A \qquad (2.13)$$

If A is the reference solution then all the terms on the right hand side are known and $\phi_B$ can be estimated. $\gamma_2$ in a single electrolyte solution can then be calculated from equation (9) (Table 2.4) This procedure, known as the isopiestic method, is widely used since the measurements can easily be made over a wide temperature and concentration range. A significant proportion of the data on electrolyte mixtures has been obtained in this way. The mathematics associated with measurements in mixtures is necessarily more complex than that shown in Table 2.4 since we must now determine the activity coefficients of *several* solutes from measurements of $\phi$. The methods take as their starting point the cross differentiation relationship,

$$v_j (d \ln \gamma_j / \partial m_i)_{m_j} = v_i (\partial \ln \gamma_i / \partial m_j)_{m_i} \qquad (2.14)$$

This form of equation was used by McKay (1952, 1955). A slightly modified version forms the basis of a more elegant and widely used treatment by McKay and Perring (1953). Details of the derivation and application of these equations are given by Lewis and Randall (1961, pp 572–576) and Robinson and Stokes (1965, pp 440–445). An alternative procedure which is more easily applied to electrolyte mixtures has been developed by Scatchard (1961, 1968, 1969). This method, (Section 2.6) is now gradually replacing the McKay procedure (see for example Rush and Johnson, 1968; Lindenbaum et al., 1972; Platford, 1968b).

Given the activities of the solutes in an electrolyte mixture it is also possible to reverse the calculations and estimate the osmotic coefficient, and hence the colligative properties, of the solution. This procedure provides a useful check on the validity of electrolyte solution models and produces information about the properties of saline waters that is of direct technological and physiological importance.

### 2.2.5. PRESSURE AND TEMPERATURE COEFFICIENTS

The pressure and temperature coefficients of the chemical potential, and hence the activity coefficient, can be defined in terms of partial molal quantities (Table 2.5). These partial molal quantities may be determined experimentally from the appropriate apparent molal parameters (Table 2.6).

The activity coefficients of a solute in a solution at any temperature $(T_1)$ may be related to the value measured at a reference temperature $(T_0)$ by the equation

$$\ln \left( \gamma_{\pm}^{(T_1)} / \gamma_{\pm}^{(T_0)} \right) = - \int_{T_0}^{T_1} (\bar{L}_2 / \nu R T^2) \, dT, \qquad (2.15)$$

TABLE 2.5

*Pressure and temperature coefficients of intensive thermodynamic functions.*

| $Y$ | $(\partial Y/\partial T)_{P, n_j}$ | $(\partial^2 Y/\partial T^2)_{P, n_j}$ | $(\partial Y/\partial P)_{T, n_j}$ | $(\partial^2 Y/\partial P^2)_{T, n_j}$ |
|---|---|---|---|---|
| $\mu_i$ | $-\bar{S}_i$ | $\bar{C}_{p,i}/T$ | $\bar{V}_i$ | $-\kappa_i$ |
| $R \ln \gamma_i^a$ | $-(\bar{H}_i - \bar{H}_i^\ominus)/T^2$ | $\bar{J}_i^b/T^2$ | $(\bar{V}_i - \bar{V}_i^\ominus)/T^c$ | $-(\kappa_i - \kappa_i^\ominus)/T$ |
| $R \ln K^a$ | $\Delta H^\ominus/T^{2d}$ | $(\Delta C_p^\ominus - 2\Delta H^\ominus/T)/T^2$ | $-\Delta V^\ominus/T^e$ | $\Delta\kappa^\ominus/T$ |

[a] by analogy with $G/T$, Table 2.1 since $R \ln m_i\gamma_i = (\mu_i - \mu_i^\ominus)/T$ and $-R \ln K = \Delta G^\ominus/T$
[b] Relative partial molal heat capacity $= \bar{C}_{p,i} - \bar{C}_{p,i}^\ominus$
[c] Variable pressure standard state, fixed pressure value $= \bar{V}_i/RT$ (Rock, 1968)
[d] $\Delta Y^\ominus = \sum_i \text{prod } n_i \bar{Y}_i^\ominus - \sum_i \text{react } n_i \bar{Y}_i^\ominus$
[e] Variable pressure standard state, fixed pressure value $= 0$ (Rock, 1968).

TABLE 2.6

*Determination of partial molal quantities*

(i) *Define apparent molal quantities* $(^\phi Y_2)$

$$^\phi Y_2 = (Y - m_w \overline{Y}_w^\ominus)/\sum_i \nu_i m_i \tag{1}$$

e.g. $\quad ^\phi V_2 = \left[\begin{array}{c}\text{Total volume of} \\ \text{solution}\end{array} - \begin{array}{c}\text{Volume of} \\ \text{pure water}\end{array}\right]\Big/ \text{moles of solute}$

(ii) *Determine partial molal quantities* $(\overline{Y}_2)$

Two component solution, $\overline{Y}_2 = {}^\phi Y_2 + m_2[\mathrm{d}^\phi Y_2/\mathrm{d}m_2]$ \qquad (2)

Multicomponent solution, $\overline{Y}_2 = {}^\phi Y_2 + \sum_i m_i[\partial^\phi Y_i/\partial m_i]_{m_j}$ \qquad (3)

(iii) *Determine standard partial molal quantities* $(\overline{Y}_2^\ominus)$

Two component solution, as $m_2 \to 0$, $^\phi Y_2 \to \overline{Y}_2^\ominus$

i.e. $\qquad\qquad ^\phi Y_2^\ominus = \overline{Y}_2^\ominus \tag{4}$

Plot $^\phi Y_2$ against the molality and extrapolate to the reference state for which $m_2 = 0$. For multicomponent systems the reference state may be either pure water or the solution containing the remaining solutes (Table 2.2).

(iv) *Molality functions*

(a) Masson equation (Masson, 1929)

$$^\phi Y_2 = \overline{Y}_2^\ominus + \mathscr{S}_Y^* m_2^{\frac{1}{2}} \tag{5}$$

where $\mathscr{S}_Y^*$ is the experimental slope.

(b) Redlich equation (Redlich and Rosenfeld, 1931a, b).

$$^\phi Y_2 = \overline{Y}_2^\ominus + \mathscr{S}_Y m_2^{\frac{1}{2}} + b_Y m_2 \tag{6}$$

where $\mathscr{S}_Y$ is a theoretical limiting slope (section 2.5) and $b_Y$ is an empirical constant.

References. Klotz (1964, pp. 253–259) and Lewis and Randall (1961, pp. 205–210) have outlined alternative methods for determining partial molal quantities. A particularly detailed discussion has been given by Harned and Owen (1958, Chapter 8), and Millero (1971a) has treated extrapolation functions for $\overline{V}_2^\ominus$ in some depth.

where $\overline{L}_2$ $(= \overline{H}_2 - \overline{H}_2^\ominus)$ is the relative standard partial molal enthalpy of the solute (Table 2.1). It is rarely a good approximation to assume that $\overline{L}_2$ is independent of temperature over more than a five degree range so that values of the relative partial molal heat capacity $(\overline{J}_2 = \overline{C}_p - \overline{C}_p^\ominus = \partial \overline{L}_2/\partial T)$ are needed for an accurate evaluation of equation (2.15). If the variation of $\overline{J}_2$ with temperature is neglected then we can write

$$\overline{L}_{2(T_1)} = \overline{L}_{2(T_0)} + \overline{J}_{2(T_0)}(T_1 - T_0). \tag{2.16}$$

Substituting this value into equation 2.15 and integrating gives (Harned and Owen, 1958, p 504)

$$\ln \gamma_\pm^{(T_1)} = (\bar{L}_{2(T_0)} - \bar{J}_{2(T_0)} T_0)/\nu RT_1 - (\bar{J}_{2(T_0)} \ln T_1)/\nu R + I_c. \qquad (2.17)$$

The integration constant $(I_c)$ may be obtained by substituting the known values of $\ln \gamma_\pm$, $\bar{L}_2$ and $\bar{J}_2$ at the reference temperature $T_0$. Unfortunately few data are available for $\bar{L}_2$ for ionic strengths greater than $0.1$ M, even at 298 K (see Parsons, 1959; Lewis and Randall, 1961; Harned and Owen, 1958 and Robinson and Stokes, 1965). The most extensive compilation of data is that given by Parker (1965) who lists tables of the apparent relative partial molal enthalpies and heat capacities of $1:1$ electrolytes at 298 K from which values of $\bar{L}_2$ and $\bar{J}_2$ may be calulated (see Table 2.6). If equation (2.15) is to be used for electrolyte mixtures then the values of $\bar{L}_2$ need to be corrected for the heats of mixing of the component single electrolyte solutions. These corrections are small, but they provide interesting information about the mixing process (see Section 2.6).

The effect of pressure on the activity coefficient may be defined in a similar manner, viz.

$$\ln (\gamma_\pm^{(P_1)}/\gamma_\pm^{(P_0)}) = \int_{P_0}^{P_1} [(\bar{V}_2 - \bar{V}_2^{\ominus})/\nu RT]\, dP \qquad (2.18)$$

where $\bar{V}_2$ and $\bar{V}_2^{\ominus}$ are the partial molal volumes of the solute in the solution and in the standard state respectively.

If the effect of pressure on the partial molal compressibility ($\bar{\kappa}_2$, Table 2.1) is neglected then equation 2.18 can be integrated to give,

$$\ln \gamma_\pm^{(P_1)} = \ln \gamma_\pm^{(P_0)} + (P_1 - P_0)(\bar{V}_2 - \bar{V}_2^{\ominus})_{P_0}/\nu RT - (P_1 - P_0)^2$$
$$\times (\bar{\kappa}_2 - \bar{\kappa}_2^{\ominus})_{P_0}/\nu RT \qquad (2.19)$$

The effect of pressure on the partial molal volume is small in the pressure range encountered in the oceans (up to 1000 atmospheres) so that it is a reasonable approximation to neglect the quadratic term in equation (2.19). Partial molal volumes are easier to measure than relative partial molal enthalpies and so data for $\bar{V}_2^{\ominus}$ are available for a large number of salts (Tables 2.7 and 2.8). The values of the partial molal volumes ($\bar{V}_2$) in single electrolyte solutions may be calculated from the apparent molal volumes (Table 2.6). The calculations are discussed in detail by Harned and Owen (1958, p 358).

The temperature and pressure coefficients of the equilibrium constant ($K^{\ominus}$, equation (2.11)) will prove to be useful in discussing sea water models (Section 2.7). The properties of the appropriate differential coefficients

TABLE 2.7

Standard partial molal volumes of salts in various ionic media at 298K and 1atm

| Salt | $\bar{V}^\ominus (10^{-6}\,m^3\,mol^{-1})$ | | | | | |
|---|---|---|---|---|---|---|
| | 0·725 M NaCl | | | 35‰ Sea water | | |
| | a | b | c | a | d | e |
| HCl | 18·0[f] | 17·8 | — | — | 19·6 | 20·1 |
| NaCl | 16·6 | 16·6 | 19·0 | 18·90 ± 0·08 | 18·9 | 18·8 |
| KCl | 26·6 | 26·9 | 29·3 | 29·20 ± 0·14 | 29·2 | 28·9 |
| MgCl$_2$ | 14·8[f] | 14·5 | 18 | 19·62 ± 0·21 | 19·6 | 19·6 |
| CaCl$_2$ | 17·8 | 17·8 | 22·6 | 22·00 ± 0·28 | 22 | 22 |
| SrCl$_2$ | — | 17·5 | — | — | — | 22·1 |
| BaCl$_2$ | 24·4[f] | 23·2 | — | — | 27·1 | 27·0 |
| NaBr | 23·5[f] | 23·5 | 25·8[h] | — | 25·9 | 25·7 |
| KBr | 33·7[f] | 33·7 | 36·2[h] | — | 36·2 | 35·8 |
| MgBr$_2$ | 28·0[g] | 28·3 | — | — | 33·6 | 33·4 |
| CaBr$_2$ | 32·4[g] | 31·6 | — | — | 36·0 | 35·8 |
| NaI | 35·1[f] | 35·0 | — | — | 37·0 | 37·3 |
| KI | 45·3[f] | 45·2 | — | — | 35·5 | 47·4 |
| NaNO$_3$ | 28·0 | 27·8 | 30 | 30·46 ± 0·17 | 30·4 | 30·0 |
| KNO$_3$ | 38·0 | 38·0 | 41 | 40·76 ± 0·21 | 40·7 | 40·1 |
| Mg(NO$_3$)$_2$ | 37·8 | 36·8 | 41 | 42·75 ± 0·52 | 42·6 | 42·0 |
| Ca(NO$_3$)$_2$ | 40·6 | 40·2 | 45 | 45·13 ± 0·64 | 45·0 | 44·4 |
| NaHCO$_3$ | 22·2 | 22·2 | 25 | 27·07 ± 0·27 | 27·0 | 23·9 |
| KHCO$_3$ | 32·2 | 32·4 | 36 | 37·38 ± 0·30 | 37·3 | 34·0 |
| Mg(HCO$_3$)$_2$ | 26·2 | 25·6 | 31 | 35·98 ± 0·67 | 35·8 | 29·8 |
| Ca(HCO$_3$)$_2$ | 29·0 | 29·0 | 35 | 38·35 ± 0·76 | 38·2 | 32·2 |
| Na$_2$CO$_3$ | −6·5[f] | −6·7 | 1·74 ± 0·23[i] | 9·0 ± 1·5[i] | 10·4[k] | −2·4 |
| K$_2$CO$_3$ | 13·1[f] | 13·7 | 23·17 ± 0·35[i] | 28·3 ± 0·8[i] | 31·0[k] | 17·8 |
| MgCO$_3$ | — | −25·5 | — | — | −7·8[k] | 1·4 |
| CaCO$_3$ | — | −22·2 | −13·9 | −6·5 ± 1·7 | −5·4[k] | 3·8 |
| Na$_2$SO$_4$ | 11·2 | 11·6 | 21 | 21·03 ± 1·4 | 20·5 | 15·4 |
| K$_2$SO$_4$ | 31·2 | 32·0 | 41 | 41·63 ± 0·37 | 41·1 | 35·6 |
| MgSO$_4$ | −7·0 | −7·2 | 1 | 2·86 ± 0·48 | 2·3 | −2·6 |
| CaSO$_4$ | −4·2 | −3·8 | 5 | 5·24 ± 0·61 | 4·7 | −0·2 |

[a] Experimental values, Duedall (1968)
[h] Calculated from single-ion values (Millero, 1972a)
[c] Owen and Brinkley (1941)
[d] Calculated from single-ion values (Millero, 1969, Table 2.4)
[e] Free ion values (see section 2.5.2.1) calculated from single-ion values (Millero, 1969, Table 2.8)
[f] Mean of experimental values tabulated by Millero (1972b)
[g] Root (1932)
[h] Quoted by Millero (1971a)
[i] Duedall (1972), 293 K
[k] Using the value of $\bar{V}^\ominus_{CO_3^{2-}}$ calculated by Millero and Berner (1972) from the experimental value of Duedall (1972)
[l] Duedall (1968) gives formulae for calculated $\bar{V}_{SALT}$ as a function of salinity and temperature.

TABLE 2.8

*Conventional[a] single-ion standard partial molal volumes at 1 atm*

| Ion | $\overline{V}_i^{\ominus}$ ($10^{-6}$ m$^3$ mol$^{-1}$) Water[b] | | | 35‰ Sea water (298 K) | |
|---|---|---|---|---|---|
| | 273 K | 298 K | 323 K | c | d |
| $H^+$ | 0 | 0 | 0 | 0 | 0 |
| $Na^+$ | −3·51 | −1·21 | −0·30 | −0·7 | −1·3 |
| $K^+$ | 7·17 | 9·02 | 9·57 | 9·6 | 8·8 |
| $Mg^{2+}$ | −21·81 | −21·17 | −20·90 | −19·6 | −20·6 |
| $Ca^{2+}$ | −19·84 | −17·85 | −18·22 | −17·2 | −18·2 |
| $Sr^{2+}$ | −20·77 | −18·16 | −17·69 | — | −18·1 |
| $Ba^{2+}$ | −15·79 | −12·47 | −11·73 | −12·1 | −13·2 |
| $NH_4^+$ | 17·47 | 17·86 | 19·20 | — | 17·7 |
| $F^-$ | −2·21 | −1·16 | −1·4 | — | 1·3 |
| $Cl^-$ | 16·45 | 17·83 | 18·0 | 19·6 | 20·1 |
| $Br^-$ | 23·06 | 24·71 | 25·49 | 26·6 | 27·0 |
| $I^-$ | 33·57 | 36·22 | 37·52 | 37·7 | 38·6 |
| $HCO_3^-$ | — | 23·4 | — | 27·7 | 25·2 |
| $NO_3^-$ | 25·6 | 29·0 | 30·3 | 31·1 | 31·3 |
| $OH^-$ | −6·8 | −4·04 | −4·35 | 1·2 | −1·6 |
| $CO_3^{2-}$ | −9·8 | −4·3 | — | 11·8[e] | 0·2 |
| $SO_4^{2-}$ | 11·1 | 13·98 | 16·0 | 21·9 | 18·0 |

[a] Absolute values may be obtained from the equation

$$\overline{V}_i^{\ominus} = \overline{V}_{i(conv)}^{\ominus} + z_i \overline{V}_{H^+}^{\ominus}$$

where $z_i$ is the charge on the ion (positive for a cation and negative for an anion). In pure water the values $-5\cdot1$ (273 K), $-5\cdot4$ (298 K) and $-5\cdot9$ (323 K) $\times 10^{-6}$ m$^3$ mol$^{-1}$ may be used for $\overline{V}_{H^+}^{\ominus}$ (Millero, 1971a). In sea water the value $-3\cdot7 \times 10^{-6}$ m$^3$ mol$^{-1}$ is recommended (Millero, 1969).

[b] Millero (1972a)

[c] Millero (1969). Values at other salinities may be calculated from the equation

$$\overline{V}_{i(SW)}^{\ominus} = \overline{V}_{i(W)}^{\ominus} + 0\cdot1058\, S'''_{00}\,(z_i^2/r_i) + 0\cdot0237\, S'''_{00}$$

where $r_i$ is the Pauling radius (Table 2.11). For the polyatomic anions the following values are used (in nanometres):

$$NO_3^- (0\cdot203),\ HCO_3^- (0\cdot192),\ CO_3^{2-} (0\cdot173),\ SO_4^{2-} (0\cdot205)$$

[d] *Free* ions only (Millero, 1969; see Section 2.6). Here use $\overline{V}_{H^+}^{\ominus} = -3\cdot2 \times 10^{-6}$ m$^3$ mol$^{-1}$

[e] Value calculated by Millero and Berner (1972) from the experimental data of Duedall (1972). Millero (1969) used the value $2\cdot9 \times 10^{-6}$ m$^3$ mol$^{-1}$.

(Table 2.5) and their integration to obtain explicit relationships between $K^{\ominus}$ and temperature or pressure are discussed by King (1965, pp 184–217) and Hamann (1963).

## 2.2.6. PREDICTION

It is possible, in principle, to provide a complete thermodynamic description of a solution by measuring the activities of all components over the relevant range of temperature, pressure and solution composition. Throughout, the activity coefficients remain as empirical correction factors that enable the properties of the real solution to be described with the same formalism as that used in the description of ideal solutions. Thermodynamics can provide no way of predicting these factors nor their variation with temperature, pressure and solution composition in systems for which there are no experimental measurements.

To predict these properties and to provide an explanation for the thermodynamic description it is necessary to create some model, be it mathematical or mechanical, of the molecular structure of the solution and of the interactions between the various components. Such a model will also provide useful clues to the behaviour of individual ions in solution so that the ambiguous thermodynamic description of electrolyte mixtures in terms of mean-ion activity coefficients may be replaced by a more useful, if less rigorous description based on conventional single-ion activity coefficients.

Although sceptics caution against reliance on structural models (e.g. Holtzer and Emerson, 1969) such models may nonetheless provide useful intuitive guidelines as long as they are not used so frequently that they become mistaken for reality. Indeed, the complexities of the real system and the primitive state of our understanding restrict the mechanical models to a qualitative description of electrolyte solution behaviour so that there is little real danger of blind adherence to a particular structural picture.

## 2.3. STRUCTURAL PATTERNS IN WATER

There is currently considerable controversy about the nature of the structural organization in liquid water (see Kavanau, 1964; Conway, 1966; Eisenberg and Kauzmann, 1969; Frank, 1972; Horne, 1972). Since the problem cannot be solved exactly (Ben-Naim and Stillinger, 1972) it is likely that there is no unique description capable of explaining completely the properties of liquid water and, at the same time, excluding all other models.

### 2.3.1. THE BASIS OF STRUCTURAL ORGANISATION

The isolated water molecule has four $sp^3$ hybrid orbitals with an approximately tetrahedral configuration (Fig. 2.4). The two hydrogen–oxygen bonds are slightly dipolar with a net positive charge on the hydrogen. The remaining two orbitals are occupied by lone-pairs of electrons, giving the molecule as

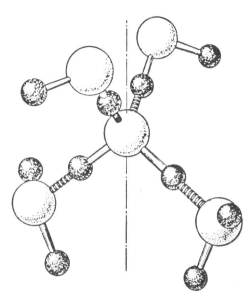

Fig. 2.4. Hydrogen-bonded structure containing five water molecules. Small spheres represent hydrogen atoms; large spheres, oxygen atoms; and discs, hydrogen bonds. Reproduced with permission from Walrafen (1964) and the American Chemical Society.

a whole a marked dipolar character. The exact size and distribution of the lone-pair orbitals on the isolated water molecule have not been uniquely defined (see Eisenberg and Kauzmann, 1969). These orbitals are probably absent in the vapour and only make their appearance when there is significant interaction between water molecules in the liquid or solid (Bader and Jones, 1963; Muirhead-Gould and Laidler, 1966).

In the liquid the net positive charge on the hydrogen atom of one water molecule will interact with the net negative charge on the lone-pair of an adjacent molecule to form a relatively weak hydrogen-bond (Fig. 2.4, see also Pimentel and McClellan, 1960; Hadzi, 1968 and Rao, 1972). As each link is formed in this way the water molecules in the cluster that do not yet have their full complement of hydrogen bonds become increasingly dipolar, and hence more likely to form more hydrogen bonds, because of the resonance of electrons throughout the linked structure (Coulson and Danielsson, 1954; Frank, 1958; Del Bene and Pople, 1970; Rao, 1972; Symons, 1972). This gives rise to *co-operative hydrogen bonding*. Water is able to maintain an extensively linked structure in the liquid phase because it has an equal number of lone pairs and hydrogen atoms. This symmetry is lacking in the neighbouring hydrides, ammonia and hydrogen fluoride, although they possess strong permanent dipoles and can form hydrogen bonds (Sanderson, 1960). The result is a very open structure in liquid water—at the temperature of maximum density a cubically close packed arrangement of water molecules would occupy $12.5 \times 10^{-6}\,m^3\,mol^{-1}$ whereas a loose tetrahedral packing would occupy $18 \times 10^{-6}\,m^3\,mol^{-1}$ (Bernal and Fowler, 1933).

All proposed structural models for water take into account the competition between collective hydrogen bonding and thermal disruption, and differ only in the patterns of organization that hydrogen bonding is supposed to promote in the liquid. Three levels of organization may be distinguished (Fig. 2.5, Eisenberg and Kauzmann, 1969, pp 150–155). If the distribution of

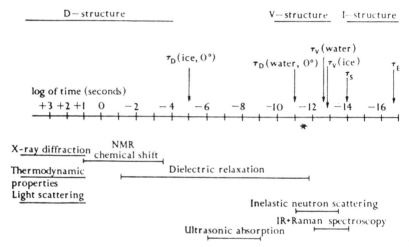

FIG. 2.5. Time scale of molecular processes in ice and liquid water. The vertical arrows indicate the periods associated with various molecular processes: $\tau_D$ and $\tau_V$ are the average periods for molecular displacement and oscillation; $\tau_S$ is the period for O–H stretching vibration and $\tau_E$ is the time required for an electron to complete one circuit in the innermost Bohr orbit.

The horizontal lines below the time scale show the time intervals for which various experimental techniques have yielded information on ice and water. Reproduced from "Structure and Properties of Water" by D. Eisenberg and W. Kauzmann (1969), with permission of the authors and the Clarendon Press.

molecules could be determined instantaneously then the *I-structure* would be observed, free from the averaging effects of molecular motion. When the disturbance caused by the experimental method is slow when compared with oscillations about a temporary equilibrium position but rapid compared to displacements from that position, then a vibrationally averaged structure will be observed (*V-structure*). When much longer time intervals are involved then the molecular pattern that is deduced will average out the effects of molecular diffusion on the organization of the system and will reveal the diffusionally averaged or *D-structure*. The distinction between the last two structures will disappear as the displacement of molecules becomes more rapid at high temperatures.

The molecular organization in liquid water is best considered in terms of the V-structure since it is simpler to visualize ordered structural patterns when the molecules are in temporary equilibrium positions than when they are constantly moving from site to site. Model makers therefore concentrate on creating structures that will be stable over a time interval of $10^{-12}$ seconds. In general, they postulate their molecular patterns on the basis of spectroscopic evidence (Walrafen, 1972) and then try to reproduce the properties of the D-structure (X-ray diffraction data, thermodynamic properties) by averaging out the properties of the V-structure over time or space.

The resulting models can be divided into two categories, (i) those postulating that, although the hydrogen-bond network in ice may be violently distorted on melting, it remains essentially intact, giving a four coordinated environment for the water molecules (continuous network models) and (ii) those assuming that a significant number of hydrogen bonds are broken giving a small number of structurally distinct environments (broken network models). The controversy between these two schools of model builders is still very active, and the main points at issue have been summarized by Eisenberg and Kauzmann (1969), Frank (1972), Kell (1972a, b) and Davis and Jarzynski (1972). The key point is the directional nature of the hydrogen bond. If it is considered to be largely covalent then it will be highly directional and little distortion will be possible. Thermal influences will, therefore, cause bond rupture as postulated in the broken network models. If, as Coulson and Danielsson (1954) suggest, there is only a small degree of covalency, then the continuous network model will be favoured. The two approaches clearly represent extreme views, but they have the advantage of being mathematically tractable and capable of giving quantitative predictions of water properties.

### 2.3.2. CONTINUOUS NETWORK MODELS

Although Bernal and Fowler (1933) proposed a continuous linked network with a few predominant structural patterns, this approach has since received little attention.

Pople (1951) and Lennard–Jones and Pople (1951) have presented a quantitative analysis of a flexible, distorted hydrogen-bond model based on four-coordinated water molecules. More recent calculations by Eisenberg and Kauzmann (1969) using this model suggest that it gives an accurate account of the dielectric constant of water, of the volume decrease on melting of Ice I and of the heat capacity and thermal energy of water. Symons (1972) has used the cooperative nature of hydrogen bonding to develop, in qualitative terms, a lucid and persuasive model of water as a continuous, linked polymeric network. A similar model, based on the concept of a random network of water molecules in which five-membered rings are a common feature was proposed by Bernal (1964).

Circumstantial evidence for the stability of such structures is provided by the low internal energy of high pressure modifications of ice which have grossly distorted hydrogen bond networks (Kamb, 1968) and by the observation of stable non-linear hydrogen bonds in other structures (see e.g. Conway 1966, p. 491). These models are frequently rejected because they apparently lack the fluidity of broken bond models. However, since the difference between a broken and a distorted hydrogen bond is more a matter of subjective definition than of physical reality, the comment has no foundation (see Frank, 1972 for a detailed discussion).

It would appear that the mass of experimental evidence is gradually beginning to favour those water models which allow a wide range of molecular environments (Eisenberg and Kauzmann, 1969; Kell, 1972b).

### 2.3.3. BROKEN NETWORK MODELS

In this category an impressive array of molecular arrangements has been invoked to explain the properties of liquid water. In many cases they have been successful—with the aid of an equally impressive array of adjustable parameters! This flexibility, together with the conceptual simplicity of a structure based upon a few discrete molecular arrangements probably explains the popularity of these models. They can trace their pedigree back to the work of Röntgen (1892) and have been extensively reviewed (see for example, Kavanau, 1964; Eisenberg and Kauzmann, 1969; Davis and Jarzynski, 1972).

The simplest models are based on a mixture of two structures only—one arrangement providing an open, bulky network of molecules and the other a denser, more degraded structure. The difference in energy between the two structures is usually 8 or 12 kJ mol$^{-1}$. Two models have proved particularly useful. The first is the flickering cluster model (Frank and Wen, 1957) which considers irregular and unstable clusters of four-coordinated water molecules interspersed with unstructured, or less structured, water molecules, and the second is the interstitial model (Forslind, 1952; Samoilov, 1965) which proposes an ice-I-like structure in the liquid with the interstices occupied by individual water molecules. Both models are able to give a good qualitative account of many of the properties of water on the basis of relatively few assumptions. The interstitial model has recently been given a more quantitative treatment by Levy and his co-workers (Danford and Levy, 1962; Narten et al, 1967; Narten and Levy, 1969) and Frank (1972) considers that a model of this kind is most likely to provide an answer to the question—"What *must* water be like?" (See also Narten and Levy, 1972.) However 'water is not expected to be a mixture of finely powdered ice and water vapour' (Desnoyers and Joliceour, 1969) and any "ice-I-like" structure must be considerably distorted to be consistent with the pronounced supercooling of water (Koefoed, 1957) and with available spectroscopic data (Eisenberg and Kauzman, 1969, p. 265). In addition, the interstitial water molecules cannot behave like free vapour molecules. Stevenson (1965) has shown that there is considerable evidence against the existence of appreciable concentrations of monomeric gas-like water molecules, and Lenzi (1972) has suggested that not more than 0·0016 mole fraction of unbonded water molecules are present in the water structure. This view is supported by the infra-red spectroscopic data collected by Symons and his co-workers (Symons, 1972).

To predict thermodynamic properties from structural models, the distribution of molecules between the various energy states must be calculated. This distribution is described mathematically by a partition function (Denbigh, 1961, pp. 331–436). Only a few models have been refined to this degree, but they have each been able to predict the thermodynamic properties of water with surprising accuracy despite their different structural bases. However, this is due in part to the large number of adjustable parameters that can be invoked to improve the fit (Table 2.9).

Némethy and Scheraga (1962a) applied the statistical thermodynamic treatment to the flickering cluster model of Frank and Wen (1957) assuming that the clusters were irregular interconnected networks of five and six membered rings. They assumed that some 70% of water molecules at any instant were involved in the formation of clusters, each containing some 50 molecules. They considered four equally spaced discrete energy levels within the cluster corresponding to the successive rupture of bonds in the tetrahedral structure (Fig. 2.4) in the region of the cluster boundary. A fifth energy level is provided by the unbonded water molecules that exist between the clusters. This model has proved popular with those mainly concerned with the effects of dissolved components on the properties of water (e.g. Kavanau, 1964; King, 1965; Horne, 1969) because it provides a simple and satisfying conceptual framework for the consideration of solute-water interactions. However, it is inconsistent with the observed X-ray diffraction data since the low angle diffraction patterns are incompatible with the existence of low density regions larger than a few molecular diameters (Narten and Levy, 1969).

Marchi and Eyring (1964) developed partition functions for a model consisting of organised, tetrahedrally bonded clusters and monomeric water molecules. As well as moving freely between the clusters, these monomers were

TABLE 2.9

*Calculation of thermodynamic properties of water from broken network models[a]*

| Authors | Number of adjustable parameters | Quality of fit of calculated properties to experiment[b] | |
|---|---|---|---|
| Némethy and Scheraga[c] (1962a) | 2 in addition to 9 vibrational frequencies | $E, S, A$ within 3·8% over range 273 K to 373 K<br>$C_v$ 18% too high at 272 K<br>28% too low at 373 K | P–V–T properties at atm. pressure fitted with several (about 6) additional parameters. $V$ shows minimum at 277 K and is within 0·5% of experimental value between 272 K and 343 K |
| Marchi and Eyring[d] (1964) | 14 | Good fit of $S$ and $A$. $C_v$ too low ($=4·19$ J mol$^{-1}$ K$^{-1}$) | Good fit of vapour pressure over range 273 K to 413 K<br>$V$ within $0·3 \times 10^{-6}$ m$^3$ mol$^{-1}$ over range 293 K to 453 K<br>Does not show minimum at 277 K |
| Vand and Senior (1965)[e] | 12 plus 6 spectroscopic constants | $E, S, A$ and $C_v$ within 1·5% over range 273 K to 373 K | |
| Jhon et al. (1966)[d] | 9 | Excellent fit of $S$ and $A$ over range 273 K to 423 K<br>$C_v$ within 12·3% over range 273 K to 373 K | $V$ within 1% over range 273 K to 423 K. Vapour pressure within 3·2% over same range |

[a] Reproduced from "Structure and Properties of Water" by D. Eisenberg and W. Kauzmann (1969), with permission of the authors and of the Clarendon Press.

[b] $C_v$ is the specific heat at constant volume and $E$ and $A$ are the analogues of $H$ and $G$ for conditions of constant temperature, *volume* and solution composition.

[c] Flickering cluster model.    [d] Interstitial model.    [e] Energy band model.

considered to occupy interstitial sites within the ordered networks themselves. Jhon *et al.* (1966) modified this model to reduce the significance of the gas-like unbonded water molecules (cf. Stevenson, 1965).

A quantitative mixture model employing a continuous range of energy levels was developed by Vand and Senior (1965). Although they considered the formation of organised clusters by cooperative hydrogen bonding, they only postulated the formation of zero, one or two such bonds with any given water molecule. Overlapping energy *bands* rather than discrete energy levels, were attributed to each bonding pattern. This model gives a good quantitative account of thermodynamic properties (Table 2.9). In particular, the model proposed by Vand and Senior (1965) was able to reproduce the rather unusual behaviour of the specific heat of water up to its boiling point—a feat which has eluded most other model makers. Development of the energy band concept will probably lead to a more realistic model for water, intermediate between the extremes of simple mixture and complex continuum.

When a single solute is introduced into water, the six dimensional statistical mechanical problem becomes a nine-dimensional one since we now need three coordinates to locate the solute molecules (Ben-Naim and Stillinger, 1972; Kell, 1972a). Also, with the addition of the solute we introduce new forms of bonding. Having observed the embarrassing multiplicity of hydogen-bonded structures we can appreciate the complexities associated with bonding around highly charged, spherical ions, intricately structured organic molecules or ions with large and highly directional electron orbitals. The range of interaction possible is in itself an indication of the intricacy and flexibility of the hydrogen-bonded water structure.

## 2.4. INFINITE DILUTION—INTERACTIONS BETWEEN WATER MOLECULES AND ISOLATED IONS

At infinite dilution the ions are so widely separated that they can be considered in isolation. By starting with this simple situation we can begin to classify ions according to their influence on the water structure. Since infinite dilution is used as the ideal reference state (section 2.2.2) the interactions observed at this level will not, in themselves, contribute to the deviations from ideal behaviour that are summarised by the activity coefficient ($\gamma$). However, the picture that emerges will provide a conceptual background for the interpretation of activity coefficients and for the estimation of single-ion properties.

### 2.4.1. CLASSIFICATION OF IONS

When an ion is added to water the water molecules re-arrange themselves to accommodate the new particle by nullifying the effect of its electrostatic charge and by packing around the intruder to mask its unfamiliar shape (Bernal and Fowler, 1933). Three factors are likely to be important here— the sign and magnitude of the charge on the ion and the ionic radius.

At infinite dilution the free energy of solution ($\Delta G_S^{\ominus}$, Fig. 2.6) can be related

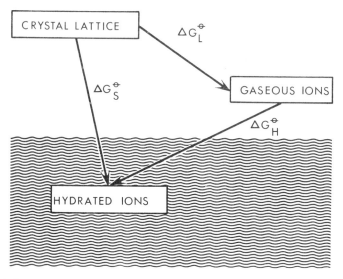

Fig. 2.6. A thermodynamic cycle representing the formation of hydrated ions at infinite dilution. The symbols are defined in the text.

to the free energies associated with lattice disruption ($\Delta G_L^{\ominus}$) and ion hydration ($\Delta G_H^{\ominus}$) by the equation,

$$\Delta G_S^{\ominus} = \Delta G_H^{\ominus} + \Delta G_L^{\ominus} \tag{4.1}$$

$\Delta G_H^{\ominus}$ and $\Delta G_L^{\ominus}$ are dominated by the enthalpy changes accompanying the disruption and formation of bonds in the condensed phase (Rosseinsky, 1965). In aqueous solutions the ion-water interactions are so intense that $\Delta H_H^{\ominus}$ and $\Delta H_L^{\ominus}$ almost cancel each other out giving rise to only a small net enthalpy of solution (Table 2.10). The re-orientation of water molecules in the presence of the ion is, therefore, able to compensate almost completely for the shift in environment from a close packed ionic lattice (Benson and Copeland, 1963; Stokes, 1963).

Since the electrostatic environment in solution is energetically similar to that in the crystal lattice, the crystallographic radii are used to describe the intrinsic size of aqueous ions (Stokes, 1963; Batsanov, 1963; Benson and Copeland, 1963). According to Pauling (1960) "Since the electron distribution

TABLE 2.10

*Enthalpies of solution ($\Delta H_S^{\ominus}$) and hydration ($\Delta H_H^{\ominus}$) for alkali metal halides*[a]

| Cation | Anion | | | |
|---|---|---|---|---|
| | $F^-$ | $Cl^-$ | $Br^-$ | $I^-$ |
| $Li^+$ | $-1035\cdot4$[b] | $-900\cdot2$ | $-868\cdot8$ | $-827\cdot3$ |
| | $4\cdot6$[c] | $-37\cdot3$ | $-49\cdot4$ | $-63\cdot2$ |
| $Na^+$ | $-919\cdot4$ | $-783\cdot8$ | $-752\cdot4$ | $-711\cdot3$ |
| | $0\cdot4$ | $3\cdot8$ | $-0\cdot8$ | $-8\cdot0$ |
| $K^+$ | $-835\cdot7$ | $-700\cdot0$ | $-668\cdot2$ | $-627\cdot2$ |
| | $-17\cdot6$ | $17\cdot2$ | $20\cdot1$ | $20\cdot5$ |
| $Rb^+$ | $-809\cdot3$ | $-673\cdot2$ | $-641\cdot8$ | $-600\cdot8$ |
| | $-26\cdot4$ | $16\cdot8$ | $21\cdot8$ | $26\cdot0$ |
| $Cs^+$ | $-791\cdot7$ | $-655\cdot7$ | $-624\cdot3$ | $-583\cdot2$ |
| | $-37\cdot7$ | $18\cdot0$ | $26\cdot0$ | $33\cdot1$ |

[a] Morris (1968)
[b] $\Delta H_H^{\ominus}$ kJ mol$^{-1}$ at 298 K
[c] $\Delta H_S^{\ominus}$ kJ mol$^{-1}$ at 298 K

function for an ion extends indefinitely it is evident that no single characteristic size can be assigned to it. Instead the apparent ionic radius will depend on the physical property under discussion and will differ for different properties". Consequently, a number of independent tabulations of ionic radii are in current use (Table 2.11). The Gourary and Adrian radii (Table 2.11, column 6) provide a simple and uniform basis for interpreting parameters associated with the intense electrostatic interactions that accompany the hydration process (Desnoyers and Joliceour, 1969; Blandamer and Symons, 1963). However, the Waddington radii (Table 2.11, column 5) give the most satis-satisfactory results in calculations involving partial molal quantities (Master-ton *et al.*, 1971; Millero, 1971a; 1972a) and will be used in the subsequent discussion.

The influence of the ionic charge on the water structure will be most intense for small, highly charged ions (i.e. for ions with a large $z/r$ factor, Table 2.12) whereas the steric effect, resulting from discrepancies in size between the intruding ion and the water molecules, will be most obvious for ions larger than the water molecule ($r = 0\cdot138$ nm, Table 2.11, Vasilëv, 1971). Smaller ions will be able to accommodate themselves quite readily in the interstices of the existing structure. These effects should be most clearly seen, from the thermodynamic point of view at least, in the standard partial molal volumes ($\bar{V}_i^{\ominus}$) and standard partial molal entropies ($\bar{S}_i^{\ominus}$) of the individual ions. These parameters will indicate, in turn, the spatial and the organizational conse-quences of the presence of the ions in solution. By considering partial molal

TABLE 2.11

*Ionic radii (nm)*[a]

| Ion | Gas-phase ions[b] | Crystallographic[c] | | | | Hydrated ions[f] |
| | | Goldschmidt (1926) | Pauling[d] (1960) | Waddington (1966) | Gourary and[e] Adrian (1960) | |
|---|---|---|---|---|---|---|
| $Li^+$ | — | (0·068) | (0·060) | 0·074 | 0·094 | 0·250 |
| $Na^+$ | 0·135 | 0·098 | 0·095 | 0·101 | 0·117 | 0·217 |
| $K^+$ | 0·167 | 0·133 | 0·133 | 0·132 | 0·149 | 0·175 |
| $Rb^+$ | 0·180 | 0·149 | 0·148 | 0·146 | 0·163 | 0·153 |
| $Cs^+$ | 0·200 | 0·165 | 0·169 | 0·170[g] | 0·186 | 0·147 |
| $Be^{2+}$ | — | — | 0·031 | — | — | — |
| $Mg^{2+}$ | 0·118 | 0·078 | 0·065 | — | — | 0·296 |
| $Ca^{2+}$ | 0·148 | 0·106 | 0·099 | — | — | 0·272 |
| $Sr^{2+}$ | 0·163 | 0·127 | 0·133 | — | — | 0·274 |
| $Ba^{2+}$ | 0·180 | 0·143 | 0·135 | — | — | 0·248 |
| $NH_4^+$ | — | 0·140 | — | 0·143[h] | — | 0·188 |
| $Fe^{2+}$ | — | — | 0·076 | — | — | — |
| $Cu^{2+}$ | — | 0·082 | — | — | — | — |
| $Zn^{2+}$ | — | 0·083 | 0·074 | — | — | 0·307 |
| $Cd^{2+}$ | — | 0·103 | 0·097 | — | — | — |
| $Pb^{2+}$ | — | 0·132 | 0·120 | — | — | — |
| $Fe^{3+}$ | — | 0·167 | 0·064 | — | — | — |
| $F^-$ | 0·191 | 0·133 | 0·136 | 0·132 | 0·116 | — |
| $Cl^-$ | 0·225 | 0·181 | 0·181 | 0·182 | 0·164 | 0·184 |
| $Br^-$ | 0·230 | 0·196 | 0·195 | 0·198 | 0·180 | 0·192 |
| $I^-$ | 0·255 | 0·220 | 0·216 | 0·224 | 0·205 | 0·216 |
| $OH^-$ | — | 0·140 | — | 0·152[h] | — | 0·246 |
| $HS^-$ | — | 0·195 | — | — | — | — |
| $NO_3^-$ | — | — | — | 0·196[h] | — | 0·203 |
| $SO_4^{2-}$ | — | — | 0·216[i] | — | — | — |

[a] 1 nm = 10 Å
[b] Guggenheim and Stokes (1969, p. 133)
[c] Compare $r_{H_2O}$ = 0·138 nm (Halliwell and Nyburg, 1963)
[d] See also Ahrens (1952)
[e] See also Morris (1968)
[f] Monk (1961)
[g] Desnoyers and Joliceour (1969)
[h] Masterton *et al.* (1971)
[i] From tetrahedral oxygen radius (0·066 nm, Wells, 1955) and crystallographic S–O distance (0·15 nm). Millero (1969) gives a value of 0·196 for the crystallographic radius of the sulphate ion.

quantities we can avoid consideration of the intense interactions associated with the hydration processes and concentrate entirely on changes occurring *within* the solution as the result of the addition of ionic components.

The appearance of small or highly charged cations in solution generally results in the promotion of organized structures (negative $\bar{S}_i^{\ominus}$) accompanied by an overall contraction of the solution (negative $\bar{V}_i^{\ominus}$) despite the finite size of the ions themselves. These ions are commonly called "structure makers" (Class I. Table 2.12, see also Kavanau, 1964). Large singly charged anions and cations normally have the opposite effect, disrupting the organized water structure (positive $\bar{S}_i^{\ominus}$) and causing an increase in the solution volume (positive $\bar{V}_i^{\ominus}$). These ions are known as "structure breakers" (Class III, Table 2.12). A number of ions do not fall readily into the categories defined in this way (e.g. fluoride and acetate and the ions in Classes II and IV).

The picture becomes a little clearer if other parameters are considered that may reflect ion-solvent interactions. Nightingale (1966) has suggested that the variation of solution viscosity ($\eta$) with salt concentration will be a sensitive indicator of ion-solvent interactions since the B-parameter of the Jones-Dole equation (Jones and Dole 1929, Table 2.12, equation (1) is characteristic of the ion in motion together with its entourage of nearest neighbour water molecules (Frank and Evans, 1945; Gurney, 1953; Stokes and Mills, 1965). Furthermore, the activation energy for viscous flow ($\Delta E_i^{\ddagger}$, equation (2), Table 2.12) provides a measure of the work required to move the ion through the mobile water network. Ions of Class I give solutions that are more viscous than pure water at the same temperature (positive $B_i$) and this effect becomes less pronounced as the temperature is increased ($\Delta E_i^{\ddagger}$ positive). Class III shows the opposite effect ($B_i$ and $\Delta E_i^{\ddagger}$ negative). Using this classification both fluoride and acetate fall clearly into Class I. Class II ions are anomalous because $B_i$ and $\Delta E_i^{\ddagger}$ have opposite signs. However, this anomaly results from the selection of 298 K as the reference temperature. Above 313 K they show positive $\Delta E_i^{\ddagger}$ values (Nightingale, 1966) and are classified as structure makers.

The behaviour summarized so far suggests that small highly charged ions gather around them a fairly large and relatively immobile sheath of water molecules to nullify their intense electrostatic charge. By contrast the large, singly charged, ions appear to disrupt the water structure because they are too big to fit neatly into the existing network and their relatively weak electrostatic fields have a confusing effect on the dipole orientation of nearest neighbour water molecules. Because of this pattern of ion-water interactions the limiting ionic conductance ($\lambda_0$) of $Li^+$ is smaller than that of $Cs^+$ (Kay and Evans, 1966). Despite its smaller intrinsic size the lithium ion has greater difficulty moving through the water network than the caesium ion because it is slowed down by relatively intense interactions with neighbouring water molecules. Broadwater and Kay (1970) have used the temperature coefficient of the $\lambda_0 \eta$ product as an indicator of the structure breaking power of ions ($\Lambda_r$, Table 2.12).

TABLE 2.12

Classification of ions according to ion–water effects at 298 K and 1 atm total pressure

| Ion | Bare ion | | | Hydrated ion | | | | | | | Class |
|---|---|---|---|---|---|---|---|---|---|---|---|
| | | | | Thermodynamic properties | | | Transport properties | | Water molecules $t_r$ | | |
| | $z/r$[a] | $\varepsilon_i$[b] | $P_i$[c] | $\bar{S}_i^{\ominus}$[d] | $\bar{V}_i^{\ominus}$[e] | $B_i$[f] | $\Delta E_i^{\ddagger}$[g] | $\Lambda_r$[h] | i | j | |
| $H^+$ | — | 3.55 | — | -23.0 | -5.4 | 0.069[k] | 210 | -0.005 | — | — | |
| $Li^+$ | 1.67 | 0.74 | 0.027 | -10.5 | -6.3 | 0.15 | 20 | 0.036 | — | — | |
| $Na^+$ | 1.05 | 0.70 | 0.21 | 36.0 | -6.6 | 0.086 | 700 | — | 2.60 | 2.88 | |
| $Be^{2+}$ | 6.45 | 1.91 | 0.008 | — | -22.8[k] | 0.392 | 420 | — | 1.27 | 1.75 | |
| $Mg^{2+}$ | 3.08 | 1.56 | 0.12 | -164.1 | -32.0 | 0.385 | — | — | $3 \times 10^{7}$[1] | 5.25 | |
| $Ca^{2+}$ | 2.02 | 1.22 | 0.9 | -101.3 | -28.6 | 0.285[k] | — | — | $10^{6}$[1] | 3.5 | I[s] |
| $Sr^{2+}$ | 1.77 | 1.10 | 1.42 | -85.4 | -29.6 | 0.265[k] | — | — | $3 \times 10^{2}$[m] | 3.13 | |
| $F^-$ | 0.74 | 5.75 | 0.99 | 9.2 | 4.2 | 0.096 | 1240 | — | $3 \times 10^{2}$[m] | 2.25 | |
| $OAc^-$ | 0.40[p] | — | — | 110.1[n] | 35[k] | 0.25[r] | (+ve) | — | — | — | |
| $(CH_3)_4N^+$ | — | — | — | — | 84.2 | 0.143 | — | — | $1.4 \times 10^{2}$[m] | 2.75 | |
| $Ba^{2+}$ | 1.48 | 1.02 | 2.4 | -33.5 | -23.3 | 0.22 | -470 | 0.043 | — | — | II[t] |
| $OH^-$ | 0.71[o] | — | — | 12.1 | 1.4 | 0.112 | -240 | — | — | — | |
| $SO_4^{2-}$ | 0.98[q] | — | — | 63.2[n] | 25.0[k] | 0.208 | -280 | — | — | 1.75 | |
| $CO_3^{2-}$ | 1.16[q] | — | — | -8.8 | 6.5 | — | -920 | — | — | 2.38 | |
| $NH_4^+$ | 0.71[o] | 1.94 | 1.94 | 90.4 | 12.4 | -0.007 | -100 | — | — | — | |
| $K^+$ | 0.75 | 0.56 | 0.87 | 79.6 | 3.6 | -0.007 | -220 | 0.122 | 0.54 | 0.88 | |
| $Rb^+$ | 0.68 | 0.53 | 1.9 | 101.3 | 8.7 | -0.030 | -490 | — | — | 0.63 | |
| $Cs^+$ | 0.59 | 0.49 | 2.9 | 110.1 | 15.9 | -0.045 | -430 | 0.150 | 0.59 | 0.5 | |
| $Cl^-$ | 0.55 | 4.93 | 3.02 | 79.6 | 23.2 | -0.007 | -220 | 0.092 | 0.70 | 0.88 | III[u] |
| $Br^-$ | 0.51 | 4.53 | 4.17 | 105.5 | 30.1 | -0.042 | -240 | 0.111 | 0.51 | 0.75 | |
| $I^-$ | 0.46 | 3.84 | 6.22 | 134.4 | 41.6 | -0.068 | -370 | 0.115 | — | 0.5 | |
| $NO_3^-$ | 0.51[p] | — | — | 169.6 | 34.4 | -0.046 | -240 | — | — | 0.75 | IV[v] |
| $(C_2H_5)_4N^+$ | 0.33 | — | — | — | 143.7 | 0.385 | 160 | — | — | — | |

[a] Charge/radius ratio (Nightingale 1966). Pauling radii unless otherwise stated.

[b] Electronegativity—ratio of average electron density to that of a hypothetical, isoelectronic inert element (Sanderson, 1960, p. 32).

[c] Polarizability [$10^{-30}$ m$^3$] (Morf and Simon, 1971; Conway, 1952).

[d] Standard partial molal entropy (J mol$^{-1}$ K$^{-1}$). Single ion parameters estimated from mean of a number of determinations of $\bar{S}_{H^+}^{\ominus}$ (Desnoyers and Joliceour, 1969; Rosseinsky, 1965; Millero, 1972a).

[e] Standard partial molal volume ($10^{-6}$ m$^3$ mol$^{-1}$), references as for d.

[f, g] Transport properties defined by the equations (Nightingale, 1966)

$$\eta/\eta_w = 1 + Ac^{\frac{1}{2}} + Bc \qquad (1)$$

$$v_+ \Delta E_+^{\ddagger} + v_- \Delta E_-^{\ddagger} = R(1 + Bc)^{-1} d(1 + Bc)/d(1/T) \qquad (2)$$

$\eta, \eta_w$ = viscosity of solution and water respectively
A, B = adjustable parameters
Single ion values obtained assuming $B_{K^+} = B_{Cl^-}$.
Values of $B$ over a range of temperatures are given by Stokes and Mills (1965). The application of equation (2) to electrolyte mixtures has been considered by Wu (1968).

[h] Defined by the equation (Broadwater and Kay 1970)

$$\Lambda_r = [(\lambda_0\eta_0)_{10} /(\lambda_0\eta_0)_{45} ] - 1 \qquad (3)$$

[i] $\lambda_0$ = limiting ionic conductance

$$t = t_i/t_0$$

[j] where $t$ represents the mean residence time of a water molecule in a particular site (Samoilov, 1965, 1972). For pure water at 298 K $t_0 = 10^{-11}$ sec (Eisenberg and Kauzmann, 1969).

$$t_r = \tau_{ri}/\tau_{r0} \quad \text{(Hertz, 1979)}$$

where $\tau_r$ represents the re-orientation time for a water molecule in a particular site. For pure water $\tau_{r0} = 0.8 \times 10^{-11}$ s. Single ion parameters for $i$ and $j$ obtained assuming $\tau_{r,K^+} = \tau_{r,Cl^-}$.

[k] Estimated from data in Millero (1972a)

[l] Akitt (1971)

[m] Eigen and Maas (1966)

[n] Estimated from data in Lewis and Randall (1961, p 400)

[o] Goldschmidt radii (Table 2.11)

[p] Radii estimated by Masterton et al. (1971)

[q] Using radii given by Millero (1969)

[r] Gurney (1953 p. 169)

[s] Class I (Nightingale,1966). Hydrophilic structure makers (Desnoyers and Joliceour, 1969); positively hydrated ions (Samoilov, 1965, 1972).

[t] Class II ions (Nightingale, 1966)

[u] Class III ions (Nightingale, 1966). Structure breakers (Desnoyers and Joliceour, 1969); negatively hydrated ions (Samoilov, 1965, 1972).

[v] Class IV ions (Nightingale, 1966). Hydrophobic structure makers (Desnoyers and Joliceour, 1969).

This general behaviour is confirmed by estimates of the residence time (Samoilov, 1965, 1972) and the re-orientation or rotation time (Hertz, 1970) of water molecules adjacent to the ion ($t_r$, Table 2.12) which indicate that water molecules next to Class I and Class II ions are less mobile than water molecules in the bulk liquid. The nearest neighbours of the Class III ions are, in contrast, *more* mobile that molecules in the bulk water. The effect of dissolved ions on the "volatility" of water molecules is nicely illustrated by the variation of the activity coefficient of water with solution concentration for the alkali halides (Fig. 2.7; Samoilov, 1965, p. 98). The classifications that

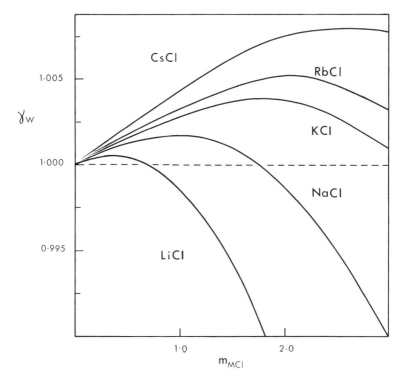

FIG. 2.7. Activity coefficient of water on the mole fraction scale in alkali metal chloride solutions at 298 K and 1 atm. The higher the value of $\gamma_w$ the more volatile is the water in the salt solution. Re-plotted from data presented in Samoilov (1965, p. 97).

result from this brief summary of solution properties are both broad and subjective and, strictly, should apply only to solutions at 298 K and 1 atmosphere. The concepts of "structure-making" and "structure-breaking" ions will, nonetheless, prove useful in the discussion of electrolyte mixtures (Section 2.7).

### 2.4.2. HYDRATION NUMBERS

Despite the complexity of the problem, numerous attempts have been made to determine the number of water molecules that are, on a time average, associated with a particular ion. The hydration numbers defined in this way are frequently used to identify the structural units in electrolyte solutions. The experimental procedures have been critically reviewed by Conway and Bockris (1954, p. 68), Robinson and Stokes (1965, p. 52) and by Conway (1970, p. 22). The water molecules "associated" with the ions are first assigned some property that differentiates them from the bulk water. For example, when compressibility measurements are being analyzed it is frequently assumed that the water immediately surrounding the ion is so rigidly orientated that it is effectively incompressible. Then experimental data for electrolyte solutions are compared with those for pure water at the same temperature and pressure, and after due allowance has been made for the presence of the unhydrated ions, the difference is attributed to the presence of a certain number of "modified" water molecules per mole of solute (i.e. to the hydration number). The number of molecules identified in this way will depend not only on the time scale of the experimental disturbance (Fig. 2.5), but also on the ability of the experimental method to differentiate between bulk water molecules and those attached to the ion (Gurney, 1953). The picture is further complicated because many of the techniques most sensitive to structural influences (X-ray diffraction, n.m.r. and i.r. spectroscopy) require relatively high concentrations ($> 1$ M) to produce meaningful results. The hydration parameters assigned therefore include the effects of competition between the ions for the attention of the water molecules.

These complications (Conway, 1970) have led to confusing literature (Hinton and Amis, 1971). For example Horne and Birkett (1967) identify a cluster of some thirty or forty water molecules as the total hydration atmosphere of NaCl at 298 K whereas Vogrin et al. (1971) give a value of 4·5 for the total effective hydration number of this salt. A survey of a more conservative spectrum of values (Table 2.13) still indicates a considerable measure of disagreement. Such discrepancies, which mainly reflect incompatible definitions of the hydration sheath, have thrown doubt on the utility of the concept, and Nightingale (1966) dismisses the hydration number as a "travesty which only serves to cloud one's understanding", echoing the sentiments of Couture and Laidler (1956) concerning the arbitrary and subjective definitions of "associated" water molecules (see also Bergqvist and Forslind, 1962).

TABLE 2.13

*Hydration numbers*[k]

| Salt | Thermodynamic | | | Structural | | | | | Adjustable | |
|---|---|---|---|---|---|---|---|---|---|---|
| | a | b | c | d | e | f | g | h | i | j |
| LiCl | 5·0 | 2·9 | 4·7 | 6 | 6·3 | 3·2 | – | 5·2 | 7·1 | 4·3 |
| LiBr | – | 2·7 | 4·0 | – | 5·6 | 4·4 | – | – | 7·6 | 4·3 |
| LiI | 2·1 | 1·8 | 3·7 | – | – | – | – | – | 9·0 | 4·3 |
| NaF | – | 7·7 | 5·7 | 4 | – | – | – | – | – | 3·6 |
| NaCl | 5·7 | 5·3 | 3·6 | 6 | 3·5 | 4·5 | 4·0 | 2·1 | 3·5 | 2·7 |
| NaBr | 4·8 | 5·1 | 2·8 | 6 | 2·8 | 4·4 | 4·4 | 0·7 | 4·2 | 2·7 |
| NaI | – | 4·2 | 2·4 | 6 | 3·0 | – | 4·6 | 0·6 | 5·5 | 2·7 |
| KF | – | 7·0 | 4·3 | – | – | – | – | – | – | 2·6 |
| KCl | 4·9 | 4·6 | 2·0 | 5 | 0·6 | 4·6 | 3·5 | 2·3 | 1·9 | 1·7 |
| KBr | 4·0 | 4·4 | 1·4 | – | 0·3 | 4·1 | 4·2 | – | 2·1 | 1·7 |
| KI | 2·2 | 3·5 | 1·0 | – | 0·3 | – | 4·3 | – | 2·5 | 1·7 |
| RbCl | – | 4·1 | 1·5 | 4 | – | 4·0 | 3·5 | – | 1·2 | 0·9 |
| RbBr | – | 3·9 | 0·9 | – | – | – | – | – | 0·9 | 0·9 |
| RbI | – | 3·0 | 0·5 | – | – | – | – | – | 0·6 | 0·9 |
| CsCl | – | 3·5 | 1·4 | – | – | 3·9 | 3·1 | 4·6 | – | 0·9 |
| CsBr | 3·0 | 3·3 | 0·8 | – | – | – | – | – | – | 0·9 |
| CsI | – | 2·4 | 0 | – | – | – | – | – | – | 0·9 |
| MgCl$_2$ | – | – | – | – | – | 8·2 | – | – | 13·7 | 6·5 |
| CaCl$_2$ | – | – | – | – | – | 9·5 | – | – | 12·0 | 5·9 |

[a] Standard partial molal compressibility, Desnoyers and Joliceour (1969)
[b] Standard partial molal volume, Desnoyers and Joliceour (1969)
[c] Standard partial molal entropy, Desnoyers and Joliceour (1969)
[d] Dielectric constant, Conway and Bockris (1954, p. 67)
[e] Diffusion, Robinson and Stokes (1965)
[f] Nuclear magnetic resonance, Vogrin *et al.* (1971), 1M
[g] Infra-red spectroscopy, McCabe and Fisher (1970), 1M
[h] Infra-red spectroscopy, Bonner and Woolsey (1968)
[i] Adjustable parameter, Stokes and Robinson (1948)
[j] Adjustable parameter, Glueckauf (1955)
[k] More extensive tables of hydration numbers are given by Hinton and Amis (1971).

Since the hydration number is clearly not a definition of the hydrated ion in molecular terms, attempts have been made to make the arbitrary nature of this parameter more explicit and thus avoid unnecessary structural speculation. Desnoyers and Joliceour (1969) treat the hydration number as a variable parameter summarizing deviations from ideal behaviour resulting from hydration effects. The hydration number therefore encompasses ion–solvent effects in the same way that the activity coefficient in dilute solution summarizes ion–ion effects. A similar approach was also employed by Glueckauf (1955) and by Stokes and Robinson (1948) in the development of theories relating activity coefficients to solution composition in concentrated electrolyte solutions (columns $i$ and $j$, Table 2.13).

Conway (1970) suggested that less confusion would arise if the adjustable hydration parameter was used not to describe the number of water molecules with given properties, but rather to describe the distribution of these properties between the water molecules of a fixed hydration sheath. Monovalent ions could be considered as four or six-coordinated, and divalent ions as six or eight-coordinated and the properties of the hydration sheath would be described on a "per-water-molecule" basis (see also Muirhead-Gould and Laidler, 1966).

Structural considerations alone give, at best, a semi-quantitative view of interactions between water molecules and the isolated ion. The treatment of concentrated solutions is necessarily more complex since the structural organisation will be disturbed even further by interactions between the ions themselves. Consequently, quantitative theories of electrolyte solutions, designed to predict activity coefficients, must make some compromise between molecular detail and computational feasibility. The extent of this compromise will be defined in the next section.

## 2.5. SOLUTIONS CONTAINING A SINGLE ELECTROLYTE

### 2.5.1. THE PARTITION FUNCTION

Although classical thermodynamics ignores the fine structure of matter, it is possible, via statistical mechanics, to relate the thermodynamic properties of a system to the energy of its individual molecular components. This approach (Denbigh, 1961; Fowler and Guggenheim, 1952; Hill, 1956) is based on the definition and determination of the partition function ($\mathcal{Q}$), which defines the distribution of the molecules between the various energy levels in the molecular assembly. For fluid phases at moderate density, the total partition function may be factorised into three mutually independent components—(i) the translational component ($\mathcal{Q}_{tr}$) related to the kinetic energy of the particles, (ii) the *intra* molecular component ($\mathcal{Q}_{intr}$) associated with rotational, vibrational etc. forces within the molecules and (iii) the *inter* molecular or configurational component ($\mathcal{Q}_c$) associated with the potential energy arising from interactions between the molecules. It is this final term ($\mathcal{Q}_c$) that contributes to the deviations from ideal behaviour which are summarised by the activity coefficient. $\mathcal{Q}_c$ is defined for conditions of constant volume and temperature. The free energy appropriate for these conditions is the Helmholtz free energy ($A$) which may be related to the Gibbs free energy $(G)$ at a given temperature and pressure and to the partition function (Denbigh, 1961, p. 356) by the equations

$$A = G - PV \tag{5.1a}$$

$$= -kT \ln \mathcal{Q}_c \tag{5.1b}$$

where $k$ is the Boltzman constant. $\mathcal{Q}_c$ can be expressed in terms of the total intermolecular potential energy $(U)$ by the equation (Vaslow, 1972)

$$\mathcal{Q}_c = \int_v \exp(-U/kT) \tag{5.2}$$

where

$$U = 0.5 \sum_i^n \sum_{j \pm i}^n u_{ij}(r) \tag{5.3}$$

Some simplification is necessary if $\mathcal{Q}_c$ is to be determined for real systems, and at least two distinct approaches have been made. The most direct of these is to consider the electrolyte solution as a granular mixture of ions in a continuous, structureless solvent. The statistical treatment of the interaction parameters $(u_{ij}(r))$ is simplified by considering only electrostatic interactions and by describing them in terms of a virial or power series (Milner, 1912, 1913). Although the fundamental problems were clearly stated by Milner and elaborated by later workers (e.g. Fowler, 1927; Kirkwood, 1934), no general quantitative treatment could be attempted until the advent of electronic computing techniques in the 1950's. The second approach, developed by Debye and Hückel (1923), dealt with a much simpler model and proved highly successful in the prediction of the properties of very dilute solutions. This theory is a special solution of equation (5.2) referring to particularly, clearly defined, boundary conditions. Any attempt to extrapolate the model beyond these conditions is purely empirical.

### 2.5.2. THE DEBYE–HÜCKEL THEORY

In this theory the molecular nature of the solution is ignored, and consideration is restricted to solutions so dilute that only long range electrostatic interactions cause deviations from ideal behaviour. The calculation involves a central ion, considered as a rigid, non-polarizable sphere, immersed in a solvent that is reduced to a dielectric continuum. The remaining ions are considered as point charges that are smeared out, on a time average, to constitute a charge cloud around the central ion. The cloud has a net charge equal in magnitude, but opposite in sign, to that of the central ion. The calculation of the electrostatic potential experienced by the central ion, and hence its activity coefficient has been considered in detail by a number of authors (Bockris and Reddy, 1970; Conway, 1970; Vaslow, 1972; Robinson and Stokes, 1965; Harned and Owen, 1958).

The distribution of point charges about the central ion is defined in statistical terms by the Boltzmann equation, and the influence of this charge cloud on the central ion is described by the Poisson equation. The two relationships give incompatible pictures of the system since the first predicts an exponential relationship between the ionic charge and the electrostatic potential whereas the second predicts a linear relationship (Harned and Owen, 1958, p. 51). However, the two equations must be combined to calculate the electrostatic potential at the central ion. The contradiction involved in making this combination may be removed by expanding the exponential Boltzmann equation and confining the analysis to very dilute solutions where only the linear term in the expansion is significant. This approximation will be most successful for ions surrounded by relatively weak electrostatic fields (i.e., large, singly charged ions, Table 2.12). The linearised Poisson–Boltzmann equation can be solved to calculate the electrostatic potential at the central ion. The activity coefficient may be related directly to the work required to charge the central ion to this potential in the presence of the charge cloud. Because of the inconsistencies associated with the linearisation of the Poisson–Boltzmann equation, different pathways for the charging process can yield different values for the electrostatic free energy (Onsager, 1933; Guggenheim and Stokes, 1969, p. 85).

Despite the crudity of the model, the resulting equations correctly predict the linear relationship between $\log \gamma_A$ and $I^{\frac{1}{2}}$ observed empirically at low concentrations (Fig. 2.8, Table 2.14) and the *slopes* of these lines can be accurately predicted. These equations however are only applicable to slightly contaminated water (electrolyte concentration less than $10^{-3}$ M). A graphic account of the breakdown of the Debye–Hückel theory at higher concentrations is given by Frank and Thompson (1959a, b). The limiting law equations (Table 2.14, equation 1) can be applied directly to mixed electrolyte solutions since only the overall ionic strength and not the specific ionic environment is significant in their derivation. The same concentration restrictions apply for mixtures as for single electrolyte solutions.

The simple theory can be extended to more interesting concentrations (up to 0·1 M in some cases) by the inclusion of an adjustable parameter ($\mathring{a}$) representing the distance of closest approach to the central ion (equations 2 and 5, Table 2.14). This avoids consideration of the region very close to the central ion where the linear form of the Poisson–Boltzmann equation may not be valid. In some instances where small, highly charged, ions are considered, very small or even negative values of $\mathring{a}$ may be required to fit the equation

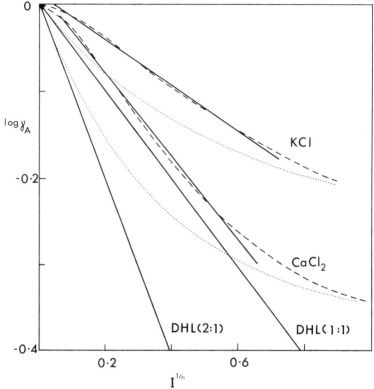

FIG. 2.8. Mean-ion activity coefficients of potassium chloride and calcium chloride solutions plotted as functions of ionic strength.
...... Experimental data versus $I^{\frac{1}{2}}$ (i.e. $n = 2$). The lines marked DHL indicate the limiting slopes predicted by the Debye-Hückel theory (Section 2.5.2).
--- Experimental data versus $I^{\frac{1}{3}}$ (i.e. $n = 3$). A solid line has been drawn through the data to indicate the approximate extent of the "cube root" behaviour discussed in Section 2.5.3.

to the experimental data. In electrolyte mixtures the cross differentiation relationships (e.g. equation 2.14) which relate the activity coefficients of the various components are obeyed only if $\mathring{a}_i = \mathring{a}_j$, etc. A number of empirical equations avoid this difficulty by giving $\mathring{a}$ a standard value for all ions (Table 2.14, equations (3) and (4)). The predictions of the various electrostatic equations begin to deviate widely at ionic strengths above 0·1 M (Fig. 2.9). A number of attempts have been made to extend the Debye–Hückel theory to give a more consistent account of the behaviour of electrolyte solutions at high ionic strengths. They may be grouped under two general headings— algebraic refinements and model extensions.

TABLE 2.14

*Electrostatic equations relating activity coefficients to solution composition.*

| Equation number | $\log \gamma_A =$ | Adjustable parameters | Reference |
|---|---|---|---|
| 1[a, b] | $-\mathscr{A}\|z_+ z_-\| I^{\frac{1}{2}}$ $= -DH$ (limiting law) | — | Debye and Hückel (1923) |
| 2[c] | $-DH/(1 + \mathscr{B}\mathring{a}I^{\frac{1}{2}})$ | $\mathring{a}$[d] | Debye and Hückel (1923) |
| 3[a, b] | $-DH/(1 + I^{\frac{1}{2}}) = -DG$ | — | Güntelberg (1926) i.e. $\mathring{a} = 0.304$ nm |
| 3(a)[a, e] | $-DH/(1 + 0.3\mathscr{B}I^{\frac{1}{2}})$ | — | Hamer (1968) |
| 4[a, b, f] | $-DH/(1 + 1.5I^{\frac{1}{2}})$ | — | Scatchard (1936) i.e. $\mathring{a} = 0.456$ nm |
| 4(a)[a, b, e] | $-DH/(1 + 0.45\mathscr{B}I^{\frac{1}{2}})$ | — | Hamer (1968) |
| 5[c] | $-DH/(1 + \mathscr{B}\mathring{a}I^{\frac{1}{2}}) + bI$ | $\mathring{a}, b$ | Hückel (1925)[g] |
| 6[b] | $-DG + \mathscr{A}\|z_+ z_-\| 0.2I$ | — | Davies (1938, 1962)[h] |

[a] Data up to 0·1 M from 273–373 K tabulated by Hamer (1968). Values of $\mathscr{A}$ and $\mathscr{B}$ as functions of temperature for aqueous solutions are tabulated by Hamer (1968), Harned and Owen (1958), Robinson and Stokes (1965) and Bates (1964). At 298 K $\mathscr{A} = 0.5108$ mole$^{-\frac{1}{2}}$ kg$^{\frac{1}{2}}$ and $\mathscr{B} = 3.287$ nm$^{-1}$ mol$^{-\frac{1}{2}}$ kg$^{\frac{1}{2}}$.

[b] For single ion activity coefficients replace $|z_+ z_-|$ by $z_i^2$.

[c] Adapted for treatment of single ion activities by the use of $a$ values tabulated by Kielland (1937) and by replacing $|z_+ z_-|$ by $z_i^2$. Bromley (1972) gives single-ion values for $b$ derived on the assumption that $\mathscr{B}\mathring{a} = 1$ nm and $b_{H+} = 0.1$ kg mol$^{-1}$.

[d] e.g. on the molal scale the equation gives a good fit for NaCl and KCl (up to 0·1 M) using the following data (298 K)

|  | $\mathring{a}$ | $\mathscr{B}\mathring{a}$ | $\mathscr{A}$ |
|---|---|---|---|
| NaCl | 0·456 | 1·5 | 0·5108 |
| KCl | 0·404 | 1·33 | 0·5108 |
|  | nm | (mol$^{-1}$ kg)$^{\frac{1}{2}}$ | (mol$^{-1}$ kg)$^{\frac{1}{2}}$ |

[e] To allow for temperature variation

[f] Equivalent to selecting NaCl as the reference electrolyte (see note [d]). Used to establish the pH scale below $I = 0.1$ M (Bates, 1964). Lewis and Randall (1961) find little basis for preferring this equation to equation 3.

[g] Values of $\mathring{a}$ and $b$ for the 1:1 halides are tabulated by Burns (1964) and, with the final term replaced by $bm_A$ for 1:2 and 2:1 electrolytes by Guggenheim and Stokes (1958)

[h] Meites (1963) tabulates values for the equation

$$\log \gamma_A = -DH/(1 + 1.5I^{\frac{1}{2}}) + 0.2Iz_i^2$$

A coefficient of 0·3 in the linear term is now preferred (Davies, 1962).

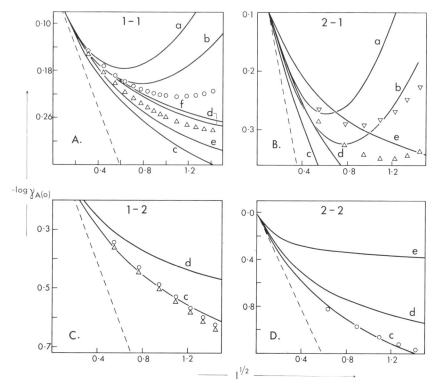

FIG. 2.9. Mean-ion activity coefficients of the major sea water components compared with those derived from calculations based on various theoretical equations. *Experimental data* (Robinson and Stokes, 1965)

A.  $\bigcirc$ = NaCl,        $\triangle$ = KCl
B.  $\triangledown$ = MgCl$_2$,      $\triangle$ = CaCl$_2$
C.  $\bigcirc$ = Na$_2$SO$_4$    $\triangle$ = K$_2$SO$_4$
D.  $\bigcirc$ = MgSO$_4$

*Theoretical equations* (equation numbers in brackets refer to Table 2.14)

a. Davies (1938, equation 6) with the coefficient of the linear term = 0·3,

b. As in a. with the coefficient of the linear term = 0·2.

c. Guntelberg (1926, equation 3).

d. Scatchard (1936, equation 4).

e. Equation (2), Table 2.14, with å set equal to the appropriate Bjerrum distance as listed by Hamer (1968).

f. Equation (2), Table 2.14, with å = 0·48 nm (Robinson and Stokes, 1965). With å = 0·456 nm (for NaCl) this equation is equivalent to d.

The broken line follows the Debye–Hückel limiting law.

The algebraic refinements of the electrostatic theory have resulted in the most elaborate and probably the least productive developments. They have included the consideration of other distribution functions to replace the Boltzmann equation (Robinson and Stokes, 1965, p. 79; Bagchi, 1950; Falkenhagen and Kelbg, 1959), the evaluation of higher terms in the expanded Poisson–Boltzmann equation (Gronwall et al., 1929) and the exact solution of the nonlinearised Poisson–Boltzmann equation (Gronwall, 1927; Müller, 1927; Guggenheim, 1959, 1962; Burley et al., 1971; Gardener and Glueckauf, 1971). The most readily accessible results of such calculations are those tabulated by Scatchard (1965) and by Gardener and Glueckauf (1971). Although these approaches do not remove the basic logical inconsistency involved in the combination of the two distribution functions, they do eliminate most of the absurdities associated with the adjustable parameter ($\mathring{a}$) in the simple Debye–Hückel equations. Furthermore, they show some improvement in the treatment of 2:2 electrolytes (Guggenheim, 1960, 1966a). Their result then is to broaden the Debye–Hückel approach to include a wider range of electrolyte solutions without significantly extending the ionic strength ranges that can be considered. Most significant advances in this direction have resulted from a re-appraisal of the *physical* model adopted by Debye and Hückel.

As a starting point an extended Debye–Hückel equation, (see Table 2.14) is assumed to give an adequate description of long range electrostatic interactions. Further terms are then added to allow for other types of interaction.

The earliest and most widely used empirical modifications of the Debye–Hückel theory (Bjerrum, 1926; Brønsted, 1922) considered that, at higher concentrations, *specific* interactions between ions of opposite charge would be superimposed on the general long-range ion cloud effect.

### 2.5.2.1. *The ion-association model*

Bjerrum assumed that ions of opposite charge to the central ion which approach within a critical distance, $q$, effectively form an ion-pair and are no longer electrostatically effective (Table 2.15). Thus, $q$ provides a natural lower limit to the integration of the Poisson–Boltzmann equation. The formation of such an ion-pair may be described formally by the equation

$$M^{z+} + A^{z-} \rightleftharpoons MA^0$$

with component concentrations, $m(1 - \theta)$, $m(1 - \theta)$ and $\theta m$ respectively.

$\theta$ is the degree of association of the ions and $m$ is the molality of the *salt MA*. In the present context ion-pairs are temporary partnerships arising between ions of opposite charge as a result of predominantly electrostatic forces. The

TABLE 2.15

*Bjerrum ion-association model*[a]

(i) *Bjerrum critical distance (q)*
Defined as distance at which

$$2kT = |z_+z_-| \, e^2/Dq \tag{1}$$

where $e$ is the electronic charge and $D$ the dielectric constant of the solvent. $q$ represents the radius at which there is a minimum probability of finding ions with a charge opposite to the central ion. It provides an energetic definition of the ion-pair.

(ii) *Degree of association ($\theta$)*
Integrate number of ions in all shells from the distance of closest approach (a)[b] to the Bjerrum critical distance (q)

$$\theta = 4\pi n_j C^3 . Q(b) \tag{2}$$

where
$$Q(b)^c = \int_2^b x^{-4} \exp(x) \, dx$$

and
$$C = |z_+z_-| \, e^2/DkT$$

$$b = C/a, \qquad x = -C/r, \qquad 2 = C/q$$

$\theta$ is given on the molar concentration scale since, $n_j$ the number of ions in unit volume of solution, is given by

$$n_j = (c_j/1000) \, N$$

---

[a] This model is illustrated because it is frequently quoted. Many alternative definitions of ion-association are in use (see for example Prue (1969) and Petrucci (1971))

[b] Not the same as $\mathring{a}$ in the Debye–Hückel theory. However, Hamer (1968) tabulates data for a limiting law expression based on the Bjerrum model where $\mathring{a}$ is replaced by $q$.

[c] Values of $Q(b)$ have been tabulated by Bjerrum (1926) and Robinson and Stokes (1965)

[d] For general discussions see Conway (1970), Robinson and Stokes (1965) and Davies (1962).

---

extent of ion-pairing will be largely controlled by the charges and the radii of the interacting ions. Most stable entities that result from the formation of largely covalent bonds between a metal ion and an electron donating ligand, which need not itself be charged, are called complexes. The stability of complexes is largely controlled by the ionisation potential of the metal ion and by the electronegativity of the ligand.

Ion-pairs are, in general, formed between "hard" (i.e. non-polarizable, Pearson, 1963) cations with an inert gas structure (e.g. $Ca^{2+}$, $Na^+$) and "hard" anions—preferably fluoride ions or ligands with an oxygen atom as the electron donor (e.g. $SO_4^{2-}$, $CO_3^{2-}$). Complexes are formed by the "softer" (more polarisable) cations of the transition metals and the B-subgroups of the periodic table with ligands containing "soft" electron donor groups (e.g. S, P, N) (see also Chapter 3).

The equilibrium constant $(K^{\ominus})$ for the formation of an ion-pair may be defined by the equation

$$
\begin{aligned}
K^{\ominus} &= a_{MA}/a_M a_A \\
&= [\theta/m(1 - \theta)^2][\gamma_{MA}/{}^F\gamma_A{}^F\gamma_M] \\
&= K^*[\gamma_{MA}/{}^F\gamma_A{}^F\gamma_M]
\end{aligned}
\tag{5.4}
$$

The prefix $F$ indicates that the activity coefficients refer to the model where free ions *and* ion-pairs are considered as the solution components (free-ion activity coefficients) in contrast to the more general model where only the ions themselves are considered as the solution components (total or stoichiometric activity coefficients, ${}^T\gamma$). The two definitions of the activity coefficient are related by the equation

$$
a_i = {}^F\gamma_i{}^F m_i = {}^T\gamma_i{}^T m_i
\tag{5.5}
$$

This gives

$$
{}^F\gamma_i = {}^T\gamma_i/(1 - \theta)
\tag{5.6}
$$

Equation 2, Table 2.14 can now be re-written using the ionic strength corrected for ion-pair formation and remembering that the Debye–Hückel treatments considers total rather than free ion activities (equation (1), Table 2.16).

In solutions of symmetrical electrolytes at moderate concentrations $\theta$-values can be calculated from the Bjerrum theory (Table 2.15). The values obtained are best considered as numerical summaries of a convenient mathematical model rather than as expressions of molecular reality (Bjerrum, 1927; Davies, 1962). This is most clearly seen by considering ion-pairs formed by 2:2 electrolytes (Robinson and Stokes, 1965). For these ions, $q$ has a value of 1·426 nm at 298 K. Since a sphere of this radius can hold approximately 400 water molecules it is unlikely that all ions approaching within the critical radius will form physically distinct ion-pairs. Fuoss (1934, 1958) developed an alternative approach in which ion-pairs are formed by direct contact between ions of opposite charge. Further attempts have been made to refine the model by considering local variations in the dielectric constant of water in the immediate vicinity of the ion (Rosseinsky, 1962; Panckhurst, 1962) and by considering the granular nature of the solvent (Levine and Wrigley, 1957; Conway, 1970).

Most applications of the ion-association model depend on the experimental determination of $\theta$. The procedures used are philosophically parallel to the methods used for the determination of hydration numbers and are

TABLE 2.16

Model extensions of the Debye–Hückel theory

| Equation number | Extra effects considered | $\log \gamma_A =$ | Adjustable parameters | Reference |
|---|---|---|---|---|
| 1. | Ion-association | $-\dfrac{\mathscr{A}\lvert z_+ z_- \rvert [(1-\theta)I]^{\frac{1}{2}}}{1 + \mathscr{B}q[(1-\theta)I]^{\frac{1}{2}}} + \log(1-\theta)$ | $\theta$ | Bjerrum (1926)[a] |
| 2. | Specific interaction between ions of opposite charge | $-DG + \underline{v}bm$ | $b$[b] | Brønsted (1922) Guggenheim (1935) |
| 3. | Entropy effects associated with ion hydration (mole fraction statistics) | $-DH/(1 + \mathscr{B}\mathring{a}I^{\frac{1}{2}})$ $+ 0.0078\,hm\phi - \log(1 + 0.018(v - h)m)$ | $\mathring{a}, h$[c] | Stokes and Robinson (1948) |
| 4. | ibid. (volume fraction statistics) | $-DH/(1 + \mathscr{B}\mathring{a}I^{\frac{1}{2}})$ $+ 0.0078\,mr(r + h - v)/[v(1 + 0.018mr)]$ $+ [(h - v)/v]\log(1 + 0.018\,mr)$ $- (h/v)\log(1 - 0.018\,mh)$ | $\mathring{a}, h$ | Glueckauf (1955)[d] |

[a] Derivation outlined by Robinson and Stokes (1965) and Bockris and Reddy (1970). A form of the Bjerrum equation in which $\mathring{a}$ in equation (2) (Table 2.14) is replaced by $q$ is discussed by Hamer (1968) (see Fig. 2.9).

[b] $$v^{-1} = 0.5(v_+^{-1} + v_-^{-1})$$

Guggenheim and Turgeon (1955) use natural logarithms and tabulate values for $\beta$ where

$$b = 2\beta/\ln 10$$

Lewis and Randall (1961) use decadic logarithms and assume, after Scatchard (1936), that $b$ is a slowly varying function of ionic strength (designated $B_{MX}$). They tabulate values of $vAB = vb_{MX} - b_{ref}$) thus combining the convenience of equation 2 and the Åkerlöf–Thomas rule (equation 5.7).

[c] $h$ is the hydration parameter. Glueckauf (1955) has shown that this is equivalent to the equation

$$\log \gamma_A = -DH/(1 + \mathscr{B}\mathring{a}I^{\frac{1}{2}}) - (h/v)\log(1 - 0.018\,hm) + [(h - v)/v]\log(1 - 0.018\,m(h - v)]$$

thus avoiding the introduction of the osmotic coefficient $\phi$.

[d] $r = {}^{\phi}Y_2/\bar{Y}_2^{\infty}$ where the volumes are expressed in molar terms. Glueckauf (1959) uses a rather more elaborate electrostatic function. The $h$ and $\mathring{a}$ parameters of the two equations therefore differ.

consequently prone to ambiguity. All the methods are based on the fundamental assumption that we have an exact knowledge of the behaviour of the free, unassociated ions (Davies, 1962, p. 4). Data for the associated system (e.g. spectroscopic properties, conductivity, activity coefficients) are then interpreted in terms of deviations from the properties of the free ions at the same total ionic strength.

At low ionic strengths ($<0.1$ M) one of the general forms of the Debye–Hückel theory (usually equation 6, Table 2.14) is used to define the properties of the free ions (Davies, 1962; Nancollas, 1966). At higher ionic strengths the activity coefficients of all electrolytes begin to deviate widely from the predictions of these simple equations (Fig. 2.9) and it is necessary either to assume ion-association in *all* solutions or to select model electrolytes in which ion-association is assumed to be negligible. Some estimate must also be made of the activity coefficient of the ion-pair.

The experimental techniques have been considered in detail by Davies (1962), Nancollas (1966) and Monk (1961, 1966). Where possible $\theta$ is determined by several methods using different properties to identify the free ions. However, Orgel and Mulliken (1957) have shown that the value of $\theta$ obtained will *necessarily* be independent of the experimental method if the law of mass action (equation 5.4) is used in reducing the data (see Friedman 1970, p. 7). Agreement on this basis is, therefore not a sufficient condition for believing that the ion-association model represents the real system.

### 2.5.2.2. *The specific interaction model*

An alternative treatment of short range interactions was proposed by Brønsted (1922) who suggested that these effects could be summarized by a linear term representing specific interactions between ions of opposite charge. Ions of the same charge were assumed to interact with each other in a uniform way so that their effects would be incorporated in the electrostatic equation. This approach was later elaborated by Guggenhein (1935) and by Scatchard (1936) who summarized the electrostatic term in an extended Debye–Hückel equation (Table 2.14, equations 3 and 4 respectively). The linear form of the additional correction factor is suggested by Hückel's treatment of short range interactions (Hückel, 1925, equation 5, Table 2.14) and is given considerable experimental support by the wide ranging validity of the Åkerlöf–Thomas rule (Åkerlöf and Thomas, 1934). According to this rule the activity or osmotic coefficients of a salt solution ($\gamma_A$, $\phi_A$) may be related to those of a reference electrolyte of the same charge type ($\gamma^0$, $\phi^0$) by the equations

$$\log \gamma_A - \log \gamma^0 = B^* m \tag{5.7}$$

$$\phi_A - \phi^0 = 0 \cdot 5B^* m \tag{5.8}$$

at a given ionic strength.

This relationship has been illustrated by Guggenheim (1966a) and by Guggenheim and Stokes (1969) and is obeyed in many cases up to ionic strengths equivalent to 2 M (see also Åkerlöf, 1934, 1937). Guggenheim (1966a, b) provided a theoretical basis for the linear relationship by considering the short range effects in terms of a virial (power) series. After cancellation of the common long-range electrostatic effects, the free energy of the solute can be treated as a virial series in molality in which only the leading quadratic term is significant. On differentiation with respect to molality this yields a linear relationship. Guggenheim (Table 2.16, equation 2) treated the interaction coefficient (b) as a constant, thus providing a particularly simple equation for treating electrolyte solutions at ionic strengths below $0 \cdot 1$ M. The Davies equation (Table 2.14, equation 6) is seen as a special case of this relationship.

Scatchard (1936) extended the Brønsted treatment to much higher concentrations by assuming that the interaction coefficient was a *slowly* varying function of ionic strength (note $b$, Table 2.16). Leyendekkers (1971b) has shown that for ions with which there is negligible complexing

$$(2v_1 v_2 / v z_1 z_2) b_{12} = A - (B/z_1 z_2) \bar{S}_1^{\ominus} \tag{5.9}$$

where $A$ and $B$ are constants and $\bar{S}_1^{\ominus}$ is the conventional standard partial molal entropy of the cation. By plotting known values of $b$ against the entropy function it is possible to construct a standard graph from which unknown $b$-values can be estimated if the value of $\bar{S}_1^{\ominus}$ is known (see Lewis and Randall, 1961, p. 400).

### 2.5.2.3. *Hydration models*

The activity coefficients of most electrolytes pass through a minimum with increasing ionic strength as the result of competition between forces tending to reduce $\gamma$ (e.g. electrostatic effects), predominant at low ionic strengths, and forces tending to increase $\gamma$, predominant at high ionic strengths. The activity coefficients of the electrolyte and of the water are related by the Gibbs–Duhem equation (Table 2.4, equation (1)) so that any effect that increases the activity coefficient of the solute decreases the activity coefficient of water (see Fig. 2.7). Therefore, attempts have been made to extend the Debye–Hückel model by attributing deviations from the electrostatic function (equation (2), Table 2.14) to the time average associations of water molecules with ions using the hydration number ($h$) as an adjustable parameter. This approach was first investigated by Bjerrum (1919, 1920) who used a cube root relationship to describe the electrostatic effect. Stokes and Robinson (1948) updated these equations, introducing the Debye–Hückel equation for the electrostatic interactions. They calculated the chemical potential of the solute by two routes —the first treating the solutions as a mixture of anhydrous ions and water in the classic manner and the second considering a mixture of hydrated ions and free water. The ideal free energy of mixing associated with the second process was considered on a statistical basis with the contributions of the individual components weighted according to their mole fractions (Robinson and Stokes, 1965). The resulting equation (Table 2.16, equation (3)) gives an accurate account of the activity coefficients of a wide range of electrolytes while using physically reasonable values for the adjustable parameters $\mathring{a}$ and $h$.

Glueckauf (1955) reconsidered the ion-hydration model following a similar derivation but weighting the contributions to the free energy of mixing according to the molar volumes of the components thus accounting for the effects of differences in ion size on the entropy of mixing (equation (4), Table 2.16). This gives values for $\mathring{a}$ and $h$ that are, according to Glueckauf, more acceptable. The $h$-parameters, for example, can be factorized into additive single-ion values that show trends with ion size in better agreement with intuitive structural models than those obtained by Stokes and Robinson (1948).

There is little difference between the fit of the two equations to the experimental data, and more extensive tables for $\mathring{a}$ and $h$ are available for the mole fraction statistics model (Table 2.17). In view of the approximations inherent in applying the Debye–Hückel equations to concentrated electrolyte solutions and the considerable uncertainties surrounding the physical interpretation of hydration numbers, it is difficult to assess the relative claims of the two equations.

### 2.5.3. ALTERNATIVE ELECTROSTATIC FUNCTIONS

These treatments (Braunstein, 1971) generally take as their starting point the suggestion of Bjerrum (1918, 1920) that the activity coefficient is related to the cube root of the concentration (Fig. 2.8). Such a relationship would be valid if the interaction effects between nearest neighbour ions were predominant and inversely proportional to the distance of separation. This quasi-lattice approach to ionic solutions was developed by Ghosh (1918), but his treatment was inconsistent. The concepts were revived by Bernal and Fowler (1933) and by Kirkwood (1936) as an alternative to the Debye–Hückel approach at higher concentrations. Frank and Thompson (1959a, b) later showed that the simple cube root equation was applicable to several 1:1 electrolytes in the concentration range 0·001 to 0·1 M. Desnoyers and Con-

TABLE 2.17

*Parameters summarizing the activity coefficients of single electrolyte solutions to high ionic strengths at 298 K*

A. 1:1 electrolytes[a]

| | $h$ | $\mathring{a}$ (nm) | | $h$ | $\mathring{a}$ (nm) |
|---|---|---|---|---|---|
| HCl | 8·0 | 0·447 | NaI | 5·5 | 0·447 |
| HBr | 8·6 | 0·518 | $NaClO_4$ | 2·1 | 0·404 |
| HI | 10·6 | 0·569 | KCl[b] | 1·9 | 0·363 |
| $HClO_4$ | 7·4 | 0·509 | KBr[b] | 2·1 | 0·385 |
| LiCl[b] | 7·1 | 0·432 | KI | 2·5 | 0·416 |
| LiBr[b] | 7·6 | 0·456 | $NH_4Cl$ | 1·6 | 0·375 |
| LiI[b] | 9·0 | 0·560 | RbCl[b] | 1·2 | 0·349 |
| $LiClO_4$ | 8·7 | 0·563 | RbBr | 0·9 | 0·348 |
| NaCl[b] | 3·5 | 0·397 | RbI | 0·6 | 0·356 |
| NaBr | 4·2 | 0·424 | | | |

A. 1:2 and 2:1 electrolytes[c]

| | $\mathring{a}$ (nm) | $b$ $mol^{-1} kg$ | Salt | $\mathring{a}$ (nm) | $b$ $mol^{-1} kg$ |
|---|---|---|---|---|---|
| $MgCl_2$[b] | 0·483 | 0·206 | $Mg(NO_3)_2$ | 0·466 | 0·208 |
| $CaCl_2$[b] | 0·467 | 0·169 | $Ca(NO_3)_2$ | 0·422 | 0·052 |
| $SrCl_2$ | 0·474 | 0·125 | $Sr(NO_3)_2$ | 0·422 | −0·043 |
| $BaCl_2$ | 0·474 | 0·066 | $Co(NO_3)_2$ | 0·483 | 0·148 |
| $MnCl_2$ | 0·474 | 0·148 | $Cu(NO_3)_2$ | 0·466 | 0·122 |
| $FeCl_2$ | 0·474 | 0·162 | $Zn(NO_3)_2$ | 0·501 | 0·172 |
| $CoCl_2$ | 0·474 | 0·188 | $Zn(ClO_4)_2$[d] | 0·618 | 20·0 |
| $NiCl_2$ | 0·474 | 0·188 | $Cd(NO_3)_2$ | 0·483 | 0·104 |
| $CuCl_2$ | 0·474 | 0·084 | $Mg(ClO_4)_2$ | 0·571 | 0·356 |
| $MgBr_2$ | 0·492 | 0·291 | $Li_2SO_4$ | 0·430 | −0·065 |
| $CaBr_2$ | 0·492 | 0·212 | $Na_2SO_4$[b] | 0·386 | −0·165 |
| $SrBr_2$ | 0·492 | 0·176 | $K_2SO_4$ | 0·325 | −0·087 |
| $BaBr_2$ | 0·474 | 0·140 | $Rb_2SO_4$ | 0·404 | −0·148 |
| $MgI_2$[d] | 0·614 | 19·0 | $Cs_2SO_4$ | 0·404 | −0·113 |
| $CaI_2$[d] | 0·569 | 17·0 | | | |
| $SrI_2$[d] | 0·558 | 15·5 | | | |
| $BaI_2$[d] | 0·544 | 15·0 | | | |

C. 2:2 Electrolytes[e]

| | $A$ | $B \times 10^2$ | $C \times 10^3$ | $D \times 10^4$ |
|---|---|---|---|---|
| $BeSO_4$ | 1·2325 | −1·54261 | 5·10411 | −0·945238 |
| $MgSO_4$ | 1·37486 | −5·42492 | 8·42636 | −1·89929 |
| $MnSO_4$ | 1·28920 | −5·47447 | 7·36518 | −1·56926 |
| $NiSO_4$ | 1·31677 | −7·20761 | 10·2081 | −2·45406 |
| $CuSO_4$ | 1·14652 | −2·25375 | −1·13297 | 6·24346 |
| $ZnSO_4$ | 1·27839 | −5·56227 | 7·60615 | −1·25189 |
| $CdSO_4$ | 1·20516 | −5·08061 | 7·00382 | −1·69009 |

way (1964) assumed that the activity coefficient could be factorized into three components representing contributions from electrostatic ($\gamma_e$), hydration ($\gamma_h$) and salting out ($\gamma_{so}$) effects. The hydration term was evaluated from the theory of Stokes and Robinson (1948) using experimental hydration parameters obtained from partial molal volume measurements. The salting out effect arises from the replacement of the polarizable water molecules by relatively unpolarizable ions, and is calculated from a classical electrostatic model. When the terms $\gamma_{so}$ and $\gamma_h$ were substituted into the expression for the activity coefficient the residual electrostatic effect ($\gamma_e$) was a linear function of $c^{\frac{1}{3}}$ for most 1:1 electrolytes at ionic strengths up to 1 M. This linearity was not observed if other sources were used for the hydration numbers nor if the Glueckauf (1955) model was used to represent the hydration effect.

Glueckauf (1969) has developed electrostatic equations based on the statistical treatment of Kirkwood (1936) that incorporate both the cube root law and the Debye–Hückel limiting law and provide a smooth transition from one to the other. For concentrations where $\mathscr{B}\mathring{a}I^{\frac{1}{2}} > 2$

$$\log \gamma_A = -0.4444 \, \mathscr{A} \, |z_+ z_-| \, (2I/\mathscr{B}\mathring{a})^{\frac{1}{3}} + b'm \tag{5.10}$$

and for concentrations where $\mathscr{B}\mathring{a}I^{\frac{1}{2}} < 2$

$$\log \gamma_A = -DH[(1 + 0.5\mathscr{B}\mathring{a}I^{\frac{1}{2}})/(1 + \mathscr{B}\mathring{a}I^{\frac{1}{2}})]^2 + b'm \tag{5.11}$$

Similar equations have been produced by Lietzke et al. (1968) and Mitra et al. (1968) who weighted the Debye–Hückel and cube-root contributions according to arbitrarily defined partition functions.

### 2.5.4. GENERAL STATISTICAL EQUATIONS

The Debye–Hückel theory ignores the specific structural influences in electrolyte solutions. Embellishments of the basic equation (Table 2.16) represent attempts to put back the structural detail that is lacking in the primitive model. The general statistical approach takes *as its starting point* the granular nature of electrolyte solutions, and attempts to break down the configurational partition function ($\mathcal{Q}_c$, equation (5.2)) into more manageable components. Two basic approaches have been considered—the first expres-

---

[a] Robinson and Stokes (1965). Equation (3) Table 2.16

[b] Parameters also available for polynomial equation (note [e])

[c] Equation 5, Table 2.14. Recalculated from the data of Guggenheim and Stokes (1958) using the value $\mathscr{A} = 0.5108 \, \text{mol}^{-\frac{1}{2}} \text{kg}^{\frac{1}{2}}$, $\mathscr{B} = 3.287 \, \text{mol}^{-\frac{1}{2}} \text{kg}^{\frac{1}{2}} \text{nm}^{-1}$ at 298 K (Hamer, 1968).

[d] Equation as in [a], second column gives $h$

[e] Lietzke and Stoughton (1962), using the equation

$$\ln \gamma_{\pm} = -2.303 \mathscr{A} I^{\frac{1}{2}}/(1 + AI^{\frac{1}{2}}) + (2B)I + (3C/2)I^2 + (4D/3)I^3$$

sing $\mathcal{Q}_c$ in terms of a virial (power) series (Mayer and Mayer, 1940; Mayer, 1950; Friedman, 1962, 1970) and the second focusing on the central ion and calculating $\mathcal{Q}_c$ via a radial distribution function which describes the disposition of neighbouring ions (Poirier, 1966; Percus and Yevick, 1958; von Leeuwen *et al.*, 1959; Ramanathan and Friedman, 1971).

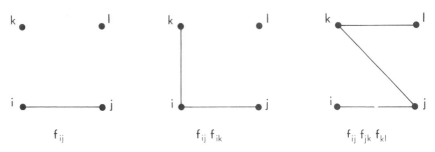

FIG. 2.10. Schematic representation of some pair-wise interactions in a four particle system.

The virial expansion equations were first extended to electrolyte solutions by Mayer (1950). The individual terms in the definition of $U$ (equation (5.3)) represent interactions between all molecules in the assembly, first as pairwise interactions, then as the resultant of pairwise interactions between three molecules, etc. until the final term involves a product representing every possible pairwise interaction. Mathematically this becomes (Fig. 2.10),

$$\exp\left(-U/kT\right) = 1 + \sum_{ij} f_{ij} + \sum_{ijk} f_{ij}f_{ik} + \sum_{ijkl} f_{ij}f_{jk}f_{kl} + \ldots \quad (5.12)$$

Each term gives an integral representing the energy contribution of that particular configuration or cluster of molecules (cluster integral). The next step is to remove the labels on the diagrams representing the clusters and to gather into separate groups all those terms that are topologically indistinguishable. The analysis at this stage becomes quite complex if there is a large number of molecules in the array (e.g. Friedman's analysis of a nine-molecule array, Friedman, 1962, p. 7). The grouped integrals are then re-investigated and the larger patterns are broken down until $\mathcal{Q}_c$ is represented by a series of irreducible cluster integrals. The number of interactions contributing to each integral is then specified (the combinatorial problem) and the series of cluster integrals is manipulated to converge rapidly so that only a limited number of terms need be considered. Stillinger and White (1971) have strongly criticized the cluster expansion approach because of the subjective nature of this stage in the calculation.

The interaction energies ($u_{ij}$) are then expressed in terms of the solution model and the partition function is evaluated. The problem can be simplified by computing the properties of the solution relative to those observed in the reference state at infinite dilution and by keeping the activity of the water constant by subjecting the solution to a pressure equal to the osmotic pressure. In this brief summary we have skated lightly over many difficult problems. For more detailed treatments, roughly in order of increasing complexity, see Vaslow (1972), Friedman (1970), Falkenhagen and Ebeling (1971) and Friedman (1962).

Although the cluster integral expansion approach has been applied to the primitive model for single electrolyte solutions with some success at low concentrations (Mayer, 1950; Poirier, 1953; Friedman, 1960a, b; 1962), the more recent emphasis has been on the evaluation of theories based on the radial distribution function ($g(r)$). This function defines the probability of finding a molecule at a specified distance ($r$) from an arbitrarily defined central molecule. In a general statistical theory of ionic solutions it will take the place of the Boltzmann distribution equation in the Debye–Hückel theory, defining the relative disposition of interacting particles. The nature and size of the interactions between particles distributed in this way can be defined by potential energy terms analogous to those introduced in the cluster integral expansion approach. However, these terms will refer here only to pairwise interactions between the central molecule and a specified neighbour. The complexity associated with the whole range of cluster interactions is summarized in $g(r)$. The equation relating the average energy of molecules in the system to $g(r)$ and $u_{ij}$ is (Vaslow, 1972).

$$\bar{U} = (\rho/2) \int_V g_{ij}(r)u_{ij}\, dr_{ij} \tag{5.13}$$

where $\rho$ is the density in molecules cm$^{-3}$. Brief summaries of the procedures for solving this integral are given by Poirier (1966) and Vaslow (1972). This approach was used by Kirkwood (1934) and by Kirkwood and Poirier (1954) for the rigorous derivation of the Debye–Hückel limiting law. More recent treatments have been based on the solutions of Percus and Yevick (1958, P–Y equations) and of von Leeuwen et al. (1959, hypernetted chain or HNC equations).

In particular, the HNC equations have been developed by Friedman and his co-workers to consider models based on charged hard spheres (Rasaiah and Friedman, 1968, 1969), on hard spheres surrounded by a penetrable hydration sheath (square mound potential model) and on charged spheres whose ion-ion pair potentials are a continuous function of interionic distance (Ramanathan and Friedman, 1971). These equations are able to give a good

E

account of the behaviour of $1:1$ electrolytes at ionic strengths equivalent to up to $1\,M$ using only a single adjustable parameter. Rasaiah et al. (1972) have completed a detailed comparison of the various integral equation theories for the primitive model and concluded that the HNC equations provide the most satisfactory treatment. Data obtained via an analytical solution of the P-Y equation (Waisman and Lebowitz, 1970) also give results that are in essential agreement with the calculations of Friedman and his associates. An interesting conclusion arising from these calculations is that the "cube-root" behaviour at moderate concentrations (Fig. 2.8) arises naturally from the forces considered in the restricted model, and no additional hypotheses concerning the formation of a "quasi-lattice" are required (Rasaiah et al., 1972).

### 2.5.5. SINGLE-ION ACTIVITY COEFFICIENTS

It is clear, structurally, that the ions and not the neutral molecules are the real components of the solution. Consequently, structural theories, such as the Debye–Hückel theory, give single-ion activity coefficients directly at low ionic strengths (see Table 2.14, footnotes $b$ and $c$). Some of the more elaborate equations (Table 2.16) can also be adapted to calculate conventional single-ion activity coefficients at higher ionic strengths. The Brønsted–Guggenheim equation (Table 2.16, equation 2) is readily adapted by replacing $|z_+ z_-|$ by $z_i^2$ and the hydration equations (Table 2.16, equations 3 and 4) can be used to calculate single ion activity coefficients by defining the hydration parameter $(h)$ for one ion in the series. Bates et al. (1970) have suggested setting $h_{Cl} = 0$ for the alkaline and alkaline earth metal chlorides. This gives for the monovalent chlorides using the mole-fraction statistics equation,

$$\log \gamma_{M^+} = \log \gamma_{\pm} + 0{\cdot}00782\,h_+ m\phi \qquad (5.14a)$$

$$\log \gamma_{Cl^-} = \log \gamma_{\pm} - 0{\cdot}00782 h_+ m\phi \qquad (5.14b)$$

The procedure has been extended to the alkali metal fluorides (Robinson et al., 1971a) by assuming that $h_{F^-} = 1{\cdot}19 = h_{K^+}$ and Elgqvist and Wedborg (1973) have proposed that comparable single-ion activities may be calculated for sulphate systems if it assumed that $h_{SO_4^{2-}} = 0$.

An alternative approach, suggested by MacInnes (1919), assumes that at a given concentration and ionic strength

$$\gamma_{K^+} = \gamma_{Cl^-} = \gamma_{\pm KCl} \qquad (5.15)$$

The two hydrated ions have about the same size and mobility and are isoelectronic with argon. The individual values of $\gamma_{K+}$ and $\gamma_{Cl-}$ are used in other salt solutions at the same concentration and ionic strength so that, for example,

$$\gamma_{Br^-} = (\gamma_{\pm KBr})^2 / \gamma_{\pm KCl}.$$

The convention has been extended (e.g. Garrels and Christ, 1965; Garrels, 1967) to cover solutions containing divalent ions by dropping the concentration restriction. This convention has come under criticism recently because calculations of single-ion activities by several different pathways do not always yield the same results, particularly at ionic strengths greater than 0·1 M (Bates and Alfanaar, 1969). In addition, calculations of $\gamma_{Cl^-}$ via the conventions of Guggenheim (1935) (see Leyendekkers, 1971a) and Bates et al. (1970) indicate that $\gamma_{Cl^-}$ varies from one chloride solution to another at a given ionic strength and is *not* constant as is implied by the MacInnes convention.

Single-ion activity coefficients may also be estimated by some arbitrary arithmetical splitting of the mean-ion activity coefficients. For 1:1 electrolytes this is simply done by defining the activity coefficients so that,

$$\gamma_+ = \gamma_- = \gamma_\pm = \gamma_2^{\frac{1}{2}}. \qquad (5.16)$$

For unsymmetrical electrolytes ambiguities arise because it is not clear whether the solute activity coefficient $(\gamma_2)$ should be split by weighting according to mole fraction or according to charge. Shatkay (1967) favours the former approach so that the $\gamma_2^-$ value is split in the ratio $v_+ : v_-$ i.e. for $CaCl_2$

$$\gamma_{\pm(CaCl_2)} = \gamma_{Ca^{2+}} = \gamma_{Cl^-} \qquad (5.17)$$

Bates and Alfenaar (1969) prefer to use charge weightings in accordance with the Debye–Hückel theory (i.e. split $\gamma_2$ in the ratio $z_+^2 : z_-^2$) so that for $CaCl_2$

$$\gamma_{Ca^{2+}} = [\gamma_{\pm(CaCl_2)}]^2 \qquad (5.18)$$

$$\gamma_{Cl^-} = [\gamma_{\pm(CaCl_2)}]^{\frac{1}{2}}$$

Both conventions must obey the relationship

$$\gamma_2 = [\gamma_{\pm(CaCl_2)}]^3 = \gamma_{Ca^{2+}}\gamma_{Cl^-}^2$$

Conventions taking the Debye–Hückel theory as their starting point will naturally deviate from the Shatkay convention, particularly at higher ionic strengths. The various conventions predict markedly different values for the single-ion activity coefficients, particularly at ionic strengths greater than 0·1 M (see Fig. 2.11). It is possible to obtain experimental single-ion activities from ion-selective electrode measurements provided that some convention is adopted for calculating the liquid junction potential (Table 2.3). However, the results of such measurements have so far failed to give unequivocal support to any one convention for defining single-ion activities (Fig. 2.11). Discussions in support of the MacInnes convention (Garrels, 1967), the Guggenheim

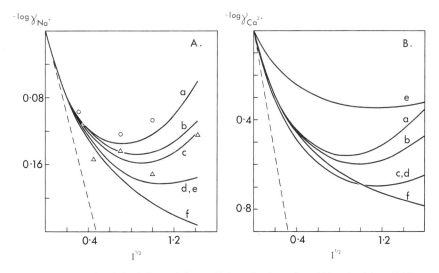

FIG. 2.11. Conventional single-ion activity coefficients for the sodium (A.) and calcium (B.) ions ions in their respective chloride solutions.

*Experimental data*

    O—estimated from Garrels (1967)

    △—data of Shatkay and Lerman as recalculated by Bates *et al.* (1970)

*Theoretical curves*

a. MacInnes convention (equation 5.15) calculated via the bromides.

b. As in a., calculated via the chlorides (cf. Garrels and Thompson, 1962).

c. Hydration convention (Bates *et al.*, 1970).

d. Total activity coefficient split according to charge weighting (equation (5.18)).

e. Total activity coefficient split according to mole fraction weighting (equation (5.17)).

f. Equation (2), Table 2.14, with å-values taken from the compilation of Kielland (1937).

The broken line follows the Debye–Hückel limiting law.

convention (Leyendekkers, 1971a) and the mole fraction convention for splitting the mean-ion activity coefficient (Shatkay, 1967) indicate the variety of approaches to the problem. Recent moves towards the establishment of ionic activity standards for use with ion-selective electrodes indicate that international conventions based on the hydration equations may eventually be adopted (Bates *et al.*, 1970; Robinson *et al.*, 1971a).

### 2.5.6. SUMMARY

Developments of the Debye–Hückel theory have provided a number of equations that give accurate summaries of the activity coefficients of electrolyte solutions using only a few adjustable parameters (Table 2.17). Since the activities of the solute and solvent are interrelated by the Gibbs–Duhem equation (Table 2.4, equation (1)), these equations can also be used to calculate the osmotic coefficient of the solution (Table 2.18).

The simplest way of applying these equations to electrolyte mixtures is to assume that the activity coefficient is affected by the ionic strength only and is impervious to specific changes in the solution composition at constant ionic strength. If the equations contain parameters that allow for the specific

TABLE 2.18

*Osmotic coefficients calculated from electrostatic equations*

| Equation number | $(1 - \phi) =$ | Adjustable parameters | Reference |
|---|---|---|---|
| 1.[a] | $(\mathscr{A}*/3)\|z_+ z_-\| I^{\frac{1}{2}} = \phi_{DH}$ | | Lewis and Randall (1961, p. 339) |
| 2.[b] | $\phi_{DH}\sigma(I^{\frac{1}{2}})^c - \underline{v}b^*m/2$ | $b^*$ | Guggenheim and Turgeon (1955) |
| 3.[d] | $\phi_{DH}\sigma(AI^{\frac{1}{2}})$ $+ BI + CI^2 + DI^3$ | $A, B, C, D$ | Lietzke and Stoughton (1962) |
| 4.[e] | $\phi_{el} + \dfrac{r + h - v}{(1 + 0{\cdot}018\,mr)}$ $- \dfrac{1}{0{\cdot}018\,mv}\ln\dfrac{(1 + 0{\cdot}018\,mr)}{(1 - 0{\cdot}018\,mh)}$ where $\phi_{el} = \phi_{DH}\sigma(\mathscr{B}\mathring{a}I^{\frac{1}{2}})$ | $\mathring{a}, h$ | Glueckauf (1955) |
| 5.[f] | $\phi_{DH}\sigma(\mathscr{B}\mathring{a}I^{\frac{1}{2}}) - b^*m/2$ | $\mathring{a}, b^*$ | Guggenheim and Stokes (1958) |

[a] From equation (1) Table 2.14 $\mathscr{A}* = \mathscr{A}\ln 10$
[b] From equation (2) Table 2.16

[c] $\sigma(x) = \dfrac{3}{x^3}\displaystyle\int_0^x \left[\dfrac{x}{1 + x}\right]^2 dx = \dfrac{3}{x^3}\left[(1 + x) - (1 + x)^{-1} - 2\ln(1 + x)\right]$

   Values are tabulated by Scatchard and Epstein (1942), Robinson and Stokes (1965, p. 460) and Harned and Owen (1958, p. 176).
[d] From equation quoted in note [e] Table 2.17
[e] From equation (4), Table 2.16. The equation for $\phi_{el}$ may be derived from equation (2), Table 2.16.
[f] From equation (5), Table 2.16. Values of $\mathring{a}$ and $b$ are given in Table 2.17. $b^* = b\ln 10$.

interaction between ions of opposite charge (e.g. equations (1) and (2), Table 2.16), then changes in the solution composition can also be taken into account. Both of these approaches would enable the behaviour of the mixture to be defined in terms of the properties of the component single electrolyte solutions. If, in mixtures, additional interactions occur that are not important in single electrolyte solutions, then more elaborate procedures will be necessary.

## 2.6. ELECTROLYTE MIXTURES

### 2.6.1. THE MIXING PROCESS

In an ideal mixing process the behaviour of the electrolyte mixture can be predicted exactly from the properties of the pure component solutions. However, for real mixtures it is necessary to define functions (excess mixing parameters) that specify how widely the actual mixing process deviates from ideal behaviour.

There are two ways of defining the ideal reference system. The first considers the system resulting from the mixing of *ideal* solutions of the component salts with the same composition as the real solutions (Fig. 2.12, Reference Mixture I). The excess free energy of the real mixture relative to this reference point ($\Delta G_m^E$, Table 2.19) will include all deviations from ideal behaviour displayed by the pure components as well as the extra effects introduced by the mixing process. The second, and more useful reference system (Fig. 2.12, Reference Mixture II) results from the ideal mixing of the *real* solutions A and B. The excess free energy relative to this reference point ($\Delta_m G^E$, Table 2.19) is *exclusively* concerned with the mixing process and will indicate how closely the properties of the mixtures may be predicted from the properties of its components. The behaviour of $\Delta_m G^E$ and its temperature and pressure derivatives ($\Delta_m H^E$ and $\Delta_m V^E$ respectively) may be summarized by a few empirical rules.

### 2.6.2. EMPIRICAL CORRELATIONS

#### 2.6.2.1. *Harned's rule.*

The excess free energy of mixing $\Delta G_m^E$ can be estimated from the chemical potentials of the solution components. The mixing effects are, in general, small so that the experimental data are confined to relatively high ionic strengths, normally greater than one molal. At constant total ionic strength (Fig. 2.12) the contribution of a particular solute (A) to $\Delta G_m^E$ is proportional to the ratio $\ln (\gamma_{A(m)}/\gamma_{A(0)})$ (Table 2.19). Harned and his co-workers deduced,

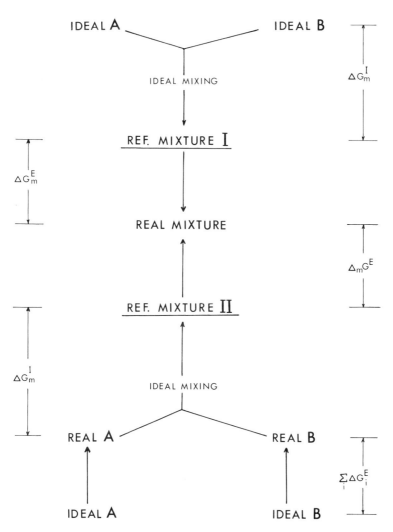

Fig. 2.12. Alternative pathways for the formation of a binary mixture from ideal components. For electrolyte solutions the mixing process is normally considered at constant total ionic strength to minimise the influence of long-range electrostatic effects. If one mole of NaCl in 500 g of water is mixed with one mole of KCl in 500 g of water at 298 K and 1 atm then the free energy parameters will have the following values (Harned and Robinson, 1968; Robinson et al., 1971b).

$$\Delta G_{NaCi}^{E} = -1 \cdot 942 \text{ kJ} \qquad \Delta G_{KCi}^{E} = -2 \cdot 328 \text{ kJ}$$

$$\Delta_{m} G^{E} = -130 \text{ J} \qquad \Delta G_{m}^{I} = -1 \cdot 720 \text{ kJ}$$

$$\Delta G_{m}^{E} = -4 \cdot 400 \text{ kJ}$$

TABLE 2.19

*Excess free energy of electrolyte mixtures*[a]

(i) *Relative to reference mixture I* (Fig. 2.12)[b]

The excess free energy of component $A$ in the mixture is given by

$$\Delta G^E_{A(m)} = RT \nu_A m_A [\ln \gamma_{A(m)} + y_A(1 - \phi_m)] \tag{1}$$

where $y_A$ is the mole fraction of $A$ in the mixture and $\phi_m$ is the osmotic coefficient of the mixture. $\gamma_{A(m)}$ is the activity coefficient of component $A$ in the mixture.

Summation over all components in the mixture gives

$$\Delta G^E_m = RT \sum_i \nu_i m_i [\ln \gamma_{i(m)} + y_i(1 - \phi_m)] \tag{2}$$

(ii) *Relative to reference mixture II* (Fig. 2.12)[c]

The excess free energy of component $A$ in the pure electrolyte solution relative to the ideal solution at the same ionic strength is given by

$$\Delta G^E_A = RT \nu_A m_A [\ln \gamma_{A(0)} + y_A(1 - \phi_{A(0)})]^d \tag{3}$$

where $\gamma_{A(0)}$ and $\phi_{A(0)}$ refer to component $A$ in a pure solution at the same total ionic strength as the mixture.

Since 

$$\Delta_m G^E = \Delta G^E_m - \sum_i \Delta G^E_i \qquad \text{(see Fig. 2.12)}$$

$$\Delta_m G^E = RT \sum_i \nu_i m_i [\ln (\gamma_{i(m)}/\gamma_{i(0)}) + y_i(\phi_{i(0)} - \phi_{(m)})] \tag{4}$$

The ideal free energy of mixing $(\Delta G^I_m)$ is given by

$$\Delta G^I_m = \sum_i \nu_i m_i \ln y_i$$

[a] Robinson *et al.* (1971b)
[b] Harned and Robinson (1968, p. 47)
[c] Scatchard (1936), Robinson and Bower (1965)
[d] Friedman (1960a).

from extensive studies of ternary common ion mixtures, that this ratio was a simple function of solution composition (equations (1) and (2), Table 2.20). In most cases only the linear term was significant.

The data on which Harned's rule is based have been surveyed by Harned and Owen (1958), Robinson and Stokes (1965), Lewis and Randall (1961) and Harned and Robinson (1968). A detailed compilation of more recent

Table 2.20 contd.
[a] Harned and Owen (1958, p. 600). Salt A contains cation 1 and anion 2 and salt B cation 3 and anion 4
[b] Harned and Owen (1958, p. 621), Harned and Robinson (1968, p. 35)
[c] Harned (1935), Harned and Gary (1954), Harned and Owen (1958, p. 622).
[d] Table 2.19
[e] Leyendekkers (1970, 1971b)
[f] Lewis and Randall (1961, p. 400).

**TABLE 2.20**

*Harned's Rule*

(i) *Statement*

For mixtures of two electrolytes $(A, B)$ prepared at constant total ionic strength[a]

$$\log\left[\gamma_{A(m)}/\gamma_{A(0)}\right] = -\alpha_{12}I_B + \xi_{12}I_B^2 \tag{1a}$$

$$\log\left[\gamma_{B(m)}/\gamma_{B(0)}\right] = -\alpha_{34}I_A + \xi_{12}I_A^2 \tag{1b}$$

or

$$\log\left[\gamma_{A(m)}/\gamma_{(0)A}\right] = \alpha_{12}I_A - \xi_{12}I_A^2 \tag{2a}$$

$$\log\left[\gamma_{B(m)}/\gamma_{(0)B}\right] = \alpha_{34}I_B - \xi_{12}I_B^2 \tag{2b}$$

where $\gamma_{(0)A}$ is the activity coefficient of a trace of $A$ in the presence of $B$ at the same overall ionic strength as the mixture; $I_B$ is the ionic strength fraction of $B$ in the mixture etc. The $\alpha$- and $\xi$- coefficients are characteristic of each salt pair and are functions of temperature, pressure and ionic strength.

(ii) *Restrictions on independent variation of parameters*
(a) Cross differentiation condition[b]

$$v_A[\partial \log \gamma_{A(m)}/\partial m_B]_{m_A} = v_B[\partial \log \gamma_{B(m)}/\partial m_A]_{m_B}$$

Differentiating equations (1a) and (1b) with respect to I

$$v_A j_B(\alpha_{12} + 2I\xi_{12}) + v_B j_A(\alpha_{34} + 2I\xi_{34}) = \text{Constant} \tag{3}$$

where $j_A = I_A/m_A$ etc.

(b) Gibbs–Duhem equation.[c]
Neglecting quadratic terms in equation (1) this gives

$$\alpha_{34}/Z_B = (\alpha_{12}/Z_A) - (2/2\cdot3031)\left[(\phi_{A(0)} - 1)/Z_A - (\phi_{B(0)} - 1)/Z_B\right] \tag{4}$$

where

$$Z_A = |z_1 z_2| \quad \text{and} \quad Z_B = |z_3 z_4|$$

(iii) *Extension to multicomponent mixtures.*[d]

$$\log\left[\gamma_{A(m)}/\gamma_{A(0)}\right] = \sum_N (-\alpha_{12}I_N + \xi_{12}I_N^2) \tag{5}$$

where $\alpha_{12}$ and $\xi_{12}$ here refer to mixtures of salts $A$ and $N$ at the same overall ionic strength as the multicomponent mixture.

(iv) *Entropy correlations.*[e]

$$\alpha_{12} = a + b\bar{S}_{12}^{\ominus} \tag{6}$$

where

$$\bar{S}_{12}^{\ominus} = Z_1\bar{S}_1^{\ominus} + Z_2\bar{S}_2^{\ominus} + Z_3\bar{S}_3^{\ominus}$$

$$Z_1 = (3z_1 - z_3)(z_2 - z_1)/4(z_3 - z_1)$$

$$Z_2 = 1/z_2^2 \quad \text{(chlorides only)}$$

$$Z_3 = 3(z_1 - z_3)(z_2 - z_1)/z_1^2(z_2 - z_3)$$

$\bar{S}_i^{\ominus}$ is the conventional standard partial molal entropy[f] of the ion $i$ and $z_i$ is its charge (negative for an anion, positive for a cation).

TABLE 2.21

*Harned's rule coefficients for major sea-water components (298 K, 1 atm.)*

| A | B | Ionic strength (molal) | | | | | |
|---|---|---|---|---|---|---|---|
| | | 0·5 | | 1·0 | | 2·0 | |
| | | $\alpha_{12}$ | $\alpha_{32}$ | $\alpha_{12}$ | $\alpha_{32}$ | $\alpha_{12}$ | $\alpha_{32}$ |
| NaCl | KCl | 0·025[a] | −0·013[a] | 0·0235[a] | −0·0095[a] | 0·023[a] | −0·0084[a] |
| NaCl | MgCl$_2$ | −0·0228[b] | −0·0061[b] | −0·019[c] | −0·012[c] | −0·014[c] | −0·015[c] |
| Na$_2$SO$_4$ | MgSO$_4$ | −0·0164[d,e] | — | −0·037[c] | 0·022[c] | −0·032[c] | 0·014[c] |
| NaCl | CaCl$_2$ | −0·005[d,f] | −0·011[d,f] | −0·004[g] | 0·00002[g] | −0·005[g] | −0·013[g] |
| KCl | MgCl$_2$ | −0·032[h] | −0·064[h] | −0·0312[i] | 0·0475[h] | −0·018[h] | 0·037[h] |
| KCl | CaCl$_2$ | −0·026[h] | 0·053[h] | −0·025[i] | 0·026[j] | −0·0185[j] | 0·0202[j] |
| CaCl$_2$ | MgCl$_2$ | | | −0·0277[k] | 0·0225[k] | | |
| NaCl | Na$_2$SO$_4$ | 0·0557[d,e] | −0·008[l] | 0·045[c] | −0·041 | 0·044[c] | −0·039[c] |
| KCl | K$_2$SO$_4$ | 0·0202[h] | −0·0412[h] | 0·029[i], 0·017[h], −0·0117[i] | −0·035[h] | 0·016[h] | −0·033[h] |
| KCl | MgSO$_4$ | | | 0·071[c] | −0·077[c] | 0·067[c] | −0·071[c] |
| MgCl$_2$ | MgSO$_4$ | 0·073[b] | −0·091[b] | | | | |
| NaCl | MgSO$_4$ | | | 0·022[m] | −0·080[m,n] | 0·024[m] | −0·083[m,n] |
| Na$_2$SO$_4$ | MgCl$_2$ | 0·0194[d,e,n] | | −0·078[m] | 0·080[m,n] | −0·068[m] | 0·073[m,n] |

[a] Robinson (1961)
[c] Wu et al. (1968)
[e] Gieskes (1966)
[g] Lanier (1965)
[i] Christensen and Gieskes (1971)
[k] Robinson and Bower (1966), $I = 0.93$

[b] Rush (1969a)
[d] $I = 0.7$ M
[f] Leyendekkers and Whitfield (1971)
[h] Calculated from cluster integral expansion equations (Table 2.31)
[j] Robinson and Covington (1968)
[l] Platford (1968a)
[m] Wu et al. (1969).
[n] $\alpha_{34}$ values in terminology used in Table 2.20.

data has been prepared by Rush (1969a) and a selection of values of interest for natural waters has been listed by Whitfield (1971). Data for ternary mixtures of the major electrolyte components in sea water are listed in table 2.21.

The form of the excess free energy function (Table 2.19, equation (4)) suggests that extensions to multi-component systems involve a simple summation of the interaction terms. This approach was suggested by Åkerlof (1934, 1937) and has been used by Lerman (1967) and by Gieskes (1966) to calculate activity coefficients in natural electrolyte mixtures.

The Åkerlof–Thomas rule (equation (5.7)) may be derived from the linear forms of equations (1) and (2) (Table 2.20) to give,

$$\log\left[\gamma_{B(0)}/\gamma_{A(0)}\right] = I(\alpha_{32} - \alpha_{12}) + \log\left[\gamma_{(0)B}/\gamma_{(0)A}\right] \qquad (6.1)$$

Where the uncommon (or hetero-) ions have similar effects on the water structure (Section 2.4.1), $\gamma_{(0)A}$ and $\gamma_{(0)B}$ are related by the equation (Harned and Robinson, 1968, p. 17),

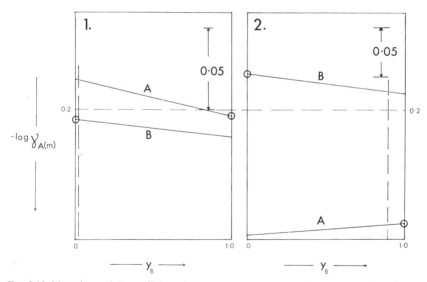

FIG. 2.13. Mean-ion activity coefficients in ternary common-ion mixtures as a function of solution composition. The vertical dashed line indicates the approximate sea-water ratio.
1. A = NaCl, B = KCl

$$\log \gamma_{(0)A} = -0{\cdot}2038, \log \gamma_{(0)B} = -0{\cdot}2082$$

2. A = MgCl$_2$, B = NaCl

$$\log \gamma_{(0)A} = -0{\cdot}3163, \ 2 \times \log \gamma_{(0)B} = -0{\cdot}3266$$

Experimental data taken from Rush (1969a).

$$|z_1 z_2| \log \gamma_{(0)B} = |z_3 z_4| \log \gamma_{(0)A} \qquad (6.2)$$

When this relationship holds (Fig. 2.13) then equation (6.1) is equivalent to the Åkerlof–Thomas rule (Robinson and Stokes, 1965, p. 446). Where the hetero-ions have markedly different effects on the water structure (e.g. HCl–CsCl, NaOH–NaCl) the conditions required by the Åkerlof–Thomas rule are not fulfilled, and frequently the additional quadratic term ($\xi$, in equations (1) and (2), Table 2.20) is required to describe the solution properties (Robinson and Stokes, 1965; Harned and Owen, 1958). The $\xi$-values are usually small ($\pm 0.002$) but have a noticeable effect on the curvature of the plots.

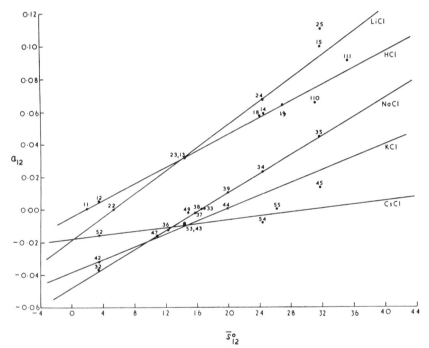

FIG. 2.14. Linear correlation between the Harned's rule coefficients ($\alpha_{12}$) and the weighted ionic entropies for alkali metal chlorides at 298 K according to equation (6), Table 2.20. The numbers $ij$ identify the cations $i$ and $j$ in the ternary common-ion mixture according to the coding,

| | | | |
|---|-----|----|-------------|
| 1 | $H^+$ | 7 | $Ca^{2+}$ |
| 2 | $Li^+$ | 8 | $Sr^{2+}$ |
| 3 | $Na^+$ | 9 | $Ba^{2+}$ |
| 4 | $K^+$ | 10 | $Al^{3+}$ |
| 5 | $Cs^+$ | 11 | $Ce^{3+}$ |
| 6 | $Mg^{2+}$ | | |

Reprinted with permission from J. V. Leyendekkers (1970) copyright of the American Chemical Society.

The relationship between the Harned's rule coefficients and ion-water interactions has been illustrated quantitatively by Leyendekkers (1970, 1971b). She has shown that, at constant ionic strength, the α-coefficients for ternary chloride mixtures are linear functions of the charge-weighted entropies of the component ions (Table 2.20, equation (6), Fig. 2.14). These relationships provide a very neat way of predicting α-coefficients. Similar linear plots may be expected with other structure sensitive parameters (e.g. the viscosity B-coefficient, Stokes and Mills, 1965).

Friedman (1960a) has shown that, for systems obeying the quadratic form of Harned's rule, the excess free energy of mixing is given by

$$\Delta_m G^E/RT = 2 \cdot 303 I^2 y_A y_B [g_0 + g_1(y_A + y_B)] \tag{6.3}$$

where

$$g_0 = -[(\alpha_{12} + \beta_{12}I)/(Z_A + (\alpha_{34} + \xi_{34}I)/Z_B] \tag{6.4a}$$

$$g_1 = -(I/3)[(\beta_{12}/Z_A) - (\xi_{34}/Z_B)] \tag{6.4b}$$

$$Z_A = |z_1 z_2| \quad \text{and} \quad Z_B = |z_3 z_4|$$

$y_A$ and $y_B$ are the mole fractions of the components $A$ and $B$ respectively. Where both the electrolytes obey the linear form of Harned's rule the equation for 1:1 electrolytes reduces to (McKay, 1957),

$$\Delta_m G^E/RT = 2 \cdot 303 I^2 y_B(1 - y_B)(\alpha_{12} + \alpha_{34}) \tag{6.5}$$

Since the final term on the right hand side of this equation normally represents the difference between two small numbers (Table 2.21) this is not a very precise method for estimating $\Delta_m G^E$. Friedman (1960a, b, 1962) indicates that, unlike the Harned's rule coefficients, the $g$-parameters defined above are not constrained by the cross-differentiation relationships (see Table 2.20). Furthermore, they are characteristic only of the mixing process and do not include the contributions from the pure component solutions that are implicit in the Harned's rule coefficients (equation (4), Table 2.20). The use of the excess mixing function and the $g$-parameters is likely to lead more directly to the elucidation of those properties arising from specific interactions within the mixture. An investigation of the pressure and temperature coefficients of $\Delta_m G^E$, $g_0$ and $g_1$ has led to the discovery of further regularities in ionic solution behaviour that are known collectively as Young's rules (Table 2.22).

### 2.6.2.2. Young's rules.

The excess heats ($\Delta_m H^E$) and volumes ($\Delta_m V^E$) of mixing, unlike $\Delta G_m^E$ can be measured directly. The data consequently do not have to undergo the exten-

TABLE 2.22

*Young's rules*[a]

(i) *General equations*[b] (mixtures of two electrolytes, $A$, $B$)

$$\Delta_m H^E / RT^2 = [\partial(\Delta_m G^E / RT)/\partial T]_P$$
$$= 2 \cdot 303 I^2 y_A y_B [h_0 + h_1(y_A - y_B)] \qquad (1)$$

where $\qquad -(\partial g_n / \partial T)_P = h_n$

$$\Delta_m V^E / RT = [\partial(\Delta_m G^E / RT)/\partial P]_T$$
$$= 2 \cdot 303 I^2 y_A y_B [v_0 + v_1(y_A - y_B)] \qquad (2)$$

where

$$(\partial g_n / \partial P)_T = v_n$$

(ii) *Young's first rule*[c]

As a first approximation assume ideal mixing so that

$$h_0 = h_1 = v_0 = v_1 = 0 \qquad (3)$$

(iii) *Young's second rule*[d] (the structure rule)

If $h_1 = v_1 = 0$ then for the mixing of two cations $(M, N)$ in the presence of a common anion $(X, Y)$ the values of $h_0$ and $v_0$ are independent of the nature of the anion

i.e.

$$h_{0\,MFX} \simeq h_{0\,MNY} \quad \text{and} \quad v_{0\,MNX} \simeq v_{0\,MNY} \qquad (4)$$

A similar rule applies for the mixing of two anions in the presence of a common cation.

(iv) *Young's third rule*[e] (the cross square rule)

For a reciprocal salt pair (e.g. $MX$, $NY$) the sum of the four heats of mixing for the systems *with* a common ion $(MY–MX, NY–NX, MY–NY, MX–NX)$ is equal to the sum of the two heats of mixing for systems *without* a common ion $(MY–NX, MX–NY)$, i.e.

$$\sum \square = \Delta_m H^E(MY–MX) + \Delta_m H^E(NY–NX) + \Delta_m H^E(MY–NY) + \Delta_m H^E(MX–NX)$$
$$= \Delta_m H^E(MY–NX) + \Delta_m H^E(MX–NY)$$
$$= \sum \times$$

This rule also applies to the excess volumes[f] and free energies[g] of mixing. The validity of Young's second rule is a necessary but not a sufficient condition for the validity of the cross-square rule[h].

---

[a] Scatchard *et al.* (1970)

[b] Friedman (1960a), Tables 2.1, 2.19. These equations are often written with $y_A = 1 - y_B$

[c] Young (1951), Young and Smith (1954), Wirth *et al.* (1963)

[d] Young *et al.* (1957), Wu *et al.* (1965a, b).

[e] Young *et al.* (1957), Krawetz (1957)

[f] Wirth and Mills (1968), Wirth and LoSurdo (1968), Wirth *et al.* (1963)

[g] Covington *et al.* (1968), Wu *et al.* (1969)

[h] Scatchard (1969)

sive manipulations associated with the determination of $\Delta_m G^E$ from activity coefficient measurements and are more sensitive to relatively small changes in solution properties on mixing. The general patterns of behaviour are summarized in Fig. 2.15. The skew terms ($h_1$ and $v_1$, Table 2.22) are in general, small so that the behaviour of most systems is adequately summarized by curve B (Fig. 2.15). The most significant deviations occur in mixtures where the mixture has no common ion. Only one example has been reported of an aqueous solution where $h_1$ is greater than $h_0$ (Na$_2$SO$_4$–KCl, Smith, 1942). The discussion that follows will be based on the assumption that $\bar{y}_1 = 0$.

As a first approximation (equation (3), Table 2.22) the formation of a ternary common-ion mixture at constant total ionic strength can be considered as

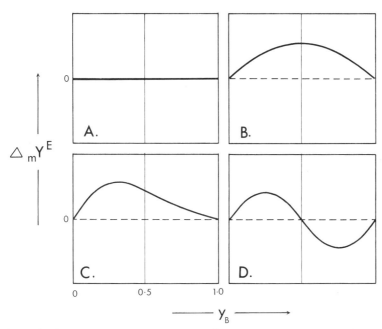

FIG. 2.15. Behaviour of excess mixing functions ($\Delta_m Y^E$) according to variations in the interaction parameters $y_0$ and $y_1$ (Wu, 1970).

A. $\bar{y}_0 = \bar{y}_1 = 0$.    Young's first rule for $\Delta_m H^E$ and $\Delta_m V^E$ and the Brønsted principle for $\Delta_m G^E$

B. $\bar{y}_1 = 0$.    Symmetrical mixing consistent with the linear form of Harned's rule when $\alpha_{12} + \alpha_{32}$ = constant e.g. $\Delta_m H^E$ for NaCl − KCl − H$_2$O.

C. $\bar{y}_0 > \bar{y}_1$.    Skew mixing consistent with the quadratic form of Harned's rule e.g. $\Delta_m H^E$ for NaCl − NaSO$_4$ − H$_2$O

D. $\bar{y}_1 > \bar{y}_0$.    Complex mixing. The curve is not necessarily symmetrical with respect to the composition $y_A = y_B = 0.5$. e.g. $\Delta_m H^E$ for KCl − Na$_2$SO$_4$ − H$_2$O

The deviations for curves B to D can be positive or negative with respect to the ideal curve A (see Table 2.24).

ideal so that $\Delta_m V^E = \Delta_m H^E = 0$ (*Young's first rule*, Young. 1951; Young and Smith, 1954). This is a much better approximation than the corresponding ionic strength principle for $\Delta_m G^E$ (Wood and Reilly, 1970) and is generally a more accurate description for volumes of mixing than for heats of mixing (e.g. Wirth *et al.*, 1963).

Millero and his co-workers have made extensive use of Young's first rule to calculate the physical properties of sea water from the properties of the component salt solutions. These calculations indicate that the isothermal compressibility (Lepple and Millero, 1971) and the apparent equivalent volume, expansibility and compressibility of sea water can be predicted quite accurately from the sums of the corresponding single electrolyte properties weighted on an equivalent basis (see Table 2.22). The procedures employed have been discussed in detail by Millero (1974).

The properties of mixtures are more accurately represented by equation (6.3) and by equations (1) and (2) (Table 2.22) with $g_1 = h_1 = v_1 = 0$ (Fig. 2.15b). Using the equations in this form and considering mixtures where $y_A = y_B = 0.5$ a further regularity can be observed (*Young's second rule*, equation (4), Table 2.22). For mixtures of two cations with a common anion (Set A, Table 2.23) or two anions with a common cation (Set B, Table 2.23) the heats of mixing are almost independent of the common ion (Fig. 2.16). This rule has also been demonstrated for the excess free energies (e.g. Covington *et al.*, 1968) and volumes (e.g. Wirth *et al.*, 1963) of mixing. It holds well down to low ionic strengths (Wood and Smith 1965) and for ions of various charge types (Wood and Anderson, 1966a; Wood *et al.*, 1969a, b) where the common ions compared have similar effects on the water structure (e.g. $Cl^-$, $Br^-$, $I^-$). Where the structural influences differ (e.g. $F^-$, $Cl^-$) or where specific interaction effects may be significant (e.g. $Na^+-NO_3^-$, $Na^+-CO_3^{2-}$) the rule breaks down (see Wood and Anderson, 1967; Joliceour *et al.*, 1969 and Fig. 2.16). Young's second rule has consequently been called the 'structure rule'.

Following the structural clues a little further, another point becomes apparent (Set D, Table 2.23). Where the hetero-ions are of the same charge type and have similar structural influences (i.e. both are structure makers (e.g. $Li^+$, $H^+$), or both structure breakers (e.g. $K^+$, $Cs^+$)) then $\Delta_m H^E$ is positive. Where these ions have differing structural effects (e.g. $H^+$. $K^+$) then $\Delta_m H^E$ is negative (Wood and Anderson, 1966a, b, 1967; Karapet'yants *et al.*, 1970; Wood and Smith, 1965). For mixtures belonging to the first group ($\Delta_m H^E$ positive) the excess heat of mixing is smaller if both ions are structure breakers than if they are structure makers (Harned and Robinson, 1968, p. 71). These effects have also been confirmed for charge asymmetric mixtures (Woods and Ghamkhar, 1969) and for ionic strengths as low as $0.1$ M. For 1:1 electrolytes, at least, similar influences may be noted on the sign of $\Delta_m V^E$ and $\Delta_m G^E$

TABLE 2.23

*Excess heats of mixing*[a] *(298 K, 1 atm)*

| Set | Components A | B | $\Delta_m H^E$ $(J\,mol^{-1})$ | Set | Components A | B | $\Delta_m H^E$ $(J\,mol^{-1})$ |
|---|---|---|---|---|---|---|---|
| A | LiCl | NaCl | 88·55[b] | B | MgCl$_2$ | MgBr$_2$ | 17·08[e, f] |
| | LiBr | NaBr | 86·54[b] | *cont.* | CaCl$_2$ | CaBr$_2$ | 18·51[e] |
| | NaCl | KCl | −40·11[b] | | | | |
| | NaBr | KBr | −39·48[b] | C | NaCl | KBr | 5·07[g] |
| | NaI | KI | −44·72[c] | | NaBr | KCl | −78·34[g] |
| | NaF | KF | −29·52[c, d] | | NaCl | KNO$_3$ | 331·6[g] |
| | NaAc | KAc | −39·15[c] | | NaNO$_3$ | KCl | −419·5[g] |
| | Na$_2$CO$_3$ | K$_2$CO$_3$ | −11·97[c] | | | | |
| | NaNO$_3$ | KNO$_3$ | −52·63[b] | D | HCl | LiCl | 54·47[b] |
| | MgCl$_2$ | CaCl$_2$ | 14·65[e] | | HCl | NaCl | 136·16[b] |
| | MgBr$_2$ | CaBr$_2$ | 11·56[e] | | HCl | KCl | −15·66[b] |
| B | LiCl | LiBr | 3·39[b] | | HCl | RbCl | −86·04[b] |
| | NaCl | NaBr | 3·31[b] | | HCl | CsCl | −142·4[b] |
| | KCl | KBr | 3·35[b] | | LiCl | KCl | −67·20[b] |
| | NaCl | NaNO$_3$ | 12·98[b] | | RbCl | KCl | 10·05[b] |
| | KCl | KNO$_3$ | 1·42[b] | | CsCl | KCl | 6·66[b] |
| | | | | | NaCl | Na$_2$SO$_4$ | −34·75[b] |

[a] Unless otherwise stated $I = 1$ M, $y_A = y_B = 0·5$
[b] Wu *et al.* (1965)
[c] Joliceour *et al.* (1969)
[d] $I = 0·9$ M
[e] Wood and Anderson (1966) $I = 3·0$ M
[f] Extrapolation to $I = 1$ M gives $\Delta H_m^E = 1·38\,J\,mol^{-1}$
Allowing for charge weighting $\Delta H_m^E(2:1) = (4/9)\,\Delta H_m^E(1:1)$
$$= 1·51\,J\,mol^{-1}$$
∴ Young's second rule obeyed
[g] Wood and Smith (1965).

ton and Lilley, 1970; see Table 2.24). It is interesting and rather unexpected that the tenuous classfication of ions discussed in Section 2.4.1 should prove to be so generally useful.

Leyendekkers (1971b) has shown that for ternary common anion mixtures a plot of $RTh_0$ (when $y_A = y_B = 0·5$) vs. the sum of the conventional cation entropies ($S_H$) yields three straight lines corresponding to mixtures of different structural types (Fig. 2.17) if Na$^+$ is considered as a structure breaker (Kaminsky, 1957; Greyson and Snell, 1969; contrast Table 2.12). The simple qualitative correlation between the structural influences of the ions and the sign of $\Delta H_m^E$ is, therefore, replaced by a more complex but more useful quantitative correlation. The three lines can be merged into a single line

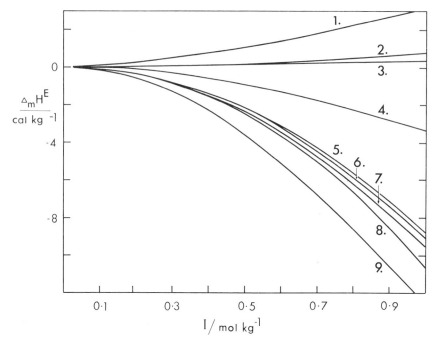

FIG. 2.16. Excess heats of mixing of ternary common-ion mixtures (1 cal = 4·187 J).

*Common cation mixtures*
1. NaCl–NaNO$_3$ (Wood and Smith, 1965).
2. NaCl–NaBr and KCl–KBr (*ibid.*)
3. KCl–KNO$_3$ (*ibid.*)

*Common anion mixtures*
4. Na$_2$CO$_3$–K$_2$CO$_3$ (Joliceour *et al.*, 1969)
5. NaF–KF (*ibid.*)
6. Na Acetate–K Acetate (*ibid.*)
7. NaCl–KCl and NaBr–KBr (Wood and Smith, 1965)
8. NaI–KI (Joliceour *et al.*, 1969)
9. NaNO$_3$–KNO$_3$ (Wood and Smith, 1965).

(E, Fig. 2.17) by correcting $S_H$ for long range structural entropy effects (Frank and Evans, 1945; Leyendekkers, 1971b).

*Young's third rule* (the cross square rule, Table 2.22) may be illustrated using the data in Table 2.23 (Wood and Smith, 1965). For the system Na$^+$–K$^+$–Cl$^-$–Br$^-$ this gives $\Sigma\square = -72\cdot93$ J kg$^{-1}$ and $\Sigma\times = -73\cdot27$ J kg$^{-1}$, and for the system Na$^+$–K$^+$–Cl$^-$–NO$_3^-$, $\Sigma\square = -78\cdot29$ J kg$^{-1}$ and $\Sigma\times = -87\cdot92$ J kg$^{-1}$. The rule does not hold so well for systems where specific ionic interactions may be large. A further example is provided by the free energy of mixing data for the system $Na^+-Mg^{2+}-Cl^--SO_4^{2-}$ (Wu *et al.*, 1969). Equations have been developed that allow for charge weighting in asymetric mixtures of this type (Reilly and Wood, 1969). However, the cross-square rule appears to apply equally well with or without charge weighting (Wood and Reilly, 1970) so that the simple rule may be retained for the present.

TABLE 2.24

*Structural influences on mixing parameters for ternary common-ion mixtures of monatomic 1:1 electrolytes (298 K, 1 atm, I = 1 M)*[a]

| Structural types | $g_0$ | $h_0^d$ | $v_0$ |
|---|---|---|---|
| SM[b] + SM | + | + | − |
| SB[c] + SB | + | + | − |
| SM + SB | − | − | + |

[a] Covington and Lilley (1970)   [c] Structure breaking
[b] Structure making   [d] $2g_0 \simeq h_0$ (Friedman 1962).

Reproduced with permission from the Chemical Society's Specialist Periodical Report, Electrochemistry Vol. I, 1970.

### 2.6.2.3. *Zdanovskii's rule.*

The rules described so far have referred to mixtures prepared at constant ionic strength (Figure 2.12) so that the long-range electrostatic effects on solute activity will cancel and thus simplify the interpretation of the mixing effects. The accent therefore has been on the effects of the mixing process on the properties of the solute.

Zdanovskii (1936) has chosen to consider the mixing process at constant water activity (i.e. with the components and the mixture at isopiestic equilibrium). The concentrations of the components making up the mixture therefore become variables, and their inter-relationship is defined by Zdanovskii's rule (equation (1), Table 2.25) for the ideal mixing process ($\Delta_m G^E = \Delta_m H^E = \Delta V_m^E = 0$). Although this equation has been applied to a wide variety of mixtures of different charge types (see e.g. Frolov *et al.*, 1971), its usefulness for *predicting* the properties of electrolyte mixtures has not yet been clearly demonstrated.

Mikhailov (1968) has suggested that the rule is inconsistent with the Debye–Hückel theory for charge asymmetric mixtures, but this is refuted by Kirgintsev (1971). Wood and Reilly (1970) have emphasized that the equations remain purely empirical and have little theoretical support. However, they should provide a useful rule of thumb for the preparation of multicomponent iso-osmotic saline solutions.

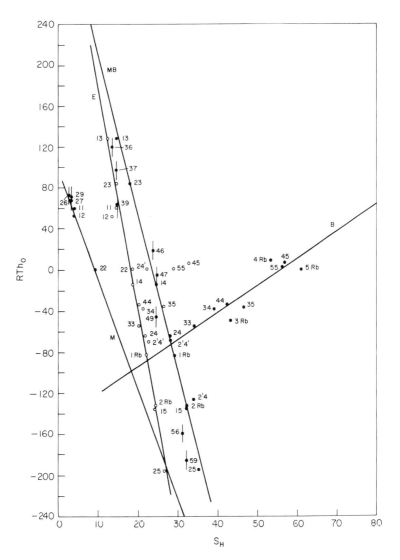

FIG. 2.17. Linear correlation between $R\,Th_0$ and the sum of the conventional cation entropies $(S_H)$ for an ionic strength of 1M at 298 K.

$$y_A = y_B = 0.5.$$

For charge symmetric mixtures $S_H = \bar{S}^0_{M^+} + \bar{S}^0_{N^+}$
For charge asymmetric mixtures $S_H = \bar{S}^0_{M^+} + (\bar{S}^0_{M^{2+}}/3)$

TABLE 2.25

*Zdanovskii's rule (mixing at isopiestic equilibrium)*

(i) *Ideal mixture*[a] (two electrolytes)

$$m^{-1} = (m_{A(0)})^{-1} y_A + (m_{B(0)})^{-1} y_B \qquad (1)^b$$

where $m_{A(0)}$ and $m_{B(0)}$ are the molalities of the pure component solutions in isopiestic equilibrium with the mixture containing $y_A m$ moles of $A$ and $y_B m$ moles of $B$.

(ii) *Non-ideal mixture*[c] (two electrolytes)

$$m^{-1} = (m_{\overline{A(0)}})^{-1} y_A + (m_{B(0)})^{-1} y_B + b y_A y_B \qquad (2)^d$$

where

$$b = A + Ba_w$$

or

$$b = k_1 + k_2 a_w + k_3 y_B$$

(iii) *Non-ideal mixture*[e] (multi-component)

$$m^{-1} = \sum_i (m_{i(0)})^{-1} y_i + \sum_{i<j} y_i y_j b_{ij} \qquad (3)$$

---

[a] Zdanovskii (1936), Zdanovskii and Deryabina (1965)
[b] A simple derivation of this equation is given by Kirginstev (1971)
[c] Kirgintsev and Luk'yanov (1963).
[d] In cases where the hetero-ions are both structure makers $b$ is positive, where both are structure breakers $b$ is approximately zero, and where one is a structure maker and one a structure breaker $b$ is negative. (Treating $Na^+$ as a structure maker). Kirgintsev and Luk'yanov (1963, 1964, 1966)
[e] Kirgintsev and Luk'yanov (1967).

Fig. 2.17 (contd).

*Curve identification*
   M.  mixture of two structure making cations
   B.  mixture of two structure breaking cations
 MB.  mixture of a structure making and a structure breaking cation
   E.  Data for curves M, B and MB corrected for long range structure entropy effects (Frank and Evans, 1945).

*Identification of data points*
The numbers $ij$ identify the cations $i$ and $j$ in the mixture according to the coding.

| | |
|---|---|
| 1. $H^+$ | 6. $Mg^{2+}$ |
| 2. $Li^+$ | 7. $Ca^{2+}$ |
| 3. $Na^+$ | 8. $Sr^{2+}$ |
| 4. $K^+$ | 9. $Ba^{2+}$ |
| 5. $Cs^+$ | |

A dashed number indicates a bromide salt, all other salts are chlorides. The units of $h_0$ are cal kg$^{-1}$. Reprinted with permission from Leyendekkers, (1971b), copyright of the American Chemical Society.

2.6.3. THEORETICAL DESCRIPTIONS

2.6.3.1. *The ion-association model.*

If the activity coefficients of the free ions and the ion-pairs are assumed to be unaffected by the ionic environment at constant ionic strength then the ion-association model (Section 2.5.2.1) can be applied directly to electrolyte mixtures. In addition to the equilibrium equations (equation (5.4), one for each ion-pair) a mass balance equation will be required to define the total concentration of each stoichiometric ionic component in terms of the free ions and the appropriate ion-pairs. These equations are then solved simultaneously by an iterative procedure to calculate the concentrations of all the solution components. A number of computer programmes are available that remove the tedium from such calculations (Ingri *et al.*, 1967; van Breemen, 1972; Morel and Morgan, 1972; Truesdell and Jones, 1973). The results are expressed as free concentrations of the various solution components. The stoichiometric activity coefficients ($^T\gamma_i$) can be calculated from these data using equation (5.5).

A characteristic feature of the ion-association model is the considerable flexibility introduced by the freedom to choose values for the activity coefficients of the free ions and the ion-pairs. This freedom, on occasion, gives rise to an embarrassing variety of models. For example, the solubility of gypsum ($CaSO_4 \, 2H_2O$) in electrolyte mixtures has been described on three occasions by ion-association models each using different definitions of the free single-ion activity coefficients (Nakayama, 1971c; Yeatts and Marshall, 1969; Gardener and Glueckauf, 1969). The concentrations of free ions in solution calculated from such models will reflect the definitions of single-ion activities, and will not necessarily represent the molecular reality in the solution. Furthermore, the ion-association model is unable to explain the changes in activity coefficient that occur when apparently unassociated halide solutions are mixed (e.g. $NaCl–CaCl_2$, Table 2.21). Indications that Harned's rule effects are in some circumstances compatible with the ion-association model (Pytkowicz and Kester, 1969) may revitalize the search for very weak ion-pairs in such mixtures. However, their behaviour is most simply described in terms of the ionic interaction theories discussed in the following sections (see Ginstrup, 1970).

The greatest problem associated with the use of the ion-association model is the assignment of activity coefficients to the ion-pairs. Since these species cannot be characterised experimentally, their properties must remain largely a matter of conjecture. Unless explicit relationships can be defined which specify the variation of the activity coefficients with temperature, pressure and composition, it is difficult to encompass a wide range of solution compositions and environmental conditions. In addition, it is not possible to calculate the osmotic coefficient from the Gibbs–Duhem equation (Table 2.4) and thus provide a comprehensive model of the solution. Where the ion-association model *has* been extended to cover a wide range of ionic environments (Yeatts and Marshall, 1969; Nakayama, 1971c) the variation of the ion-pair activity coefficient with ionic strength has been used as an adjustable parameter to fit the model to the experimental data.

Despite these shortcomings, the ion-association model has proved both popular and useful in marine chemistry because it provides a clear mechanistic picture of the interactions in solution. In addition, the effects of these interactions, resulting largely from electrostatic forces, can be treated by simple equilibrium equations in the same way as more familar examples of complex formation which result from predominantly covalent forces. The resulting equilibrium picture has a unity to it that so far appears to be lacking in the more physically oriented electrolyte solution theories.

### 2.6.3.2. *Specific interaction model.*

The Brønsted–Guggenheim equation (Table 2.16, equation (2)) is readily extended to electrolyte mixtures if the interaction coefficients are independent of the ionic environment at a given ionic strength (Table 2.26). A ternary common-ion mixture of two 1:1 electrolytes is used to illustrate the application of these equations in detail (Table 2.27).

The equations predict the properties of the mixture solely on the basis of the properties of the component single electrolyte solutions. This is equivalent to the assumption of Young's first rule for the mixing process (Table 2.27). By combining equations for $\log \gamma_{A(m)}$ and $\log \gamma_{A(0)}$ according to the linear form of Harned's rule, it is possible to predict the $\alpha$-coefficients (Table 2.26). A number of authors have compared these predicted values with experimental Harned's rule coefficients (e.g. Lakshmanan and Rangarajan, 1970; Rosseinsky and Hill, 1971; Lewis and Randall, 1961; Boyd *et al.*, 1971) and a good general correlation is observed (Fig. 2.18).

The cross differentiation relationship between the two Harned coefficients (Table 2.20, equation (3)) will be obeyed so long as the interaction coefficients (*b*) are slowly varying functions of ionic strength and the linear correlation noted between the $\alpha$-coefficients and the weighted ionic entropies (Leyendekkers, 1970, 1971b; Table 2.20, equation 6) is consistent with a similar relationship noted for the *b*-coefficients (Section 2.5.2.2). In addition, the relationship between the trace activity coefficients of the two components (equation (6.2)) can be predicted (Table 2.27, equation (6)) since both this relationship and the Brønsted–Guggenheim theory stem from the Åkerlöf–Thomas rule (equation (5.7)).

TABLE 2.26

*Brønsted specific interaction model.*

(i) *Excess free energy of mixture*[a] (Table 2.19)

$$\Delta G_m^E/RT = 0.5 \sum_j \sum_k [n_j n_k \bar{\beta}_{jk}/(n_w + \sum_i n_i)] \tag{1}$$

(ii) *Activity coefficients*
(a) Single ion (for the ions 1 and 2 of component $A$)

$$\log \gamma_{1(m)} = -|z_1/z_2| DG^b + \sum_k b_{1k}[k]_T \tag{2a}$$

$$\log \gamma_{2(m)} = -|z_2/z_1| DG + \sum_j b_{j2}[j]_T \tag{2b}$$

$b_{jk}$ is defined in Table 2.16 and $[k]_T$ is the stoichiometric concentration of component $k$ etc.

(b) Mean ion (component $A$)

$$\log \gamma_{A(m)} = -DG + (v_1/v_A) \sum_k b_{1k}[k]_T + (v_2/v_A) \sum_j b_{j2}[j]_T \tag{2c}$$

where

$$v_A = v_1 + v_2$$

(iii) *Osmotic coefficient*[a]

$$(\sum_i v_i m_i)(\phi_m - 1) = I\phi^{DG^c} + 2.303 \sum_j \sum_k m_j m_k b_{jk} \tag{3}$$

(iv) *Harned's rule coefficients* $(\alpha)$[d]

Combine equations (2a)–(2c) of this table with equation (1), Table 2.20 and equation (2), Table 2.16. For a reciprocal salt mixture $A(1, 2)$, $B(3, 4)$ (e.g. NaCl, MgSO$_4$) single-ion activity coefficients

$$\alpha_1/2 = (v_2/w_A) b_{12} - (v_4/w_B) b_{14} \tag{4a}$$

$$\alpha_2/2 = (v_1/w_A) b_{12} - (v_3/w_B) b_{32} \tag{4b}$$

mean-ion activity coefficients

$$\alpha_{12}/2 = (2v_1v_2/w_Av_A) b_{12} - (v_1v_4/w_Bv_B) b_{14} - (v_2v_3/w_Bv_A) b_{32} \tag{4c}$$

where

$$w_A = |z_1z_2| v_A \quad \text{and} \quad w_B = |z_3z_4| v_B$$

To calculate other α-values replace subscript $x$ in equations (4a)–(4c) by subscript $y$ as shown below

*To calculate other α-values replace subscript x in equations (4a)–(4c) by subscript y as shown below and on facing page.*

| α-value | $\alpha_{34}$[e] | | | | $\alpha_{14}$[e] | | $\alpha_{32}$[e] | | $\alpha_{12}$[f] | $\alpha_{32}$[f] | | | $\alpha_{14}$[g] | | | $\alpha_{12}$[g] |
|---------|---|---|---|---|---|---|---|---|---|---|---|---|---|---|---|---|
| $x$ | 1 | 2 | 3 | 4 | 2 | 4 | 1 | 3 | 4 | 1 | 3 | 4 | 3 | 2 | 4 | 3 |
| $y$ | 3 | 4 | 1 | 2 | 4 | 2 | 3 | 1 | 2 | 3 | 1 | 2 | 1 | 4 | 2 | 1 |

TABLE 2.26—*continued*

| α-value | $\alpha_3^{\,e}$ | $\alpha_4^{\,e}$ | $\alpha_1^{\,f}$ | $\alpha_2^{\,f}$ | $\alpha_3^{\,f}$ | $\alpha_1^{\,g}$ | $\alpha_2^{\,g}$ | $\alpha_4^{\,g}$ |
|---|---|---|---|---|---|---|---|---|
| $x$ | 1  3 | 2  4 | 4 | 4 | 1  3  4 | 3 | 3 | 3  2  4 |
| $y$ | 3  1 | 4  2 | 2 | 2 | 3  1  2 | 1 | 1 | 1  4  2 |

(v) *Interaction between ions of like charge*[d, h]
Equations (2a)–(2c) become

$$\log \gamma_{1(m)} = BG(2a)^{ij} + \sum_j \delta_{1j'}^i [j']_T \tag{5a}$$

$$\log \gamma_{2(m)} = BG(2b) + \sum_{k'} \delta_{2k'}[k']_T \tag{5b}$$

$$\log \gamma_{A(m)} = BG(2c) + (v_1/v_A)\sum_{j'} \delta_{1j'}[j']_T + (\cancel{v}_2/v_A)\sum_{k'} \delta_{2k'}[k']_T \tag{5c}$$

and equations (4a)–(4c) become

$$\alpha_{12}/2 = BG(4a) - (v_1 v_3/w_B v_A)\,\delta_{13} - (v_2 v_4/w_B v_A)\,\delta_{24} \tag{6a}$$

$$\alpha_1/2 = BG(4b) - (v_3/w_B)\,\delta_{13} \tag{6b}$$

$$\alpha_2/2 = BG(4c) - (v_4/w_B)\,\delta_{24} \tag{6c}$$

These equations can be used to deduce Young's rules (Table 2.27).

---

[a] Harned and Robinson (1968), $\bar{\beta}_{jk} = \ln 10 . b_{jk}$
[b] Electrostatic term, Table 2.14. equation (3)
[c] Given by the first term of equation (2) Table 2.18
[d] Harned and Robinson (1968), Leyendekkers (1971a, b)
[e] Reciprocal salt mixtures $A(12)$, $B(34)$, e.g. NaCl, MgSO$_4$
[f] Ternary common anion mixtures $A(12)$, $B(32)$, e.g. NaCl, MgCl$_2$
[g] Ternary common cation mixtures $A(12)$, $B(14)$, e.g. NaCl, Na$_2$SO$_4$
[h] Guggenheim (1966a, b)
[i] $BG(x)$ refers to the terms given on the right hand side of equation $x$ of this table
[j] $\delta_{ii'}$ represents interactions between the ions $i$ and $i'$ with like charge (e.g. Na$^+$ − K$^+$, Cl$^-$ − Br$^-$). When $i$ and $i'$ are identical $\delta_{ii'} = 0$ (Harned and Robinson, 1968, p. 23). The table following equation (4) is also valid here.

The specific interaction theory is, however, only an approximation. The interrelationship between the α-coefficients predicted by equation (4c) (Table 2.26) is rarely, if ever, accurately obeyed. For example, for 1:1 electrolytes the theory predicts that $\alpha_{12} = -\alpha_{32}$ whereas Table 2.21 gives for the NaCl–KCl system $\alpha_{12} = 0{\cdot}0235$, $\alpha_{32} = -0{\cdot}0095$ (I = 1·0 M), and for the CaCl$_2$–MgCl$_2$ system $\alpha_{12} = -0{\cdot}0277$, $\alpha_{32} = 0{\cdot}0225$ ($I = 1{\cdot}0$ M). In addition, careful measurements on electrolyte mixtures indicate that the mixing process is not ideal and that there are finite changes in volume and enthalpy when the mixture is formed (Section 2.6.2.2). These effects can be encompassed by introducing into the equations additional parameters characteristic of

TABLE 2.27
*Specific interaction model for a ternary common-ion mixture of* $1:1$ *electrolytes.*[a]

---

(i) *Activity coefficients* (at ionic strength $I$)
   From equation (2), Table 2.16

$$\log \gamma_{A(0)} = -DG + b_{12}m \tag{1}$$

From equation (2c), Table 2.26

$$\log \gamma_{A(m)} = -DG + b_{12}m/2 + b_{12}m_A/2 + b_{32}m_B/2 \tag{2}$$

where $m_A = y_A I$, $m_B = y_B I$, $m = m_A + m_B = I$

(ii) *Harned's rule coefficients*
   Combine equations (1) and (2) of this table with equation (1a), Table 2.20

$$-\alpha_{12}Iy_B = \log \left[\gamma_{A(m)}/\gamma_{A(0)}\right] = -0{\cdot}51[b_{12}(y_A - 1) + b_{32}y_B]$$
$$= -0{\cdot}51y_B[b_{32} - b_{12}] \tag{3}$$
$$\therefore \quad \alpha_{12} = 0{\cdot}5[b_{32} - b_{12}] \tag{4}$$

similarly

$$\alpha_{32} = 0{\cdot}5[b_{12} - b_{32}] = -\alpha_{12} \tag{5}$$

(iii) *Trace activities*
   Combining equation (2a), Table 2.20 with equations (1) and (4) of this table gives[b]

$$\log \gamma_{(0)A} = -DG + 0{\cdot}51(b_{12} + b_{32})$$
$$= \log \gamma_{(0)B} \tag{6}$$

(iv) *Young's rules*[c]
(a) Young's first rule
   Combine equation 6.5 with equation (5) of this table.
(b) Young's second rule
   Derive equation (3) of this table from equation (5c) of Table 2.26 to give

$$2\log \left[\gamma_{A(m)}/\gamma_{A(0)}\right] = -Iy_B(b_{32} - b_{12} + \delta_{13})$$
$$\therefore \quad \alpha_{12} = b_{32} - b_{12} + \delta_{13} \tag{7}$$

Similarly

$$\alpha_{21} = b_{12} - b_{32} + \delta_{13} \tag{8}$$

From equation 6.5

$$\Delta_m G^E / RT = 2{\cdot}3031^2 y_B(1 - y_B)(2\delta_{13}) \tag{9}$$

The excess mixing functions are therefore independent of the common ion.
(c) Young's third rule
   From equation (9) of this table for ternary common ion mixtures,

$$\Delta_m G^E(12, 32) + \Delta_m G^E(14, 34) + \Delta_m G^E(12, 14) + \Delta_m G^E(32, 34)$$
$$= \text{const}^d(\delta_{13} + \delta_{13} + \delta_{24} + \delta_{24}) = 2\times\text{const}\,(\delta_{13} + \delta_{24})$$

and for reciprocal salt pairs,

$$\Delta_m G^E(12, 34) + \Delta_m G^E(14, 32) = \text{const}\left[(\delta_{13} + \delta_{24}) + (\delta_{13} + \delta_{24})\right]$$
$$= 2\times\text{const}\,(\delta_{13} + \delta_{24})$$

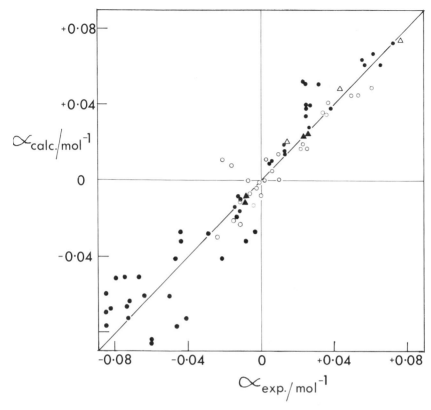

FIG. 2.18. Comparison of experimental Harned's rule coefficients with values calculated from equation (4c). Table 2.26. Data points are too closely crowded to be labelled individually but the corresponding mixtures may be identified by consulting the references listed below.

○, Lakshmanan and Rangarajan (1970)
●, Rosseinsky and Hill (1971)
△, Lewis and Randall (1961)
▲, Harned and Robinson (1968).

(Reproduced with permission of Elsevier Press).

Table 2.27 (contd).

[a] Guggenheim (1966a, b), Harned and Robinson (1968). There is a difference in notation between Guggenheim (1966a) and Guggenheim (1966b). The treatment in the second reference is fuller and is consistent with the tabulated values of the interaction parameters.

[b] Calculate $\log\left[\gamma_{A(0)}/\gamma_{(0)A}\right]$ and $\log\left[\gamma_{B(0)}/\gamma_{(0)B}\right]$ from equations (1) and (2), Table 2.20. Substitute for pure electrolyte values and for $\alpha_{12}$ and $\alpha_{32}$ from equations (1), (4) and (5) of the present table.

[c] See Table 2.22

[d] $\text{Const} = 4 \cdot 605 I^2 RT y_B (1 - y_B)$.

the interactions between ions of like charge (Table 2.26). Empirical observations suggest that such interactions are less pronounced between anions than between cations and that the higher the charge on the ions the less they will interact with ions of the same sign (Wood and Anderson, 1966a). Using the extended equations, Young's second and third rules may be deduced (Table 2.27, Guggenheim, 1966b).

Values for the new interaction parameters $(\delta_{jj'}, \delta_{kk'})$ may be estimated from the experimental Harned's rule coefficients. For charge symmetric systems with a common anion this gives (Leyendekkers, 1971a)

$$\delta_{13} = -(\alpha_{12} + \alpha_{32})(4v_1v_3/w_Av_B)^{-1}$$

and for a $1:1/2:1$ mixture with a common anion (see Table 2.26),

$$\delta_{13} = -2(\alpha_{12} + \alpha_{32}) - b_{32}/9$$

Guggenheim (1966a, p. 175) also indicates a procedure for estimating $\delta$-values graphically using deviations from the Åkerlöf–Thomas rule. As the solution becomes more concentrated, the $\delta$-terms will become more important and multiple interactions between three or maybe four ions might become significant. These more complex effects will be reflected in the $g_1$-term (equation (6.3)) and its derivatives.

The specific interaction theory provides a good qualitative basis for the discussion of the regularities observed in electrolyte mixtures (see Rosseinsky, 1971b) and it has featured prominently in the discussion of electrolyte mixtures in the standard text books (Robinson and Stokes, 1965; Harned and Owen, 1958; Lewis and Randall, 1961; Harned and Robinson, 1968). Various applications of the model have been discussed including the prediction of activity coefficients in saturated salt solutions (Glueckauf, 1949) and in multi-component systems (Meissner and Kusik, 1972) and as the basis for the calculation of complex formation constants (Chan and Panckhurst, 1972).

### 2.6.3.3. *Scatchard's equations.*

Scatchard (1961) suggested that the free energy of the mixture could be expressed as a power series in the ionic strength (Table 2.28, equation (2)). The electrostatic contributions to the free energy are described by an extended Debye–Hückel equation. They are assumed to be functions of the ionic strength only and are estimated from the properties of the component single electrolyte solutions. Any errors resulting from this assumption will be absorbed into the $b$-coefficients of the power series representing the interactions in the two-salt mixtures (equations (10a) (10d), Table 2.28).

TABLE 2.28

*Scatchard's theory. General equations for the neutral electrolyte approach.*

(i) *Excess free energy of mixture*[a]

If concentrations are expressed in ionic strength fractions ($y_A = I_A/I$) then $\Delta G_m^E$ can be written as a power series in ionic strength (cf. Table 2.17, footnote [e])

$$(\Delta G_m^E/RT)/\sum_T n_I^* = \sum_J A_J y_J + \sum_J \sum_{K>J} y_J y_K \sum_{t \geqslant 0} B_{JK}^{(t)} (y_J - y_K)^t \tag{1}^b$$

where

$$B_{KJ}^{(t)} = -B_{JK}^{(t)} (-1)^t$$

$$n_J^* = n_J \sum_i \nu_{iJ} z_i^2/2, \qquad y_J = n_J^*/\sum_I n_I^*$$

$\nu_{iJ}$ = No. of moles of ion $i$ in 1 mole of component $J$. $A_J$ and $B_{JK}^{(t)}$ are adjustable parameters.

(ii) *Activity coefficients* (for the component $Q$)

$$(\partial G_m^E/RT)/\partial n_Q^* = (\nu_Q n_Q/n_Q^*) \ln \gamma_{Q(m)}$$

$$= A_Q + \alpha_Q + \sum_J (\alpha_J - \alpha_Q) y_J + \sum_J y_J \sum_{t \geqslant 0} B_{QJ}^{(t)} \left[ (y_Q - y_J)^t + t y_Q (y_Q - y_J)^{(t-1)} \right]$$

$$+ \sum_J \sum_{K>J} y_J y_K \sum_{t \geqslant 0} \left[ \beta_{JK}^{(t)} - (t+1) B_{JK}^{(t)} \right] (y_J - y_K)^t \tag{2}^b$$

where

$$\alpha_J = \partial A_J/\partial \ln I, \qquad \beta_{JK}^{(t)} = \partial B_{JK}^{(t)}/\partial \ln I \tag{3}^b$$

(iii) *Osmotic coefficient*

$$\partial(\Delta G_m^E/RT \sum_I n_I^*)/\partial(n_w w_w) = (\sum_I \nu_I m_I/\sum_I n_I^*)(\phi - 1)$$

$$= \sum_J \alpha_J y_J + \sum_J \sum_{K>J} y_J y_K \sum_{t \geqslant 0} \beta_{JK}^{(t)} (y_J - y_K)^t \tag{4}^b$$

where

$$w_w = M_w/1000 = 0 \cdot 018$$

(iv) *Determination of adjustable parameters*

(a) $A_J$ and $\alpha_J$

From equation (2) for a single electrolyte solution

$$(\nu_J m_J/I_J) \ln \gamma_{J(0)} = A_J + \alpha_J \tag{5}^{c, d}$$

From equation (4)

$$(\nu_J m_J/I_J)(\phi - 1) = \alpha_J \tag{6}^{c, d}$$

$A_J$ and $\alpha_J$ may be obtained directly from these equations at each ionic strength or they may be summarized as polynomial functions of ionic strength, i.e.

$$\alpha_J = 2\mathscr{A}^* I^{\frac{1}{2}} \sigma(\mathring{a}_J I^{\frac{1}{2}}) + a_J^{(1)} I + a_J^{(2)} I^2 + a_J^{(3)} I^3 \tag{7}^e$$

and

$$(A_J + \alpha_J) = 2\mathscr{A}^* I^{\frac{1}{2}}/(1 + \mathring{a}_J I^{\frac{1}{2}}) + 2a_J^{(1)} I + (3/2) a_J^{(2)} I^2 + (4/3) a_J^{(3)} I^3 \tag{8}$$

TABLE 2.28—*continued*

The terms $\hat{a}_J$ and $a_J^{(t)}$ are treated as adjustable parameters. Values for the major sea salt components are listed in Table 2.30[d].

(b)  $B_{JK}^{(t)}$      and      $\beta_{JK}^{(t)}$

Obtained by curve-fitting to data for mixtures of two salts. The equations are considered in detail in Table 2.29. The parameters are defined so that,

$$\beta_{IJ}^{(0)} = \sum_{t \geq 1} b_{IJ}^{(0, t)} I^t \tag{9a}$$

$$\beta_{IJ}^{(1)} = \sum_{t \geq 2} b_{IJ}^{(1, t)} I^t \tag{9b}$$

$$B_{IJ}^{(0)} = \sum_{t \geq 1} (b_{IJ}^{(0, t)} I^t / t) \tag{10a}$$

$$B_{IJ}^{(1)} = \sum_{t \geq 2} (b_{IJ}^{(1, t)} I^t / t) \tag{10b}$$

The terms $b_{IJ}^{(0, t)}$ and $b_{IJ}^{(1, t)}$ are treated as adjustable parameters (Table 2.30).

---

[a]  Scatchard *et al.* (1970). Their symbol $\Delta G^E$ has been replaced here by $\Delta G_m^E$ to be consistent with Table 2.19

[b]  Scatchard (1961) including corrections given by Rush and Johnson (1968)

[c]  Wu *et al.* (1968) including corrections given by Rush (1969b)

[d]  Rush (1969a)

[e]  Rush and Johnson (1968). The functions $\sigma(x)$ and $\mathscr{A}^*$ are defined in Table 2.18. $\hat{a}_J$ can be used as an additional adjustable parameter, usually it is set equal to $1 \cdot 5^d$ (equation (4), Table 2.14). The $a_J^{(t)}$ parameters for some electrolytes may be obtained from the data of Lietzke and Stoughton (1962) by multiplying the B, C, D parameters by $-2/|z_+ z_-|$ (Table 2.17)[c]

To illustrate the application of the method the equations for a two-salt mixture are shown in Table 2.29. Although the equations appear quite formidable, the procedure for using them is straightforward (Tables 2.28 and 2.29; Rush, 1969a).

Because there are so many adjustable parameters available a good fit to the data is inevitable. To minimize the arbitrary nature of the fit the *b*-parameters are chosen by initially allowing *all* the parameters to vary to give the best fit. The parameters are then tested one by one until the *smallest* set is found that satisfies the data without a significant increase in the standard deviation (Wu *et al.*, 1968).

Once the parameters have been determined they can be used to predict the osmotic and activity coefficients in any mixture containing these components (Table 2.30). The tabulated data (Rush 1969a) therefore provide a convenient way for storing with a high precision information on the properties of electrolyte solutions. The deviation function

$$\delta_J = \log \gamma_{J(m)} - \log \gamma_{J(m)\alpha}$$

TABLE 2.29

*Scatchard's theory. Equations for mixtures of two salts*
*(neutral electrolyte approach)*

---

(i) *Excess free energy of mixture*

$$(\Delta G_m^E/RT)/(n_A^* + n_B^*) = A_A + (A_B - A_A)y_B + B_{AB}^{(0)}y_B(1 - y_B)$$
$$+ B_{AB}^{(1)}y_B(1 - y_B)(1 - 2y_B) \qquad (1)^{a, b}$$

(ii) *Activity coefficients*

$$\ln \gamma_{A(m)} = (I_A/v_A m_A)\left[A_A + \alpha_A + (\alpha_B - \alpha_A)y_B + \beta_{AB}^{(0)}y_B\right.$$
$$\left. + (B_{AB}^{(0)} - \beta_{AB}^{(0)})y_B^2 + \beta_{AB}^{(1)}y_B + 3(B_{AB}^{(1)} - \beta_{AB}^{(1)})y_B^2 - 2(2B_{AB}^{(1)} - \beta_{AB}^{(1)})y_B^3\right] \qquad (2)^{a, b}$$

$$\ln \gamma_{B(m)} = (I_B/v_B m_B)\left[A_B + \alpha_B + (\alpha_A - \alpha_B)y_A + \beta_{AB}^{(0)}y_A\right.$$
$$\left. + (B_{AB}^{(0)} - \beta_{AB}^{(0)})y_A^2 - \beta_{AB}^{(1)}y_A - 3(B_{AB}^{(1)} - \beta_{AB}^{(1)})y_A^2 + 2(2B_{AB}^{(1)}) - \beta_{AB}^{(1)})y_A^3\right] \qquad (3)^{a, b}$$

(iii) *Osmotic coefficient*

$$\phi = 1 + [I/(v_A m_A + v_B m_B)]\left[\alpha_A + (\alpha_B - \alpha_A)y_B + \beta_{AB}^{(0)}y_B(1 - y_B)\right.$$
$$\left. + \beta_{AB}^{(1)}y_B(1 - y_B)(1 - 2y_B)\right] \qquad (4)^{a, b}$$

(iv) *Determination of adjustable parameters*
(a) $A_J$ and $\alpha_J$—see Table 2.28
(b) $B_{AB}^{(i)}$ and $\beta_{AB}^{(i)}$—define in terms of the parameters $b_{AB}^{(i, j)}$

$$\beta_{AB}^{(0)} = b_{AB}^{(0, 1)}I + b_{AB}^{(0, 2)}I^2 + b_{AB}^{(0, 3)}I^3 \qquad (5a)^{c, d}$$

$$\beta_{AB}^{(1)} = b_{AB}^{(1, 2)}I^2 + b_{AB}^{(1, 3)}I^3 \qquad (5b)^{c, d}$$

$$B_{AB}^{(0)} = b_{AB}^{(0, 1)}I + (\tfrac{1}{2})b_{AB}^{(0, 2)}I^2 + (\tfrac{1}{3})b_{AB}^{(0, 3)}I^3 \qquad (6a)^c$$

$$B_{AB}^{(1)} = (\tfrac{1}{2})b_{AB}^{(1, 2)}I^2 + (\tfrac{1}{3}) b_{AB}^{(1, 3)}I^3 \qquad (6b)^c$$

---

[a] For most purposes it is sufficient to consider only terms as high as $B_{JK}^{(1)}$ in the expansion (Scatchard, 1961). Including corrections given by Rush and Johnson (1968)
[b] Wu *et al.* (1968) including corrections given by Rush (1969b)
[c] Rush and Johnson (1968)
[d] Different notations are used in the references quoted above; notably

$$b_{AB}^{(j, i)}(\text{ref. b}) = \beta_{AB}^{(j, i)}/I^i (\text{ref. a})$$

has been used extensively to assess the applicability of the Brønsted principle and Harned's rule (Wu *et al.*, 1968; 1969; Lindenbaum *et al.*, 1972). $\log \gamma_{J(m)\alpha}$ corresponds to the value of $\log \gamma_{J(m)}$ estimated from the properties of the component single electrolyte solutions only. If the Brønsted hypothesis is valid then $\delta$ is zero so that only the $\alpha_J$ and $A_J$ parameters are required. For the perchlorate mixtures studied by Rush and Johnson (1968, $H^+-Na^+-Li^+-ClO_4^-$) the Brønsted principle gave predictions that were accurate to $\pm7\%$

TABLE 2.30

Scatchard's theory. Experimental a- and b-coefficients[a] (neutral electrolyte approach)

A.

| Parameters \ Salt[b] | 1 NaCl | 2 NaCl | 3 KCl | 4 MgCl$_2$ | 5 CaCl$_2$ | 6 BaCl$_2$ | 7 Na$_2$SO$_4$ | 8 MgSO$_4$ |
|---|---|---|---|---|---|---|---|---|
| $\mathring{a}_J$ | 1·5 | 1·45397 | 1·5 | 1·60067 | 1·5 | 1·5 | 1·24072 | 1·37486 |
| $a^{(1)}$ | 0·03684 | 0·04472 | −0·06408 | 0·06633 | 0·05848 | 0·03166 | −0·06580 | −0·02712 |
| $a^{(2)}$ | 0·02108 | 0·018616 | 0·05244 | 0·009003 | 0·005532 | 0 | 0·007263 | 0·004213 |
| $a^{(3)}$ | −0·00134 | −0·0010724 | −0·011124 | −0·0002545 | −0·000068 | 0·000912 | −0·0001945 | −0·000095 |
| $a^{(4)}$ | 0 | 0 | 0·000918 | 0 | 0 | −0·0001006 | 0 | 0 |
| $\mathscr{A}^*$ | −1·17082 | −1·17202 | −1·17082 | −1·17202 | −1·17082 | −1·17082 | −1·17202 | −1·17202 |
| $I^c_{max}$ | 5·4 | 5·9 | 5·4 | 5·9 | 6·3 | 4·8 | 9·4 | 7·8 |

B.

| Parameters \ Mixture[d] | (1, 3) | (4, 2) | (1, 5) | (6, 1) | (2, 7) | (2, 8) | (7, 8) | (7, 4) |
|---|---|---|---|---|---|---|---|---|
| $b^{(0,1)}_{AB}$ | −0·0253 | 0·0654 | 0·0703 | 0·03674 | −0·05821 | 0 | 0·03178 | 0 |
| $b^{(0,2)}_{AB}$ | −0·00299 | −0·0176 | −0·0225 | −0·02512 | 0 | −0·00798 | −0·003055 | −0·00657 |
| $b^{(0,3)}_{AB}$ | 0 | 0·00191 | 0·00235 | 0·00304 | 0·000439 | 0·000855 | 0 | 0·000231 |
| $b^{(1,2)}_{AB}$ | 0 | 0 | 0·00405 | 0·0095 | 0 | 0 | 0 | 0 |
| $b^{(1,3)}_{AB}$ | 0 | 0 | −0·0050 | −0·00257 | 0 | 0 | 0 | 0 |
| $I^c_{max}$ | 5·4 | 5·9 | 6·3 | 4·8 | 9·4 | 7·8 | 8·8 | 8·2 |

TABLE 2.30—continued

B—continued

| Parameters \ Mixture[d] | (3, 5) | (6, 3) |
|---|---|---|
| $b_{AB}^{(0,1)}$ | 0·03332 | 0·0197 |
| $b_{AB}^{(0,2)}$ | −0·01036 | −0·0295 |
| $b_{AB}^{(0,3)}$ | 0 | 0·00451 |
| $b_{AB}^{(1,2)}$ | 0 | 0·00288 |
| $b_{AB}^{(1,3)}$ | 0 | −0·00166 |
| $I_{max}^{c}$ | 5·0 | 4·7 |

[a] Rush (1969a)

[b] Substitute $a$-values into equations (7) and (8) of Table 2.28 to calculate $\alpha_J$ and $A_J$. The activity and osmotic coefficients of the single electrolyte solutions can then be obtained directly from equations (5) and (6) of the same table.

[c] The parameters were obtained by curve fitting up to this ionic strength. The behaviour of the polynomial function may be erratic at higher ionic strengths.

[d] The numbers in parentheses refer to the pure electrolyte components listed in part A. To calculate the properties of two salt mixtures substitute the $b$-values into equations (5) and (6) Table 2.29 and obtain $\beta_{AB}$ and $B_{AB}$. These values, together with the appropriate $\alpha_J$ and $A_J$ values are then substituted into equations (1) to (4) of the same table to calculate activity and osmotic coefficients in the mixture.

To calculate the properties of a multicomponent mixture combine the values of the $A$, $B$, $\alpha$ and $\beta$ terms of all the component salts according to equations (2) and (4) of Table 2.28.

F

TABLE 2.31

*Cluster integral expansion theory of electrolyte mixtures*

---

(i) *Free energy of mixture*

For a solution containing $m_i^M$ moles of cation $M_i$ with charge $z_i^M$, $m_j^X$ moles of anion $X_j$ with charge $z_j^X$, etc.

$$G/RT = (2IRT) \sum_{l=1}^{l=i} \sum_{m=1}^{m=j} E_A Z_{lm} G_{M_l X_m}^0$$

$$+ \sum_{k=2}^{k=i} \sum_{l=1}^{l=(k-1)} \sum_{m=1}^{m=j} (E_A/4) E_k^M Z_{km} Z_{lm} g_{M_k M_l X_m}$$

$$+ \sum_{k=1}^{k=i} \sum_{k=2}^{l=j} \sum_{m=1}^{m=(l-1)} E_B E_m^X Z_{kl} Z_{km} g_{X_l X_m^M k} \qquad (1)^a$$

where the equivalent weighting factors are given by

$$E_k^M = z_k^M m_k^M; \quad E_m^X = -z_m^X m_m^X$$

$$E = \sum_{k=1}^{k=i} E_k^M = \sum_{m=1}^{m=j} E_m^X$$

$$E_A = E_l^M E_m^X / E; \qquad E_B = E_k^M E_l^X / 4E$$

the charge weighting terms by

$$Z_{km} = z_k^M - z_m^X \text{ etc.}$$

and the ionic strength by

$$2I = \sum_{k=1}^{k=i} z_k^M E_k^M - \sum_{l=1}^{l=j} z_l^X E_l^X$$

(ii) *Osmotic coefficient*

$$m(1 - \phi) = - \sum_{l=1}^{l=i} \sum_{m=1}^{m=j} (E_A Z_{lm}/z_l^M z_m^X)(l - \phi_{M_l X m}^0)$$

$$- \sum_{k=2}^{k=i} \sum_{l=1}^{l=(k-1)} \sum_{m=1}^{m=j} (E_A/4) E_k^M Z_{km} Z_{lm} g^M$$

$$- \sum_{k=1}^{k=i} \sum_{l=2}^{l=j} \sum_{m=1}^{m=(l-1)} E_B E_m^X Z_{kl} Z_{km} g^X \qquad (2)^b$$

where

$$g^M = g_{M_k M_l X_m} + I \partial (g_{M_k M_l X_m})/\partial I$$

$$g^X = g_{X_l X_m M_k} + I \partial (g_{X_l X_m M_k})/\partial I$$

(iii) *Activity coefficient*

For the salt $M_p X_q$

$$(Z_{pq}/z_p^M z_q^M) \ln \gamma_{\pm}^{M_p X_q} = \sum_{l=1}^{l=i} \sum_{m=1}^{m=j} (E_A/E)(A_1 + A_2 - A_3)$$

$$- \sum_{l=1}^{l=i} \sum_{m=1}^{m=j} (E_A/4) Z_{pm} Z_{lm} g_{M_p M_l X_m}$$

TABLE 2.31—continued

$$+ \sum_{k=2}^{k=i} \sum_{l=1}^{l=(k-1)} \sum_{m=1}^{m=j} (E_A/4E) E_k^M B$$

$$+ \sum_{k=1}^{k=i} \sum_{l=1}^{l=j} E_B Z_{kl} z_{kq} g_{X_l X_q} M_k$$

$$+ \sum_{k=1}^{k=i} \sum_{l=2}^{l=j} \sum_{m=1}^{m=(l-1)} (E_B/E) E_m^X C \qquad (3)^c$$

where
$$A_1 = (Z_{pm}/z_p^M z_m^X)(1 - \phi_{M_p X_m}^0 + \ln \gamma_{M_p X_m}^0)$$

$$A_2 = (Z_{lq}/z_l^M z_q^X)(1 - \phi_{M_l X_q}^0 + \ln \gamma_{M_l X_q}^0)$$

$$A_3 = (Z_{lm}/z_l^M z_m^X)([(1 + Z_{pq}E/2I)(1 - \phi_{M_l X_m}^0) + \ln \gamma_{M_l X_m}^0]$$

$$B = Z_{km}Z_{lm}[g_{M_k M_l X \, m} - (Z_{pq}E/2)\partial(g_{M_k M_l X \, m})/\partial I] - Z_{kq}Z_{lq}g_{M_k M_l X q}$$

$$C = Z_{kl}Z_{km}[g_{X_l X_m M_k} - (Z_{pq}E/2)\partial(g_{X_l X_m M_k})/\partial I] - Z_{pl}Z_{pm}g_{X_l X_m M p}$$

$$g_{M_p M_p} X_m = g_{X_q X_q M_k} = 0$$

(iv) *Estimation of interaction terms*
(a) From the osmotic coefficient of the corresponding ternary common ion mixture (e.g. $MX, NX$)
Equation (3) simplifies to

$$\frac{z^M - (z^M - z^M)y}{z^M z^N z^X}(1 - \phi) = \frac{y}{z^M z^X}(1 - \phi_{MX}^0) + \frac{(1-y)}{z^N z^X}(1 - \phi_{FX}^0)$$

$$+ \frac{y(1-y)I}{2}\left[g_{MN}X + \frac{I\partial}{\partial I}(g_{MN}X)\right] \qquad (4)$$

where $y$ is the ionic strength fraction of component $MX$.

The final term in parentheses can be estimated by comparing the experimental value of $\phi$ with that calculated on the basis of the weighted contributions of the single electrolyte solutions (Reilly et al., 1971).

(b) From the equation derived by Friedman (1960a).

Since only pairwise interactions are considered $g_{MNX}$ may be equated to $g_0$ of equation (6.3) for the mixing of $MX$ and $NX$ at the same total ionic strength as the multicomponent mixture[d]. Consequently the interaction parameters may be calculated from Harned's rule coefficients using equation (6.4a.)

---

[a] Reilly and Wood (1969). The second two terms in equation (1) correspond to $\Delta_m G^E$ in Friedman's equations. $G_{M_l X_m}^\circ$ is the free energy of a pure solution of $M_l X_m$ at the same total ionic strength as the mixture. $g_{M_k M_l X_m}$ is a parameter representing the interaction between cation $M_k$ and cation $M_l$ in the presence of anion $X_m$. The estimation of these parameters from experimental data is considered in part (iv) of this table.
[b] Reilly et al. (1971). $\phi_{M_l X_m}^\circ$ is the osmotic coefficient of a pure solution of $M_l X_m$ at the same total ionic strength as the mixture.
[c] Reilly et al. (1971). $\gamma_{M_l X_m}^\circ$ is the activity coefficient of a pure solution of $M_l X_m$ at the same total ionic strength as the mixture.
[d] Reilly and Wood (1969).

in $\gamma_\pm$ up to an ionic strength of 5 M. On the same basis, the properties of common-ion mixtures in the system $Na^+-Mg^{2+}-Cl^--SO_4^{2-}$ can be predicted with a maximum uncertainty of $\pm 8\%$ in $\gamma_\pm$ at an ionic strength of 1 M but at an ionic strength of 6 M the error may be as much as 56% (Wu et al., 1968). Predictions for mixtures *without* a common ion in the same system are better, the error varying from 1% at $I = 1$ M to a maximum of 18% at $I = 6$ M (Wu et al., 1969). The full equations will naturally give a better account of the properties of the system. However, in view of its simplicity, the use of the Brønsted hypothesis is worth investigating at ionic strengths around 1 M.

In the Scatchard equations the neutral electrolytes are treated as the solution components. Consequently, for the mixture $Na^+-Mg^{2+}-Cl^--SO_4^{2-}$ it is necessary to decide whether the solution is to be prepared by mixing $NaCl$ and $MgSO_4$ or $Na_2SO_4$ and $MgCl_2$ (or any combination of these salts) before the properties of the mixture can be calculated. This is an artificial situation which could give rise to ambiguities in a complex mixture such as sea water. Scatchard therefore developed an alternative procedure in which the ions are treated as the solution components. The resulting equations have a daunting complexity (Scatchard, 1968; Scatchard et al., 1970). Although they remove the ambiguities in the definition of the solution composition, these equations do not necessarily result in more accurate predictions of the properties of electrolyte mixtures (Rush and Johnson, 1970).

The Scatchard model can be used to calculate the excess mixing functions ($\Delta_m Y^E$). The equation for $\Delta_m G^E$ gives *directly* the form deduced by Friedman (1961) and the appropriate equations for the other mixing functions can be obtained by differentiation of $\Delta_m G^E$ with respect to temperature and pressure (Table 2.22, equations (1) and (2)). The conditions necessary for the validity of Young's rules are discussed by Scatchard (1968, 1969) and by Scatchard et al. (1970) using the ionic component approach.

### 2.6.3.4. Cluster integral expansion equations.

Wood and his co-workers have developed equations from Friedman's version of the cluster integral expansion theory (Friedman, 1960a, b; 1962) to describe the properties of mixed electrolyte solutions. They take as their starting point the general equation for the free energy of the multi-component mixture (e.g. equation (1), Table 2.28) and they express the excess mixing function ($\Delta G_m^E$) in terms of the interaction products defined by Friedman's theory (Reilly and Wood, 1969). These equations account correctly for all pairwise interactions. In mixtures containing electrolytes of a single charge type, all triplet interactions (except those between ions of the same sign) are accurately accounted for since their effects will cancel out. However, in

mixtures containing electrolytes of different charge types the triplet interactions are not accurately considered and the equations become less accurate
at high ionic strengths.

Although the equations (Table 2.31) appear complex at first sight, they
are actually simpler to use than the Scatchard equations (Table 2.28) because
the input data required (activity and osmotic coefficients) are more readily
accessible than the b-coefficients of the Scatchard polynomials. In addition,
the equations deal with various permutations of ion concentrations (Reilly
and Wood, 1969) so that the composition of the mixture can be expressed
using the ions as components. The ambiguities associated with Scatchard's
neutral electrolyte approach are, therefore, avoided.

The equations have been used for the prediction of activity and osmotic
coefficients in multi-component electrolyte solutions (Reilly et al., 1971).

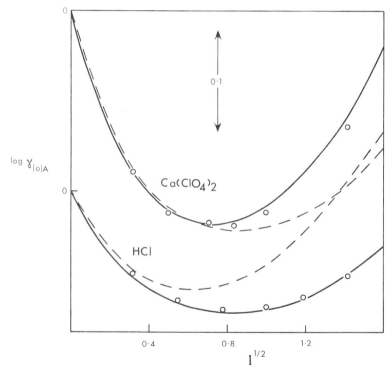

FIG. 2.19. Comparison of predicted and experimental trace activity coefficients (molal) in the
mixture $Ca^{2+}-H^+-Cl^--ClO_4^--H_2O$ at 298 K.

The solid line indicates experimental data and the dashed line predictions based on the
ionic strength principle. The zero values have been displaced for clarity. Values calculated from
the cluster integral expansion equation (Reilly et al., 1971) are shown as open circles.

The prediction of activity coefficients was given a severe test by comparing calculated and experimental values in reciprocal salt mixtures of the type $M^{2+} - H^+ - Cl^- - ClO_4^-$. As noted above, the present theory, based on pairwise interactions, is expected to break down in such asymmetric mixtures at high ionic strengths. In general, the activity coefficients can be predicted within experimental error using only the properties of the single electrolyte components (Fig. 2.19). The most significant errors were noted with the system $Mg^{2+} - H^+ - Cl^- - ClO_4^-$ (Reilly et al., 1971), but even in this case the predictions of the cluster expansion theory are within 6% in $\gamma_\pm$ at $I = 1$ M and within 15% at $I = 3$ M for $\gamma_{(0)HCl}$. Accurate predictions were also obtained for the osmotic coefficients of mixtures of three chlorides (Table 2.32), In this case the simple Brønsted theory predictions show significant deviations from the experimental values, but the use of the full equation (Table 2.31, equation (3)) gives values within a few tenths of a percent of the experimental data.

The extension of the model to consider Young's rules arises naturally since the equations used here have the same root as equation (6.3) which was used in the construction of Table 2.22. The detailed derivation of equations describing the excess mixing functions will be found in the papers by Wood and his co-workers referred to in Section 2.6.2.2 and in the paper by Reilly

TABLE 2.32

*The prediction of osmotic coefficients by the cluster integral expansion equations*[a]

| Composition (molalities) | | | Ionic Strength | Osmotic Coefficient | | |
|---|---|---|---|---|---|---|
| LiCl | NaCl | CsCl | | Exptl. | Calc. 1[b] | Calc. 2[c] |
| 1·5336 | 1·5280 | 1·6486 | 4·7102 | 1·1045 | 1·2117 | 1·1073 |
| 1·7037 | 1·6974 | 1·8314 | 5·2325 | 1·1359 | 1·2586 | 1·1397 |
| LiCl | NaCl | BaCl$_2$ | | | | |
| 0·7136 | 0·7112 | 0·2374 | 2·1371 | 1·0295 | 1·0383 | 1·0308 |
| 1·0647 | 1·0612 | 0·3543 | 3·1888 | 1·1215 | 1·1383 | 1·1217 |
| 1·1010 | 1·0973 | 0·3663 | 3·2973 | 1·1337 | 1·1494 | 1·1315 |
| KCl | NaCl | BaCl$_2$ | | | | |
| 0·4708 | 0·4708 | 0·3559 | 2·0093 | 0·9179 | 0·9255 | 0·9197 |
| 0·1740 | 0·1740 | 0·7449 | 2·5827 | 0·9205 | 0·9269 | 0·9224 |
| 1·1980 | 1·1980 | 0·2126 | 3·0338 | 0·9718 | 0·9859 | 0·9713 |
| 1·0281 | 1·0281 | 0·4292 | 3·3438 | 0·9772 | 0·9944 | 0·9768 |
| 0·5509 | 0·5509 | 1·0022 | 4·1084 | 0·9976 | 1·0175 | 0·9980 |

[a] Reilly et al. (1971)
[b] Calculated assuming $g$-parameters $= 0$ (i.e. equivalent to the Brønsted specific interaction principle)
[c] Calculated using the full theory (equation 3, Table 2.31).

and Wood (1969). The temperature derivative of equation (1) (Table 2.31) has been used to predict the excess heat of mixing of three chlorides (Wood et al., 1969a). The experimental results are predicted with a surprising accuracy considering the very small heats of mixing that are involved (Fig. 2.20).

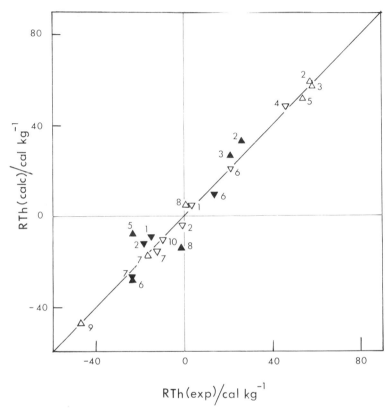

FIG. 2.20. Comparison between experimental heats of mixing at constant total ionic strength and values calculated from the cluster integral expansion theory (Wood et al., 1969a)

$$\triangle, \blacktriangle M = Mg \qquad \triangledown, \blacktriangledown M = Ba$$

Open symbols I = 1M, solid symbols I = 3M

*Identification of mixtures*

| | |
|---|---|
| 1 = LiCl/NaCl + MCl$_2$ | 6 = LiCl/MCl$_2$ + NaCl/MCl$_2$ |
| 2 = LiCl/KCl + MCl$_2$ | 7 = LiCl/MCl$_2$ + KCl/MCl$_2$ |
| 3 = LiCl/CsCl + MCl$_2$ | 8 = KCl/MCl$_2$ + CsCl/MCl$_2$ |
| 4 = NaCl/KCl + MCl$_2$ | 9 = LiCl/MCl$_2$ + CsCl/MCl$_2$ |
| 5 = KCl/CsCl + MCl$_2$ | 10 = NaCl/MCl$_2$ + KCl/MCl$_2$ |
| (1 cal = 4·187 J). | |

## 2.6.3.5. *Hydration models.*

Equations which take into account short range interactions between ions and water molecules provide excellent curve-fitting models for the activity coefficients of single electrolyte solutions (Table 2.17). They have not however been used to any great extent for the prediction of the properties of

TABLE 2.33

*Hydration models (mole fraction statistics)*

(i) *Osmotic coefficients*

$$\phi = \left(\sum_N v_N \left| z_N^+ z_N^- \right| m_N \Big/ \sum_N v_N m_N \right) \phi_{EL} + (L_A - L_B)/0.018 \sum_N v_N m_N \qquad (1)^a$$

where ·

$$L_A = \ln \left(1 - 0.018 \sum_N (h_N - v_N) m_N \right)^b$$

$$L_B = \ln \left(1 - 0.018 \sum_N h_N m_N \right)$$

and $\phi_{EL}$ is defined in Table 2.18 (equation (4)$^c$)

(ii) *Activity coefficient*

$$\ln \gamma_{J(m)} = \left| z_J^+ z_J^- \right| \ln \gamma_{EL} + (h_J/v_J)(L_A - L_B) - L_A \qquad (2)^a$$

where in $\gamma_{EL}$ is defined by equation (2), Table 2.14$^c$. Expanding the logarithms and selecting only the linear terms

$$\ln \gamma_{J(m)} = \left| z_J^+ z_J^- \right| \ln \gamma_{EL} + (l_J/2v_J) \sum_K v_K m_K + 0.5 \sum_K l_K m_K \qquad (3)^a$$

where $l_K = 0.018 (2h_K - v_K)$ etc.

(iii) *Harned's rule coefficients* (charge symmetric mixtures)$^d$

$$\log \gamma_{A(0)} = \log \gamma_{EL} - (h_A/v_A) \log a_{w(A)} - \log \left[1 + 0.018(v_A - h_A) m_A/y_A \right] \qquad (4)$$

$$\log \gamma_{(0)A} = \log \gamma_{EL} - (h_B/v_B) \log a_{w(B)} - \log \left[1 + 0.018(v_B - h_B) m_B/y_B \right] \qquad (5)$$

Substituting for $a_w$ using the osmotic coefficient (Table 2.4) and combining equations (4) and (5) gives

$$\alpha_{12} I = \log (\gamma_{A(0)}/\gamma_{(0)A})^e = 0.0078 [(m_A h_A/y_A) \phi_A - (m_B h_B/y_B) \phi_B]$$
$$- \log \left[1 + 0.018(v_A - h_A) m_A/y_A \right] + \log \left[1 + 0.018(v_B - h_B) m_B/y_B \right] \qquad (6)^{f, g}$$

---

$^a$ Gardener and Glueckauf (1969)

$^b$ The single electrolyte equations are considered in Table 2.16

$^c$ Gardener and Glueckauf (1969) consider the electrostatic term defined by equations (5.10) and (5.11). The original equations (Glueckauf, 1955; Stokes and Robinson, 1948) used equation (2) (Table 2.14). Adjustable parameters determined on this basis are listed in Table 2.17

$^d$ Robinson and Stokes (1965, p. 452)

$^e$ Table 2.20, equations (1a) and (2a)

$^f$ Robinson and Stokes (1965) derive the special case for mixtures of 1:1 electrolytes

$^g$ Analogous equations derived from the volume fraction statistics approach (Section 2.5.2.3) are discussed by Leyendekkers (1971b) and Leyendekkers and Whitfield (1971).

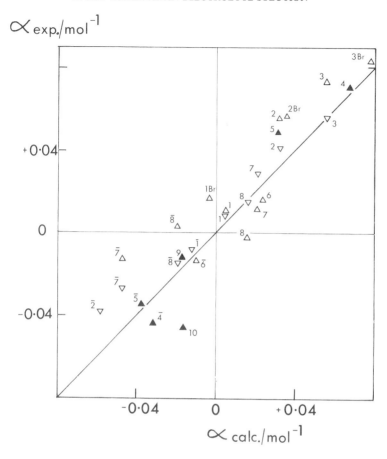

Fig. 2.21. Comparison of experimental Harned's rule coefficients with values calculated from hydration equations.

△, ▲ Volume fraction statistics (Leyendekkers, 1971b)
▽ Mole fraction statistics (Robinson and Stokes, 1965, p. 453)

For open symbols I = 1 M and for solid symbols I = 2 M.

*Identification of mixtures*

| | |
|---|---|
| 1 = HCl/LiCl | 6 = NaCl/KCl |
| 2 = HCl/NaCl | 7 = NaCl/CsCl |
| 3 = HCl/KCl | 8 = KCl/CsCl |
| 4 = LiCl/KCl | 9 = NaCl/NaNO₃ |
| 5 = NaCl/LiCl | 10 = CsCl/LiCl |

The $\alpha$-coefficient usually refers to the mean-ion activity coefficient of the first electrolyte listed. A barred symbol indicates that the $\alpha$-coefficient refers to the second electrolyte listed, e.g. 2 refers to $\alpha_{HCl}$ and $\bar{2}$ to $\alpha_{NaCl}$ in the mixture HCl/NaCl/H₂O

Symbols followed by Br indicate that bromide has replaced chloride in the mixture listed.

electrolyte mixtures. The most convenient equations for multi-component systems have been developed by Gardener and Glueckauf (1969) from the mole fraction statistics model (Table 2.33, equations (1)–(4)). These equations consider only the properties of the component single electrolyte solutions and are restricted in their scope to systems that can be adequately treated by the Brønsted specific interaction theory.

Robinson and Stokes (1965) and Leyendekkers (1971b) have derived specific equations for calculating Harned's rule coefficients in ternary common-ion mixtures (equations (7) and (8), Table 2.33). When the calculated values are compared with those obtained experimentally (Fig. 2.21) it is clear that the predictions are poorest where interactions between ions of like charge are expected to be most intense (e.g. $Li^+$–$Cs^+$–$Cl^-$). This is consistent with the basic assumptions of the model.

## 2.7. SEA WATER—THE ELECTROLYTE COMPONENT

### 2.7.1. ION-ASSOCIATION MODELS

#### 2.7.1.1. *The MacInnes convention.*

Garrels and Thompson (1962) used the ion-association concept to prepare the first quantitative model for standard sea water ($S = 35‰$) at 298 K and 1 atmosphere pressure. In addition to the major electrolytes (Table 2.34) they considered the components of the carbon dioxide system in their calculations. The metal chloride solutions were considered as the model unassociated electrolytes (Section 2.5.2.1).

Single-ion activity coefficients were calculated from the MacInnes convention (equation (5.15)). The free, single-ion activity coefficients of the carbonate and bicarbonate ions were calculated from the mean-ion activity coefficients of the corresponding potassium salts which were assumed to be unassociated. The activity coefficients of the singly charged ion-pairs were set equal to the value for $\gamma_{HCO_3^-}$—and the values for the uncharged species to the activity coefficient of carbon dioxide in sodium chloride solutions at the same ionic strength (i.e. $\gamma_{MX^0} = 1·13$ at 298 K and $I = 0·7$ M). The free activity coefficient for the sulphate ion was obtained from the mean-ion activity coefficient of potassium sulphate using the calculated values for $\gamma_{K^+}$ and $\gamma_{KSO_4^-}$ and the association constant for the $KSO_4^-$ ion-pair. Ten ion-pairs were considered in all (footnote b, Table 2.35). The results of the calculations were originally presented as tables showing the mole percentage of each species present in the mixture. Since the fundamental purpose of any sea water model is to calculate the stoichiometric activity coefficients ($^T\gamma_i$) these values have been recalculated from the original data (Garrels and

Thompson, 1962) via equation (5.5). Later calculations using the same model (Berner, 1971; van Breemen, 1972; Truesdell and Jones, 1973) have benefited from the use of more reliable activity coefficients for the metal chloride solutions (Garrels and Christ, 1965; Robinson and Stokes, 1965) and from later estimates of the ion-pair formation constants (Table 2.35). In addition, the mathematical solutions are more exact because the calculations were carried out rigorously on a computer rather than by approximate iterative

TABLE 2.34

*Standard sea water composition*

| Ion | Na$^+$ | K$^+$ | Mg$^{2+}$ | Ca$^{2+}$ | Cl$^-$ | SO$_4^{2-}$ |
|---|---|---|---|---|---|---|
| Molecular weight | 22·99 | 39·10 | 24·31 | 40·08 | 35·453 | 96·06 |
| Δ | 41·39 | 1898 | 363·4 | 1885 | 35·49 | 686·2 |

Formal concentration $(F_i) = 10^{-3}$ kg ion kg$^{-1}$ sea water

$$= Cl‰/\Delta = S‰/1·80655\Delta$$

Molality $(m_i) = 10^{-3}$ kg ion kg$^{-1}$ water

$$= 1000 \, F_i/[1000 - (0·073 + 1·811 \, Cl‰)]$$

$$= 1000 \, F_i/[1000 - (0·073 + 1·0025 \, S‰)]$$

Molarity $(c_i) = 10^{-3}$ kg ion dm$^{-3}$ sea water

$$= F_i d_{sw}, \text{ where } d_{sw} \text{ is the density of sea water}$$

Molal ionic strength $(I) = 35·9997 \, Cl‰/(1000 - 1·81578 \, Cl‰)$

$$= 19·9273 \, S‰/(1000 - 1·005109 \, S‰)$$

methods (Garrels and Thompson, 1962). The various models (Table 2.36) show a tolerable agreement with one another and, for the more recent calculations at least, are in good agreement with the available "experimental" evidence. The "experimental" single-ion activity coefficients are themselves conventional, being based mainly on assumptions about the liquid junction potential in ion-selective electrode cells (Table 2.3). The agreement between the several conventions, even under these restricted environmental conditions, is encouraging.

The various models can be compared directly with experimental data without recourse to conventional definitions via the mean-ion activity coefficients of the component salts. These values may be calculated from

TABLE 2.35

*Ion-pair formation constants[a] for a sea water model (298 K, 1 atm)*

|        | $SO_4^{2-}$ | $CO_3^{2-}$ | $HCO_3^-$ |
|--------|-------------|-------------|-----------|
| $Na^+$ | $0.72^b$, $1.06^{c, q}$ $0.98^d$, $0.65^e$ $0.7^f$ | $1.27^b$, $0.68^d$ $0.55^{g, h}$ | $-0.25^b$, $0.161^{g, h}$ |
| $K^+$  | $0.96^b$, $0.85^{d, q}$ $0.77^e$, $1.0^f$ | — | — |
| $Mg^{2+}$ | $2.36^b$, $2.40^d$ $2.29^i$, $2.23^f$ | $3.4^b$, $2.4^c$ $3.24^j$ | $1.16^b$, $1.00^{d, q}$ $1.23^j$, $1.28^k$ |
| $Ca^{2+}$ | $2.31^b$, $2.57^l$ $2.03^m$, $2.28^f$ | $3.2^b$, $3.1^{n, q}$ | $1.26^b$, $1.32$ |
| $Sr^{2+}$ | $3.10^o$, $2.55^l$ $2.1^l$ | — | $1.24^p$ |
| $Ba^{2+}$ | $2.30^o$ | — | $1.52^p$ |

[a] Expressed as $\log_{10} K^{\leftrightarrow}$ (equation 2.11)
[b] Garrels and Thompson (1962)
[c] Berner (1971), other values as in ref. [b]
[d] Truesdell and Jones (1969)
[e] Izatt *et al.* (1969)
[f] Davies (1962)
[g] Nakayama (1971b), also gives the temperature dependence.
[h] Butler and Huston (1970)
[i] Larson (1970)
[j] Nakayama (1971a)
[k] Nakayama (1968)
[l] Sillén and Martell (1971)
[m] Yeatts and Marshall (1969), also give values from 273 K to 623 K.
[n] Lafon (1970)
[o] Hanor (1969)
[n] Nakayama and Rasnick (1969)
[q] Used by Lafon (1969), other values were taken from reference b.

the single-ion values in Table 2.36 using the appropriate version of equation (2.6). The results (Table 2.37) are in reasonable agreement with the available experimental data.

The concept of ion-pair formation in sea water has been given circumstantial support by the differential chromatographic studies carried out by Mangelsdorf and Wilson (1971) and by the interpretation of ultrasonic absorption phenomena in terms of the formation of magnesium sulphate ion-pairs (Fisher, 1965, 1967). Although recent investigations indicate that ion-pair formation may be observed indirectly in solution by Raman spectroscopy (Daly et al., 1972) the ion-pairs cannot yet be characterized experimentally. Consequently, their thermodynamic properties can only be

TABLE 2.36

*Ion association models for sea water (298 K, 1 atm 35‰ S).*
*Single-ion activity coefficients*

| Ion | $F_{\gamma_i}$ | | | $T_{\gamma_i}$ | | | | | $T_{\gamma_i}$ |
|-----|------|------|------|-------|-------|-------|-------|-------|----------------|
|     | a | b | c | a | b | c | d | e | (experimental) |
| $Na^+$ | 0·76 | 0·71 | 0·69 | 0·753 | 0·703 | 0·675 | 0·697 | 0·694 | 0·68[f], 0·70[g] |
| $K^+$ | 0·64 | 0·63 | 0·63 | 0·634 | 0·624 | 0·617 | 0·630 | 0·620 | 0·614[h], 0·60[a] |
| $Mg^{2+}$ | 0·36 | 0·29 | 0·32 | 0·313 | 0·252 | 0·280 | 0·257 | 0·254 | 0·26[i, j] |
| $Ca^{2+}$ | 0·28 | 0·26 | 0·25 | 0·255 | 0·237 | 0·225 | 0·241 | 0·228 | 0·203[k], 0·21[l] |
| $Cl^-$ | 0·64 | 0·63 | 0·63 | 0·640 | 0·630 | 0·625 | 0·637 | 0·630 | 0·65[f], 0·68–0·73[m] |
| $SO_4^{2-}$ | 0·12 | 0·17 | 0·17 | 0·065 | 0·068 | 0·068 | 0·081 | 0·090 | 0·11[n] |
| $HCO_3^-$ | 0·68 | 0·68 | 0·67 | 0·471 | 0·504 | 0·527 | 0·444 | 0·521 | 0·45[o], 0·55[k] |
| $CO_3^{2-}$ | 0·20 | 0·20 | 0·19 | 0·018 | 0·021 | 0·020 | 0·029 | — | 0·021[k] |

[a] Garrels and Thompson (1962)
[b] Berner (1971). $F_{\gamma_i}$ values for references d and e are identical, except that van Breemen uses $F_{\gamma_{SO_4}} = 0·21$
[c] Lafon (1969)
[d] Truesdell and Jones (1973)
[e] van Breemen (1972)
[f] Platford (1965a)
[g] Garrels (1967)
[h] Mangelsdorf and Wilson (1971)
[i] Thompson (1966)
[j] Pytkowicz and Gates (1968)
[k] Berner (1965)
[l] Thompson and Ross (1966)
[m] Platford (1965b)
[n] Platford and Dafoe (1965)
[o] Berner and Wilde (1972)
Note: the experimental uncertainties for $T_{\gamma_i}$ are $\pm 0·01$ to $\pm 0·02$ units.

assigned on the basis of sensible guesses. Garrels and Thompson (1962) treated the uncharged ion-pairs as neutral molecules when estimating their activity coefficients and this assumption has been questioned by later workers. Yeatts and Marshall (1969), in their studies of the solubility of gypsum ($CaSO_4 2H_2O$) in electrolyte solutions suggested that the calcium sulphate ion-pair should behave more like a dipolar ion than a neutral molecule. Using an ion-association model that neglected the formation of sodium sulphate ion-pairs they related the activity coefficient of the calcium sulphate ion-pair to the ionic strength by the equation,

$$\log \gamma_{CaSO_4^0} = -S^* I^{\frac{1}{2}} \tag{7.1}$$

At 298 K, $S^*$ has a value of $0.17 \, \text{mol}^{-\frac{1}{2}} \, \text{kg}^{\frac{1}{2}}$ giving $\gamma_{CaSO_4^0} = 0.72$ at an ionic strength of $0.7 \, \text{M}$. At 273 K, $S^* = 0.19 \, \text{mol}^{-\frac{1}{2}} \, \text{kg}^{\frac{1}{2}}$. The variation of $\gamma_{CaSO_4^0}$ with temperature is closely similar to that observed for 1:1 electrolytes at high dilution. Kester (1969) considered the magnesium sulphate ion-pair to be dipolar and suggested a value of $0.8$ for $\gamma_{MgSO_4^0}$ at 298 K. Similar effects

TABLE 2.37

*Ion-association models for sea water (298 K, 1 atm, 35‰ S).*
*Total mean-ion activity coefficients*

| Salt | $\gamma_\pm$ (calc.) | | | | | $\gamma_\pm$ (expt.) | $\gamma_{\pm 0}$[j] |
| | a | b | c | d | e | | |
|---|---|---|---|---|---|---|---|
| NaCl | 0.694 | 0.666 | 0.662 | 0.666 | 0.650 | 0.668 ± 0.003[f] | 0.667 |
| | | | | | | 0.672 ± 0.007[g] | |
| KCl | 0.637 | 0.627 | 0.625 | 0.634 | 0.621 | 0.645 ± 0.008[h] | 0.626 |
| MgCl$_2$ | 0.504 | 0.464 | 0.465 | 0.471 | 0.478 | — | 0.482 |
| CaCl$_2$ | 0.471 | 0.455 | 0.449 | 0.461 | 0.447 | — | 0.465 |
| Na$_2$SO$_4$ | 0.333 | 0.323 | 0.352 | 0.340 | 0.314 | 0.378 ± 0.016[i] | 0.353 |
| K$_2$SO$_4$ | 0.297 | 0.299 | 0.326 | 0.318 | 0.296 | 0.352 ± 0.018[h] | 0.340 |
| MgSO$_4$ | 0.143 | 0.131 | 0.151 | 0.144 | 0.138 | — | 0.124[k] |
| CaSO$_4$ | 0.129 | 0.127 | 0.143 | 0.140 | 0.124 | — | — |

[a] Garrels and Thompson (1962)     [b] Berner (1971)
[c] van Breemen (1972)              [d] Truesdell and Jones (1973)
[e] Lafon (1969)                    [f] Gieskes (1966)
[g] Platford (1965b)
[h] Whitfield (unpublished results), with ion-selective electrode cells of type I (Table 2.3)
[i] Platford and Dafoe (1965)
[j] Values observed in single electrolyte solutions at the same overall ionic strength (Robinson and Stokes, 1965)
[k] Pitzer (1972)

might be expected for the corresponding carbonate ion-pairs. If Berner's ion-association model (Berner, 1971) is recalculated using these values, a small but significant change is noted in the calculated distribution of the dissolved species (Table 2.38). The consistent extension of the ion-association model to cover a wide range of natural aqueous systems, particularly at high ionic strengths, will depend on the development of a clear and reasonably based procedure for expressing the activity coefficients of the ion-pairs as a function of solution composition. Similarly, the use of the model over a reasonable temperature (273 K to 313 K) and pressure (to 1000 atm) range will depend on the definition of the partial molal enthalpies and volumes of

TABLE 2.38

*Ion-association models for sea water[a] (298 K, 1 atm, 35‰ S).*
*Distribution of ion-pairs*

| Anion | Ion-pair | % anion as ion-pair[b] | | | | | | | |
|-------|----------|---|---|---|---|---|---|---|---|
| | | Theoretical constants | | | | | Stoichiometric constants | | |
| | | c | d | e | f | g | h | i | j |
| $SO_4^{2-}$ | $NaSO_4^-$ | 18·3 | 21·9 | 19·9 | 37·6 | 24·7 | 37·0 | 39·0 | 43·2 |
| | $KSO_4^-$ | 0·6 | 0·7 | 0·6 | 0·4 | 0·8 | 0·4 | 0·4 | — |
| | $MgSO_4^0$ | 22·7 | 23·5 | 29·0 | 19·6 | 24·2 | 19·3 | 14·5 | 17·5 |
| | $CaSO_4^0$ | 3·2 | 3·7 | 5·1 | 2·7 | 3·9 | 3·9 | 4·1 | 3·4 |
| | Free | 55·2 | 50·2 | 54·5 | 39·5 | 46·4 | 39·4 | 42·0 | 35·9 |
| $CO_3^{2-}$ | $NaCO_3^-$ | 17·5 | 19·1 | 15·1 | 17·8 | 21·4 | 17·1 | 16·7 | — |
| | $MgCO_3^0$ | 67·0 | 63·3 | 68·2 | 63·0 | 59·0 | 67·8 | 68·4 | 54·8 |
| | $CaCO_3^0$ | 6·6 | 7·1 | 8·4 | 8·8 | 7·3 | 6·3 | 6·2 | 10·6 |
| | Free | 8·9 | 10·5 | 8·3 | 10·4 | 12·3 | 8·8 | 8·7 | 34·6 |
| $HCO_3^-$ | $NaHCO_3^0$ | 8·6 | 8·3 | 8·4 | 8·0 | 8·2 | 8·5 | 8·5 | — |
| | $MgHCO_3^+$ | 17·1 | 14·5 | 14·1 | 9·5 | 12·4 | 17·6 | 18·1 | 4·8 |
| | $CaHCO_3^+$ | 3·4 | 3·2 | 3·1 | 3·4 | 2·8 | 3·3 | 3·3 | 0·9 |
| | Free | 70·9 | 74·0 | 74·4 | 79·1 | 76·6 | 70·6 | 70·1 | 94·3 |

[a] Sea water composition (molalities, Garrels and Thompson, 1962). $Na^+$ (0·4752), $K^+$ (0·0100), $Mg^{2+}$ (0·0540), $Ca^{2+}$ (0·0104), $Cl^-$ (0·5543), $SO_4^{2-}$ (0·0284), $CO_3^{2-}$ (0·000269), $HCO_3$ (0·00238).
[b] Recalculated for the composition shown in note a using the computer programme HALTAFALL (Ingri *et al.*, 1967) and the appropriate parameters from Tables 2.35, 2.36 and 2.39.
[c] Garrels and Thompson (1962)      [d] Berner (1971)
[e] Berner (1971) with $\gamma_{MgX^0} = 0·8$, (Kester, 1969) and $\gamma_{CaX^0} = 0·72$, (Yeatts and Marshall, 1969)
[f] Lafon (1969)
[g] As for c but $\gamma_{\pm(MCl)}$ values used to calculate $^F\gamma_i$ were calculated from a specific interaction model for the mixture $Na^+-K^+-Mg^{2+}-Ca^{2+}-Cl^-$ (Table 2.41, note b)
[h] Pytkowicz and Kester (1969)
[i] Elgqvist and Wedborg (1973)      [j] Dyrssen and Hansson (1973)

the ion-pairs. Some idea of the uncertainties associated with the definitions of these parameters may be gained from the recent discussion concerning the partial molal volume of the magnesium sulphate ion-pair (Kester and Pytkowicz, 1970; Millero, 1971b; Pytkowicz, 1972).

### 2.7.1.2. The supporting electrolyte convention

Lafon (1969) constructed an ion-association model for sea water on the basis of the "supporting electrolyte" procedure of Helgeson (1969). This convention assumes that, in an electrolyte mixture in which one component predominates (e.g. sodium chloride in sea water), the free individual ion activities are given by the equation

$$\log{}^F\gamma_i = \frac{-z_i^2 \mathscr{A} I^{\frac{1}{2}}}{1 + \mathscr{B}\mathring{a}_i I^{\frac{1}{2}}} + B^{\cdot}I \tag{7.2}$$

(cf. Table 2.14, equation (5))

where $\mathring{a}_i$ may be obtained from Kielland's tables (Kielland, 1937). For sea water, the deviation function $B^{\cdot}I$ is obtained by substituting the experimental *mean-ion* activity coefficients for sodium chloride into equation (7.2) together with the appropriate values for $\mathring{a}_i$ ($\mathring{a}(NaCl) = 0.397$ nm at 298 K). The resulting value of $B^{\cdot}$ ($= 0.038$ mol$^{-1}$ kg for $I = 0.67$) together with the appropriate values of $\mathring{a}_i$ are then used to calculate the conventional single-ion activity coefficients of the ionic components. The $\mathring{a}_i$-values for the charged ion-pairs are assumed to be the same as those for similar ionic species (e.g. $\mathring{a}(NaCO_3^-) = \mathring{a}(HCO_3^-)$). The activity coefficients of the uncharged ion-pairs are equated with the values for carbon dioxide as in the Garrels and Thompson model. Variations of the $^F\gamma_i$ values over the temperature (273 K to 333 K), pressure (1 to 1000 atm) and ionic strength (0.67 to 2.01 M) ranges were estimated but assumed to be negligible in comparison with the other errors involved.

The effect of pressure on the equilibrium constants was calculated using a value of $7.7 \times 10^{-6}$ m$^3$ mol$^{-1}$ for the standard partial molal volume change accompanying the formation of the $MgSO_4^0$ ion-pair and an arbitrary value of $20 \times 10^{-6}$ m$^3$ mol$^{-1}$ for all other ion-associations. The effect of temperature was estimated from the equation (Helgeson, 1967),

$$-RT \ln K^{\ominus} = \Delta \bar{H}_r^{\ominus} - \Delta \bar{S}_r^{\ominus} \left\{ T_r - \frac{\theta}{\omega} \left[ \exp \left( \exp \left( b + aT \right) \right. \right. \right.$$
$$\left. \left. \left. - C + (T - T_r)/\theta \right) \right] \right\} \tag{7.3}$$

where $\theta$, $\omega$, $a$, $b$ and $c$ are constants characteristic of the solvent with values of 219.0, 1.00322, 0.01875, $-12.741$ and $7.84 \times 10^{-4}$ respectively for water. $\Delta \bar{H}_r^{\ominus}$ and $\Delta \bar{S}_r^{\ominus}$ are the standard partial molal enthalpy and entropy changes accompanying the ion-association reaction at the reference temperature $T_r$. Values of $\Delta \bar{H}_r^{\ominus}$ at 298 K are available for most of the ion-association reactions involved in the sea water model. Values of $\Delta \bar{S}_r^{\ominus}$ may be estimated by combining the conventional single-ion entropies (see e.g. Lewis and Randall, 1961, p. 400) with the ion-pair entropies calculated using the "corrected entropy" concept used by Cobble (1953). This gives for singly charged ion-pairs,

$$\bar{S}^{\ominus}(MX) = \bar{S}^{\ominus}_{H_2O} + 49 - 99\,|z|\,(r_{1-2}/0.65) \qquad (7.4)$$

and for neutral ion-pairs

$$\bar{S}^{\ominus}(MX) = \bar{S}^{\ominus}_{H_2O} - 116 - 302/r_{1-2} \qquad (7.5)$$

The appropriate value for the distance parameters $r_{1-2}$ are listed by Lafon (1969).

The total single-ion activity coefficients calculated from Lafon's model at 298 K and 1 atm (Table 2.36) show significant variations from those predicted from the models based on the MacInnes convention. The corresponding mean-ion activity coefficients are in rather poor agreement with the available experimental evidence (Table 2.37).

### 2.7.1.3. Stoichiometric constants

The difficulties associated with the characterization of the ion-pairs may be avoided by determining the stoichiometric ion-association constants ($K^*$, equation (5.4)) directly in sea water or similar ionic media. The experimental work carried out so far (Kester and Pytkowicz, 1969; Elgqvist and Wedborg, 1974) is consistent in suggesting that the extent of sulphate ion-pairing in sea-water is greater than that predicted from chemical models based on the MacInnes convention (Table 2.38). In this respect the model proposed by

TABLE 2.39

Stoichiometric ion-pair formation constants[a] for a sea water model
(298 K, 1 atm, I = 0.7 M)

| | $SO_4^{2-}$ | | $CO_3^{2-}$ | | $HCO_3^{-}$ | |
|---|---|---|---|---|---|---|
| Na$^+$ | $-0.15^b$, | $-0.03^c$ | $0.62^b$, | $0.59^c$ | $-0.59^b$, | $-0.62^c$ |
| | $0.30^d$, | $0.31^e$ | $0.56^d$ | | $-0.66^d$ | |
| | $0.42^f$, | $\underline{0.30}^g$ | | | | |
| K$^+$ | $0.13^b$, | $0.16^c$ | — | | — | |
| | $0.49^d$, | $\underline{1.01}^e$ | | | | |
| Mg$^{2+}$ | $0.94^b$, | $1.00^c$ | $2.20^b$, | $2.11^c$ | $0.71^b$, | $0.62^c$ |
| | $1.01^d$, | $\underline{1.01}^e$ | $2.10^d$, | $\underline{1.51}^f$ | $0.40^d$, | $\underline{0.02}^f$ |
| | $1.00^f$, | $0.85^g$ | | | | |
| Ca$^{2+}$ | $0.78^b$, | $0.90^c$ | $1.90^b$, | $1.86^c$ | $0.71^b$, | $0.68^c$ |
| | $0.85^d$, | $\underline{1.03}^e$ | $1.94^d$, | $\underline{1.51}^f$ | $0.65^d$, | $\underline{0.02}^f$ |
| | $1.00^f$ | | | | | |

[a] Expressed as $\log_{10} K^*$. The underlined values are experimental. Other values are calculated from the appropriate parameters in Tables 2.35 and 2.36
[b] Garrels and Thompson (1962)  [c] Berner (1971)
[d] Lafon (1969)  [e] Kester and Pytkowicz (1969)
[f] Dyrssen and Hanson (1973)  [g] Elgqvist and Wedborg (1974).

Lafon (1969) gives a better account of the experimental data (Table 2.38, column $d$). The stoichiometric ion-pair formation constants used in the various models are summarized in Table 2.39. Kester and Pytkowicz (1970) have extended their experimental data to include conditions typical of the abyssal depths (275 K, 1000 atm, see Table 2.40). The influence of temperature and pressure on the distribution of the sulphate species predicted by Lafon's model is not in agreement with the experimental observations (Table 2.40). In particular, the marked decrease in concentration of free sulphate ions predicted by Lafon as the temperature rises from 273 K to 298 K contrasts sharply with the experimentally observed *increase* over a similar temperature range. This difference may be due in part to Lafon's neglect of the influence

TABLE 2.40

*Ion-association models for sea water[a] as a function of temperature and pressure. Distribution of ion-pairs*

| | | | % as ion-pair | | | | | | | | | | |
|---|---|---|---|---|---|---|---|---|---|---|---|---|---|
| | | T/K | 273 | | | | | | 298 | | | | 333 |
| Anion | Ion-pair | P/atm | 1 | | 400 | 1000 | | | 1 | | 400 | 1000 | 1 |
| | | | b | c | b | b | c | d | b | c | b | b | b |
| $SO_4^{2-}$ | $NaSO_4^-$ | | 26·5 | 47 | 20·9 | 14·5 | 32 | 35 | 36·4 | 38 | 30·5 | 22·3 | 48·4 |
| | $KSO_4^-$ | | 0·4 | – | 0·3 | 0·2 | – | – | 0·4 | – | 0·4 | 0·3 | 0·4 |
| | $MgSO_4^0$ | | 15·4 | 21 | 15·3 | 14·5 | 24 | 19 | 20·6 | 19 | 21·1 | 21·0 | 26·2 |
| | $CaSO_4^0$ | | 3·0 | 4 | 2·4 | 1·7 | 5 | 4 | 2·7 | 4 | 2·2 | 1·7 | 2·0 |
| | Free | | 54·7 | 28 | 61·1 | 69·1 | 39 | 42 | 39·9 | 39 | 45·8 | 54·7 | 23·0 |
| $CO_3^{2-}$ | $NaCO_3^-$ | | 5·7 | – | 5·2 | 4·7 | – | – | 17·5 | – | 16·9 | 15·5 | 52·7 |
| | $MgCO_3^0$ | | 79·1 | – | 75·9 | 69·0 | – | – | 67·1 | – | 64·4 | 59·0 | 36·3 |
| | $CaCO_3^0$ | | 3·5 | – | 3·4 | 3·1 | – | – | 4·8 | – | 4·7 | 4·4 | 5·0 |
| | Free | | 11·7 | – | 15·5 | 23·2 | – | – | 10·6 | – | 14·0 | 21·1 | 6·0 |
| $HCO_3^-$ | $NaHCO_3^0$ | | 8·3 | – | 6·6 | 4·2 | – | – | 7·9 | – | 6·1 | 4·0 | 4·9 |
| | $MgHCO_3^+$ | | 2·6 | – | 1·9 | 1·2 | | | 10·1 | | 7·7 | 5·0 | 41·8 |
| | $CaHCO_3^+$ | | 1·5 | – | 1·1 | 0·7 | – | – | 3·3 | – | 2·6 | 1·7 | 6·5 |
| | Free | | 87·6 | – | 90·2 | 93·9 | – | – | 78·7 | – | 83·6 | 89·3 | 46·8 |

[a] Sea water composition (molal), $Na^+$ (0·4733), $K^+$ (0·0101), $Mg^{2+}$ (0·05752), $Ca^{2+}$ (0·0103), $SO_4^{2-}$ (0·0286), $Cl^-$ (0·5589), $HCO_3^-$ (0·00226), $CO_3^{2-}$ (0·000228), $'H_2CO_3'$ (1·43 × 10$^{-5}$) at 298 K and 1 atm and pH = 8·15

[b] Lafon (1969). Data for 273 K at pH 7·7, other data at pH 8·15. Unknown volume changes of association set at $20 × 10^{-6}$ $m^3$ $mol^{-1}$

[c] Kester and Pytkowicz (1970). Experimental data at 275 K and 298 K

[d] Millero (1971b), 275 K

of temperature and pressure on the activity coefficients of the individual components.

The stoichiometric constants provide experimental data that are essential for checking the predictions of the detailed solution models. However, suggestions that they should be considered as *replacements* for the chemical models (Kester and Pytkowicz, 1970) turn the predictive concepts full circle by re-introducing the ionic medium method, and thus requiring separate determinations of the stoichiometric constants for each ionic strength and solution composition.

### 2.7.1.4. *Extended models*

The ion-association model has been extended to include strontium and barium (Hanor, 1969) and a range of trace metals (zinc, copper, cadmium and lead, Zirino and Yamamoto, 1972). The model employed by Zirino and Yamamoto involves the use of two separate conventions to characterise the free ions—the MacInnes convention for the major components and the Davies equation (Table 2.14, equation (6)) for the trace components. Since the stoichiometric constants and hence the distribution of dissolved species will depend on the assumptions made about the behaviour of the free ions, the resulting tables showing the concentrations of the various ions and ion-pairs should be viewed as conventional definitions of the chemical system rather than as expressions of analytical reality. A number of electrostatic equations are available that could be used to describe the behaviour of the free ions (see Table 2.14) and it would be worthwhile to explore the possibility of adopting a uniform convention for the major and minor components.

### 2.7.1.5. *The carbon dioxide system*

The original model suggested by Garrels and Thompson (1962) has been widely used to study ionic equilibria in sea water. It has been used for example in the elucidation of precipitation equilibria of carbonate (Broecker and Takahashi, 1966) and sulphate (Hanor, 1969) and in the development of an evolutionary model for sea water itself (Kramer, 1965). These and other applications have been reviewed by Garrels and Christ (1965) and by Berner (1971) and will be considered in detail in later chapters.

### 2.7.2. SPECIFIC INTERACTION MODELS

The concept of specific interaction in electrolyte solutions (Sections 2.5.2.2 and 2.6.3.2) enables the total single-ion and mean-ion activity coefficients to be calculated directly. No new chemical species are postulated, and the variable parameters are calculated from the mean-ion activity coefficients of the pure component electrolyte solutions at the same overall ionic strength as the mixture (equation (2), Table 2.16). The data for the single electrolyte solutions are then combined according to equations (2a) to (2c), Table 2.26. Calculations on the sea water model are, at present, restricted to the major anions and cations. The effects of the carbon dioxide system cannot be considered because data on the interaction coefficients are incomplete. Since the carbonate and bicarbonate complexes in the ion-association model never represent more than 1% of the total concentration of a given *cation* this omission should have little effect on the model for the major components. The mean-ion activity coefficients of the neutral salt components can be taken from the compilation given by Robinson and Stokes (1965). More recent data for magnesium sulphate are given by Pitzer (1972). It is assumed that $\gamma_{\pm(CaSO_4)} = \gamma_{\pm(MgSO_4)}$ at the same ionic strength since the insolubility of gypsum ($CaSO_4 \cdot 2H_2O$) prevents the direct determination of this activity coefficient. A more accurate estimate of $\gamma_{\pm(CaSO_4)}$ may be made using the entropy correlations suggested by Leyendekkers (1971b, 1973, see equation (5.9)).

The total single-ion activity coefficients calculated on this basis (Table 2.41, column *c*) are in reasonable agreement with both the predictions of the ion-association model and the available experimental evidence (columns *e* and *f* respectively). It is interesting, and re-assuring, that three independent conventions for splitting the mean-ion activity coefficients into their individual ionic components should result in similar descriptions of such a complex mixture. The greatest discrepancy is noted in the value of $^T\gamma_{SO_4}$. The estimate provided by the specific interaction theory is given some support by the recent calculations made by Elgqvist and Wedborg (1974). Using the hydration convention (Section 2.5.5) they suggested that for sulphates of monovalent cations,

$$\log \gamma_{SO_4} = 2 \log \gamma_{\pm} - 0.0156 \, hm\phi + \log [1 + 0.018 (3 - h)m]$$

and for divalent cations,

$$\log \gamma_{SO_4} = \log \gamma_{\pm} - 0.00782 \, hm\phi \qquad (7.7)$$

where $h$ is the hydration number of the cation (3.5, 1.9 and 13.7 for $Na^+$, $K^+$ and $Mg^{2+}$ respectively). The values of $^T\gamma_{SO_4}$ calculated from these equations is practically independent of the cation for sodium, potassium and magnesium sulphates for ionic strengths up to 3 M. At an ionic strength of 0.7 M the equations give, $^T\gamma_{SO_4}(K_2SO_4) = 0.115$, $^T\gamma_{SO_4}(Na_2SO_4) = 0.122$ and $^T\gamma_{SO_4}(MgSO_4) = 0.122$. In the absence of strong interactions between the chloride and sulphate ions this would suggest a value of 0.12 for $^T\gamma_{SO_4}$ in sea water at this ionic strength, in agreement with the predictions of the specific interaction model.

Leyendekkers (1973) used a more elaborate specific interaction model, incorporating interactions between ions of like charge (equation (5c), Table 2.26), calculate the mean-ion activity coefficients of the neutral electrolyte components in sea water. She then calculated the total single-ion

TABLE 2.41

*Specific interaction models for sea water[a] (298 K, 1 atm.).*
*Total single-ion activity coefficients*

| Ion | $^T\gamma_i$ | | | | | | |
|---|---|---|---|---|---|---|---|
| | | | $I = 0.7$ M | | | $I = 0.5$ M | |
| | b | c | d | e | f | c | d |
| $Na^+$ | 0.659 | 0.650 | 0.680 | 0.703 | 0.68 | 0.669 | 0.686 |
| $K^+$ | 0.621 | 0.617 | 0.630 | 0.624 | 0.61 | 0.643 | 0.649 |
| $Mg^{2+}$ | 0.240 | 0.217 | 0.234 | 0.252 | 0.26 | 0.232 | 0.251 |
| $Ca^{2+}$ | 0.218 | 0.203 | 0.214 | 0.237 | 0.21 | 0.221 | 0.235 |
| $Cl^-$ | 0.691 | 0.686 | 0.658 | 0.630 | 0.66 | 0.694 | 0.676 |
| $SO_4^{2-}$ | — | 0.122 | 0.108 | 0.09 | 0.11 | 0.150 | 0.142 |

[a] Using the Güntelberg electrostatic equation (Table 2.14, equation (3))
[b] Composition (molalities), $Na^+$ (0.4944), $K^+$ (0.0101), $Mg^{2+}$ (0.0547), $Ca^{2+}$ (0.0105), $Cl^-$ (0.6348)
[c] Composition (molalities), $Na^+$ (0.4699), $K^+$ (0.0101), $Mg^{2+}$ (0.0536). $Ca^{2+}$ (0.0104), $Cl^-$ (0.5518), $SO_4^{2-}$ (0.0281). Whitfield (1973a), using equations (2a) and (2b). Table 2.26
[d] Leyendekkers (1973) using equation (5c), Table 2.26 and the hydration convention (equation 5.14a) for $\gamma_{Na^+}$. $I = 0.74$ M and $0.51$ M respectively
[e] Ion-association model as calculated by Berner (1971), see Table 2.36.
[f] Arithmetic mean of experimental values listed in Table 2.36.

activity coefficient of the sodium ion using the hydration convention (equation (5.9)) and used this value to calculate the activity coefficients of the other ionic components. The resulting model which is based on a fourth independent convention for estimating single-ion activities (column d, Table 2.41) is in excellent agreement with the experimental data. The value calculated for $^T\gamma_{SO_4}$ is in agreement with the predictions of the specific interaction model. A value of 0.11 to 0.12 for $^T\gamma_{SO_4}$ when combined with Berner's estimate of 0.17 for $^F\gamma_{SO_4}$ would suggest the presence of 30 to 40% free sulphate in an ion-association model. This is in much closer agreement with the experimental evidence (Table 2.38) than the 50 to 60% free sulphate derived from ion-association models based on the MacInnes convention. Data for a sulphate-free sea water (Table 2.41, column b) can be used to provide a more realistic basis for an ion-association model based on the MacInnes convention since they take into account the interactions observed in mixed chloride solutions that were disregarded in the original models (Section 2.7.1.1). Calculations based on the stability constants used by Garrels and Thompson (1962) give a model that is intermediate in its predictions between those of Berner (1971) and Lafon (1969) (see Table 2.38).

TABLE 2.42

*Specific interaction models for sea water[a] (298 K, 1 atm). Total mean-ion activity coefficients*

| Salt | $^T\gamma_\pm$ | | | | | | | |
|------|------|------|------|------|------|------|------|------|
|      | $I = 0.7$ M | | | | | $I = 0.5$ M | | |
|      | b | c | d | e | f | c | d | f |
| NaCl | 0·675 | 0·668 | 0·666 | 0·666 | 0·670 | 0·681 | 0·681 | 0·690 |
| KCl | 0·655 | 0·650 | 0·644 | 0·627 | 0·645 | 0·668 | 0·663 | — |
| $MgCl_2$ | 0·486 | 0·467 | 0·466 | 0·464 | — | 0·482 | 0·486 | — |
| $CaCl_2$ | 0·470 | 0·457 | 0·453 | 0·455 | — | 0·474 | 0·475 | — |
| $Na_2SO_4$ | — | 0·372 | 0·368 | 0·323 | 0·378 | 0·406 | 0·406 | 0·405 |
| $K_2SO_4$ | — | 0·360 | 0·355 | 0·299 | 0·352 | 0·396 | 0·392 | — |
| $MgSO_4$ | — | 0·163 | 0·151 | 0·131 | — | 0·187 | 0·189 | — |
| $CaSO_4$ | — | 0·157 | — | 0·127 | — | 0·182 | 0·183 | — |

[a-f] See corresponding notes on Table 2.41. Data for columns e and f were taken from Table 2.37.

The mean-ion activity coefficients of the component salts calculated from the specific interaction model are in close agreement with the experimental values where data are available (Table 2.42). The inclusion of interactions between ions of like sign (Leyendekkers, 1973) has only a small influence on the calculated values. The agreement with the mean-ion activity coefficients calculated from the ion-association model is reasonable, although the latter model yields consistently lower values for the sulphates because it over-estimates the extent of sulphate association (Table 2.38). Clearly, the behaviour of this particular electrolyte mixture can be adequately accounted for without the introduction of ion-pairs.

The specific interaction model is mathematically simple, involving only the summation of linear terms, and it enables both the mean-ion and single-ion activity coefficients to be calculated directly from readily available data. In its basic form the model requires only the mean-ion activity coefficients of the component single electrolyte solutions. No new chemical species are introduced and explicit equations are used to relate the activity coefficient of each component to the solution composition. The equations are readily differentiated with respect to temperature and pressure (Whitfield, 1973c) and the temperature and pressure coefficients are expressed in terms of readily accessible thermodynamic parameters (Table 2.5). The equation can also be differentiated with respect to solution composition (Whitfield, 1973b) so that the osmotic coefficient of sea water may be calculated (see Section 2.8). The specific interaction model should, therefore, provide a simple procedure for extending the sea water model to cover a wide temperature, pressure and composition range.

TABLE 2.43

*Values of stoichiometric single-ion activity coefficients[a] that provide a self-consistent model for the major components in sea water ($\mathring{a} = 0.394\,nm$)*

| Ion | Ionic strength | | | | | | | |
|-----|------|------|------|------|------|------|------|------|
|      | 0·5  | 0·7  | 0·8  | 0·9  | 1·0  | 2·0  | 3·0  | 4·0  |
| $Na^+$     | 0·670 | 0·652 | 0·645 | 0·640 | 0·636 | 0·628 | 0·651 | 0·692 |
| $K^+$      | 0·643 | 0·618 | 0·609 | 0·600 | 0·594 | 0·555 | 0·543 | —[b]  |
| $Mg^{2+}$  | 0·231 | 0·215 | 0·209 | 0·206 | 0·203 | 0·214 | 0·259 | 0·339 |
| $Ca^{2+}$  | 0·220 | 0·201 | 0·195 | 0·190 | 0·186 | 0·184 | 0·206 | 0·250 |
| $Cl^-$     | 0·693 | 0·681 | 0·678 | 0·675 | 0·674 | 0·693 | 0·746 | 0·822 |
| $SO_4^{2-}$ | 0·149 | 0·121 | 0·111 | 0·103 | 0·099 | 0·060 | 0·042 | 0·035 |

[a] Reproduced with permission from Whitfield (1973).
[b] Data for potassium sulphate are restricted to ionic strengths 3 M and below.

The use of this model to investigate many important equilibria in sea water is, at present, precluded because the appropriate interaction coefficients are unknown. For example, the carbon dioxide system cannot be treated because the coefficients representing interactions between the alkaline earth cations and the carbonate and bicarbonate ions are not available. In some instances it may be possible to obtain these values by retracing the steps followed originally in the calculation of the appropriate ion-pair formation constants, and in others it should be feasible to estimate the $b$-values using entropy correlations (Leyendekkers, 1971b).

Calculations of the osmotic coefficient of sea water (Section 2.8) indicate that a value of 0·394 nm for $\mathring{a}$ in the electrostatic equation (Table 2.14, equation (2)) is more appropriate in sea water than the value of 0·304 nm suggested by Guggenheim (1935). This alteration has only a minor effect on the calculated activity coefficients (Table 2.43).

### 2.7.3. THE ÅKERLÖF–THOMAS RULE

Åkerlöf and Thomas (1934) suggested that the rule summarized by equation (5.7) could be extended to electrolyte mixtures by the simple summation of the linear interaction terms. Åkerlöf (1934, 1937) used these relationships to calculate the solubility of evaporite minerals (e.g. halite, gypsum) in concentrated brines. The approach was revived by Lerman (1967) to prepare a model for the chemical evolution of the Dead Sea (see also Berner, 1971). Lerman expressed the activity coefficients in a more convenient form using the generalized form of Harned's rule (Table 2.20, equation (5)). It is only necessary to consider the contributions of the component ternary common-ion mixtures at the same total ionic strength. For sodium chloride in sea water we can write

$$\log \gamma_{NaCl(m)} = \log \gamma_{NaCl(0)} - (\alpha I)_{KCl} - (\alpha I)_{MgCl_2} - (\alpha I)_{CaCl_2} - (\alpha I)_{Na_2SO_4} \quad (7.8)$$

where $(\alpha I_{MX} = \alpha_{FaCl-MX}.I_{MX}$ at the same total ionic strength as the mixture. A simplified version of this equation was used by Gieskes (1966). Using the Harned rule coefficients listed in Table 2.21 and the sea water recipe (molal), NaCl (0·416), KCl (0·0101), $MgCl_2$ (0·054), $CaCl_2$ (0·0103), $Na_2SO_4$ (0·0286), the following mean-ion activity coefficients are predicted:

$$NaCl, 0·666; \quad KCl, 0·640; \quad MgCl_2, 0·479; \quad Na_2SO_4, 0·378$$

These values are in reasonable agreement with the results of the more elaborate calculations considered in earlier sections.

This model provides a simple way of calculating mean-ion activity coefficients in complex mixtures. It is particularly well suited to the treatment of concentrated electrolyte solutions (see Glueckauf, 1949) since the fit of the Åkerlöf–Thomas and Harned rules actually improves at high ionic strengths. In contrast, the assumptions of the other models become less tenable as the concentration increases since strong interactions between aggregates larger than ion-pairs become significant. The major drawback of the model is that it treats the neutral electrolytes as the solution components and this introduces some ambiguity into the expression describing the ionic interactions. This ambiguity could be removed by expressing Harned's rule in terms of conventional single-ion activity coefficients (see e.g. Leyendekkers 1971a).

### 2.7.4. CLUSTER INTEGRAL EXPANSION MODEL

These equations (Section 2.6.3.4) have been used to calculate mean-ion activity coefficients in synthetic sea water solutions (Robinson and Wood, 1972; Whitfield, 1973a). The simplest equation (Table 2.31, equation (3), line 1) considers interactions between ions of opposite sign and requires only the properties of the component single electrolyte solutions. Calculations on this basis (Whitfield, 1973a) give mean-ion activity coefficients that are in good agreement with the experimental data and with the predictions of the specific interaction model (Table 2.44, cf. Table 2.42). The use of the more complex equation, allowing for interactions between ions of like sign, results in only a slight alteration in the predictions at an ionic strength of 0·7 M. The two procedures for estimating the contributions of interactions in ternary common ion mixtures (Table 2.31, part iv) yield slightly different estimates of the mean ion activity coefficients at an ionic strength of 1 M (Table 2.44, columns $f$ and $d$). Although the procedure based on the Harned's rule co-efficients is the simpler to use in conjunction with the published data, it

TABLE 2.44

*Cluster integral expansion model for sea water* (298 K, 1 atm). *Total mean-ion activity coefficients*

| Salt | $I = 0.7$ M | | | | | $I = 1.0$ M | | |
|------|------|------|------|------|------|------|------|------|
|      | a | b | c | d | e | b | f | d |
| NaCl | 0·669 | 0·667 | 0·674 | 0·669 | 0·668, 0·672 | 0·654 | 0·659 | 0·657 |
|      |       | 0·682[g] |       | 0·684[g] | 0·690[g] |       |       |       |
| KCl | 0·646 | 0·644 | 0·654 | 0·639 | 0·645 | 0·627 | 0·630 | 0·618 |
| MgCl$_2$ | 0·470 | 0·460 | 0·479 | 0·467 | — | 0·448 | 0·470 | 0·462 |
| CaCl$_2$ | 0·461 | 0·448 | 0·464 | — | — | 0·437 | 0·449 | — |
| Na$_2$SO$_4$ | — | 0·381 | 0·385 | 0·366 | 0·378 | 0·350 | 0·342 | 0·334 |
|      |   | 0·412[g] |   | 0·399[g] | 0·405[g] |   |   |   |
| K$_2$SO$_4$ | — | 0·364 | 0·368 | 0·345 | 0·352 | 0·331 | 0·323 | 0·308 |
| MgSO$_4$ | — | 0·165 | 0·168 | 0·158 | — | 0·145 | 0·140 | 0·141 |
| CaSO$_4$ | — | 0·159 | 0·164 | — | — | 0·140 | 0·135 | — |

[a] Composition (molal) Na$^+$ (0·4944), K$^+$ (0·0101), Mg$^{2+}$ (0·0547), Ca$^{2+}$ (0·0105), Cl$^-$ (0·6348), $I = 0.700$

[b] Composition (molal) Na$^+$ (0·4702), K$^+$ (0·0101), Mg$^{2+}$ (0·0536), Ca$^{2+}$ (0·0104), Cl$^-$ (0·5516), SO$_4^{2-}$ (0·0281), $I = 0.700$ M. Considering only interactions between ions of opposite charge (Whitfield, 1973a)

[c] Recipe as in b. Cation–cation interactions included using equation (6.3) and experimental Harned's rule coefficients (Table 2.21).

[d] Composition (molal), Na$^+$ (0·4697), K$^+$ (0·0103), Mg$^{2+}$ (0·0639), Cl$^-$ (0·5510), SO$_4^{2-}$ (0·0284), $I = 0.7$ M
Using the full equation (Table 2.31, equation (3)) with interaction terms estimated from the osmotic coefficients of the component ternary common-ion mixtures (Robinson and Wood, 1972)

[e] Experimental values, sources as in Table 2.37

[f] As in c but anion–anion interactions also included

[g] $I = 0.5$ M.

TABLE 2.45

*Total mean-ion activity coefficients of sea water components as a function of ionic strength (298 K, 1 atm).*

| $I$ | NaCl | KCl | $MgCl_2$ | $CaCl_2$ | $Na_2SO_4$ | $K_2SO_4$ | $MgSO_4$ | $CaSO_4$ |
|---|---|---|---|---|---|---|---|---|
| 0·4[a] | 0·691 | 0·678 | 0·478 | 0·487 | 0·427 | 0·416 | 0·193 | 0·198 |
| | 0·694 | 0·679 | 0·490 | 0·488 | 0·433 | 0·421 | 0·203 | 0·202 |
| | — | — | — | — | — | — | — | — |
| 0·5 | 0·681 | 0·668 | 0·481 | 0·473 | 0·406 | 0·395 | 0·186 | 0·181 |
| | 0·682 | 0·665 | 0·477 | 0·471 | 0·412 | 0·399 | 0·188 | 0·184 |
| | 0·684 | 0·661 | 0·483 | — | 0·399 | 0·381 | 0·181 | — |
| 0·6 | 0·668 | 0·652 | 0·470 | 0·460 | 0·379 | 0·367 | 0·173 | 0·164 |
| | 0·673 | 0·654 | 0·466 | 0·459 | 0·395 | 0·380 | 0·175 | 0·170 |
| | 0·675 | 0·649 | 0·474 | — | 0·381 | 0·361 | 0·168 | — |
| 0·7 | 0·666 | 0·649 | 0·464 | 0·453 | 0·372 | 0·359 | 0·161 | 0·156 |
| | 0·667 | 0·644 | 0·460 | 0·448 | 0·381 | 0·364 | 0·165 | 0·159 |
| | 0·669 | 0·639 | 0·467 | — | 0·366 | 0·345 | 0·158 | — |
| 0·8 | 0·661 | 0·643 | 0·458 | 0·448 | 0·359 | 0·345 | 0·152 | 0·147 |
| | 0·661 | 0·638 | 0·455 | 0·446 | 0·369 | 0·352 | 0·158 | 0·153 |
| | 0·664 | 0·631 | 0·464 | — | 0·354 | 0·331 | 0·151 | — |
| 0·9 | 0·657 | 0·636 | 0·455 | 0·442 | 0·348 | 0·334 | 0·146 | 0·140 |
| | — | — | — | — | — | — | — | — |
| | 0·660 | 0·624 | 0·463 | — | 0·344 | 0·319 | 0·146 | — |
| 1·0 | 0·655 | 0·633 | 0·452 | 0·384 | 0·342 | 0·327 | 0·142 | 0·136 |
| | 0·654 | 0·627 | 0·448 | 0·437 | 0·350 | 0·331 | 0·145 | 0·140 |
| | 0·657 | 0·618 | 0·462 | — | 0·334 | 0·308 | 0·141 | — |
| 2·0 | 0·659 | 0·620 | 0·468 | 0·445 | 0·287 | 0·264 | 0·113 | 0·105 |
| | — | — | — | — | — | — | — | — |
| | 0·662 | 0·593 | 0·488 | — | 0·280 | 0·241 | 0·115 | — |
| 3·0 | 0·697 | 0·637 | 0·524 | 0·486 | 0·261 | 0·231 | 0·104 | 0·093 |
| | — | — | — | — | — | — | — | — |
| | 0·700 | 0·598 | 0·558 | — | 0·258 | — | 0·112 | — |

[a] Three models are considered at each ionic strength. The first row gives values calculated from the simple specific interaction model as summarized in Table 2.43. The second row gives values calculated from the cluster integral expansion equations considering only interactions between ions of opposite sign. The third and final row shows values calculated from the *full* cluster integral expansion equations (Robinson and Wood, 1972).

would be expected to yield less accurate results than the more direct procedure used by Robinson and Wood (1972). Over a wide range of ionic strengths (Table 2.45), the mean-ion activity coefficients calculated from the full cluster integral expansion equation (Table 2.31, equation (3)) are in essential agreement with the predictions of the much simpler specific interaction equation (Table 2.26, equation (2c)).

## 2.8. SEA WATER—THE OSMOTIC COEFFICIENT

The activity of water in a complex salt solution is most conveniently expressed by the osmotic coefficient ($\phi$, Table 2.4). From $\phi$ it is possible to calculate the boiling point, freezing point, osmotic pressure and vapour pressure of the salt solution. This parameter is of central importance in processes as diverse as osmoregulation and flash evaporation, and has consequently attracted considerable experimental and theoretical attention. The equations developed may be divided between those that provide adequate summaries or descriptions of the experimental data and those that allow the experimental values to be predicted from the properties of the solution components. The former are restricted to the solutions that they were designed to describe, and the latter can be extended to cover a range of related systems.

### 2.8.1. DESCRIPTIVE EQUATIONS

Robinson (1954) took advantage of the precise isopiestic data available for sodium chloride solutions to establish a relationship (Table 2.46) between the molalities of sodium chloride solutions and the chlorinities of sea water samples at isopiestic equilibrium (i.e., with the same water activity). The osmotic coefficient of sea water could be calculated from this relationship using the tabulated osmotic coefficients of sodium chloride solutions (see e.g. Robinson and Stokes, 1965). Rush and Johnson (1966) extended these calculations to higher concentrations and derived an explicit relationship between the osmotic coefficient and the concentration of sodium chloride solutions that is useful for interpolation (Table 2.46). Their results may be summarized by the equation

$$\phi_{sw} = \phi_{EL} + 0{\cdot}02212I + 0{\cdot}00852I^2 - 0{\cdot}000403I^3 \qquad (8.1)$$

where $\phi_{EL}$ is defined in Table 2.46 (equation (2b)). Small variations in the sea water recipe (e.g. replacing $Ca^{2+}$ by $Mg^{2+}$) have only a slight effect on the osmotic coefficient. These equations are restricted to 298 K for which abundant experimental data are available.

Stoughton *et al.* (1964) treated sea water as a slightly perturbed sodium chloride solution and assumed that the osmotic coefficient could be given by the equation

$$\phi_{sw} = \phi_{EL} + \sum_n C_n I^n \qquad (8.2)$$

which is similar in form to that derived by Lietzke and Stoughton (1962, see Table 2.18). The first term represents the electrostatic contribution to the osmotic coefficient and is given by equation (4a) (Table 2.46). The mean charge

TABLE 2.46

*Empirical procedures for calculating the osmotic coefficient of sea water.*

---

(i) *At 298 K*

   (a) Calculate the molality of the sodium chloride solution at isopiestic equilibrium with sea water using one of the equations

$$m_{NaCl} = 0.02782\,(Cl\permil) + 0.000079\,(Cl\permil)^2 \tag{1a[a]}$$

$$m_{NaCl} = 0.01540\,(S\permil) + 2.4206 \times 10^{-5}\,(S\permil)^2 \tag{1b[a]}$$

$$m_{NaCl} = 0.9700\,(m_{sw}) + 0.02006\,(m_{sw})^2 - 0.00083\,(m_{sw})^3 \tag{1c[b]}$$

   where $m_{sw} = 0.5 \sum_i m_i$

   (b) Obtain $\phi_{NaCl}$ for this molality from tables (e.g. Robinson and Stokes, (1965) or calculate using the equation[b]

$$\phi_{NaCl} = \phi_{EL} + 0.01842\,(m_{NaCl}) + 0.0154\,(m_{NaCl})^2 - 0.000652\,(n_{NaCl})^3 \tag{2a}$$

   where $\phi_{EL} = 1 - \phi_{DH}\sigma(1.5I^{\frac{1}{2}})$[c]. \hfill (2b)

   (c) Since the sea water and the sodium chloride solutions have the same water activity

$$\phi_{sw} = \phi_{NaCl}m_{NaCl}/m_{sw} \tag{3}$$

(ii) *At other temperatures*[d]

   Use the equation,

$$\phi_{sw} = \phi_{EL} + BI + CI^2 \tag{4}$$

   where

$$\phi_{EL} = 1 - (\mathscr{A}^*Z_{sw}I^{\frac{1}{2}}/3)\sigma(1.5I^{\frac{1}{2}}) \tag{4a[c]}$$

$$Z_{sw} = 1/0.5 \sum_i m_i = 1.2457$$

$$B = -348.662\,T^{-1} + 6.72817 - 0.971307 \ln T$$

$$C = 40.5016\,T^{-1} - 0.721404 + 0.103915 \ln T$$

---

[a] Robinson (1954), 9–22‰ Cl.

[b] Rush and Johnson (1966) for $m_{sw}$ up to 6 M. The two sets of equations give slightly different results as they are fitted to different experimental data.

[c] $\sigma(x)$ is defined in Table 2.18

[d] Stoughton *et al.* (1964), Stoughton and Lietzke (1965). For ionic strengths up to 4 M at 373 K and 3 M at 643 K

[e] Using the sea water recipe of Stoughton and Lietzke (1965). Although the original papers use $I'\,(=0.5 \sum_i m_i)$ in equation (4) the experimental osmotic coefficients are more accurately reproduced if the ionic strength is used as shown (Rush and Johnson, 1966).

of the composite sea salt is used instead of the $|z_+ z_-|$ factor in the Debye–Hückel equation. This however is the only concession to the complexity of sea water since the parameters in the power series in ionic strength are obtained by curve fitting to the data for sodium chloride solutions alone. Stoughton and Lietzke (1965) revised the equations in the light of more recent experimental data and gave parameters for calculating the osmotic coefficient of sea water at temperatures up to 523 K.

Gibbard and Scatchard (1972) have measured the osmotic coefficients of sea water solutions (ionic strengths 1·0, 2·8 and 5·8 M) at temperatures up to 373 K. They confirmed the validity of the 'perturbed sodium chloride' approach by comparing their experimental data with the predictions of an equation similar to that given by Stoughton et al. (1964), i.e.

$$\phi_{sw} = \phi_{EL} + 0.5 \sum_j B^{(j)} I^j \qquad (8.3)$$

where $\phi_{EL}$ is defined by equation (4a) (Table 2.46). The Debye–Hückel limiting slope ($\mathscr{A}$) and the interaction parameters ($B^{(j)}$) were expressed as functions of temperature. The $B^{(j)}$ terms were derived from experimental measurements on sodium chloride solutions. The results were in reasonable agreement with the experimental values and with the calculations by Rush and Johnson (1966) at 298 K. The maximum deviation at 373 K and an ionic strength of 5·8 M did not exceed 1·5%.

### 2.8.2. PREDICTIVE EQUATIONS

As a first approximation, the properties of an electrolyte mixture may be predicted from the properties of the component single electrolyte solutions at the same overall ionic strength. The simplest approach is to consider a mole fraction weighting of the osmotic coefficients of the component salts.

If the osmotic coefficients are estimated at the same total ionic strength as the mixture then a simple adaptation of the ionic strength principle will give,

$$\phi_{sw} = \sum_I v_I m_I \phi_I \Big/ \sum_I v_I m_I \qquad (8.4)$$

where the summation is made over the $I$ salts used in the sea water recipe. This calculation gives osmotic coefficients that are within 0·4% of the experimental values (see Table 2.47, column b).

The success of this naïve approach and the even closer predictions of the "perturbed sodium chloride" theory suggest either that interactions in electrolyte mixtures are unimportant or that the various effects cancel each other out in sea water. Gibbard and Scatchard (1972) attempted to assess the contributions of the various interactions by predicting the properties of

TABLE 2.47

*Calculated values for the osmotic coefficient of sea water at 298 K and 1 atm.*

| $I$ | Additivity rule[a] | Perturbed NaCl[b] | Specific interaction | | Cluster integral expansion | | | Exptl.[h] |
|---|---|---|---|---|---|---|---|---|
| | | | c | d | e | f | g | |
| 0·5 | 0·899 | 0·902 | 0·902 | 0·903 | 0·900 | 0·902 | 0·901 | 0·902 |
| 0·6 | — | 0·904 | — | — | 0·902 | 0·904 | 0·903 | 0·904 |
| 0·7 | 0·903 | 0·906 | 0·907 | 0·908 | 0·904 | 0·906 | 0·906 | 0·906 |
| 0·8 | 0·906 | 0·909 | 0·910 | 0·911 | 0·907 | 0·909 | 0·908 | 0·909 |
| 0·9 | 0·909 | 0·912 | 0·913 | 0·915 | 0·910 | 0·912 | 0·911 | 0·912 |
| 1·0 | 0·912 | 0·915 | 0·917 | 0·919 | 0·913 | 0·915 | 0·914 | 0·916 |
| 2·0 | 0·959 | 0·962 | 0·962 | 0·965 | 0·962 | 0·964 | 0·962 | 0·962 |
| 3·0 | 1·021 | 1·023 | 1·016 | 1·020 | 1·025 | 1·026 | 1·023 | 1·023 |
| 4·0 | 1·092 | 1·097 | 1·078 | 1·081 | 1·099 | 1·098 | 1·095 | 1·094 |
| 5·0 | — | 1·184 | — | — | 1·179 | 1·179 | 1·174 | 1·172 |
| 6·0 | — | 1·283 | — | — | 1·264 | 1·266 | 1·260 | 1·254 |

[a] Equation (8.4) (Whitfield, 1973c). Composition (molal), NaCl (0·4699), KCl (0·0101), $MgCl_2$ (0·0255), $CaCl_2$ (0·0104), $MgSO_4$ (0·0281) at I = 0·7 M

[b] Equation (4), Table 2.46

[c] Equation (8.6) with $\mathring{a} = 0·394$ nm (Whitfield, 1973b). Composition (molal), $Na^+$ (0·4699), $K^+$ (0·0101), $Mg^{2+}$ (0·0536), $Ca^{2+}$ (0·0104), $Cl^-$ (0·5518), $SO_4^{2-}$ (0·0281) at I = 0·7 M

[d] As in note c but using only the Na–Cl, Mg–Cl and Mg–$SO_4$ interactions (Whitfield, 1973b)

[e] Robinson and Wood (1972). Data from component single electrolyte solutions only. Composition (molal), $Na^+$ (0·4800), $Mg^{2+}$ (0·0639), $Cl^-$ (0·5510), $SO_4^{2-}$ (0·0284) at I = 0·7 M

[f] As in note e but including interaction parameters calculated from component ternary common-ion mixtures

[g] As for note f with sea water composition (molal), $Na^+$ (0·4786), $K^+$ (0·0105), $Mg^{2+}$ (0·0651), $Cl^-$ (0·5349), $SO_4^{2-}$ (0·0289) at I = 0·7 M. Recalculation using the sea water recipe shown in note c has a negligible effect on the calculated osmotic coefficients

[h] Data of Rush and Johnson (1966) as interpolated by Robinson and Wood (1972).

sea water on the basis of Scatchard's ionic component theory (Scatchard, 1968; see also Section 2.6.3.3). Considering only the contributions of the component single electrolyte solutions their equation may be written as

$$\phi_{sw} = \phi_{EL} + (\sum_i x_i/z_i)^{-1} \sum_k \sum_j x_k x_j (B_{kj}^{(1)}m + B_{kj}^{(2)}m^2 + B_{kj}^{(3)}m^3) \quad (8.5)$$

where

$$\phi_{EL} = 1 - \mathscr{A}^*(\sum_i x_i/z_i)^{-1} \sum_k \sum_j x_k x_j (z_k + z_j)(y - y^{-1} - 2\ln y)/(\mathscr{B}\mathring{a}_{kj})^3 I$$

$$y = 1 + \mathscr{B}\mathring{a}_{kj}I^{\frac{1}{2}} \qquad m = 0·5 \sum_i m_i$$

and $x_i = E_i/m$ (see Table 2.31, equation (1)).

Using a restricted model (containing only $Na^+$, $Mg^{2+}$, $Cl^-$ and $SO_4^{2-}$) they were able to predict the osmotic coefficient of sea water to within $\pm 0.005$ units at temperatures up to 373 K and ionic strengths up to 2·8 M. Forty adjustable parameters were required to specify the sea water model over this range. Their detailed analysis revealed that the success of the simpler approach of Stoughton *et al.* (1964) may be due in part to the fortuitous cancellation of the large non-Debye–Hückel contributions of sodium sulphate and magnesium chloride in the region of 323 K.

The specific interaction model (Section 2.6.3.2) may be used to calculate the osmotic coefficients of sea water by re-arranging equation (3) (Table 2.26) to give (treating the ions as components),

$$\phi_{sw} - 1 = \phi_{EL} + 2·303 \sum_j \sum_k m_j m_k b_{jk} / \sum_i m_i \qquad (8.6)$$

where $\phi_{EL} = (\mathscr{A}^*/3)Z_{sw}I^{\frac{1}{2}}\sigma(\mathscr{B}\hat{a}I^{\frac{1}{2}})$. The function $\sigma(x)$ is defined in Table 2.18 and $Z_{sw}$ is defined in Table 2.46.

Rather unexpectedly, the Guggenheim model (with $\mathscr{B}\hat{a} = 1\ \text{mol}^{-\frac{1}{2}}\text{kg}^{\frac{1}{2}}$) which has been used to calculate the solute activity coefficients in sea water (Section 2.7.2) predicts values for the osmotic coefficient of sea water that are much too high. On the other hand the $\hat{a}$ value suggested by Scatchard (1936, $\hat{a} = 0.456$ nm) results in values that are much too low (Whitfield 1973b). An alternative approach to the definition of $\hat{a}$ based on the single-ion values tabulated by Kielland (1937) and by Bonino (e.g. Bonino and Centola, 1933) has been suggested by Bianucci and de Stefani (1962). If the single-ion parameters are weighted according to the mole fractions of the individual ionic components then $\hat{a}$ can be defined by the equation

$$\hat{a} = 0·5 \sum_k x_k \hat{a}_k + 0·5 \sum_j x_j \hat{a}_j \qquad (8.7)$$

For sea water this gives $\hat{a} = 0.395$ nm using Kielland's $a_i$-values and $\hat{a} = 0.394$ using Bonino's values. The difference between the two results is negligible. Using $\hat{a} = 0.394$ nm, equation 8·6 gives osmotic coefficients that are in excellent agreement with the experimental values for ionic strengths up to 2 M (Table 2.47, column $c$). Detailed calculations (Whitfield, 1973b) suggest that the Na–Cl and Mg Cl interactions make predominant contributions to the statistical term throughout the concentration range and are counterbalanced by a small negative contribution from the Mg–$SO_4$ interaction. The other terms effectively cancel each other out. Calculations using only these three interaction terms give a good account of the experimental osmotic coefficients (Table 2.47, column $d$).

The calculations can be extended to consider a wider range of temperature and pressure (Whitfield, 1973c). The electrostatic term $\phi_{EL}$ may be calculated over the range 273 to 373 K and 1 to 1000 atm using the tabulated values of $\mathscr{A}$ and $\mathscr{B}$ (Hamer, 1968; Hills and Ovenden, 1966) if $å$ is treated as a constant. The statistical term at a reference temperature $T_0 (b_{k,j}^{(T_0)})$ may be related to the value at $T_1 (b_{k,j}^{(T_1)})$ by the equation

$$v_{kj}m_{kj}[b_{k,j}^{(T_1)} - b_{k,j}^{(T_0)}] = \log [\gamma_{kj}^{(T_1)}/\gamma_{kj}^{(T_0)}] - \log [\gamma_{EL}^{(T_1)}/\gamma_{EL}^{(T_0)}] \quad (8.7)$$

The first term on the right hand side is defined by equation (2.15) and the second term may be calculated if values of $\mathscr{A}$ and $\mathscr{B}$ are available since $\log \gamma_{EL} = -\mathscr{A}|z_+ z_-| I^{\frac{1}{2}}/(1 + 0.394\mathscr{B}I^{\frac{1}{2}})$.

Because few data are available for the relative partial molal enthalpies and heat capacities of electrolyte solutions at high ionic strengths, equation (8.7) can be fully evaluated for sodium chloride only over the range 298 K to 373 K. If the electrostatic term $\phi_{EL}$ and the interaction term $b_{Na,Cl}$ are corrected for changes in temperature using the activity coefficients tabulated by Harned and Owen (1958, p. 726) and the Debye–Hückel parameters tabulated by Hamer (1968) then the experimental data of Gibbard and Scatchard (1972) can be reproduced at an ionic strength of 1 M up to 373 K (Table 2.48). A similar procedure may be used to calculate the effect of pressure on the osmotic coefficient (Whitfield, 1973c).

Robinson and Wood (1972) have calculated the osmotic coefficient of sea water at 298 K over a wide concentration range using the cluster integral expansion equation (Table 2.31, equation (3)). Considering the first term, describing interactions between ions of opposite charge, the calculations on a restricted sea water model (containing $Na^+$, $Mg^{2+}$, $Cl^-$ and $SO_4^{2-}$)

TABLE 2.48

*Osmotic coefficient of sea water at 1 M ionic strength from 298 to 373 K.*

| T/K | Specific interaction model | | Experimental | |
|-----|-----|-----|-----|-----|
| | a | b | c | d |
| 298 | — | 0·917 | 0·918 | 0·916 |
| 323 | 0·912 | 0·919 | 0·920 | 0·919 |
| 333 | 0·910 | 0·918 | (0·915)[e] | 0·919 |
| 348 | 0·907 | 0·915 | 0·917 | 0·917 |
| 373 | 0·900 | 0·906 | 0·903 | 0·908 |

[a] 298 K data corrected for changes in $\phi_{EL}$ only
[b] Values listed under a corrected for variations in $b_{Na,Cl}$
[c] Gibbard and Scatchard (1972)
[d] Stoughton and Lietzke (1965)
[e] This value refers to 335·5 K and is anomalously low when compared with other data at this ionic strength (Gibbard and Scatchard 1972).

give osmotic coefficients that are within $\pm 0.002$ units of the experimental values (Table 2.47, column $e$). The agreement is only slightly improved by the use of the full equation which also considers interactions between ions of like charge. The inclusion of potassium in the sea water results in an almost exact prediction of the experimental data to ionic strengths as high as 6 M (Table 2.47, column $g$).

## 2.9. SUMMARY

Regardless of whether chemical processes in the oceans are controlled by equilibrium forces or by a delicate balance of dynamic systems they cannot be treated quantitatively without due consideration of the interactions associated with the presence of the major electrolyte component. The consequences of such interactions are most conveniently summarized by the solute activity coefficients which are directly related to the partial molal free energies of the various solution components.

Despite the structural complexity of aqueous electrolyte solutions (Section 2.4) predictive equations have been derived for the activity coefficients of the solutes at low concentrations (millimolar) using only fundamental constants (Section 2.5). A number of descriptive equations have been developed from these relationships that enable the solute activity coefficients to be calculated over a wide concentration range using only a few adjustable parameters (Table 2.17). These parameters are estimated by curve fitting to the experimental data.

The existence of regular patterns in the behaviour of ternary common-ion mixtures, together with statistical treatments of the mixing process result in the development of some tediously long, but computationally reasonable, equations describing the properties of electrolyte mixtures (Section 2.6). Four such models have been extended to calculate the activity coefficients of the major electrolyte components in sea water. The results of the calculations are in good general agreement with the available experimental evidence (Section 2.7). Two of the models are also able to give accurate predictions of the osmotic coefficient of sea water (Section 2.8) and thus provide comprehensive models covering the solutes *and* the solvent. These models must now be extended to allow quantitative predictions to be made about the behaviour of trace components in sea water.

It is hoped that the background presented in this section and the numerous examples provided in the chapters that follow will encourage others to help to provide not merely a coherent approach to marine chemistry, but a consistent attitude towards the chemistry of all natural electrolyte solutions.

G

162     M. WHITFIELD

## References

Ahrens, L. H. (1952). *Geochim. Cosmochim. Acta*, **2**, 155.
Åkerlöf, G. (1934). *J. Amer. Chem. Soc.* **56**, 1539.
Åkerlöf, G. (1937). *J. Phys. Chem.* **41**, 1053.
Åkerlöf, G. and Thomas, H. C. (1934). *J. Amer. Chem. Soc.* **56**, 593.
Akitt, J. W. (1971). *J. Chem. Soc.* (A) **14**, 2347.
Anderson, G. M. (1970). *J. Chem. Ed.* **47**, 676.
Bader, R. F. W. and Jones, G. (1963). *Can. J. Chem.* **41**, 586.
Bagchi, S. N. (1950). *J. Indian Chem. Soc.* **27**, 199.
Bates, R. G. (1964). "Determination of pH. Theory and Practice", 435 pp. John Wiley & Sons, New York.
Bates, R. G. and Alfenaar, M. (1969). *In* "Ion selective electrodes" (R. A. Durst, ed.), pp. 191–214. National Bureau of Standards Special Publication No. 314. U.S. Govt. Printing Office, Washington D.C.
Bates, R. G., Staples, B. R. and Robinson, R. A. (1970). *Anal. Chem.* **42**, 867.
Batsanov, S. S. (1963). *J. Struct. Chem. (USSR)*, **4**, 158.
Ben-Naim, A. and Stillinger, F. H. (1972). *In* "Water and aqueous solutions" (R. A. Horne, ed.), pp. 295–330. Wiley Interscience, New York.
Benson, S. W. and Copeland, C. S. (1963). *J. Phys. Chem.* **67**, 1194.
Bergqvist, M. S. and Forslind, E. (1962). *Acta. Chem. Scand.* **16**, 2069.
Bernal, J. D. (1964). *Proc. R. Soc. Lond.* **A280**, 299.
Bernal, J. D. and Fowler, R. H. (1933). *J. Chem. Phys.* **1**, 515.
Berner, R. A. (1965). *Geochim. Cosmochim. Acta*, **29**, 947.
Berner, R. A. (1971). "Principles of Chemical Sedimentology", pp. 44–48. McGraw-Hill, New York.
Berner, R. A. and Wilde, P. (1972). *Amer. J. Sci.* **272**, 826.
Bianucci, G. and de Stefani, G. (1962). *Acqua Industriale*, **4**, 14.
Bjerrum, N. (1918). *Z. Electrochem.* **24**, 321.
Bjerrum, N. (1919). *Medd. vetensk. Acad. Nobelinst.* **5**, No. 16.
Bjerrum, N. (1920). *Z. Anorg. Chem.* **109**, 275.
Bjerrum, N. (1926). *Kgl. Danske Videnskab. Selskab.* 7 No. 9. See also "Selected Papers", p. 108. Einar Munksgaard, Copenhagen, 1949.
Bjerrum, N. (1927). *Trans. Faraday Soc.* **23**, 433.
Blandamer, M. J. and Symons, M. C. R. (1963). *J. Phys. Chem.* **67**, 1304.
Bockris, J. O'M. and Reddy, A. K. N. (1970). "Modern Electrochemistry" Vol. I, 622 pp. McDonald, London.
Bonino, G. B. and Centola, G. (1933). *Atti. R. Acad. Nazionale Lincei.* **18**, 145.
Bonner, O. D. and Woolsey, G. B. (1968). *J. Phys. Chem.* **72**, 899.
Boyd, G. E., Lindebaum, S. and Robinson, R. A. (1971). *J. Phys. Chem.* **75**, 3153.
Braunstein, J. (1971). *In* "Ionic Interactions from Dilute Solutions to Fused Salts" (S. Petrucci, ed.) Vol. I, pp. 187–190. Academic Press, London.
van Breemen, N. (1972). *Geochim. Cosmochim. Acta*, **37**, 101.
Broadwater, T. L. and Kay, R. L. (1970). *J. Phys. Chem.* **74**, 3802.
Broecker, W. S. and Takahashi, T. (1966). *J. Geophys. Res.* **71**, 1575.
Bromley, L. A. (1972). *J. Chem. Thermodynam.* **4**, 669–673.
Brønsted, J. N. (1922). *J. Am. Chem. Soc.* **44**, 877, 938.
Burley, D. M., Hutson, V. C. L. and Outhwaite, C. W. (1971). *Chem. Phys. Lett.* **9**, 109.

Burns, D. T. (1964). *Electrochim. Acta*, **9**, 1545.

Butler, J. N. (1969). *In* "Ion Selective Electrodes" (R. A. Durst, ed.), pp. 143–189. N.B.S. Special Publication No. 314. U.S. Govt. Printing Office, Washington, D.C.

Butler, J. N. and Huston, R. (1970). *J. Phys. Chem.* **74**, 2976.

Chan, C. Y. and Panckhurst, M. H. (1972). *Aust. J. Chem.* **25**, 317.

Christenson, P. G. and Gieskes, J. M. (1971). *J. Chem. Eng. Data.* **16**, 398.

Cobble, J. W. (1953). *J. Chem. Phys.* **21**, 1446.

Conway, B. E. (1952). "Electrochemical Data". Elsevier, New York, 374 pp.

Conway, B. E. (1966). *Ann. Rev. Phys. Chem.* **17**, 491.

Conway, B. E. (1970). *In* "Physical Chemistry. An Advanced Treatise" (H. Eyring, D. Henderson and W. Jost, eds) Vol. IXA, pp 1–166. Academic Press, New York.

Conway, B. E. and Bockris, J. O'.M. (1954). *In* "Modern Aspects of Electrochemistry" (J. O'M. Bockris, ed.) Vol. 1, pp. 47–102. Butterworths, London.

Coulson, C. A. and Danielsson, U. (1954). *Ark. Fys.* **8**, 239, 245.

Couture, A. M. and Laidler, K. J. (1956). *Can. J. Chem.* **34**, 1209.

Covington, A. K. and Lilley, T. H. (1970). "Specialist Periodical Reports. Electrochemistry" Vol. I, pp. 1–30. Chemical Society, London.

Covington, A. K., Lilley, T. H. and Robinson, R. A. (1968). *J. Phys. Chem.* **72**, 2759.

Daly, F. P., Brown, C. W. and Kester, D. R. (1972). *J. Phys. Chem.* **76**, 3664.

Danford, M. D. and Levy, H. A. (1962). *J. Amer. Chem. Soc.* **84**, 3965.

Davies, C. W. (1938). *J. Chem. Soc.* 2093.

Davies, C. W. (1962). "Ion association", 190 pp. Butterworths, London.

Davis, C. M. and Jarzynski, J. (1972). *In* "Water and Aqueous Solutions" (R. A. Horne, ed.). Wiley-Interscience, New York.

Debye, P. and Hückel, E. (1923). *Physik Z.* **24**, 185, 305.

Del Bene, J. and Pople, J. A. (1970). *J. Chem. Phys.* **52**, 4858.

Denbigh, K. (1961). "The Principles of Chemical Equilibrium". 491 pp. The University Press, Cambridge.

Desnoyers, J. E. and Conway, B. E. (1964). *J. Phys. Chem.* **68**, 2305.

Desnoyers, J. E. and Jolicoeur, C. (1969). *In* "Modern Aspects of Electrochemistry" (J. O'M. Bockris and B. E. Conway, eds) Vol. 5, pp. 1–89. Butterworths, London.

Dittmar, W. (1884). "Report on the Scientific Results of the Exploring Voyage of H.M.S. *Challenger*. Physics and Chemistry", Vol. I. H.M. Stationery Office, London.

Duedall, I. W. (1968). *Environ. Sci. Technol.* **2**, 706.

Duedall, I. W. (1972). *Geochim. Cosmochin. Acta,* **36**, 729.

Durst, R. A. (1969), ed. "Ion Selective Electrodes". N.B.S. Special Publication No. 314. 452 pp. U.S. Govt. Printing Office, Washington, D.C.

Dyrssen, D. and Hansson, I. (1973). *Mar. Chem.* **1**, 137.

Edmond, J. M. (1972). *Proc. Roy. Soc. Edinburgh* **B72**, 371.

Eigen, M. and Maas, G. (1966). *Z. Phys. Chem. N.F.* **49**, 163.

Eisenberg, D. and Kauzmann, W. (1969). "The structure and Properties of Water", 296 pp. Clarendon Press, Oxford.

Elgqvist, B. and Wedborg, M. (1974). *Mar. Chem.* **2**, 1.

Falkenhagen, H. and Ebeling, W. (1971). *In* "Ionic Interactions" (S. Petrucci, ed.) Vol. I. pp. 1–59, Academic Press, New York.

Falkenhagen, H. and Kelbg, G. (1959). *In* "Modern Aspects of Electrochemistry" (J. O'M. Bockris, ed.) Vol. 2. Butterworths, London.

Forch, C., Knudsen, M. and Sørensen, S. P. L. (1902). *K. Danske Vidensk. Selsk. Skr.* **12**, 1.

Forslind, E. (1952). *Acta Polytechnica,* **115**, 9.

Forslind, E. (1953). *Proc. 2nd Int. Congr. Rheol.* Butterworths, London.

Fowler, R. H. (1927). *Trans. Faraday Soc.* **23**, 434.

Fowler, R. H. and Guggenheim, E. A. (1952). "Statistical Thermodynamics", 701 pp. Cambridge University Press, London.

Frank, H. S. (1958). *Proc. R. Soc. Lond.* **A247**, 481.

Frank, H. S. (1972). *In* "Water. A Comprehensive treatise" (F. Franks, ed.) Vol. 1. Plenum Press, New York.

Frank, H. S. and Evans, M. W. (1945). *J. Chem. Phys.* **13**, 507.

Frank, H. S. and Thompson, P. T. (1959a). *In* "Structure of Electrolyte Solutions" (W. Hamer, ed.), pp. 113–134. Wiley, New York.

Frank, H. S. and Thompson, P. T. (1959b). *J. Chem. Phys.* **31**, 1086.

Frank, H. S. and Wen, W. Y. (1957). *Disc. Faraday Soc.* **24**, 133.

Friedman, H. L. (1960a). *J. Chem. Phys.* **32**, 1351.

Friedman, H. L. (1960b). *J. Chem. Phys.* **32**, 1134.

Friedman, H. L. (1962. "Ionic Solution Theory", 265 pp. Interscience, New York.

Friedman, H. L. (1970). *In* "Modern Aspects of Electrochemistry" (J. O'M. Bockris and B. E. Conway, eds) Vol. 6. Plenum Press, New York.

Frolov, Yu. G., Nikolaev, V. P., Karapet'yants, M. Kh. and Vlasenko, K. K. (1971). *Russ. J. Phys. Chem.* **45**, 1054.

Fuoss, R. M. (1934). *Trans. Faraday Soc.* **30**, 967.

Fuoss, R. M. (1958). *J. Amer. Chem. Soc.* **80**, 5059.

Gardener, A. W. and Glueckauf, E. (1969). *Proc. R. Soc. London.* **A313**, 131.

Gardener, A. W. and Glueckauf, E. (1971). *Proc. R. Soc. Lond.* **A321**, 515.

Garrels, R. M. (1967). *In* "Glass Electrodes for Hydrogen and Other Cations" (G. Eisenman ed.) Marcel Dekker, New York.

Garrels, R. M. and Christ, C. (1965). "Solutions, Minerals and Equilibria", 450 pp. Harper & Row, New York.

Garrels, R. M. and Thompson, M. E. (1962). *Amer. J. Sci.* **260**, 57.

Ghosh, J. C. (1918). *J. Chem. Soc.* **113**, 449, 707.

Gibbard, H. F. and Scatchard, G. (1972). *J. Chem. Eng. Data,* **17**, 498.

Gieskes, J. M. (1966). *Z. Phys. Chem.* (*Frankfurt*), **50**, 78.

Ginstrup, O. (1970). *Acta Chem. Scand.* **24**, 875.

Glueckauf, E. (1949). *Nature, Lond.* **163**, 414.

Glueckauf, E. (1955). *Trans. Faraday Soc.* **51**, 1235.

Glueckauf, E. (1959). *In* "Structure of Electrolyte Solutions" (W. Hamer, ed.), pp. 97–112. Wiley, New York.

Glueckauf, E. (1969). *Proc. R. Soc. Lond.* **A310**, 449.

Goldschmidt, V. M. (1926). *Skr. Norske Vidensk. Akad., Oslo Math. Nat. Kl.* **1**, 21.

Gourary, B. S. and Adrian, F. J. (1960). *Solid State Phys.* **10**, 127.

Greyson, J. and Snell, H. (1969). *J. Phys. Chem.* **73**, 3208.

Gronwall, T. H. (1927). *Proc. Nat. Acad. Sci. USA,* **13**, 198.

Gronwall, T. H., LaMer, V. K. and Sandved, K. (1929). *Physik. Z.* **29**, 358.

Guggenheim, E. A. (1935). *Phil. Mag.* **19**, 588.

Guggenheim, E. A. (1956). "Thermodynamics", 3rd Ed., 476 pp. North-Holland, Amsterdam.

Guggenheim, E. A. (1959). *Trans. Faraday Soc.* **55**, 1714.

Guggenheim, E. A. (1960). *Trans. Faraday Soc.* **56**, 1152.
Guggenheim, E. A. (1962). *In* "Electrolytes" (B. Pesce, ed.). Pergamon Press, London.
Guggenheim, E. A. (1966a). "Applications of Statistical Mechanics", 211 pp. Oxford University Press, London.
Guggenehim, E. A. (1966b). *Trans. Faraday Soc.* **62**, 3446.
Guggenheim, E. A. and Stokes, R. H. (1969). "Equilibrium Properties of Aqueous Solutions of Single Strong Electrolytes", 148 pp. Pergamon Press, London.
Guggenheim, E. A. and Turgeon, J. C. (1955). *Trans. Faraday Soc.* **51**, 747.
Güntelberg, E. (1926). *Z. Physik. Chem.* **123**, 199.
Gurney, R. W. (1953). "Ionic Processes in Solutions", 275 pp. McGraw-Hill, New York.
Haase, R. (1969). "Thermodynamics of Irreversible Processes", 509 pp. Addison-Wesley, London.
Hadzi, D., ed. (1968). "The Hydrogen Bond". Pergamon Press, London.
Halliwell, H. F. and Nyburg, S. C. (1963). *Trans. Faraday Soc.* **58**, 1126.
Hamann, S. D. (1963). *In* "High Pressure Physics and Chemistry" (R. S. Bradley, ed.) Vol. 2. Academic Press, New York.
Hamer, W. J. (1968). "Theoretical mean activity coefficients of strong electrolytes in aqueous solutions from 0 to 100°C". National Standard reference data series— National Bureau of Standards 24 (NSRDS-NBS24). U.S. Govt. Printing Office, Washington, D.C.
Hanor, J. S. (1969). *Geochim. Cosmochim. Acta,* **33**, 894.
Hansson, I. (1973). *Deep-Sea Res.* **20**, 461.
Harned, H. S. (1935). *J. Amer. Chem. Soc.* **57**, 1865.
Harned, H. S. and Gary, R. (1954). *J. Amer. Chem. Soc.* **76**, 5924.
Harned, H. S. and Owen, B. B. (1958). "The Physical Chemistry of Electrolytic Solutions", 803 pp. Amer. Chem. Soc. Monograph No. 137. Reinhold, New York.
Harned, H. S. and Robinson, R. A. (1968). "Multicomponent Electrolyte Solutions", 110 pp. Pergamon Press, Oxford.
Helgeson, H. C. (1967). *J. Phys. Chem.* **71**, 3121.
Helgeson, H. C. (1969). *Amer. J. Sci.* **267**, 729.
Hertz, H. G. (1970). *Angew. Chem., internat. ed.* **9**, 124.
Hill, T. L. (1956). "Statistical Mechanics", 432 pp. McGraw-Hill, New York.
Hills, G. J. and Ovenden, P. J. (1966). *Adv. Electrochem.* **4**, 185.
Hinton, J. F. and Amis, E. S. (1971). *Chem. Rev.* **71**, 627.
Holtzer, A. and Emerson, M. F. (1969). *J. Phys. Chem.* **73**, 26.
Horne, R. A. (1969). "Marine Chemistry", 568 pp. Wiley-Interscience, New York.
Horne, R. A. ed. (1972). "Water and Aqueous Solutions". Wiley-Interscience, New York.
Horne, R. A. and Birkett, J. D. (1967). *Electrochim. Acta,* **12**, 1153.
Hückel, E. (1925). *Physik. Z.* **26**, 93.
Ingri, N., Kakolowicz, W., Sillén, L. G. and Warnqvist, B. (1967). *Talanta,* **14**, 1261.
Ives, D. J. G. and Janz, G. J. (1961). "Reference Electrodes", 651 pp. Academic Press, New York.
Izatt, R. M., Eatough, D., Christensen, J. J. and Bartholemew, C. H. (1969). *J. Chem. Soc. A,* 45.
Jhon, M. S., Grosh, J., Ree, T. and Eyring, H. (1966). *J. Chem. Phys.* **44**, 1465.
Joliceour, C., Picker, P. and Desnoyers, J. E. (1969). *J. Chem. Thermodynam.* **1**, 485.
Jones, G. and Dole, M. (1929). *J. Amer. Chem. Soc.* **51**, 2950.

Kamb, B. (1968). *In* "Structural Chemistry and Molecular Biology" (A. Rich and N. Davidson, eds). Freeman, San Fransisco.
Kaminsky, M., (1957). *Disc. Faraday Soc.* **24**, 171.
Karapet'yants, M. Kh., Vlasenko, K. K. and Solov'eva, S. G. (1970). *Russ. J. Phys. Chem.* **44**, 305.
Kavanau, J. L. (1964). "Water and Solute-water Interactions", 101 pp. Holden-Day, San Francisco.
Kay, R. L. and Evans, D. F. (1966). *J. Phys. Chem.* **70**, 2325.
Kell, G. S. (1972a). *In* "Water. A Comprehensive Treatise" (F. Franks, ed.) Vol. I, pp. 363–413. Plenum Press, New York.
Kell, G. S. (1972b). *In* "Water and Aqueous Solutions" (R. A. Horne, ed.). Wiley, New York.
Kester, D. R. (1969). Ph.D. Thesis, 116 pp. Oregon State University, Corvallis.
Kester, D. R. and Pytkowicz, R. M. (1969). *Limnol. Oceanogr.* **14**. 586.
Kester, D. R. and Pytkowicz, R. M. (1970). *Geochim. Cosmochim. Acta*, **34**, 1039.
Kielland, J. (1937). *J. Amer. Chem. Soc.* **59**, 1675.
King, E. J. (1965). "Acid-Base Equilibria", 341 pp. Pergamon Press, New York.
Kirgintsev, A. N. (1971). *Russ. J. Phys. Chem.* **45**, 74.
Kirgintsev, A. N. and Luk'yanov, A. V. (1963). *Russ. J. Phys. Chem.* **37**, 1501.
Kirgintsev, A. N. and Luk'yanov, A. V. (1964). *Russ. J. Phys. Chem.* **38**, 867.
Kirgintsev, A. N. and Luk'yanov, A. V. (1966). *Russ. J. Phys. Chem.* **40**, 686.
Kirgintsev, A. N. and Luk'yanov, A. V. (1967). *Russ. J. Phys. Chem.* **41**, 54.
Kirkwood, J. G. (1934). *J. Chem. Phys.* **2**, 767.
Kirkwood, J. G. (1936). *Chem. Rev.* **19**, 275.
Kirkwood, J. G. and Poirier, J. C. (1954). *J. Phys. Chem.* **58**, 591.
Klotz, I. M. (1964). "Chemical Thermodynamics. Basic Theory and Methods", 468 pp. W. A. Benjamin, New York.
Koefoed, J. (1957). *Disc. Faraday Soc.* **24**, 216.
Kramer, J. R. (1965). *Geochim. Cosmochim. Acta*, **29**, 921.
Krawetz, A. A. (1957). *Disc. Faraday Soc.* **24**, 77.
Lafon, G. M. (1969). Ph.D. Thesis, Northwestern University, Evanston, Illinois.
Lafon, G. M. (1970). *Geochim. Cosmochim. Acta*, **34**, 935.
Lakshmanan, S, and Rangarajan, S. K. (1970). *J. Electroanal. Chem.* **27**, 170.
Lanier, R. D. (1965). *J. Phys. Chem.* **69**, 3992.
Larson, J. W. (1970). *J. Phys. Chem.* **74**, 3392.
Lee, T. S. (1959). *In* "Treatise on Analytical Chemistry" (I. M. Kolthoff and P. J. Elving, eds) Part I, Vol. 1, pp. 187–275. Interscience, New York.
von Leeuwen, J. M. J., Groeneveld, J. and DeBoer, I (1959). *Physica*, **25**, 792.
Lennard-Jones, J. and Pople, J. A. (1951). *Proc. R. Soc. Lond.* **A205**, 163.
Lenzi, F. (1972). *Can. J. Chem.* **50**, 1008.
Lepple, F. K. and Millero, F. J. (1971). *Deep-Sea Res.* **18**, 1233.
Lerman, A. (1967). *Geochim. Cosmochim. Acta*, **31**, 2309.
Levine, S. and Wrigley, H. E. (1957). *Disc. Faraday Soc.* **24**, 43, 73.
Lewis, G. N. and Randall, M. (1961). "Thermodynamics" 2nd Ed., (revised by K. S. Pitzer and L. Brewer), 723 pp. McGraw-Hill, New York.
Leyendekkers, J. V. (1970). *J. Phys. Chem.* **74**, 2225.
Leyendekkers, J. V. (1971a). *Anal. Chem.* **43**, 1835.
Leyendekkers, J. V. (1971b). *J. Phys. Chem.* **75**, 946.
Leyendekkers, J. V. (1973). *Mar. Chem.* **1**, 75.

Leyendekkers, J. V. and Whitfield, M. (1971). *J. Phys. Chem.* **75**, 957.
Lietzke, M. H. and Stoughton, R. W. (1962). *J. Phys. Chem.* **66**, 508.
Lietzke, M. H., Stoughton, R. W. and Fuoss, R. M. (1968). *Proc. Nat. Acad. Sci. U.S.A.* **58**, 39.
Lindenbaum, S., Rush, R. M. and Robinson, R. A. (1972). *J. Chem. Thermodynam.* **4**, 381.
McCabe, W. C. and Fisher, H. F. (1970). *J. Phys. Chem.* **74**, 2990.
MacInnes, D. A. (1919). *J. Am. Chem. Soc.* **41**, 1086.
McKay, H. A. C. (1952). *Nature, Lond.* **169**, 464.
McKay, H. A. C. (1955). *Trans. Faraday Soc.* **51**, 903.
McKay, H. A. C. (1957). *Disc. Faraday Soc.* **24**, 76.
McKay, H. A. C. and Perring, J. K. (1953). *Trans. Faraday Soc.* **49**, 163.
Mangelsdorf, P. C., Jr. and Wilson, T. R. S. (1971). *J. Phys. Chem.* **75**, 1418.
Marchi, R. P. and Eyring, H. (1964). *J. Phys. Chem.* **68**, 221.
Masson, D. O. (1929). *Phil. Mag.* **8**, 218.
Masterton, W. L., Bolocofsky, D. and Lee, T. P. (1971). *J. Phys. Chem.* **75**, 2809.
Mayer, J. E. (1950). *J. Chem. Phys.* **18**, 1426.
Mayer, J. E. and Mayer, M. G. (1940). "Statistical Mechanics", 495 pp. Wiley, New York.
Meissner, H. P. and Kusik, C. L. (1972). *Amer. Inst. Chem. Eng. J.* **18**, 294.
Meites, L. (1963). *In* "Handbook of Analytical Chemistry" (L. Meites, ed.), p. 1–8. McGraw-Hill, New York.
Mikhailov, V. A. (1968). *Russ. J. Phys. Chem.* **42**, 1414.
Millero, F. J. (1969). *Limnol. Oceanogr.* **14**, 376.
Millero, F. J. (1971a). *Chem. Rev.* **71**, 147.
Millero, F. J. (1971b). *Geochim. Cosmochim. Acta,* **35**, 1089.
Millero, F. J. (1972a). *In* "Water and Aqueous Solutions" (R. A. Horne, ed.), pp. 519–564. Wiley-Interscience, New York.
Millero, F. J. (1972b). *In* "Water and Aqueous Solutions" (R. A. Horne, ed.), pp. 565–595. Wiley-Interscience, New York.
Millero, F. J. (1974). *In* "The Sea" (E. D. Goldberg, ed.) Vol. 5. Interscience, New York.
Millero, F. J. and Berner, R. A. (1972). *Geochim. Cosmochim. Acta,* **36**, 92.
Milner, S. R. (1912). *Phil. Mag.* **23**, 551.
Milner, S. R. (1913). *Phil. Mag.* **25**, 742.
Mitra, P., Jain, D. V. S. and Kapoor, M. H. (1968). *Indian J. Chem.* **6**, 391.
Monk, C. B. (1961). "Electrolytic Dissociation", 320 pp. Academic Press, New York.
Monk, C. B. (1966). *In* "Chemical Physics of Ionic Solutions" (B. E. Conway and R. G. Barradas, eds), pp, 175–195. Wiley, New York.
Moody, G. J. and Thomas, J. D. R. (1972). "Selective Ion Sensitive Electrodes", 148 pp. Merrow Publishing Co., Watford, England.
Morel, F. and Morgan, J. (1972). *Environ. Sci. Technol.* **6**, 58.
Morf, W. E. and Simon, W. (1971). *Helv. Chim. Acta,* **54**, 794.
Morris, D. F. C. (1968). *In* "Structure and Bonding" Vol. 4, p. 63. Springer, New York.
Muirhead-Gould, J. S. and Laidler, K. J. (1966). *In* "Chemical Physics of Ionic Solutions" (B. E. Conway and R. G. Barradas, eds) J. Wiley & Sons, New York.
Müller, E. (1927). *Physik. Z.* **28**, 324.
Nakayama, F. S. (1968). *Soil. Sci.* **103**, 213.
Nakayama, F. S. (1971a). *J. Chem. Eng. Data,* **16**, 178.

Nakayama, F. S. (1971b). *J. Inorg. Nucl. Chem.* **33**, 1287.

Nakayama, F. S. (1971c). *Soil. Sci. Amer. Proc.* **35**, 881.

Nakayama, F. S, and Rasnick, B. A. (1969). *J. Inorg. Nucl. Chem.* **31**, 3491.

Nancollas, G. H. (1966). "Interactions in Electrolyte Solutions", 214 pp. Elsevier, Amsterdam.

Narten, A. and Levy, H. A. (1969). *Science, N.Y.* **165**, 447.

Narten, A. H. and Levy, H. A. (1972). *In* "Water. A Comprehensive Treatise" (F. Franks, ed.) Vol. I, pp. 311–332. Plenum Press, New York.

Narten, A. H., Danford, M. D. and Levy, H. A. (1967). *Disc. Faraday Soc.* **43**, 97.

Némethy, G. and Scheraga, H. A. (1962a). *J. Chem. Phys.* **36**, 3382.

Némethy, G. and Scheraga, H. A. (1962b). *J. Chem. Phys.* **36**, 3401.

Nightingale, E. R. Jr. (1966). *In* "Chemical Physics of Ionic Solutions" (B. E. Conway and R. G. Barradas, eds), pp. 87–100. John Wiley & Sons, Inc., New York.

Onsager, L. (1933). *Chem. Rev.* **13**, 72.

Orgel, L. E. and Mulliken, R. S. (1957). *J. Amer. Chem. Soc.* **79**, 4839.

Owen, B. B. and Brinkley, S. R. J. (1941). *Chem. Rev.* **29**, 461.

Panckhurst, M. H. (1962). *Aust. J. Chem.* **15**, 383.

Parker, V. B. (1965). "Thermal Properties of Aqueous Uni-univalent Electrolytes". NSRDS–NBS2. U.S. Govt. Printing Office, Washington D.C.

Parsons, R. (1959). "Handbook of Electrochemical Constants", 113 pp. Butterworths, London.

Pauling, L. (1960). "The Nature of the Chemical Bond", 3rd Ed., 644 pp. Cornell University Press, New York.

Pearson, R. G. (1963). *J. Amer. Chem. Soc.* **85**, 3533.

Percus, J. K. and Yevick, E. J. (1958). *Phys. Rev.* **110**, 1.

Petrucci, S. (1971). *In* "Ionic Interactions from Dilute Solutions to Fused Salts" ((S. Petrucci, ed.), pp. 117–177. Academic Press, London.

Pimentel, G. C. and McClellan, A. L. (1960). "The Hydrogen Bond", 475 pp. Freeman, San Francisco.

Pitzer, K. S. (1972). *J. Chem. Soc. Faraday Trans.* II, **1**, 101.

Platford, R. F. (1965a). *J. Fish. Res. Bd. Can.* **22**, 885.

Platford, R. F. (1965b). *J. Mar. Res.* **23**, 55.

Platford, R. F. (1968a). *J. Chem. Eng. Data*, **13**, 46.

Platford, R. F. (1968b). *J. Phys. Chem.* **72**, 4053.

Platford, R. F. and Dafoe, T. (1965). *J. Mar. Res.* **23**, 63.

Poirier, J. C. (1953). *J. Chem. Phys.* **21**, 972.

Poirier, J. C. (1966). *In* "Chemical Physics of Ionic Solutions" (B. E. Conway and R. G. Barradas, eds), pp. 9–27. Wiley, New York.

Pople, J. A. (1951). *Proc. R. Soc. Lond.* **A205**, 163.

Prausnitz, J. M. (1960). *Amer. Inst. Chem. Eng. J.* **6**, 78.

Prue, J. E. (1969). *J. Chem. Ed.* **46**, 12.

Pytkowicz, R. M. (1968). *Oceanogr. Mar. Biol. Ann. Rev.* **6**, 83

Pytkowicz, R. M. (1972). *Geochim. Cosmochim. Acta,* **36**, 631.

Pytkowicz, R. M. and Gates, R. (1968). *Science, N.Y.* **161**, 690.

Pytkowicz, R. M. and Kester, D. R. (1969). *Amer. J. Sci.* **267**, 217.

Ramanathan, P. S. and Friedman, H. L. (1971). *J. Chem. Phys.* **54**, 1086.

Rao, C. N. (1972). *In* "Water. A Comprehensive Treatise" (F. Franks, ed.) Vol. 1, pp. 93–114. Plenum Press, New York.

Rasaiah, J. C. and Friedman, H. L. (1968). *J. Chem. Phys.* **48**, 2742.

Rasaiah, J. C. and Friedman, H. L. (1969). *J. Chem. Phys.* **50**, 3965.
Rasaiah, J. C., Card, D. N. and Valleau, J. P. (1972). *J. Chem. Phys.* **56**, 248.
Redlich, O, and Rosenfeld, P. (1931a). *Z. Electrochem.* **37**, 705.
Redlich, O. and Rosenfeld, P. (1931b). *Z. Phys. Chem.* **A155**, 65.
Reilly, P. J. and Wood, R. H. (1969). *J. Phys. Chem.* **73**, 4292.
Reilly, P. J., Wood, R. H. and Robinson, R. A. (1971). *J. Phys. Chem.* **75**, 1305.
Robinson, R. A. (1954). *J. Mar. Biol. Ass. U.K.* **33**, 449.
Robinson, R. A. (1961). *J. Phys. Chem.* **65**, 662.
Robinson, R. A. and Bower, V. E. (1965). *J. Res. Nat. Bur. Std.* **69A**, 439.
Robinson, R. A. and Bower, V. E. (1966). *J. Res. Nat. Bur. Std.* **70A**, 305.
Robinson, R. A. and Covington, A. K. (1968). *J. Res. Nat. Bur. Std.* **72A**, 239.
Robinson, R. A. and Stokes, R. H. (1965). "Electrolyte Solutions", 571 pp. Butterworths, London.
Robinson, R. A. and Wood, R. H. (1972). *J. Solution Chem.* **1**, 481.
Robinson, R. A., Duer, W. C. and Bates, R. G. (1971a). *Anal. Chem.* **43**, 1862.
Robinson, R. A., Wood, R. H. and Reilly, P. J. (1971b). *J. Chem. Thermodynam.* **3**, 461.
Rock, P. A. (1967). *J. Chem. Ed.* **44**, 104.
Röntgen, W. C. (1892). *Ann. Phys. Chem. (Wien)*, **45**, 91.
Root, C. W. (1932). Ph.D. Thesis, Harvard University, Cambridge, Massachusetts.
Rosseinsky, D. R. (1962). *J. Chem. Soc.* 785.
Rosseinsky, D. R. (1965). *Chem. Rev.* **65**, 467.
Rosseinsky, D. R. (1971). *Ann. Repts Chem. Soc.* **68A**, 81.
Rosseinsky, D. R. and Hill, R. J. (1971). *J. Electroanal. Chem.* **30**, App. 7.
Rush, R. M. (1969a). Oak Ridge National Laboratory Report No. ORNL-4402. Oak Ridge, Tennessee.
Rush, R. M. (1969b). *J. Phys. Chem.* **73**, 4433.
Rush, R. M. and Johnson, J. S. (1966). *J. Chem. Eng. Data*, **11**, 590.
Rush, R. M. and Johnson, J. S. (1968). *J. Phys. Chem.* **72**, 767.
Rush, R. M. and Johnson, J. S. (1970). *J. Chem. Thermodynam.* **3**, 779.
Samoilov, O. Ya. (1965). "Structure of Aqueous Electrolyte Solutions and the Hydration of Ions", 185 pp. Consultants Bureau, New York.
Samoilov, O. Ya (1972). *In* "Water and Aqueous Solutions" (R. A. Horne, ed.), 837 pp. Wiley, New York.
Sanderson, R. T. (1960). "Chemical Periodicity", Ch. 8, 330 pp. Reinhold, New York.
Scatchard, G. (1936). *Chem. Rev.* **19**, 309.
Scatchard, G. (1961). *J. Amer. Chem. Soc.* **83**, 2636.
Scatchard, G. (1965). *Z. Phys. Chem.* **228**, 354.
Scatchard, G. (1968). *J. Amer. Chem. Soc.* **90**, 3124.
Scatchard, G. (1969). *J. Amer. Chem. Soc.* **91**, 2410.
Scatchard, G. and Epstein, L. F. (1942). *Chem. Rev.* **30**, 211.
Scatchard, G., Rush, R. M. and Johnson, J. S. (1970). *J. Phys. Chem.* **74**, 3786.
Shatkay, A. (1967). *Anal. Chem.* **39**, 1056.
Sillén, L. G. (1967). *In* "Equilibrium Concepts in Natural Water Systems" (W. Stumm, ed.), pp. 47–50. Amer. Chem. Soc., Washington D.C.
Sillén, L. G. and Martell, A. E. (1964). "Stability Constants of Metal-ion Complexes", 754 pp. Special Publication No. 17. The Chemical Society, London.

Sillén, L. G. and Martell, A. E. (1971). "Stability Constants of Metal-ion Complexes", 865 pp. Special Publication No. 25. The Chemical Society, London.

Skirrow, G. (1965). *In* "Chemical Oceanography" (J. P. Riley and G. Skirrow, eds) Vol. I, pp. 262–263. Academic Press, London.

Smith, M. B. (1942). Ph.D. Thesis, University of Chicago.

Stevenson, D. P. (1965). *J. Phys. Chem.* **69**, 2145.

Stillinger, F. H. and White, R. J. (1971). *J. Chem. Phys.* **54**, 3405.

Stokes, R. H. (1963). *J. Amer. Chem. Soc.* **86**, 979, 2332.

Stokes, R. H. and Mills, R. (1965). "Viscosity of Electrolytes and Related Properties", 151 pp. Pergamon Press, New York.

Stokes, R. H. and Robinson, R. A. (1948). *J. Amer. Chem. Soc.* **70**, 1870.

Stoughton, R. W. and Lietzke, M. H. (1965). *J. Chem. Eng. Data*, **10**, 254.

Stoughton, R. W., Lietzke, M. H. and White, R. J. (1964). *J. Tennessee Acad. Sci.* **39**, 109.

Symons, M. C. R. (1972). *Nature, Lond.* **239**, 257.

Thompson, M. E. (1966). *Science, N.Y.* **153**, 866.

Thompson, M. E. and Ross, J. W. (1966). *Science, N.Y.* **154**, 1643.

Truesdell, A. H. and Jones, B. F. (1969). *Chem. Geol.* **4**, 51.

Truesdell, A. H. and Jones, B. F. (1973). *U.S. Geol. Surv. J. Res.* (In press).

Vand, V. and Senior, W. A. (1965). *J. Chem. Phys.* **43**, 1878.

Vasilëv, V. A. (1971). *Russ. J. Phys. Chem.* **45**, 834.

Vaslow, F. (1972). *In* "Water and Aqueous Solutions" (R. A. Horne ed.), pp. 465–518. Wiley-Interscience, New York.

Vogrin, F., Knapp, P. S., Flint, W. L., Anton, A., Highberger, G. and Malinowski, E. R. (1971). *J. Chem. Phys.* **54**, 178.

Waddington, T. C. (1966). *Trans. Faraday Soc.* **62**, 1482.

Waisman, E. and Lebowitz, J. L. (1970). *J. Chem. Phys.* **52**, 4307.

Walrafen, G. E. (1964). *J. Chem. Phys.* **40**, 3249.

Walrafen, G. E. (1972). *In* "Water. A Comprehensive Treatise" (F. Franks, ed.) Vol. 1. Plenum Press, New York.

Wells, A. F. (1955). "Structural Inorganic Chemistry", 727 pp. Clarendon Press, Oxford.

Whitfield, M. (1971). "Ion-selective Electrodes for the Analysis of Natural Waters", 130 pp. AMSA Handbook No. 2. Australian Marine Sciences Association, Sydney.

Whitfield, M. (1972). *Proc. R. Soc. Edinburg.* **B72**, 389.

Whitfield, M. (1973a). *Mar. Chem.* **1**, 251.

Whitfield, M. (1973b). *Deep-Sea Res.* **21**, 57.

Whitfield, M. (1973c). *J. Mar. Biol. Assoc.* **53**, 685.

Wirth, H. E. and LoSurdo, A. (1968). *J. Chem. Eng. Data*, **13**, 226.

Wirth, H. E. and Mills, W. L. (1968). *J. Chem. Eng. Data*, **13**, 102.

Wirth, H. E., Lindstron, R. E. and Johnson, R. E. (1963). *J. Phys. Chem.* **67**, 2339.

Wood, R. H. and Anderson, H. L. (1966a). *J. Phys. Chem.* **70**, 992.

Wood, R. H. and Anderson, H. L. (1966b). *J. Phys. Chem.* **70**, 1877.

Wood, R. H. and Anderson, H. L. (1967). *J. Phys. Chem.* **71**, 1869.

Wood, R. H. and Ghamkhar, M. (1969). *J. Phys. Chem.* **73**, 3959.

Wood, R. H. and Reilly, P. J. (1970). *Ann. Rev. Phys. Chem.* **21**, 387.

Wood, R. H. and Smith, R. W. (1965). *J. Phys. Chem.* **69**, 2943.

Wood, R. H., Ghamkhar, M. and Patton, J. D. (1969a). *J. Phys. Chem.* **73**, 4298.

Wood, R. H., Patton, J. D. and Ghamkhar, M. (1969b). *J. Phys. Chem.* **73**, 346.
Wu, Y. C. (1970). *J. Phys. Chem.* **74**, 3781.
Wu, Y. C. and Friedman, H. L. (1966). *J. Phys. Chem.* **70**, 2020.
Wu, Y. C., Smith, M. B. and Young, T. F. (1965a). *J. Phys. Chem.* **69**, 1868.
Wu, Y. C., Smith, M. B. and Young, T. F. (1965b). *J. Phys. Chem.* **69**, 1873.
Wu, Y. C., Rush, R. M. and Scatchard, G. (1968). *J. Phys. Chem.* **72**, 4048.
Wu, Y. C., Rush, R. M. and Scatchard, G. (1969). *J. Phys. Chem.* **73**, 2047.
Yeatts, L. B. and Marshall, W. L. (1969). *J. Phys. Chem.* **73**, 81.
Young, T. F. (1951). *Rec. Chem. Progr.* **12**, 81.
Young, T. F. and Smith, M. B. (1954). *J. Phys. Chem.* **58**, 716.
Young, T. F., Wu, Y. C. and Krawetz, A. A. (1957). *Disc. Faraday Soc.* **24**, 37, 77, 80.
Zdanovskii, A. B. (1936). *Tr. Solyanoi Lab. Akad. Nauk SSSR*, Vol. 6.
Zdanovskii, A. B. and Deryabina, L. D. (1965). *Russ. J. Phys. Chem.* **39**, 357, 485, 774.
Zirino, A. and Yamamoto, S. (1972). *Limnol. Oceanogr.* **17**, 661.

Chapter 3

# Chemical Speciation

## WERNER STUMM

*Swiss Federal Institute for Aquatic Science and Water Pollution Control*
*(EAWAG), Zürich, Switzerland*

and

## PHYLLIS A. BRAUNER

*Chemistry Department, Simmons College, Boston, U.S.A.*

## 3.1 INTRODUCTION

When man first began to study the chemistry of the sea, his interest lay primarily in the elemental composition. It has become increasingly obvious that information about this alone is inadequate for the identification of the mechanisms that control the composition of the oceans. The oceanographer needs to know the species in which the element is present in order to gain insight into the role of the elements in the sedimentary cycles, into the physical chemistry of the sea (e.g. the colligative properties, sound absorption, surface properties, conductivity), and into the nature of pollutant interactions as well as into complexities of the biochemical cycle.

The term *species* refers to the actual form in which a molecule or ion is present in solution. For example, iodine in aqueous solution may conceivably exist as one or more of the species: $I_2$, $I^-$, $I_3^-$, $HIO$, $IO^-$, $IO_3^-$ or as an ion pair or complex, or in the form of organic iodo compounds. Figure 3.1 shows the various forms in which metals are thought to occur in sea water. It is operationally difficult to distinguish between dissolved and colloidally dispersed substances. Colloidal metal ion precipitates, such as $Fe(OH)_3(s)$ or $FeOOH(s)$ may occasionally have particle sizes smaller than 100 Å— sufficiently small to pass through a membrane filter. Organic substances can assist markedly in the formation of stable colloidal dispersions. Information on the types of species encountered under different chemical conditions (types of complexes, their stabilities and rates of formation) is a prerequisite to a better understanding of the distribution and functions of trace elements in natural waters.

Although the importance of species distribution has long been recognized, especially for acids (e.g. $H_2CO_3$, $B(OH)_3$, $Si(OH)_4$, $H_2O$, $NH_4^+$, $H_2S$) and their conjugate bases, it is only during the last two decades that general awareness of the relevance of chemical speciation in sea water systems has been aroused. Goldberg (1954) and Krauskopf (1956) were among the first to consider the forms of the reacting metal species. Sillén (1961) further developed the application of equilibrium models to the portrayal of many aspects of the species composition of sea water, and Garrels (1960) demonstrated how chemical relations could be interpreted from mineral equilibria (Garrels and Christ, 1965). Identification of species was considered to be one of the most urgent oceanographic analytical problems at the 1971 24th Summer Symposium on Analytical Chemistry (Carpenter, 1972).

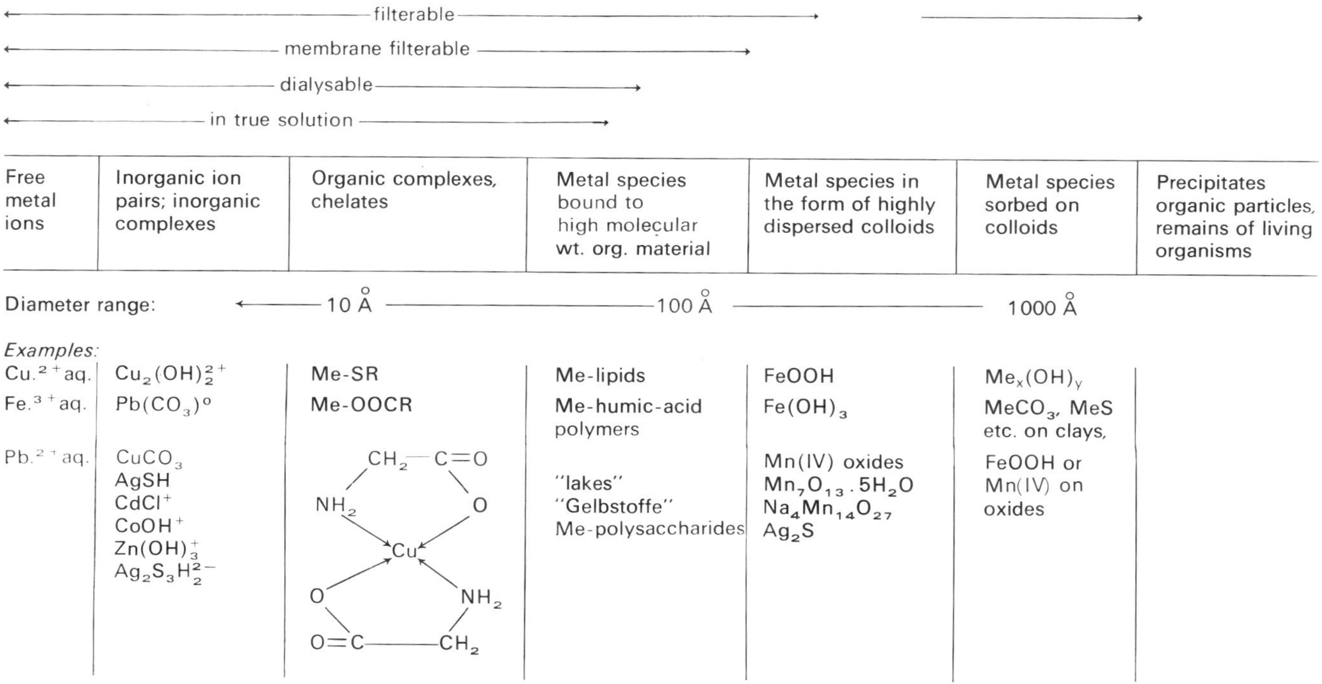

Diameter range: ——— 10 Å ——— 100 Å ——— 1000 Å ———

| Free metal ions | Inorganic ion pairs; inorganic complexes | Organic complexes, chelates | Metal species bound to high molecular wt. org. material | Metal species in the form of highly dispersed colloids | Metal species sorbed on colloids | Precipitates organic particles, remains of living organisms |
|---|---|---|---|---|---|---|
| *Examples:* | | | | | | |
| $Cu^{2+}$ aq. | $Cu_2(OH)_2^{2+}$ | Me-SR | Me-lipids | FeOOH | $Me_x(OH)_y$ | |
| $Fe^{3+}$ aq. | $Pb(CO_3)^0$ | Me-OOCR | Me-humic-acid polymers | $Fe(OH)_3$ | $MeCO_3$, MeS etc. on clays, | |
| $Pb^{2+}$ aq. | $CuCO_3$ | | "lakes" | Mn(IV) oxides | FeOOH or | |
| | AgSH | | "Gelbstoffe" | $Mn_7O_{13} \cdot 5H_2O$ | Mn(IV) on | |
| | $CdCl^+$ | | Me-polysaccharides | $Na_4Mn_{14}O_{27}$ | oxides | |
| | $CoOH^+$ | | | $Ag_2S$ | | |
| | $Zn(OH)_3^+$ | | | | | |
| | $Ag_2S_3H_2^{2-}$ | | | | | |

Top-of-figure ranges (arrows):
— filterable —
— membrane filterable —
— dialysable —
— in true solution —

Organic complexes, chelates (Cu chelate structure):

$$CH_2\!-\!C\!=\!O$$
$$NH_2 \quad O$$
$$Cu$$
$$O\!=\!C \quad CH_2$$
$$O \quad NH_2$$

Fig. 3.1. Forms of occurrence of metal species

## 3.2. CRITERIA FOR SPECIES PREDOMINANCE

Natural aquatic environments show a complexity seldom encountered in the laboratory. Often it is difficult either to detect individual species in sea water by any direct method or to isolate them from the water. Because of these difficulties it is necessary to consider *kinetic and thermodynamic information* together with the *analytical data* in order to infer the physical and chemical forms present, and to distinguish between oxidized vs. reduced, complexed or chelated vs. non-complexed, dissolved vs. colloidal, and monomeric vs. polymeric species.

A system at thermodynamic equilibrium is in its most stable state, i.e. its species are in their most stable distribution. Real systems will tend towards equilibrium but to attain this state they need to surmount activation energy barriers. The magnitude of these often determine the rate and therefore the time scales of the reaction. In ocean waters time scales may range from milliseconds or less to geological periods. Many reactions which appear slow in the laboratory proceed sufficiently fast to reach equilibrium in the ocean. Reactions feasible from a thermodynamic point of view may be kinetically unfavourable. Metastable conditions may prevail when a given substance undergoes a particular reaction only slowly while at the same time participating in other reactions which attain equilibrium. For example, in aerobic waters, the concentrations of organic substances should be negligible at equilibrium; although their oxidative decomposition may be very slow, these compounds often undergo rapid acid-base and complex formation equilibria (Blumer, 1967; Stumm and Morgan, 1970; Dayhoff, 1971).

*Kinetic Considerations.* Although high energy bonds are broken during complex formation reactions, these processes are often very fast. In addition to acid-base and simple precipitation and dissolution reactions, other fast processes include ligand exchanges with alkali earth and $3+$ rare earth cations and with Ag(I) and Hg(II). However, Rh(III), Ir(III), Pt(II, IV), Au(III), Pd(II), Cr(III) and Co(III) tend to form kinetically "inert" complexes (i.e. complexes that form or dissociate only slowly).

Many complexes containing macrocyclic organic molecules (e.g. porphyrin) are also "inert". Among reactions that are "slow" on a time scale relevant to the oceans, are many redox processes (e.g. oxidation of certain metal ions, sulphides, $N_2$, etc. and reduction of sulphate, etc.), various metal oxy or hydroxy ion polymerizations and depolymerizations (e.g. vanadium, aluminium), and precipitation of silicates and carbonates (e.g. dolomite). Some of these reactions can be greatly accelerated by biological catalysis (microbial mediation).

Photosynthesis and the maintenance of life are major causes of non-equilibrium conditions. Above all, many organic compounds, although thermo-dynamically unstable, are long-lived and may participate in complexation, acid-base and redox equilibria (Blumer, 1967). In most marine chemical systems there exist at least localized regions in which equilibrium conditions are closely approached, notwithstanding the fact that spatial and temporal gradients exist throughout the system as a whole. For this reason it is worth-while attempting to predict what the stable components would be for parti-cular environmental conditions (viz. pH, redox potential, pressure and temperature).

### 3.2.1. EQUILIBRIUM CONSTANTS AND ACTIVITY CONVENTIONS

Procedures for finding the *equilibrium distribution of species* are based upon the principle that, at equilibrium, the total free energy of the system is a minimum. This total free energy is the sum of the contributions from each of the constituent chemical species in the system; the contribution of each species depends on its standard free energy of formation, its concentration (activity) and the temperature and pressure of the system. Two approaches can be adopted for determining the equilibrium distribution. Firstly, the equilibrium constant approach—solution of the set of non-linear equations based on the mass law and the mole balance equations. Secondly, minimiza-tion of the Gibbs free energy function bearing in mind the constraints of mole balance requirements.

Tables of free energies and related thermodynamic properties are contained in the compilations by Latimer (1952), the National Bureau of Standards (Wagman *et al.*, 1965, 1966) and Robie and Waldbaum (1968). The most extensive compilation of equilibrium constants is that assembled by Sillén and Martell (1964, 1971). Because one of the "ligands" considered is the electron, this compilation contains, in addition to formation constants of metal ion complexes and solubility products, equilibrium constants for redox reactions, e.g.

$$IO_3^- + 3H_2O + 6e^- = I^- + 6\,OH^- :$$

$$\log K = 26 \cdot 1\,(25°C, I = 0)\,(0 \cdot 26\,V) \qquad (3.1)$$

which means

$$\{I^-\}\,\{OH^-\}^6/\{IO_3^-\}\,\{e^-\}^6 = 10^{26 \cdot 1}; \qquad (3.2)$$

a reaction for which the standard electrode potential on the hydrogen scale (IUPAC convention) is 0·26 V. Feitknecht and Schindler (1963) have critically selected solubility product data for metal oxides and metal hydroxides and metal ion hydrolysis equilibrium.

Three types of constants are in common use:

(1) thermodynamic constants based on activities (rather than concentrations), the activity scale being based on the infinite dilution scale (i.e. $I = 0$),
(2) apparent equilibrium constants expressed as concentration quotients and valid for a medium of given ionic strength,
(3) conditional constants that hold only under specified experimental conditions (e.g. at a given pH).

As pointed out by Sillén (1967), equilibrium constants in the form of concentration quotients are just as thermodynamically valid as the traditional "thermodynamic" constants, the main difference between the two being the choice of activity scale. On the infinitely dilute solution scale the activity coefficient is defined in such a way that the activity coefficient of a species approaches unity as the concentrations of all the solutes approach zero. However, an alternative increasingly used convention is defined such that the activity coefficient of a species becomes unity as its concentration approaches zero in the medium of the given ionic strength (i.e. salinity); activity coefficients of a species on this scale are close to unity as long as its concentration is small in comparison to those of the medium ions. Since the oceanic composition is relatively constant the ionic medium approach is well suited to the needs of marine chemists for the study of the equilibria of minor components (Sillén 1967).

For computational purposes it is possible to use thermodynamic equilibrium constants in conjunction with activities. In order to do this and to relate the constants to concentrations the values of single ion activity coefficients must be known. Alternatively, apparent equilibrium constants valid for the medium of particular interest (or a closely similar one) or constants that have been corrected for the medium under consideration can be used in conjunction with concentrations. Non-thermodynamic assumptions are involved in either case.

### 3.2.2. pH, p$\varepsilon$ AND pX AS MASTER VARIABLES

Proton exchange and electron exchange processes (acid-base and redox processes respectively), are of prime importance in the aqueous environment. The equilibrium redox potential, $E_H$ (hydrogen scale) is an indication of the oxidizing intensity of the system. It is conveniently expressed in terms of the parameter pε, used extensively by Sillén. pε is the negative logarithm of a relative electron activity (cf. pH) viz.

$$p\varepsilon = -\log \{e^-\} = E_H/(2\cdot303RT/F), \quad (3.3)$$

where $(2\cdot303RT/F) = 59\cdot15$ mV at 25°C. Like pH, it is a convenient device for facilitating the application of thermodynamic data to environmental problems. The species present at equilibrium are controlled by pH, pε and $\Sigma pX_i$, where $X_1$, $X_2 \ldots$, are the activities of the various reacting species.

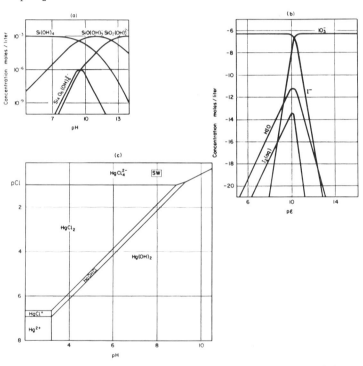

FIG. 3.2. Examples of dependence of species predominance on pH, pε or pCl (1 atm, 25°C). (a) Dissolved silica: $Si(OH)_4$ predominates at pH = 8 ($Si_T = 10^{-3}$ M). (b) Dissolved iodine: pH = 8, $I_T = 5 \times 10^{-7}$ M. Under aerobic conditions (pε = 12·7) the iodate ion is the thermodynamically stable iodine species. (c) Hg(II)—$H_2O$—Cl⁻ : in sea water (SW) $HgCl_4^{2-}$ predominates.

These quantities (pH, pε and pX) are termed Master Variables and can be used in the rapid numerical and graphical representation of equilibrium data even in complicated systems (Bjerrum, 1914; Sillén, 1961; Garrels, 1960; Butler, 1964; Garrels and Christ, 1965; Stumm and Morgan, 1970; Blackburn, 1969; Berner, 1971). Examples are given in Figs. 3.2(a–c). Figs. 3.2a and 3.2b indicate that at equilibrium, orthosilicic acid is the predominant soluble silica species at the pH of sea water (pH = 8), and that at $p\varepsilon = 12.7$ (the value for air-saturated sea water) $IO_3^-$ should be the predominant iodine containing species. The thermodynamic "prediction" regarding the silica system is consistent with experimental observation. However, the simplified picture implied by Fig. 3.2b fails to properly represent the known facts for the iodine system, presumably because of biological activity. Analytically, iodide forms 20–30% of the total iodine present (Riley and Chester, 1971). The simple equilibrium calculation is deficient in that it does not take account of complexes formed between Ag(I), Hg(II), Cd(II), Zn(II) and Cu(II) and iodide ion. In Fig. 3.2c, the predominance area diagram with pH and pCl as variables, suggests in sea water that Hg(II) occurs as $HgCl_4^{2-}$. It also suggests that in brackish waters $HgCl_2$ predominates and that in river waters $HgCl_2$, HgOHCl or $Hg(OH)_2$ species predominate according to the pH and pCl (Anfält *et al.*, 1968).

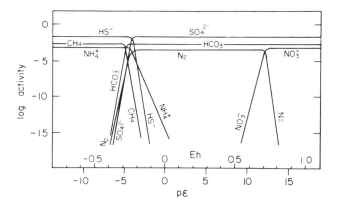

FIG. 3.3. Equilibrium distribution of dissolved C, N and S. Species as a function of $p\varepsilon$ at pH = 7.5, (1 atm, 25°C). $C_T = 3 \times 10^{-3}$ M, $N_T = 10^{-3}$ M, $S_T = 3 \times 10^{-2}$ M. These concentrations approximate to those in sea water (Thorstenson, 1970)

Figure 3.3 gives the equilibrium distribution of those compounds that participate in the biochemical cycles (Thorstenson, 1970) of the oceans. It appears that, with partial exception of reactions involving $N_2(g)$ and C(s), microbial mediation tends to catalyze the redox processes involving C, N, and S species toward redox equilibria. The fact that nitrogen gas is not converted to any great extent into $NO_3^-$ under prevailing aerobic conditions indicates that kinetic factors are important in this system.

### 3.2.2.1. *Activity ratio diagrams*

In order to elucidate the relative proportions of the various possible equilibrium species it is convenient to plot relative activity or concentration ratios (e.g. $[MeX]/[Me]$) as a function of the appropriate master variables ($p\varepsilon$,

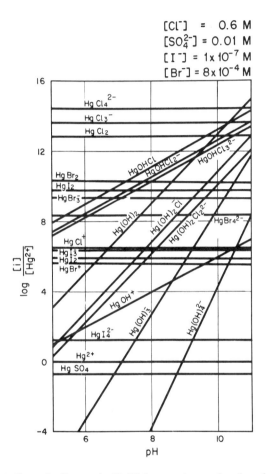

$$[Cl^-] = 0.6 \text{ M}$$
$$[SO_4^{2-}] = 0.01 \text{ M}$$
$$[I^-] = 1 \times 10^{-7} \text{ M}$$
$$[Br^-] = 8 \times 10^{-4} \text{ M}$$

FIG. 3.4. Concentration ratio diagram for Hg(II) in sea water as a function of pH. This diagram gives the equilibrium concentrations of Hg(II)-complexes relative to the concentration of $Hg^{2+}$.

pH or pCl). Such diagrams immediately indicate the prevalent species; furthermore, relative proportions of the various species can be readily derived from these diagrams. Figure 3.4 shows a concentration ratio diagram for

Hg(II). For sea water ($[Cl^-] = 0.6$ M, $[SO_4^{2-}] = 0.01$ M, $[I^-] = 10^{-7}$ M, $[Br^-] = 8 \times 10^{-4}$ M, $25°C, 1$ atm) of pH = 8, the equilibrium model suggests that chloro complexes, especially $HgCl_4^{2-}$ prevail. Other concentration ratio diagrams are given in Figs. 3.15a and 3.15b.

### 3.2.2.2. *Dynamic reaction progress diagrams*

These diagrams show not only the distribution of aqueous species at equilibrium, but also allow comment to be made on the amount of mass transfer resulting from irreversible reactions. Consequently, they are useful for showing the effects of initial conditions, on the reaction path, on the appearance or disappearance of stable or metastable solid phases and on the redistribution of aqueous species (Helgeson, 1968; Thorstenson, 1970; Gardner, 1973).

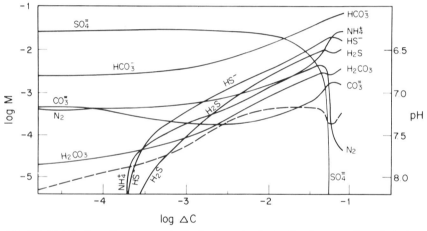

FIG. 3.5. Distribution of predominant dissolved species during the decomposition of alanine ($C_3H_7O_2N$) in sea water at $25°C$, as a function of $\Delta C$ (the number of gram-atoms of organic carbon reacted per litre of solution). The dashed line shows pH (Thorstenson, 1970).

Such a reaction path calculation has been used by Thorstenson (1970) to predict the compositional changes in the aqueous phase as a function of the amount of organic matter decomposed. Figure 3.5 shows the distribution of predominant dissolved species at hypothetical equilibrium as a function of $\Delta C$ (the amount of organic matter—in this example alanine ($C_3H_7O_2N$)— which had decomposed. Berner (1971), Thorstenson (1968), and Whitfield (1971) have shown that the predominant dissolved species in reducing marine environments frequently tend to approach equilibrium.

3.2.3. COMPUTER CALCULATIONS OF EQUILIBRIA

Typically, equilibrium models require the computation of the composition of systems containing numerous species which may be distributed between aqueous, gaseous and solid phases. Although specific numerical programs can be formulated for each individual problem, there are now available general-purpose computer programs for this purpose. A general method based on the minimization of the Gibbs free energy function (White *et al.*, 1958) was applied to large multiphase systems by Shapiro (1964). The equilibrium constant approach (simultaneous solution of the appropriate mass law and mole balance equations) was pioneered by Brinkley (1947). This has found one of its best known applications to aqueous systems through the HALTAFALL program developed by Sillén and his co-workers (Ingri *et al.*, 1967; Dyrssen *et al.*, 1968).

Morel and Morgan (1972) have developed a general-purpose computer program especially well adapted to the study of coordinative interactions and dissolution and precipitation in aqueous systems. The program uses the stability constant approach and the Newton–Raphson method for digital computation of equilibria. It can be used to solve numerous equations simultaneously in order to determine the composition of the aqueous phase and the corresponding set of solids at equilibrium; for example 56 iterations were necessary to compute (within 60 s) at given pH, pε, temperature, pressure and ionic strength, the equilibrium composition involving 20 metals ($+H^+$), 31 ligands ($+OH^-$), 738 complexes, 83 possible solids, and one gas phase component. The program can also handle redox equilibria. As has been shown by many authors, (Perrin, 1967; Helgeson, 1968; Thorstenson, 1970; Stumm and Morgan, 1970; Childs, 1971; Zirino and Yamamoto, 1972; Paces, 1972), computer programs are invaluable for the study of the interdependence of the different chemical constituents of a system. It has been demonstrated by Morel *et al.* (1973—see 3.4.5.4) that metals and ligands in sea water, participate in an interrelated and an interdependent network of chemical interactions.

3.2.4. EXPERIMENTAL METHODS FOR SPECIES IDENTIFICATION

No single method at present available permits an unequivocal identification of a species. Some of the principal methods in use are listed in Table 3.1. Usually, the evidence for a particular form of occurrence is circumstantial and is based on some complementary evaluations together with kinetic and thermodynamic considerations. Much progress has been made in extending our knowledge of the major species in sea water; this will be discussed in a subsequent section. The identification of minor species has

TABLE 3.1

*Methods for assisting in the specific identification of individual species*

| Method and principle | Examples |
|---|---|
| *Physical–mechanical separation* Separation based on size (molecular weight), density, or charge | Membrane filtration, dialysis, electrodialysis, centrifugation, chromatography, gel filtration. |
| *Auxiliary equilibria* A familiar equilibrium system (e.g. a colour forming reaction or an ion exchange system) is introduced to provide indication for the species. | Effect of complex-formation on acid-base equilibrium, adsorption, ion exchange or redox reaction, or solubility equilibrium. Solvent extraction. |
| *Equilibrium potentiometric methods* Evaluation of an electrical potential difference related to the chemical potential (activity) of certain species. | Redox-electrodes. Ion-selective electrodes (metal-, glass-, hydrogen-, solid state- and membrane electrodes). Electrodes of the second kind (e.g. Ag/AgCl). |
| *Electrode kinetics* Interdependence of current, potential, and time for a given electrode process. Depends on the species participating. | Polarography (square-wave-, pulse-, inverse- = anodic stripping), Chrono-potentiometry, chrono-amperometry. |
| *Direct detection of electronic or atomic structure* Measurement of properties based on electronic or atomic structure. | Optical methods (spectrophotometry). Magnetic properties (electron spin resonance). Sound absorption. |
| *Catalytic effects and bioassays* Many species, especially metal ions act as catalysts. Growth (or inhibition) of organisms or rate of enzyme processes depend on species. | Initiation of coordination- or electron transfer reactions. Batch or continuous culture experiments with organisms. Enzymatic reactions. |

proved to be very difficult as such species occur at concentrations smaller than $10^{-6}$ M, in the presence of large excesses of substances that often interfere with specific *in situ* sensing methods. Because of this, investigations using synthetic solutions in which the variables are known and can be controlled may frequently provide valuable clues to the types of species that exist in real sea water. A few examples are given to illustrate the various identification methods.

### 3.2.4.1. Size fractionation

The concept of chemical species as "dissolved" or "particulate" as defined by the pore size of a membrane filter can no longer be considered adequate. Among the size fractionation methods that seem to be particularly promising for characterizing the molecule-size distribution of soluble organic macro-molecular material and of the metal ions associated with it is gel filtration (filtration and elution from columns containing gels of dextran, silica or other molecular sieves). It has been used so far mostly for the fractionation of humic compounds and other coloured components of natural waters (Gjessing, 1967; Christman, 1970; de Haan, 1972).

### 3.2.4.2. Competition between complex formation and adsorption equilibria

Duursma (1970) has reported investigations of the effect of chelate formers in reducing the adsorption of trace metals on sediments. Experimental results were compared with those computed from stability constants. For example, the addition of leucine at a concentration similar to that found in sea water ($2 \cdot 10^{-8}$ M; $1.5 \, \mu g \, l^{-1}$ C) to water of pH = 8.2 ($[Ca^{2+}] = 1.3 \times 10^{-3}$ M, $[Mg^{2+}] = 4.5 \times 10^{-3}$ M) that contained as trace metals Co(II) ($1.5 \times 10^{-8}$ M) and Zn(II) ($1.7 \times 10^{-6}$ M) would lead, according to the calculation, to the formation of a negligible amount of Co-leucine chelates ($7.4 \times 10^{-10}$ M, = 5% of Co(II)) and Zn-leucine chelates ($8 \times 10^{-10}$ M, = 1% of Zn(II)) because 98% of the leucine is bound to $Ca^{2+}$ and $Mg^{2+}$. Duursma demonstrated that there was an excellent agreement between the experimental and calculated results. In order for leucine to produce any effect on the sorption of Zn(II) and Co(II), concentrations of leucine at least $10^4$ times the natural concentrations had to be added.

Inorganic complex formation has also a pronounced effect on adsorption equilibria. Hydroxo, sulphato, carbonato and especially uncharged inorganic complexes, tend to be sorbed much more strongly at an interface than are free metal ions. This enhancement of adsorption by complex formation is caused primarily by the reduction in the solvation energy resulting from the lowering of the ionic charge (James and Healy, 1972).* The observation that many trace metals are sorbed onto suspended materials (Turekian, 1971), or become adsorbed at the walls of the sampling bottle is a strong indication

---

* Adsorption of metal ions on hydrated oxide surfaces or other surfaces containing hydroxo groups can be interpreted as pH-dependent surface complex formation, viz. $- R - OH + Me^+ \rightarrow$ ROMe $+ H^+$. Because both reactions may be similarly pH dependent it is difficult to distinguish between coordinative bending of metal ions and the adsorption of hydroxo metal complexes (see also Chapter 4).

that these metal ions may be present, at least partially, as uncharged inorganic complex species. Partial desorption of trace metals, initially adsorbed onto the detrital load of streams (cf. Turekian, 1971), may result from changes in the distribution of complex species when the stream enters the sea.

### 3.2.4.3. *Equilibrium potentials*

The use of potentiometric methods for the investigation of natural aqueous environments is fraught with practical problems and difficulties of theoretical interpretation which belie the apparent simplicity of the electrochemical technique. The point of zero-applied current is not necessarily the equilibrium potential; hence, the measured electrochemical potential with inert metallic electrodes is not a reliable indication of the $E_h$ level in many sea water systems (Morris and Stumm, 1967). Other potentiometric electrodes may well serve for the potentiometric determination of individual solutes. The glass-electrode and many other metal-, solid state- and membrane electrodes are sufficiently specific for this purpose (Whitfield 1971; see also Chapter 20). However, at the low concentration levels encountered, most of the trace metals and trace constituents cannot establish electrochemical reversible equilibrium with the electrode system. The sulphide electrode is particularly sensitive to $S^{2-}$ (Berner, 1971).

### 3.2.4.4. *Electrode kinetic measurements* (*polarography, inverse polarography*)

Electroanalytical methods lend themselves to a variety of modifications. The kinetics of an electrode process depend on the species involved and this is reflected in the readily measurable interrelationship of faradaic current, electrode potential and time. The discussion will be restricted to polarography and inverse polarography (anodic stripping) in which the morphology of the current–potential curve may reveal the nature of the species present in solution. For electrochemical reactions in which the rate controlling step is the transport to the electrode surface (e.g. dropping Hg-electrode), the half wave potential, $E_{1/2}$, depends on the depolarizer species. Complex formation with a ligand $L'''$ , e.g.

$$M^{n+} + pL^{m-} \rightleftarrows ML_p^{(n-pm)},$$

shifts $E_{1/2}$ to more negative values:

$$\Delta E_{1/2} = -\frac{RT}{nF} \ln \beta_p - \left( p \frac{RT}{nF} \ln L^{m-} \right) \tag{3.4}$$

where $\beta_p$ is the stability constant of the complex $ML_p^{(n-pn)}$ and $n$ in $RT/nF$ is the number of electrons involved in the redox process ($2 \cdot 303 \, RTF^{-1} =$

exist in real sea water. A few examples are given to illustrate the various identification methods.

### 3.2.4.1. *Size fractionation*

The concept of chemical species as "dissolved" or "particulate" as defined by the pore size of a membrane filter can no longer be considered adequate. Among the size fractionation methods that seem to be particularly promising for characterizing the molecule-size distribution of soluble organic macro-molecular material and of the metal ions associated with it is gel filtration (filtration and elution from columns containing gels of dextran, silica or other molecular sieves). It has been used so far mostly for the fractionation of humic compounds and other coloured components of natural waters (Gjessing, 1967; Christman, 1970; de Haan, 1972).

### 3.2.4.2. *Competition between complex formation and adsorption equilibria*

Duursma (1970) has reported investigations of the effect of chelate formers in reducing the adsorption of trace metals on sediments. Experimental results were compared with those computed from stability constants. For example, the addition of leucine at a concentration similar to that found in sea water ($2 \cdot 10^{-8}$ M; $1 \cdot 5 \, \mu g \, l^{-1}$ C) to water of pH = 8·2 ($[Ca^{2+}] = 1 \cdot 3 \times 10^{-3}$ M, $[Mg^{2+}] = 4 \cdot 5 \times 10^{-3}$ M) that contained as trace metals Co(II) ($1 \cdot 5 \times 10^{-8}$ M) and Zn(II) ($1 \cdot 7 \times 10^{-6}$ M) would lead, according to the calculation, to the formation of a negligible amount of Co-leucine chelates ($7 \cdot 4 \times 10^{-10}$ M, = 5% of Co(II)) and Zn-leucine chelates ($8 \times 10^{-10}$ M, = 1% of Zn(II)) because 98% of the leucine is bound to $Ca^{2+}$ and $Mg^{2+}$. Duursma demonstrated that there was an excellent agreement between the experimental and calculated results. In order for leucine to produce any effect on the sorption of Zn(II) and Co(II), concentrations of leucine at least $10^4$ times the natural concentrations had to be added.

Inorganic complex formation has also a pronounced effect on adsorption equilibria. Hydroxo, sulphato, carbonato and especially uncharged inorganic complexes, tend to be sorbed much more strongly at an interface than are free metal ions. This enhancement of adsorption by complex formation is caused primarily by the reduction in the solvation energy resulting from the lowering of the ionic charge (James and Healy, 1972).* The observation that many trace metals are sorbed onto suspended materials (Turekian, 1971), or become adsorbed at the walls of the sampling bottle is a strong indication

* Adsorption of metal ions on hydrated oxide surfaces or other surfaces containing hydroxo groups can be interpreted as pH-dependent surface complex formation, viz. $- R - OH + Me^+ \rightarrow$ ROMe $+ H^+$. Because both reactions may be similarly pH dependent it is difficult to distinguish between coordinative bending of metal ions and the adsorption of hydroxo metal complexes (see also Chapter 4).

that these metal ions may be present, at least partially, as uncharged inorganic complex species. Partial desorption of trace metals, initially adsorbed onto the detrital load of streams (cf. Turekian, 1971), may result from changes in the distribution of complex species when the stream enters the sea.

### 3.2.4.3. *Equilibrium potentials*

The use of potentiometric methods for the investigation of natural aqueous environments is fraught with practical problems and difficulties of theoretical interpretation which belie the apparent simplicity of the electrochemical technique. The point of zero-applied current is not necessarily the equilibrium potential; hence, the measured electrochemical potential with inert metallic electrodes is not a reliable indication of the $E_h$ level in many sea water systems (Morris and Stumm, 1967). Other potentiometric electrodes may well serve for the potentiometric determination of individual solutes. The glass-electrode and many other metal-, solid state- and membrane electrodes are sufficiently specific for this purpose (Whitfield 1971; see also Chapter 20). However, at the low concentration levels encountered, most of the trace metals and trace constituents cannot establish electrochemical reversible equilibrium with the electrode system. The sulphide electrode is particularly sensitive to $S^{2-}$ (Berner, 1971).

### 3.2.4.4. *Electrode kinetic measurements* (*polarography, inverse polarography*)

Electroanalytical methods lend themselves to a variety of modifications. The kinetics of an electrode process depend on the species involved and this is reflected in the readily measurable interrelationship of faradaic current, electrode potential and time. The discussion will be restricted to polarography and inverse polarography (anodic stripping) in which the morphology of the current–potential curve may reveal the nature of the species present in solution. For electrochemical reactions in which the rate controlling step is the transport to the electrode surface (e.g. dropping Hg-electrode), the half wave potential, $E_{1/2}$, depends on the depolarizer species. Complex formation with a ligand $L^{m-}$, e.g.

$$M^{n+} + pL^{m-} \rightleftarrows ML_p^{(n-pm)},$$

shifts $E_{1/2}$ to more negative values:

$$\Delta E_{1/2} = -\frac{RT}{nF} \ln \beta_p - \left( p \frac{RT}{nF} \ln L^{m-} \right) \tag{3.4}$$

where $\beta_p$ is the stability constant of the complex $ML_p^{(n-pm)}$ and $n$ in $RT/nF$ is the number of electrons involved in the redox process ($2 \cdot 303 \, RTF^{-1} =$

0·059 V/eq.) Thus, the complexing ability of suitable metal ions with sea water constituents can be studied polarographically by plotting $\Delta E_{1/2}$ vs. the log of ligand concentrations. Barić and Branica (1967) investigated the effect of changes in pH and pCl in sea water on the polarographic processes of Cd(II) and Zn(II).

Conventional polarography gives reliable data only at concentrations greater than ca. $10^{-5}$ M. In order to circumvent this difficulty, Barić and Branica used zinc and cadmium amalgams and measured the I-E relationship for the oxidation waves. They suggested that zinc in sea water is present as the hydrated $Zn^{2+}$ ion, and possibly, to a smaller extent as $ZnOH^+$. Cadmium on the other hand exists, according to their interpretation, as $CdCl^+$; its hydrolysis does not occur at pH values below 9·5.

*Inverse polarography* (IP) (anodic stripping voltammetry, ASV) square wave and pulse polarography are very sensitive methods of trace metal characterization.* As with normal polarography, several elements may be detected and determined during one analysis. In addition, the useful concentration range can be extended to $10^{-6}$–$10^{-10}$ M. Because the method, in principle, can be used to elucidate the type of metal ion species encountered in the solution, IP is very promising for both identifying and distinguishing the various forms (ionic, complex, particulate) of metal ions actually present in sea water. In IP, two consecutive electrochemical steps are involved: (a) electrolytic separation and concentration of metals to form a deposit or an amalgam on the working electrode, and (b) the dissolution (stripping) of the deposit. The separation step may be done quantitatively or arranged so that a reproducible fraction of the electroactive species is plated out. This can be achieved by controlling the potential and the time of electrolysis and using reproducible mixing conditions in the test solution. The stripping step is usually carried out in an oxygen-free, unstirred solution, most often by applying a linearly increasing potential. The current–potential curve provides the analytical information of interest (Kemula and Kubik, 1958; Shain, 1963; Fitzgerald, 1970; Matson, *et al.*, 1967; Mancy, 1972). Elements which can be characterized by IP include Zn, Cd, Pb, Cu, Bi, Ni and Co. Figure 3.6a (from Mancy, 1972) shows an IP curve for sea water from the Mediterranean. The heights of the current peaks are related to the concentrations of the metal species and the peak potential, $E_p$, is characteristic of the species. Several different types of working electrodes may be used, e.g. hanging mercury drop, thin Hg-films on solid (e.g. graphite) electrodes. Many modifications of inverse polarography are in use. At present, IP has proved most successful with reversible metal ion systems. In such systems the reduction potential, $E_p$, lies near the polarographic half-wave potential $E_{1/2}$. As in

* (see Chapter 20).

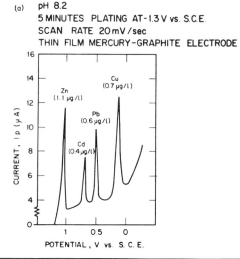

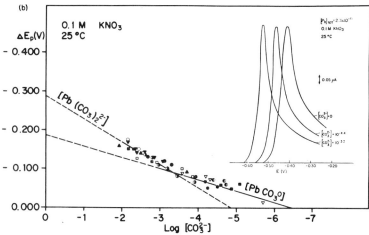

FIG. 3.6. Anodic stripping voltammetry. (a) Typical ASV voltammogram of Mediterranean sea water (Mancy, 1972). (b) Effect of increasing free $CO_3^{2-}$ upon peak potential. Insert: graphical evaluation of complex formation constant from Eq. 3.4 (using $\Delta E_p$ instead of $\Delta E_{1/2}$ where $\Delta E_p = E_p M^{n+} - E_p M^{(n-pn)}$ (from Bilinski and Stumm, 1973).

polarography (Eq. 3.4), the peak potential is shifted under ideal conditions towards a more negative value if the solution contains a complexing ligand. Figure 3.6b, from Bilinski and Stumm (1973), illustrates how the increase in pH shifts the anodic curves of Pb(II) in $HCO_3^-$-solution towards more negative potential values because of the formation of hydroxo and carbonato complexes.

*Experimental difficulties.* The electrochemical irreversibility of a metal complex depolarizer system (electron transfer or chemical steps prior to, or after, the discharge are slow in comparison to the transportation steps) may prevent the metal from being fully deposited in the plating step and thus cause it to remain undetected or to be underestimated in the subsequent stripping step. Often such complexes can be dissociated by acidifying the solution. Results obtained in acidified solutions do not provide the investigator with any information about the species, but differences between the voltammograms of acidified and unacidified samples may provide valuable indications that the metal is present in a "non-reactive" form. However, without additional information it is not possible to interpret differences in peak heights of acidified vs. unacidified samples in terms of (i) "free" vs. complex (or total), (ii) free vs. organically bound or (iii) "labile" vs. non-labile complexes. Nor is it legitimate to infer the presence of complex species from double peaks.

Two factors other than complex formation that may affect the response of the electrode system illustrate the difficulties involved: (1) many trace metals (especially their hydroxo and carbonato complexes) have a strong tendency to become adsorbed on suspended matter and on glass walls; acidification shifts the adsorption equilibria and increases the concentration of metal ion "available" to the electrode. (2) organic substances, even in trace quantities, can exert strong, and often insidious, effects on electrode processes. Such effects are known in polarography. However, because the surfaces of the hanging drop or solid electrodes are not continuously renewed as are those of the dropping electrode in polarography, these adsorption effects are much more pronounced in IP. An adsorbed layer on the electrode surface can act either by hindering the approach of the reacting substance to the electrode surface, or by creating conditions at the interface that are unfavourable either for chemical reactions prior to, or after, the cathodic or anodic discharge, or for the electron transfer itself. Many depolarization reactions that are polarographically reversible become irreversible on stationary electrodes because of surface contamination of the electrode. Other artefacts, (obliterated waves, double peaks etc.) may also result from the specific adsorption of metal complexes at the electrode surface.

Because of these difficulties, anodic stripping voltammetry cannot be used routinely for *in situ* species identification, but the method provides valuable diagnostic clues for the characterization of metals. Matson (1968) and Fitzgerald (1969) have made valuable suggestions concerning the experimental estimation of amounts of metal bound by organic material in natural waters.

### 3.3. Aspects of Coordination Chemistry

Atoms, molecules and ions tend to increase the stability of their outer shell electron configuration by undergoing changes of coordinative relations. Such changes arise if the coordination partner or the coordination number (number of nearest ligand atoms neighbours of a particular atom) is changed. Acid-base, precipitation, and complex formation reactions are all coordination reactions and, hence are phenomenologically and conceptually similar.

*Definitions.* In the treatment which follows, any combination of cations with molecules or anions containing unshared electron pairs (bases) is called coordination (or complex formation) and can be either electrostatic or covalent, or a mixture of both. The metal cation will be called the "central atom", and the anions or molecules with which it forms a coordination compound will be referred to as ligands. If the ligand is composed of several atoms, the one responsible for the basic or nucleophilic nature of the ligand is called the ligand atom. If a base contains more than one ligand atom, and thus can occupy more than one coordination position in the complex, it is referred to as a "multidentate" complex former. Ligands occupying one, two, three, etc. positions are referred to as unidentate, bidentate, tridentate etc. Typical examples are oxalate and ethylenediamine as bidentate ligands, citrate as a tridentate ligand, ethylenediamine tetraacetate (EDTA) as a hexadentate ligand. Complex formation with multidentate ligands is called "chelation", and the complexes are called chelates. The most conspicious feature of a chelate is the formation of a ring. For example, in the reaction between glycine and $Cu.aq^{2+}$, a chelate with two rings, each of five members, is formed, (see Fig. 3.1). Glycine is a bidentate ligand; O and N are the donor atoms. If there is more than one metal atom (central atom) in a complex, we speak about multi- or polynuclear complexes. One also speaks of polycations (e.g. $Hg_2^{2+}$, $Mo_6Cl_8^{4+}$) and of polyanions (e.g., $Si_4O_6(OH)_6^{2-}$, $Mo_7O_{24}^{6-}$).

One essential distinction between a proton complex and a metal complex is that the coordination number of protons is different from that of metal ions. The usual coordination number of the proton is 1 (although in hydrogen bonding, $H^+$ can also exhibit a coordination number of 2). Most metal cations exhibit an even coordination number of 2, 4, 6 and occasionally 8. In complexes of coordination number 2, the ligands and the central ion are linearly arranged. If the coordination number is 4, the ligand atoms surround the central ion either in a square planar or in a tetrahedral configuration. If the coordination number is 6, the ligands occupy the corners of an octahedron, in the centre of which stands the central atom. An example is given in Fig. 3.7a, where the oxygen ligand atoms of trihydroxamic acid surround

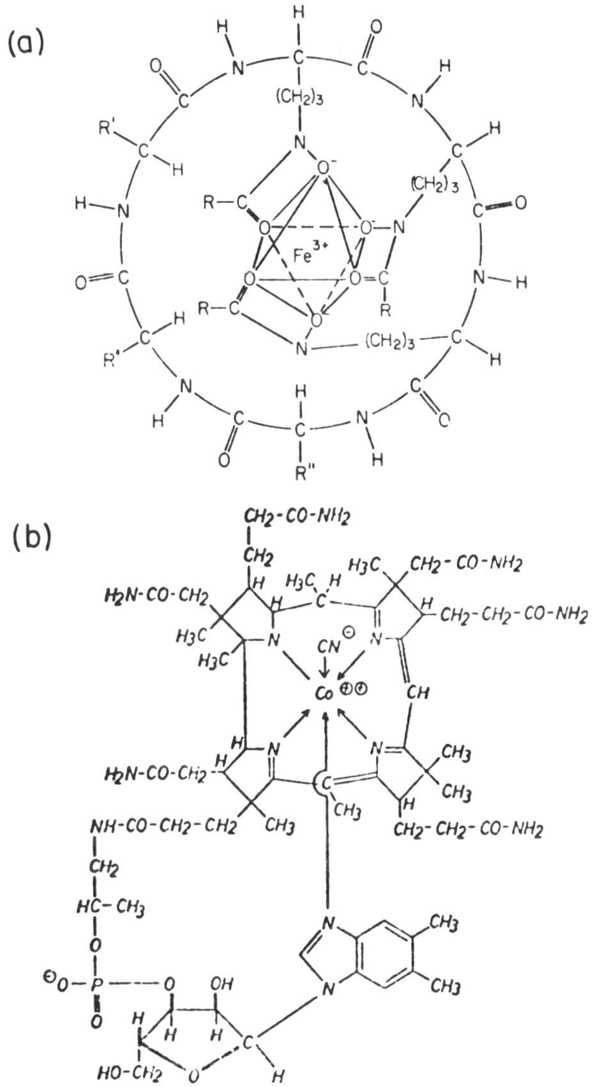

Fig. 3.7. Macrocyclic complex formers. (a) Structure of a ferrichrome (desferri–ferrichrome). One of the strongest complex formers presently known for Fe(III). The iron-binding centre is an octahedral arrangement of six oxygen donor atoms of trihydroxamate. It has been suggested that such naturally occurring ferrichromes play an important role in the biosynthetic pathways involving iron (Neilands, 1964). Reproduced with permission from Birkhäuser Verlag, Basel, Switzerland. (b) The vitamin $B_{12}$ (or cyano cobalamin) as a macrocyclic multidentate complex of cobalt in the porphyrin-resembling part. The cyanide group $CN^-$ can be exchanged for $Cl^-$ or $OH^-$. Vitamin $B_{12}$ is an essential growth factor for several bacteria and auxotrophic phytoplankton.

$Fe^{3+}$ in an octahedral arrangement, thus satisfying all the coordinative requirements of $Fe^{3+}$. The complex depicted in Fig. 3.7a belongs to a group of natural compounds called ferrichromes consisting of heteromeric peptides containing polyhydroxamate as the Fe(III)-binding centre (Neilands, 1964). Complexes of polydentate ligands frequently occur in nature where they are important in enzymatic and biological processes (Anderegg, 1971).

The macrocyclic ligands constitute a very interesting category of complexing agents. There are many natural macrocyclic compounds such as the porphyrins (e.g. vitamin $B_{12}$, a cobalt complex; haem, an iron porphyrin molecule, and chlorophyll, a magnesium complex) and certain antibiotics, but recently synthetic macrocyclic compounds have been prepared (Busch, 1967; Christenson et al., 1971), Fig. 3.7b. These macrocyclic molecules contain central hydrophilic cavities ringed with either electronegative or electropositive ligand atoms and the external framework exhibits hydrophobic behaviour. These substances function as multidentate complexing agents in living matter or at the interface between living matter and water.

*Stability constants.* The term *stable* (or unstable) is applied only in reference to thermodynamic stability (free energy of formation; formation constant). The term *inert* or *robust* is used when a complex is *kinetically* slow in dissociation (or formation). Table 3.2 illustrates the principles of formulating stability constants (= reciprocal of dissociation constants) of metal-ion complexes.

*Mixed complexes.* Additionally, there is the possibility that mixed complexes may form, e.g. BeOHF, HgOHCl, $Fe(OH)_2SO_4^-$, $MgCaCO_3^{2+}$, CuOH, So far, few stability constants for mixed complex formation have been experimentally determined. Often it is possible to estimate a stability constant on the basis of statistical arguments (Dyrssen et al., 1968). For example, the stability constant of the species $Hg(OH)_nCl_m$ can be calculated from the stability of $HgOH_{m+n}$ and $HgCl_{m+n}$ where $m + n \leqslant 4$, by the equation

$$\beta_{mn} = \frac{(m + n)!}{m!\,n!} \left[ \beta_{HgOH_{m+n}}^m \beta_{HgCl_{m+n}}^n \right]^{1/(m+n)} \tag{3.5}$$

Thus, for the formation of HgOHCl, $\beta_{11} = [HgOHCl]/[Hg^{2+}][OH^-][Cl^-]$ is given by

$$\beta_{11} = 2\sqrt{\beta_{Hg(OH)_2}\beta_{HgCl_2}} \tag{3.6}$$

For statistical reasons, such mixed complexes usually predominate only under very restricted conditions (see e.g. Fig. 3.2c).

### 3.3.1. ION PAIRING AND ION ASSOCIATION

In sea water the dissolved ions are in close proximity, the distance of

TABLE 3.2

*Formulation of stability constants†*

---

I. Mononuclear complexes
   (a) Addition of ligand

$$M \xrightarrow[K_1]{L} ML \xrightarrow[K_2]{L} ML_2 \cdots \xrightarrow[K_i]{L} ML_i \cdots \xrightarrow[K_n]{L} ML_n$$

$$\xrightarrow{\quad\quad \beta_2 \quad\quad}$$

$$\xrightarrow{\quad\quad \beta_i \quad\quad\quad\quad}$$

$$\xrightarrow{\quad\quad \beta_n \quad\quad\quad\quad\quad\quad}$$

$$K_i = \frac{[ML_i]}{[ML_{(i-1)}][L]} \tag{1}$$

$$\beta_i = \frac{[ML_i]}{[M][L]^i} \tag{2}$$

   (b) Addition of protonated ligands

$$M \xrightarrow[*K_1]{HL} ML \xrightarrow[*K_2]{HL} ML_2 \cdots \xrightarrow[*K_3]{HL} ML_i \cdots \xrightarrow[*K_n]{HL} ML_n$$

$$\xrightarrow{\quad\quad *\beta_2 \quad\quad}$$

$$\xrightarrow{\quad\quad *\beta_i \quad\quad\quad\quad}$$

$$\xrightarrow{\quad\quad *\beta_n \quad\quad\quad\quad\quad\quad}$$

$$*K_i = \frac{[ML_i][H^+]}{[ML_{(i-1)}][HL]} \tag{3}$$

$$*\beta_i = \frac{[ML_i][H^+]^i}{[M][HL]^i} \tag{4}$$

II. Polynuclear complexes

   In $\beta_{nm}$ and $*\beta_{nm}$ the subscripts $n$ and $m$ denote the composition of the complex $M_m L_n$ formed. [If $m = 1$, the second subscript $(=1)$ is omitted.]

$$\beta_{nm} = \frac{[M_m L_n]}{[M]^m [L]^n} \tag{5}$$

$$*\beta_{nm} = \frac{[M_m L_n][H^+]^n}{[M]^m [HL]^n} \tag{6}$$

---

† Notation as used in Sillén and Martell (1971).

separation being 10 Å or less. Under these conditions the approximations underlying the Debye-Hückel ionic interaction theory are no longer valid, and the mutual attractive energy of oppositely charged ions will be comparable with or greater than the thermal energy which tends to maintain a random distribution. Although there is a continual interchange of ions in concentrated electrolyte solutions, the high ionic concentration and increased interaction potentials lead to the formation of what amounts to a new entity in the solution, the *ion pair*, capable of persisting through a number of collisions with water molecules (Nancollas, 1966). These pairs which may be ions or uncharged molecules (Horne, 1969) speaks of the ion pair as a "liaison but not a marriage") may behave quite differently from the constituent ions.

Solution properties are significantly affected by ion associations, and in order to understand the properties of sea water, information on both the nature of the species and the extent of such associations are needed. Many difficulties are involved in the evaluation of the degree of association between a cation and a ligand even in electrolyte solutions that are less complicated than sea water. Because there is no method by which unequivocal structures of the species present may be obtained, ion association is a phenomeno-logical concept. Ion pair formation is invoked to explain deviations from "normal" behaviour. However, the impossibility of knowing unam-biguously the relevant activity coefficients in sea water, implies that the con-cept of normal behaviour is not clearly defined (see Hindman and Sullivan, 1972, p. 409, see also Chapter 2).

### 3.3.1.1. *Ion pairs and complexes*

Two types of complex species can be distinguished:

1. the metal ion or the ligand or both retain the coordinated water when the complex compound is formed; that is, the metal ion and the base are separated by one or more water molecules,
2. the interacting ligand is immediately adjacent to the metal cation.

The first of these is described as an outer-sphere species or an ion pair; the second type is an inner-sphere species or complex. Hence, in ion pairs, the association is caused mainly by long range electrostatic forces; in com-plexes, short range or covalent forces contribute to the bonding.

During the formation of a true inner-sphere complex, a dehydration step must precede the association reaction. Complex formation is often accom-panied by changes in the absorbance in the visible region, whereas for ion pair formation absorption changes often occur in the u.v. region.

Estimates of stability constants of ion pairs can be made on the basis of simple electrostatic models which consider *coulombic* interactions between the ions. Calculations made in this way indicate the following ranges of stability

constants (25°C) (Stumm and Morgan, 1970):
Ion pairs formed between ions of opposite charge of 1:

$$\log K \simeq 0\text{--}1\,(I = 0); \qquad \log K = -0.5\text{--}0.5\,(\text{SW})$$

Ion pairs formed between ions of opposite charge of 2:

$$\log K \simeq 1.5\text{--}2.2\,(I = 0); \qquad \log K = 0.1\text{--}1.2\,(\text{SW})$$

Ion pairs formed between ions of opposite charge of 3:

$$\log K \simeq 2.8\text{--}4.0\,(I = 0)$$

The range for sea water (SW) was estimated on the basis of assumed single ion activity coefficients for a medium of ionic strength 0·7.

*The $MgSO_4$-system.* An interesting example is given by the interpretation obtained from ultrasonic absorption measurements on 2–2 electrolyte systems. Eigen and Tamm (1962) proposed a three-step process, e.g.

$$\text{Mg}^{2+}\cdot\text{aq.} + \text{SO}_4^{2-}\cdot\text{aq.} \underset{k_{2,1}}{\overset{k_{1,2}}{\rightleftharpoons}} \left[\text{Mg}^{2+}\text{O}\begin{matrix}\text{H H}\\\text{H H}\end{matrix}\text{OSO}_4^{2-}\right]\text{aq.}$$

State 1                          State 2

$$\overset{k_{2,3}}{\underset{k_{3,2}}{\diagup}}$$

$$\text{Mg}^{2+}\text{O}\begin{matrix}\text{H}\\\text{H}\end{matrix}\text{SO}_4^{2-}\ \text{aq.} \underset{k_{4,3}}{\overset{k_{3,4}}{\rightleftharpoons}} [\text{MgSO}_4]\ \text{aq.}$$

State 3                          State 4                          (3.7)

On the basis of this kinetic evidence, three species are supposed to co-exist at equilibrium, although equilibrium measurements do not allow these various species to be distinguished. Conductance and e.m.f. measurements lead to estimates of association constants which are composites for the formation of several species, viz.

$$K^c = \frac{[\text{MgSO}_4] + [\text{Mg(H}_2\text{O})\text{SO}_4] + [\text{Mg(H}_2\text{O})_2\text{SO}_4]}{[\text{Mg}^{2+}][\text{SO}_4^{2-}]} \tag{3.8}$$

Equilibrium constants computed from the kinetic data (Atkinson and Petrucci, 1966) are consistent with those derived from conductivity measurements. Fischer (1967) compared sound absorption data of sea water with those of $MgSO_4$ solutions and calculated that 9% of the total Mg ions are associated as $MgSO_4$ ion pairs in sea water (20°C, 1 atm). These results agree well with those obtained by Garrels and Thompson (1962) and Thompson (1966), and those from Pytkowicz's laboratory (Kester and Pytkowicz, 1968; Pytkowicz and Hawley, 1974; see also Section 2.7.1).

### 3.3.2. METAL IONS AND LIGANDS

In aqueous solutions cations are coordinated with a more or less definite number of water molecules and, in principle at any rate, will interact with a ligand to replace one or more coordination positions in accordance with the mass law, the extent of replacement being governed by the relative stabilities of the aquo and complex species. The barest of the metal cations is the free hydrogen ion, the proton.

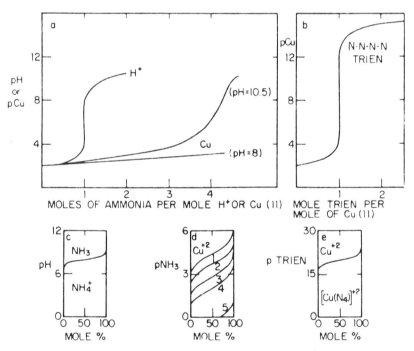

FIG. 3.8. Phenomenological similarity between the "neutralization" of $H^+$ with bases and that of metal ions with complex formers. Titration of $H^+$.aq. and $Cu.aq.^{2+}$ with ammonia (a) and with tetramine (trien) (b). Equilibrium diagrams for the distribution of $NH_3$—$NH_4^+$ (c) of the amino copper (II) complexes (d), and of $Cu^{2+}$, Cu-trien (e). If, four $NH_3$ molecules are packaged together into one single molecule such as trien (triethylenetetramine) a 1:1 Cu-trien complex is formed. The Cu-trien equilibrium (e) is as simple as the $H^+$—$NH_3$ equilibrium (c). From Stumm and Morgan (1970); reproduced with permission from Wiley Interscience.

*Brönsted acidity and Lewis acidity.* Because there is little difference in principle between a free metal ion and a proton, there is a phenomenological and conceptual similarity between the "neutralization" of $H^+$ with bases and that of metal ions with complex formers. In Fig. 3.8, the titration of $H^+$(aq.) with ammonia is compared with that of $Cu.aq.^{2+}$ with ammonia. The bases, molecules or ions which complex $H^+$ or metal ions possess free electron pairs. According to Brönsted, acids are proton donors although Lewis has proposed a much more generalized definition of an acid in the sense that acidity is not attributed to a particular element but to a unique electronic arrangement, i.e. the availability of an empty orbital for the acceptance of an electron pair. Such acid-analogue properties are possessed by $H^+$, metal ions and other Lewis acids (e.g. $SOCl_2$, $AlCl_3$, $SO_2$, $BF_3$). In aqueous solutions, protons and metal ions are in competition with each other for the available bases.

### 3.3.2.1. *Classification of metal ions and ligands*

Inorganic and organic ligands contain the following possible donor atoms in the 4th, 5th, 6th and 7th vertical column of the periodic table:

$$
\begin{array}{cccc}
C & N & O & F \\
 & P & S & Cl \\
 & As & Se & Br \\
 & & Te & I
\end{array}
\tag{3.9}
$$

In water, the halogens are effective complexing agents only as anions, but not if bonded to carbon. For special reasons cyanide ion is a particularly strong complex former. The more important donor atoms include nitrogen, oxygen and sulphur (Anderegg, 1971).

In aqueous solution, preference of a cation for one type of ligand as opposed to another depends on the cation. Ahrland *et al.* (1958) divided metal ions into two categories, depending on whether the metal ions formed their most stable complexes with the first ligand atom of each periodic group (i.e. F, O, N) or with a later member of the group (e.g. I, S, P). As Table 3.3 shows, this classification into A- and B-type metal cations is governed by the number of electrons in the outer shell. Class A metal cations have the inert gas type (d°) electron configuration. These ions may be visualized as being of spherical symmetry; their electron sheaths are not readily deformed under the influence of electric fields, such as those produced by adjacent charged ions. They are, as it were, *hard spheres* whereas Class B metal cations have a more readily deformable electron sheath (higher polarizability) than A-type metals and may be visualized as *soft spheres* ($nd^{10}$ and $nd^{10}(n+1)s^2$ configurations.

TABLE 3.3

*Classification of metal ions*

| A-type metal cations | Transition metal cations | B-type metal cations |
|---|---|---|
| Electron configuration of inert gas. Low polarizability. "Hard spheres" $(H^+)$, $Li^+$, $Na^+$, $K^+$, $Be^{2+}$, $Mg^{2+}$, $Ca^{2+}$, $Sr^{2+}$, $Al^{3+}$, $Sc^{3+}$, $La^{3+}$, $Si^{4+}$, $Ti^{4+}$, $Zr^{4+}$, $Th^{4+}$ | 1–9 outer shell electrons; not spherically symmetrical $V^{2+}$, $Cr^{2+}$, $Mn^{2+}$, $Fe^{2+}$, $Co^{2+}$, $Ni^{2+}$, $Cu^{2+}$, $Ti^{3+}$, $V^{3+}$, $Cr^{3+}$, $Mn^{3+}$, $Fe^{3+}$, $Co^{3+}$ | Electron number corresponds to $Ni^\circ$, $Pd^\circ$ and $Pt^\circ$ (10 or 12 outer shell electrons). Low electronegativity, high polarizability. "Soft spheres" $Cu^+$, $Ag^+$, $Au^+$, $Tl^+$, $Ga^+$, $Zn^{2+}$, $Cd^{2+}$, $Hg^{2+}$, $Pb^{2+}$, $Sn^{2+}$, $Tl^{3+}$, $Au^{3+}$, $In^{3+}$, $Bi^{3+}$ |

According to Pearson's (1968) Hard and Soft Acids

| Hard acids | Borderline | Soft acids |
|---|---|---|
| All A-metal cations plus $Cr^{3+}$, $Mn^{3+}$, $Fe^{3+}$, $Co^{3+}$, $UO^{2+}$, $VO^{2+}$ | All bivalent transition metal cations plus $Zn^{2+}$, $Pb^{2+}$, $Bi^{3+}$ | All B-metal cations minus $Zn^{2+}$, $Pb^{2+}$, $Bi^{3+}$ |
| as well as species like $BF_3$, $BCl_3$, $SO_3$, $RSO_2^+$, $RPO_2^+$, $CO_2$, $RCO^+$, $R_3C^+$ | $SO_2$, $NO^+$, $B(CH_3)_3$ | All metal atoms, bulk metals $I_2$, $Br_2$, $ICN$, $I^+$, $Br^+$ |

Preference for ligand atom

| | | |
|---|---|---|
| $N \gg P$ | | $P \gg N$ |
| $O \gg S$ | | $S \gg O$ |
| $F \gg Cl$ | | $I \gg F$ |

Qualitative generalizations on stability sequence

| Cations | Cations | |
|---|---|---|
| Stability $\simeq$ prop. $\dfrac{charge}{radius}$ | Irving–Williams order $Mn^{2+} < Fe^{2+} < Co^{2+} < Ni^{2+} < Cu^{2+} > Zn^{2+}$ | |
| Ligands $F > O > N$ $= Cl > Br > I > S$ $OH^- > RO^- > RCO_2^-$ $CO_3^{2-} \gg NO_3^-$ $PO_4^{3-} \gg SO_4^{2-} \gg ClO_4^-$ | Ligands $S > I > Br > Cl$ $= N > O > F$ | |

Metal cations in class A form complexes preferentially with the fluoride ion and ligands having oxygen as donor atom. Water is more strongly attracted to these metals than are ammonia or cyanide. No sulphides (precipitates or complexes) are formed by these ions in aqueous solution, since $OH^-$ ions readily displace $HS^-$ or $S^{2-}$. Chloro or iodo complexes are weak and occur most readily in acid solutions under which conditions competition with $OH^-$ is minimal. The univalent alkali ions form only relatively unstable ion pairs with some anions; some weak complexes of $Li^+$ and $Na^+$ with chelating agents, macrocyclic ligands and polyphosphates are known. Chelating agents containing only nitrogen of sulphur as ligand atoms do not coordinate with A-type cations to form complexes of appreciable stability. Class A metal cations tend to form difficultly soluble precipitates with $OH^-$, $CO_3^{2-}$ and $PO_4^{3-}$; no reaction occurs with sulphur and nitrogen donors (addition of $NH_3$, alkali sulphides or alkali cyanides produces solid hydroxides). With class A metals a simple electrostatic picture of the binding of cation and ligand gives a satisfactory first approximation explanation of complex stability. For example, the stability increases rapidly with an increase in charge on the metal ion and those ions with the smallest radii form the most stable complexes. Some stability sequences are indicated in Table 3.3.

In contrast, class B metal ions coordinate preferentially with bases containing I, S, or N as donor atoms. Thus, metal ions in this class may bind ammonia more strongly than water, $CN^-$ in preference to $OH^-$ and form more stable $I^-$ or $Cl^-$ complexes than $F^-$ complexes. These metal cations, as well as transition metal cations (Table 3.3), forms insoluble sulphides and soluble complexes with $S^{2-}$ and $HS^-$. In this group, electrostatic forces do not appear to be of primary importance because neither the charge nor the size of the interacting ions is entirely decisive for the stability sequence. Non-coloured components often yield a coloured compound (charge transfer bands), thus indicating a significant deformation of the electron orbital overlap. Hence, in addition to coulombic forces, types of interactions other than simple electrostatic forces must be considered. These other types of interactions can be interpreted in terms of quantum mechanics, and in a somewhat oversimplified picture the bond is regarded as resulting from the sharing of an electron pair between the central atom and the ligand (*covalent bond*). The tendency toward complex formation increases with the capability of the cation to take up electrons (increasing ionization potential of the metal) and with decreasing electronegativity of the ligand, (increasing tendency of the ligand to donate electrons). In the series F, O, N, Cl, Br, I, S, the electronegativity decreases from left to right, whereas the stability of complexes with B-type cations increases. However, other factors such as steric hindrance and entropy effects distort the picture, and for this reason stability sequences with cations of the B group are often irregular.

*Transition metal cations* have between zero and ten d-electrons ($nd^q$ configuration where $0 < q < 10$) (Table 3.3). For these cations, a reasonably well-established rule on the sequence of complex stability, the *Irving-Williams* (1953) order is valid. According to this rule the stability of complexes increases in the series

$$Mn^{2+} < Fe^{2+} < Co^{2+} < Ni^{2+} < Cu^{2+} > Zn^{2+}.$$

An example is given in Fig. 3.9. As a first approximation it might be argued qualitatively that the electrovalent behaviour of the bivalent transition metal cations remains almost constant (A-type character), but that the non-electrovalent behaviour (B-type character) changes markedly, in going from $3d^5$ to the $3d^{10}$ configuration.

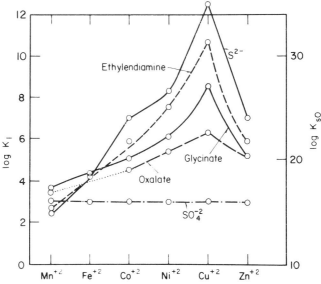

FIG. 3.9. Stability constants of 1:1 complexes of transition metals and solubility products of their sulphides (Irvings–Williams series).

The *Irving-Williams* order is usually explained in terms of ligand field or simple-crystal theory, but other explanations can also account for this order (Schwarzenbach, 1973). For other transition elements, the generalization may be made that the B-type character increases slowly with $q$ and markedly with $n$.

3.3.2.2. *Soft and hard acids and bases*

It is possible to recognize the two following categories of reactions:

1. Reactions of A-type cations with ligands which attach preferentially to A-type ions ($F^- \gg Cl^-$; $OH^- > NH_3$).

2. Reaction of B-type cations with ligands that have a strong tendency to share electrons with B ions ($I^- > Cl^- > F^-$; $NH_3 > OH^- > H_2O$).

In Pearson's concept of soft and hard acids and bases (SHAB concept), reactions listed under (1) are classified as hard acid-hard base interactions, whereas reactions under (2) involving B cations fall into the category of soft acid–soft base interactions. Some useful trends, usable for the qualitative prediction of chemical reactions and relative sequences of compound stability are indicated in Table 3.3. Typically, reactions listed under (1) are characterized by a small $\Delta H$ and a positive and, usually, large $T\Delta S$, whereas those given under (2) above show an appreciable negative $\Delta H$ and a small positive or negative $T\Delta S$ term. That is, the entropy increase is the primary "drive" behind hard acid–hard base reactions whereas the free energy change of soft acid–soft base reactions is dominated by a negative enthalpy change.

The principle of hard and soft acids and bases may also be applied to the rates of nucleophilic and electrophilic substitution reactions. Water is a very hard solvent, with respect to both its acidic and basic functions. It is the ideal solvent for hard acids, hard bases and hard complexes (Pearson, 1968). Modern theories on coordination chemistry are reviewed by Schläfer and Gliemann (1967), Burns and Fyfe (1967), Jørgensen (1963, 1971) and by Cotton and Wilkinson (1972).

### 3.3.3. THE CHELATE EFFECT

Complexes with monodentate ligands are usually less stable than are those with multidentate ligands. More important is the fact that the degree of complex formation (as measured by $\Delta pM$—see below) decreases more markedly with dilution for monodentate complexes than for multidentate complexes (chelates). This is illustrated in Fig. 3.10, in which the dependence of $\Delta pM$ on concentration is shown for uni-, bi- and tetradentate copper (II) amine complexes. In this figure, $\Delta pM \equiv \Delta pCu = \log [Cu_T]/[Cu^{2+}]$). Clearly, the complexing effect of $NH_3$ on $Cu^{2+}$ is negligible under the conditions existing in sea water. However, chelate formation is still appreciable even at great dilution.

Chelates are generally much more stable than the corresponding complexes formed with unidentate ligands. The enthalpy changes associated with the formation of either type of complex are often about the same; ($\Delta H$ for the

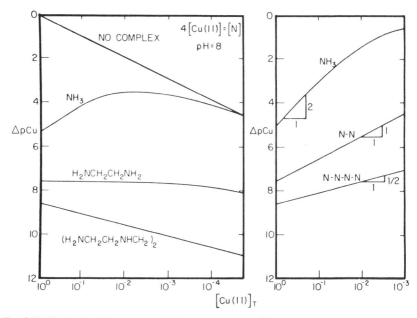

FIG. 3.10. The chelate effect on complex formation of Cu. aq.$^{2+}$ with monodentate, bidentate and tetradentate amines. (a) pCu as a function of concentration. (b) The degree of complex formation ($\Delta$p[Cu] as a function of concentration. $\Delta$p[Cu] is larger with chelate complex formers than with chelate complex formers than with corresponding unidentate ligands. As a result, unidentate complexes are dissociated in dilute solutions whereas chelates remain essentially undissociated at great dilutions. From Stumm and Morgan (1970), with permission from Wiley Interscience.

formation of an ethylenediamine complex is similar to that of an amine complex because a similar type of bond occurs in these complexes). Since the enthalpy changes are similar, the much greater negative free energy change associated with the chelate formation must arise from the more positive entropy change associated with ring formation compared with complexation with a monodentate ligand.

### 3.3.4. PROBLEMS OF SELECTIVITY

Compounds containing F and O donor atoms tend to be non-specific in the type of cations with which they form complexes. However, the soft bases (for example, ligands with N, Cl, Br, I and S) are more selective, the selectivity increasing with decreasing electronegativity of the element. Polyamines and sulphur containing compounds in which the link is formed through sulphur are much more selective for B-type cations and transition metals than are polycarboxylic acids. Although the bond strengths for different ions may differ, inspection of tables of stability constants of metal-ion complexes shows that it is not possible to achieve complete specificity for a single metal ion with synthetic chelating agents. Thus, it is not surprising that although many valuable metals (e.g. $7 \times 10^7$ tons of gold and $7 \times 10^9$ tons of copper) are known to be dissolved in sea water, their commercial extraction with the help of chelating agents is not yet possible (Bayer 1964). However, although simple chelating agents cannot concentrate a single metal exclusively, living organisms often take up metals extremely selectively.

The introduction of bulky substituents into a chelate ligand usually does not aid complex formation and may have an an adverse effect on the kinetics of complex formation. However, specificity can be improved in this way. It is conceivable that protein structures contain enclaves into which only certain metal ions can fit, and that specificity in enzymic catalysis by metals arises in this way (Gurd, 1954). Consistent with this are the observations made by Bayer *et al.* (1964) on the haemocyanin derived from *Octopus vulgaris*. This haemocyanin has a molecular weight of $2 \cdot 7 \times 10^6$. Its composition is that of a protein containing no prosthetic groups, that is, the complex-forming function is part of the molecular protein. Apparently, in addition to the donor atoms (e.g. mercapto groups of the cystein linked by peptide bonds), the sequence of the links and the steric structure must be responsible for the specificity. Bayer has been able to synthesize macromolecular chelating agents possessing high selectivity for rare metal ions.

The ferrichrome depicted in Fig. 3.7 which, it has been suggested, is a cofactor in microbial iron metabolism (Neilands, 1964) is a relatively specific ferric ion chelator. This ligand may serve the double function of dissolving iron and specifically donating it to biosynthetic pathways at the point at which it is required.

## 3.4. Inorganic Complexes and Ion Pairs in Sea Water

Some stability constants for the association of representative cations and ligands are summarised in Table 3.4 which illustrates some of the features of the classification of cations described in Table 3.3. The first nine cations listed are those of A-type metals or are classified as hard acids and tend to form stronger complexes with ligands containing oxygen and $F^-$. At the bottom of the list are B-type metal cations which tend to form chloro and ammonia complexes. These cations form very insoluble sulphides and stable thio complexes. The transition metal cations are intermediate in their association behaviour. The formation constants of hydroxo complexes, especially of class A cations, are large compared with those of most other inorganic ligands. This explains why, with most tri- and tetravalent metal ions, the hydroxides or hydrous oxides are the only stable precipitates in the pH range of natural waters. However, for many bivalent cations, $CO_3^{2-}$, $S^{2-}$, $S_2^{2-}$ and $PO_4^{3-}$ may successfully compete with $OH^-$ ions to satisfy their coordinative requirements.

TABLE 3.4

*Stability constants of inorganic association (expressed by log $K_1$ at 25° C). The format of the Table has been adopted in part from Riley and Chester (1971). Values have been selected with the help of "Stability Constants" (Sillén and Martell 1964, 1971) and are valid in a non-complexing medium of ionic strength similar to that of sea water. A dash indicates that there is no or little evidence for complex formation; + indicates that there is evidence for complex formation. The value given is underlined when a significant part of the cation is present in that form. ~ means order of magnitude of constant. \* indicates that these metals tend to form polynuclear hydroxo complexes—especially in concentrated solutions.*

| Ligand | $F^-$ | $OH^-$ | $CO_3^{2-}$ | $SO_4^{2-}$ | $Cl^-$ | $NH_3$ | $HS^-$ |
|---|---|---|---|---|---|---|---|
| −log concentration of the ligand in sea water | 4·4 | 5·8 | 4·9 | 1·9 | 0·3 | — | — |
| $Be^{2+}$ | <u>5</u> | ~7* | — | 0·7 | ~1 | — | — |
| $Mg^{2+}$ | 1·3 | 1·5 | 2·2 | 1·0 | 0·2 | −0·1 | — |
| $Ca^{2+}$ | 0·6 | 0·8 | 1·9 | 1·0 | — | — | — |
| $Al^{3+}$ | 6·4 | ~9* | — | 1·1 | — | — | — |
| $Sc^{3+}$ | <u>6·2</u> | ~8·6* | — | 2·6 | 1·1 | + | — |
| $Y^{3+}$ | 3·6 | <u>5·4*</u> | — | ~2 | 0·4 | — | — |
| $Th^{4+}$ | <u>7·6</u> | <u>10·5*</u> | — | 3·3 | 0·2 | — | — |
| $UO_2^{2+}$ | 4·5 | <u>8·0</u> | ± | 1·8 | ~0·3 | — | — |
| $Fe^{3+}$ | 5·0 | <u>11</u> | — | 2·3 | 0·5 | — | — |
| $Mn^{2+}$ | — | 3 | — | ~1 | — | — | + |
| $Co^{2+}$ | — | 3·9 | — | 1·2 | ~0 | 2 | + |
| $Cu^{2+}$ | 0·7 | ~7* | <u>5·5</u> | 1·2 | ~0·7 | 4·2 | + |
| $Pb^{2+}$ | 0·3 | <u>~6·8</u> | <u>6·2</u> | | ~0·8 | — | + |
| $Zn^{2+}$ | 0·7 | <u>~4·7</u> | <u>~4</u> | 1·2 | ~0 | ~2 | + |
| $Ag^+$ | −0·2 | 1·8 | — | 0·4 | <u>3·5</u> | 3·6 | 13·3 |
| $Cd^{2+}$ | 0·5 | ~4 | ~4 | 1·2 | <u>1·5</u> | 2·7 | 7·6 |
| $Hg^{2+}$ | 1·0 | 10 | + | 1·3 | <u>6·7</u> | ~8·0 | + |
| $H^+$ | 2·9 | 13·7 | 9·1 | 1·4 | −7 | 9·4 | 6·9 |

The distribution of ion pairs and complexes can readily be calculated if all pertinent equilibria can be identified and equilibrium constants valid for sea water or for solutions of equivalent ionic strength are available.

3.4.1. THE MAJOR SPECIES IN SEA WATER (see also Section 2.7)

In discussing the major specific chemical species in sea water, we will consider first the constituents that consist of, or are combined with, the ligands $H_2O$ (and $OH^-$), $SO_4^{2-}$, $CO_3^{2-}$, $HCO_3^-$, $Cl^-$, $F^-$, $SiO_2(OH)^{2-}$, $B(OH)_4^-$, $PO_4^{3-}$ and the cations $Ca^{2+}$, $Mg^{2+}$, $Na^+$, $K^+$ and $H^+$.

### 3.4.1.1. *Water*

The predominant species in the sea is water. As has already been mentioned, much of this solvent is to some extent associated with the ions and the concentration of "free" water is appreciably less than that of total water. Christenson and Gieskes (1973) have estimated that only about 2·5 moles of water per kg are "free". On an activity scale so defined that the activity of a pure liquid (at 1 atm pressure and at specified temperature) is unity, the activity of $H_2O$ ($a_{H_2O}$) in sea water is given by the ratio of the vapour pressure (strictly, fugacity) of sea water ($p$) to that of pure water ($p^0$) at the same temperature, viz

$$a_{H_2O} = p/p^0 \qquad (3.10)$$

In sea water of 19·4‰ chlorinity at 25°C, the activity of $H_2O$ is 0·981.

Aquo complexes of A-type metal cations are more stable than would be expected on the basis of their electrostatic ion-dipole interaction. This enhanced stability has been accounted for (Jørgensen, 1954) by assuming that interconnecting hydrogen bridges reinforce the first hydration sheath. $F^-$- and O-donors are able to react with these aquo-A-metal cations, but halogen ions, $CN^-$ and $S^{2-}$ cannot compete with, and successfully displace, the $H_2O$ dipoles. For example, $Al(H_2O)_5F^{2+}$ has a formation constant of $\sim 10^5$, but even in concentrated HCl no corresponding chloro complex can be formed; apparently, $F^-$ is able to replace an $H_2O$ of the coordinated water sheath; $Cl^-$ cannot do this, partly for structural reasons, but also because it has little tendency to form hydrogen bridges (Schwarzenbach, 1973). The role of water structure in coordination chemistry has been reviewed by Krindel and Eliezen (1972).

Measurements of partial molar volumes of dissolved electrolytes give information on structural interactions between ions and water and on ion–ion interactions. A dissolved salt appears to cause a similar effect on the water to that produced by a high external pressure (Millero, 1971a, 1972). Duedall and Weyl (1967) have measured the partial molar volumes, $\bar{V}^0$, (at infinite dilution in the ionic medium) for a number of salts in synthetic sea water. Millero (1969) has separated the $\bar{V}^0$ values in sea water and 0·725 M NaCl into their ionic components and analyzed these ionic $\bar{V}^0$ components using a simple model for ion–water interactions. To a first approximation the transfer of an ion to sea water results in a volume change, $\Delta \overline{V}^0$ (trans) that is proportional to $z^2/r$ where $r$ = crystal radius and $z$ is the charge of the ion. The ions of $OH^-$, $HCO_3^-$, $CO_3^{2-}$ and $SO_4^{2-}$, however, give much larger values of $\Delta \overline{V}^0$ (trans). This appears to be the result of ion pairing effects that occur in sea water. When an ion pair is formed, part of the electrostricted water of the individual ions is liberated and the volume of the system is increased (Spiro et al., 1968). This forms the basis of one method for the estimation of the amount of ion pair formation.

The formation of ion pairs in dilute solutions is usually characterized by a positive $\Delta H$ and a positive $\Delta S$. If the process was simply the association of two bare ions (loss of a single particle only), a negative entropy change would be expected. The observed entropy increase results from changes in the co-ordinated water spheres of both partners (breakdown of the structure of the coordinated water molecules) accompanying the ion pair formation. It might be expected that the entropy gain for ion association in sea water would be smaller than that in dilute solutions. For the formation of sulphato complexes with $Na^+$, $Ca^{2+}$ and $Mg^{2+}$ in sea water a decrease in entropy and in enthalpy was observed (Kester and Pytkowicz, 1970).

Comparison of the coefficient of thermal expansion of solutions of salts in sea water with those for solutions in pure water led Millero (1969) to infer that the major difference between the hydration of simple ions (those that do not form contact ion pairs) in sea water and pure water is the disappearance of the disordered structural region in sea water.

### 3.4.2. ION ASSOCIATION WITH $HCO_3^-$, $CO_3^{2-}$ AND $SO_4^{2-}$ (see also Section 2.7)

Because, as yet, the fundamental theory of electrolyte solutions is inadequate to explain the interactions in sea water, *models* have been developed to account for the observed chemical properties. Garrels and Thompson (1962) made the first comprehensive interpretation of the major chemical species present in sea water in terms of such a model (Table 3.5). For the computation they used stability constants (determined in simple electrolytes and corrected or extrapolated to $I = 0$) and estimated activity coefficients of the individual ionic species in sea water. The mean ionic activity coefficients were assumed to be the same as those which would apply to a pure solution of the salt at the same ionic strength as sea water. Gieskes (1966) and others have provided phenomenological support for this assumption.

Since none of the major cations of sea water ($Na,^+ K^+$, $Ca^{2+}$, $Mg^{2+}$) interact significantly with chloride to form ion pairs, synthetic solutions of these chlorides can be used to provide reference solutions in obtaining activity coefficients of the cations. There are two main approaches which

TABLE 3.5

Chemical sea water models

Major species at 25°C (1 atm), chlorinity 19·0‰ (G, H) or 19·375‰ (H + P). (Garrels and Thompson, 1962[G]; Hanor, 1969[H]; and Pytkowicz and Hawley 1973[P+H]).

|  | Na(I) | | | Mg(II) | | | Ca(II) | | | K(I) | | |
|---|---|---|---|---|---|---|---|---|---|---|---|---|
|  | G | H | P + H | G | H | P + H | G | H | P + H | G | H | P + H |
| molality | 0·4752 | 0·4823 | 0·4822 | 0·0540 | 0·05485 | 0·05489 | 0·0104 | 0·01062 | 0·01063 | 0·0100 | 0·01020 | 0·01052 |
| % free ion | 99 | 99·0 | 97·7 | 87 | 89·9 | 89·2 | 91 | 91·5 | 88·5 | 99 | 98·5 | 98·9 |
| % MSO$_4$ | 1·2 | 1·0 | 2·2 | 11 | 9·2 | 10·3 | 8 | 7·6 | 10·8 | 1 | 1·5 | 1·1 |
| % MHCO$_3$ | 0·01 | 0·0 | 0·1 | 1 | 0·6 | 0·1 | 1 | 0·7 | 0·3 | — | 0·0 | — |
| % MCO$_3$ | — | 0·0 | 0·0 | 0·3 | 0·3 | 0·1 | 0·2 | 0·2 | 0·3 | — | 0·0 | — |
| % Mg$_2$CO$_3$ | — | — | — | — | — | 0·0 | — | — | — | — | — | — |
| % MgCaCO$_3$ | — | — | — | — | — | 0·0 | — | — | 0·1 | — | — | — |

|  | SO$_4^{2-}$ | | | HCO$_3^-$ | | | CO$_3^{2-}$ | | | F$^-$ |
|---|---|---|---|---|---|---|---|---|---|---|
|  | G | H | P + H | G | H | P + H | G | H | P + H | P + H |
| molality | 0·0284 | 0·02909 | 0·02906 | 0·00238 | 0·00186 | 0·00213 | 0·000269 | 0·00011 | 0·000171 | 0·000080 |
| % free ion | 54 | 62·9 | 39·0 | 69 | 74·1 | 81·3 | 9 | 10·2 | 8·0 | 51·0 |
| % NaX | 21 | 16·4 | 37·1 | 8 | 8·3 | 10·7 | 17 | 19·4 | 16·0 | — |
| % MgX | 21·5 | 17·4 | 19·5 | 19 | 14·4 | 6·5 | 67 | 63·2 | 43·9 | 47·0 |
| % CaX | 3 | 2·8 | 4·0 | 4 | 3·2 | 4·0 | 7 | 7·1 | 21·0 | 2·0 |
| % KX | 1 | 0·5 | 0·4 | — | 0·0 | 0·4 | — | 0·0 | — | — |
| % Mg$_2$CO$_3$ | — | — | — | — | — | — | — | — | 7·4 | — |
| % MgCaCO$_3$ | — | — | — | — | — | — | — | — | 3·8 | — |

TABLE 3.6

*Stability constants for ion association between the major inorganic ions of sea water.*

| I | HCO₃⁻ (Temp) | HCO₃⁻ Log $K_1$ | HCO₃⁻ Ref. | HCO₃⁻ I | CO₃²⁻ (Temp) | CO₃²⁻ Log $K_1$ | CO₃²⁻ Ref. | CO₃²⁻ I | SO₄²⁻ (Temp) | SO₄²⁻ Log $K_1$ | SO₄²⁻ Ref. | SO₄²⁻ I | F⁻ (Temp) | F⁻ Log $K_1$ | F⁻ Ref. |
|---|---|---|---|---|---|---|---|---|---|---|---|---|---|---|---|
| **Na⁺** →0 | 25° | −0·25 | 62G | →0 | 25° | 1·27 | 61G | →0 | 25° | 0·72 | 50J | →0 | | | |
| →SW | 25° | −0·585 | 62G/P69 | →SW | 25° | 0·62 | 62G/P69 | SW | 25° | 0·34 | 69P | SW | | | |
| 0·5(NaCl) | 25° | −0·41 | 70B | 0·5(NaCl) | 25° | 0·14 | 70B | SW | 2·4° | 0·53 | 70K | | | | |
| 1(NaCl) | 25° | −0·67 | 70B | 1(NaCl) | 25° | 0·27 | 70B | SW | 1·5° | 0·55 | 70K | | | | |
| →0 | 25° | −0·08 | 70B | 3(NaCl) | 25° | 0·37 | 70B | | | | | | | | |
| | | −0·30 | 70B | →0 | | 0·77 | 70B | | | | | | | | |
| 0·72~SW | 25° | −0·55 | 73H | 0·72 | 25° | 0·97 | 73H | | | | | | | | |
| **K⁺** →0 | | | | | | | | | 25° | 0·96 | 50J | →0 | | | |
| →SW | | | | | | | | | 25° | 0·13 | 69P | SW | | | |
| **Mg²⁺** →0 | 25° | 1·16 | 41G | →0 | 25° | 3·40 | 61G | →0 | 25° | 2·36 | 52J | →0 | 07NaCl~SW 25° | 1·27 | 68B |
| →SW | 25° | 0·95 | 63H | →SW | 25° | 2·20 | 61G/P69 | SW | 25° | 1·01 | 68K | SW | | 1·27 | 70E |
| | 25° | 0·72 | 41G/P69 | | | | | | 1·7° | 1·18 | 70K | SW | | | |
| 0·72~SW | 25° | 0·21 | 73H | 0·72~SW | | | | | | | | | | | |
| **Ca²⁺** →0 | 25° | 1·26 | 41G | →0 | 25° | 3·2 | 62G | →0 | 25° | 1·53 | 69D | 0·7NaCl | 25° | 0·62 | 70E |
| →SW | 25° | 0·71 | 41G/P69 | →SW | 25° | 1·89 | 62G/P69 | 0·2/SW | 25° | 1·03 | 69P | SW | | | |
| 0·72~SW | 25° | 0·29 | 73H | 0·72~SW | | | | | | | | | | | |
| **H⁺** →0 | 25° | 6·35 | 67H | →0 | 25° | 10·32 | 67H | →SW25° | 25° | 1·38 | 69D | 1(NaNO₃) | 25° | 2·9 | 68S |
| SW | 5° | 5·89* | 56L | SW | 5° | 9·33* | 56L | SW | | | | | | | |
| | 5° | 5·88* | 73M | | 5° | 9·35* | 73M | | | | | | | | |
| | 25° | 6·00* | 56L | | 25° | 9·09* | 56L | | | | | | | | |
| | 25° | 6·035* | 73M | | 25° | 9·09* | 73M | | | | | | | | |

(1) $I \to 0$, indicates constants extrapolated or corrected to zero ionic strength.

1 (NaCl) indicates ionic strength held constant at the value stated (molality) by addition of salt shown in parenthesis.

$\to SW$ indicates constant extrapolated to be valid for seawater ($35\%_0$ salinity).

$0.72 \sim SW$ indicates constants evaluated at $I$ given in salt mixtures or synthetic sea waters.

* The constants are defined by

$$K_1^1 = [CO_{2T}^*]/a_{H^+}[HCO_{3T}^-], \qquad K_2^1 = [HCO_{3T}^-]/a_{H^+}[CO_{3T}^{2-}]$$

where $a_{H^+}$ is the hydrogen ion activity according to the pH definition of the NBS (National Bureau of Standards); and $[CO_{2T}^*]$ is the total dissolved $CO_2$ (hydrated and unhydrated) and $[HCO_{3T}^-]$ and $[CO_{3T}^{2-}]$ are the concentrations of free $HCO_3^-$ plus the same of the concentrations of the complexes formed between it and the medium ions, (e.g. $[HCO_{3T}^-] + [NaHCO_3^0] + [CaHCO_3^+] + [MgHCO_3^+]$, and free $CO_3^{2-}$ and its complexes with medium ions, respectively. The constants determined by M73 have been corrected by these authors to the same pH-scale as the values of L56.

*References*

41G. Greenwald. I.. (1941). 50J. Jenkins and Monk (1950). 61G. Garrels *et al.* (1961). 62G. Garrels and Thompson (1962). Data of 62G corrected with activity coefficients to sea water conditions by 69P. Pytkowicz and Kester (1969). 62G. Garrels and Thompson (1962). Data of 62G corrected with activity coefficients to sea water conditions by 69P. Hostetler (1963). 67H. Helgeson (1967). 68B. Brewer (1968). 68K. Kester and Pytkowicz (1968). 68S. Srinivasan and Rechnitz (1968). 69D. Dyrssen (1969). 69D. Dyrssen *et al.* (1969). 69P. Pytkowicz and Kester (1969). 70B. Butler and Huston (1970). 70E. Elgqvist (1970). 70K. Kester and Pytkowicz (1970). 73H. Pytkowicz and Hawley (1974). 73M. Mehrbach *et al.* (1973).

have been used to derive the thermodynamic composition of mixed electrolytes or of sea water.

(1) the mean activities of salts (NaCl, KCl etc.) are measured in mixed
electrolytic solutions using ion-selective glass-, membrane- or amalgam
electrodes (e.g. Lanier, 1965; Gieskes, 1966; Thompson, 1966; Platford,
1965a, b; Kester and Pytkowicz, 1968; Pytkowicz and Kester, 1969a, b;
Butler and Huston, 1970; Christenson and Gieskes, 1971, 1973). A decrease
in activity caused by the addition of another electrolyte at given ionic strength
(e.g. a marked decrease of the activity of NaCl consequent on the addition of
carbonate or bicarbonate) may be explained in terms of a model in which
ion pairs ($NaHCO_3$ or $NaCO_3^-$) are formed. Such a model is self-consistent
when the formation constants are relatively independent of composition at
constant ionic strength (Butler and Huston, 1970).

(2) the effects of different ionic media on equilibria (such as carbonate
protonation) can be determined. For example, the acidity constant of
$HCO_3^-$ is increased by the presence of $Mg^{2+}$ ions. The lowering of the pK
value in such solutions and in sea water is caused by the formation of carbonate
ion pairs (Garrels, et al., 1961; Culberson and Pytkowicz, 1968; Butler and
Huston, 1970; Pytkowicz and Hawley, 1974).

In both these procedures it is difficult to separate the effects of ionic
strength and of ion association. Either the ion association is known from
other experiments, or the activity coefficient effect is known from other
experiments. Both involve non-thermodynamic assumptions (Butler and
Huston, 1970).

Table 3.6 gives a summary of stability constants that have been used in
calculating species distribution in sea water. The notation employed in
"Stability Constants" (Sillén and Martell, 1971) has been used. Table 3.7
gives a brief comparison of activity coefficients valid for sea water conditions.
For sea water of salinity 34·8‰, the total ionic strength when formally
calculated from the elemental composition, assuming no ion association, is
0·72; however, because of association, the *effective* ionic strength is 0·67.

In Table 3.7, the values listed under A are "total" activity coefficients
which have been measured in seawater:

$$\delta_A = a_A/[A_{total}] \qquad (3.11)$$

The values listed under B are "free" activity coefficients:

$$\gamma_A = a_A/[A_{free}] \qquad (3.12)$$

The difference between $\delta_A$ and $\gamma_A$ is the distribution coefficient

$$\alpha_A = \frac{[A_{free}]}{[A_{total}]} = \frac{\gamma_A}{\delta_A} \qquad (3.13)$$

which can be explained by the association model. For example, $\alpha_{Na}$ in sea water is given by

$$\alpha_{Na} = \frac{[Na^+_{free}]}{[Na^+_{free}] + [NaHCO_3^0] + [NaCO_3^-] + [NaSO_4^-]} \quad (3.14)$$

$$\alpha_{Na} = (1 + K_{NaHCO_3}[HCO_3^-] + K_{NaCO_3}[CO_3^{2-}] + K_{NaSO_4}[SO_4^{2-}])^{-1} \quad (3.15)$$

The distribution of the species in sea water according to the model suggested by Garrels and Thompson and as calculated on the basis of more recent information by Hanor and by Pytkowicz and Hawley is given in Table 3.5. According to these models, the major cations in sea water are predominantly present as free aquo metal ions. This is understandable because they are A-type metals and the concentration of major cations is much greater than the concentration of associating anions. Hence, a significant fraction of the anions $CO_3^{2-}$, $SO_4^{2-}$, and $HCO_3^-$ are associated with metal ions.

TABLE 3.7

*Comparison of activity coefficients at 25°C and effective ionic strength 0·67 measured in sea water (A), in single-salt solutions (B) and calculated from the association model (C). Taken from Kester and Pytkowicz (1969).*

| Constituent | A | B | C |
|---|---|---|---|
| $Na^+$ | 0·67* | 0·71 | 0·69 |
| $Ca^{2+}$ | 0·20 ± 0·01† | | |
| | 0·22 ± 0·02† | 0·26 | 0·23 |
| $HCO_3^-$ | 0·55 ± 0·01† | | |
| | 0·56 ± 0·01† | 0·68 | 0·48 |
| $CO_3^{2-}$ | 0·021 ± 0·004† | | |
| | 0·024 ± 0·004† | 0·20 | 0·018 |
| NaCl | 0·67 ± 0·01‡ | | |
| | 0·668 ± 0·003§ | 0·67 | 0·66 |
| $Na_2SO_4$ | 0·38 ± 0·03‖ | 0·31 | 0·34 |

* Platford (1965a).   † Berner (1965.   ‡ Platford (1965b).   § Gieskes (1966).   ‖ Platford and Dafoe (1965).

*Note:* Garrels and Thompson (1962) have used the following activity coefficients of individual species in seawater: $Na^+$, 0·76; $K^+$, 0·64; $Mg^{2+}$, 0·36; $Ca^{2+}$, 0·28; $Cl^-$, 0·64; $CO_3^{2-}$, 0·20; $SO_4^{2-}$, 0·12; $NaCO_3^- = NaSO_4^- = MgHCO_3^+$ etc., 0·68; $NaHCO_3^0 = MgCO_3^0 = CaCO_3^0 = MgSO_4^0 = CaSO_4^0$, 1·13. Pytkowicz and Kester (1971) have pointed out that uncharged ion pairs are dipolar ions and they consider an activity coefficient of about 0·8 for $MgSO_4^0$ more appropriate.

The use of recent values of activity coefficients and stability constants has modified views on the distribution of ligands more than of metal ions, but the pertinent qualitative features of the Garrels and Thompson model are still valid. Garrels and Christ (1965) show that the $Mg^{2+}$ and $Na^+$ contents of sea water are nearly as important as $H^+$ in controlling $CaCO_3$ deposition. They point out that, for example, an organism that excludes $Mg^{2+}$ from its cell fluids would tend to cause precipitation of $CaCO_3$ because of the the loss of the association tendency of $Mg^{2+}$ for $CO_3^{2-}$.

Using the stability constants for $HSO_4^-$ from Table 3.6, it is possible to show that only a small fraction of the total sulphate in sea water is in the form of $HSO_4^-$, ($[HSO_4^-] \approx 2 \cdot 7 \times 10^{-9}$ mol $kg^{-1}$); this affects the $SO_4^{2-}$ balance only negligibly but has an appreciable effect on the proton balance.

*3.4.2.1. Effect of Ion Association on pH-buffer capacity.* Wangersky (1972) and others have called attention to the question of the extent to which the whole set of ion-association equilibria influences the control of sea water pH. Because some of the cations present compete with $H^+$ for the basic ligands $HCO_3^-$ and $CO_3^{2-}$, an alkalimetric titration curve (pH vs. equivalent fraction of base added) becomes somewhat flatter; because the buffer capacity

$$\beta = \frac{d[\text{Alk}]}{dpH} \qquad (3.16)$$

is inversely proportional to the slope of this curve (Stumm and Morgan, 1970), the capacity is slightly enhanced by ion association. Here $\beta$ is the buffer capacity for addition of an increment of a strong base (d[Alk]), to a system of constant total $CO_2$ (in any form). The effect of ion association upon buffering can be assessed by comparing the slopes of titration curves of the carbonate system in the presence or absence of associating ions or by numerically calculating the difference in buffer capacity. The enhancement of buffer capacity by ion association is very small (McDuff *et al.*, 1973).

*3.4.2.2. Pressure and Temperature Dependence*

In dilute electrolyte solutions, ion association is usually accompanied by a positive change in entropy. Thus, in dilute solutions, a decrease in the extent of ion association with decreased temperature is to be expected. The data given by Kester and Pytkowicz (1970) for $SO_4^{2-}$ association with $Na^+$, $Mg^{2+}$ and $Ca^{2+}$ (Table 3.6), show that at high I or in sea water the tendency to form $MSO_4$ increases as the temperature decreases. These workers have calculated that a decrease in the water temperature causes a decrease in the free $[SO_4^{2-}]$, a marked increase in $[NaSO_4^-]$, a slight increase in $[MgSO_4^0]$ and no change in $[CaSO_4^0]$.

The effect of pressure on ion association can be estimated from partial molar volume data (Millero, 1971a). Various assumptions are involved in these calculations and until direct measurements are made, the effect of pressure on ion pair formation is uncertain. Pytkowicz and Kester have determined the pressure dependence of the $NaSO_4^-$ association. At $1.5\,°C$ the following dependence of log $K_1$ ($I = 0.72$) was found:

$$\ln K_1 = 1.26 - 0.7 \times 10^{-3} P(\text{atm}) \tag{3.17}$$

TABLE 3.8

*The effect of temperature and pressure on sulphate speciation in sea water*†

| $T(°C)$ | $P(\text{atm})$ | %Free $SO_4^{2-}$ | %$NaSO_4^-$ | %$MgSO_4^0$ | %$CaSO_4^0$ |
|---|---|---|---|---|---|
| 25 | 1 | 39·0 | 38·0 | 19·0 | 4·0 |
| 2 | 1 | 28·0 | 47·0 | 21·0 | 4·0 |
| 2 | 1000 | 39·0 | 32·0 | | |
| | | 42·0‡ | 35·0‡ | 19·0‡ | 4·0‡ |

† Given as % of the total sulphate as given species; results at 25° and 2° (1 atm) are taken from Kester and Pytkowicz (1970)
‡ Estimated theoretically by Millero (1971(b)).

Hence association of $NaSO_4^-$ increases with pressure. Millero (1971b) has estimated theoretically the effect of pressure on $CaSO_4^0$ and $MgSO_4^0$ association (Table 3.8). With the exception of $NaSO_4^-$ at 1000 atmospheres, little is known about the formation of other ion pairs in sea water at high pressures.

### 3.4.3. FLUORIDE, BORATE, SILICATE AND PHOSPHATE

F, B and Si are usually present at concentrations larger than 1 mg per litre and may, therefore, be included among the major components of sea water. Fluorine has a high electronegativity and is the only one of the three bases that can form relatively stable complexes with $Ca^{2+}$ and $Mg^{2+}$. Using the constants given in Table 3.6 the following distribution of total fluoride in sea water can be computed—($[F_T] = 8 \times 10^{-5}\,\text{mol kg}^{-1}$):

$$[F^-] = 4.1 \times 10^{-5}\,M\,(51\%), \qquad [MgF^+] = 3.7 \times 10^{-5}\,M\,(47\%),$$

$$[CaF^+] = 1.6 \times 10^{-6}\,M\,(2\%), \qquad [HF] = 3 \times 10^{-10}\,M,$$

$$[HF_2^-] = 7.2 \times 10^{-14}\,M.$$

Borate ($B(OH)_4^-$) and silicate $H_3SiO_4^-$ form weak complexes with major cations. Santschi and Schindler (1973) have determined the complex formation between $Ca^{2+}$, $Mg^{2+}$ and silicate:

$$M^{2+} + H_3SiO_4^- \rightleftarrows MH_3SiO_4^+; Ca^{2+}, \log K_1 = 0.39;$$

$$Mg^{2+} \log K_1 = 0.63; [25°C, I = 1\,(NaClO_4)] \tag{3.18}$$

$$M^{2+} + 2H_3SiO_4^- \rightleftarrows M(H_3SiO_4)_2; Ca^{2+}, \log \beta_2 = 2.89;$$
$$Mg^{2+} \log \beta_2 = 3.82; [25°C, I = 1\,(NaClO_4)] \qquad (3.19)$$

$$H^+ + H_3SiO_4^- \rightleftarrows H_4SiO_4; \log K = -9.47; [25°C, I = 1\,(NaClO_4)] \quad (3.20)$$

These are quite small stability constants. Because the concentration of major cations is much larger than that of silicate, complexing of cations by silicate is negligible, but complexing of silicate by metal ions may be significant. For the complex formation $M + A \rightleftarrows MA$, the ratio of complex bound to unbound anion A is

$$[MA]/[A] = K_1[M] \qquad (3.21)$$

Because $[M]$ can be as large as $0.5$ or $0.05$, even a small $K_1$ value (e.g. $K_1 = 1$) indicates a marked degree of complexing of A.

*Phosphate.* The same considerations apply to phosphate where $[P] \ll [M]$. By comparing thermodynamic acidity constants of phosphoric acids with apparent constants determined in sea water (Kester and Pytkowicz, 1967), it is possible to distinguish between non-specific interactions (effect of

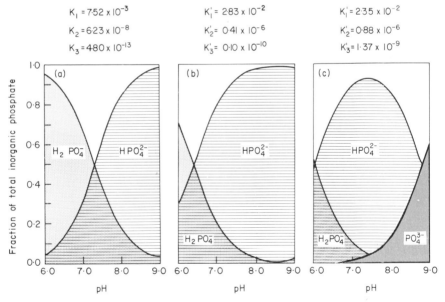

FIG. 3.11. Distribution of phosphate species at 20°C. (A) Pure water. (B) 0.68 M NaCl. (C) Artificial sea water, 33‰ salinity. The shift in the distribution of phosphate species between A and B is due mainly to ionic strength effects; the shift between B and C is caused predominantly by ion association. From Kester and Pytkowicz (1967). Reproduced with permission from Limnology and Oceanography.

ionic strength) and specific interactions, i.e. interactions, such as ion pair and complex formation, which depend on the specific constituents of the medium. This distinction is made very apparent by comparing the differences in distribution of phosphate species in pure water, in 0·68M NaCl, and in artificial sea water (Fig. 3.11, cf. Kester and Pytkowicz, 1967). The shift of the peaks from A to B is caused mainly by ionic strength effects whereas the shift from C to B is due to specific interactions. In sea water of pH = 8, 12% of the inorganic phosphate exists as $PO_{4T}^{3-}$, 87% as $HPO_{4T}^{2-}$ and 1% as $H_2PO_{4T}^{-}$ ($T$ for total). Kester and Pytkowicz estimate that 99·6% of the $PO_4^{3-}$ species and 44% of the $HPO_4^{2-}$ species are complexed with cations other than $Na^+$ in sea water.

### 3.4.4. HYDROXO COMPLEXES

The formation constants of hydroxo complexes, especially of class A cations, are large in comparison to those of most other inorganic ligands. Most tri- and tetravalent cations form insoluble hydroxides or hydrous oxides and are present in sea water as mainly hydroxo complexes. Baran (1972) has recently reviewed the literature on the hydroxide ion as a ligand.

#### 3.4.4.1 The Acidity of the Metal Ions

The term hydrolysis is still used, especially in connection with metal ions, but is no longer necessary to describe the proton transfer from an acid to water or the proton transfer from water to a base. Multivalent metal ions participate in a series of consecutive proton transfers:

$$Fe(H_2O)_6^{3+} \rightarrow Fe(H_2O)_5OH^{2+} + H^+ \rightarrow Fe(H_2O)_4(OH)_2^+ + 2H^+$$
$$\rightarrow Fe(OH)_3(H_2O)_3(s) + 3H^+ \rightarrow Fe(OH)_4(H_2O)_2^- + 4H^+ \quad (3.22)$$

All hydrated ions can, in principle, donate a larger number of protons than that corresponding to their charge and can form anionic hydroxo-metal complexes, but, because of the limited pH range of aqueous solutions, not all elements can exist as anionic hydroxo or oxo complexes.

The acidity of aquo metal ions is expected to increase with decreasing radius and increasing charge of the central ion. Figure 3.12 is an attempt to illustrate how the oxidation state of the central atom determines the predominant species (aquo, hydroxo, hydroxo–oxo and oxo complexes) in the pH range of aqueous solutions.

At the pH value of sea water most cations with $z = +1$ and many cations with $z = +2$ are coordinated solely by $H_2O$ molecules. Most trivalent metal ions are already coordinated with $OH^-$ ions. For $z = +4$ the aquo ions have become too acidic and are out of the accessible pH range of aqueous

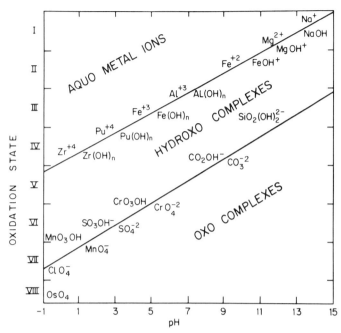

FIG. 3.12. Predominant pH range for the occurrence of aquo, hydroxo, hydroxo-oxo and oxo complexes for various oxidation states. The scheme attempts to show a useful generalization, but many elements cannot be properly placed in this simplified diagram because other factors, such as ionic radius and those related to electron distribution, have to be considered in interpreting the acidity of metal ions. A similar diagram has been given by Jørgensen (1963).

solutions, and $O^{2-}$ appears as a ligand. Thus, for example, for C(IV) and Si(IV) the oxo-hydroxo complexes, $H_2CO_3$ $(=CO(OH)_2)$ and $HCO_3^-$ $(=CO_2(OH)^-)$ are important. With even higher oxidation states of the central atom, hydroxo complexes can only occur at very low pH values. However, the scheme given in Fig. 3.12 is an oversimplification. For each oxidation state, a distribution of acidity according to the ionic radius exists; thus, for $Ba^{2+}$, $Ca^{2+}$, $Mg^{2+}$ and $Be^{2+}$ the acidity, as indicated by the $p^*K_1$ values given in parentheses, increases in the sequence

$$Ba^{2+} (14 \cdot 10), \qquad Ca^{2+} (13 \cdot 3), \qquad Mg^{2+} (12 \cdot 2), \qquad Be^{2+} (5 \cdot 7) \qquad (3.23)$$

The establishment of hydrolysis equilibria is generally rapid, provided that simple species are involved in the hydrolysis reactions. However, some metal ions such as Cr(III), Co(III), Pt(II) and ruthenium form their complexes very slowly. Hydrolysis equilibria can be interpreted in a meaningful way if the solutions are not oversaturated with respect to the solid hydroxide or oxide. Most hydrolysis equilibrium constants have been determined in the presence of a swamping "inert" electrolyte of constant ionic strength ($I = 0.1$, 1 or 3). The formation of hydroxo species can be formulated in terms of acid-base equilibria (see Table 3.2).

The following rules concerning hydrolysis can be recognized—

1 the tendency of metal ion solutions to protolyze (hydrolyze) increases with dilution of metal and with increasing pH,

2. the fraction of polynuclear complexes in a solution decreases on dilution. The first rule can be illustrated by comparing the equilibria

$$Mg^{2+} + H_2O = MgOH^+ + H^+; \qquad \log {}^*K_1 = -11.4;$$
$$I = 0, \quad 25°C \qquad (3.24)$$

$$Cu^{2+} + H_2O = CuOH^+ + H^+; \qquad \log {}^*K_1 = -6.0;$$
$$I = 0, \quad 25°C \qquad (3.25)$$

At great dilution (pH $\approx$ 7), most of the Cu(II) of a pure Cu-salt solution [e.g. $Cu(ClO_4)_2$] will occur as a hydroxo complex

$$\alpha_{CuOH^+} = \frac{[CuOH^+]}{Cu_T} = \left(1 + \frac{[H^+]}{{}^*K_1}\right)^{-1} = 0.91 \qquad (3.26)$$

In sea water, in which $\log {}^*K_1 \approx -6.4$, $\alpha_{CuOH} = 0.88$ and $\alpha_{CuCO_3} = \dfrac{[CuCO_3^0]}{Cu_T}$ is 0.09. On the other hand, because of the low acidity of $Mg^{2+}$, even at infinite dilution the fraction of hydrolyzed $Mg^{2+}$ ions in a solution of a $Mg^{2+}$-salt is very small:

$$\alpha_{MgOH^+} = \frac{[MgOH^+]}{Mg_T} = \left(1 + \frac{[H^+]}{{}^*K_1}\right)^{-1} = 0.0001 \qquad (3.27)$$

In sea water, where $\log {}^*K_1 \approx -11.8$, $\alpha_{MgOH^+} = 1.6 \times 10^{-4}$ and $\alpha_{MgCO_3}$ is $2 \times 10^{-3}$.

Only salt solutions of sufficiently acidic metal ions which fulfil the condition

$$p^*K_1 < 1/2pK_w \qquad \text{or} \qquad p^*\beta_n < (n/2) pK_w \qquad (3.28)$$

(where $^*\beta_n$ is the cumulative acidity constant (see Table 3.2)), undergo substantial hydrolysis upon dilution.

### 3.4.4.2. Polynuclear Hydroxo Complexes

The concept of consecutive stepwise hydroxide complex formation is an oversimplification. Polymeric hydrolysis species (isopolycations) have been reported for most metal ions and metalloids. A sequence of such hydrolytic and condensation reactions, sometimes called olation and oxolation, leads, under conditions of oversaturation with respect to the (usually very insoluble) metal

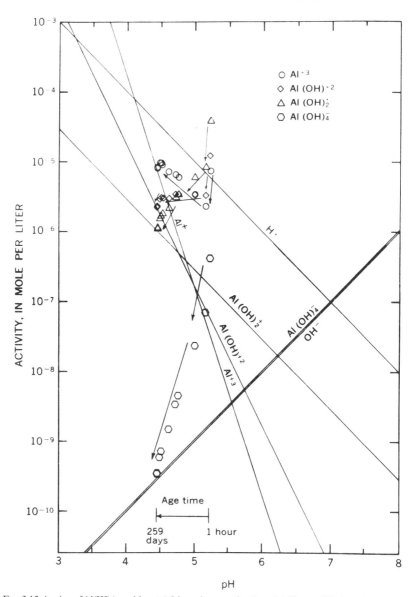

FIG. 3.13. Ageing of Al(III) (perchlorate). Lines characterize the solubility equilibrium of α-Al(OH)$_3$. Experimental points illustrate ageing time effects on activity of monomeric species. These species move toward the equilibrium lines with increasing ageing time. $4.5 \times 10^{-4}$ molar aluminium perchlorate was adjusted to an OH:Al ratio of 2·76. During the ageing, the concentration of monomeric Al(III) species remained relatively constant ($2 \times 10^{-5}$ M). From Smith and Hem (1972) (with permission of U.S. Geological Survey).

hydroxide, to the formation of colloidal hydroxo polymers and ultimately to the production of precipitates. In the pH range lower than the zero point of charge of the metal hydroxide precipitate, positively charged metal hydroxo polymers prevail. In solutions more alkaline than the zero point of charge, anionic hydroxo complexes (isopolyanions) and negatively charged colloids exist. Although multinuclear complexes have been recognized for many years for a few hydrolysis systems such as Cr(III) and Be(II) and for anions of Cr(VI), Si(IV), Mo(VI) and V(V), more recent studies have shown that multinuclear hydrolysis products of metallic cations are of almost universal occurrence in the water solvent system.

Polynuclear complexes are frequently formed rather slowly; they are kinetic intermediates in the slow transition from free metal ions to solid precipitates and are thus thermodynamically unstable. Some metal ion solutions "age", that is, they change their composition over periods of months or years because of slow structural transformations of the iso-polyions. For many metals the polynuclear species are formed only under conditions of oversaturation with respect to the metal hydroxide or metal oxide and are thus not thermodynamically stable, for example,

$$Cu_2(OH)_2^{2+} = Cu^{2+}(aq.) + Cu(OH)_2(s), \quad \Delta G^\circ = -2 \cdot 6 \, \text{kcal mol}^{-1} \quad (3.29)$$

Multinuclear hydrolysis species are not usually observed during dissolution of the most stable modification of the solid hydroxide or oxide; however, they are formed by oversaturating a solution with respect to the solid phase. Such polynuclear species, even if thermodynamically unstable, are of significance in natural water systems, e.g. those polluted with metal bearing wastes. Many multinuclear hydroxo complexes may persist as metastable species for years.

Interesting illustrations are provided, for example, by the hydroxo aluminium system. Smith and Hem (1972) have shown that equilibrium is achieved only very slowly; the time required depends on the formal OH:Al-ratio and on the mode of the preparation of the solution. Figure 3.13 (from Smith and Hem, 1972) depicts the distribution of monomeric species as a function of pH in equilibrium with $\alpha$-Al(OH)$_3$ and the ageing time effect on the activity of monomeric species in a solution incipiently oversaturated with respect to $\alpha$-Al(OH)$_3$. They showed that for each OH:Al ratio the concentration of monomeric species remains constant with ageing time whereas soluble polymeric species decrease and the amount of solid Al(III) (with the structure of gibbsite) increases.

### 3.4.4.3. Mono-nuclear Wall

If hydrolysis leads to mono-nuclear and poly-nuclear hydroxo complexes, it can be shown that mono-nuclear species prevail beyond a certain dilution. If, for example, the dimerization of $CuOH^+$, viz.

$$2CuOH^+ = Cu_2(OH)_2^{2+}; \quad \log *K_{22} = 1\cdot5 \qquad (3.30)$$

is considered, it is apparent from the equilibrium constant that the dimerization is concentration dependent. Thus, for a Cu(II) system where $Cu_T = [Cu^{2+}] + [Cu(OH)^+] + 2[Cu_2(OH)_2^{2+}]$

$$\frac{[Cu_2(OH)_2^{2+}]}{[CuOH^+]^2} = \frac{[Cu_2(OH)_2^{2+}]}{(Cu_T - [Cu^{2+}] - 2[Cu_2(OH)_2^{2+}])^2} = *K_{22} \qquad (3.31)$$

and it is obvious that $[Cu_2(OH)_2^{2+}]$ is dependent upon $[Cu_T]$. For each pH, the mono-nuclear wall (e.g., $Cu_T$ for $[Cu_{dimer}] = 1/100[Cu_{monomer}]$ can be calculated.

### 3.4.5. COORDINATION WITH SULPHIDE

Metal ions that form insoluble sulphides have a tendency to form thio complexes. Experimental evidence for the existence of these complexes and qualitative data on their stability are available for the species $AgSH$, $Ag(SH)_2^-$, $Ag_2S_3H_2^{2-}$, $AuS^-$, $CdSH^+$, $Cd(SH)_2$, $ZnHS_2^-$, $HgS_2^{2-}$, $CH_3HgS^-$, $(CH_3Hg)_2S$, $(CH_3Hg)_3S^+$, $Tl_2SH^+$ and $TlSH$. The stabilities of the mono-nuclear SH complexes of many transition metals have not yet been determined. Figure 3.9 illustrates that the tendency to coordinate with $S^{2-}$ increases from Mn(II) to Co(II); comparison of the sulphide solubility with the tendency to co-

TABLE 3.9

*Thio complex equilibria of $Hg^{2+}$ and $Ag^+$*

|  | $I = 1\,(NaClO_4)\,(20°C)$ | |
|---|---|---|
| $Ag^+ + SH^- = AgSH$ | $\log K_1$ | $13.3$ |
| $AgSH + SH^- = Ag(SH)_2^-$ | $\log K_2$ | $3\cdot9$ |
| $Ag_2S(s) = 2Ag^+ + S^{2-}$ | $\log K_{s0}$ | $-49\cdot7$ |
| $Ag_2S(s) + 2HS^- = Ag_2S_3H_2^{2-}$ | $\log K_s$ | $-4\cdot8$ |
|  | $I = 1\,(KCl)\,(20°C)$ | |
| $Hg^{2+} + 2HS^- = Hg(SH)_2$ | $\log \beta_2 =$ | $37\cdot7$ |
| $Hg(SH)_2 = HgS_2H^- + H^+$ | $\log K_{a1}^1 =$ | $-6\cdot2$ |
| $HgS_2H^- = HgS_2^{2-} + H^+$ | $\log K_{a2}^1 =$ | $-8\cdot3$ |
| $Hg^{2+} + 2S^{2-} = HgS_2^{2-}$ | $\log \beta_2 =$ | $51\cdot5$ |
| $HgS(s) = Hg^{2+} + S^{2-}$ | $\log K_{s0} =$ | $-50\cdot96$ |
| $H_2S = H^+ + HS^-$ | $\log K_{a1} =$ | $-6\cdot88$ |
| $HS^- = H^+ + S^{2-}$ | $\log K_{a2} =$ | $-14\cdot15$ |

* Schwarzenbach and Widmer (1963, 1966).

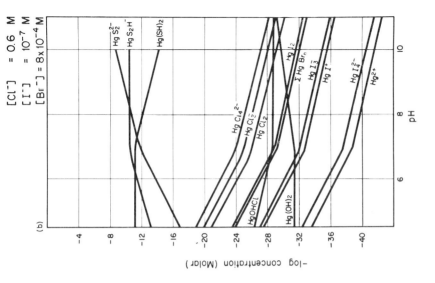

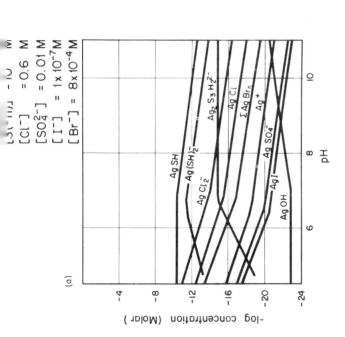

FIG. 3.14. Solubility of $Ag_2S$ and $HgS$ in sea water ($25°C$, 1 atm pressure). (a) Species in equilibrium with solid $Ag_2S$. (b) Species in equilibrium with solid $HgS$.

ordinate with O-groups illustrates why, in nature, Mn and Fe occur primarily in oxygen bearing ores, whereas Co, Ni, Cu, Zn are found as the sulphides. Conceivably, the geochemical deposition of the sulphide ores of Co, Ni, Cu and Zn occurred from solutions containing excess hydrogen sulphide, the metals being in the form of soluble $M(SH)_n^{(2-n)}$ complexes (Schwarzenbach, 1973).

The tendency to form thio complexes is particularly strong with B-type metals (Table 3.9); as Fig. 3.14 shows, traces of sulphide readily displace OH and Cl groups from the coordination sheaths of Ag(I) and Hg(II).

### 3.4.6. COMPLEX FORMATION AND THE SOLUBILITY OF SOLIDS

The solubility of a compound cannot be calculated solely from the solubility product and acid-base equilibria. The solid may, simultaneously, be in equilibrium with complexes of $OH^-$, and various other ligands, including those comprising the lattice constituents. The solubility of a compound is given by the sum of the concentrations of the free metal ion, and of the various metal complexes:

$$[M_T] = [M]_{free} + \Sigma[M_m H_k L_n (OH)_i]$$

where L represents the different ligand types and where all values of $m$, $n$, $i$, or $k \geqq 0$ have to be considered in the summation. For example, the solubilities in marine waters of such "insoluble" substances as $Ag_2S$ ($K_{s0} \approx 10^{-50}$), HgS ($10^{-52}$), FeOOH ($10^{-38}$), CuO ($10^{-20}$), $Al_2O_3$ ($10^{-34}$) are most likely to be determined by the presence of, respectively, AgSH, $HgS_2^{2-}$ or $HgS_2H^-$, $Fe(OH)_2^+$, $CuCO_3$ and $Al(OH)_4^-$.

It has been pointed out by Schwarzenbach and Widmer (1963), that if a compound has a very small solubility product $K_{s0}$, it must be expected that molecular associates between the metal ion and the ligand exist in solution as stable complexes. The presence of covalent bonds is frequently evident when the colour of the solid (pink MnS, yellow AgI, black HgS) is not a composite of the colours of the ions. Thus, the observation by Barton (1959) that those metal sulphides having the smallest solubility products behave as if they were the most soluble is not incompatible with the interpretation that the smaller the solubility product, the stronger is the tendency to form soluble complexes. Figs. 3.14a and 3.14b illustrate that, in the presence of sulphide, the solubility of HgS and $Ag_2S$ is determined by the concentrations of soluble S-complexes; these are many orders of magnitude larger than the concentrations of the free metal ions.

### 3.4.7. INORGANIC SPECIES OF TRACE METALS AND TRACE ELEMENTS

Despite the biochemical importance of many of the trace elements, the chemical forms of these components are seldom known. At concentrations of less than $10^{-6}$ M, analytical techniques usually cannot provide information concerning the speciation. Table 3.10 gives a tentative list of probable dissolved inorganic sea water species. Complexation by organic material needs to be considered in addition to inorganic complex formation.

One may approach the thermodynamics of the problem by extending the chemical model for the equilibrium distribution of the major species—such as the Garrels–Thompson model—to minor components (Morel and Morgan, 1972; Zirino and Yamamoto, 1972). The relative predominance of various possible species can also be elucidated from activity or concentration ratio diagrams from which relative proportions of the various species can be derived. The concentrations of the free ligands, i.e. residual concentrations remaining after complexation with major cations (Table 3.5)—enter into the computation of the complexation equilibria of the trace metals. For inorganic ligands $[L_{free}] \gg [M_{trace}]$ and there is negligible competition in the complexing between individual trace metals. Furthermore, the quantitative degree of complex formation, $\Delta pM$ is independent of the total metal ion concentration and hence inorganic complex formation systems of the trace metals may be considered individually. The concentration ratio diagrams for Zn(II) and Cd(II) (Fig. 3.15) may be compared with that given earlier for Hg(II) (Fig. 3.4). These computations indicate that chloro complexes ($HgCl_4^{2-}$) and ($CdCl_2^0$) are the most stable species in the Hg(II) and Cd(II) systems, whereas $ZnOH^+$ and $Zn(OH)_2^0$ predominate in the Zn(II) system. Determination of the complexing of Cd(II) and Zn(II) is difficult as there is some uncertainty concerning the stability of carbonato complexes; the lines given in Fig. 3.15 have been computed by assuming stability constants of log $K_1(ZnCO_3^0) = 4.1$ and log $K_1$ $(CdCO_3^0) = 4.2$ valid for $I = 0.7$ and 25°C. These are the constants estimated by Zirino and Yamamoto (1972) from studies of the dependence of stability constants on electronegativity and corrected for sea water ionic strength conditions. (Note that the concentration of free $CO_3^{2-}$ at pH = 8 has been taken from Table 3.5 as $1.4 \times 10^{-5}$ M; between pH 7 and 8.5 the line for $[MCO_3]/[M]$ can be constructed by considering that d log $([MCO_3]/[M])/dpH \approx +1$).

By using similar types of computations, one can establish that in sea water, from a thermodynamic point of view, chloro complexes should predominate not only for Hg(II) and Cd(II) but also for Ag ($AgCl_2^-$) and Au ($AuCl_2^-$). Carbonato complexes appear to be the most stable species in the Pb(II) (Fig. 3.16), $UO_2^{2+}$ and, perhaps, Cu(II) systems. Soluble hydroxo-species are

TABLE 3.10

*Probable main dissolved inorganic species in sea-water (aerobic conditions)*

| Element | Probable main species | Element | Probable main species |
|---|---|---|---|
| H | $H_2O$ | Rb | $Rb^+$ |
| Li | $Li^+$ | Sr | $Sr^{2+}$ |
| Be | $BeOH^+, (?)$ | Y | $Y(OH)_n^{3-n}$ |
| B | $H_3BO_3, B(OH)_4^-$ | Zr | $Zr(OH)_n^{4-n} (12)$ |
| C | $HCO_3^-$ | Mo | $MoO_4^{2-}$ |
| N | $N_2, NO_3^-$ | Ag | $AgCl_2^-$ |
| O | $H_2O$ | Cd | $CdCl_2^0 (6,13)$ |
| F | $F^-, MgF^+$ | Sn | $SnO(OH)_3^- (?)$ |
| Na | $Na^+$ | Sb | $Sb(OH)_6^- (?)$ |
| Mg | $Mg^{2+}$ | Te | $HTeO_3$ |
| Al | $Al(OH)_4^-$ | I | $I^-, IO_3^- (14, 15)$ |
| Si | $Si(OH)_4, MgH_3SiO_4^+ (?) (1)$ | Cs | $Cs^+$ |
| | | Ba | $Ba^{2+}$ |
| P | $HPO_4^{2-}, MgPO_4^-$ | La | $La^{3+}$ |
| S | $SO_4^{2-}, NaSO_4^+$ | Ce | $Ce^{3+} (16)$ |
| Cl | $Cl^-$ | Pr | $Pr^{3+}$ |
| K | $K^+$ | Nd | $Nd^{3+}, NdOH^{2+}$ |
| Ca | $Ca^{2+}$ | other | |
| Sc | $Sc(OH)_3^0$ | rare | $Me^{3+}, MeOH^{2+} (16, 17, 18)$ |
| Ti | ? | earths | |
| V | $H_2VO_4^-, HVO_4^{2-} (2)$ | Lu | $LuOH^{2+}$ |
| Cr | $Cr(OH)_3^0, CrO_4^{2-} (3, 4, 5)$ | W | $WO_4^{2-}$ |
| Mn | $Mn^{2+}, MnCl^+$ | | |
| Fe | $Fe(OH)_2^+$ | Re | $ReO_4^-$ |
| Co | $Co^{2+}, CoCO_3^0 (?)$ | Au | $AuCl_2^- (19)$ |
| Ni | $Ni^{2+}, NiCO_3^0 (?)$ | Hg | $HgCl_4^{2-} (20)$ |
| Cu | $CuCO_3^0, CuOH^+ (6)$ | Tl | ? |
| Zn | $ZnOH^+, Zn^{2+}, ZnCO_3^0 (6-9)$ | Pb | $PbCO_3^0, Pb(CO_3)_2^{2-} (6. 21)$ |
| Ga | $Ga(OH)_4^-$ | Bi | $BiO^+, Bi(OH)_2^+, Bi_6(OH)_{12}^{6+} ?(22)$ |
| Ge | $GeO(OH)_3^-, GeO_2(OH)_2^{2-} (?)$ | Ra | $Ra^{2+}$ |
| As | $HAsO_4^{2-}$ | Th | $Th(OH)_n^{4-n} Th(CO_3)_n^{4-2n} (?)$ |
| Se | $SeO_3^{2-} (10, 11)$ | U | $UO_2(CO_3)_3^{4-} (23)$ |
| Br | $Br^-$ | | |

References: (1) Santschi and Schindler (1973); (2) Kolk (1963); (3) Curl et al. (1965); (4) Fukai (1969); (5) Elderfield (1970); (6) Zirino and Yamamato (1972); (7) Hood (1966); (8) Barić and Branica (1967); (9) Zirino and Healy (1970); (10) Tischendorf and Ungethüm (1964); (11) Chau and Riley (1965); (12) Branica et al (1969); (13) Barić and Branica (1967); (14) Sugarawa and Terada (1958); (15) Johannesson (1958); (16) Carpenter and Grant (1967); (17) Høgdahl et al. (1968); (18) Popov (1970); (19) Peshchevitskiy et al. (1965); (20) Anfält et al. (1968); (21) Stumm and Bilinski (1972); (22) Vinogradov (1967); (23) Takai and Yamabe (1971).

prevalent species in the systems: Be(II) $(Be_3(OH)_3^{3+}(?))$; Al(Al(OH)$_4^-$); Sc(III)-
$(Sc(OH)_3^0)$ and some rare earth ions $(Ln(OH)_3^0)$; Fe(III) $(Fe(OH)_2^+, Fe(OH)_3^0(?))$;
$Zn(ZnOH^+)$; Th(IV). A few trace elements occur in an inorganic sea water
medium as "free" aquo ions: $Rb^+$, $Cs^+$, $Ba^{2+}$, $Ra^{2+}$, some of the rare
earth ions and probably $Mn^{2+}$, $Ni^{2+}$ and $Co^{2+}$ (see Table 3.12). Under

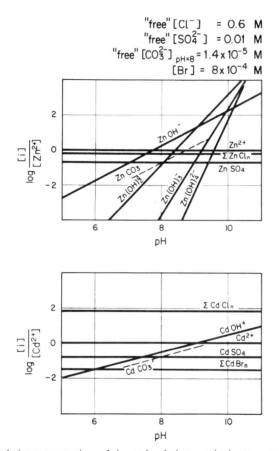

FIG. 3.15. Relative concentrations of zinc and cadmium species in sea water. (25°C, 1 atm).

anaerobic conditions in the presence of S(-II), thio complexes become the
thermodynamically most stable species for all B-type cations and most
transition elements.

Clearly, the conclusions reached above, depend on the values of the
equilibrium constants used in the computations; for some of the species
here are large variations between values reported by various authors and

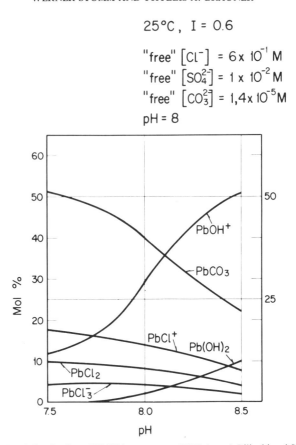

$25°C, \ I = 0.6$

"free" $[Cl^-] = 6 \times 10^{-1}$ M
"free" $[SO_4^{2-}] = 1 \times 10^{-2}$ M
"free" $[CO_3^{2-}] = 1,4 \times 10^{-5}$ M
pH = 8

FIG. 3.16. Calculated distribution of Pb(II) in sea water. (25°C, 1 atm). Bilinski and Stumm (1973).

for some of the species no thermodynamic data exist. Although calculated equilibrium species distributions remain tentative and are subject to corrections when better data become available, these calculations constitute a useful general framework that permits a better understanding of the observed behaviour and functions of trace metals and of how their chemical forms are influenced by solution variables.

## 3.5. ORGANIC COMPLEXES IN SEA WATER

The history of chemical oceanography has been primarily a history of inorganic and analytical chemistry. However, sea water is not simply a solution of inorganic material (Richards, 1970), and although it contains only very small amounts of organic compounds (Table 3.11), these exert a pronounced influence on the inorganic constitutents.

Metal ions in sea water often appear not to behave as predicted from the known chemistry of the metals in question. Frequently sea water shows an apparent supersaturation with respect to many inorganic compounds. A substantial proportion of the metal species present are not dialysable or extractable with reagents known to be suitable for simpler aqueous systems.

TABLE 3.11

*Dissolved organic compounds in the Northeast Pacific Ocean. From Williams (1971). Reproduced with permission from M. Dekker, Inc.*

| | | $\mu g\, C\, l^{-1}$ (mean) | |
|---|---|---|---|
| | Depth | 0–300 m | 300–3000 m |
| Total organic carbon | | 1000 | 500 |
| Amino acids (free and combined) | | 25 | 25 |
| Sugars (free) | | 10 | 10 |
| Fatty acids (free and combined) | | 40 | 10 |
| Urea (free) | | 20 | $<2$ |
| Aromatics (substituted phenols) | | 1 | – |
| Vitamins ($B_{12}$, $B_1$, biotin) | | $10^{-2}$ | $10^{-2}$ |
| | | $\Sigma \approx 100$ | 50 |
| | % Identified of total $\approx$ | 10 | 10 |

The addition of organic substances to water or their exudation by algae may increase or decrease biological activity, presumably because of increased or decreased rates of metal ion uptake by organisms. Conversely, the reactions of the organic substances present in sea water are influenced by the presence of metal ions. Complex formation frequently renders the organic molecule more reactive toward nucleophilic reagents. Obviously, interactions between inorganic and organic substances must account for at least some of the observed phenomena enumerated above. In this section the evidence for organic complex-formation or chelation in sea water is considered and an examination is made of whether knowledge of the degree of combination of organic compounds with metal ions can resolve some of the discrepancies between observed and predicted behaviour. A recent symposium on "Organic Matter in Natural Waters" (Hood, 1970) contains information concerning the distribution of dissolved and particulate organic matter and on inorganic–organic associations, as well as a summary of the circumstantial evidence found by many investigators for the apparent occurrence of metal organic complexes in sea water.

### 3.5.1. THE COMPLEX FORMING PROPERTIES OF ORGANIC MATTER

In sea water, the concentration of dissolved organic carbon is typically 0·5 to 1 mg $Cl^{-1}$ (0·4–0·8 × $10^{-5}$ M), or even less; in contrast the total concentration of inorganic material is some $10^5$ times greater. Concentrations of dissolved organic carbon may be higher for waters over the continental shelf and especially so for regions where nutrients are returned to the surface by up-welling. Sediments are the principal depositories of posthumous organic debris, and in interstitial waters dissolved organic matter may reach concentrations of 100 to 150 mg $Cl^{-1}$ (Brown et al., 1972). The term "dissolved" organic matter is generally considered to comprise that which passes a 0·45 μm pore membrane filter; this material therefore includes colloidal as well as truly dissolved material.

Table 3.11 (from Williams, 1971) illustrates that relatively little is known concerning the dissolved organic carbon. Fatty acids, amino acids and sugars have been identified. Duursma (1965) and Williams (1971) have reviewed the information available (see also Chapter 12). As Williams (1971) points out, identification of the molecular nature of dissolved oceanic organic matter (especially that of the deep sea) is a problem, similar to, but more difficult than, the identification of "humus" in soils. As Table 3.11 shows, only about 10% of the total dissolved organic carbon in surface and subsurface waters has been identified, leaving some 90% unidentified, some of which has been termed "Gelbstoffe".

The *interstitial water* of sediments of Saanich Inlet (British Columbia), a reducing fjord (Brooks et al., 1968; Brown et al., 1972; Presley et al., 1972) contains relatively high concentrations of organic matter (30–800 mg $Cl^{-1}$) containing polymerized sugars and amino acids; this high-molecular weight material failed to pass through a dialysis bag. According to Brown et al. (1972) the relatively high concentration of polymerized sugars and amino acids in interstitial water suggests that the pathway for the transfer of soluble oxygenated and amino-rich groups is through complexing (i.e. condensation and polymerization reactions) in solution. As the high molecular-weight components grow in complexity, a point is reached at which, because of the decreased number of hydrophilic groups, the organic polymers can no longer remain soluble, and become converted into *fulvic acids* and precipitate eventually as *humic acids*. Humic acids appear to form largely from plankton-derived material and not from terrigenous plant or soil sources (Nissenbaum et al., 1971, 1972).

In most particulate organic matter found in sea water, fatty acids, amino acids (upon hydrolysis) and sugars have been identified (see Chapter 13). Marine organisms excrete organic compounds into the water. Most of these substances are polymeric and often contain both polypeptides and poly-saccharides (Fogg, 1966; Hellebust, 1970) and the total concentration of amino acids may reach $100 \mu g \, l^{-1}$. Although some exudation products of biota may have special steric arrangements of donor atoms which make them relatively selective toward individual trace metals (see for example Fig. 3.7a), most organic matter may not be present in the form of selective complex formers; the organic functional groups compete for association with inorganic cations and protons; cations and protons can satisfy their coordinative requirements with inorganic anions including $OH^-$ as well as with organic ligands. Most organic ligands in sea water appear to have, at best, complex forming tendencies similar to acetate, citrate, amino acids, phthalate, salicylate and quinoline-carboxylate. An inspection of "Stability Constants" shows that these classes of compounds are able to form moderately stable complexes with most multivalent cations. Ethylenediaminetetraacetate and nitrilotri-acetate, both of which are able to form particularly stable 1:1 complexes, are probably not representative models for the complex forming tendency of the organic matter normally encountered in sea water.

### 3.5.2. SOLUBLE, POLYMERIC AND COLLOIDAL COMPLEXES

When the coordination of metal ions by organic matter is discussed it is convenient to distinguish the following three categories of organic material
(1) low molecular weight organic substances present in true solution,
(2) polymeric organic substances which contain a sufficient number of hydrophilic functional groups, ($-COO^-$, $-NH_2$, $-R_2NH$, $-RS^-$, $ROH$, $RO^-$) to allow them to remain in solution despite their molecular size,
(3) colloidal organic material, either high-molecular weight compounds or organic substances, sorbed or chemically bound to inorganic colloids.

In sea water very few individual organic substances will be present at concentrations as high as $10^{-6} M$. An unknown, but probably significant, fraction of organic material is present in macro-molecular or colloidal form. Polypeptides, certain lipids, polysaccharides, and many of the substances classified as humic acids, fulvic acids and Gelbstoffe belong to category (2).

Despite the high electrolyte content of sea water, most colloids present are resistant to agglomeration because they behave as hydrophilic colloids. This implies that many of them consist of inorganic colloids which are coated with "colloid-protective" organic substances (Hahn and Stumm, 1970).

All three categories of organic material may interact chemically with, and play important roles in, the transformation of metal ions. Coordination of metal ions at the interfaces of colloids (like ion exchange at other solid surfaces) is not readily amenable to generalized quantitative description and treatment. Polymeric bases combine coordinatively with metal ions; even alkali metal ions can be quite strongly bonded to functional groups. Very few metal complex equilibria involving macromolecules have been studied, so far; furthermore, the procedures used for the evaluation of such equilibria are varied and differ from those used for small organic ligands. The tendency of a given metal ion to interact with a particular ligand group of the macromolecule varies with the electrostatic free energy needed to change the electric charge of the macromolecule; this free energy is, itself, a function of all the ionic equilibria in which the various types of ligand groups in the macromolecule may take part (Sillén and Martell, 1971). Even for reasonably defined polyelectrolytes such as polyacrylic acids or casein, semi-empirical expressions must be used to account for the variation of complex formation tendency with pH, because the acidity constant of these molecules depends on the macromolecular pH-dependent charge, the constant can be interpreted as being composed of two factors; an intrinsic constant (independent of the degree of neutralization) and one which represents the electrostatic interaction (Gregor *et al.*, 1955; Tanford, 1961). Silten and Martell (1971) give information concerning a few macromolecules. More recently, Jellinek and Sangal (1972) and Jellinek and Chen (in press) have characterized the complex formation of polygalacturonic and alginic acids with Zn, Cd, Cu, Ni and Cr(III). The literature abounds with empirical evaluations of the complex forming tendency of humic (or fulvic) acids; however, much of this information cannot be generalized, because the acid-base characteristics of the ligand are not yet fully understood and because metal hydrolysis is not considered. Furthermore, account should be taken of whether the system is in true solution or in a colloidal state. The formation of organic complexes with boric acid (as boron-polyhydroxy complexes) and silicic acid have been suggested, and relatively high concentrations of Si in interstitial organic polymer has been found (Nissenbaum *et al.*. 1971). but the complexes have not been isolated. Williams and Strack (1966) have shown that boron complexes are not significant in sea water.

3.5.3. A HYPOTHETICAL EXPERIMENT: TITRATION OF INORGANIC SEA WATER
        WITH ORGANIC MATERIAL

In this section an attempt is made to gain some insight into the interaction of common organic material with metal ions. This is done by computing the effects of increments of organic matter added to an inorganic sea water

TABLE 3.12

*Equilibrium model: effect of complex formation on distribution of metals*
(All concentrations are given as −log (mol⁻¹)). (Charges of species are omitted)

**Inorganic sea water**, pH = 8.0, 25°C — free ligands: $pSO_4$ 1.95; $pHCO_3$ 2.76; $pCO_3$ 4.86; $pCl$ 0.25

**Inorganic sea water plus soluble organic matter (2.3 mg C l⁻¹)[1]**, pH = 8.0, 25° — $pSO_4$ 1.95; $pHCO_3$ 2.76; $pCO_3$ 4.86; $pCl$ 0.25. Organic complexes with ligands[2] (free ligand concentrations shown in column headers).

| M | $M_T$[3] | free M | Major species | | free M | Major inorg. species | | Acet. 5.21 | Citr. 14.7 | Tartr. 5.41 | Glyc. 6.96 | Glut. 6.89 | Phthal. 5.2 |
|---|---|---|---|---|---|---|---|---|---|---|---|---|---|
| Ca | 1.97 | 2.03 | $CaSO_4$ 2.94 | $CaCO_3$ 3.50 | 2.03 | $CaSO_4$ 2.94 | $CaCO_3$ 3.50 | 7.41 | 5.90 | 6.41 | 9.06 | 8.19 | 6.28 |
| Mg | 1.26 | 1.31 | $MgSO_4$ 2.25 | $MgCO_3$ 3.3 | 1.31 | $MgSO_4$ 2.25 | $MgCO_3$ 3.3 | 6.06 | 5.25 | 5.56 | 7.31 | 6.34 | —[4] |
| Na | 0.32 | 0.33 | $NaSO_4$ 1.97 | $NaHCO_3$ 3.3 | 0.33 | $NaSO_4$ 1.97 | $NaHCO_3$ 3.3 | — | — | — | — | — | — |
| K | 1.97 | 1.98 | $KSO_4$ 3.93 | — | 1.98 | $KSO_4$ 3.93 | — | — | — | — | — | — | — |
| Fe(III) | 8.0 | 18.9 | $Fe(OH)_2$ 8.3 | $FeSO_4$ 18.5 | 18.9 | $Fe(OH)_2$ 8.3 | $FeSO_4$ 18.5 | 20.7 | 8.6 | — | 15.9 | 13.7 | — |
| Mn(II) | 7.5 | 8.1 | $MnCl$ 7.8[5] | $MnCl_2$ 8.3[5] | 8.1 | $MnCl$ 7.8 | $MnCl_2$ 8.3 | 12.8 | 11.4 | — | 13.1 | 12.2 | — |
| Cu(II) | 7.7 | 9.2 | $CuCO_3$ 7.7 | $Cu(CO_3)_2$ 9.1 | 10.8 | $CuCO_3$ 9.4 | $Cu(CO_3)_2$ 10.5 | 14.3 | 7.7 | 16.7 | 9.6 | 10.6 | 13.0 |
| Cd | 8.5 | 10.9 | $CdCl_2$ 8.7 | $CdCl$ 9.2 | 10.9 | $CdCl_2$ 8.7 | $CdCl$ 8.7 | 15.1 | 13.1 | 13.5 | 13.5 | 13.4 | 13.6 |
| Ni | 7.7 | 7.9 | $NiSO_4$ 8.3 | $NiCl$ 8.7 | 8.0 | $NiSO_4$ 8.5 | $NiCl$ 8.8 | 12.5 | 8.4 | — | 9.2 | 9.4 | 11.1 |
| Pb | 8.2 | 9.9 | $PbCO_3$ 8.6 | $PbOH$ 8.7 | 9.9 | $PbCO_3$ 8.6 | $PbOH$ 8.7 | 13.2 | 11.34 | 11.5 | 11.8 | — | 11.7 |
| Co(II) | 8.3 | 8.5 | $CoCl$ 9.0 | $CoSO_4$ 9.1 | 8.5 | $CoCl$ 9.0 | $CoSO_4$ 9.0 | 12.7 | 26.5 | 11.9 | 10.8 | 10.8 | 14.9 |
| Ag | 8.7 | 13.1 | $AgCl_2$ 8.7 | $AgCl$ 10.0 | 8.7 | $AgCl_2$ 8.7 | $AgCl$ 10.0 | 17.9 | 26.5 | — | 16.7 | — | — |
| Zn | 7.2 | 7.8 | $ZnOH$ 7.4 | $ZnCl$ 8.0 | 7.8 | $ZnOH$ 7.4 | $ZnCl$ 8.0 | 11.7 | 11.3 | 10.9 | 8.8 | 9.7 | 10.9 |
| %[6] | | | | | | | | 13.0 | 98.6 | 44.9 | 0.7 | 6.6 | 7.5 |

(1) Organic matter of approx. composition $C_{13}H_{17}O_{12}N$ consists of a mixture of acetate, citrate, tartrate, glycine, glutamic acid and phthalate, each present at $7 \cdot 10^{-6}$ mol l⁻¹ (11 millimoles donor groups per litre).

(2) The concentrations given refer to the sum of all complexes, e.g., CuCit, CuHCit, CuCit₂.

(3) Total concentration of metal species; note that Fe(III) is slightly oversaturated with respect to $Fe(OH)_3(s)$; Cu(II) is oversaturated with respect to malachite but because formation is slow, precipitation of this solid has not been allowed. All other metals are thermodynamically soluble at the concentrations specified.

(4) − means that no stability constants for such complexes are available.

(5) There is some uncertainty regarding the validity of the stability constants of chloro complexes of Mn; according to other computations $Mn^{2+}$ is a major inorganic species.

(6) Percentage of total ligand bound to metal ions.

medium; after each addition, the distribution of all the species is computed by considering all the equilibria involved*. The assumed inorganic medium contains major inorganic cations and anions according to the tabulation by Pytkowicz and Hawley (1974) (Table 3.5). Total concentrations of trace metals are taken to be similar to those reported for sea water, as listed in the second column of Table 3.12. Organic matter of the approximate composition $C_{13}H_{17}O_{12}N$ is made up by mixing equal molar quantities of acetic acid, citric acid, tartaric acid, glycine, glutamic acid, and phthalic acid. It thus contains the functional groups hydroxy, carboxylic and amino which are believed to be principally responsible for the complex and chelate forming properties of organic sea water constituents. A few representative results of the calculations are given in Table 3.12; the effect of organic matter on the distribution of two metals is exemplified in Figs. 3.17a and 3.17b.

Some significant implications can be derived from the results of these calculations as displayed in Table 3.12. (1) "Conventional" complex forming organic ligands at concentrations much larger than those encountered in waters of the open sea affect the distribution of trace metal species only to a limited extent; for the example taken only the distribution of Cu species is affected markedly by organic complex formation. (2) About one-third of the organic complex forming donor groups become bound to cations, mostly to Ca and Mg. (3) Concentrations of individual amino acids as high as $100\,\mu g\,l^{-1}$ are not sufficient to cause significant interactions with trace metals. (4) At the concentration levels considered, monodentate complexing agents like acetate are less efficient than chelate formers in binding trace metals (e.g. glycine with Zn, citrate with Cu); B-type cations are relatively more strongly bound by molecules containing mixed N and O donor groups (amino acids) than A-type cations.

Figures 3.17a and 3.17b illustrate the different effects of organic matter upon the distributions of Cd and of Cu. Under the conditions assumed for this model, the distribution of inorganic Cd-species (and that of most other trace metals) is not affected by the presence of organic functional groups at the concentration levels normally found in sea water. The trend of the curves in Fig. 3.17a indicates, and computations confirm, that higher concentrations of organic matter (e.g. $5 \times 10^{-4}$ equiv. $l^{-1}$ and above) will influence the distribution of most trace metals. For Cu(II) however, organic ligands are able to compete successfully with inorganic ligands for the available coordinative sites of the Cu atom even at low concentrations.

The model presented suggests that many soluble trace metals in waters

---

* The computer program developed by Morel and Morgan (1972) was used. For help with the calculations we are indebted to K. Baumann, J. Davis and J. J. Morgan.

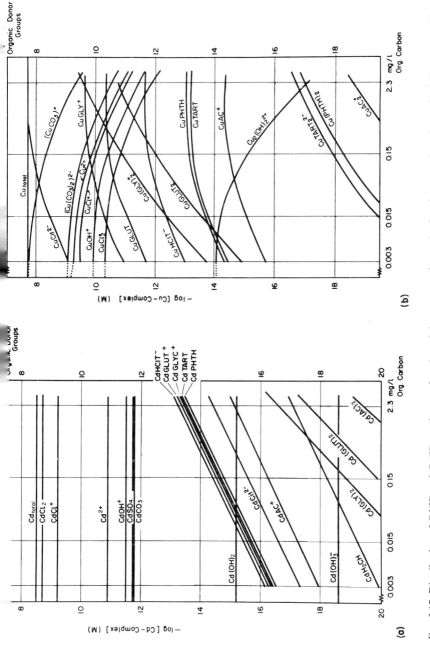

FIG. 3.17. Distribution of Cd(II) and Cu(II) species as a function of the amount of complex forming organic carbon added. For composition of organic material see Table 3.12. Whereas organic complex formation affects the distribution of Cu species (b), there is little effect on the distribution of inorganic species of most other trace metals, e.g. distribution of Cd species (a). [Note. citric acid behaves as a tetraprotic acid, $H_4L$].

of the open sea are primarily present as inorganic species. Only if the concentrations of metals and of organic complex forming components are higher, as they may be in interstitial waters (Nissenbaum *et al.*, 1971), or if stronger and much more selective complex formers are present, will there be interactions with trace metals normally found in sea water.

### 3.5.4. INTERDEPENDENCE OF COMPLEX FORMATION

In the example discussed in the previous section (Fig. 3.17b), one notes that even at low concentration, organic matter becomes bound to Cu(II). This indicates that Cu(II), by tying up organic ligands, may regulate the chelation of other metals. A computation for example shows that increasing the total Cu(II) concentration (e.g. as with coastal pollution) by a factor of 100 will decrease the concentration of citrato complexes of all other metal ions by a factor of 10; this affects in turn—although to a lesser extent—the equilibrium concentrations of other metal complexes. Similar effects can be caused by the addition of other elements or by the addition of individual ligands. As Fig. 3.17b suggests, metals and ligands form a complicated network of interactions; because each cation interacts and equilibrates with all ligands and each ligand similarly equilibrates with all cations, the free concentration of metal ions and the distributions of both cations and ligands depend on the total concentrations of all the other constituents of the system (Duursma 1970; Morel *et al.*, 1973). The addition of Fe(III) (or of any other metal) to a sea water medium, for example in a productivity experiment, produces significant reverberations in the interdependent "web" of metals and ligands, and may lead to a redistribution of all trace metals; (hence an observed change in productivity is not necessarily caused by a change in the availability of Fe to the cell). Morel *et al.* (1973) have illustrated how organic complex forming substances mediate the many pathways of interdependence among the constituents in natural water systems and have elaborated on methods for quantifying such interactions.

### REFERENCES

Anderegg, G. (1971). *In* "Coordination Chemistry" (A. E. Martell, ed.). Van Nostrand Reinhold, New York.
Anfält, T., Dyrssen, D., Ivanova, E. and Jagner, D. (1968). *Sv. Kem. Tidskr.* **80**, 10.
Atkinson, G. and Petrucci, S. (1966). *J. Phys. Chem.* **79**, 3122.
Baran, V. (1972). *Coord. Chem. Rev.* **6**, 65.
Barić, A. and Branica, M. (1967). *J. Polarogr. Soc.* **13**, 4.
Barton, P. B. (1959). *In* "Researches in Geochemistry" (P. H. Abelson, ed.). Wiley-Interscience, New York.

Bayer, E. (1964). *Angew. Chem.* **3**, 325.

Berner, R. A. (1963). *Geochim. Cosmochim. Acta*, **27**, 563.

Berner, R. A. (1965). *Geochim. Cosmochim. Acta*, **29**, 947.

Berner, R. A. (1971). "Principles of Sedimentation". McGraw-Hill, New York.

Bilinski, H. and Stumm, W. (1973). *EAWAG News*, **2**, 3.

Bjerrum, N. (1914). Samml. Chem. Chem.-Tech. Vor. **21**, 575.

Blackburn, T. R. (1969). "Equilibrium, A Chemistry of Solutions". Holt, Rinehart & Winston, N.Y.

Blumer, M. (1967). *Advan. Chem. Ser.* **67**, 312.

Branica, M., Bilinski, H. and Pokric, B. (1969). *Thalassia Jugosl.* **5**, 17.

Brewer, P. G. (1968). Quoted by D. Dryssen (1969).

Brinkley, S. R. (1947). *J. Chem. Phys.* **15**, 107.

Brown, F. S., Baedecker, M. J., Kaplan, I. R. and Nissenbaum, A. (1972). *Geochim. Cosmochim. Acta*, **36**, 1185.

Brown, R. R., Kaplan, I. R. and Presley, B. J. (1968). *Geochim. Cosmochim. Acta*, **32**, 397.

Burns, R. G. and Fyfe, W. S. (1967). *In* "Researches in Geochemistry" (P. H. Abelson, ed.). Wiley and Sons, New York.

Busch, D. H. (1967). *Helv. Chim. Acta*, **50**, 174.

Butler, J. N. (1964). "Ionic Equilibrium, A Mathematical Approach". Addison-Wesley, Reading, Mass.

Butler, J. N. and Huston, R. (1970). *J. Phys. Chem.* **24**, 2976.

Carpenter, J. H. (1972). *In* "Analytical Chemistry: Key to Problems" (W. W. Meinke and J. K. Taybr, eds.), pp. 393–419. National Bureau of Standards, Special Publication 351, Washington D.C.

Carpenter, J. H. and Grant, V. E. (1967). *J. Mar. Res.* **25**, 228.

Chau, Y. K. and Riley, J. P. (1965). *Anal. Chim. Acta*, **33**, 36.

Childs, C. W. (1971). Proc. 14th Conf. Great Lakes Res. (1971), 198.

Christenson, J. J., Hill, J. O. and Izatt, R. M. (1971). *Science, N.Y.* **174**, 459.

Christenson, P. G. (1973). *J. Chem. Eng. Data.* **18**, 286.

Christenson, P. G. and Gieskes, J. M. (1971). *J. Chem. Eng. Data*, **16**, 398.

Christman, R. F. (1970). In *In* "Organic Matter in Natural Waters" (D. W. Hood, ed.). University of Alaska College, Alaska.

Cotton, F. A. and Wilkinson, G. (1972). "Advanced Inorganic Chemistry" 3rd Edn. Wiley Interscience, New York.

Culberson, C. H. and Pytkowicz, R. (1968). *Limnol. Oceanogr.* **13**, 403.

Curl, H., Cutshall, N. and Oschenberg, C. (1965). *Nature, Lond.* **205**, 275.

Dayhoff, M. O. (1971). *In* "Organic Compounds in Aquatic Environments" (S. Faust and J. Hunter, eds.). Dekker, New York.

De Haan, H. (1972). *Freshwater Biol.* **2**, 235.

Duedall, I. W. and Weyl, P. (1967). *Limnol. Oceanogr.* **12**, 521.

Duursma, E. K. (1965). *In* "Chemical Oceanography" (J. P. Riley and G. Skirrow, eds.). Academic Press, London.

Duursma, E. K. (1970). *In* "Organic Matter in Natural Waters" (D. W. Hood, ed.). University of Alaska, College, Alaska.

Duursma, E. K. (1972). *Oceanogr. Mar. Biol. Ann. Rev.* **10**, 137.

Dyrssen, D. (1969). *In* "Chemical Oceanography" (R. Lange, ed.). Universitets Forlaget, Oslo.

Dyrssen, D., Jagner, D. and Wengelin, F. (1968). "Computer Calculations of Ionic

Equilibria and Titration Procedures". Almqvist and Wiksell, Stockholm.

Dyrssen, D., Ivanova, E. and Aren, K. (1969). *J. Chem. Ed.* **46**, 252.

Eigen, M. and Tamm, K. (1962). *Z. Elektrochem.* **66**, 93.

Elderfield, M. (1970). *Earth Planet. Sci. Lett.* **9**, 10.

Elgqvist, B. (1970). *J. Inorg. Nucl. Chem.* **32**, 437.

Feitnecht, W. and Schindler, P. (1963). "Solubility Constants of Metal Oxides, Metal Hydroxides and Metal Hydroxide Salts in Aqueous Solution". Butterworths, London.

Fischer, F. H. (1967). *Science, N.Y.* **157**, 823.

Fitzgerald, W. F. (1969). "A Study of Certain Trace Metals in Sea Water Using Anodic Stripping Voltammetry". Ph.D. Thesis. Mass. Inst. Tech. 130 pp.

Fogg, G. E. (1966). *Oceanogr. Mar. Biol. Ann. Rev.* **4**, 195.

Fukai, R. (1969). *J. Oceanogr. Soc. Jap.* **25**, 47.

Fukai, R. and Huynh-Ngoc, L. (1968). *In* "Radioactivity in the Sea". Internatl. Atomic Energy Agency, Publ. No. 22, Vienna.

Gardner, L. R. (1973). *Geochim. Cosmochim. Acta* **37**, 53.

Garrels, R. M. (1960). "Mineral Equilibria at Low Temperature and Pressure". Harper, New York.

Garrels, R. M. and Christ, C. L. (1965). "Solutions, Minerals and Equilibria". Harper & Row, New York.

Garrels, R. M. and Thompson, M. (1962). *Amer. J. Sci.* **260**, 57.

Garrels, R. M., Thompson, M. E. and Siever, R. (1961a). *Amer. J. Sci.* **259**, 24.

Garrels, R. M., Thompson, M. E. and Siever, R. (1961b). *Amer. J. Sci.* **259**, 43.

Gieskes, J. M. (1966). *Z. Phys. Chem.* N.F. **50**, 78.

Gjessing, E. T. (1967). *In* "Chemical Environment in the Aquatic Habitat" (H. L. Golterman and R. S. Clyro, eds.). Noord-Hollandische Uitgevens, Amsterdam.

Goldberg, E. D. (1954). *J. Geol.* **62**, 249–265.

Greenwald, I. (1941). *J. Biol. Chem.* **141**, 789.

Gregor, H. P., Loebl, E. M. and Luftinger, L. B, (1955). *J. Phys. Chem.* **59**, 34.

Gurd, F. R. N. (1954). "Chemical Specificity in Biological Interactions". Academic Press, New York.

Hahn, H. H. and Stumm, W. (1970). *Amer. J. Sci.* **268**, 354.

Hanor, J. S. (1969). *Geochim. Cosmochim. Acta*, **33**, 894.

Hawley, J. E. (1973). PhD. Thesis, Oregon State University, Corvallis, Oregon.

Helgeson, H. C. (1967). *J. Phys. Chem.* **71**, 3121.

Helgeson, H. C. (1968). *Geochim. Cosmochim. Acta*, **32**, 853.

Hellebust, J. A. (1970). *In* "Organic Matter in Natural Waters" (D. W. Hood ed.). University of Alaska, College, Alaska.

Hindman, J. C. and Sullivan J. C. (1971). *In* "Coordination Chemistry" (A. E. Martell, ed.), Vol. 1. Van Nostrand Reinhold, New York.

Høgdahl, O. T., Bowen, V. T. and Melsom, S. (1968). *Adv. Chem. Ser.* **73**, 308.

Hood, D. W. (1966). "The Chemistry and Analysis of Trace Metals in Sea Water". A & M Project 276, Ref. 66-2F AEC Contract AT (40–1) 2799, 105 p.

Hood, D. W. (ed.) (1970). "Organic Matter in Natural Waters". University of Alaska, College, Alaska.

Horne, R. A. (1969). "Marine Chemistry", p. 166. Wiley-Interscience, New York.

Hostettler, P. B. (1963). *J. Phys. Chem.* **67**, 720.

Ingri, N., Kakolovicz, W., Sillén, L. G. and Warnqvist, D. (1967). *Talanta*, **14**, 1261.

Irving, H. and Williams, R. J. P. (1953). *J. Chem. Soc.* 3192.

James, R. O. and Healy, T. W. (1972). *J. Colloid Interface Sci.* **40**, 42, 53, 65.

Jelinek, H. H. G. and Chen, P. A. (1973). *J. Polym. Sci.* In press.

Jelinek, H. H. G. and Sangal, S. P. (1972). *Water Res.* **6**, 305.

Jenkins, I. L. and Monk, C. B. (1950). *J. Amer. Chem. Soc.* **72**, 2695.

Johannesson, J. K. (1958). *Nature, Lond.* **182**, 251.

Jørgensen, C. K. (1963). "Inorganic Complexes". Academic Press, New York.

Jørgensen, C. K. (1971). "Modern Aspects of Ligand Field Theory". North-Holland, Amsterdam.

Kemula, W. (1960). "Advances in Polarography". Pergamon Press, Oxford.

Kemula, W. and Kubik, Z. (1958). *Anal. Chim. Acta*, **18**, 104.

Kester, D. R. and Pytkovicz, R. M. (1967). *Limnol. Oceanogr.* **12**, 243.

Kester, D. R. and Pytkovicz, R. M. (1968). *Limnol Oceanogr.* **13**, 670.

Kester, D. R. and Pytkovicz, R. M. (1969). *Limnol. Oceanogr.* **14**, 686.

Kester, D. R. and Pytkovicz, R. M. (1970). *Geochim. Cosmochim. Acta*, **34**, 1039.

Kolk, M. (1963). *Nature, Lond.* **198**, 1010.

Krauskopf, K. B. (1956). *Geochim. Cosmochim. Acta*, **9**, 1.

Krindel, P. and Eliezen, I. (1972). *Coordin. Chem. Rev.* **6**, 217.

Lanier, R. D. (1965). *J. Phys. Chem.* **69**, 3992.

Latimer, W. M. (1954). "Oxidation Potentials", 2nd. Edn. Prentice-Hall, Eaglewood Cliffs, N.J.

Mancy, K. H. (1972). 6th Internatl. Water Poll. Research Conf. Pergamon Press, London.

Matson, W. R. (1968). PhD. Dissertation. MIT, Cambridge, Mass.

Matson, W. R., Carritt, D. E. and Roe, D. K. (1967). *Anal. Chem.* **37**, 1594.

McDuff, R. E., Morel, F. and Morgan, J. J. (1973). Personal communication.

Mehrbach, C., Culberson, C. H., Hawley, J. E. and Pytkowicz, R. M. (1973). *Limnol. Oceanogr.* **18**, 897.

Millero, F. J. (1969). *Limnol. Oceanogr.* **14**, 376.

Millero, F. J. (1971a). *Chem. Rev.* **71**, 147.

Millero, F. J. (1971b). *Geochim. Cosmochim. Acta* **35** 1089.

Millero, F. J. (1972). *In* "Water and Aqueous Solutions" (R. A. Horne, ed.). Wiley-Interscience, New York.

Morel, F. and Morgan, J. J. (1972). *Environ. Sci. Technol.* **6**, 58.

Morel, F., McDuff, M. F. and Morgan, J. J. (1973). California Institute of Technology, Pasadena. In preparation.

Morris, J. C. and Stumm, W. (1967). *Adv. Chem. Ser.* **67**, 270. Amer. Soc. Washington.

Nancollas, G. H. (1966). "Interactions in Electrolyte Solutions", p. 1. Elsevier, London.

Neilands, J. B. (1964). *In* "Essays in Coordination Chemistry" (W. Schneider, R. Gut and G. Anderegg, eds.). Birkhäuser, Basel.

Nissenbaum, A., Baedecker, M. J. and Kaplan, I. R. (1971). *In* "Advances in Organic Geochemistry" (H. R. Gaertner and H. Wehner, eds.), 427 pp. Pergamon Pres, Oxford.

Nissenbaum, A., Baedecker, M. J. and Kaplan, I. R. (1972). *Geochim. Cosmochim. Acta.* **36**, 709.

Paces, T. (1972). *Geochim. Cosmochim. Acta* **36**, 217.

Pearson, R. G. (1968). *J. Chem. Ed.* **45**, 581–643.

Perrin, D. D. and Sayce, I. G. (1967). *Talanta*, **14**, 833.

Peshchevitskiy, B. I., Anoshin, G. N. and Erenburg, A. M. (1965), *Dokl. Akad.*

*Nauk USSR.* **162**, 205.

Platford, R. F. (1965a). *J. Fish. Res. Bd. Can.* **22**, 885.

Platford, R. F. (1965b). *J. Mar. Res.* **23**, 55.

Platford, R. F. and Dafoe, T. (1965c). *J. Mar. Res.* **23**, 63.

Popov, N. I. (1970). *Tech. Rep. Ser.* No. 118. Internatl. Atomic Energy Assoc., Vienna, p. 213.

Presley, B. J., Kolodny, Y., Nissenbaum, A. and Kaplan, I. R. (1972). *Geochim. Cosmochim. Acta*, **36**, 1073.

Pytkovicz, R. M. and Hawley, J. E. (1974). *Limnol. Oceanogr.* **19**, 223.

Pytkowicz, R. M. and Kester, D. R. (1969a). *Limnol. Oceanogr.* **14**, 686.

Pytkowicz, R. M. and Kester, D. R. (1969b). *Amer. J. Sci.* **267**, 217.

Pytkowicz, R. M. and Kester, D. R. (1971). *Oceanogr. Mar. Biol. Ann. Rev.* **9**, 11.

Richards, F. A. (1970). *In* "Organic Matter in Natural Waters" (D. W. Hood, ed.). University of Alaska, College, Alaska.

Riley, J. P. and Chester, R. (1971). "Introduction to Marine Chemistry". Academic Press, London. p. 74.

Robie, R. A. and Waldbaum, D. W. (1968). Thermodynamic Properties of minerals and related substances at 298·15°K (25°C) and one atmosphere pressure and at higher temperatures. *Geol. Surv. Bull.* 1259. Washington.

Santschi, P. and Schindler, P. W. (1974). *J. Chem. Soc. Dalton Trans.* 181.

Schläfer, H. L. and Gliemann, G. (1967). "Einführung in die Ligandfeld-theorie". Verlagsgesetllschaft, Frankfurt.

Schwarzenbach, G. (1973). *Chimia*, **27**, 1.

Schwarzenbach, G. and Widmer, M. (1963, 1966). *Helv. Chim. Acta*, **46**, 2613; **49**, 111.

Shain, I. (1963). *In* "Treatise on Analytical Chemistry" (I. M. Kolthoff and P. J. Elving, eds.). Wiley-Interscience, New York. pp. 2533–2568.

Shapiro, N. Z. (1964). Rand Corp. Memo 4205-PR, AD695316.

Sillén, L. G. (1961). *In* "Oceanography" (M. Sears, ed.). *Am. Ass. Adv. Sci.* Washington, D.C. 549.

Sillén, L. G. (1967). *Adv. Chem. Ser.* **67**, 45.

Sillén, L. G. and Martell, A. E. (1964, 1971). "Stability Constants". Chemical Society, London. Special Publications No. 17 and No. 25. The Chemical Society, London.

Smith, R. W. and Hem, J. D. (1972). "Chemistry of Aluminum in Natural Waters". U.S. Geol. Survey, Washington D.C.

Spiro, T. G., Lee, J. and Revesz, A. (1968). *J. Amer. Chem. Soc.* **90**, 4000.

Srinivasan, K. and Rechnitz, S. A. (1968). *Anal. Chem.* **40**, 509.

Stumm, W. and Bilinski, H. (1972). Proc. Water Poll. Res. Conf. Israel. Pergamon Press, Oxford.

Stumm, W. and Morgan, J. J. (1970). "Aquatic Chemistry". Wiley-Interscience, New York.

Sugawara, K. and Terada, K. (1958). *Nature, Lond.* **182**, 250, 251

Takai, N. and Yamabe, T. (1971). *Mizi Shori Gijutsu*, **12**, 3.

Tanford, C. (1961). "Physical Chemistry of Macromolecules". Wiley and Sons, New York.

Thompson, M. (1966). *Science, N.Y.* **153**, 966.

Thorstenson, D. C. (1970). *Geochim. Cosmochion. Acta*, **34**, 745.

Tischendorf, S. and Ungethüm, H. (1964). *Chem. Erde*, **23**, 279.

Turekian, K. K. (1971). *In* "Impingement of Man on the Oceans" (D. W. Hood, ed.). Wiley-Interscience, New York.

Vinogradov, A. P. (1967). *In* "Introduction to Geochemistry of the Oceans". Science Publ. House Moskow; cited in "Handbook of Geochemistry" (K. H. Wedepohl, ed.). Springer (1969).

Wagman, D. D., Evans, W. H., Helow, I., Parker, V. B., Bailey, S. M. and Schumm, R. H. (1965, 1966, 1968). "Selected Values of Chemical Thermodynamic Properties, Part I and II." Natl. Bureau of Standards, Techn. Notes 270–1 and 270–2 and 270–3.

Wangersky, P. J. (1972). *Limnol. Oceanogr.* **17**, 1.

White, W. B., Johnson, S. M. and Danzig, G. B. (1958). *J. Chem. Phys.* **28**, 751.

Whitfield, M. (1971). "Ion Selective Electrodes for the Analysis of Natural Waters". Australian Marine Sciences Associ., Sydney.

Williams, P. M. (1971). *In* "Organic Compounds in Aquatic Environments". (S. J. Faust and J. V. Hunter, eds.). Dekker, New York.

Williams, P. M. and Strack, P. M. (1966). *Limnol. Oceanogr.* **11**, 401.

Zirino, A. and Healy, M. L. (1970). *Limnol. Oceanogr.* **15**, 956.

Zirino, A. and Yamamoto, T. (1972). *Limnol. Oceanogr.* **17**, 661.

Chapter 4

# Adsorption in the Marine Environment

## GEORGE A. PARKS

*Department of Applied Earth Sciences, Stanford University,*
*Stanford, California 94305, U.S.A.*

## 4.1. INTRODUCTION

### 4.1.1. ADSORPTION IN THE MARINE ENVIRONMENT

For several heavy metals sea water is considerably undersaturated with respect to any likely solid phase, and Krauskopf (1956) has shown experimentally that adsorption on suspended solids is probably responsible for reducing the concentration levels of these metals to those observed. The use of hydrous iron (III) oxide as a collector of trace elements in the analysis of sea water illustrates the way in which adsorption may bind trace elements to suspended solids in marine systems (Kim and Zeitlin, 1971a, b, 1972).

Several trace metals (Turekian, 1965), including mercury (Cranston and Buckley, 1972), are more strongly concentrated in the finer size fractions of sediments than in the coarser ones. Further, mercury is notably more concentrated in suspended solids than in the water or bottom sediments at the same station (Cranston and Buckley, 1972). Adsorption is almost certainly responsible for this, since if the adsorption density were constant, the higher specific surface area associated with the finer particles would mean that on a weight basis the finer fraction would contain the greatest amount of the adsorbate.

Kharkar *et al.* (1968) have studied trace elements associated with river-borne suspended solids and have proved by experiment that some, but not all, of them can be expected to desorb when the solids enter the sea. Chester (Riley and Chester, 1971, p. 412) cites evidence that cobalt is adsorbed onto clays from a sea water and that under other circumstances the reverse process may occur.

### 4.1.2. PROBLEMS AND OBJECTIVES

Adsorption, or some kind of surface limited association, must play an important role in the distribution of trace constituents between water and solids and between different solids. Clearly, the partition between the various phases will vary with the identity of the trace component and that of the solids, the chemical environment (e.g. pH and salinity) and the suite of inorganic and organic ligands present. If we are to be able to interpret many of the chemical processes occurring, a proper appreciation of the physical chemistry of adsorption is essential.

Adsorption is often used to explain observations such as those cited in the previous section; it is also frequently held that adsorption of ionic species occurs in response to attraction by solids of opposite electrical charge. For example, it is often claimed that $Fe_2O_3$ is a positively charged solid and that $SiO_2$ and $MnO_2$ are negatively charged and that this difference perhaps explains the observed selectivity in their adsorptive properties. It does not. These generalizations are dangerous oversimplifications since they fail to explain adsorption of non-electrolytes, selectivity between ions of like charge, adsorption of ionic species on solids of like charge and the reversal of charge which occurs when an excess of certain ionic species is adsorbed. These phenomena which are commonly observed in fresh water laboratory systems must also be expected to occur in natural systems as well; they stem from the fact that the behaviour of solids is complex and that it is not possible dogmatically to characterize one as positive and another as negative.

Adsorption occurs as a result of a variety of binding mechanisms. Most solids in aquatic environments are electrically charged, and some adsorption of ionic species would be expected to occur through electrostatic attraction. Electrostatic adsorption does take place, of course. It is rapid and readily reversible, and electrostatically adsorbed ions undergo equilibrium exchange.

The charge on the solid is sensitive to the composition of the aqueous phase and frequently it is possible to identify certain ionic solutes as being potential determining, i.e. primarily responsible for the surface charge. There is usually a particular concentration of the potential determining species at which the charge is zero. This is the point of zero charge, PZC. The charge becomes positive at higher concentrations of the positive potential determining species. If the solid is positively charged, anion adsorption will occur and it is an anion exchanger. Anions can be desorbed by reversing the charge, by, for example, adding an excess of a negative potential determining ion. It will be shown below that other adsorption mechanisms exist. Nevertheless, the assumption that surface charge controls adsorption is fairly widespread (see e.g. Krauskopf, 1967; Pravdić, 1970), and this concept has been used as the basis for selective leaching methods for identifying the mode of binding of trace metals in sediments and suspended solids.

There are many causes of adsorption other than electrostatic attraction. Changes in hydration state of the solid or the adsorbate, interactions between the adsorbate molecules or ions themselves, and covalent, Van der Waals or hydrogen bonding between the adsorbate and the solid are examples. When contributions to the free energy of adsorption from these sources are large, adsorption is said to be *specific* and an ion can be adsorbed against electrostatic repulsion, even to the extent that the total adsorbed charge exceeds the original charge on the solid thus causing charge reversal. Clearly, purely electrostatic adsorption cannot lead to charge reversal. Electrostatic attraction or repulsion may play an insignificant role in specific adsorption. Unlike electrostatic adsorption, specific adsorption may be slow and is less readily reversible in the sense that equilibrium is not achieved in a reasonable time.

Specific and electrostatic adsorption occur simultaneously. Regardless of the original charge, specific adsorption may increase, reduce, neutralize or reverse the effective charge on the solid. If a net charge remains, the solid-specific adsorbate complex will still be capable of adsorbing more ionic species electrostatically. Ions thus adsorbed mày have the same sign as the original charge on the solid. Hydrous iron (III) oxide is slightly positive in pure water or in NaCl solutions up to $1\cdot0$ M. However, it specifically adsorbs sulphate ions from dilute sulphate solutions, and becomes negatively charged (Healy and Kavanagh, 1972). Thus, in fresh water or NaCl brines it is an anion exchanger or adsorbent, but in sea water, owing to the $SO_4^{2-}$ present, it will probably adsorb cations.

The purpose of this chapter is to aquaint the reader with the importance of surface charge, to briefly review the consequences of the existence of this charge (i.e. the development of an electrical double layer—see below), to discuss the origins of surface charge and the fact and origins of the extreme variability of the surface charge of real solids. On the basis of this background the reasons for adsorption will be discussed and the phenomena examined quantitatively. Whenever possible data appropriate to marine systems will be cited.

## 4.2. Electrical Double Layer, Non-specific Adsorption and Surface Charge

### 4.2.1. introduction

#### 4.2.1.1. *Structure of the double layer*

An electric charge on a suspended solid generates an ionic atmosphere of opposite charge, part of which is diffuse. The surface charge and the counter charge comprise an electrical double layer (EDL) which is described in great detail in the literature of colloid chemistry (Overbeek, 1952; Van Olphen, 1963; and Haydon, 1964) and electrochemistry (e.g. Grahame, 1947; Parsons, 1954). For the present purpose, the EDL is assumed to consist of the charge on the solid which for the most part is assumed to be limited (or fixed) to the surface (the *surface charge*), a layer of ions in contact with the surface (the *Stern layer*) and a diffuse layer of ions in the aqueous phase near the surface (the *Gouy* or *diffuse layer*). The distributions of charge and potential in the double layer are illustrated schematically in Fig. 4.1 in which the nomenclature and symbolism are also summarized.

The simplest models (Case I, Figure 4.1) of the EDL envisage just two zones of charge, the surface charge and the Gouy layer charge. These models are approximately valid if adsorption bonding is purely electrostatic, i.e., in the absence of any strong, specific, or chemical bond. The counter ions in the Gouy layer are fully hydrated and none approach the surface more closely than their hydrated radii. A plane through the centres of the counter ions closest to the surface is the *Stern Plane*. That portion of the total Gouy-layer charge localized in the Stern Plane is the *Stern Layer Charge*. The distance from the Stern Plane to the surface is denoted by $d$.

Specifically adsorbed species are thought to approach more closely to the surface than do non-specifically adsorbed species. Multivalent cations are considered to be adsorbed strongly enough to expel all water of hydration except for the inner hydration sheath (James and Healy, 1972c). It is conceivable that adsorption bonding might be strong enough to expel all coordinating water.

If specific adsorption occurs, the EDL is considered to comprise three zones of charge (Case II, Figure 4.1), except in the unlikely event that the surface charge is exactly neutralized by the specifically adsorbed charge; the three

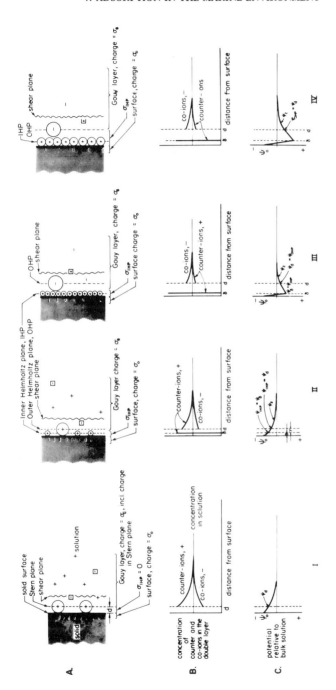

FIG. 4.1. Distribution of Charge (A), Counter and Co-Ions (B), and Potential (C) in the Electrical Double Layer. Double Layer Nomenclature and Symbolism. Case I: No specific adsorption, $\sigma_{IHP} = 0$, $\sigma_G = -\sigma_0 \neq 0$. Case II: Finite specific adsorption, $|\sigma_{IHP}| < |\sigma_0|$, $\sigma_G = -(\sigma_0 + \sigma_{IHP}) \neq 0$. Case III: Superequivalent specific adsorption, Stern model with adsorbate ions not hydrated, $|\sigma_{IHP}| > |\sigma_0|$, $\sigma_G = -(\sigma_0 + \sigma_{IHP}) \neq 0$. Case IV: Superequivalent adsorption, James–Healy model with adsorbate ions hydrated, $|\sigma_{IHP}| > |\sigma_0|$, $\sigma_G = -(\sigma_0 + \sigma_{IHP}) \neq 0$.

zones are the surface charge, the specifically adsorbed charge and the Gouy-layer charge. The locus of the centres of specifically adsorbed species is known as the *Inner Helmholtz Plane* (IHP). The distance from the IHP to the surface is designated $\delta$. There is no charge between the IHP and the surface. The locus of the centres of the Gouy-layer ions closest to the surface is the *Outer Helmholtz Plane* (OHP). The distance from the OHP to the surface is denoted by d. The IHP and OHP together replace the Stern Plane. The Gouy-layer starts at the OHP. Charge localized in the OHP is part of the Gouy layer charge; charge localized in the IHP is not.

The "surface charge" which is effective in accumulating the Gouy Layer is the charge inside the OHP; this corresponds with the true surface charge of the solid, $\sigma_0$, only if there is no specific adsorption. If specific adsorption does not occur, $\sigma_{IHP} = 0$, the effective surface charge is $\sigma_0$, and the sign of the Gouy-layer charge is opposite to that of the surface, (*ie* $\sigma_G = -\sigma_0$). The Gouy Layer charge accumulates by attraction of counterions (ions with charges opposite to $\sigma_0$). Counter-ions are positively adsorbed in the Gouy Layer and co-ions negatively adsorbed. The net charge accumulated in the Gouy Layer results from an excess of counterions and a deficit of co-ions (an excess of co-ion vacancies). At low potentials ($\Psi_d < 25$ mV) both contribute equally, i.e.

$$\sigma_G = \sigma_{G+} + \sigma_{G-}$$

and $\sigma_{G+}$ is approximately equal to $-\sigma_{G-}$. The algebraic sign must be included when substituting numerical values for $\sigma$. At larger potentials, the charge due to an excess of counterions exceeds that due to co-ion vacancies (DeBruyn and Agar, 1962).

If specific adsorption does occur and the specifically adsorbed species absorb in the IHP, $\sigma_{IHP}$ is not zero and

$$\sigma_0 = -(\sigma_{IHP} + \sigma_G)$$

If specific adsorption is strong, $\sigma_{IHP}$ may exceed $\sigma_0$, either increasing the effective surface charge if the specifically adsorbing species has the same sign as $\sigma_0$, or reversing the effective charge if the adsorbing species has the opposite sign to $\sigma_0$. The sign of $\sigma_G$ is determined by the sign of $(\sigma_{IHP} + \sigma_0)$.

$$\sigma_G = -(\sigma_{IHP} + \sigma_0)$$

If $\sigma_{IHP} > \sigma_0$, adsorption is said to be *superequivalent*.

### 4.2.1.2. *Evidence for surface charge*

If the solid moves with respect to the aqueous phase the double layer is split at a shear plane between the hydrodynamic layer on the solid and the bulk of the solution. A variety of methods based on the relative motion between solid and liquid have been evolved for measuring the "surface charge" or "surface potential"; they are known collectively as electrokinetic methods. The potential (the zeta potential, $\Psi_z$) at the shear plane, or the net charge within it, is sensed, not the surface potential or $\Psi_o$. The shear plane is near the OHP at low potentials in low ionic strength media (Li and deBruyn, 1966). At high potentials and high ionic strength it is outside the OHP, probably because the water viscosity near the surface increases (Li and deBruyn, 1966). The relationships between $\Psi_z$ and $\Psi_0$ or $\Psi_d$ (the potentials at the surface and OHP respectively) remains ambiguous (e.g. Haydon, 1964), in part because of the variable dimensions of the boundary layer.

Electrokinetic methods involve either observation of the motion of charged particles in response to an applied potential (electrophoresis), or observation of a potential (streaming potential) generated by forcing a solution past the solid. At the ionic strength of sea water, the conductivity of the solution is so high that thermal convection and electrolysis interfere with measurements; furthermore, $\Psi_z$ is small and difficult to sense (see equation 4.3 below). Neihof (1960, 1969) has devised a way of preventing interference from electrolysis during electrophoresis so that measurements are possible in a thermostatically controlled system (Neihof and Loeb, 1972).

### 4.2.1.3. Non-specific and specific adsorption

Adsorption in the Gouy Layer is electrostatic and, to a first approximation, obeys the Gouy–Chapman Theory of the EDL. It is relatively non-selective. Specific adsorption occurs in response to a variety of binding energies, is more selective, and is best described by a combination of the Langmuir adsorption isotherm and EDL theory.

### 4.2.2. NON-SPECIFIC ADSORPTION

Non-specific adsorption refers to adsorption in the Gouy Layer, whether specifically adsorbed species are present or not.

### 4.2.2.1. Counter and co-ions

The concentrations of counter and co-ions, mainly $Na^+$ and $Cl^-$ in sea water, change with distance from the OHP, approaching the bulk solution concentration as distance increases. The concentration and potential at any distance from the OHP are related by the expression

$$C_{xi} = C_{bi} \exp\left[\frac{-z_i F \Psi_x}{RT}\right] \tag{4.1}$$

in which

$x$ = distance from OHP or (Case I, Fig. 4.1) from the Stern Plane. Distance from surface $= x + d$,

$C_{xi}$ = concentration of ion, $i$, at $x$,

$C_{bi}$ = concentration of ion, $i$, in the bulk solution far from the surface,

$z_i$ = ionic charge on species, $i$,

$F$ = The Faraday, 96,500 coulombs equiv.$^{-1}$ or joules volt$^{-1}$ equiv,$^{-1}$

$\Psi_x$ = Gouy layer potential at $x$,

$R$ = Gas constant,

$T$ = Absolute temperature.

In substituting for $z_i$ and $\Psi_x$ the algebraic sign should be included.

The adsorption density of species $i$ is the (positive or negative) excess of $i$ over the amount that would be present if the concentration of $i$ was unaffected by the surface. Thus, the total adsorption density is given by

$$\Gamma_{Gi} = \int_{x=d}^{\infty} (C_{xi} - C_{bi})dx$$

(Grahame, 1947; Stumm and Morgan, 1970, p. 466). The integrated adsorption isotherm is

$$\Gamma_{Gi} = \left(\frac{2\varepsilon kT}{z_i^2 e^2 N_a} \times 10^3 C_{bi}\right)^{\frac{1}{2}} \left(\exp\left[\frac{-z_i F \Psi_d}{2RT}\right] - 1\right) \tag{4.2}$$

in which

$\Psi_d = \psi_x$ evaluated at $x = d$,

$\varepsilon$ = permittivity of water,

$k$ = Boltzmann constant,

$e$ = electronic charge,

$N_a$ = Avogadro's number,

The algebraic sign should be included when substituting for $z$ and $\Psi_d$ as (4.2) applies to both the counter and co-ions. If $C_{bi}$ has the units mol l$^{-1}$ and all other units are SI, $\Gamma_{Gi}$ is in mol m$^{-2}$. $\Gamma_{Gi}$ includes the ions in the Stern plane (Grahame, 1947).

Adsorption in the double layer is more often expressed in terms of charge than of adsorption densities (Grahame, 1947; or Shaw, 1970). Thus, metrical the concentration of the counter ion must be used.

$$\sigma_G = zF(\Gamma_{G+} - \Gamma_{G-})$$

$$= -(8 \times 10^3 \times N_a \varepsilon kTC_{bi})^{\frac{1}{2}} \sinh \frac{zF\Psi_d}{2RT} \tag{4.3}$$

$C_{bi}$ is in mole $l^{-1}$ if all else is in SI units.

At low potentials, $\sigma_G \sim -\varepsilon\kappa\Psi_d$.

Use of equation (4.2) requires a knowledge of the variation of potential with distance. Within the Gouy Layer, $\Psi$ and $x$ are related as follows,

$$\Psi_x = \frac{2RT}{zF} \ln\left(\frac{1 + y\exp(-\kappa x)}{1 - y\exp(-\kappa x)}\right) \tag{4.4}$$

in which

$$y = \frac{\exp\left[zF\Psi_d/(2RT)\right] - 1}{\exp\left[zF\Psi_d/(2RT)\right] + 1}$$

and

$$\kappa = \left(\frac{10^3 \times e^2 N_a Cz^2}{\varepsilon kT}\right)^{\frac{1}{2}} = \left(\frac{2 \times 10^3 \times e^2 N_a I}{\varepsilon kT}\right)^{\frac{1}{2}} \tag{4.5a}$$

if $I$ = ionic strength (mol $l^{-1}$) if the electrolyte is symmetrical

$C$ = concentration of electrolyte (mol $l^{-1}$). If the electrolyte is not symmetrical the concentration of the counter ion must be used.

*Validity of non-specific adsorption isotherm.* The non-specific adsorption isotherm derived from electrical double layer theory, equation (4.2), has been successfully tested. Li and de Bruyn (1966) made careful measurements of $Na^+$ adsorption densities and streaming potentials on quartz. They used the streaming potentials to calculate $\Psi_z$ and compared the experimental values of $\Psi_z$ with $\Psi_d$ calculated from adsorption densities using equation (4.2). At potentials below 100 mV and with an ionic strength of $10^{-3}$ or $10^{-4}$ molar, agreement is good. At higher ionic strengths or potentials, $\Psi_d$ is larger than $\Psi_z$; this deviation can be explained in terms of changes in thickness of the hydrodynamic boundary layer which arise as a result of changes in electrolyte viscosity.

#### 4.2.2.2. Minor solutes.

When considering a complex solution such as sea water, in which many ionic species are present, but one set, $Na^+$ and $Cl^-$, predominates, equation (4.1) applies only to the predominant set. Equation (4.2) does not apply directly to the adsorption of ions present in trace concentrations because the surface field ($\Psi_d$ in equation (4.2)) is fixed by total electrolyte concentration whereas the adsorption of a minor ion depends on field strength and the concentration of the minor ion (MacNaughton and James, 1973). The adsorption density of the dilute species, $j$, can be expressed in terms of that of the dominant ion, $i$, of the same charge, thus

$$\Gamma_j = \frac{C_j}{C_i} \Gamma_i. \tag{4.5b}$$

Alternatively a direct isotherm can be derived by substituting for $\Gamma_i$. (see Section 4.2.3.7). Note that the effect of a change in $C_i$ on ionic strength is not obvious from equation (4.3) because $\Gamma_i$ changes with $C_i$ through equation (4.2). The net effect of an increase in $C_i$ is a decrease in $\Gamma_j$.

#### 4.2.2.3. Gouy Layer thickness

$\Psi_x$ decays to $\Psi_d/e$ in a distance $1/\kappa$. The distance $1/\kappa$ is a measure of the "thickness" of the Gouy Layer. The thickness of the EDL is $1/\kappa + d$.

In NaCl solution, $1/\kappa$ is 10 nm if $C$ is $10^{-3}$ mol l$^{-1}$ and 0·36 nm if $C$ is 0·7 mol l$^{-1}$. In sea water, the Gouy Layer is very thin. Most of the "diffuse" layer lies within the first layer of water molecules, i.e. most of the surface charge is masked by the counter charge which is confined within the first monolayer of water on the particle surface and will not be sensed by electrokinetic methods.

#### 4.2.2.4. Evaluation of $\Psi_0$ and $\Psi_d$

The surface potential, $\Psi_0$, can be related to the solution composition. However, use of the equations given above to calculate $\Gamma_{Gi}$ requires a knowledge of $\Psi_d$. Calculation of $\Psi_d$ from $\Psi_0$ requires a knowledge of $d$ and $\varepsilon$, the permittivity in the Stern layer. These quantities are not well known. Several approaches have been used in obtaining approximations of $\Psi_d$ for use in equation (4.4). These have been reviewed briefly by James and Healy (1972c). The two best methods are probably to determine $\Psi_z$ (Li and de Bruyn, 1966), empirically, and to calculate $\Psi_x$ using equation (4.4) after replacing $\Psi_d$ with $\Psi_0$ and setting $x = d$. However, the latter is equivalent to assuming that the potential drop across the Stern plane obeys the same law as does the potential decay in the Gouy Layer, and is clearly not correct. Methods of calculating $\Psi_d$ are available (see e.g. Hunter and Wright, 1971), but no account has been published of direct optimization of approximation methods; the determination of $\Psi_0$ from the properties of the solid will be discussed in Section 4.23.

### 4.2.2.5. Surface concentrations and ion exchange in the Gouy Layer

Equation (4.1) can be used to calculate the "surface concentrations" of non-specific ions. Stumm and Morgan (1970) have defined a surface pH in this way, and Gaudin and Fuerstenau (1955) have used the relationship to show that surface concentrations of non-specifically adsorbing surfactants can approach the critical micelle concentration before the bulk concentration does so.

The relative concentrations of various species in the Gouy Layer can be derived from equations (4.1) and (4.2). The result for total adsorption densities of ions of the same charge is given by equation (4.3). The derivation of ion exchange equations has been discussed by Stumm and Morgan (1970), Van Olphen (1963) and Overbeek (1952) as well as in many more detailed papers to which these will serve as a guide. If used as they stand, equations (4.1) and (4.2) lead to the conclusion that there is no selectivity among ions of like charge. In fact, slight selectivity is observed and in more detailed treatments corrections are applied in the equations (see references in Stumm and Morgan (1970) and Van Olphen (1963)).

According to equation (4.2) cations alone adsorb positively when $\Psi_d$ is negative. Hence the solid is a cation exchanger when $\Psi_d$ is negative and conversely it is an anion exchanger when $\Psi_d$ is positive.

### 4.2.3. ORIGINS OF CHARGE AND POTENTIAL

#### 4.2.3.1. Solids present in marine waters

Suspended matter in sea water ranges in size from colloidal ($10^{-5}$–$10^{-7}$ cm) to large flora and fauna. If living organisms are excluded, the key differences between colloidal and larger particulates are the longer residence time and the much larger specific surface area of colloidal materials.

The composition of the suspended solids in the sea varies markedly with location and depth (see Ch. 27 and 35). On average it resembles that of surface sediment (Riley and Chester, 1971; Horne, 1969; Manheim et al., 1972), although the proportion of land derived materials other than clays, decreases in the finer sizes (Jacobs and Ewing, 1969). Direct observation and inference suggest that the inorganic and organic components are present in variable proportions with the inorganic fraction usually slightly predominating (Riley and Chester, 1971, p. 210; Horne, 1969 p. 229). In inshore waters the suspended solids may be mainly inorganic; seaward of the shelf break, the organic components predominate in the upper few hundred metres of the water column (Manheim et al., 1972). Land-derived material in deep sea suspended matter is largely colloidal (Riley and Chester, 1971, p. 235).

The organic fraction is largely detrital (Riley and Chester, 1971, p. 200; Horne, 1969, p. 229; see also Chapter 13) and consists of carbohydrates, such as cellulose (Parsons, 1963; Riley and Chester, 1971, p. 204), starch and chitin (Degens, 1965), proteins (such as collagen), lipids and perhaps hydrocarbons.

The inorganic fraction includes detrital terrigenous minerals (such as quartz, feldspars, carbonates and clays), halmyrogenous or hydrogenous minerals (such as the precipitated silicates glauconite and phillipsite, carbonates, sulphates and hydrous oxides), and biogeneous materials such as carbonate, phosphate and silica skeletal debris) (see Chapters 28 and 29).

The proportion of the suspended solid load which is of colloidal dimensions is unknown. Particulate material is customarily defined as the fraction which is retained by a filter having a pore size of 0·5 to 1·0 μm. Everything that passes the filter is considered to be dissolved (see e.g. Parsons, 1963). Since colloidal solids are generally finer than 1 μm, they are included as part of the dissolved fraction. We can infer that colloidal material is present among the suspended solids because the < 2 μm fraction comprises the major proportion of most deep sea sediments (Chester and Johnson, 1971); this contains the same minerals that are present in the coarser fractions of the same sediment. Hydrolysis products and authigenic minerals may also be expected to occur in the colloidal fraction.

### 4.2.3.2. Introduction

The charge on a suspended solid can have many origins. It may be intrinsic to the structure and composition of the solid or can arise from the interaction of the solid with the water itself or its $H^+$ and $OH^-$ ions. Alternatively, it may be extrinsic to the solid, and arise from specific adsorption of a solute foreign to the solid. Some examples of intrinsic charge development are barite ($BaSO_4$) which develops a surface charge by non-stoichiometric dissolution, proteins and insoluble oxides which develop charges by electrolytic dissociation of bound ionogenic functional groups, and clays in suspension in which the silicate structure may carry a net charge which is not completely neutralized by exchangeable ions. In contrast, much of the suspended matter in sea water, regardless of the solid composition, is negatively charged, apparently by strong specific adsorption of ubiquitous anionic organic solutes (Neihof and Loeb, 1972). (See Section 4.2.3.11.)

Unfortunately for simple presentation, real solids defy rigid categorization and their properties cover a wide range. Even so, it is useful to recognize the extremes of behaviour. Table 4.1 gives a classification of aqueous colloidal systems; this scheme has been compiled from those used by Bungenberg de Jong (1949), Kruyt (1952), and Van Olphen (1963). It has been expanded to clarify use of the word hydrophobic (see Vold and Vold, 1964; and Shaw, 1970) and to take account of the various mechanisms of charge development (Haydon, 1964 (p. 126); Parks, 1967; and Fuerstenau, 1971). Hydrophobic colloids are largely non-biogenic, whereas hydrophilic colloids are mainly bio-colloids. In the paragraphs which follow discussion will centre on several of the modes of charge development and how $\Psi_0$ is related to the composition of both solution and solid.

### 4.2.3.3. Salts of simple ions

(a) *Origin of charge.* Salts of non-hydrolyzing ions such as $BaSO_4$ or $AgCl$ are largely ionic and they interact with water principally through dissolution. Surface charge appears to develop as part of the dissolution process. As normally written the reaction

$$BaSO_4(s) \rightleftarrows Ba^{2+} (aq) + SO_4^{2-}(aq) \tag{4.6}$$

implies stoichiometry and electroneutrality in solid and solution. However, the existence of an observable charge on the solid is evidence that electroneutrality is not preserved. Apparently the dissolution of anions and cations is not strictly coupled. The dissolution reactions should be written separately,

$$\left. \begin{array}{l} Ba^{2+} \text{ (in } BaSO_4(s)) \rightleftarrows Ba^{2+} (aq) \\ SO_4^{2-} \text{(in } BaSO_4(s)) \rightleftarrows SO_4^{2-} (aq) \end{array} \right\} \tag{4.7}$$

and if one reaction proceeds to a greater extent than the other an excess of one ion over the other will remain on the solid. An excess of one ion could equally well arise through stoichiometric dissolution and selective resorption. In either case the surface charge, $\sigma_0$, and surface excess charge density $\Gamma$, are related by,

$$\sigma_0 = z_+ F\Gamma_+ + z_- F\Gamma_-$$

$z$ should take account of the algebraic sign.

TABLE 4.1

*A behavioural classification of colloidal systems*

---

A. *Irreversible Colloids*

The colloidal or sol state is unstable with respect to coagulation, flocculation or sedimentation. It is a non-equilibrium state, a suspension, not a solution. The solid is hydrophobic in the sense that formation of the colloidal state from the solid or coagulated state is not spontaneous but requires mechanical energy.

1. *Electrolytic*

Solid is hydrophilic relative to water wetting. It is likely to be at least partly ionic.

a. *Intrinsic*

Charge originates from electrolytic interaction between solid and water.
  i.   Salts of simple ions, e.g. $BaSO_4$
  ii.  Salts of weak acids or bases, e.g. $CaCO_3$, apatites, calcium phosphates.
  iii. Oxides and non-clay silicates

b. *Extrinsic*

Charge originates in defect structure of solid.
  i.  Clays
  ii. Defect solids, e.g. $\delta$-$MnO_2$

2. *Hybrids*

AgI is electrolytic in the sense of 1a (Overbeek, 1952) but hydrophobic in the sense that water wetting is incomplete (Fuerstenau, 1971).

3. *Non-Electrolytic*

Solid is hydrophobic relative to water wetting. It is likely to be non-polar (e.g. hydrocarbons and air bubbles).

B. *Reversible Colloids*

The colloid or sol state is stable with respect to coagulation, flocculation and sedimentation. It is an equilibrium state approximating to true solution. The solid is hydrophilic in the sense that formation of the colloidal state from the solid or coagulated state is spontaneous.

1. *Electrolytic*

Solid is hydrophilic relative to water wetting; charge originates in electrolytic interaction between ionogenic groups on the solid and water. Examples include proteins, cells, and any macromolecule incorporating the requisite ionogenic functional groups.

2. *Non-Electrolytic*

Solid is hydrophilic with respect to water wetting. Examples include polysaccharides, such as starch, with abundant polar but non-ionogenic groups, e.g. —OH and —O—.

---

*Solid space charge.* Because it is known that equilibrium concentrations of vacancies and interstitial ions exist in solids such as the silver halides, it has been suggested that the equilibration of solid with aqueous phases involves the transfer of ions between the interior of the solid and the solution; this will result in a space charge or diffuse layer charge both on the solid and the solution side of the interface. Haydon (1964) has reviewed this model and the evidence for its validity. If the solid is a semi-conductor, an electronic space charge could develop in the solid (Parks, 1965).

(b) *The point of zero charge, PZC.* The algebraic sign of $\sigma_0$ depends on the relative magnitudes of $\Gamma_+$ and $\Gamma_-$. These, in turn, depend on the equilibrium constants for reactions (4.7) and the activities of the anion and cation in both the solid and the solution. Obviously, increasing $a_+(aq)$ (the activity of $Ba^{2+}$ in the $BaSO_4$ example) increases the surface concentration of the cation, and the excess $a_+$ leads to a positively charged surface. There will be some combination of $a_+(aq)$ and $a_-(aq)$ at which $\sigma_0$ is zero; this is the point of zero charge, PZC. Since $a_+$ and $a_-$ are coupled through the solubility product, the PZC is fully defined by specifying either $a_+$ or $a_-$.

(c) *Surface potential and potential determining ions.* The potential difference between the solid and the solution is determined by the activities of the dissolving ions (see e.g. Overbeek and Lijklema, 1959; de Bruyn and Agar, 1962). The chemical potential of each species must be the same in all parts of the system. At equilibrium, the total chemical potential is

$$\mu_{i,\,p} = \mu^0_{i,\,p} + z_i F \Psi_p + RT \ln a_{i,\,p} \tag{4.8}$$

where $\eta_{i,\,p}$ = chemical potential of species $i$ in phase (or position) $p$.

$\mu^0_{i,\,p}$ = standard chemical potential of $i$ in $p$.

$z_i$ = charge on ionic species $i$.

$\Psi_p$ = absolute potential in phase (or position) $p$.

$a_{i,\,p}$ = thermodynamic activity of $i$ in $p$.

At equilibrium,

$$\mu_{i,\,aq} = \mu_{i,\,s} \tag{4.9}$$

where "aq" refers to the bulk solution phase and "s" to the solid surface which is assumed to be the seat of the charge on the solid. Expanding and separating variables yields

$$\Psi_0 = \Psi_s - \Psi_{aq} = \frac{\mu^0_{i,\,aq} - \mu^0_{i,\,s}}{z_i F} + \frac{RT}{z_i F} \ln \frac{a_{i,\,aq}}{a_{i,\,s}} \tag{4.10}$$

$\Psi_0$ is the surface potential relative to the bulk solution potential.

(i) *Potential determining ions.* Ions for which equation (4.10) holds are called potential determining ions (PDI) of the first kind (James, 1971). For barite, the PDI are $Ba^{2+}$ and $SO_4^{2-}$. For AgI, they are $Ag^+$ and $I^-$.

(ii) $\Psi_0$ *and the PZC, the Nernst equation.* If no polar molecules are adsorbed, $\Psi_0$ is zero when $\sigma_0$ is zero, i.e. $\sigma_0$ and $\Psi_0$ are both zero at the PZC, thus

$$\Psi_0(PZC) = 0 = \frac{\mu_{i,aq}^0 - \mu_{i,s}^0}{z_i F} + \frac{RT}{z_i F} \ln \frac{a_{i,aq}(\text{at PZC})}{a_{i,s}(\text{at PZC})}$$

$\Psi_0$ at any other PDI activity is $\Psi_0(a_{i,aq}) - \Psi_0(PZC)$, or

$$\Psi_0 = \frac{RT}{z_i F} \ln \frac{a_{i,aq}}{a_{i,aq}(\text{at PZC})} - \frac{RT}{z_i F} \ln \frac{a_{i,s}}{a_{i,s}(\text{at PZC})} \tag{11}$$

It is not known how $a_{i,s}$ can be evaluated. It is usually assumed to be a constant, independent of $\Gamma_i$ or $\sigma_0$ (de Bruyn and Agar, 1962). This is tantamount to assuming that the change in concentration of PDI at the surface accompanying changes in $\Psi_0$ is negligible with respect to the total concentration, or that ions at the surface do not interact. Berubé and de Bruyn (1968) and Hunter and Wright (1971) are doubtful about the general validity of this assumption, but concede that it is probably true for ionic crystals in which the surface concentration of PDI is large.

On the assumption that $a_{i,s}$ is constant, equation (4.11) takes the simple form

$$\Psi_0 = \frac{RT}{z_i F} \ln \frac{a_{i,aq}}{a_{i,aq}(\text{at PZC})} \tag{4.12a}$$

or, because the PZC is usually given as the logarithm of $a_i$ at $\sigma_0 = 0$,

$$\Psi_0 = \frac{2 \cdot 3 RT}{z_i F} (\log a_{i,aq} - PZC) \tag{4.12b}$$

Equation (4.12a) resembles the Nernst equation, and is often referred to by that name. It is obeyed well only close to the PZC.

(iii) *Adsorption of potential determining ions.* The adsorption densities of positive and negative PDI determine the surface charge, $\sigma_0$, thus

$$\sigma_0 = zF(\Gamma_{PDI+} - \Gamma_{PDI-}) \tag{4.13}$$

In the absence of specific adsorption, $\sigma_0 = -\sigma_G$, and the Gouy–Chapman equation (4.3) allows calculation of $(\Gamma_{PDI+} - \Gamma_{PDI-})$. If it is assumed that adsorption of both PDIs is zero at the PZC, it is possible to obtain an adsorption isotherm for the PDI. This treatment applies reasonably well for AgI (Kruyt, 1952) and for $SiO_2$ (Li and de Bruyn, 1968), but is reliable only close to the PZC where the Nernst equation holds at least to a first approximation. Other, better, isotherms are being developed. They are based on an approach used first with proteins, and will be discussed briefly in Section 4.2.3.8.

(iv) *Potential determining ions of the second kind.* Any solutes which react with a PDI, and alter its activity, will obviously affect $\Psi_0$, and thus will indirectly assist in determining the potential. Such ions may be called PDI of the second kind (James, 1971).

(v) *The nomenclature of zero charge.* The following definitions are based on terminology adopted by the IUPAC Commission on Colloid and Surface Chemistry in July, 1971 (Everett, 1972).

It has been noted above that some sources of charge involve adsorption of unique potential determining ions which are directly related to the composition of the solid and produce a true surface charge, $\sigma_0$, which is *on* the solid, ie. part of its surface structure. The Point of Zero Charge, PZC, is defined as the negative logarithm of the PDI activity corresponding to zero true surface charge, $\sigma_0 = 0$. It is best determined by direct measurement of the adsorption densities of PDI. The measurement must be made in the presence of an indifferent electrolyte sufficiently concentrated to ensure that the PDI do not populate the Gouy Layer. If this precaution is neglected, adsorption will be observed, but the PDI adsorption density will not represent the Stern surface alone.

Attempts to use electrokinetic methods for the direct measurement of surface charge yield only the net charge, $\sigma_z$, or the Zeta potential $\Psi_z$ effective at the slipping or shear plane, and not the true surface charge. If $\sigma_0$ is zero but a specifically adsorbed ion is present, $\sigma_z$ will not be zero. Even in the absence of specifically adsorbed ionic species there may be sources of charge or potential, such as ion exclusion or oriented water molecules between the surface and the shear plane. Thus, there is no guarantee that the true surface charge or potential will be zero when electrokinetic methods indicate zero charge or potential even if specific adsorption is absent.

The negative logarithm of the PDI activity corresponding to zero electrokinetic potential under the hypothetical restriction that no electrolyte other than the PDI is present, is called the isoionic point, IP. The IP has been defined for proteins and in this instance $H^+$ and $OH^-$ are the only ions allowed to be present. The corresponding term if an indifferent electrolyte is present is the Isoelectric Point, IEP. The PZC and IP should be unique characteristics of the solid. To a rough approximation, the IEP, IP, and PZC should be the same if only indifferent electrolytes are present, but oriented adsorption of water or ion exclusion may lead to a difference.

K

These definitions take no account of the zero-charge condition of solids which do not develop charge by PDI adsorption. With clays, the usual practice seems to be to adopt $H^+$ and $OH^-$ as potential determining and to use the above definitions even though the principal charge on clays is structural and is independent of pH. If the charge is determined by adsorption either of potential determining ions of the second kind, or of specifically adsorbed species unrelated to the solid (e.g. if the variation of charge with concentration of such species is being studied at constant PDI activity) the electrolyte concentration corresponding to zero electrokinetic potential ($\Psi_z = 0$) is termed the Reversal of Charge Concentration, RCC (Ottewill, personal communication). The negative logarithm of the RCC can be referred to as the Point of Zeta Reversal, PZR (Fuerstenau, 1971). The exhaustive review of reversal of charge data presented by Bungenberg de Jong (1949) consists largely of RCC data for systems described in Section 4.2.3.12. Presumably, these terms can also be used when dealing with solids such as hydrocarbons for which there are no PD species. Apparently, no term has been defined to describe the PDI activity corresponding to zero charge when specifically adsorbed species contribute charge and the PDI must compensate to achieve $\sigma_z = 0$. It can be referred to as an *apparent-PZC*.

There is much confusion in the use of these terms within each of the fields which employ them, and still more when usage is transferred from one field to another. It is quite common to refer to any condition corresponding to zero electrokinetic potential as an IEP. The same term is sometimes used to describe the pH at which the net proton charge on proteins is zero (see Section 4.2.3.9), and the pH at which the concentrations of all species in a suite of related positive and negative complexes of the same ligand result in zero net charge being carried by the suite when no solid is present, i.e. the IEP of a suite of hydroxo complexes of a metal, M, is the pH at which

$$\sum_n (n - z) [M(OH)_n^{z-n}] = 0,$$

or the IEP of a suite of chloro complexes of M is the pCl at which

$$\sum_m (m - z) [M(Cl)_m^{z-m}] = 0,$$

This latter usage will be distinguished as IEP(aq).

In the preceding and following discussions it has been assumed that the PZC, IP, and IEP are identical in the absence of specific adsorption.

(vi) *Empirical PZC's and IEP's for salts of simple ions.* The PZC's and IEP's of salts are highly variable. This is illustrated by the data in Table 4.2 which

TABLE 4.2

*PZC's and IEP's for salts of simple ions*[a]

| Salt | IEP | PZC | Charge at equiv. pt, |
|------|-----|-----|----------------------|
| AgCl | pAg 4·1–4·6 | nd | |
| AgBr | pAg 4·2–5·4[b] | 5·6–5·9 | |
| AgI | pAg 5·1–6·2[c] | 4·8–6·05 | |
| $CaF_2$ | pCa 2·6–7·7 | nd | |
| $BaSO_4$ | pBa 3·9–7·0 | nd | $-(9), +(7)^d$ |
| $PbSO_4$ | pPb 5·1 | 4·5 | $+(4)^d$ |
| $CaWO_4$ | pCa 4·8 | | |

[a] References: Parks (1967), Honig and Hengst (1969), Fuerstenau and Healy (1972).
[b] IEP's as low as 1·3 have been observed.
[c] IEP's as low as 2 have been observed.
[d] Number in parentheses indicates number of investigators reporting observed charge at equivalence point.

is based on reviews by Parks (1967), Honig and Hengst (1969) and Fuerstenau and Healy (1972). Honig and Hengst have attributed the variability to variable non-stoichiometry and variable impurity content, i.e. to variable intrinsic and/or extrinsic lattice defect concentrations and types.

#### 4.2.3.4. *Oxides and hydroxides*

(a) *Origins of charge.* Interaction with water hydroxylates the surfaces of oxides, i.e. oxides such as MgO, $Al_2O_3$, $Fe_2O_3$ and $SiO_2$, have hydroxide surfaces, as must solid hydroxides themselves (see e.g. Parks, 1967). The reaction of cations in fresh fracture surfaces with water to produce the hydroxylated surface is equivalent to a surface hydrolysis. A surface MOH groups have Brønsted acidity, and charge can develop by amphoteric dissociation or hydrolysis (equivalent to desorption and adsorption of $H^+$),

$$\left.\begin{array}{l} MOH \rightleftarrows MO^- + H^+ \\ H^+ + MOH \rightleftarrows MOH_2^+ \end{array}\right\} \tag{14}$$

$MOH_2^+$ is equivalent to a hydrated $M^+$. This is probably the principal mechanism of charge development on insoluble oxides such as crystalline $SiO_2$, $Al_2O_3$ and $Fe_2O_3$ (Healy and Fuerstenau, 1965; Parks, 1965).

Oxides and hydroxides must also establish equilibrium with respect to dissolution. For the more soluble hydroxides such as $Mg(OH)_2$ or $Zn(OH)_2$, charge may also develop by non-stoichiometric dissolution as with $BaSO_4$. The dissolved cation hydrolyzes, producing a suite of hydroxo complexes which may re-sorb, offering still another mechanism of charge development (Parks and de Bruyn, 1962). Matijevic (cited in Parks, 1967) has suggested that the polymeric species among the suite of hydroxo complexes should adsorb more strongly than the others and dominate surface charge. However, James and Healy (1972) contest this claim.

*Solid space charge.* Onoda and de Bruyn (1966) have suggested that $H^+$ may diffuse into oxides producing a positive interior space charge, and that $H^+$ may diffuse out of hydroxides and hydrous oxides leaving a negative interior space charge. Obviously these processes would alter the potential and charge of the surface.

(b) *Potential determining ions and the PZC.* $H^+$ is probably the principal sorbed species controlling the $\sigma_x$ of insoluble oxides. If PZC is expressed as a pH, equations (4.11) and (4.12) can be written

$$\Psi_0 = 2 \cdot 3 \frac{RT}{F} (\text{PZC} - \text{pH})$$

or, at 25°C,                                                                                                     (4.5)

$$\Psi_0 = 59 \, (\text{PZC} - \text{pH}) \, \text{mV}.$$

If the cation, $M^{z+}$, of a metal having a relatively soluble hydroxide $(M(OH)_z)$ is potential determining, $\Psi_0$ is still uniquely determined by pH if the solid is the only source of $M^{z+}$ in the system, thus

$$\Psi_0 = \frac{RT}{zF} \ln \frac{a_M}{a_M \, (\text{at PZC})}$$

but

$$a_M = K_{s0}(\text{OH}^-)^z = K_{s0}K_w^{-z}(\text{H}^+)^z$$

and

$$\Psi_0 = 2 \cdot 3 \frac{RT}{zF} \log \frac{K_{s0}K_w^{-z}(\text{H}^+)^z}{K_{s0}K_w^{-z}(\text{H}^+)^z_{\text{PZC}}} = 2 \cdot 3 \frac{RT}{F} (\text{PZC} - \text{pH})$$

In general, it is probably appropriate and simplest to regard $H^+$ as the PDI and to treat adsorption of metal hydroxo complexes as specific adsorption in response to $\Psi_0$ (Healy and Jellett, 1967). This approach does not contradict Matijevic's viewpoint since the specifically adsorbed hydroxo complexes are in the Stern Layer, and hence contribute to, and may indeed dominate, the effective or observable surface charge—the charge responsible for electrokinetic and coagulation behaviour.

(i) *Failure of the Nernst equation.* The activity of the PDI on the solid surface, $a_{i,s}$, is unlikely to be independent of $\sigma_0$ on insoluble oxides since the PDI is present on the solid only on the surface (Berubé and de Bruyn, 1968). The Nernst equation is to be considered approximately valid near the PZC, but the calculated $\Psi_0$ rises more rapidly than $\Psi_0$ determined empirically by measuring $\Gamma_{H+}$ directly, or derived from measured $\Psi_z$ (Hunter and Wright, 1971). Hunter and Wright (1971) and Ahmed (1972) have attempted to refine the Nernst equation itself. Lai and Fuerstenau (1972) have offered a direct relationship between $\Gamma_{H+}$ or $\sigma_0$ and $H^+$ activity. A promising approach to the derivation of direct relationships between $\sigma_0$ and $H^+$ activity, based on the existing theory of protein sorption is described briefly in Section 4.2.3.9.

(ii) *Empirical PZC's and IEP's for oxides and hydroxides.* Empirical data for many oxides and hydroxides, in forms ranging from freshly precipitated amorphous gels to well crystallized macrocrystalline minerals, have been summarized by Parks (1965). In general, the PZC's of oxides range from pH $\sim 0$ for some silica gels to pH $> 12$ for $Mg(OH)_2$, and correlate roughly with the ratio of cationic charge to radius after correction for coordination number and crystal field stabilization energy. The ranges of PZC's observed are illustrated in Table 4.3 which emphasizes the extreme variability mentioned earlier. Amorphous hydrous materials are, in general, likely to have more basic PZC's than well crystallized, anhydrous materials.

TABLE 4.3

*Ranges of PZC's for oxides and hydroxides*

| $M_2O$, MOH | PZC $>$ pH 11·5 | |
|---|---|---|
| MO, $M(OH)_2$ | 8·5 $<$ PZC $<$ | 12·5 |
| $M_2O_3$, $M(OH)_3$ | 2 $<$ PZC $<$ | 11 |
| $MO_2$, $MO_2 \cdot nH_2O$ | 0 $<$ PZC $<$ | 7·5 |
| $M_2O_5$, $MO_3$ | $<$ PZC $<$ | 0·5 |

Among the more common oxides and hydrous oxides in marine environments are silica, manganese oxides and ferric oxides.

*Silica.* The PZC of silica ($SiO_2$) in all forms is invariably more acid than pH 3·7 (Parks, 1965, 1967) and, unless altered by specific sorption, the surface charge will be negative in all natural waters.

*$MnO_x$.* The PZC of $MnO_x$ varies from pH 1·5 $\pm$ 0·5 to pH 7·3 $\pm$ 0·2 according to composition and structure (Healy, *et al.*, 1966) and perhaps ion exchange capacity (Parks, 1967). Observed PZC's and IEP's are listed in Table 4.4.

TABLE 4.4

*PZC's and IEP's of manganese oxides*

| | PZC or IEP, pH | Method (see footnote) | Ionic Str. mol l⁻¹ | Comments | Source |
|---|---|---|---|---|---|
| δ-MnO₂ (birnessite) | PZC, 2.8 ± 0.3 | Ti | $I \gtrsim 0.01$, Na⁺, H⁺, Cl⁻ | Synth. $KMnO_4 + Mn^{2+} = MnO_x$ $x = 1.9$ to $1.95$ | Morgan and Stumm (1964) |
| | IEP, 1.5 ± 0.5 | Coag. | $I < 0.001$, NaNO₃ | Synth. $KMnO_4 + HCl$. IEP increases with I above 0.001. | Healy *et al.* (1966) |
| | PZC, ~4 | Ti | $I \leqslant 0.1$ | | Jenkins and Stumm (1971) |
| | PZC, 2.2 ± 0.1 IEP, <2.5 | $\Gamma$; Na⁺, K⁺ mep. | $I \leqslant 0.01$ NaCl or KCl | Synth. $MnO_4^- + Mn^{2+} = MnO_x$ $x = 1.92$ | Murray (1972) |
| Mn(II)-manganite | IEP, 1.8 ± 0.5 | Coag. mep. | $I < 0.001$, NaNO₃ | Synth. $KMnO_4 + HCl$. IEP increases with I above 0.001 | Healy, *et al.* (1966) |
| | PZC, 1.8 ± 0.5 | $\Gamma$; Li⁺, Na⁺, K⁺ | $I < 0.01$, M⁺, Cl⁻ | $\Gamma$ indep. pH if $I > 0.1$ | Murray, *et al.* (1968) |
| α-MnO₂ (cryptomelane) | IEP, 4.5 ± 0.5 | Coag. mep. | $I < 0.001$, NaNO₃ | Synth. by heating Mn(II) manganite in aq. KCl. IEP increase with I above 0.001 | Healy, *et al.* (1966) |
| β-MnO₂ (pyrolusite) | IEP, 7.3 ± 0.2 | Coag. mep. | $I < 1.0$, NaNO₃ | Baker Analyzed Reagent IEP indep. of I for $I < 1.0$ | Healy, *et al.* (1966) |
| | IEP, 4.6 ± 0.2 | mep. | $I < 0.01$, Na⁺, Cl⁻ or NO₃⁻ | — | Malati, *et al.* (1969) |
| γ-MnO₂ | IEP, 5.5 ± 0.2 | Coag. mep. | $I < 1.0$ NaNO₃ | Synth. electrolytic, IEP indep. I up to 1.0 | Healy, *et al.* (1966) |

*Coag:* IEP estimated as pH of maximum rate of coagulation

*Ti:* $(\Gamma_+ - \Gamma_-)$ estimated by potentiometric titration

*Mep:* microelectrophoresis

$\Gamma$: $\sigma_o$ assumed zero when $\Gamma$ of indifferent ion drops to zero

In sea water, unless the high salinity affects its behaviour drastically, $MnO_x$ will be negative, but its capacity for non-specific adsorption will be extremely variable, reflecting the variability of the PZC ($\Gamma = f(\Psi_0)$; $\Psi_0 = f$(pH and PZC)).

The surface charge density at pH 8 ($I < 0.01$ mol l$^{-1}$) ($\sigma_0$ as estimated by adsorption of Na$^+$) on $\delta$-MnO$_2$ is much higher than that on SiO$_2$, which has about the same PZC (Murray, 1972). $\delta$-MnO$_2$ sorbs Mn$^{2+}$ with a capacity of 0.5 mol Mn(II) per mol MnO$_2$ at pH 7.5 (Morgan and Stumm, 1964).

As would be expected, adsorption of Na$^+$, K$^+$, and Li$^+$ onto Mn(II) manganite is pH dependent at ionic strengths below 0.01 mol l$^{-1}$, but becomes independent of pH above 0.1 mol l$^{-1}$. Ni$^{2+}$, Co$^{2+}$, and Cu$^{2+}$ are specifically sorbed by Mn(II) manganite (Murray et al., 1968).

The negative surface charge on $\beta$-MnO$_2$ is reversed by specific sorption of Ba$^{2+}$. The RCC at pH 10 is $6.6 \times 10^{-4}$ mol l$^{-1}$ in $10^{-3}$ mol l$^{-1}$ NaCl. Mn$^{2+}$ reverses the charge also, but at higher concentration (Malati et al., 1969).

*Fe(III) oxides and hydroxides.* Observed PZC's and IEP's of Fe(III) oxides and hydroxides are summarized in Table 4.5. If specifically adsorbed anions, such as silicate, sulphate (and perhaps carbonate) and organic species, are absent, the charge on precipitated iron oxides should be positive in sea water whereas that of crystalline haematite should be negative; the charge on goethite should be zero or negative. However, coprecipitation of anions shifts the IEP. Thus, an amorphous precipitate containing Cl$^-$ (mol ratio Cl$^-$/Fe$_2$O$_3$ = 0.005) has an IEP of 7.2, and a precipitate containing SO$_4^{2-}$ (SO$_4^{2-}$/Fe$_2$O$_3$ = 0.014) has an IEP of 7.0 (see references cited in Parks, 1965). Furthermore, specific adsorption of silicate at $1.2 \mu$ mol l$^{-1}$ shifts the PZC of goethite from 8.5 to 7.5 (Hingston et al., 1972). Specific adsorption of sulphate also shifts the IEP of goethite dramatically. A goethite with an IEP (in KCl solution) of 7.5 has an IEP of 6 in the presence of $10^{-6}$ M sulphate ($I = 10^{-2}$) and of 2.5 in the presence of $3 \times 10^{-2}$ M sulphate! (Kavanagh and Healy, 1972). Information on the influence of impurities has been summarized by Parks (1965).

Hydrous Fe(III) oxide precipitated from sea water coprecipitates several metals effectively enough to serve as an analytical collector (see e.g. Kim and Zeitlin, 1972, 1971a, b and Chapter 19). Some examples of metals collected and the optimum pH for the hydrolysis are:

Mo, present as MoO$_4^{2-}$         : pH 4.0

U, present as (UO$_2$(CO$_3$)$_3$)$^{4-}$ : pH 6.7

Zn(II)                              : pH 7.6

Cu(II)                              : pH 7.6

TABLE 4.5

*PZC's and IEP's of Fe(III) oxides and hydroxides*

| | PZC or IEP, (pH) | Method (see footnote) | Ionic Strength mol l$^{-1}$ | Comments | Source |
|---|---|---|---|---|---|
| Amorphous Fe$_2$O$_3 \cdot$nH$_2$O or Fe(III) "hydroxide" | 8·5 to 8·8 | Various | Various | Range of precipitated materials free from Cl$^-$ and SO$_4^{2-}$ and never dried | Parks (1965) |
| α-FeOOH (Goethite) | 5·9 to 6·7 PZC, 7·8 to 8·3 | Various Titrn[a] | Various $I \leqslant 1 \cdot 0$, NaCl | Synthetic and natural PZC depends on preparation history | Parks (1965) Hingston et al. (1972) |
| α-Fe$_2$O$_3$ | 8·0 to 9·03 | Various | Various including 1·0, NaCl | Synthetic, never dried | Parks (1965) |
| | PZC, 8·4 to 9·3 | Titrn[a] | $I \leqslant 1 \cdot 0$, NaCl | Synthetic | Atkinson et al. (1967) |
| | 4·2 to 6·9 | Various | Various | Natural | Parks (1965) |

[a] $(\Gamma_H - \Gamma_{OH})$ estimated from potentiometric titration

It might be inferred from these data that the PZC of the Fe(III) precipitate lies between 6·7 and 7·6. A PZC in that range is possible either through coprecipitation or specific sorption of the $Cl^-$ or $SO_4^{2-}$ from sea water. However, specific sorption of the metals is more likely since silicate (Hingston et al., 1972), sulphate (Kavanagh and Healy, 1972) and all hydrolyzable metal ions (James and Healy, 1972a, b, c) are adsorbed specifically. It should be recalled that a knowledge of surface charge is useful for predicting *non-specific* adsorption, but may be irrelevant for specific adsorption.

### 4.2.3.5. *Salts of weak acids*

(a) *Origins of charge.* If the solid is a salt of a weak acid or base, such as calcite or apatite, hydrolysis complicates the chemistry of charge development. Charge may develop by non-stoichiometric dissolution, but the dissolution products hydrolyze, and resorption of hydrolysis products is another candidate mechanism, as has been suggested for $CaCO_3$ by Samasundaran and Agar (1967) and for apatite by the same authors (1972). Hydrolysis may also occur on the surface, leading to charge development by $H^+$ adsorption (Fuerstenau and Miller, 1967). Regardless of the mechanism, $Ca^{2+}$ and $CO_3^{2-}$ will be present in the solution and on the surface, and equation (4.10) and its derivatives can be used to relate $\Psi_0$ to solution composition, i.e. regardless of the mechanism of charge formation, the principal anion and cation of the solid must be potential determining. However, because these ions hydrolyze their activities change with the concentrations of $H^+$, $OH^-$ or any of the hydrolysis products; thus all of these species are potential determining ions of the second kind. The concept of a PZC is complicated in systems with several PDI. This can be illustrated by two examples.

(b) *PDI and pH dependence of $\Psi_0$.* The two PDI of the first kind for $CaCO_3$ are $Ca^{2+}$ and $CO_3^{2-}$. If the activity of one is fixed, that of the other is also fixed by virtue of the solubility product relationship. Thus, the PZC is determined by the $CO_3^{2-}$ activity. If the solid itself is the only source or sink for calcium and carbonate in the system, there should be a unique pH corresponding to the zero charge condition, a "pH(PZC)". The $CO_3^{2-}$ activity is determined under these conditions by the activity of $H^+$, the solubility product, $K_{s0}$, and the first and second dissociation constants of carbonic acid, $K_1$ and $K_2$, as follows:

$$(CO_3^{2-})_{PZC} = \left[ \frac{K_{s0}}{1 + \dfrac{(H^+)_{PZC}}{K_2} + \dfrac{(H^+)_{PZC}^2}{K_1 K_2}} \right]^{\frac{1}{2}} \qquad (4.16)$$

Substitution of this expression for $(CO_3^{2-})_{PZC}$ into the appropriate form of equation (4.10) and its derivatives, enables $\Psi_0$ to be obtained as a function of $H^+$. Because the carbonate concentration corresponding to the true PZC is unique, the $(H^+)_{PZC}$ appearing in equation (4.16), is also unique, and in terms of pH is written pH(PZC).

If there are other sources of calcium or carbonate in the system there is no unique PZC on the pH scale; instead, there is a different pH(PZC) for each solution composition. If it is assumed that an independent source of carbonate produced C mol $1^{-1}$ of carbonate in addition to that derived from dissolution of calcite and that a source of calcium contributed M mol $1^{-1}$, the relationship between $(CO_3^{2-})_{PZC}$ and $(H^+)_{PZC}$ becomes

$$0 = \left\{ (CO_3^{2-})_{PZC}^2 \left[ 1 + \frac{(H^+)_{PZC}}{K_2} + \frac{(H^+)_{PZC}^2}{K_1 K_2} \right] + (CO_3^{2-})_{PZC}[M - C] - K_{s0} \right\}$$

and the PZC on the pH scale is a function of the amounts of extra calcium and carbonate in the system.

If $M$ is large, i.e. if an excess of the cation is added, equation (4.17) reduces to $(CO_3^{2-})M = K_{s0}$ and $CO_3^{-2}$ is fixed; thus, an excess of the cation removes the pH dependence of $\Psi_0$. This has been observed for scheelite, $CaWO_4$, (Choi, 1963; O'Connor, 1957). An excess of anion does not remove the pH dependence of $(CO_3^{2-})$ or $\Psi_0$.

Interdependence of $\Psi_z$, pH and pCa have been demonstrated for fluorite, $CaF_2$, (Choi and Han, 1963). $H^+$, $OH^-$, and $H_2PO_4^-$ have been shown to play potential determining roles for apatite (Somasundaran and Agar, 1972). The apatites have more than two PDI of the first kind. For example, $Ca^{2+}$, $PO_4^{3-}$ and $OH^-$ should be PD for hydroxyapatite, $Ca_5(PO_4)_3OH$. No quantitative study of the validity of analyses such as these is yet available.

(c) *Empirical surface charge, PZC's and IEP's of phosphates and carbonates.* The interpretation given above requires that the surface charges ($\sigma_z$) of phosphates and carbonates are pH dependent and vary with the concentration of $Ca^{2+}$ or $Mg^{2+}$ and $CO_3^{2-}$ or $PO_4^{3-}$. PZC's and IEP's for several salts estimated in water solutions of the solid alone, or in solutions of an indifferent electrolyte are collected in Table 4.6.

(i) *Apatites.* The effects of added $CaCl_2$ and $KH_2PO_4$ on the surface charge (actually $\Psi_z$) of a natural crystalline fluorapatite at pH 8, and on the PZC of a synthetic hydroxyapatite are illustrated in Fig. 4.2. The solid is negative under most conditions, but a PZC is observed at pCa = 3·2. At higher $Ca^{2+}$ concentrations $\Psi_z$ is positive.

Bell *et al.* (1972) attributed the fact that the pH(PZC) observed by them is greater than that reported by Somasundaran (1968) and by Somasundaran

TABLE ? ... *of phosphates and carbonates*

| | PZC or IEP, (pH) | Method (see footnote) | Ionic Strength mol l$^{-1}$ | Comments | Source |
|---|---|---|---|---|---|
| Eggonite, AlPO$_4\cdot$2H$_2$O Strengite, FePO$_4\cdot$2H$_2$O Monazite, (Ce, La, Th)PO$_4$ | PZC, 4 PZC, 2·8 IEP, 3·4 | Dr Dr Mep | | Synthetic Synthetic Natural | From review by Parks (1967) |
| Fluorapatite, Ca$_5$(PO$_4$)$_3$(F, OH) | IEP, 5·6 | Sp | 0·01, KNO$_3$ | Nat'l Crystalline, aged hours | Somasundaran (1968) |
| | IEP, ~4 | Sp | 0·01, KNO$_3$ | Nat'l Crystalline, aged hours | Somasundaran and Agar (1972) |
| | IEP, ~6 | Sp | 0·01, KNO$_3$ | Nat'l Crystalline, aged days | Somasundaran and Agar (1972) |
| Fluorapatite, Ca$_5$(PO$_4$)$_3$(F, OH) | PZC, 6·9 ± 0·2 | Ti | 0·01–1·0, KCl | Synthetic, aged 504 hours | Bell et al. (1972) |
| Hydroxyapatite | PZC, 8·5 ± 0·2 | Ti | 0·01–1·0, KCl | Synthetic, aged 504 hours | Bell et al. (1972) |
| Calcite, CaCO$_3$ | IEP, 10·8 | Sp | | Natural, aged minutes | Somasundaran and Agar (1967) |
| | IEP, 9·5 | Sp | | Natural, aged 2 months log ΣCa = −3·5; log ΣC$_3$ = −3 | |
| Dolomite, CaMg(CO$_3$)$_2$ | IEP, <8·5 | Sp | 0 to 0·01, KCl | Natural, aged 30 minutes | Predali (1970) |
| Magnesite, MgCO$_3$ | IEP, 5·2 ± 0·1 | Sp | 0 to 0·01, KCl | Natural, aged 30 minutes | Predali (1970) |

*Dr*: PZC is the pH at which salt can be synthesized and aging produces no pH drift    *Ti*: (Γ$_H$ − Γ$_{OH}$) determined by titration
*Mep*: microelectrophoresis    *Sp*: streaming potential

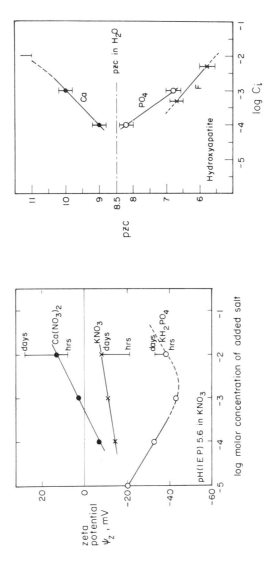

Fig. 4.2. Effect of Electrolyte Additions at pH 8 on the Zeta Potential, $\Psi_z$, of a fluorapatite and on the PZC of a synthetic hydroxyapatite. Data from Somasundaran and Agar (1968, 1972) and Bell et al. (1973).

and Agar (1972), to possible specific adsorption of $CO_3^{2-}$ or $HCO_3^-$ (from dissolution of atmospheric $CO_2$) or to impurities in natural apatites. Sorption of Al(III) or Fe(III) masks the inherent $\Psi_z$ of the apatite (Aleinikov and German, 1967). The IEP's of coprecipitates of hydrous oxides of Fe(III), Al(III) and Si(IV) incorporating P(V) vary roughly linearly with the mole fraction of phosphorus (Parks, 1967).

If these data can be extrapolated to sea water, the charge on fluorapatite should be negative, and that on hydroxyapatite should be slightly positive in this medium.

(ii) *Carbonates*. Dodecyl sulphate and dodecylamine ions sorb onto calcite, but do not shift the IEP significantly, and thus are not specifically adsorbed (Somasundaran and Agar, 1967).

$Ca^{2+}$ and $Mg^{2+}$ decrease the negative charge on dolomite and magnesite (Predali, 1970). The charge on both minerals reverses at pMg $\sim 2 \cdot 7$ (pH = 10). $CO_3^{2-}$ increases the negative charge on these minerals. If these results can be extrapolated to sea water conditions the charge on calcite should be positive and that on magnesite (and probably also dolomite) negative.

### 4.2.3.6. *Non-Clay silicates*

The PZC's of silicates are determined by the composition of the exposed surface of the mineral. It has been shown (see e.g. Smolik *et al.*, 1966; Parks, 1967; Smith and Trivedi, 1971; Parks and Luce, 1972) that weathering selectively and progressively depletes the silicate surfaces of cations of lower charge. Since oxides of higher valent cations have the more acid PZC's, the PZC's of silicates become more acidic, or the surface charge becomes more negative as weathering proceeds. A maximum PZC can be predicted from the composition (Parks, 1967). The physical form or habit of the mineral particle also affects the PZC; chrysotile ($Mg_3Si_2O_5(OH)_8$) should have, according to its bulk composition, a PZC near pH 9. However, the observed values for samples exposed to water for only a short time are between pH 10 and 11·8 and apparently reflect the fibrous habit which exposes a magnesium rich surface (Parks, 1967).

No investigation in sea water media of the IEP's of minerals common in marine sediments or suspended solids, has been reported. Empirical IEP's of a few silicates related to some of marine significance are presented in Table 4.7.

Co-precipitated hydrous oxides of Al(III) with Si(IV), Fe(III) with Si(IV), and some complex hydrous $M_2O_3$–$P_2O_5$–$SiO_2$ coprecipitates have IEP's which vary roughly linearly with composition between the IEP's of the component hydrous oxides (Parks, 1967).

TABLE 4.7

*Isoelectric points (IEP's) of silicates*

|  |  | Source[a] |
|---|---|---|
| Aluminum silicates (see also 4.2.3.7) |  |  |
| $Al_2SiO_5$ (kyanite, andalusite, sillimanite) | 5·2–7·9[b] | Smolik *et al.* (1966) |
| Feldspars |  |  |
| $NaAlSi_3O_8$–$KAlSi_3O_8$ (albite–orthoclase) | 2·0–2·4 |  |
| microcline ($KAlSi_3O_8$, in 0·01 m KCl, aged 3 hours) | <2 | Warren and Kitchener (1972) |
| Magnesium Silicates ($Mg_3Si_2O_5(OH)_8$) |  |  |
| chrysotile | 10–11·8 |  |
| chrysotile: freshly immersed | >12 | Smith and Trivedi (1971) |
|               aged 70 days | ~9 | Smith and Trivedi (1971) |
| lizardite | 9·6 |  |
| serpentine (aged 1 to 50 days) | 6·6–8·9 | Smith and Trivedi (1971) |
| Forsterite ($Mg_2SiO_4$) | ⩽8·4[b] | Luce and Parks (1972) |
| Cummingtonite ($(Fe, Mg)_7(Si_4O_{11})_2(OH)_2$) | 5·2 |  |

[a] If no source is given, data have been selected from the review by Parks (1967).
[b] Range reflects differences in history. IEP becomes more acidic as sample equilibrates in water.

Synthetic calcium silicates and calcium silicate hydrates (Stein, 1968) are negative (i.e., $\Psi_z$ is negative) in 0·005 to 0·02 mol $1^{-1}$ NaOH solutions. $Ca^{2+}$ adsorbs and reverses the surface charge. RCC's (see p 258) in 0·01 mol $1^{-1}$ NaOH are:

| | |
|---|---|
| $\alpha$-$CaSiO_3$, pseudowollastonite | 1·3 to 3·1 $\times 10^{-3}$ mol $1^{-1}$ $Ca^{2+}$ |
| $\beta$-$CaSiO_3$, wollastonite | 2·5 $\times 10^{-3}$ |
| amorphous $CaSiO_3$ | 3·3 $\times 10^{-3}$ |
| $\alpha$-$Ca_3Si_2O_7$, rankinite | 1·3 $\times 10^{-3}$ |
| $\gamma$-$Ca_2SiO_4$ | 1·8 $\times 10^{-3}$ |
| Xonotlite | 4·3 $\times 10^{-3}$ |
| Tobermorite, e.g. $5CaO \cdot 6SiO_2 \cdot 9H_2O$ | 4·4 $\times 10^{-3}$ |

The effect of $Ca^{2+}$ at lower pH values is difficult to predict. If specific adsorption of $Ca^{2+}$ is responsible for charge reversal, a lower concentration of $Ca^{2+}$ should suffice at lower pH. However, the activity product $[Ca^{2+}][OH^-]^2$ at the RCC's listed is close to the solubility product of

$Ca(OH)_2(s)$, so the RCC may be a manifestation of $Ca(OH)_2$ precipitation on the silicate surface.

### 4.2.3.7. Clays

Most clay minerals are layered structures consisting of sheets of interbonded $SiO_4$ tetrahedra linked via shared oxygen ions with sheets of $AlO_6$ octahedra. Their charge originates mainly by isomorphous substitution of Al(III) for Si(IV) at tetrahedral sites or of Mg(II) or Fe(II) for Al(III) at octahedral sites. The resulting negative charge is internal to the structure; it is not a surface charge. It might be called a structural or intrinsic charge. Unlike the flat surfaces, however, the edges of layers terminate in broken or unsatisfied Si—O or Al—O bonds. These hydroxylate in water, producing SiOH and AlOH sites capable of amphoteric dissociation. Like the surfaces of oxides and non-clay silicates, the edges of clay particles develop a pH-dependent true "surface" charge ($\sigma_0$ (edge)) which may be positive or negative and has a PZC.

The structural charge on clays is fixed by the composition of the clay and is independent of solution composition. As it is not determined by adsorption of ions from solution there are no particular PD species. The charge is constant and consequently there is no PZC associated with the structural charge. It is distributed throughout the particle volume, hence it interacts with the surroundings mainly through the flat layer surfaces.

The net charge on the clay particle is the algebraic sum of the structural and edge charges. When the edge charge is negative the particle charge is negative. When the edge charge is sufficiently positive to neutralize the face charge (on the acid side of the edge's PZC) the particle charge is zero. The pH corresponding to zero particle charge is not a PZC for two reasons. The PZC is defined in terms of the PDI, and the particle charge is not determined exclusively by PDI adsorption. The surface potential should be zero at the PZC (equations (4.12a), (4.12b)). The potentials between the particle faces and the solution and between the particle edges and the solution are certainly not zero when particle charge is zero since a net edge charge is required to neutralize the permanent structural charge.

The pH at which the net particle charge is zero in indifferent electrolytes is best called an Isoelectric Point or pH(IEP) in recognition of the fact that electrophoretic mobility, for example, will be zero at that pH. If the maximum edge charge is small relative to the structural charge, it may not be possible to acquire zero net charge without specific adsorption of a positive species. The pH(IEP) of kaolinite and montmorillonite are given in Table 4.8.

When the edge charge is positive, electrostatic interaction between the positive edges and negative faces of clay particles leads to interesting proper-

ties. The gelling and high viscosity of very dilute suspensions of mont-
morillonites, for example, is attributable to edge-to-face coagulation leading
to a cellular structure enclosing large amounts of water (Van Olphen, 1963).

TABLE 4.8

*pH(IEP) for kaolinite and montmorillonite*

| Mineral | pH(IEP) | Source |
|---------|---------|--------|
| Kaolinite | 3·3 to 4·6 | Parks (1967) |
| | ~3·3 | Buchanan and Oppenheim (1968) |
| | <2 | Vestier (1969) |
| Montmorillonite | ≤2·5 | Parks (1967) |
| | <2 | Touret (1970) |

Vestier found IEP < 2 for kaolin in $Li^+$, $NH_4^+$, $H^+$, $Ca^{+2}$, and $Mg^{+2}$ forms at an ionic strength of 0·01 mol $l^{-1}$ (KCl).
Touret found IEP < 2 for montmorillonite in NaCl, $CaCl_2$ and $UO_2(NO_3)_2$ solutions from 0 to 0·01 mol $l^{-1}$.

When the edge charge is positive, the edge should adsorb anions whereas
the face of the same particle should adsorb cations. The clay should be
simultaneously an anion and a cation exchanger.

(a) *Constant potential and constant charge systems.* When a solid develops
charge by adsorption of PDI, its potential is fixed by the activities of the
PDI and it is referred to as a *constant potential system.* Its charge adjusts to
the potential and the solution composition. Thus, *in the absence of specific
adsorption,*

$$\sigma_0 = -\sigma_G = -F(\Sigma z_+ \Gamma_{G_+} + \Sigma z_- \Gamma_{G_-}) \tag{4.18}$$

in which $\Gamma_{G_+}$ and $\Gamma_{G_-}$ are the adsorption densities of positive and negative
species in the Gouy Layer and are given by equation (4.2), $\sigma_0$ is given more
directly by equation (4.3). At constant surface potential, charge increases
as the square root of the electrolyte concentration.

The structural charge of a clay leads to a potential difference between the
solid and liquid phases, but it is the potential which must adjust to the charge
and solution composition (Van Olphen, 1963) since the charge is constant.
If the total charge on a clay particle is dominated by structural charge, the
clay is referred to as a *constant charge system.* The potential and charge are
still related through equations (4.18) and (4.2) or (4.3) and the rest of the
properties of the double layer are unchanged.

(b) *Ion exchange.* The ion exchange capacity of a constant potential surface is $\sigma_G$ and depends on both the activities of the PDI (or $\Psi_0$) and the solution's electrolyte concentration. The ion exchange capacity of a clay has two components, the constant cation exchange capacity determined by the structural charge, and a variable capacity for anion or cation exchange associated with the edge charge.

### 4.2.3.8. *Hydrocarbons and air bubbles*

(a) *Simple electrolytes.* Hydrocarbons and air bubbles are generally thought to be negatively charged in water and simple electrolyte solutions (Davies and Rideal, 1963; Haydon, 1964) although not all authors agree (Haydon, 1964; MacKenzie, 1969; MacKenzie and O'Brien, 1969). However, the simple interpretation offered by Davies and Rideal is appealing. They suggested, when discussing the pH dependence of the charge (which becomes increasingly negative as the pH increases), that $H^+(aq)$ cannot approach the surface as closely as $OH^-(aq)$ because $H^+$ is more strongly hydrated and hence its hydrated radius is larger than that of $OH^-$. This is equivalent to negative sorption of $H^+$ or positive sorption of $OH^-$. The charge is pH dependent; it is not a true surface charge ($\sigma_0$), but resides somewhere in the Stern Layer. Apparently there is no quantitative theory relating potential or charge to solution composition.

Some authors prefer to interpret the exclusion of cations from the interface in terms of the differences in water structure induced by the hydrocarbon or air bubble and by cations (see Haydon, 1964; and Horne, 1969). Haydon (1964) has suggested that the same mechanisms should result in charge separation at all non-ionogenic surfaces.

(b) *Surfactant solutions.* Adsorption of surface active ions, such as alkyl sulphates, at the air/water or hydrocarbon/water interfaces results in a surface charge and attracts a diffuse layer of counter ions in the aqueous phase. The theory of the resulting diffuse double layer has been explored in detail, and qualitative agreement with experiment is good (Davies and Rideal, 1963; Haydon, 1964; MacKenzie, 1969).

Explicit isotherms relating adsorption density of surfactant ion, and hence surface charge, to solution concentration have been derived for limited conditions. The magnitude of the free energy of adsorption is known only for a few long chain sulphates (Davies and Rideal, 1963) and interpretation of the isotherms is not yet satisfactory (Haydon, 1964).

Adsorption of surface active weak electrolytes such as fatty acids and alkyl amines leads to a pH dependent surface charge on hydrocarbon droplets. Good qualitative correlation is observed between surface charge (electrophoretic mobility) and the degree of dissociation calculated from bulk solution pH and the dissociation constant of the surfactant. Nujol droplets are negative at all pH values in both pure water and Na-dodecyl sulphate solutions and positive at all pH values in cetyl trimethyl ammonium bromide medium. The charge is highly pH dependent in oleic acid and dodecylamine-hydrochloride solutions (Mackenzie and O'Brien, 1969).

If the adsorption density of a surfactant ion or the surface potential (or charge) is known, the theory of the diffuse double layer, modified to take account of penetration of counter ions into the surface charge "plane", describes adsorption of counter ions reasonably well (Haydon, 1964).

### 4.2.3.9. Proteins

(a) *Origin of charge.* The surface charge ($\sigma_0$) on macromolecular or particulate proteins unquestionably develops through the electrolytic dissociation or hydrolysis of functional groups which are part of the amino acid components of the protein. The most common functional groups, in neutral and charged form, together with intrinsic proton dissociation constants (as $pK_i$ values) are (Tanford, 1962);

$$-COOH \rightleftharpoons -COO^- + H^+ \qquad\qquad 4\cdot0 \text{ to } 4\cdot8$$

$$-\langle\bigcirc\rangle-OH \rightleftharpoons -\langle\bigcirc\rangle-O^- + H^+ \qquad 9\cdot4 \text{ to } 10\cdot8$$

$$-SH \rightleftharpoons -S^- + H^+ \qquad\qquad 9\cdot1 \text{ to } 9\cdot5 \text{ (est.)}$$

$$-NH_3^+ \rightleftharpoons -NH_2 + H^+ \qquad\qquad 9\cdot6 \text{ to } 10\cdot4$$

$$-NH-C\overset{NH_2^+}{\underset{NH_2}{\diagup\diagdown}} \rightleftharpoons -NH-C\overset{NH}{\underset{NH_2}{\diagup\diagdown}} + H^+ \qquad 11\cdot9 \text{ to } 12$$

$$-\underset{\underset{C}{HN\diagdown\;\diagup NH^+}}{C}=CH \rightleftharpoons -\underset{\underset{C}{HN\diagdown\;\diagup N}}{C}=CH + H^+ \qquad 6\cdot3 \text{ to } 7\cdot4$$

(b) *Potential and pH.* $H^+$ and $OH^-$ are the logical choice for PDI and, in principle, the Nernst equation should relate $\Psi_0$ and pH. Although the number of functional groups and hence the maximum charge, is fixed, the fraction ionized changes with pH, and proteins are, therefore, not fixed charge systems. However, as with the oxides, the activity of $H^+$ on the surface,

$a_{i,s}$, is unlikely to be constant. Since it is known that the charge originates in acid-base dissociation, it is more appropriate to use the theory of electrolytic dissociation to relate charge and pH (Overbeek and Bungenberg de Jong, 1949). To do this it is necessary to know the relative numbers of each functional group present and their dissociation constants.

Each functional group has an intrinsic dissociation constant, $K_i$, which is the dissociation constant observed for that group in simple molecules in which only one dissociable group is present. If several groups are present, $K_i$ is modified by adjacent polar groups or by the charge derived from dissociation of adjacent groups. Hydrogen ions will be less strongly bound to all sites on a positively charged molecule or surface. The theory required to correct $K_i$ for these effects is well developed as it is used in interpreting hydrogen ion titration curves for the purpose of extracting both the $pK_i$'s and the relative numbers of functional groups (Tanford, 1962; Nozaki and Tanford, 1967; Steinhardt and Reynolds, 1969). Alternatively, the net proton charge on colloidal proteins (average molecular charge due to bound protons or to groups which have lost protons) can be calculated for any pH if the titration curve is known (Tanford, 1962). If the functional group composition (types present and relative numbers of each type), the $pK_i$'s of each group and a valid theory for correcting the $K_i$ values are available, then of course, the titration curve and surface charge can be predicted. The pH values corresponding to zero net proton charge, (and to PZC), is referred to by Tanford (1962) as the "point of zero net proton charge". The pH of a suspension containing only the protein, $H^+$, $OH^-$ and water is called the isoionic pH (IP) and is very close to the point of zero net proton charge or PZC.

If the surface charge is known, $\Psi_0$ can be estimated from equation (4.3) as for constant charge surfaces provided that specific adsorption is absent and some relationship between $\Psi_0$ and $\Psi_d$ is assumed. If it is possible to assume a value for the thickness of the slipping plane, $\Psi_z$ can also be estimated. Good qualitative agreement has been found between the values of $\Psi_z$ derived from titration curves and from electrophoresis data (Overbeek and Bungenberg de Jong, 1949). Cann (1969) has concluded that electrical double layer and electromigration theory provide a satisfactory explanation. Failure to obtain quantitative agreement is attributable to the binding of ions from the solution (Cann, 1969; Nozaki and Tanford, 1967; Overbeek and Bungenberg de Jong, 1949), to oriented adsorption of dipoles such as water (Overbeek and Bungenberg de Jong, 1949), or to location of the ionizable groups at some depth within the surface (Haydon, 1964). In addition to these possibilities, there is also the classic problem of locating the slipping plane relative to the surface; this prevents unambiguous calculation of $\Psi_z$ from $\Psi_0$ and vice versa.

*Intrinsic dissociation constants of hydroxide groups on oxide surfaces.* Schindler and Kamber (1968), Stumm *et al.* (1970) and Schindler and Gamsjäger (1972) have applied similar approaches i.e. the definition of intrinsic acidic and basic dissociation constants, which vary with surface charge, to the problem of modelling the pH dependence of $\Gamma_{H^+}$ on metal oxides.

c. *Effect of salts.* If specifically adsorbed ions, i.e. ions capable of combining (binding or complexing) with the functional groups are present, the effective $K_i$ and therefore the titration curve is altered and the net surface charge is not the net proton charge. Steinhardt and Reynolds (1969) have given a detailed account of the multiple equilibria involved in simultaneous binding or adsorption of $H^+$, other ions and neutral molecules.

d. *Charges on natural proteins.* Most natural proteins will be negatively charged at the pH of sea water. There are two reasons for this. Acidic and amphoteric proteins are known, but basic proteins are uncommon (Overbeek and Bungenberg de Jong, 1949). The PZC's of amphoteric proteins lie in the pH range 4 to 7, judging from observed isoelectric points summarized by Houwink (1949) and from IP's reported by Tanford (1962) and Steinhardt and Reynolds (1969).

The charges on macromolecular proteins which are too small to observe microscopically can be studied by observing the electrophoretic behaviour of solid particles upon which the protein has specifically adsorbed. At a sufficiently high protein concentration the influence of the substrate is negligible (Overbeek and Bungenberg de Jong, 1949; Brinton and Lauffer, 1959).

e. *Viruses and cells.* Viruses and cells, like proteins, are negatively charged at high pH, at least in the buffers used to simulate body fluids. Biological materials are usually complex, consisting of several substances; for this reason the surface charge sensed by electrokinetic methods is an average, in the same way as is that of clays. Zero electrophoretic mobility, for instance, signifies zero net particle charge, not zero charge at all points on the particle. Apparent IEPs of viruses (in biochemical buffers of ionic strength 0·01 to 0·02) fall in the range of pH 3–6. Bacterial cells are more negative, having IEP's in the pH range 2–4. Brinton and Lauffer (1959) have attributed this difference to the presence of lipids (which might behave similarly to hydrocarbons) and glutamic acid in the cell capsule, and also to the relative paucity of basic amino acids in bacterial cells.

### 4.2.3.10. *Carbohydrates*

The commonest carbohydrates likely to be found among marine colloids are polysaccharides and their derivatives, e.g. starch, cellulose and chitin, which are all hydrophilic.

Starch:

Cellulose:

Chitin:

Starch and cellulose form hydrates. Starch, at least, alters the structure of water beyond the first monolayer (Tait, 1972). The abundant OH groups in these compounds are responsible for their hydrophilic behaviour.

(a) *Origin of charge.* Starches are negatively charged over a wide pH range (Banajee and Iwasaki, 1969). The charge changes with pH and is at a maximum at pH 8–9. The PZC, if any, is below pH 2. Banajee and Iwasaki have implicity assumed that the hydroxyl groups are not responsible for the charge. They attribute the charge to ionized anionic groups, perhaps polymer end groups. Houwink (1949) and Hermans (1949 also considered $-OH$ groups on proteins and cellulose to be incapable of dissociation.

The acidity of $-OH$ groups is a matter of degree. The hydroxide groups on alcohols and simple sugars are very weak acids. Rochester (1971) has reported the following ranges of dissociation constants;

$$\text{alcohols;} \quad ROH = RO^- + H^+ \qquad pK_a \sim 15.5$$

$$\text{sugars;} \quad R'OH = R'O^- + H^+ \qquad pK_a \sim 11.7\text{–}13.1$$

Phenol, the alcohols, and glycol, at least, can bind an extra proton in acid solutions, thus (Rochester, 1971):

$$CH_3OH + H^+ = CH_3OH_2^+ \qquad pK_b \sim 2.2$$

It is conceivable that the hydroxyl groups of polysaccharides are also very weak acids and bases and that their charge orginates from their dissociation, as with oxides. If a polysaccharide had $-OH$ groups with $pK_a$ and $pK_b$ values close to those of the sugars and alcohols respectively, its PZC should be below pH 5.5, $(\frac{1}{2}(pK_a - pK_b))$. This is consistent with its usual negative charge. Because the hydroxyl group is so much weaker as an acid than is the

—COOH groups of proteins, the magnitude of charge from this source must be small and indeed Weilaud (1959) has stated that sugars and poly-alcohols are uncharged unless complexed with boric acid.

Since the polysaccharides bind water and alter its structure, cation exclusion may also contribute to the negative charge at the air–water surface and on hydrocarbons (Haydon, 1964; Horne, 1969) as it may on silica. Of course any chemical modification which introduces dissociable functional groups also introduces sources of charge.

### 4.2.3.11. Specific adsorption

Specific adsorption of ionic species can lead to the generation of an apparent charge on otherwise uncharged solids. This is exemplified by the development of a charge on oil droplets by adsorption of ionic surfactants. Specific adsorption of ionic species onto solids which do have an intrinsic source of charge can significantly alter their apparent charge. The intrinsic mechanism fixes $\sigma_0$, the specifically adsorbed ions comprise $\sigma_{IHP}$, and the apparent charge sensed by electrokinetic means or counter ion adsorption is close to the algebraic sum of $\sigma_0$ and $\sigma_{IHP}$.

If $\sigma_{IHP}$ is much greater than $\sigma_0$ the properties of the solids may be irrelevant to the behaviour of the coated particle. Thus, the use of a carrier-solid to study the ionization of proteins (Section 4.2.3.4) would fail unless the protein completely masked the properties of the substrate. Neihof and Loeb (1972) in a most significant study, observed that solids as different as $CaCO_3$, a wax, a polysaccharide, and a glass, exhibiting both positive and negative charges in artificial sea water, all became negatively charged in natural sea water. They have shown that this is due to sorption of organic material. If this occurs, it is unimportant whether the solid is a constant potential or constant charge system; the solid itself may respond to changes in solution composition or pH by adjusting its charge or potential, but its effect goes unnoticed as long as the $\sigma_{IHP}$ due to specific adsorption swamps the $\sigma_0$.

(a) *Charge reversal.* If $\sigma_{IHP}$ is greater than $\sigma_0$, specific adsorption reverses the apparent charge. There are many examples of charge reversal. In a classic paper, Bungenberg de Jong (1949) has given an account of reversal of charge in many colloidal systems. Many metal ions and organic cations, including some alkaloids and some aliphatic and aromatic amines, were found to be capable of reversing the originally negative charge of several biocolloids characterized by phosphate, carboxylate, or sulphate ionogenic groups. Charge reversal was also observed on clupein, gelatin, and casein, at pH values at which they are positive, by short chain alkyl sulphates, long chain aliphatic fatty acids, and aromatic carboxylic acids. The relative effectiveness of a series of specifically adsorbed ions in reversing charge on a colloid is characteristic of the functional group present in the colloid. Bungenberg de Jong pioneered the use of electrophoresis and charge reversal spectra in the characterization of biocolloid surfaces. The subject has been reviewed more recently by Brinton and Lauffer (1959), Curtis (1967) and others.

We have already seen (Section 4.2.3.4) that many anions are capable of reversing charge on oxides. James and Healy (1972b) showed that metal ions are effective in reversing the charge on negative $SiO_2$, $TiO_2$, AgI and polystyrene latexes only in the pH range in which they are hydrolyzed and present as hydroxo complexes of low ionic charge. They believe that the reversal of charge is caused by precipitation of a hydroxide of the added metal ion on the solid particle surface and that the charge after reversal is that characteristic of a hydroxide of the added metal ion. The effect of the metal ion

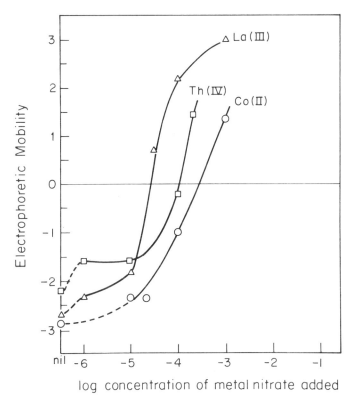

Fig. 4.3. Effect of added hydrolyzable metal ions on surface charge of quartz at pH 8 (after James and Healy, 1972b).

concentration (concentration *added* not equilibrium concentration) on the surface of quartz is illustrated with their data in Figure 4.3. It must be emphasized that the metal adsorption density and hence its effect on surface charge, depends on the amount of metal ion added and the amount of available surface present; i.e. the PZR or RCC will be independent of these parameters only if it is defined in terms of *equilibrium* adsorbate concentrations.

Fuerstenau (1971) and Han *et al.* (1972) have reviewed several papers devoted to the adsorption of surfactants and have shown that most long chain surfactants, including aliphatic amines, fatty acids, sulphonates and sulphates are capable of reversing the charges on many mineral surfaces, AgI and polystyrene latexes. The effect of three surfactants on the charge of $Al_2O_3$ and $Fe_2O_3$ are illustrated in Fig. 4.4. Nicol and Hunter (1970) have observed

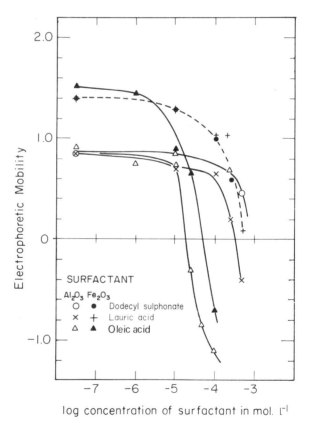

FIG. 4.4. Effect of sodium dodecylsulphonate lauric acid and oleic acid concentration on particle charge for $Fe_2O_3$ and $Al_2O_3$ at pH 8 (Han *et al.*, 1972).

a reversal of charge on kaolin by quaternary ammonium bromides with chain lengths from $C_{10}$ to $C_{16}$ (PZC $\sim 5$ for $C_{16}$).

Wilkins and Meyers (1970) have shown that proteins and protein derivatives are adsorbed onto and are capable of reversing the charge on silica, oil droplets, red blood cells, and polystyrene latexes.

Species capable of being adsorbed onto an oppositely charged surface sufficiently strongly to reverse the charge may not adsorb onto uncharged surfaces. Han et al. (1972) have shown that surfactants with very simple functional groups, such as $SO_4^{2-}$, require the electrostatic attractive free energy of adsorption provided by an oppositely charged surface, in addition to hydrophobic bonding, in order for adsorption to be sufficiently strong to cause charge reversal; they do not adsorb at the PZC. Of course, the total free energy of adsorption and the concentration of the adsorbate in solution determine whether or not it will absorb, and if so, the extent of adsorption.

Judging from the data in Figs. 4.3 and 4.4 and that in Section 4.3.2.4, specifically adsorbed solutes which are simple ions can be expected to dominate the surface behaviour of solids present, if they are present at concentrations in the range of $10^{-3}$ to $10^{-6}$ M. Presumably macromolecular solutes with many ionogenic groups would be adsorbed and dominate surface behaviour at lower concentrations.

## 4.3. Specific Adsorption

### 4.3.1. THEORETICAL ISOTHERMS

If the models of Fig. 4.1 (Cases II, III, IV) are correct, adsorption into the IHP involves replacement of water (which is assumed to be present on the surface) by either a partially dehydrated ion (Cases II, III) or an ion and its inner hydration sheath (Case IV). The adsorption reaction should be, approximately,

$$X(aq) + S(H_2O)_n = SX + nH_2O \qquad (4.19)$$

where $X$ is the adsorbing ion

$S$ is an adsorption site and $SX$ an occupied site

The Stern–Grahame derivation of an adsorption isotherm assumes $n = 1\cdot0$. An equilibrium constant can be written for (4.19) (with $n = 1\cdot0$) and related to the standard free energy of adsorption, $\Delta G_{ads}^{\circ}$ by

$$\left.\begin{aligned} K &= \frac{a_w a_{sx}}{a_{x\,aq}\, a_{sw}} \\[2mm] \Delta G_{ads}^{\circ} &= -RT\log K \end{aligned}\right\} \qquad (4.20)$$

In order to relate $K$ and $\Delta G^\circ_{ads}$ to experimentally measurable quantities, the activities

$a_w$ = activity of water in the aqueous phase
$a_{sx}$ = activity of adsorbed $X$
$a_{sw}$ = activity of water on adsorption site $S$
$a_{x\,aq}$ = activity of adsorbate $x$ in the aqueous phase

must be defined. The specific adsorption isotherms derived by Stern and Grahame (Grahame, 1947) results if the following definitions of activities are used (these choices ensure that $K$ is dimensionless);

$$a_{sx} = k_{sx} n_{sx}$$
$$a_{sw} = k_{sw} n_{sw}$$
$$a_x = k_x n_x$$
$$a_w = k_w n_w = k_w \frac{N_a \rho}{M}$$

$n_w$ = concentration of water, molecules per cm$^3$
$n_x$ = concentration of adsorbate, ions per cm$^3$
$n_{sx}$ = surface concentration of adsorbate, ions per cm$^2$
$n_{sw}$ = surface concentration of hydrated adsorption sites, or of surface bound water, molecules per cm$^2$
$N_a$ = Avogadro's number
$\rho$ = density of water
$M$ = molecular weight of water

The $k$ values are proportionality constants or activity coefficients and will be assumed to be constant. The surface is assumed to have a finite adsorption capacity which is always satisfied so that $n_{sw} + n_{sx} = n_{max} = n_m$. The resulting equilibrium constant is

$$K = n_w \frac{n_{sx}}{n_x(n_m - n_{sx})} = \left(\frac{N_a \rho}{M}\right) \frac{n_{sx}}{n_x(n_m - n_{sx})} = \exp\left(\frac{-\Delta G^\circ_{ads}}{RT}\right) \quad (4.21)$$

Solving for $n_{sx}$ yields one form of the Stern specific adsorption isotherm,

$$\frac{n_{sx}}{n_m - n_{sx}} = n_x\left(\frac{M}{N_a \rho}\right)\exp\left(\frac{-\Delta G^\circ_{ads}}{RT}\right) \quad (4.22)$$

The $\Delta G^\circ_{ads}$ has the form of a standard free energy of adsorption, but is not strictly comparable to classical free energies because the activity scales and standard states have not been reconciled. Perhaps $\Delta G^\circ_{ads}$ should be called a Stern free energy of adsorption.

The fundamental Stern isotherm may be cast in many forms, which differ mainly in the units in which various terms are expressed. Some of these are summarized in Table 4.9.

TABLE 4.9

*Relationships among common forms of the Stern specific adsorption isotherm*

| Concentrations in solution | Adsorption density | | |
|---|---|---|---|
| ions cm⁻³, $n$ | ions cm⁻², $n$ | $\dfrac{n_{sx}}{n_m - n_{sx}} = \dfrac{\theta_x}{1-\theta_x} = n_x\left(\dfrac{M}{N_a\rho}\right)K$ | $n_{sx} = \dfrac{n_m}{1 + (1/Kn_x)(N_a\rho/M)}$<br><br>$= \dfrac{n_m}{1+(1/KX_x)} = \dfrac{n_m KX_x}{1+KX_x}$ |
| mol dm⁻³, $m$ | ions cm⁻², $n$ | $\dfrac{n_{sx}}{n_m - n_{sx}} = \dfrac{m_x M}{10^3}K = \dfrac{m_x}{55.5}H$ | $n_{sx} = \dfrac{n_m}{1 + (55.5/Km_x)} = \dfrac{n_m Km_x}{55.5 + Km_x}$ |
| mol dm⁻³, $m$ | mol cm⁻², $\Gamma$ | $\dfrac{\Gamma_{sx}}{\Gamma_m - \Gamma_{sx}} = \dfrac{n_{sx}}{n_m - n_{sx}}$ | $\Gamma_{sx} = \dfrac{\Gamma_m}{1 + (55.5/Km_x)} = \dfrac{\Gamma_m Km_x}{55.5 + Km_x}$ |

$$K = \exp\left(\frac{-\Delta G^\circ_{ads}}{RT}\right)$$

$X_x \equiv$ mole fraction of $x$ in solution
$\theta_x \equiv$ fraction of adsorption capacity filled by species $x$.

If the adsorption density is small, i.e. $n_{sx} \ll n_m$,

$$n_{sx} \simeq n_x n_m \left(\frac{M}{N_a \rho}\right) \exp\left(\frac{-\Delta G^\circ_{ads}}{RT}\right) \qquad (4.23)$$

The quantity $n_m(M/N_a \rho)$ is the volume occupied by a monolayer of water per square centimetre of surface. Grahame (1947) derived an equation of this form, the Stern–Grahame isotherm,

$$n_{sx} = 2r n_x \exp\left(\frac{-\Delta G^\circ_{ads}}{RT}\right) \qquad (4.24)$$

in which $r$ is the appropriate radius of the adsorbing species. The Stern–Grahame equation can be derived by considering the species in the IHP to occupy a surface phase of thickness equal to the diameter ($2r$) of the adsorbate species. The term $n_m(M/N_a \rho)$ also has the units of length, and Equation (4.23), defines the thickness of the surface phase as specified by Grahame as the thickness of a monolayer of water. Equation (4.24) is identical to equation (4.1) if $\Delta G^\circ_{ads}$ is electrostatic and $r = \delta$.

Free energies derived from the Stern–Grahame equation (4.24) will be on a different scale from those derived from the Stern equation (4.23), etc.; i.e.

$$\Delta G^\circ_{Stern} = \Delta G^\circ_{Stern-Grahame} + \ln\left(\frac{n_m(M/N_a\rho)}{2r}\right)$$

$$= \Delta G^\circ_{Stern-Grahame} + \ln\left(\frac{2r_w}{2r_{ion}}\right) \qquad (4.25)$$

where $r_w$ = water molecule radius
$\quad r_{ion}$ = radius of adsorbing species
The adsorption reaction can be written as

$$X(aq) + S = SX \qquad (4.26)$$

if displacement of water is neglected and equation (4.21) then takes the form,

$$K = \frac{n_{sx}}{n_x(n_m - n_{sx})} = \exp\left(\frac{\Delta G^\circ_{ads}}{RT}\right) \qquad (4.27)$$

which is equivalent to the Langmuir adsorption isotherm. The $\Delta G^\circ$ and K values derived from equation (4.27) differ from the corresponding quantities in equations (4.21) or (4.24), as the former are based on Langmuir isotherms and the latter pair are based on Stern and Stern–Graham isotherms; thus using appropriate subscripts for $\Delta G^\circ$ and K

$$\Delta G_L^\circ = -RT \ln \frac{n_{sx}}{n_x(n_m - n_{sx})} = -RT \ln K_L \qquad (4.28)$$

$$\Delta G_{SG}^\circ = -RT \ln \frac{n_{sx} n_w}{n_x(n_m - n_{sw})} = -RT \ln K_L - RT \ln n_w \qquad (4.29)$$

Thus $\Delta G_{SG}^\circ = \Delta G_L^\circ - RT \ln n_w$.
If concentrations are molar, $n_w$ is 55·5 (Table 4.9).

### 4.3.2. THE FREE ENERGY OF ADSORPTION

#### 4.3.2.1. *Introduction. Contributions to* $\Delta G_{ads}^\circ$

The free energy of adsorption, $\Delta G_{ads}^\circ$, is the standard free energy change associated with the total change in state of the system when adsorption occurs and not simply the energy of the adsorption bond. Recalling equation (4.19),

$$X(aq) + S(H_2O)_n = SX + nH_2O(aq)$$

It is conceivable that $X(aq)$ adsorbs as the hydrated ion $X(H_2O)_h$. If so, the process involves vacation of a surface site by removal of sufficient $H_2O$ to accommodate $X(H_2O)_h$, the movement of $X(H_2O)_h$ from zero potential in solution to $\Psi_S$ at the Stern plane, and the formation of the bond responsible for adsorption. Alternatively, the hydration numbers, $n$ or $h$, may change upon adsorption so that the process involves, in addition to the other steps, partial dehydration of $X$. All of these processes, including the change of hydration number, require or yield energy and contribute to $\Delta G_{ads}^\circ$. In general, then, changes in the extent of hydration of the adsorbate and the surface and changes in the electrical environment of the water of hydration and of the adsorbate must be expected to accompany adsorption and to contribute to $\Delta G_{ads}^\circ$.

The adsorption process and the contribution to $\Delta G_{ads}^\circ$ may be further complicated if the adsorption site is already occupied, e.g.,

$$X(aq) + SY \rightleftarrows SX + Y(aq) \qquad (4.30)$$

or if the adsorbent reacts with the adsorbate, e.g.,

$$ROH(aq) + SOH \rightleftarrows SOR + H_2O(aq) \qquad (4.31)$$

Contributions to $\Delta G_{ads}^\circ$ from processes such as the formation of water in equation (4.31) may be referred to as $\Delta G_{rxn}^\circ$.

No tables are available for the free energies of formation of surface species which are needed to calculate $\Delta G^{\circ}_{ads}$ and reliance must be placed on the estimation of the contributions from individual steps or groups of steps and from the adsorption bond energy unless $\Delta G^{\circ}_{ads}$ can be determined empirically. Many authors have pointed out that $\Delta G^{\circ}_{ads}$ comprises several individual contributions; thus, $\Delta G^{\circ}_{ads} = \Delta G^{\circ}(ads\ bond) + \Delta G^{\circ}_{hyd} + \Delta G^{\circ}_{rxn} + etc.$ Recent reviews have been given by Fuerstenau (1970, 1971), James and Healy (1972c) and Healy (1971).

The magnitude of a $\Delta G^{\circ}_{ads}$ derived from adsorption data is likely to be on a different scale from that of contributions calculated from, for example, standard free energies of formation; a term must be added to the absolute magnitude of $\Delta G^{\circ}_{ads}$ to adjust free energy contributions to consistent scales.

The adsorption bond is a chemical bond. Like the chemical bond, it may be mainly ionic (or electrostatic), covalent, hydrogen bonding or Van der Waals bonding. Like many chemical bonds, it may be a composite of these. The free energy change associated with formation of the adsorption bond (exclusive of other contributions) can be broken down into contributions from each bond type (see, e.g., Fuerstenau, 1970; Healy 1971; James and Healy, 1972c). Thus,

$$\Delta G^{\circ}(ads\ bond) = \Delta G^{\circ}_{elec} + \Delta G^{\circ}_{cov} + \Delta G^{\circ}_{vw} + \Delta G^{\circ}_{hb} + etc.$$

in which the subscripts are abbreviations for the bond types listed above.

### 4.3.2.2. Hydration terms in $\Delta G^{\circ}_{ads}$

(a) *Inorganic adsorbates.* It has been noted above that a variety of solvation or hydration-related changes may accompany adsorption. These include: changes in the number of water molecules in the hydration sheaths of either adsorbate or surface, changes in the polarizing field to which the water in the solvation sheath is exposed, and substitution of the solid surface for all, or part, of the hydration environment of the adsorbate. These changes may include, or be manifested in, changes in the structure of the water near the surface. The sum of the energies involved can be combined in a $\Delta G^{\circ}_{hyd}$ term which contributes to $\Delta G^{\circ}_{ads}$. Very little attention has been paid to "hydration effects", although a few useful results have been reported.

James and Healy (1972c) have shown that metal ions and their hydroxo complexes probably retain their primary hydration sheaths upon adsorption, but they have pointed out that adsorption does change the electrical environment of the hydrated ion drastically. In bulk solution, the hydrated ion is enveloped in bulk water at zero potential. At the surface, the ion and its hydration sheath are subject to a potential field gradient ($d\Psi_s/dx$), and the enveloping environment is asymetric with the solid on one side and water, polarized by the surface field, on the other.

James (1972, cf. James and Healy, 1972c) used an electrostatic model to calculate the work or free energy ($\Delta G^\circ_{hyd}$) involved in these changes. He found that, on solids of low dielectric constant such as silica, silicates or carbonates, $\Delta G^\circ_{hyd}$ is dominated by the dielectric constant of the solid, and is positive, thus hindering adsorption. $\Delta G^\circ_{hyd}$ increases rapidly with ionic charge, but changes little with pH or ionic strength on low dielectric solids. (On silica, $\Delta G^\circ_{hyd}$ for nickel (II) hydroxo complexes is 11 to 12 kJ mol$^{-1}$ for monovalent species and 44 to 47 kJ for divalent species). For small di- and tri-valent cations, $\Delta G^\circ_{hyd}$ is larger than $\Delta G^\circ_{elec}$, and adsorption is prevented even if there is a negative $\Delta G^\circ_{chem}$ amounting to a few tens of kJ mol$^{-1}$. If the sum of $\Delta G^\circ_{hyd}$, $\Delta G^\circ_{elec}$, and $\Delta G^\circ_{chem}$ is positive, adsorption should be negative; the ion should actually be excluded from the surface. Horne (1969) has cited evidence of cation exclusion from silica gel and from the water/air interface, and interprets it on the basis of incompatibility of the structures induced in adjacent water by the surface and the cation.

Adsorption of hydrolyzable metal ions is characterized by low adsorption at low pH and an abrupt increase at a pH characteristic of the adsorbing metal (but not dependent on the solid). The pH at which adsorption increases (pH$_{ads}$) is close to the pH of initial hydrolysis (James, 1971). This is explained by considering all hydroxo complexes and the metal ion itself as potential adsorbates. $M^{+z}$ ($z > 1$) does not absorb because $\Delta G^\circ_{hyd}$ is large and positive. As the pH increases, the relative concentrations of hydroxo complexes increase and ionic charge decreases, thus, $\Delta G^\circ_{hyd}$ decreases. However, $\Delta G^\circ_{hyd}$ is still large enough to largely counteract $\Delta G^\circ_{elec}$, and adsorption occurs through $\Delta G^\circ_{chem}$.

This phenomenon and the fact that pH$_{ads}$ is in the pH range 5 to 9 for most divalent metal ions has significant consequences for the behaviour of these metals in natural waters if they are not complexed. A summary of pH$_{ads}$ values collected by James (1971) is given in Table 4.10.

Quadrivalent and trivalent metals, unless complexed by a ligand other than OH$^-$, are likely to be strongly sorbed in most natural waters whereas divalent metals, unless complexed, etc., are unlikely to be sorbed on oxides or silicates in fresh-water media of pH $< 8$. However, these divalent metals should be more strongly adsorbed in marine or saline waters (pH $\geqslant 8$). Whether any metal is or is not adsorbed will depend on its pH$_{ads}$, its concentration, other species competing for the adsorption surfaces, and the form in which the metal is present. For example, the author has recently found preliminary evidence that chlorocomplexes are unlikely to adsorb.

(b) *Organic adsorbates.* For the adsorption of neutral organic molecules, it is

TABLE 4.10

*Empirical adsorption edges of hydrolyzable metal ions (after James, 1971)*

| Metal | $pH_{ads}$ | Substrates Represented |
|-------|-----------|------------------------|
| Hf(IV) | 1·9–3·6 | glass, PVC latex, AgI |
| Th(IV) | 3–6·1 | glass, $SiO_2$ gel, AgI |
| Fe(III) | 2·1–4 | glass, $SiO_2$ gel, $\alpha$-$SiO_2$, $\alpha$-$Al_2O_3$ |
| Cr(III) | 5–6 | $SiO_2$ gel, $\alpha$-$SiO_2$ |
| Al(III) | 5 | $SiO_2$ gel |
| La(III) | 6 | AgI |
| Cu(II) | 6–7 | $SiO_2$ gel, $\alpha$-$SiO_2$ |
| Pb(II) | 5–6·2 | $\alpha$-$SiO_2$, $\alpha$-$Al_2O_3$ |
| Zn(II) | 6·5–8 | $SiO_2$ gel, "river silt" |
| Co(II) | 5·4–7·8 | $SiO_2$ gel, $\alpha$-$SiO_2$, $\alpha$-$TiO_2$, $Fe_2O_3$ |
| Ni(II) | 7 | $SiO_2$ gel |
| Mn(II) | 6·9–9·1 | $SiO_2$ gel, glass |
| Mg(II) | 9·7 | $\alpha$-$SiO_2$ |
| Ca(II) | 10–11·5 | $\alpha$-$SiO_2$ (7·0 on $TiO_2$) |

commonly found that non-polar solutes are adsorbed preferentially from polar solvents onto non-polar solids and that polar solutes prefer polar solids. For a homologous series adsorption onto charcoal increases with decreasing solubility, or with increasing molecular weight (Traube's rule), but the trend is reversed on silica gel (Adamson, 1967). These observations can be interpreted in terms of the interplay of the free energies of solvation and adsorption.

(i) *Hydration and solubility.* The solubility of an uncharged organic molecule in water is determined largely by hydration, which in turn reflects its ability to interact with water through H-bonding. If the molecule is either *non-polar*, (i.e. both uncharged and unable to engage in either H-bonding or dipole-dipole interactions), or unable to hydrate for steric reasons it will probably be insoluble (Kavanau, 1964). Generally, polar molecules are more strongly hydrated and more soluble in water than non-polar ones. In keeping with their low solubility, the free energy of hydration of non-polar molecules is either positive or very slightly negative.

Analogously, the free energy of hydration of polar surfaces should be negative and that of non-polar surfaces small or positive.

(ii) *Hydrophobic bonding.* The tendency of non-polar parts of organic molecules (such as proteins) to adhere to one another in an aqueous environment is known as hydrophobic bonding (Kauzmann, 1959; Nemethy and Scheraga,

1962). This tendency results in micelle formation among organic solutes, adsorption of non-polar species onto non-polar surfaces (Kauzmann, 1959) and oriented adsorption of amphiphilic solutes on non-polar surfaces (polar groups directed away from the surface) (Fuerstenau, 1970, 1971).

The free energy change responsible for hydrophobic bonding, $\Delta G^{\circ}_{hydroph}$, includes Van der Waals bonding between non-polar groups and the surface. However, hydrophobic bonding requires withdrawal of the non-polar group (surface or solute) from its aqueous environment. Since non-polar solutes alter the structure of the adjacent water (Kavanau, 1964), withdrawal allows the water structure to relax, leading to a gain of entropy. This entropy change dominates $\Delta G^{\circ}_{hydroph}$ (Kauzmann, 1959). Healy (1971) has discussed recent work on the influence of molecular configuration on $\Delta G^{\circ}_{hydroph}$.

The binding of organic molecules and ions to proteins involves a significant, and for neutral molecules often dominant, contribution from $\Delta G^{\circ}_{hydroph}$. Hydrocarbons (e.g., butane), many steroids, and alkyl sulphates and sulphonates are bound with a $\Delta G^{\circ}_{ads}$ in the range 20 to 45 kJ mol$^{-1}$ which is thought to be dominated by $\Delta G^{\circ}_{hydroph}$ (Steinhardt and Reynolds, 1969). At least eight carbon atoms are required for strong binding in these systems as also in the adsorption of simple surfactants on oxides (Fuerstenau, 1971). Since the affinities of several adsorbates of the same hydrocarbon chain length depend on the functional groups it is likely that for binding to proteins both $\Delta G^{\circ}_{hydroph}$ and $\Delta G^{\circ}_{chem}$ are involved. On bovine serum albumin the binding affinity increases in the order

$$ROH < RCOO^- < RSO_3^- < RSO_4^-$$

For dodecyl-compounds there is a large change in $\Delta G^{\circ}_{ads}$ in this series. Thus,

$$-8\cdot7 < -32 \approx -31 < -35 \, (kJ \, mol^{-1})$$

The difference must be attributed to covalent or hydrogen bonding because it is too large to be attributed to $\Delta G^{\circ}_{elec}$.

Some use of the concept of hydrophobic bonding has been made in interpreting the adsorption of organics onto inorganic solids. Somasundaran *et al.* (1964) and Wakamatsu and Fuerstenau (1968) observed increased adsorption of monofunctional ionic surfactants (e.g., alkyl-NH$_3^+$) at solution concentrations above a critical concentration (the h.m.c.) which is roughly one hundredth of the critical micelle concentration of the surfactant (see also Fuerstenau, 1964, 1970, 1971; Healy, 1971). This can be attributed to two-dimensional aggregation of pre-adsorbed ions into surface or hemi-micelles (Gaudin and Fuerstenau, 1955). Purely electrostatic adsorption in the Gouy layer occurs below the h.m.c.; hemi-micelle formation occurs

L

only when the ionic charge of the adsorbate and the surface charge are opposite in sign. Apparently, binding to the surface is predominantly electrostatic, whereas hydrophobic bonding is responsible for aggregation of the adsorbed ions. The two processes should probably be treated separately.

$$RNH_3^+(aq) + surface \rightleftarrows RNH_3^+ \text{ (in Gouy layer on surface)} \quad (4.32)$$

$$n\,RNH_3^+ \text{ (Gouy layer)} \rightleftarrows (RNH_3^+)_n \text{ (hemi-micelle on surface)} \quad (4.33)$$

$\Delta G_{ads}^\circ$ for reaction (4.32) is purely electrical, $\Delta G_{elec}^\circ$. $\Delta G_{ads}^\circ$ for reaction (4.33) amounts to $-2 \cdot 4\,kJ\,mol^{-1}$ of adsorbate per $CH_2$ group in the alkyl chain.

Repulsion between the charged $NH_3^+$ groups should contribute a positive $\Delta G_{elec}^\circ$ to $\Delta G_{ads}^\circ$ for reaction 4.33 and limit the size of the hemi-micelles. In support of this supposition are the observations that increased adsorption results if a significant fraction of the surfactant is present as the neutral molecule, $RNH_2$ (Li and de Bruyn, 1966) or is replaced by an alcohol of the same chain length (Smith, 1963).

Hydrophobic bonding between the non-polar part of ionic surfactants and non-polar surfaces results in oriented adsorption of the surfactant and in the polar or ionic group being directed away from the surface (Fuerstenau, 1971).

### 4.3.2.3. $\Delta G_{elec}^\circ$

Negative surfaces attract (adsorb) positive ions and vice versa. $\Delta G_{elec}^\circ$ is the energy gained or spent in bringing an ionic adsorbate from the bulk solution to an adsorption site on or near a charged surface where the potential is $\Psi_x$.

$$\Delta G_{elec}^\circ = zF\Psi_x. \quad (4.34)$$

$\Delta G_{elec}^\circ$ can be negative or positive. Electrostatic attraction can lead to enough positive adsorption to neutralize $\sigma_0$, but no more. Further adsorption reverses the effective surface charge and $\Delta G_{elec}^\circ$ for the same ion becomes positive, and adsorption is thereby hindered. Adsorption in response to $\Delta G_{elec}^\circ$ *alone* is relatively non-specific, the adsorbate is in the Gouy layer and the adsorption density is best described by the double layer equations described earlier.

Non-electrical negative contributions to $\Delta G_{ads}^\circ$ and $\Delta G_{elec}^\circ$ lead to specific adsorption in the Stern layer where $\Delta G_{elec}^\circ = zF\Psi_d$ ($\Psi_d$ is the potential in the plane of the adsorbing ion; see Section 4.2.2.4). If negative contributions to $\Delta G_{ads}^\circ$ are larger than $\Delta G_{elec}^\circ$, adsorption can occur in spite of electrostatic repulsion, and charge adsorption densities can exceed $\sigma_0$, leading to effective surface charge reversal.

$\Delta G^\circ_{e\,lec}$ varies from zero at the PZC to 11 kJ mol$^{-1}$ at $\Psi_d = 0\cdot12$ V for a monovalent ion, or 34 kJ mol$^{-1}$ at $\Psi_d = 0\cdot12$ V for a trivalent ion.

(a) *Activated adsorption.* Normally, non-specific ions with the same sign as $\sigma_0$ will not adsorb appreciably. They can be induced to adsorb by changing the sign of the effective surface charge through specific adsorption of another species. This is called "activated" adsorption of the non-specific ion (Fuerstenau 1971).

Alkyl carboxylates, sulphonates and sulphates will adsorb onto negative oxides in the presence of hydrolyzable metal ions in the pH range at which the metal ion is partially hydrolyzed. (see e.g., Fuerstenau and Bhappu, 1963). Alkylammonium ions will adsorb onto positively charged solids (see examples in Fuerstenau and Healy, 1972).

(b) *Selectivity in simple ion adsorption: interplay between* $\Delta G_{elec}$ *and* $\Delta G_{hyd}$. Adsorption dominated by $\Delta G^\circ_{elec}$ is characterized by a change from cation adsorption on the negative side of the PZC to anion adsorption on the positive side. The solid is a cation exchanger when $\sigma_0$ is negative and an anion exchanger when $\sigma_0$ is positive. Adsorption of simple monovalent anions and cations and non-hydrolyzing bivalent cations on many solids shows this behaviour, but this cannot lead to the reversal of the sign of $\sigma_0$.

(i) *Cation adsorption.* As has been seen, purely electrostatic adsorption, according to Equation (4.2) should be non-specific. There should be no preference for one ion over any other of the same charge. In fact, however, at least two classes of behaviour are observed. Most clays and a Mn(IV) oxide (Amphlett, 1964), $SiO_2$, AgI, $As_2S_3$ (Overbeek, 1952c), a hydrous zirconium phosphate and at least two zeolites (chabazite and faujasite) (Amphlett, 1964) show decreasing affinity for monovalent cations, when the solid is negative, in the order.

$$Ag^+ > Cs^+ > Rb^+ > NH_4^+, K^+ > Na^+ > Li^+$$

Some of the same solids show decreasing affinity for non-hydrolyzed bivalent ions in the order,

$$UO_2^{2+}, Ca^{2+} > Ba^{2+} > Mg^{2+}$$

But the order for clays and faujasite is,

$$Ba^{2+} > Sr^{2+} > Ca^{2+} > Mg^{2+}$$

The order of decreasing adsorption affinity is reversed for these mono- and bi-valent ions on some zeolites (see e.g., Amphlett, 1964), $TiO_2$ (Berubé and deBruyn, 1968) and some glasses (Stumm and Morgan, 1971).

Selectivities among members of these series can be measured by ion exchange.

$$M_1(aq) + M_2(ads) = M_2(aq) + M_1(ads)$$

$$K_1 = \frac{(\underline{M}_1)(M_2)}{(\underline{M}_2)(M_1)}$$

(4.35)

where $(\underline{M}_1)$ and $(\underline{M}_2)$ indicate the activities of the adsorbed species and $(M_1)$ and $(M_2)$ are the activities in the solution. Values of $K$ range from about unity to a few hundred. Amphlett (1964) has reported

| Mineral | $M_1(aq)$ | $M_2(ads)$ | $K$ |
|---------|-----------|------------|-----|
| montmorillonite | $Cs^+$ | $Na^+$ | 13·2 |
| chabazite | $Ag^+$ | $Na^+$ | 11·4 |
| basic sodalite | $Ag^+$ | $Na^+$ | 335 |
| chabazite | $K^+$ | $Na^+$ | 0·86 |

These exchange constants vary with the relative adsorption densities of the two species involved in the exchange.

Of course, in sea water the high concentrations of some ions forces the equilibrium toward $Na^+$ adsorption. River-borne clays lose $K^+$ by exchange for $Na^+$ on exposure to sea water (Berner, 1971), although phillipsite concentrates $K^+$ relative to $Na^+$, and hence must have a relatively large selectivity coefficient or K/Na exchange constant (Amphlett, 1964).

The order of increasing adsorption affinity is also the order of decreasing hydrated-ion radius (Amphlett, 1964) for the "normal" affinity series (i.e. normal for, clays for example). Selectivity is attributable to a variety of properties of the adsorbing ions (see, e.g., Van Olphen, 1963; Stumm and Morgan, 1971; Berubé and de Bruyn, 1968). The most successful interpretations introduce selectivity through a variable $\Delta G°_{ads}$. Berubé and de Bruyn have emphasized hydration of both surface and ion. These affect the structure of water, and adsorption preference would be expected for ions which induce the same kind of structural changes in water as does the surface—i.e., those which "fit" at the surface-altered water structure.

(ii) *Anion adsorption.* Gouy layer adsorption of simple anions might also be expected to be non-specific, but this is not found to be so. Some proteins (albumins) (Steinhardt and Reynolds, 1969), a positively charged carbon (Overbeek, 1952), anion exchange resins and chrysotile asbestos (Naumann and Dresher, 1966) adsorb anions as counter ions with the following order of decreasing affinity;

$$SCN^- > I^- > Cl^-, Br^- > SO_4^{2-} \text{ (carbon)}$$

$$CCl_3COO^- > SCN^- > I^-, F^- > Cl^- \text{ (proteins)}$$

$$MnO_4^- > SCN^- > I^- > NO_3^- > Br^- \approx IO_3^- \approx ClO_4^- \approx NO_2^- \approx Cl^-$$
$$\text{(asbestos)}$$

Air bubbles made positive through adsorption of cationic surfactants behave similarly (Shinoda and Fujihira, 1968)

$$SO_4^{2-} > NO_3^- > ClO_3^- > Br^- > Cl > CH_3COO^-$$

whereas positive magnetite (Amphlett, 1964), positive $TiO_2$ (Berubé and deBruyn, 1968), positive $Al_2O_3$ and $Fe_2O_3$ (Overbeek, 1952) exhibit the opposite order of affinities.

Naumann and Dresher (1966) found that the order of increasing affinity for the surface was the same as the order of increasing ionic size and attributed differences in behaviour to differences in polarizability. The $\Delta G_{ads}^\circ$ contribution has been treated in this case as a $\Delta G_{vw}^\circ$ in Section 4.2.2.5. Bungenberg de Jong (1949) has offered a more detailed interpretation of selectivity in terms of ionic size and polarizability.

(c) *Organic solute adsorption: interplay of* $\Delta G_{elec}^\circ$ *and* $\Delta G_{hydroph}^\circ$. Fuerstenau and his co-workers (e.g., Fuerstenau, 1971; Han, et al., 1972; and Wakamatsu and Fuerstenau, 1968) have explored the adsorption of monofunctional, long-chain alkyl-surfactants onto simple oxides in great detail. They concluded that adsorption occurred only on oppositely charged surfaces, thus that $\Delta G_{ads}^\circ$ is dominated by $\Delta G_{elec}^\circ$. At low adsorption densities the surfactant is in the Gouy layer, and adsorption occurs by ion exchange with other counter ions available. At higher concentrations, hydrophobic bonding between the hydrocarbon chains of the adsorbate ions leads to a negative $\Delta G_{hydroph}^\circ$ and to increased concentration dependence of adsorption. This has been discussed in Section 4.3.2.2. and will be brought up again in discussion of the shapes of isotherms, (Section 4.3.2.7).

$\Delta G_{hydroph}^\circ$ is dominant in the binding of neutral molecules, such as hydrocarbons to proteins, with exposed non-polar sites. $\Delta G_{elec}^\circ$ is relatively unimportant with respect to $\Delta G_{cov}^\circ$, $\Delta G_{hb}^\circ$, or $\Delta G_{hydroph}^\circ$ in the binding of organic anions to proteins (Steinhardt and Reynolds, 1969). Thus, $\Delta G_{ads}^\circ$ often differs by 8 kJ mol$^{-1}$ or less between neutral molecule adsorbates such as ROH and corresponding ionic species such as RCOO$^-$ when the total $\Delta G_{ads}^\circ$ amounts to tens of kJ mol$^{-1}$. This should be contrasted with the fact that ROH fails to adsorb from water solution onto oxides unless a charged long-chain ion is also present (see Section 4.3.2.6).

The relative unimportance of $\Delta G^{\circ}_{elec}$ in the determination of $\Delta G^{\circ}_{ads}$ for proteins does not mean the pH is an unimportant variable. The availability of adsorption sites is highly dependent on the conformation of the protein and this, in turn, is pH dependent (Steinhardt and Reynolds, 1969).

### 4.3.2.4. $\Delta G^{\circ}_{cov}$ and $\Delta G^{\circ}_{chem}$

Covalent bonding in adsorption is often suspected but not widely documented. A $\Delta G^{\circ}_{chem}$ has been defined in such a way as to include predominantly covalent bonding (Fuerstenau, 1971), or to include all contributions to $\Delta G^{\circ}_{ads}$, except $\Delta G^{\circ}_{elec}$ and certain solvation terms (James and Healy, 1972c). In the latter case, covalent bonding is not necessarily involved. $\Delta G^{\circ}_{chem}$ is often used to embrace all these possibilities.

(a) $\Delta G^{\circ}_{chem}$ *in metal ion adsorption.* In their study of the adsorption of hydrolyzable metal ions, James and Healy have reviewed other work and have also derived new empirical estimates of $\Delta G^{\circ}_{chem}$ for two substrates, $SiO_2$ and $TiO_2$. $\Delta G^{\circ}_{chem}$ ranges from $-26$ to $30 \text{ kJ mol}^{-1}$ for divalent ions and their hydrolysis products on $SiO_2$ and is in the vicinity of $-35 \text{ kJ mol}^{-1}$ for trivalent ions and their hydrolysis products on $SiO_2$. $\Delta G^{\circ}_{chem}$ for $Co^{+2}$, $CoOH^+$, etc., on $TiO_2$ is about $-17 \text{ kJ mol}^{-1}$ (James and Healy, 1972c).

(b) $\Delta G^{\circ}_{chem}$ *in adsorption on proteins.* Steinhardt and Reynolds (1969) report association constants for metal ion binding to proteins, and refer to the bond as covalent. As they have pointed out, it is difficult to make generalizations which apply to all proteins since proteins compositions differ so greatly. It is probably safe to say that the association constant of a metal ion to a protein will not be smaller than that of the same metal ion to a single amino acid closely approximating to the composition of the most abundant amino acid in the protein. $\Delta G^{\circ}_{chem}$ calculated from the association constants ranges from $-2 \cdot 5$ to $-30 \text{ kJ mol}^{-1}$ for $Zn^{2+}$ on several proteins and is as high as $-17 \text{ kJ}$ for $Ca^{2+}$ on one protein. $\Delta G^{\circ}_{chem}$ for $Cu^{2+}$ on a single protein changes with pH between $-10$ and $-30 \text{ kJ mol}^{-1}$. As has been noted in Section 4.3.2.2b, there is probably a $\Delta G^{\circ}_{chem}$ amounting to a few tens of kJ $\text{mol}^{-1}$ in the $\Delta G^{\circ}_{ads}$ responsible for binding organic ions to proteins.

(c) *Relative magnitudes of* $\Delta G^{\circ}_{chem}$. The same forces responsible for complexation or compound formation are responsible for adsorption. There are no others available. This observation provides a means of comparing the expected magnitudes of $\Delta G^{\circ}_{chem}$ among various systems. Solute species which are capable of forming strong complexes or insoluble compounds with some component of the solid are likely to adsorb more strongly than those which are not. The actual complex may, or may not, form. There seems to be evidence of compound formation in some systems (see e.g., Peck 1963; Wadsworth, 1965; Peck *et al.*, 1966, 1967), but for the present purpose the question need not be answered; it is enough to know that the abundant tables of complex stability constants and solubility products can be used to anticipate adsorption behaviour. As an example, Fe(III) forms insoluble soaps but its sulphates are quite soluble; alkyl carboxylates chemisorb on $Fe_2O_3$ whereas alkyl sulphates do not.

### 4.3.2.5. $\Delta G^{\circ}_{vw}$

Many approximate expressions for the Van der Waal's binding energy between atoms have appeared (see, e.g., Overbeek, 1952; Von Hippel, 1954). All depend on the electronic polarizability of the species bound, which in turn depends on the ionic or atomic volume (Von Hippel, 1954). Overbeek has pointed out that London's formulation of Van der Waals binding energy is additive, so that the Van der Waals binding energy of a group of atoms to a surface should be the sum of the individual binding energies. Van der Waal's binding will affect adsorption in several ways. It will contribute to the bond between adsorbate and surface and to the interaction between adsorbates.

Nauman and Dresher (1966) have presented evidence that in any series of monovalent inorganic anions (e.g., $Cl^-$, $NO_3^-$, and $MnO_4^-$) the strength of adsorption increases as the ionic volume increases. This could be interpreted as evidence of increasing $\Delta G^{\circ}_{vw}$, but the conjecture has not been verified. In contrast, Davies and Rideal (1963) have assigned a value of zero to $\Delta G^{\circ}_{vw}$ for simple ions, including $Cl^-$, $F^-$, $SO_4^{2-}$, $K^+$, $Ca^{2+}$, $Mg^{2+}$, $Cu^{2+}$ and $Th^{4+}$.

Consideration of the results obtained by Nauman and Dresher (1966) and the estimated $\Delta G^{\circ}_{vw}$ tabulated for other species by Davies and Rideal (1963) leads to the generalization that $\Delta G^{\circ}_{vw}$ will be larger the larger the adsorbing species. This is supported by the observation that polystyrene sulphonate adsorbs strongly on a negative silica surface whereas toluene sulphonate does not (Morgan and Stumm, 1970). $\Delta G^{\circ}_{vw}$ ranges from close to zero for small simple ions to tens of kJ $mol^{-1}$ for large molecules.

### 4.3.2.6. $\Delta G^{\circ}_{hb}$

Hydrogen bonding requires two electronegative atoms, one in the adsorbate and one in the surface. One, the "A" group, must have a hydrogen atom bound to it forming an acidic or proton donor group. The other, the "B" group, may carry a bonded hydrogen atom although this is not essential. The common hydrogen bonding atoms are O, N, and F. Pimentel and McClelland (1960) also include the carbon of the acetylenic —CH group, halogen activated —CH groups, and sulphur. Giles (1959) considers that carbon in a —C≡O activated —CH group should also be included. Single hydrogen bond energies range from $-8$ to $-40\,\mathrm{kJ\,mol^{-1}}$.

Giles (1959) has gathered evidence that most pairs of the functional groups

alkyl-OH          alkyl-NH$_2$

aryl-OH           tri-alkyl-N

alkyl-COOH

can be expected to form hydrogen bonds in aqueous systems if one is present in an organic surface and the other in a solute. The phenols are especially strong hydrogen bonding solutes. They adsorb from water solution, presumably by hydrogen bonding, onto cellulose acetate (not cellulose), chitin, nylon, silk and wool and also onto alumina and silica (Giles, 1959).

Undissociated carboxylic acids are thought (Giles, 1959) to adsorb onto wool by hydrogen bonding whereas the carboxylate ion adsorbs by anion exchange. Degens and Matheja (1971) have suggested that $HCO_3^-$ bonds to proteins with hydrogen bonds involving the amide hydrogen and carboxyl oxygen, and that amino acids adsorbing onto clays form hydrogen bonds between the amino group of the acid and oxygen ions in the clay structure. They have postulated that adsorption can catalyze polymerization of the amino acids.

Wadsworth and co-workers (Peck, 1963; Peck and Wadsworth, 1965; Peck et al., 1966, 1967) have attributed physical adsorption of oleic acid onto the HF-treated surface of beryl $(Be_3Al_2Si_6O_{18})$ and onto fluorite $(CaF_2)$ to hydrogen bonding between the carboxyl oxygen and $F^-$ in the solid surface. Similarly, they and others (see Kipling, 1965) have attributed physical adsorption of oleic (Ol) and stearic acids onto phenacite $(Be_2SiO_4)$, haematite $(Fe_2O_3)$, and corundum $(Al_2O_3)$ to hydrogen bonding between surface —OH or oxygen and the carboxyl group of the acid. On beryl, adsorption occurs by H-bonding alone. In all other instances, the carboxylic acid is present in two states; as the H-bonded neutral molecule and as what appears to be a metal carboxylate. It is though that the H-bonded state may be an intermediate in the chemisorption process. The total reaction is, e.g.

$$
\left.
\begin{aligned}
&(Fe_2O_3)\!-\!OH + HOl \rightleftarrows (Fe_2O_3)\!-\!O\!\!\begin{array}{c} H \\ \diagup \diagdown \\[-2pt] \diagdown \diagup \\ H \end{array}\!\!Ol \\[10pt]
&(Fe_2O_3)\!-\!O\!\!\begin{array}{c} H \\ \diagup \diagdown \\[-2pt] \diagdown \diagup \\ H \end{array}\!\!Ol \rightleftarrows (Fe_2O_3)\!-\!Ol + H_2O
\end{aligned}
\right\} \quad (4.36)
$$

Thus, the net reaction is related to the adsorption of carboxylate ions by wool cited above.

MacKenzie and O'Brien (1969) have suggested that metal hydroxo complexes adsorbed onto silica are hydrogen bonded to surface —OH (silanol) groups. Colombera *et al.* (1971) offer evidence of H-bonding between Al(III) hydroxo-complexes and illite. More recently, James and Healy (1972c) have cited evidence that metal hydroxo complexes adsorbed on oxides remained hydrated, and hence are *unlikely* to hydrogen bond directly to the surface. Giles (1959) has pointed out that strong hydration of either solute or surface may suppress adsorption by hydrogen bonding. Strong hydration would introduce a large positive contribution to $\Delta G_{ads}^{\circ}$, and thus would reduce the negative contribution of a hydrogen bond to the adsorption bond. Examples are the failure of normal alcohols and glucose to adsorb from water onto cellulose acetate or wool (in contrast to phenol which does so), presumably because solute hydration is stronger than the solute-surface interaction. Cellulose, and carbohydrate —OH surfaces in general, do not adsorb hydrogen bonding solutes, presumably, in this case, because surface hydration is stronger than the solute-surface interaction.

Healy (1971) and Fuerstenau (1971) reviewed evidence that ionic surfactants in which $-NH_3^+$ is the functional group adsorbed onto oxides solely electrostatically. If this is so, they do not hydrogen bond as might be expected. Even the $RNH_2$ molecule appears not to adsorb onto oxides unless $RNH_3^+$ ions are also present and are electrostatically bound (Smith, 1963).

### 4.3.2.7. $\Delta G_{rxn}^{\circ}$ (p.285) *adsorption of hydrolyzable anions*

According to Hingston *et al.* (1972), adsorption of fully dissociated hydrolyzable anions (e.g., $PO_4^{3-}$) on oxides occurs only on surfaces of positive charge. (Whether such highly charged species can be expected to adsorb in response to $\Delta G_{elec}^{\circ}$ alone depends on the magnitude, if any, of a positive $\Delta G_{hyd}^{\circ}$, if the model proposed by James and Healy for cation uptake is generally applicable.) When such adsorption occurs it obeys a Langmuir equation with $\sigma_0 = -z_i F\Gamma_{max,i}$ and $\Delta G_{ads}^{\circ}$ is dominated by $\Delta G_{elec}^{\circ}$. In the pH range at which the anion is hydrolyzed and the surface contains MOH or $MOH_2$ groups, adsorption is aided by a $\Delta G_{rxn}^{\circ}$ arising in the reaction of a proton from the adsorbing species and an OH on the surface; thus,

$$HF(aq) + FeOH = FeF + H_2O \qquad (4.37)$$

The authors have shown that $\Gamma_{max}$ varies with pH in a manner influenced by hydrolysis of the adsorbate and by surface charge. They have studied adsorption of several phosphates, arsenate, selenite, silicate and fluoride ions onto goethite ($\alpha$-FeOOH) and gibbsite ($\alpha$-Al(OH)$_3$).

Cook and his co-workers (1950, 1969) have offered a similar explanation for
the adsorption of xanthates, $RC\overset{\diagup S}{\underset{\diagdown SH}{}}$ , symbolized HX, on sulphide minerals.
Presumably the metal ions in the sulphide surface hydroxylate and

$$FeOH \text{ (in FeS surface)} + HX = FeX + H_2O \qquad (4.38)$$

### 4.3.3. EMPIRICAL ADSORPTION ISOTHERMS

If adsorptive behaviour is viewed in terms of the solute-concentration
dependence of the adsorption density it can be divided into two classes with
many variants in each class. Giles *et al.* (1960) have developed a comprehensive classification from which that shown in Fig. 4.5 has been derived (see
also Kipling, 1965). The difference between the two classes lies in the shape
of the initial low-concentration portion of the curve. The $S$-family isotherms
involve a $\Delta G^{\circ}_{ads}$ which increases with increasing adsorption, at least for a
short range of concentrations and these isotherms are initially concave
upward. The $L$- and $H$-family isotherms are initially convex upward and
may involve a $\Delta G^{\circ}_{ads}$ which is constant or decreases with increasing adsorption
density. The $H$-family is a special case of the $L$-family in which initial adsorption is so strong that it *appears* to start at zero concentration.

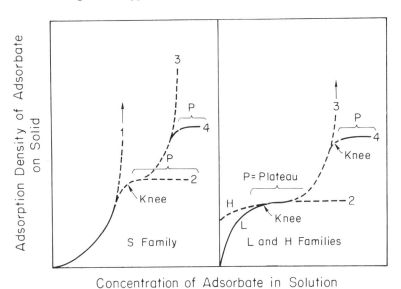

FIG. 4.5. Types of adsorption isotherms (adapted from Giles, *et al.*, 1960).

The $S$-family curves are obtained (Giles $et$ $al.$, 1960; Kipling, 1965) if (a) the solvent is adsorbed very strongly, competing with the adsorbate for surface sites (i.e., a $\Delta G^\circ_{hyd}$, as might be expected of water on hydroxyl surfaces), (b) there is strong inter-adsorbate interaction, such as a $\Delta G^\circ_{hydroph}$ and (c) the adsorbate is mono-functional. An example is adsorption of a monohydric phenol from water onto polar solids (Kipling, 1965). Competition between the adsorbate and water molecules for surface sites (see equation 4.32) is a special case of exchange adsorption. $S$-family isotherms might also be expected where ionic adsorption is competitive (see equation 4.30). Adsorption of $Tl^+$ onto an $Ag^+$-loaded zeolite exhibits an $S$-isotherm (Amphlett, 1964).

$L$ or Langmuir-isotherms often characterize adsorption in the absence of competition from the solvent (or a pre-adsorbed competitor) and adsorption of linear or planar molecules which lie flat on the surface (Kipling, 1965; Giles $et$ $al.$, 1960). Kitchener (1965), and Kipling (1965) have discussed the reasons for variation in the forms of $L$-isotherms. If $\Delta G^\circ_{ads}$ is very large and negative, as for strong chemisorption or adsorption of polymers, the initial slope of the $L$-isotherm is very steep and it may appear that adsorption occurs at essentially zero equilibrium concentration. This is the $H$-family of isotherms. Chemisorption of oleate on $CaF_2$ is an example of $H$-2 isotherm (Peck and Wadsworth, 1965).

### 4.3.3.1. *Variants of the basic isotherm*

The short plateaux in $S$-3 and $H$ or $L$-3 isotherms and the $\Gamma_{max}$ plateaux in $S$-2, $S$-4, $H$ or $L$-2 and $H$ or $L$-4 isotherms have several alternative explanations. Many examples are given in the review papers cited in the next few sections. (a) *Effects of adsorbate activity.* As the concentration or activity of the solute (adsorbate) increases, several limitations may be reached. The solute concentrations may approach saturation with respect to capillary condensation or even to homogeneous nucleation and precipitation. In both cases *apparent* adsorption (measured as a loss from solution) will increase without limit (either before any other phenomena are observed ($S$-1) or after ($S$-3 and $L$ or $H$-3)). Well below saturation with respect to precipitation many long chain heteropolar solutes, such as detergents, aggregate into micelles. This aggregation limits the activity of the solute since further increases in the total concentration increase the concentration of micelles but not the activity of the solute. Unless the micelles themselves "sorb", the adsorption density reaches a maximum and the isotherm develops an·ultimate plateau ($S$-2, $L$ or $H$-4).

(b) $\Gamma_{max}$ *and plateaux: limited adsorption capacity.* If adsorption is dominated by an adsorbate-surface bond which is non-localized, adsorption should cease when the surface is covered. $\Gamma_{max}$ corresponds to that for monolayer coverage and a type $S$, $L$, or $H$-2 isotherm should result. If the adsorbate is heteropolar (e.g., a monofunctional alkyl surfactants) adsorption densities may exceed that corresponding to a monolayer. The first layer is bound by interaction between the polar functional group and the surface; the second is bound by hydrophobic bonding between hydrocarbon chains and results in a polar-group outward orientation of the second layer. The capacity of the second layer might be limited if the interaction between the now outward oriented polar groups is repulsive.

If adsorption occurs on localized sites (as for example, with metal ions on proteins), the maximum adsorption density is limited by the site density when $\Delta G_{ads}$ on the site is large and attractive interaction among adsorbates is small. Again a type $L$ or $H$-2 isotherm should result, but the plateau no longer corresponds to that for monolayer coverage. Specific site density may be pH sensitive. Moderate interadsorbate interaction could lead to a further increase in adsorption, perhaps limited to close-packed monolayer coverage, at higher concentration and a type $S$, $H$, or $L$-4 isotherm could result.

### 4.3.3.2. Complex isotherms. Examples

(a) *Ionic surfactants on charged solids.* An idealized isotherm for adsorption of simple monofunctional alkyl (having eight or more carbon atoms) surfactants with no tendency to chemisorb ($\Delta G^{\circ}_{chem} \to 0$) onto charged surfaces (oxides, carbons with charged surfaces, carboxylated latexes) is shown in Fig. 4.6. It is a composite of behaviours similar to those observed in several systems as summarized by Fuerstenau (1971) and Tamamushi (1963). Adsorption of n-alkyltrimethylammonium ions onto polystyrene is similar (Connor and Ottewill, 1971). On the basis of the shape of its enlarged, low concentration extreme, it is classified as an $L$-4 isotherm.

According to Fuerstenau (1971) (see also, Wakamatsu and Fuerstenau, 1968) the isotherm is linear in three ranges on a log–log scale. In range I, adsorption is coulombic, with $\Delta G^{\circ}_{elec}$ dominating. The surfactant is in the Gouy layer. In range II, adsorption increases more rapidly with concentration because hemi-micelle aggregation of the adsorbate has commenced so that $\Delta G^{\circ}_{ads} = \Delta G^{\circ}_{elec} + \Delta G^{\circ}_{hydroph}$. In range III, the adsorption density has increased to the point where the charge adsorbed exceeds $\sigma_0$ and a positive $\Delta G^{\circ}_{elec}$ reduces adsorption although $\Delta G^{\circ}_{hydroph}$ is still negative. According to Tamamushi (1963), adsorption becomes constant for adsorbate concentrations higher than the critical micelle concentration (CMC), but $\Gamma_{max}$ is pH dependent.

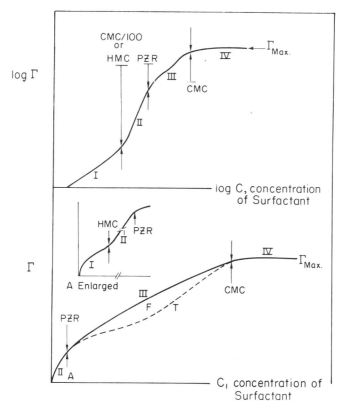

Fig. 4.6. Idealized adsorption isotherms for simple monofunctional surfactants on charged surfaces.

A few predictive generalizations are possible. If the CMC of the adsorbate is known, coulombic adsorption would be expected at concentrations below CMC/100 and constant adsorption above CMC. The effect of changes in molecular configuration on CMC can be predicted to some extent (Shimomuta, 1963). Since $\Delta G^{\circ}_{\text{hydroph}}$ is linear with carbon chain length, adsorption should increase with chain length in ranges II and III and perhaps IV but not in I. Since a PZC is observed, $\Gamma_{\text{max}}$ cannot be equated with surface charge. Tamamushi (1963) has pointed out that the $\Gamma_{\text{max}}$ is insensitive to ionic strength, but that adsorption densities in regions II and III, at least, increase as the ionic strength increases. The adsorption isotherms of members of a homologous series are all the same if $\Gamma$ is expressed as a function of surfactant activity or $C/CMC$ rather than concentration, $C$.

(b) *Oleate on silicates.* Adsorption of oleate on forsterite, $Mg_2SiO_4$, at pH 8·5 follows a type $S$ isotherm, but reaches a maximum at $10^{-4}$ molar and decreases thereafter. This is a type $S$-5 isotherm in the Giles classification—a type which has not been discussed. Oleate on tephroite, $Mg_2SiO_4$ and fayalite $(Fe_2SiO_4)$ follows type $L$-3 isotherms with adsorption at the knee amounting to about $5 \times 10^{-3}$ mol m$^{-2}$ at about $2 \times 10^{-5}$ molar equilibrium oleate concentration. The isotherms have been interpreted in terms of chemisorption (ion exchange of oleate for OH$^-$), hydrolysis and hydration of the adsorbate, and surface polarizability (Read and Manser, 1972).

(c) *Reviews.* Interpretative reviews have been published by Kipling (1965), Healy (1971), Fuerstenau (1971), Tamamushi (1963) and Giles *et al.* (1960).

## 4.4. CLOSING COMMENTS

It has been shown that the reasons for adsorption are very complex. Except in the simplest systems, surface charge is unlikely to be a reliable guide to adsorption behaviour. The origins of charge are many and for this reason, if for no other, many systems in which surface charge controls adsorption, often show unexpected behaviour. Just as dissolved ions and molecules form complexes which greatly change their thermodynamic activities in multicomponent solutions, so do surfaces as a result of adsorption. It is realistic to think of both the surface and the potential adsorbate as being subject to complexation and to recognise that their behaviour will be strongly influenced by this. The best documented example is probably the observation made by Neihof and Loeb (1972) that all solids regardless of whether they are positively or negatively charged in organic-free sea water become negatively charged in natural sea water. If we were unaware that the solid is "complexed" by anionic organic solutes (or colloids) we would predict some Gouy-layer anion adsorption or anion exchange capacity for the positively charged solid which, in reality, is a cation exchanger. An understanding of any system, especially one so complex as sea water containing as it does colloidal suspended solids, will require a detailed knowledge of the composition of both the solids and the solution. This knowledge must comprise not only the elemental analysis but also the speciation, including "complexation", that is resurfacing of solids by sorption.

The variety of sources of surface charge and the extreme sensitivity of the PZC or intrinsic charge (on e.g., clays, $\delta$-$MnO_2$ etc) towards coprecipitated or specifically adsorbed impurities lead to an extreme variability in the surface charge, unless the high concentration and favourable $\Delta G^\circ_{ads}$ of certain ubiquitous solutes leads to greater uniformity through their own strong but non-selective adsorption.

Adsorption can be considered to be a reaction proceeding in response to a suitable $\Delta G^\circ_{ads}$. The net free energy of adsorption comprises contributions from many sources; of these contributions, the coulombic (or electrostatic) terms are relatively small for the inorganic ionic species (charge $\pm 2$ or $\pm 1$) common in sea water. Contributions from solvation effects, hydrophobic, covalent (chemical), and hydrogen bonding or from several of these are often larger. For this reason, adsorption is often non-coulombic and unresponsive to original surface charge.

Effective surface charge, either intrinsic or extrinsic to the solid, does play a controlling role in the adsorption of inorganic or small (fewer than $\sim 8$ carbon atoms) organic–monovalent ions which do not hydrolyze and have no tendency to form strong complexes or compounds with any component of the solid *or with any component of specifically adsorbed* species. The term "controlling role" means that an attractive surface charge is required for ion adsorption to occur. The adsorption density may be controlled by charge and by factors which contribute to $\Delta G^\circ_{ads}$ (adsorption of long-chain alkyl-ammonium and sulphonate surfactants).

Strong, non-coulombic bonding to the solid surface can be expected if there is reason to believe that strong complexing occurs between the solute and a component of the surface. Strong non-coulombic bonding of metal ions to a pre-adsorbed ligand would be expected if the ligand is multi-functional, i.e., capable of binding both surface and metal ion. Strong bonding by virtue of a favourable $\Delta G^\circ_{hydroph}$ can be expected for non-polar molecules or the non-polar parts of heteropolar molecules if molecular structure allows contact with the solid.

## 4.5. ACKNOWLEDGEMENTS

The author particularly wishes to thank Dr. R. O. James for much helpful discussion. He is also greatly indebted to three of his former students, M. G. MacNaughton, Rita P. Ewing and P. V. Avotins for their help. In addition thanks are due to the many authors, particularly including D. W. Fuerstenau and T. W. Healy, who provided advance copies of many of their papers, to P. M. Williams, Roy Carpenter and G. I. Loeb for their continued interest. The U.S. National Science Foundation, through its support of past (GA 1451, Geochemistry) and present (GI 32943, RANN) research projects of the Stanford Center for Materials Research made it possible to take enough time to write this paper.

REFERENCES

Ahmed, A. (1971). Ph.D. Dissertation, Department of Applied Earth Sciences, Stanford University, Stanford, Calif., U.S.A.

Aleinikov, N. A. and German, T. P. (1965). Tr. Nauch.-Tekh. Sess. Inst. "Mekhanobr" (Vses. Nauch. Issled. Proekt. Inst. Mekh. Obrab. Polez. Iskop.) 5th., 1, 500 (in Russian, cf. C.A.69–99743d).

Amphlett, C. B. (1964). "Inorganic Ion Exchangers." Elsevier, London.

Atkinson, R. J., Posner, A. M. and Quirk, J. P. (1967). J. Phys. Chem. 71, 550.

Banajee, S. R. and Iwasaki, I. (1969). Trans. Amer. Inst. Mining Metal. Eng. 244, 401.

Bell, L. C., Posner, A. M. and Quirk, J. P. (1972). Nature, 239, 515.

Bell, L. C., Posner, A. M. and Quirk, J. P. (1973). J. Coll. Interf. Sci. 42, 250.

Berubé, Y. G. and de Bruyn, P. L. (1968) J. Coll. Interf. Sci. 28, 92.

Brinton, C. C., Jr. and Lauffer, M. A. (1959). In "Electrophoresis" (M. Bier, ed.). Academic Press, New York and London.

Buchanan, A. S. and Heymann, E. E. (1948). Proc. Roy. Soc. A 195, 150

Buchanan, A. S. and Oppenheim, R. C. (1968). Aust. J. Chem. 21, 2367.

Bungenberg de Jong, H. G. (1949). In "Colloid Science" (H. R. Kruyt, ed.) Vol. II, pp. 1–18. Elsevier, London.

Cann, J. R. (1969). In "Physical Principles and Techniques of Protein Chemistry" (S. J. Leach, ed.), Part A. Academic Press, New York and London.

Chester, R. and Johnson, L. R. (1971). Nature, 227, 105.

Colombera, P. M., Posner, A. M. and Quirk, J. P. (1971). J. Soil Sci. 22, 118.

Connor, P. and Ottewill, R. H. (1971). J. Coll. Interf. Sci. 37, 642.

Cook, M. A. and Nixon, J. L. (1960). J. Phys. Coll. Chem. 54, 445.

Cook, M. A. (1969). In "Hydrophobic Surfaces" (F. M. Fowkes, ed.), Academic Press, London and New York.

Cranston, R. E. and Buckley, D. E. (1972). Envir. Sci. Tech. 6, 274.

Curtis, A. S. G. (1967). "The Cell Surface: Its Molecular Role in Morphogenesis." Academic Press, New York and London.

Davies, J. T. and Rideal, E. K. (1963). "Interfacial Phenomena." Academic Press, London and New York.

deBruyn, P. L. and Agar, G. E. (1962). In "Froth Flotation" (D. W. Fuerstenau, ed.), Amer. Inst. Mining and Met. Eng., New York.

Degens, E. T. (1965). "Geochemistry of Sediments." Prentice-Hall, Englewood Cliffs, New Jersey.

Degens, E. T. and Matheja, J. (1971). In "Organic Compounds in Aquatic Environments" (S. J. Faust and J. V. Hunter, eds.), Marcel Dekker, New York.

Everett, D. H. (1972). Pure Appl. Chem. 31, 579.

Eyring, E. M. and Wadsworth, M. E. (1956). Trans. Amer. Inst. Mining Metal. Eng. 205, 531.

Fuerstenau, D. W. (1971). In "The Chemistry of Biosurfaces" (M. L. Hair, ed.), Marcel Dekker, New York.

Fuerstenau, D. W. (1970). Pure Appl. Chem. 24, 135.

Fuerstenau, D. W. and Healy, T. W. (1972). In "Adsorptive Bubble Separation Techniques" (R. Lemlich, ed.), Academic Press, London and New York.

Fuerstenau, D. W., Healy, T. W. and Somasundaran, P. (1964). Trans. Amer. Inst. Mining Metal. Eng. 229, 321.

Fuerstenau, M. C., Martin, C. C. and Bhappu, R. B. (1963). *Trans. Amer. Inst. Mining Metal. Eng.* **226**, 449.

Fuerstenau, M. C. and Miller, J. D. (1967). *Trans. Amer. Inst. Mining Metal. Eng.* **238**, 153.

Gaudin, A. M. and Fuerstenau, D. W. (1955). *Trans. Amer. Inst. Mining Metal. Eng.* **202**, 958

Giles, C. H. (1959). *In* "Hydrogen Bonding" (D. Hadzé, ed.), Pergamon Press, New York.

Giles, C. H., MacEwan, T. H., Nakhwa, S. N. and Smith, D. (1960). *J. Chem. Soc.* 3973.

Grahame, D. C. (1947). *Chem. Rev.* **41**, 441.

Han, K. M., Healy, T. W. and Fuerstenau, D. W. (1972). *J. Coll. Interf. Sci.* in the press.

Haydon, D. A. (1964). *In* "Recent Progress in Surface Science" (J. F. Danielli, K. G. A. Pankhurst and A. C. Riddiford, eds) Vol. I, Academic Press, London and New York.

Healy, J. W. (1971). *In* "Organic Compounds in Aquatic Environments" (S. J. Faust and J. V. Hunter, eds.), Marcel Dekker, New York.

Healy, T. W. and Fuerstenau, D. W. (1965). *J. Coll. Interf. Sci.* **20**, 376.

Healy, T. W., Herring, A. P. and Fuerstenau, D. W. (1966). *J. Coll. Interf. Sci.* **21**, 435.

Healy, T. W. and Jellett, V. R. (1967). *J. Coll. Interf. Sci.* **24**, 41.

Healy, T. W. and Kavanagh, B. V. (1972). Written personal communication of unpublished work.

Hermans, P. J. (1949). *In* "Colloid Science", (H. R. Kruyt, ed.), Vol. II, Elsevier, London.

Hingston, F. J., Posner, A. M. and Quirk, J. P. (1972). *J. Soil Sci.* **23**, 177.

Honig, E. P., and Hengst, J. H. T. (1969). *J. Coll. Interf. Sci.* **29**, 510.

Horne, R. A. (1969). "Marine Chemistry." Wiley-Interscience.

Houwink, R. (1949). *In* "Colloid Science" (H. R. Kruyt, ed.), Vol. II, Elsevier, London.

Hunter, R. J. and Wright, H. J. L. (1971). *J. Coll. Interf. Sci.* **37**, 564.

Jacobs, M. B. and Ewing, M. (1969). *Science*, **163**, 805.

James, R. O. (1971). "The Adsorption of Hydrolyzable Metal Ions at the Oxide–Water Interface," Ph.D. Dissertation, Department of Physical Chemistry, University of Melbourne, Parkville, Australia.

James, R. O. and Healy, T. W. (1972a). *J. Coll. Interf. Sci.* **40**, 42.

James, R. O. and Healy, T. W. (1972b). *J. Coll. Interf. Sci.* **40**, 53.

James, R. O. and Healy, T. W. (1972c). *J. Coll. Interf. Sci.* **40**, 65.

James, R. O. and Healy, T. W. (1972d). In preparation.

Jenkins, S. R. and Stumm, W. (1971). Presented before the Division of Air, Water and Waste Chem. Amer. Chem. Soc.

Kavanagh, B. and Healy, T. W. (1972). Written personal communication.

Kavanau, J. L. (1964), "Water and Solute-Water Interactions." Holden-Day, London.

Kauzmann, W. (1959). *Adv. Protein Chem.* **14**, 1.

Kharkar, D. P., Turekian, K. K. and Bertine, K. K. (1968). *Geochim. Cosmochim. Acta*, **32**, 285.

Kim, Y. S. and Zeitlin, H. (1971a). *Chem. Commun.* 672.

Kim, Y. S. and Zeiltin, H. (1971b). *Analyt. Chem.* **43**, 1390.

Kim, Y. S. and Zeitlin, H. (1972). *Separation Sci.* **7**, 1.

Kipling, J. J. (1965). "Adsorption from Solutions of Non-Electrolytes." Academic Press, London and New York.

Kitchener, J. A. (1965). *J. Photo. Sci.* **13**, 152.

Krauskopf, K. B. (1956). *Geochim. Cosmochim. Acta* **9**, 1.

Krauskopf, K. B. (1967). "Introduction to Geochemistry." McGraw-Hill, New York.

Kruyt, H. R. (1952). *In* "Colloid Science" (H. R. Kruyt, ed.), Vol. I, Elsevier, London.

Lai, R. W. M. and Fuerstenau, D. W. (1972). Manuscript in preparation.

Lauzon, R. V. and Matijevic, E. (1972). *J. Coll. Interf. Sci.* **38**, 440.

Li, H. C. and deBruyn, P. L. (1966). *Surface. Sci.* **5**, 203.

Luce, R. W. and Parks, G. A. (1973). *Chem. Geol.* **12**, 147

MacKenzie, J. M. W. (1969). *Trans. Amer. Inst. Mining Metal Eng.* **224**, 393

MacKenzie, J. M. W. and O'Brien, R. T. (1969). *Trans. Amer. Inst. Mining Metal. Eng.* **244**, 168.

MacNaughton, M. G. and James, R. O. (1973). Personal communication.

Malati, M. A., Yousef, A. A. and Arafa, M. A. (1969). *Chemical Ind.* 459.

Manheim, F. T., Hathaway, J. C., Uchupi, E. (1972). *Limnol. Oceanogr.* **17**, 17.

Morgan, J. J. and Stumm, W. (1964). *J. Coll. Sci.* **19**, 347.

Murray, J. W. (1972). *J. Coll. Interf. Sci.* in the press.

Murray, D. J., Healy, T. W. and Fuerstenau, D. W. (1968). *In* "Adsorption from Aqueous Solution" (W. J. Weber, Jr. and E. Matijević, eds.), American Chemical Society, Washington, D.C.

Naumann, A. W. and Dresher, W. H. (1966). *J. Phys. Chem.* **70**, 288.

Neihof, R. A. and Schuldiner, S. (1960), *Nature*, **185**, 526.

Neihof, R. A. (1969). *J. Coll. Interf. Sci.* **30**, 128.

Neihof, R. A. and Loeb, G. I. (1972). *Limnol. Oceanogr.* **17**, 7.

Némethy, G. and Scheraga, H. A. (1962). *J. Phys. Chem.* **62**, 1773.

Nicol, S. K. and Hunter, R. J. (1970). *Aust. J. Chem.* **23**, 2177.

Nozaki, Y. and Tanford, C. (1967). *In* "Methods in Enzymology" (C. H. W. Hirs, ed.), Vol. XI, Academic Press, London and New York.

Onoda, G. Y. and de Bruyn, P. L. (1966). *Surface Sci.* **4**, 48.

Overbeek, J. Th. G. (1952). *In* "Colloid Science" (H. R. Kruyt, ed.), Vol. I, "Irreversible Systems," Elsevier, London.

Overbeek, J. Th. G. and Lijklema, J. (1959). *In* "Electrophoresis" (M. Bier, ed.), Academic Press, New York and London.

Parks, G. A. (1965). *Chem. Rev.* **65**, 177.

Parks, G. A. (1967). *In* "Equilibrium Concepts in Natural Water Systems" (W. Stumm, ed.), American Chemical Society, New York.

Parks, G. A. and deBruyn, P. L. (1962). *J. Phys. Chem.* **66**, 967.

Parsons, R. (1954). *In* "Modern Aspects of Electrochemistry" (J. O'M. Bockris, ed.), Vol. I, Butterworths Scientific Publications, London.

Parsons, T. R. (1963). *In* "Progress in Oceanography" (M. Sears, ed.), Vol. I, MacMillan Co., New York.

Peck, A. S. (1963). *U.S. Bur. Mines, Rept. Invest.* **6202**.

Peck, A. S. and Wadsworth, M. E. (1965). *In* "Proc. 7th Intl. Mineral Processing Congress" (N. Arbiter, ed.), Gordon and Breach, New York.

Peck, A. S., Raby, L. H. and Wadsworth, M. E. (1966). *Trans. Amer. Inst. Mining Metal. Eng.* **235**, 301.

Peck, A. S., Raby, L. H. and Wadsworth, M. E. (1967a). *Trans. Amer. Inst. Mining Met. Eng.* **238**, 245.

Peck, A. S., Raby, L. H. and Wadsworth, M. E. (1967b). *Trans. Amer. Inst. Mining Mining and Met. Eng.* **238**, 264.

Pimentel, G. C. and McClellan, A. L. (1960). "The Hydrogen Bond." W. H. Freeman, London.

Pravdic, V. (1970). *Limnol. Oceanogr.* **15**, 230.

Predali, J. J. (1970). Proc. 9th Intl. Min. Proc. Congr. Prague, 1970, pp. 241–249.

Read, A. D. and Manser, R. M. (1972). *Trans. I.M.M.* **81**, C69–C78.

Riley, J. P. and Chester, R. (1971). "Introduction to Marine Chemistry." Academic Press, London and New York.

Rochester, C. H. (1971). *In* "The Chemistry of the Hydroxyl Group" (S. Patai, ed.), Part I, Interscience, New York.

Schindler, P. and Kamber, H. R. (1968). *Helv. Chim. Acta.* **51**, 1781.

Schindler, P. W. and Gamsjäger, A. (1972). *Kolloid Z. Z. Polym.* **250**, 759.

Shaw, D. J. (1970). "Introduction to Colloid and Surface Chemistry." 2nd Ed., Butterworths Scientific Publications, London.

Shergold, H. L. (1972). *Trans. Inst. Mining Metal. Eng.* **81**, C148.

Shinoda, K. (1969). *In* "Colloidal Surfactants" (K. Shinoda, T. Nakagawa, B.-I. Tamamushi and T. Isemura, authors), Academic Press, New York and London.

Shinoda, K. and Fujihira, M. (1968). *In* "Adsorption from Aqueous Solution" (W. J. Weber, Jr. and E. Matijević, eds.).

Smith, R. W. (1963). *Trans. Amer. Inst. Mining and Metal. Eng.* **229**, 427.

Smith, R. W. and Trivedi, N. (1971). Presented at The 1971 Annual Mtg. Amer. Inst. Mining and Met. Eng., San Francisco.

Smolik, T. J., Harman and Fuerstenau, D. W. (1966). *Trans. Amer. Inst. Mining and Metal. Eng.* **235**, 367.

Somasundaran, P. (1968). *J. Coll. Interf. Sci.* **27**, 659.

Somasundaran, P. and Agar, G. E. (1967). *J. Coll. Interf. Sci.* **24**, 433.

Somasundaran, P. and Agar, G. E. (1972). *Trans. Amer. Inst. Mining Metal. Eng.* **254**, 348.

Somasundaran, P., Healy, T. W. and Fuerstenau, D. W. (1964). *J. Phys. Chem.* **68**, 3562.

Stein, H. N. (1968). *J. Coll. Interf. Sci.* **28**, 203.

Steinhardt, J. and Reynolds, J. A. (1969). "Multiple Equilibria in Proteins", Academic Press, London.

Stumm, W. and Morgan, J. J. (1970). "Aquatic Chemistry." Wiley-Interscience, London.

Stumm, W., Huang, C. P. and Jenkins, S. R. (1970). *Croat. Chem. Acta.* **42**, 223.

Tait, M. J., Ablett, S. and Franks, F. (1972). *In* "Water Structure at the Water–Polymer Interface" (H. H. G. Jellinek, ed.), Plenum Press, London.

Tamamushi, Bun-Ichi (1963). *In* "Colloidal Surfactants" (K. Shinoda, T. Nakagawa, B.-I. Tamamushi and T. Isemura, authors), Academic Press, New York and London.

Tanford, C. (1962). *In* "Advances in Protein Chemistry", C. B. Anfinsen, Jr., M. L. Anson, K. Bailey, J. T. Edsall, eds.), Academic Press, New York and London.

Touret, C. (1969). *C. R. Acad. Sci. Paris D.* **269**, 1591.

Turekian, K. K. (1965). *In* "Chemical Oceanography" (J. P. Riley and G. Skirrow, eds.), Academic Press, London and New York.

Van Olphen, H. (1963). "Introduction to Clay Colloid Chemistry." Interscience Publishers, London.

Vestier, D. (1969). *Sci. Terre*, **14**, 289 –300.

Vold, M. J. and Vold, R. D. (1964). "Colloid Chemistry." Van Nostrand-Reinhold, London.

Von Hippel, A. R. (1954). "Dielectrics and Waves." Chapman and Hall, London.

Wakamatsu, T. and Fuerstenau, D. W. (1968). *In* "Adsorption from Aqueous Solution" (W. J. Weber, Jr. and E. Matijević, eds.), American Chemical Society, Washington, D.C.

Warren, L. J. and Kitchener, J. A. (1972). *Trans. Inst. Mining Metal. Eng.* **81**, C137.

Wieland, T. (1959). *In* "Electrophoresis" (M. Bier, ed.), Academic Press, New York.

Wilkins, D. J. and Myers, P. A. (1970). *In* "Surface Chemistry of Biological Systems" (M. Blank, ed.), Plenum Press, New York.

Chapter 5

# Sedimentary Cycling and the Evolution of Sea Water

FRED T. MACKENZIE

*Department of Geological Sciences, Northwestern University,*
*Evanston, Illinois 60201, U.S.A.*

## 5.1. INTRODUCTION

Any interpretation of the chemical history of sea water is fraught with many difficulties. First and foremost of these is the fact that there is no direct means of assessing the composition of sea water which existed in the past. It is unlikely that solutions found in the pores of sedimentary rocks or in inclusions in sedimentary minerals correspond in composition to that of the

309

sea water at the time of their deposition since these fluids undergo chemical changes soon after incorporation into sediments. Second, to a significant degree any interpretation of sea water composition during geological time depends on sedimentological data. During the last two decades some of these data, including chemical, mineralogical, and isotopic compositions of sedimentary rocks as a function of rock age, have become available for rocks of Phanerozoic age (< 600 million years). However, much less information is available for the preceding 3 aeons of earth history represented by preserved sedimentary rocks. Third, any model of the origin and evolution of sea water depends on answers to problems in many areas of research at the forefront of earth science today (e.g. degassing history of the earth, sea floor spreading and the rate at which the mantle contributes materials to the earth's surface). Without answers to these questions it is difficult to construct an acceptable unique model for sea water history.

Recognizing these limitations, an attempt will be made below to construct a model for sea water history based principally on the distribution of sedimentary rock features as a function of rock age. In the first section, chemical, mineralogical, isotopic, and rock type distributions as a function of geological age are presented and interpreted as being the result of secondary processes in the rock mass occurring even as long ago as 1·5 aeons. This interpretation implies that no major evolutionary change has occurred in sea water composition for 600 and perhaps for 1500 million years. A brief resumé of models of sea water composition will be presented in the second section. A steady-state model will be developed and this forms the basis for discussion of cycling of the elements within the atmosphere-biosphere-ocean-sediments system. Finally, an attempt will be made in the last section to deduce the evolutionary changes in sea water composition which occurred between 3·5 and $1·5 \times 10^9$ years ago. This section is certainly the most speculative of all.

Much of my research on the chemical history of sea water has been stimulated by the works of E. J. Conway, W. W. Rubey, H. D. Holland, A. B. Ronov, and the late L. G. Sillén and by discussions with my colleague and friend, R. M. Garrels.

## 5.2. SEDIMENTARY CHEMISTRY AND MINERALOGY

### 5.2.1. INTRODUCTION

The chemistry and mineralogy of existing sedimentary rocks as a function of their ages provide major clues for the decipherment of the history of the earth's surface environment. Indeed, because of lack of direct evidence (e.g., compositions of fluid inclusions and connate waters) of the composition of sea water in times past, data gleaned from the distributions, chemistry, and mineralogy of sedimentary rocks form the major basis for the interpretation of the chemical history of sea water.

It has been observed that the masses and proportions of sedimentary rock types, the chemical and mineralogical compositions of shales and carbonates, and the isotopic compositions of limestones, cherts and evaporites are functions of geological age (see summaries by Ronov, 1964 and also by Garrels and Mackenzie, 1971). These trends, assuming that they are statistically significant, may be viewed as being either primary or secondary in origin. If primary, the trends reflect progressive changes in the compositions and rock-type ratios of sediments deposited, and record changes in the types and/or magnitudes of processes operating at the earth's surface during geological time, and perhaps, in sea water composition. If secondary, these trends reflect alterations of the sediments after deposition, and do not directly record the nature of the depositional environment.

The purpose of this section is to show that the long-term trends observed for sedimentary rocks of Phanerozoic age, and perhaps, for the past 1·5 aeons are consistent with a model in which (1) the sediments deposited have nearly always been the same in their proportions of type, (2) the trends observed are a result of differential cycling of the chemical components of the sedimentary mass and, additionally, the effects of long-term diagenesis, and (3) there is little cogent evidence from the sedimentary record of a major continuous change in ocean composition over the last $1·5 \times 10^9$ years. It should be emphasized that this model does not argue against chemical changes in sea water composition, only against a significant continuous evolutionary change with time.

### 5.2.2. SEDIMENTARY MASS AND GEOLOGIC AGE

#### 5.2.2.1. *Mass distribution*

The only direct information we have about the distribution of sedimentary mass as a function of the ages of the rocks covers the interval from Devonian to Jurassic. Various estimates (Ronov, 1959; Gregor, 1968a; Garrels and Mackenzie, 1969) have been made of the Phanerozoic mass distribution, but little information is available about the total mass-age relations for Precambrian sedimentary rocks. Figure 5.1 shows a schematic representation of the mass distribution. The Precambrian mass of $14\,000 \times 10^{20}$ g (including volcanic sediments) is spread over a time span of $3 \times 10^9$ years whereas the last 600 million years of geological time are represented by an approximately equal mass ($18\,000 \times 10^{20}$ g) of preserved rock (Garrels and

Mackenzie, 1971). It can be seen from Fig. 5.1 that the mass per unit time diminishes exponentially from the present through the Precambrian. In other words, the sedimentary rock mass is crowded toward the front of geological time.

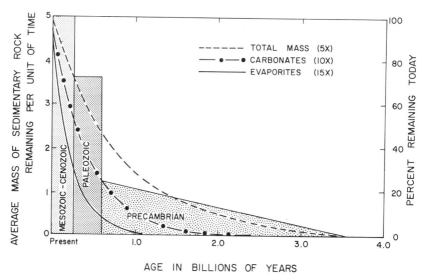

FIG. 5.1. Schematic representation of existing sedimentary rock mass. Calculated constant mass models for the distribution of the total sedimentary mass (– – – –), carbonate rocks (– ·– ·–), and evaporites (———) with rock age are also shown (Garrels and Mackenzie, 1971).

Two explanations have been proposed for this distribution (Garrels and Mackenzie, 1971; Garrels et al., 1972). In the first, it is assumed that a mass of sedimentary rock equal to the existing mass was formed very early from primary igneous rock, and that this has been continuously recycled by erosion and destroyed by metamorphism and crustal plate resorption into the mantle with concomitant recycling of $CO_2$ and HCl. This model entails an early degassing of the earth with the emission of all the water to the hydrosphere and the liberation of all the $CO_2$, HCl, and other acid gases ("excess volatiles", see Rubey, 1951) which reacted with primary igneous rock to form sedimentary rocks. The second explanation involves a model in which the sedimentary mass grows linearly through geological time as a result of the progressive and linear degassing of water and acid gases from the earth's interior. This model of linear accumulation represents an extreme alternative explanation to the constant mass model, and geophysical evidence (Higgins,

1968; Birch 1965; Fanale, 1971) suggests that it is a less likely model than one of nearly constant mass.

The dashed exponential curve in Fig. 5.1 shows the best fit to observations based on a constant mass model in which the total sedimentary mass has been turned over about 5 times; that is, the total mass of sediments deposited during geological time was $5 \times 32000 \times 10^{20} = 160000 \times 10^{20}$ g. This mass is compatible with an average rate of deposition during the last $3 \cdot 5 \times 10^9$ years of about $50 \times 10^{14}$ g per year. Details of the calculation are given in Garrels and Mackenzie (1971).

### 5.2.2.2. *Rock type distribution*

Information on the distribution of rock types within the total sedimentary rock mass has been assembled by Ronov (1964) and Fig. 5.2 has been adapted from his work. He expressed his results in terms of the relative volume percentages of the various rock types as functions of age. In assessing the trends shown on the diagram it must be remembered that the total rock mass

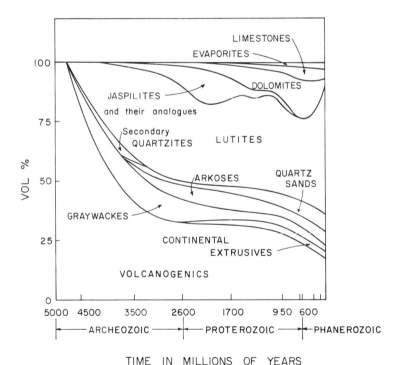

TIME  IN  MILLIONS  OF  YEARS

FIG. 5.2. Volume percent of rock types as a function of age (Ronov, 1964).

diminishes with increasing age, and that a given volume percentage of rock $2 \times 10^9$ years old represents much less rock than the same volume percentage for rocks only 200 million years old.

It can be seen from Fig. 5.2 that the percentages of lutites (shales) and sandstones are apparently relatively constant through time. However, Ronov suggested that graywackes increase at the expense of quartz sandstones and arkoses in the middle and early Precambrian, and that the percentage of volcanogenic sediments increases with increasing geological age. Carbonate rocks increase strikingly in their relative percentages from almost zero $3 \times 10^9$ years ago to a significant fraction of the total column in Phanerozoic time. Dolomites dominated the carbonates in the Precambrian, and limestones apparently are restricted almost entirely to the post-Precambrian. Evaporites behave like carbonate rocks, being almost entirely restricted to the Phanerozoic. Halite deposits are absent from Precambrian rocks (Lotze, 1964).

An interesting feature revealed by the diagram is that jaspilites were a relatively important component of the rocks in the middle Precambrian but were almost absent from younger rocks. Jaspilites are massively layered cherty rocks, commonly associated with iron deposits. The chert-iron association is unique to the rocks of the Precambrian.

The half-life or "half-mass age" of the total sedimentary mass is about 430 million years; that is, it would take 430 million years to deposit a sedimentary mass equal to that of today. It can be seen from Fig. 5.2, however, that the half-mass ages of carbonate rocks and evaporites are less, being 300 and 200 million years, respectively. The calculated constant mass model curves for these rock types and the half-mass ages are shown in Fig. 5.1. Shales and sandstones have half-mass ages of about 500 million years, slightly greater than that of the total sedimentary mass. Therefore, to account for the present mass distributions, it would have been necessary to deposit 15 times the present mass of evaporites and 10 times the present mass of carbonate rocks during geological time. In other words, evaporites are turned over $1\frac{1}{2}$ times faster than carbonate rocks which, in turn, cycle twice as fast as shales and sandstones. *Thus, the present percentage distribution of evaporites, carbonates, sandstones and shales as a function of geological age could reasonably be the result of differential cycling of these rock types rather than a reflection of progressive changes in the earth's surface environment or sea water composition.*

### 5.2.3. COMPOSITIONAL TRENDS IN SHALES

The chemical and mineralogical compositions of shales vary progressively with the geological ages of the rocks. In Fig. 5.3, four composite chemical analyses of shaly rocks are plotted as a function of time. The analyses shown represent averaged estimates of the chemical composition of modern argillaceous sediments (Garrels and Mackenzie, 1971), of Mesozoic and Cenozoic shales (Clarke, 1924; Vinogradov and Ronov, 1956a), of Palaeozoic shales (Clarke, 1924; Vinogradov and Ronov, 1956a), and of Precambrian slates,

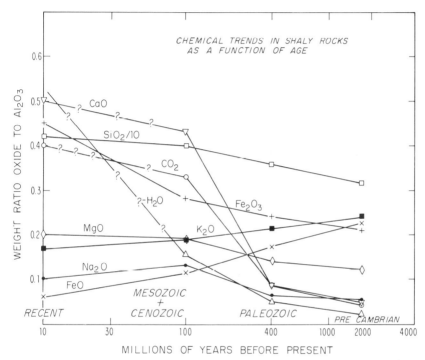

FIG. 5.3. Chemical trends in average shales and similar rocks as a function of age (Garrels, et al., 1972).

schists, and phyllites (Nanz, 1953). Because of the immobility of alumina in the exogenic cycle, the ratios of the metal oxides to alumina were chosen to illustrate compositional trends. All of the oxides but one ($Na_2O$) exhibit consistent trends with geological age. The decrease in the $Na_2O/Al_2O_3$ ratio between Mesozoic-Cenozoic and Recent is the only reversal in trend, and because of the difficulties inherent in obtaining an estimate of Recent argillaceous sediment composition, this change in trend may be spurious.

All of the oxide ratios, except those of $K_2O$ and FeO, diminish with increasing rock age. The amounts of water, CaO, and $CO_2$ decrease at a more rapid rate than do those of $SiO_2$, $Na_2O$, and $MgO$. There is a reciprocal relationship between the $FeO$ and $Fe_2O_3$ contents; the amount of total iron remains nearly constant.

The chemical trends are reflected in mineralogical differences. Figure 5.4 is a schematic diagram illustrating the mineralogical trends in shaly rocks as a function of geological age. The post-Palaeozoic shales contain a more diverse mineral assemblage than do older shales. The older shales are characterized by the clay minerals illite and chlorite (Weaver, 1967) and contain little admixed $CaCO_3$. The younger shales contain an average of up to 50%

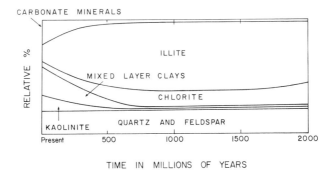

FIG. 5.4. Schematic diagram illustrating mineralogical trends in shales and similar rocks as a function of age (Garrels and Mackenzie, 1971).

of mixed-layer clays (including montmorillonite) and kaolinite together with significant proportions of illite and a little chlorite; they are relatively rich in $CaCO_3$. The increase in the percentage of illite in older rocks correlates with their relative enrichment in $K_2O$, whereas increased percentages of mixed-layer clays in younger rocks account for increases in $SiO_2$, $MgO$, and $Na_2O$, the three oxides that behave similarly in the plot of shaly rock chemical compositions. The increase in the percentage of calcium carbonate in younger rocks is reflected in the higher $CaO$ and $CO_2$ content of these shales.

Similar chemical and mineralogical trends are observed for shales of the Gulf Coast as a function of depth of burial. Burst (1959) reports that in the Eocene Wilcox formation, montmorillonite, which is a common clay mineral found in Wilcox outcrop sediments, is diagenetically transformed to illite, and perhaps chlorite, with increasing depth of burial. However, from a detailed subsurface chemical and mineralogical study of Pleistocene to Eocene shales in the Gulf Coast, Perry and Hower (1970) showed that mixed-layer illite/montmorillonite undergoes a monotonic decrease in expandability with increasing depth from about 80% expandable layers to 20%.

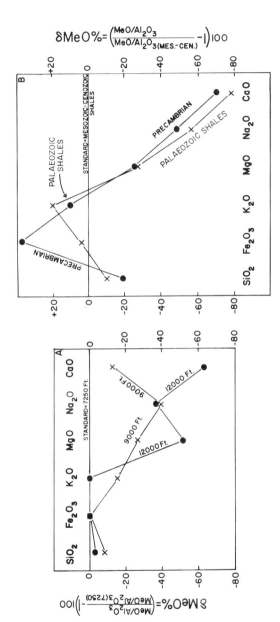

FIG. 5.5. Comparison of chemical trends with depth in Gulf Coast shales with trends with age for average shales. A. Changes with depth shown as percentage deviation from the average composition at 7250 ft. (data from Perry and Hower, 1970). B. Changes with age portrayed as the percentage difference from the composition of average Mesozoic–Cenozoic shale (data from Garrels and Mackenzie, 1971).

Since the proportion of illite increases with depth, they attributed this change in mineralogy to the breakdown of detrital mica to give potassium which is fixed in the illite/montmorillonite mixed layer. The rate of decrease of the expandable layers with depth is proportional to the geothermal gradient and not to the depth of burial; consequently, they believed temperature is more important than pressure in governing the transformation reaction. Garrels and Mackenzie (1973) have pointed out that the crystallization of quartz may drive the conversion reaction of mixed-layer clay to illite. Perry and Hower also observed that kaolinite or chlorite apparently did not undergo any transformations with depth; however, they reported that calcium carbonate, and perhaps potassium feldspar, is lost owing to diagenesis following burial.

Chemical analyses of the whole rock and clay-size fraction of the shales were obtained by Perry and Hower (*loc. cit.*). The whole rock compositions as a function of depth in Well E are plotted in Fig. 5.5A and compared with the time-dependent trends of Fig. 5.5B for the oxides analyzed. In Gulf Coast shales, the $Fe_2O_3$, $K_2O$, and $SiO_2$ contents change little with depth, but those of CaO, MgO, and $Na_2O$ diminish markedly. Perry and Hower pointed out that the percentage of $K_2O$ in the clay-size fraction increases monotonically with depth. The similarity between the chemical changes with depth of the Gulf Coast shales and the chemical trends in shales with age are obvious. Mesozoic shales of Papua, New Guinea, show mineralogical changes with depth similar to those of the Gulf Coast shales (van Moort, 1971).

Burial diagenesis apparently also leads to the disappearance of kaolinite and to the formation of chlorite with increasing depth, as reported for Upper Cretaceous shales in the African Cameroun (de Segonzac, 1965). This trend is also compatible with the mineralogical trends observed in shales as a function of geological age.

The chemical and mineralogical trends of shaly rocks, at least for Phanerozoic time, and perhaps, as far back as $1.5 \times 10^9$ years, are compatible with a model in which these trends are due to secondary processes. The carbonate (CaO, $CO_2$, and some MgO) is lost from the older shales during burial and uplift owing to a gentle long term leaching of the shales by waters. Water, which correlates with CaO and $CO_2$ in the composition-age diagram for shaly rocks, is expressed from the shales during burial and carries calcium and bicarbonate in solution. CaO and $CO_2$ continue to be leached as waters percolate through the shales later in their history, reflecting the differential cycling rate of calcium carbonate. Long-term diagenetic transformations account for the other chemical and mineralogical trends. One generalized reaction suggested by Garrels and Mackenzie (1973) is:

montmorillonite-rich mixed layer

$$2 \cdot 5X^+_{0 \cdot 05}K_{0 \cdot 35}[(Mg_{0 \cdot 34}Fe^{+3}_{0 \cdot 17}Fe^{+2}_{0 \cdot 04}Al_{1 \cdot 50})(Al_{0 \cdot 17}Si_{3 \cdot 83})O_{10}(OH)_2]$$
$$+ \, K^+ + AlO_2^- =$$

illite-rich mixed layer

$$2 \cdot 5X^+_{0 \cdot 05}K_{0 \cdot 75}[(Mg_{0 \cdot 34}Fe^{+3}_{0 \cdot 17}Fe^{+2}_{0 \cdot 04}Al_{1 \cdot 50})(Al_{0 \cdot 57}Si_{3 \cdot 43})O_{10}(OH)_2]$$
$$+ \, SiO_2, \qquad (5.1)$$

where $X^+$ represents exchangeable univalent cations. This transformation has been discussed in connection with the studies of Tertiary and Cretaceous rocks mentioned above.

Supporting evidence for the occurrence of the chemical and mineralogical transformations noted both in the Gulf Coast and in shales as a function of geological age has been obtained from K–Ar ages of bulk samples of the shales. Weaver and Wampler (1970) showed that the apparent K–Ar ages of mud samples from a well in the Mississippi Delta area decrease with depth by about 100 million years in 3700 m. They attributed this decrease to release of potassium from the potassium feldspar and mica in the muds and its incorporation in the mixed-layer illite/montmorillonite/chlorite phase with concomitant formation of illite layers and lowering of the apparent age of the clay fraction. Hurley (1965) had previously cited examples in which both the clay fraction and the whole rock K–Ar ages of shales were less than their respective sedimentation ages. It is likely that redistribution of potassium in the manner suggested by Perry and Hower and also by Weaver and Wampler accounts for these observations. It is also probable that, on a longer term basis, potassium addition to the clays will occur as a result of migration of potassium in the fluid phase into the clays, this being particularly intense in regions of moderately high geothermal gradients.

The reciprocal relationship between $Fe_2O_3$ and $FeO$ observed in the composition-age diagram for shales could also be viewed as a secondary feature of the rocks, and attributed to the reduction of ferric iron by carbonaceous material in the shales during geological time.

Thus, the chemical and mineralogical trends observed in shaly rocks during the last $1 \cdot 5 \times 10^9$ years are compatible with a model in which these trends are secondary in nature and result from differential cycling of the more easily erodible components in the shale and from long-term diagenesis. Approximately 15–20% of the shale mass is lost during these long-term processes. It is probable that time alone plays a role in the diagenetic stabilization process as well as in processes accompanying burial diagenesis. The half-life for conversion of an orginally mixed-layer clay shale to illite is about 50 million years. *The chemical and mineralogical trends in shaly rocks do not necessarily imply a marked evolutionary change in sea water composition during the past* $1\cdot5 \times 10^9$ *years.*

### 5.2.4. COMPOSITIONAL TRENDS IN CARBONATE ROCKS

Carbonate rocks become more dolomitic and enriched in magnesium with increasing rock age. Fig. 5.6 illustrates the Ca/Mg ratio in Phanerozoic carbonates as a function of geological age; the Period averages represent composite analyses of sediments from North America and the Russian Platform. Recent carbonate rocks are nearly pure calcium carbonate with Ca/Mg weight ratios of about 40 whereas Cambrian carbonates approach dolomite in composition. The scanty analyses of Precambrian carbonates give a mean Ca/Mg ratio of 2, suggesting that the trend shown in Fig. 5.6 extends into the Precambrian.

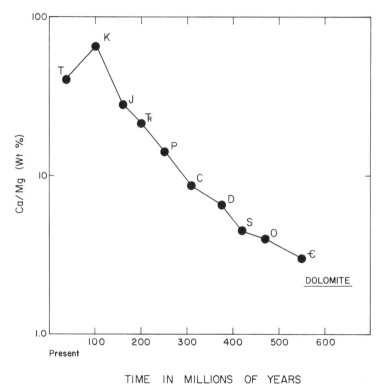

FIG. 5.6. Period averages of the calcium/magnesium ratio of carbonate rocks plotted against age (data from Vinogradov and Ronov, 1956a, b; Chilingar, 1956).

Fairbridge (1964) has discussed the four explanations which have been proposed for the observed overall trend: (1) higher Palaeozoic water temperatures, (2) widespread growth of magnesium-rich algal limestones in the Palaeozoic and Precambrian, (3) higher $P_{CO_2}$ in the Palaeozoic and Precambrian resulting in greater formation of primary dolomites, and (4) increasing probability of older rocks being dolomitized.

The oxygen isotopic composition of Phanerozoic limestones as a function of their ages provides a clue to the origin of the Ca/Mg ratio trend in carbonate rocks. Figure 5.7 shows a plot of $\delta^{18}O$ values of average limestones and cherts for various Periods. The limestone data represent composite analyses of measurements made by Degens and Epstein (1962) and Keith and Weber (1964). The chert analyses were carried out by Degens and Epstein on samples intimately associated with the limestones analyzed.

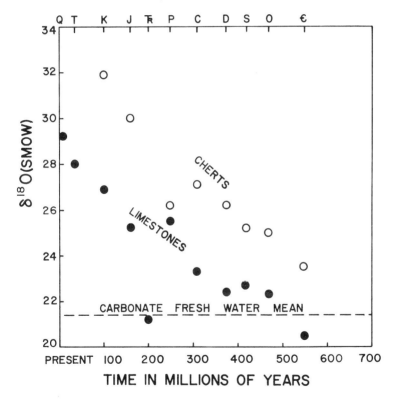

FIG. 5.7. Period averages of oxygen isotopic composition of carbonate rocks and cherts plotted against age (data from Degens and Epstein, 1962).

There is a systematic trend in $\delta^{18}O$ values for both limestones and cherts. Quaternary limestones have values of about $+29$, whereas Cambrian rocks average about $+21$. Cherts show a similar decrease in $\delta^{18}O$ with increasing age. In addition, with increasing age the $\delta^{18}O$ values for both rock types approach the $\delta^{18}O$ values of their respective fresh water analogues. Although the deficiency of $\delta^{18}O$ in progressively older limestones could in part reflect higher sea water temperatures in ancient oceans or a progressive increase in the $^{18}O/^{16}O$ ratio of sea water from the Cambrian to the present, it is much more likely that the observed trends are the result of partial equilibration of the limestones and cherts with isotopically lighter continental surface and sub-surface waters (cf. Degens and Epstein, 1962; Keith and Weber, 1964; Weber, 1965). The older the rock the greater the probability that it was bathed in meteoric waters.

There is a moderately good relationship between the $\delta^{18}O$ values of limestones, their Ca/Mg ratio and geological age. Figure 5.8 illustrates the average $\delta^{18}O$ values of Phanerozoic limestones plotted against their

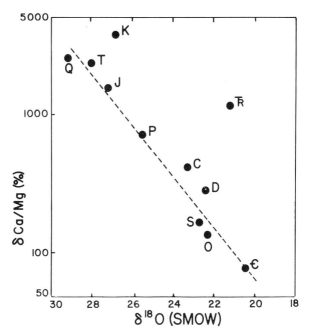

FIG. 5.8. Period averages of the calcium/magnesium ratios of carbonate rocks expressed as the percentage difference from dolomite versus Period averages of the oxygen isotopic composition of the rocks (data from Vinogradov and Ronov, 1956a, b; Chilingar, 1956; Degens and Epstein, 1962).

respective Ca/Mg ratios expressed in percent differences from dolomite, i.e.

$$\delta Ca/Mg = \frac{Ca/Mg_{(period)} - Ca/Mg_{(dolomite)}}{Ca/Mg_{(dolomite)}} \times 100.$$

Younger limestones have a relatively high Ca/Mg ratio and are isotopically heavy, whereas older limestones have a low Ca/Mg ratio nearly equivalent to that of dolomite, and are isotopically light. The progressive overall trend (Fig. 5.8) with increasing age for the Ca/Mg and $\delta^{18}O$ values of limestones suggests that the individual Ca/Mg and $\delta^{18}O$ trends (Figs. 5.6 and 5.7) have similar causes. Consequently, despite the fact that dolomite can form in a variety of ways and that the data for both the Ca/Mg ratios of carbonates and the $\delta^{18}O$ values of limestones do not necessarily pertain to rocks having the same mode of formation, it is likely that the progressive decrease of the Ca/Mg ratio of limestone represents an increased opportunity for the older rocks to be dolomitized by continental, isotopically-light, meteoric waters. It appears that long-term diagenetic processes are quantitatively more important in affecting the composition of the total mass of carbonate rocks than are penecontemporaneous and early diagenetic dolomitization. It is possible that the magnesium derived from shales during their stabilization was transferred to contemporaneous limestones by subsurface water movement and this led to their dolomitization.

The Sr/Ca ratio of carbonate rocks provides additional evidence that the chemical and mineralogical trends observed in carbonates are secondary. Figure 5.9 shows that the Sr/Ca ratios of carbonate rocks from North America and the Russian Platform decrease irregularly with increasing age. This trend corresponds to that of the Ca/Mg ratio of carbonate rocks and most likely reflects the progressive loss of Sr from the limestones by ground water leaching during inversion of aragonite to calcite, and long-term dolomitization and recrystallization.

Unlike the oxygen isotopic composition, the carbon isotopic composition of limestone shows no systematic long-term trend of $\delta^{13}C$ with geological age (Fig. 5.10). The $\delta^{13}C$ values of Precambrian carbonates are indistinguishable from those of younger rocks (Keith and Weber, 1964; Galimov, et al., 1968). These observations suggest that ocean water bicarbonate has been of nearly constant carbon isotopic composition for much of geological time. Furthermore, the absence of any pronounced trend in both the organic carbon content and the C/N ratio of sedimentary rocks with age (Ronov, 1958) suggests that input of organic carbon into the sedimentary rock mass has been nearly constant with time (Holland, 1965; Garrels and Perry, 1974).

The chemical, mineralogical and oxygen isotopic changes in carbonate rocks with increasing age reflect long-term diagenetic process. Most of the available data are for Phanerozoic carbonates; however, the Ca/Mg trend extends into the Precambrian (Vinogradov and Ronov, 1956b; Chilingar,

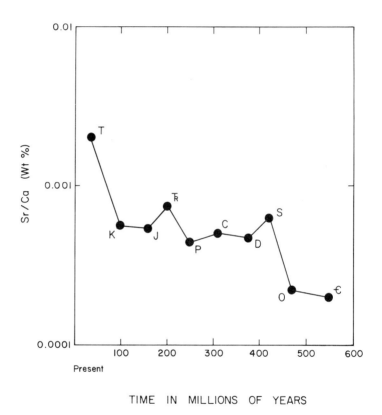

FIG. 5.9. Period averages of the strontium/calcium ratio of carbonate rocks as a function of age (data from Kulp, *et al.* 1952; Vinogradov and Ronov, 1956b).

1956; Ronov, 1968). *These data do not preclude changes in Phanerozoic ocean composition, but are not incompatible with a model in which no marked evolutionary change in sea water composition occurred during Phanerozoic time, or even perhaps, for the last* $1\cdot5 \times 10^9$ *years.*

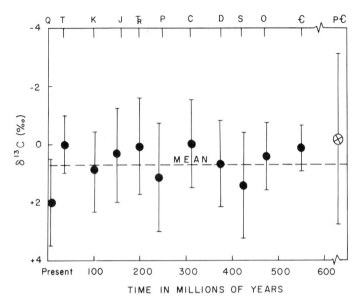

FIG. 5.10. Period averages of the carbon isotopic composition of carbonate rocks as a function of age (data from Keith and Weber, 1964; Galimov, *et al.*, 1968).

### 5.2.5. COMPOSITIONAL TRENDS IN EVAPORITES

Calcium sulphate deposits first appear in rocks of late Precambrian age; however, casts of gypsum modules are found in rocks as old as 1·8 billion years. Collapsed breccias believed to be caused by solution of calcium sulphate are found in the dolomite succession of the Transvaal Series, South Africa; these rocks are approximately $2·7 \times 10^9$ years old. Deposits of NaCl are restricted to the Phanerozoic. The distribution with age of evaporites in the sedimentary rock mass is compatible with cycling of components of evaporites at a rate faster than those of other rock types, and does not necessarily imply major changes in sediment ratios at the time of deposition or in the chemistry of sea water.

The variation of the sulphur isotopic composition of marine evaporites as a function of geological age has been used as evidence for changes in the total sulphate content of sea water (Fig. 5.11; Holser and Kaplan, 1966). Holser and Kaplan postulated that the deposition of oceanic sulphate in evaporites would have little effect on the sulphur isotopic composition of sea water because of the small isotopic fractionation involved. Consequently, changes in the isotopic composition of sea water would reflect primarily the balance between light sulphur addition to the oceans by rivers and light sulphur removal into marine sediments. If the inflow of light sulphur exceeds the outflow, the oceans would become isotopically light; the converse would be true if outflow exceeded inflow. This model implies that changes in the isotopic composition of sea water reflect changes in the sulphate concentration in sea water. Thus, according to this model, the sulphate content of sea water could have varied by as much as a factor of 2 during the Phanerozoic.

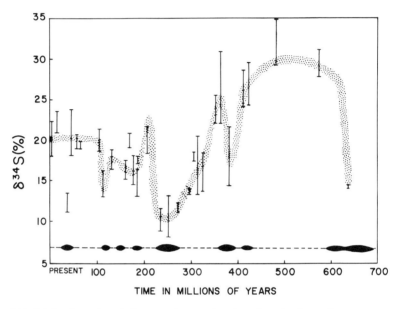

FIG. 5.11. Sulphur isotopic ratio in evaporites as a function of age (Holser and Kaplan, 1966). Dark areas represent times of major evaporite deposition (data from Lotze, 1964).

However, it has been shown recently, (Rees, 1970; McKenzie, 1972; Garrels and Perry, 1974) that the isotopic composition of ocean water depends not only on outflow and inflow of light sulphur, but also on the ratio of the rates of formation of evaporite and sulphide deposits if the sulphate content of ocean water is constant. At a steady state, high isotopic values represent times of minor evaporite formation and vice versa. A variation in the ratio of evaporite to sulphide formation rates of an order of magnitude is necessary if the oceans were to have had a constant sulphate content during the Phanerozoic, and for evaporite deposits to show the isotopic variability illustrated in Fig. 5.11, provided that the isotopic compositions of stream sulphate and sulphide sulphur have remained constant. This ratio variation is certainly not large, and is not precluded by the scanty data available for the distribution of evaporites with geological age. Isotopically light evaporites are approximately contemporary with times of major evaporite formation (Fig. 5.11).

To a first approximation, the distribution of evaporites probably reflects differential cycling of their components; however, the sulphur isotopic data for sedimentary sulphates shows that the isotopic composition of sulphur in sea water varied during the Phanerozoic and late Precambrian. Times of maximum evaporite deposition closely correspond to periods of isotopically light sedimentary sulphate, suggesting that the sulphur isotopic composition of ocean water does not reflect major changes in the total sulphur content of sea water, but to a significant degree is caused by variations in the sulphate-sulphur/sulphide-sulphur depositional ratio. *Nevertheless, the distribution does not require progressive or long-term (hundreds of millions of years) changes in sea water composition since late Precambrian time.*

### 5.2.6. SUMMARY

An attempt has been made above to show that the trends with age of the chemical, mineralogical, and isotopic compositions of sedimentary rocks of Phanerozoic age are compatible with a model in which these trends are principally the result of secondary processes, and thus, do not reflect rock-type ratios or the chemistry and mineralogy at the time of deposition. As a consequence, there are no cogent reasons for suggesting that major evolutionary change in sea water composition took place during the Phanerozoic. Because many of these trends extend back into the late Precambrian, it is likely that the same conclusion can be drawn for sea water composition as far back as $1.5 \times 10^9$ years. For rocks of greater age, rock-type ratios, chemistry and mineralogy appear to be sufficiently different to warrant other conclusions; these are discussed in a following section concerned with the history of sea water. Again, it should be emphasized that the model does not preclude periodic changes in sea water composition, but only major evolutionary change.

## 5.3. SEA WATER AND MODELLING

### 5.3.1. SYNOPSIS

Present-day sea water is an aqueous solution of ten principal ions (Culkin, 1965, see also Chapter 6), seven of which constitute more than 99 % of the total dissolved load ($Na^+$, $K^+$, $Mg^{2+}$, $Ca^{2+}$, $Cl^-$, $SO_4^{2-}$, $HCO_3^-$). The relative proportions of major components, other than $CO_2$, are nearly constant except in areas of major river runoff or in semi-enclosed basins. Variations of the total concentration of dissolved carbon are small ($<10\%$; Koczy, 1956). The pH of sea water usually ranges from about 7·8 to 8·3. These conditions must have existed for at least a few thousands of years, the age of the oldest water found in the oceans. However, as mentioned previously, it is necessary to turn to indirect evidence for sea water composition when we consider greater ages.

The residence times of all the dissolved constituents of sea water are much smaller than the age of the earth (Table 5.1). Dissolved species such as $SiO_2$ and $Ca^{2+}$ would rapidly accumulate in the oceans over short periods of

TABLE 5.1

*The residence times of elements in sea water calculated from river input and sedimentation.* (After Goldberg, 1965).

| Element | Mass in ocean (units of $10^{20}$ g) | Residence time in millions of years River input | Sedimentation |
|---|---|---|---|
| Na | 147·8 | 210 | 260 |
| Mg | 17·8 | 22 | 45 |
| Ca | 5·6 | 1 | 8 |
| K | 5·3 | 10 | 11 |
| Sr | 0·11 | 10 | 19 |
| Si | 0·052 | 0·035 | 0·01 |
| Li | 0·0023 | 12 | 19 |
| Rb | 0·00165 | 6·1 | 0·27 |
| Ba | 0·00041 | 0·05 | 0·084 |
| Al | 0·00014 | 0·0031 | 0·0001 |
| Mo | 0·00015 | 2·15 | 0·5 |
| Cu | 0·000041 | 0·043 | 0·05 |
| Ni | 0·000027 | 0·015 | 0·018 |
| Ag | 0·0000041 | 0·25 | 2·1 |
| Pb | 0·00000041 | 0·00056 | 0·002 |

geological time, if no removal processes operated. The concentrations of $Na^+$, $K^+$, $Mg^{2+}$, $Cl^-$, and $SO_4^{2-}$ would also increase, and the solubility products of evaporite minerals would soon be exceeded. The sedimentary and palaeontologic records offer no evidence that such extreme conditions persisted on a world-wide basis for extended periods of time. Indeed, possible excursions from the present composition of sea water are severely constrained by the probable precipitation of minerals such as gypsum, brucite, sepiolite, and others.

Holland (1972) has pointed out that the absence of minerals typical of non-marine evaporites and the depositional sequences of minerals in marine evaporite basins can be used as criteria to place limits on the composition of sea water during the Phanerozoic. For example, absence of bicarbonate minerals, other than those of Ca, in marine evaporite deposits suggests that the calcium concentration of sea water has never dropped below one-half that of bicarbonate. If the calcium concentration in sea water had reached such a minimum, late-stage marine evaporites should contain bicarbonate salts of sodium and magnesium. These salts are characteristic of non-marine evaporites, but not of marine ones. Using similar arguments, Holland concluded that Phanerozoic sea water major element concentrations were rarely, if ever, greater than twice or less than half of the present concentrations.

Because of the shortness of element residence times and the fact that for some elements the residence times calculated from both river input and sedimentation output data are equal, many geologists believe that in sea water a steady state exists. In the system which includes the atmospheric, oceanic, biospheric and sedimentary reservoirs sea water can be regarded as an aqueous solution having an invariant composition which is determined by the balance between the rate at which the elements enter the oceans and that at which they are removed. It is true, however, that calcium is the only major element for which it can be shown that this is true (Turekian, 1965; Pytkowicz, 1967).

Pytkowicz (1972) has recently outlined the mathematical treatment necessary to describe cyclic systems consisting of several reservoirs in which both equilibrium and steady states coexist. Unfortunately, because of lack of data for variables such as element fluxes, reservoir sizes, and rate constants, it is difficult at present to formulate an *unique* model for the apparent chemical stability of the ocean. Indeed, Broecker (1971) questions the validity of the deduction that oceanic composition is stable, and contends that steady or equilibrium state models provide "little incentive to those who search the sedimentary record for evidence of changes in ocean chemistry" (Broecker, 1971, p. 188). He, as well as others, feels that it is also very important to consider kinetic factors.

The work of L. G. Sillén, published during the past decade (Sillén, 1961, 1967a, b), has provided the stimulus for much of the recent controversy concerning the state of sea water. Sillén proposed a multi-phase equilibrium model for the ocean in which the composition of the aqueous solution closely resembled that of sea water and was determined by the choice of temperature, chlorinity, and of seven solid phases, mostly aluminosilicates, chosen from those minerals that are commonly found in marine sediments. He emphasized that, although chemical equilibrium is obviously not attained in the oceans, the equilibrium model provides a useful approximation to the real system.

He also stressed more particularly that reactions involving fine-grained silicates may exert an important control on sea water composition.

Further refinements of Sillén's model (Helgeson et al., 1969; Helgeson and Mackenzie, 1970) have taken cognizance of the fact that there are two categories of dissolved constituents in sea water. Major components, when expressed as chloride and sulphate salts, have practically constant chemical potentials in an equilibrium model. In contrast, the chemical potentials of minor components, such as HCl, $H_2SO_4$, $H_4SiO_4$, $AlCl_3$, $FeCl_3$, and $FeCl_2$, vary at the present time according to location, temperature, pressure, biological activity, and other factors. If we confine our attention to substances and components of major geological interest, phase relationships among sedimentary minerals coexisting with sea water can be expressed in terms of T, P, pH, log $a_{H_4SiO_4}$ and log $f_{CO_2}$. Figure 5.12, taken from Helgeson and Mackenzie (1970), portrays these relationships at 25°C, and 1 atmosphere total pressure, assuming equilibrium with calcite. It should be emphasized that the relationships depicted are an approximation to the real system. Important unknown factors are the mechanisms by which dissolved constituents entering the ocean are removed, and whether such minerals as can be identified, are able to maintain their components close to their saturation values.

Coexistence of magnesian chlorite, gibbsite, montmorillonite-illite solid solution and calcite, for example, suffices to fix the composition of sea water, including its pH and dissolved silica concentration. It is not necessary, however, that all the minerals considered in Fig. 5.12 should coexist together, but only that an assemblage of three suitable aluminosilicates and one additional carbonate phase be present at equilibrium with the aqueous phase. Different conditions of pH and silica activity can be accommodated if the oceans are regarded as responding to a mosaic of local equilibrium states, in each of which these two variables may be controlled by differing mineralogies and/or other parameters such as biological activity, decay of organic matter, and circulation. Conversely, steady state concentrations of dissolved Na, K, Ca, Mg, and Cl are seen to be compatible with a variety of mineral assemblages commonly found today in marine sediments under the conditions of T, P, pH, $a_{CO_2}$ and $a_{H_4SiO_4}$ characteristic of present day sea water.

It is difficult at present to determine whether or not the oceans are chemically stable. The fossil and evaporite rock record for the past 600 million years is consistent with little change having occurred in sea water composition. Indeed, the oxygen isotopic content and the Sr/Ca, and Mg/Ca ratios of fossil brachiopods in Mississippian age rocks (Lowenstam, 1961) suggest that sea water of that time was like that of today. It is unlikely that drastic

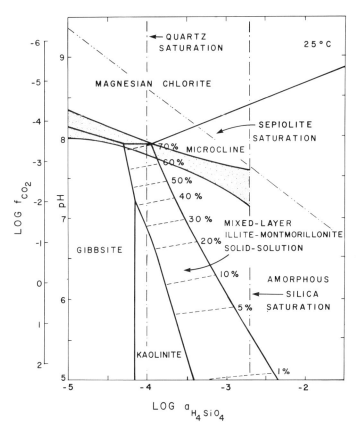

FIG. 5.12. Activity diagram illustrating phase relations in an idealized nine-component ocean system at 25°C, 1 atm, and unit fugacity of $H_2O$. The activities of $Na^+$, $K^+$, $Ca^{2+}$ and $Mg^{2+}$ are fixed at $10^{-0.50}$, $10^{-2.21}$, $10^{-2.64}$ and $10^{-1.79}$, respectively. Log $f_{CO_2}$ denotes the fugacity of $CO_2$ for calcite saturation. Contours in illite-montmorillonite field represent percent of illite component. Speckled area shows the approximate range of sea water composition (Helgeson and Mackenzie, 1970).

changes in sea water composition could have occurred during the Phanerozoic because it would have been difficult for a variety of marine organisms to have flourished and survived from Cambrian times until today. Consequently, it is likely that the composition of sea water has been reasonably constant for at least the past 600 million years. Documentation based on the sedimentary rock record for this stability has been provided in the previous section, but the model presented did not preclude periodic or short-term changes.

It is not possible at this time to arrive at a quantitative evaluation of the relative importance of equilibrium and of steady state kinetics in determining the chemical composition of sea water. In the final analysis, it is likely that these viewpoints will be complementary. Indeed, our views concerning the oceans as a chemical system have changed dramatically in the past century; it is now well established that the ocean is not simply a storehouse of the salts delivered to it. However, it is true that the chemical composition of sea water must be viewed as a variable parameter of the oceanic reservoir in a system that includes the reservoirs of atmosphere, biosphere, ocean and sedimentary lithosphere. In this system, shown schematically in Fig. 5.13, each

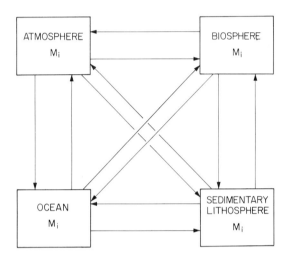

FIG. 5.13. Schematic representation of sedimentary system showing paths of transfer among reservoirs. $M_i$ represents mass of $i$th element in reservoir.

element has a finite concentration in one or more reservoirs and two or more pathways of transfer (fluxes). The reaction rates (rate coefficients for transfer) are defined by

$$K_{A_iB_i} = \frac{F_{A_iB_i}}{M_{A_i}},\qquad (5.2)$$

where $K_{A_iB_i}$ is the reaction rate (per year) for the $i$th element being transferred from reservoir $A$ to $B$, $F_{A_iB_i}$ is the flux in units of g yr$^{-1}$ for the $i$th element between reservoirs $A$ and $B$, and $M_{A_i}$ is the mass in grams of that element in reservoir $A$. In a system of this nature, the chemical composition of sea water is dependent on the stability of the fluxes of elements entering the ocean reservoir. Further conditions may be imposed on the ocean reservoir by specifying equilibrium or steady states for particular chemical components in sea water.

The system shown in Fig. 5.13 is the most general case, and solutions of the array of equations involving the rate of change of the $i$th element in the various reservoirs can lead to reasonable estimates of the maximum deviation of sea water composition in the past from that of today. It is likely that for a chemical component of sea water not controlled by an equilibrium state, the concentration of the constituent in sea water oscillated throughout geological time; the amplitude and wavelength of the oscillation being proportional to the difference between the input and output rates of the component for the ocean reservoir and to the relative masses of the component in other reservoirs.

Sodium chloride may be considered as an example. The mass of NaCl as halite in evaporite deposits is about one-half of that in the ocean (Garrels and Mackenzie, 1971). Salt deposition requires tectonic and climatic conditions that lead to circulatory isolation of a marine basin in a region in which evaporation is greater than rainfall plus runoff. When these conditions are met the NaCl entering the basin via streams will accumulate, except insofar as it is removed in sea aerosol, pore waters, and minerals other than halite. The rate of accumulation will depend on the difference between the influx rate of NaCl via streams and its efflux rate via aerosol, pore water and solids. The influx rate, if first order, will be directly proportional to the mass of NaCl exposed to weathering, i.e. the greater the mass, the higher the influx rate. The total efflux rate will depend on the mass of NaCl in the basin and the rate constants for the various mechanisms removing NaCl. If there were no removal mechanisms for NaCl other than halite deposition, it is possible that NaCl could accumulate in the ocean if evaporite deposition ceased; if this were to occur, NaCl accumulation would lead to a maximum increase in sea water salinity of 45%.

Our conception of the ocean as a chemical system has progressed from the stage of viewing sea water as an accumulation of salts to that in which sea water is considered as a solution, the composition of which is controlled by equilibria, or by kinetic factors which can lead to either a steady or a non-steady state. The viewpoint presented here is that the concentrations of individual sea water constituents are variables in the ocean reservoir which is part of a system which also includes the reservoirs of atmosphere, biosphere, and sedimentary lithosphere. The concentration of a species in sea water may be constant if either a steady state or an equilibrium state prevails; however, in certain circumstances the concentration of an element may be oscillatory. Evidently, the ocean can be regarded as being analogous to a capacitor in an electrical system.

Because we are only just beginning to appreciate the consequences resulting from oscillatory changes in the concentrations of elements in sea water, the next section will be devoted to the concept of sea water as part of a system in a steady state. The steady state approach will permit examination of the sources, pathways and removal mechanisms of major elements in sea water and allow a preliminary assessment of deviations from steady state to be made.

### 5.3.2. STEADY-STATE SEDIMENTARY SYSTEM

#### 5.3.2.1. *General relationships*

Perhaps one of the most significant advances in our understanding of the ocean in the last two decades has been the recognition that its composition cannot be viewed as a separate entity but must be considered in relation to the major sedimentary cycle of the elements. This cycle involves the reservoirs of atmosphere, biosphere, ocean, and sedimentary lithosphere, with the land being an ephemeral source of the products of weathering. The residence time of an element in a particular reservoir is dictated by the element mass in the reservoir and the input or output rate of the element; at the steady state, the input and output rates are equal. If the individual reservoirs are at a steady state with respect to an element, then the system as a whole is in a steady state for that element. In a system of this nature, the ocean is viewed as a chemical reactor, or reflux condenser, in which materials are added, react and leave. This view contrasts sharply with that held in the early part of this century in which the ocean was considered to be effectively a storage bin or accumulator of elements transported to it principally by streams.

The concept of the ocean as a reservoir in the sedimentary rock cycle provides a means for assessing the fate of the elements after entering the ocean. Garrels and Mackenzie (1972) have recently developed a steady state quantitative model for the sedimentary cycle. This model is shown in Fig. 5.14 and will be used to discuss the sources, rates of addition, and reactions involving elements in sea water. The model was developed on the basis of several assumptions which are outlined in the paper. For the present purposes, the most important of these is that the material balance in the model was accomplished assuming that sea water composition is time independent. The internal consistency of the model with a number of known variables, such as the chemical composition of the dissolved and suspended load of rivers, suggests that this assumption is applicable at least to Phanerozoic time, and perhaps, as far back as $1.5 \times 10^9$ years—the period of time for which we have sedimentary rock types similar to those of today (see Section 5.2). The rest of the discussion is keyed to Fig. 5.14 and Table 5.2.

The model is constructed with four reservoirs: biosphere, atmosphere,

TABLE 5.2
*Element reservoir sizes and residence times (T)*
*(units of $10^{20}$ g or moles and $10^6$ years).* (After Garrels and Mackenzie, 1972).

| Element | Ocean (g) | Ocean (mol) | $T^a$ | Sedimentary lithosphere (g) | Sedimentary lithosphere (mol) | $T^c$ |
|---|---|---|---|---|---|---|
| Na | 162·38 | 7·06 | 193 | 345 | 15·00 | 288 |
| K | 5·85 | 0·15 | 8·1 | 507·13 | 12·97 | |
| Mg | 19·68 | 0·82 | 14·9 | 610·08 | 25·42 | 381 |
| Ca | 6·00 | 0·15 | 1·2 | 1986·40 | 49·66 | 351 |
| Si | 0·0392 | 0·0014 | 0·016 | 6160·84 | 220·03 | 450 |
| Al | — | — | — | 1527·66 | 56·58 | 504 |
| Fe | — | — | — | 961·52 | 17·17 | 481 |
| Ti | — | — | — | 66·95 | 1·39 | 515 |
| S | 13·44 | 0·42 | 22·4 | 157·12 | 4·91 | 242 |
| Cl | 288·40 | 8·24 | 300 | 209·45 | 5·90 | 218 |
| $C_{inorg}$ | 0·396 | 0·033 | 0·1[b] | 610·08 | 50·84 | 381 |
| $C_{org}$ | 0·007 | 0·00058 | 0·0025 | 125·04 | 10·42 | 417 |

| | Atmosphere | | | Biosphere | | |
|---|---|---|---|---|---|---|
| $C_{inorg}$ | 0·00648 | 0·00054 | 0·0026 | — | — | — |
| $C_{org}$ | — | — | — | 0·0504 | 0·0042 | 0·0144 |

[a] Corrected for atmosphere cycling.
[b] Includes C input from atmospheric $CO_2$ used in weathering.
[c] Includes input flux of both dissolved and suspended stream loads.

ocean, and sedimentary lithosphere. The sediments are divided into New Rocks ($\sim$Cenozoic and Mesozoic) and Old Rocks ($\sim$Palaeozoic and Pre-cambrian). Weathering and diagenetic changes as discussed in Section 5.2 result in the conversion of New Rocks to Old and the transfer of much dissolved material from the land to the ocean. The admixture of metastable phases of detrital minerals, biochemical precipitates, and authigenic minerals is slowly converted to stable mineral assemblages of Old Rocks; that is, shales develop a monotonous mineralogy of chlorite, illite and quartz and lose $CaCO_3$, whereas limestones become high in magnesium. The mineralogical changes undergone by muddy sediments during stabilization are shown diagrammatically in Fig. 5.15. The triangle connects the 3-phase assemblage quartz-illite-chlorite; this assemblage tends to be the final product of the post-depositional change of the initial multiphase assemblages of minerals in muddy sediments.

The ratio of dissolved load derived from New Rocks to that from Old

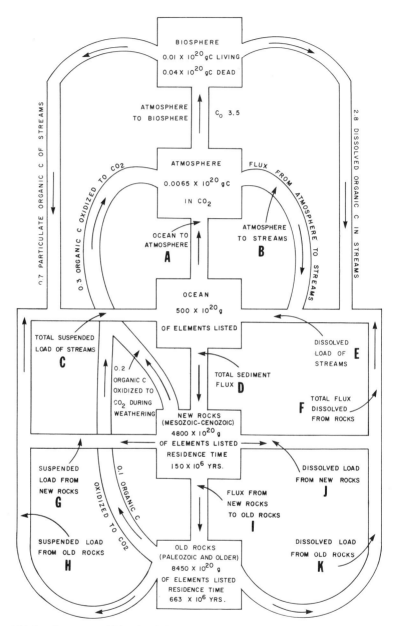

FIG. 5.14. Steady-state model for circulation of elements in the sedimentary cycle. Fluxes are in units of $10^{14}$ g per year. $C_0$ denotes organic carbon and $C_i$ inorganic carbon (Garrels and Mackenzie, 1972).

FIG. 5.14 (*cont.*)

| **A** | |
|---|---|
| Si | 0.00 |
| Al | 0.00 |
| Fe | 0.00 |
| Mg | 0.30 |
| Ca | 0.90 |
| Na | 1.68 |
| K | 0.18 |
| Ti | 0.00 |
| S | 0.84 |
| Cl | 2.10 |
| $C_i$ | 2.50 |
| $C_0$ | 3.20 |
| | 11.70 |

| **B** | |
|---|---|
| Si | 0.00 |
| Al | 0.00 |
| Fe | 0.00 |
| Mg | 0.30 |
| Ca | 0.90 |
| Na | 1.68 |
| K | 0.18 |
| Ti | 0.00 |
| S | 0.84 |
| Cl | 2.10 |
| $C_i$ | 2.50 |
| $C_0$ | 0.00 |
| | 8.50 |

| **C** | |
|---|---|
| Si | 11.16 |
| Al | 3.01 |
| Fe | 2.00 |
| Mg | 0.28 |
| Ca | 0.61 |
| Na | 0.36 |
| K | 0.58 |
| Ti | 0.13 |
| S | 0.05 |
| Cl | 0.00 |
| $C_i$ | 0.20 |
| $C_0$ | 0.70 |
| | 19.08 |

| **D** | |
|---|---|
| Si | 13.68 |
| Al | 3.03 |
| Fe | 2.00 |
| Mg | 1.60 |
| Ca | 5.66 |
| Na | 1.20 |
| K | 1.30 |
| Ti | 0.13 |
| S | 0.65 |
| Cl | 0.96 |
| $C_i$ | 1.60 |
| $C_0$ | 0.30 |
| | 32.11 |

| **E** | |
|---|---|
| Si | 2.52 |
| Al | 0.00 |
| Fe | 0.00 |
| Mg | 1.62 |
| Ca | 5.94 |
| Na | 2.52 |
| K | 0.90 |
| Ti | 0.00 |
| S | 1.44 |
| Cl | 3.06 |
| $C_i$ | 3.90 |
| $C_0$ | 2.80 |
| | 24.70 |

| **F** | |
|---|---|
| Si | 2.52 |
| Al | 0.00 |
| Fe | 0.00 |
| Mg | 1.32 |
| Ca | 5.04 |
| Na | 0.84 |
| K | 0.72 |
| Ti | 0.00 |
| S | 0.60 |
| Cl | 0.96 |
| $C_i$ | 1.40 |
| $C_0$ | 0.00 |
| | 13.40 |

| **G** | |
|---|---|
| Si | 5.58 |
| Al | 1.39 |
| Fe | 1.00 |
| Mg | 0.14 |
| Ca | 0.43 |
| Na | 0.23 |
| K | 0.20 |
| Ti | 0.06 |
| S | 0.05 |
| Cl | 0.00 |
| $C_i$ | 0.13 |
| $C_0$ | 0.00 |
| | 9.21 |

| **H** | |
|---|---|
| Si | 5.58 |
| Al | 1.62 |
| Fe | 1.00 |
| Mg | 0.14 |
| Ca | 0.18 |
| Na | 0.14 |
| K | 0.38 |
| Ti | 0.07 |
| S | 0.00 |
| Cl | 0.00 |
| $C_i$ | 0.07 |
| $C_0$ | 0.00 |
| | 9.18 |

| **I** | |
|---|---|
| Si | 6.21 |
| Al | 1.62 |
| Fe | 1.00 |
| Mg | 0.56 |
| Ca | 1.72 |
| Na | 0.25 |
| K | 0.47 |
| Ti | 0.07 |
| S | 0.09 |
| Cl | 0.10 |
| $C_i$ | 0.56 |
| $C_0$ | 0.10 |
| | 12.75 |

| **J** | |
|---|---|
| Si | 1.89 |
| Al | 0.00 |
| Fe | 0.00 |
| Mg | 0.90 |
| Ca | 3.49 |
| Na | 0.72 |
| K | 0.63 |
| Ti | 0.00 |
| S | 0.51 |
| Cl | 0.85 |
| $C_i$ | 0.91 |
| $C_0$ | 0.00 |
| | 9.90 |

| **K** | |
|---|---|
| Si | 0.63 |
| Al | 0.00 |
| Fe | 0.00 |
| Mg | 0.42 |
| Ca | 1.55 |
| Na | 0.12 |
| K | 0.09 |
| Ti | 0.00 |
| S | 0.09 |
| Cl | 0.11 |
| $C_i$ | 0.49 |
| $C_0$ | 0.00 |
| | 3.50 |

Rocks is high showing that much of the stream load entering today's ocean is dominated by materials from New Rocks. Bicarbonate, $Ca^{2+}$, $SiO_2$, $SO_4^{2-}$, $Mg^{2+}$, $Na^+$, and $Cl^-$ are the principal dissolved species in rivers, and these are the ones lost from New Rocks as they are converted to Old.

The most recently uplifted rocks are being destroyed by mechanical erosion and selectively leached by chemical erosion to form a new sedimentary mass very similar to them in composition. In the course of time the remaining New Rocks should take on the characteristics of today's Old Rocks, the transformation being accompanied by the formation of a new sedimentary mass with a chemical and mineralogical composition like that of the present

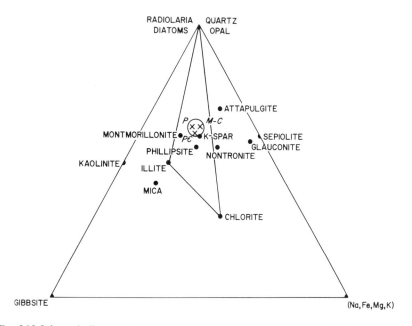

FIG. 5.15. Schematic diagram illustrating the variety of phases found in muddy sediments plotted as a function of atomic percentages of Al, Si and Na + Fe + Mg + K. Circled area encloses compositions of average Precambrian (P∈), Palaeozoic (P) and Mesozoic–Cenozoic (M–C) shales. The triangle connects phases thought to be the final phase assemblage in the post-depositional alteration of muddy sediments to shaly rocks (Garrels and Mackenzie, 1973).

New Rocks. The rocks of this new "Era", to judge from the chemistry of the dissolved load of present-day streams, will be enriched in their proportions of limestone and evaporites over the original New Rock deposits. If this picture is accurate in its broad long-term relationships, no matter how much it is modified by shorter term Period to Period relationships, the kinds of detrital and dissolved materials being brought to the oceans will not have changed greatly since the Late Precambrian, and a mean sea water composition approximately in equilibrium with calcite, K-feldspar, illite-montmorillonite solid solution and chlorite is implied (Helgeson and Mackenzie, 1970).

5.3.2.2. *Element cycles*

5.3.2.2.1. *Silicon.* To begin the discussion of element cycles, it is convenient to start with silicon. During the past decade silicon has been the focus of considerable attention in debates on the factors which control sea water composition (cf. Sillén, 1961; Holland, 1965; Mackenzie and Garrels, 1966a; Harris, 1966; Pytkowicz, 1967; Burton and Liss, 1968; Calvert, 1968; Gregor, 1968b; Siever, 1968a; Helgeson and Mackenzie, 1970; Liss and Spencer, 1970; Broecker, 1971; Lisitzin, 1972; Heath, 1973; Wollast, 1974). This debate was initiated by L. G. Sillén in his address at the first International Oceanographic Congress in 1959 at which he proposed an equilibrium model for the ocean. In this model, silicate minerals played a major role in determining the pH and cation concentrations of sea water. Prior to this it was generally thought that biological activity alone controlled the distribution of dissolved silica in the ocean, although previous attention had been paid to silicate interactions with sea water (see e.g. Grim *et al.*, 1949).

Following Sillén's lead, Mackenzie and Garrels (1966a, b) attempted to draw up a mass balance between river waters and sea water, in a manner similar to that proposed by Conway (1942, 1943) and Heck (1964). To accomplish the balance, they found it necessary to synthesize new clay minerals by reactions involving the suspended sediment load of streams and dissolved silica and cations as they entered the ocean. At that time, it was felt that the silicate synthesis reactions occurred very early, perhaps in the water column itself. These reactions also entailed the return to the atmosphere of $CO_2$ used during weathering of rock minerals, a requirement necessary to prevent the draining of $CO_2$ from the atmosphere. A schematic "reversing weathering" reaction is:

$$\text{degraded silicate} + H_4SiO_4 + HCO_3^- + \text{cations} \rightarrow$$

$$\text{new clay mineral} + CO_2 + H_2O. \quad (5.3)$$

Experimental data on the reaction of silicates with sea water (Mackenzie and Garrels, 1965; Mackenzie *et al.*, 1967; Siever, 1968b) show that silicate minerals can release or take up dissolved silica in silica-deficient and silica-spiked sea water, respectively. Experimental and field observations, however, have not supported the magnitude of very early silicate synthesis suggested originally by Mackenzie and Garrels (loc. cit.).

The problem can be further developed by reference to Fig. 5.14. Silicon delivered to the oceans by streams is principally found in the silicate mineral and quartz content of the suspended load, but about 20% is carried as $H_4SiO_4$ in the dissolved load. The smaller residence time of Si than those of Fe, Al and Ti in the sedimentary rock reservoir (Table 5.2) reflects transport of

Si in the dissolved load. The dissolved silica is derived from weathering reactions of the general type (Garrels and Mackenzie, 1971):

silicate $+ CO_2 + H_2O \rightarrow$

$$\text{clay mineral} + HCO_3^- + 2H_4SiO_4 + \text{cations.} \qquad (5.4)$$

At today's rate of transfer of dissolved silica to the ocean, silicate weathering would require $3·5 \times 10^{12}$ mol $CO_2$ per year. At this rate, it would take only 15 000 years to remove all the $5·4 \times 10^{16}$ mol of $CO_2$ at present in the atmosphere by silicate weathering alone. This implies that there must be reactions that return $CO_2$ to the atmosphere. Indeed, it is likely that these reactions, instead of occurring in the water column, take place during diagenesis and involve silica (Wollast and DeBroeu, 1971; Wollast, 1974). Mackenzie and Garrels (1966b) have pointed out that if all the silica entering the ocean today via streams accumulated as amorphous silica or its diagenetic products, chalcedony and quartz, the sedimentary rock mass would contain a greater proportion of silicon stored in these minerals than that which is observed.

Recent interstitial water analyses of modern deep-sea and shallow-water marine sediments (cf. Pressley and Kaplan, 1970; Mannheim, *et al.*, 1970; Bischoff and Ku, 1970, 1971; Bischoff and Sayles, 1972; Drever, 1971a; Ristvet *et al.*, 1973; Nissenbaum *et al.*, 1972; Sayles *et al.*, 1972; see Chapter 33) have shed some light on the site of silicate regeneration. In interstitial waters that are chemically different for one or more elements from normal sea water, $Mg^{2+}$ is generally depleted. Both $K^+$ and $Ca^{2+}$ enrichments and depletions have been observed, whereas $Na^+$ either decreases or does not change significantly. In general, $HCO_3^-$ is enriched and $SO_4^{2-}$ depleted with respect to normal sea water; particularly in waters from anoxic zones of sediments. Dissolved silica is enriched in interstitial waters; Heath (1973) obtained an average interstitial $SiO_2$ concentration for 800 determinations of $0·4$ mol $l^{-1}$ (24 ppm), a value less than that at saturation with amorphous silica at *in situ* sea bottom temperatures.

It is true, however, that not all interstitial waters have been extracted at *in situ* temperatures. Because dissolved silica and $K^+$ concentrations in interstitial waters appear to be dependent on the temperature of extraction of the waters (Fanning and Pilson, 1971; Mangelsdorf *et al.*, 1969), some of the data for these species are particularly suspect (see p. 395).

Nevertheless, various reactions have been proposed to account for pore water chemical changes. Drever (1971a, b) has suggested that depletions of $Mg^{2+}$ and $SO_4^{2-}$ in the interstitial waters of anoxic sediments may be due to the replacement of Fe by Mg in clay minerals, and to the precipitation of the Fe as iron sulphide. $K^+$ and $Mg^{2+}$ removal has been attributed to the formation of illite and chlorite or sepiolite, respectively (Pressley and Kaplan, 1970; Bischoff and Sayles, 1972; Wollast, 1974).

Wollast (1974) has recently proposed a reaction that has direct bearing on the silicate regeneration problem. From an examination of the species concentration profiles in the interstitial water of Hole 149 of the Deep Sea Drilling Project, he suggested the following overall reaction to account for the observed $Mg^{2+}$ decrease and $Ca^{2+}$ and dissolved silica increase with depth;

$$6SiO_{2(amorph)} + 4CaCO_3 + 4Mg^{2+} + 7H_2O =$$

$$2Mg_2Si_{3(sepiolite)}O_6(OH)_4(1.5H_2O) + 4CO_2 + 4Ca^{2+}. \qquad (5.5)$$

This is a diagenetic reaction which results in the consumption of $SiO_2$ and the eventual return of $CO_2$ to the atmosphere. Indeed, Wollast (1973) has calculated that $3.5 \times 10^{14}$ g yr$^{-1}$ of silica reacts during diagenesis to make new silicate minerals; if all this silica reacted as in reaction (5.5), this would result in a flux of $CO_2$ out of the ocean nearly equivalent to that necessary to compensate for its removal by silicate weathering reactions.

5.3.2.2.2. *Iron, titanium, and aluminium.* Iron, titanium and aluminium are transported to the ocean principally in solids of the suspended load of rivers. Their similar residence times in the sedimentary mass (Table 5.2) show that the composition of today's suspended load is not greatly different from that which existed in the past (Garrels and Mackenzie, 1972).

Iron deserves special attention. It is transported in the stream suspended load as discrete particles of iron(III) oxide, as iron(III) oxide coatings on detrital grains, and in the structures of silicate minerals. It is remobilized after deposition in anaerobic sediments where the ferric iron is reduced to iron(II) and precipitated as the iron(II) monosulphides greigite and mackinawite, to be later transformed to pyrite. The oxidation of buried organic matter by oxygen derived from the bacterical reduction of $SO_4^{2-}$ provides the electrons for the reduction of iron(III). The mass of iron(II) sulphide formed in modern anaerobic environments appears to be principally controlled by the presence of organic matter that can be metabolized by sulphate-reducing bacteria, although the availability of iron and the sedimentation rate are also important factors (Berner, 1973). As mentioned previously, iron(II) for sulphide mineral formation may also be derived from the structure of clay minerals. There is some likelihood that the $Fe^{2+}:Fe^{3+}$ ratio of shales decreases with increasing age pointing to continuous reduction of iron(III) by carbonaceous material even during the later stages in the post-depositional alteration of muddy sediments.

5.3.2.2.3. *Magnesium and potassium.* The residence times of magnesium and potassium in the sedimentary rock mass are less than those of Fe, Ti, Al and Si, but greater than those of Ca or Na, reflecting the tendency of Mg and K to be transported to the ocean in the dissolved, as well as, the suspended load of streams. In the steady state system, approximately 20% of the potassium and magnesium entering the ocean in dissolved form is recycled through the atmosphere back to the continents.

Dissolved magnesium is derived from the weathering of carbonate minerals (e.g. dolomite, magnesian calcites), and a variety of silicates (e.g. olivine, biotite). Typical weathering reactions are:

$$MgCO_3 + CO_2 + H_2O \rightleftharpoons Mg^{2+} + 2HCO_3^- \tag{5.6}$$

and

$$MgSiO_3 + 2CO_2 + 3H_2O \rightleftharpoons Mg^{2+} + 2HCO_3^- + H_4SiO_4. \tag{5.7}$$

It should be noted that in both reactions $CO_2$ is consumed; in a steady state system, these reactions must be reversed during deposition and burial for the ocean–atmosphere to remain constant with respect to $CO_2$. However, it is known that a large percentage of the dissolved magnesium in streams is derived from weathering of carbonate minerals, but that only a very small portion of this magnesium is being removed from the ocean in magnesian calcites and dolomite. Thus, reaction (5.6) is not reversed significantly in the ocean or during the early stages after burial of sediments.

In the model presented here the magnesium balance is maintained by transfer of magnesium from shales to carbonates during advanced diagenesis. The higher Mg to Ca ratio in Old Rocks reflects the tendencies for New Rocks to lose their calcium by selective leaching and for Mg to leave the shales of New Rocks during diagenesis and to be incorporated in the carbonate rocks of Old Rocks.

Drever (1974) has reviewed the magnesium balance in detail and has concluded that present removal mechanisms of Mg from the ocean could account for only 50% of that supplied by the river flux. These oceanic removal mechanisms involve carbonate mineral formation, ion exchange, glauconite formation, Mg–Fe exchange during burial of sediments, and burial of interstitial water. However, it is interesting to note that complete removal of the river flux of magnesium *via* the interstitial waters of the sediments would require a concentration gradient in pore waters of only $0.2 \, \text{mg} \, 1^{-1} \, \text{cm}^{-1}$ over the entire sea floor. Gradients of this magnitude are not unusual for interstitial waters of sediments. In addition, if Wollast's (1974) value of $3.5 \times 10^{14} \, \text{g} \, \text{yr}^{-1}$ for the rate of silica removal by the neo-formation of clay minerals is correct, and if only 1/3 of this silica enters minerals such as sepiolite, then all of the Mg entering the ocean could be removed by reaction (5.5).

However, as Drever has pointed out, if all the Mg enters new mineral phases, the MgO content of marine sediments should be 1–2% greater than that of river sediments. Although data are sparse, the magnesium content of river sediments does not appear to be significantly lower than that of marine sediments; thus, at present, there is no satisfactory explanation for removal of all the river-derived Mg flux from the ocean.

Sepiolite, which is effectively a sink for magnesium, may react with kaolinite to form chlorite during later stages of diagenesis according to:

$$5\,Mg_2Si_3O_6(OH)_4 + 2\,Al_2Si_2O_5(OH)_4 + 20\,H_2O =$$

$$2\,Mg_5Al_2Si_3O_{10}(OH)_8 + 13\,H_4SiO_4. \qquad (5.8)$$

The silica generated in this reaction may be lost from the shale via fluid movement and account in part for the decrease of $SiO_2$ in shales with increasing age (Fig. 5.3).

Potassium is derived principally from the weathering of potassium silicates. The weathering process involves loss of $CO_2$ from the atmosphere. The present rate of removal of potassium from the ocean by mechanisms such as zeolite formation or clay mineral exchange is insufficient to account for all the potassium entering via surface run-off. Hart (1973) has recently suggested that a significant fraction of the river-derived potassium may be removed by the formation of K-rich smectite during low temperature weathering of basalts in the deep sea.

Potassium, like magnesium, tends to be retained in sediments; however, with increasing time the potassium present in the micas and feldspars of shales of New Rocks migrates into the inter-layers of mixed-layer illite/montmorillonite. In addition, it appears that the increase in the potassium/aluminium ratio of shaly rocks with time suggests that the potassium content of shales is enhanced by additions of potassium from carbonate rocks and sandstones via subsurface waters, the converse of the pathway suggested for magnesium.

5.3.2.2.4. *Sodium and chlorine.* The mineral halite dissolves to release $Na^+$ and $Cl^-$ to streams. Sodium is also derived from the weathering of feldspars and other silicates. The residence time of sodium in rocks is about 70 million years longer than that of chlorine, reflecting the fact that part of the sodium is transported in the suspended load of streams. About 2/3 of the sodium in the sedimentary cycle circulates with chlorine. In a steady-state system, an amount of sodium in streams which is balanced by the chloride content would precipitate as halite in evaporite basins or be recycled from the sea back to the atmosphere. On a global basis, the values for the fluxes of sodium and chlorine from the sea to the continents are poorly known.

The remaining $\frac{1}{3}$ of the sodium circulates through the sedimentary cycle in the silicate minerals of shales. This sodium is found in the suspended load of streams, particularly as plagioclase feldspar, and in the dissolved load, derived from the weathering of feldspars, and is balanced by bicarbonate and sulphate. Weathering of Na-silicates causes a drain on the $CO_2$ of the atmosphere; maintenance of a $CO_2$ balance necessitates deposition of silicates accompanied by release of $CO_2$ equal in amount to that used during weathering. Silicate reactions involving sodium in the ocean today are precipitation of zeolites, and clay-mineral ion exchange. Hart (1973) has suggested that the total flux of stream-derived sodium to the ocean can be accounted for by greenschist grade metamorphism of basalts at ridge crests. It is true, however, that most of the sodium in the sedimentary cycle circulates with chloride, and halite deposits are distributed sporadically throughout the Phanerozoic rock column. Thus, at times such as the present when very little evaporite deposition takes place, it is possible that sodium and chlorine could be accumulating in the ocean.

*4.3.2.3.5. Sulphur.* The cycling of sulphur through the sedimentary system is, in reality, much more complex than is suggested by the model. Estimates of total sulphur in sedimentary rocks have increased markedly as major evaporite sequences have been discovered in unsuspected locations. For example, the Joint Oceanographic Institute's Deep Earth Sampling (JOIDES) programme has penetrated into evaporites in the deeps of the Gulf of Mexico and in the Mediterranean Sea.

No indication is given in Fig. 5.14 that sulphur moves directly from the land into the atmosphere. In the model sulphur moving in this way is combined with that transferred to the oceans via the dissolved load of streams. However, there is, in fact, a significant flux of $H_2S$ and $SO_2$ from land directly to the atmosphere and thence to the ocean. It is derived from the partial oxidation of pyrite and of the sulphur of terrestrial organisms. In the model, this flux is shown as being from rocks directly into streams, before being passed into the ocean.

The relatively short residence time of 245 million years for sulphur gives an indication of its overall mobility. Evidence is accumulating that there is a correlation between the mobilization of sulphur and that of magnesium in the sedimentary cycle. There has been a tendency to assume that sulphur which is present in streams and which originates from rocks is mainly balanced by calcium derived from the minerals gypsum and anhydrite. It now appears that magnesium has as much influence on the movement of sulphur as does calcium. It has been mentioned above that Drever (1971b) associated fixation of magnesium in clay minerals with sulphate reduction and pyrite crystallization. The reverse of this process, the oxidation of pyrite to sulphuric acid and iron oxide, with release of magnesium and other cations from shales during the subsequent acid attack on the clay minerals, may be an important cause of the correlation between magnesium and sulphur in streams.

Although the total concentration of sulphur in sedimentary rocks may well be nearly constant with time, the ratio of oxidized to reduced sulphur has changed as is indicated by the marked temporal variations of $\delta^{34}S$ in evaporite deposits (Fig. 5.11). Most investigators (cf. Li, 1972) have calculated that when the total amount of sulphur present in the ocean and in sedimentary rocks is considered, the sulphide/sulphate–sulphur ratio is about unity at the present time. McKenzie (1972) has estimated that this ratio may, in the past, have ranged by as much as 4/1 to 1/4. It is likely, but not yet proved, that periods of major evaporite deposition in the Phanerozoic correlate with low $\delta^{34}S$ values in the evaporite deposits and with low sulphate concentrations in sea water.

Garrels and Perry (1974) have recognized a major feedback loop in the sedimentary cycle involving sulphur-bearing minerals and atmospheric $O_2$ and $CO_2$. Sulphide weathering involves an $O_2$ drain on the atmosphere, resulting in the production of both iron(III) and sulphate. When these oxidized constituents are deposited in the sea, they react with organic matter originally derived from photosynthesis to regenerate iron sulphide. Carbon dioxide is released, and through photosynthesis can react to restore $O_2$ to the atmosphere.

Today, sulphur is either accumulating as sulphate in the ocean or being removed chiefly as reduced sulphur by bacterial reduction on the sea floor (Berner, 1972). The rate of formation of sulphate evaporite minerals is negligible. In contrast, because the $\delta^{34}S$ value of evaporite deposits has not changed markedly since the Cretaceous (Holser and Kaplan, 1966), it is probable that the present situation has not endured for more than a few million years.

5.3.2.2.6. *Calcium and carbon.* Calcium and carbon are intimately related. Calcium cycles through the system dominantly in the solution flux and is balanced by bicarbonate ion, but about 20% of the calcium circulates in the suspended load of streams as calcite or dolomite. Calcium is separated from magnesium when it is precipitated from the oceans; the skeletal materials of carbonate-secreting organisms are made up of nearly pure calcite, aragonite, or magnesian calcites that range up to 20% (wt) of magnesium carbonate. The material that is eventually buried in new sediments contains approximately 6% (wt) of magnesium carbonate (Chave, 1954). This implies that, as mentioned previously, a large part of the magnesium flux into new sediments must find mineral sinks other than carbonates.

The carbon cycle has been discussed in detail by many authors (see, Bolin, 1970; Pytkowicz, 1967, 1972; Johnson, 1970). In the model depicted in Fig. 5.14, attention is focused on the oceanic, atmospheric, and biospheric net fluxes necessary to maintain the inorganic–organic carbon balances between streams and rocks. The concentrations of dissolved organic carbon and of particulate organic carbon in streams are not well known (Stumm and Morgan, 1970), but the fluxes given are probably underestimates. They are certainly much larger than the organic carbon flux into sediments from the ocean, implying that oxidation significantly exceeds photosynthesis in the oceans, and that the converse is true for the land. The flux of organic carbon into the atmosphere from the oxidation of "old" carbon in the weathering of rocks must almost exactly balance the flux of organic carbon into New Rocks, because the organic carbon content of rocks is nearly independent of rock age (Broecker, 1970). The average overall land and sea excess of photosynthesis over oxidation is only about 0·1% of the total mass of C photosynthesized, as determined from the average organic content of present-day sediments and the mass of sediments deposited each year. The restoration of this tiny deficiency through oxidation of "old" carbon that has passed through the sedimentary cycle and is currently being weathered shows that over geological time intervals, the huge amounts of carbon fixed and oxidized may have differed by even less than 0·1%.

In the balance shown among streams, ocean, and atmosphere, there is an interesting compensation between the fluxes of inorganic and organic carbon which are required for steady-state conditions. About 80% of the inorganic carbon flux into streams is involved in the weathering of carbonate minerals, and about 20% in the weathering of silicates (Garrels and Mackenzie, 1971; Li, 1972). Furthermore, the average carbonate mineral weathered contains several times as much magnesium as that deposited as carbonates in the ocean. Therefore, only the $CO_2$ abstracted from the atmosphere in the weathering of the calcium component of carbonates is restored to the atmosphere when carbonates are precipitated from the ocean, leaving more than 20% of the $CO_2$ converted to bicarbonate in streams (that used for silicate minerals plus that required for $MgCO_3$) to accumulate in the oceans. If there were no compensatory mechanism, carbon dioxide would be drained out of the atmosphere in a few tens of thousand years. In the model the drain is balanced by $CO_2$ restored to the atmosphere from part of the excess of oxidation over photosynthesis in the oceans. Clearly, the sea is an important reservoir in which overlap between the organic and inorganic carbon cycles

occurs. The total flux of organic carbon out of the sea as carbon dioxide
($3.2 \times 10^{14}$ g yr$^{-1}$) is of the same order of magnitude as that calculated by
Kroopnick (1971) from his studies of the oxidation of carbon in the deep sea,
(about $2.5 \times 10^{14}$ g yr$^{-1}$).

The residence time of inorganic carbon in rocks (381 million years) is
consistent with the ready solubility of carbonate minerals, which leads to a
high flux of inorganic carbon out of New Rocks, as a consequence of both
diagenesis and weathering (Perry and Hower, 1970).

## 5.4. EARLY HISTORY OF SEA WATER

### 5.4.1. INTRODUCTION

The oldest rocks for which there are firm radiometric dates have an age of
$3.5$ to $3.6 \times 10^9$ years (Goldich et al., 1970). Thick and geographically
extensive sedimentary deposits found in South Africa and Swaziland are
nearly as old as these rocks, and provide the first clues in the unravelling of

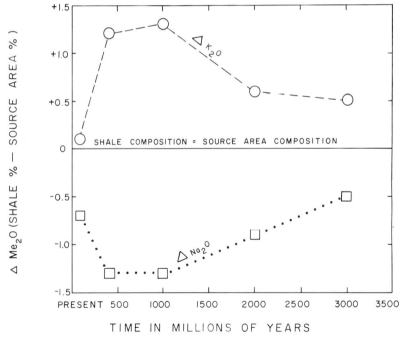

FIG. 5.16. Potash and soda contents of average shales expressed as the percentage difference
from estimates of contemporaneous source rock compositions as a function of age (data from
Ronov, 1972).

the mystery of the nature of the earth's early surface environment. It is, perhaps, remarkable that these rocks are not substantially different in mineralogy and chemistry from younger sedimentary deposits. There are, however, trends in sedimentary rock composition with geological age; in Section 5.2, a model was presented that accounted for these trends as secondary features not only of sedimentary rocks of Phanerozoic age, but perhaps also of rocks as old as $1 \cdot 5 \times 10^9$ years. Although data are sparse, there are trends in sedimentary rock types, chemistry, mineralogy and biology between 3·5 and 1·5 aeons which imply progressive changes in the nature of the earth's surface environment and sea water composition. The Russian geochemical school, especially Ronov, Migdisov, Yaroshevsky, and their colleagues (cf. Ronov, 1959, 1964, 1968, 1972; Ronov and Migdisov, 1971; Ronov et al., 1969; Vinogradov and Ronov, 1956a, b), have been particularly active in obtaining sedimentological data from Phanerozoic and older rocks and in deducing from these data the nature of the earth's surface environment in earliest times. The consensus of their views is that these trends are primary and represent an evolutionary pattern in rock composition, depositional environments, and intensity and relative importance of geochemical processes. This opinion is shared by many other geologists (cf. Engel and Engel, 1964; Rutten, 1966; Cloud, 1972a). In a recent article, Ronov (1972) concluded "the truth may lie in a synthesis that takes into account both theories: the evolution of chemical processes, and the secondary alteration of rocks" (p. 169).

Figure 5.16 illustrates the types of compositional inflections in sedimentologic data which can be observed in rocks $1–3 \cdot 5 \times 10^9$ years old. It indicates the $Na_2O$ and $K_2O$ contents of shales and their contemporaneous source rocks and is plotted as the percentage difference in composition between the shale and source rock as a function of composition against geological age. The increasing potash and decreasing soda contents of shales with increasing rock age up to 1–1·5 aeons ago reflect stabilization of the chemistry and mineralogy of shales with time. Sedimentary cycling and diagenetic reactions result in loss of sodium and gain of potassium by the shales. Superimposed on the trends produced by secondary processes are compositional trends reflecting, among other factors, changes in source rock composition with time (Ronov, 1971). The similarity between source rock and shale compositions prior to $1–1·5 \times 10^9$ years ago may reflect the more mafic composition of the crust 3·5 aeons ago than that later in earth history.

In this section a model for the evolution of sea water composition will be presented based on the hypothesis that the trends observed in sedimentary rock types, composition and mineralogy over the past $1·5 \times 10^9$ years are secondary, as documented in Section 5.4.2, and that trends before this time are to a large extent primary, and however modified by secondary processes, suggest an evolutionary trend in sea water composition up to 1·5 aeons ago.

5.4.2. ORIGIN OF SEA WATER

Comparison of the abundances of rare gases in the earth with cosmic abundances strongly suggests that the atmosphere and hydrosphere are of secondary origin. The method of calculation of geochemical balances between sedimentary and igneous rocks (e.g. Goldschmidt, 1933; Rubey, 1951; Horn and Adams, 1966; Garrels and Mackenzie, 1971) demonstrates that some elements present at the earth's surface, and termed "excess volatiles" by Rubey (1951), cannot have been derived solely from the weathering of igneous rocks and must have been brought to the earth's surface primarily by hot spring and volcanic activity. These elements, H and O (as combined in $H_2O$), C, Cl, N, S, B, Br, F, and As, are particularly abundant in the present atmosphere and oceans. The above observations have led to the commonly accepted theo, that the atmosphere and oceans were generated by degassing of the earth's interior. No firm method of estimating the rate of the earth's degassing as a function of time is available, and widely differing models have been proposed by various investigators. As was pointed out by Rubey, one school of thought prefers rapid degassing followed by large-scale reaction between a primordial ocean and the juvenile crust, whereas the other advocates slow and progressive degassing, stretching the generation of the hydrosphere over most of the history of the earth.

It is likely that chemically the early atmosphere was in a comparatively reduced state, no free oxygen being present. Volcanic gases do not, at present, contain free oxygen and there is no evidence of a progressive change in their composition over considerable periods of geological time (Holland, 1964). Equilibrium models involving basalt melts and volcanic gases (Holland, 1962, 1964) also suggest that the fugacity of oxygen must have been very small. In addition, the inorganic synthesis of primitive life required that very little oxygen be present at the surface of the earth and, in any event, free oxygen would not have become important until photosynthetic processes became established.

The rate of production of free oxygen at the surface of the earth is a source of disagreement. In a series of well known papers Berkner and Marshall (1964, 1965, 1966) proposed that photodissociation of water vapour by solar radiation was a negligible source of $O_2$ because of the absorption of ultraviolet photons by $O_2$ itself in the upper atmosphere. In their model, establishment of a significant amount of $O_2$ in the atmosphere must have awaited the development of large-scale photosynthesis. However, Brinkmann (1969), showed that more precise calculations based on a similar physical model for the structure of the present atmosphere can lead to production of significant amounts of free $O_2$ by the photodissociation mechanism over times of a few hundred million years.

The composition of sea water is continuously affected by active processes such as river discharge, ion-exchange, diffusion of dissolved material, and reactions with the suspended load and sediments. Compositional changes continue in waters trapped as pore fluids in marine sediments and, therefore, there is little hope of ever finding direct evidence bearing on the evolution of the oceans. One viable avenue of investigation is the comparison of the compositions of primary igneous rocks with those of sediments, sea water and the atmosphere, a method used often in the past (Clarke, 1924; Goldschmidt, 1933; Kuenen, 1946; Rubey, 1951; Garrels and Mackenzie, 1971). This method of geochemical balances expresses the fact that sedimentary rocks are the product of a long-term titration of primary igneous rock minerals by acids contained in the excess volatiles, a reaction summarized by the equation

Primary igneous rock minerals + Excess volatiles

$$\rightarrow \text{Sedimentary rocks} + \text{Oceans} + \text{Atmosphere.} \qquad (5.9)$$

This reaction can be used in two ways: (1) balancing of elemental abundances permits an evaluation of the mass of the excess volatiles; (2) physicochemical modelling of the reaction permits prediction of the types and masses of sedimentary materials derived from interactions among chosen initial reactants. Comparison of these results with the known sedimentary record may impose restrictions on the possible reactants and provide some information about the early history of the earth's surface.

To obtain a better picture of the evolution of the ocean as reaction (5.9) proceeds and to quantify further Rubey's arguments, Lafon and Mackenzie (1973) simulated on a high-speed computer the irreversible attack of average igneous rock by water and acid volatiles. The composition of the aqueous phase as the reaction proceeded is shown in Fig. 5.17.

The important aspects of this figure to consider here are: (1) the predicted composition of ocean water resembles that of present sea water. The differences are that the hypothetical ocean is saturated with amorphous silica and its salinity is about twice that of today because of the non-removal of NaCl in evaporite deposits; (2) the pH of the solution increases as the reaction proceeds, representing the titration of the minerals (bases) in igneous rock with the acid solution to produce as neutralization products sea water and sedimentary minerals.

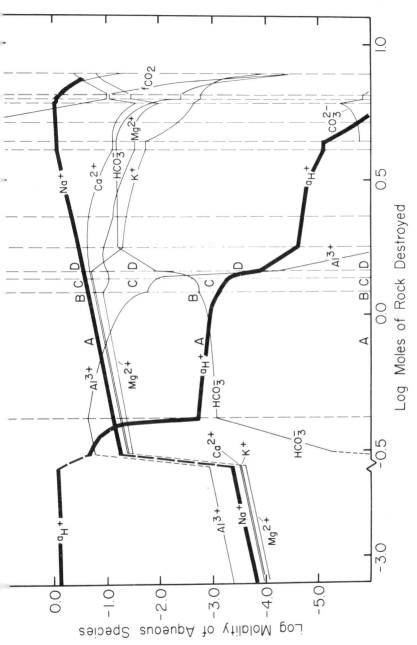

FIG. 5.17. Computer simulation of the variation in the composition of the aqueous phase as the reaction igneous rocks + excess volatiles = sedimentary rocks + oceans + atmosphere proceeds (Lafon and Mackenzie, 1973).

This calculation supports Rubey's contention that a solution having the composition of sea water can be produced by reaction (5.9). However, it sheds no light on the problem of whether degassing was early and complete, or whether the earth has degassed water and acid volatiles at a constant or episodic rate throughout geological time. Geophysical and geological data at present, however, appear to be more consistent with a model in which formation of much of the hydrosphere and atmosphere took place early in the earth's history (cf. Fanale, 1971).

### 5.4.3. SEA WATER BETWEEN 1·5 AND 3·5 AEONS AGO

#### 5.4.3.1. *Data and observations*
A number of observations suggest that sea water composition from the time of the first preserved sedimentary rocks ($3·5 \times 10^9$ years old) until about 1·5 aeons ago underwent continuous chemical change from a moderately acid solution, resulting from reactions between the early crust and acid volatiles, to a sea water of present day composition (Garrels and Mackenzie, 1971). This conclusion can be supported by the following facts:

(1) Source rocks of sediments $3 \times 10^9$ years ago were more basaltic than later ones (Ronov, 1968, 1972; Engel, 1963; Engel and Engel, 1964).

(2) Iron formations were important rock types $1·5–3·0 \times 10^9$ years ago, whereas their importance decreases significantly in younger deposits (Ronov, 1964; see also Fig. 5.2). Iron formations exhibit the facies chert ($SiO_2$)–haematite ($Fe_2O_3$), chert–magnetite ($Fe_3O_4$), chert–greenalite ($Fe_3SiO_5$-$(OH)_4$), chert–siderite ($FeCO_3$), and chert–pyrite ($FeS_2$) (James, 1954). Iron(II) is important in these facies.

(3) Detrital siderite is found in rocks of Animikie age (1·8 to $2·2 \times 10^9$ years B.P.), and detrital grains of uraninite ($UO_2$) and pebbles of pyrite are found in the 2·5 aeons old Witwatersrand conglomerates of Africa.

(4) Haematite of Archean age is restricted to iron formations; thick and geographically extensive clastic red beds did not appear until about 1·9 aeons ago.

(5) The organic carbon content of Archean and early Proterozoic shales is greater than that of Phanerozoic shales. The reduced sulphur content of pre-$1·5 \times 10^9$ year old shales is higher than that of younger shales, whereas the oxidized sulphur content of pre-Phanerozoic rocks is small (Fig. 5.18). Calcium sulphate deposits are not found in rocks older than 1·9 aeons and are rare in younger Precambrian deposits. It is true, however, that collapsed breccias apparently resulting from solution of $CaSO_4$ are found in rocks approximately $2·7 \times 10^9$ years old.

(6) Although the organic carbon content of Phanerozoic sedimentary

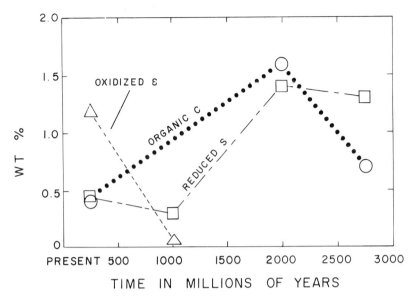

FIG. 5.18. Organic carbon and reduced sulphur content of shales, and oxidized sulphur content of sedimentary rocks as a function of age (data from Ronov and Migdisov, 1971; Cameron and Jonasson, 1972).

rocks fluctuates with age, the variation does not appear to be progressive. Indeed, the slight decrease with increasing age (Ronov, 1958) may represent post-depositional loss (Fig. 5.19C). The C/N ratio of Phanerozoic rocks is remarkably constant (Trask and Patnode, 1942; see also Fig. 5.19B).

(8) The carbon isotopic composition of marine limestones from the Precambrian to the present is remarkably constant (Fig. 5.10).

(9) The oldest eukaryotic organisms known are green and perhaps yellow-brown algae from the 1·3 aeons old Beck Spring Dolomite of California. Prokaryotic blue-green algae and bacteria are found in older deposits (Cloud, 1972a). Indeed, structures known as stromatolites and presumably produced by the sediment-binding and carbonate-precipitating activities of prokaryotic blue-green algae, and to a lesser extent, bacteria are the most common organic structures found in pre-1·5 × 10⁹ year old rocks.

(10) The upper temperature limit of eukaryotic organisms is near 60°C (Tansey and Brock, 1972), whereas prokaryotic photosynthesizing algae can exist at temperatures up to about 75°C. Some bacteria can live and grow near the boiling point of water (Brock, 1967).

(11) Blue-green algae apparently cannot survive in habitats in which the pH is less than about 5 (Brock, 1973; Shapiro, 1973). Lowering of the pH will stimulate a shift from blue-green to green algae in a particular habitat.

N

(12) Most micro-organisms can utilize $NH_3$ directly as a source of $N_2$; however, prokaryotic blue-green algae are largely $N_2$-fixers (Stanier et al., 1963). Phosphate and $CO_2$ uptake kinetics apparently favour blue-green over green algae (King, 1970; Shapiro, 1973); that is, blue-green algae are more efficient at obtaining $CO_2$ and phosphate from waters containing low concentrations of these constituents.

(13) $\delta^{18}O$ values for cherts suggest that temperatures in the Precambrian for the central and western United States may have been higher than 50°C $1\cdot3 \times 10^9$ years ago (Knauth, 1972).

(14) Thermophilic bacteria can oxidize reduced sulphur under conditions of low pH and high temperature. *Thiobacillus* is common in moderately hot habitats (40–55°C) and *Sulpholobus* is found in habitats above 60–65°C and over a pH range of 0·9 to 5·8 (Brock et al., 1972).

### 5.4.3.2. Interpretations

It is likely that the surface of the earth was hot during an early stage in its evolution. Calculations assuming high temperature equilibrium between the gases and the mafic crust suggest that at that time, the atmosphere was highly reducing and contained substantial quantities of hydrogen, methane, ammonia, and carbon monoxide (cf. Holland, 1962; French, 1966). Water vapour, hydrogen sulphide, nitrogen, argon and hydrogen chloride were also present. As Garrels and Perry (1973) so succinctly point out, to obtain surface conditions like those of today, which presumably have existed for 600, if not 1500, million years, it is necessary that the original reduced system be slowly titrated with oxygen.

Although the sequence of events is still a matter for speculation, it appears that as the original crust cooled, the reduced gases may have reacted with partly reduced ferric compounds, such as $Fe_3O_4$, in the crust leading to their reduction in the crust and to the oxidation of the gases. Photodissociation of water vapour in the upper atmosphere resulting in oxygen production and hydrogen escape, as well as photosynthesis, could also lead to oxidation of these early reduced gases. The acid gases, $H_2S$, HCl, and $CO_2$ derived from the oxidation of $CH_4$ and CO, would also react with the crust to produce sedimentary minerals. Under equilibrium conditions CO would be the first to be oxidized to $CO_2$, followed progressively by $H_2$ to $H_2O$, $NH_3$ to $N_2$, and $CH_4$ to $CO_2$.

Whatever the actual path, it is likely that by $3–3\cdot5 \times 10^9$ years ago, the original reduced gases had been nearly completely oxidized and the atmosphere was composed principally of nitrogen and carbon dioxide. The presence of minerals such as pyrite, siderite, calcite and dolomite in sedimentary rocks of this age supports this conclusion. The absence of clastic red beds and of evidence for extensive $CaSO_4$ deposition suggest vanishingly low oxygen levels at this time, about $10^{-70}$ atm. In addition, in all the experiments in which organic compounds have been produced that could lead to evolution of a living cell, no free oxygen was present (e.g. Abelson, 1966). The oldest sedimentary rocks contain fossilized bacteria, and most probably, prokaryotic blue-green algae. These organisms could not be greatly separated in time from the advent of the first living cells which were anaerobic heterotrophs (Cloud, 1972a); these organisms evolved and lived in an oxygen-free environment and fed on non-biologically produced organic matter or each other.

Reduced gases probably still existed in the 3·5 aeons old atmosphere; their concentrations were small and they may have been in non-equilibrium states. Indeed, parts per million levels of ammonia, which is an effective absorber of thermal energy, may have been necessary in the early and middle Precambrian atmosphere. The principal contributors to the greenhouse effect of the earth today are the gases $CO_2$ and $H_2O$. Sagan and Mullen (1972) have calculated that if these constituents were the only absorbers present in the Precambrian atmosphere, the global mean temperature of the earth would be less than the freezing point of sea water 2·3 aeons ago. However, similar calculations based on a Precambrian atmosphere containing a few parts per million of $NH_3$ lead to a global temperature of the earth at this time above the freezing point, a result in accord with geological and palaeontological evidence. Based on the rate of deamination of aspartic acid and the assumption that aspartic acid is necessary for life, it has been estimated by Bada and Miller (1968) that the concentration of $NH_4^+$ in the primitive Precambrian ocean was $1 \times 10^{-3}$ M, a value compatible with an atmosphere containing several parts per million of ammonia.

Furthermore, it is possible that the surface of the earth 3·5 aeons ago was warmer than in later times. Prokaryotic blue-green algae and bacteria can live at temperatures above those at which eukaryotic forms can exist. The prokaryotes are found fossilized in the oldest sedimentary rocks, whereas, eukaryotes did not appear until 1·3 to 1·6 aeons ago (Licari and Cloud, 1972). It is speculated that the evolution of eukaryotic forms had to await a general cooling off of the earth's surface environment. Perhaps the oxidation of most of the remaining $NH_3$ in the atmosphere between $3 \times 10^9$ and $1·6 \times 10^9$ years ago led to a general decrease in temperature of the earth's surface. The finding that the earth's surface temperatures in some regions may have been as hot as 50°C 1·3 aeons ago supports this conclusion to some extent. In addition, the greater efficiency of blue-green algae compared with green algae in obtaining phosphate from waters with low concentrations of phosphate would favour development of the former at this time because of

the possiblity that phosphorus, and not nitrogen, was the limiting nutrient.
It is difficult to obtain an estimate of the composition of sea water of 3·5
aeons ago because of innumerable variables for which there are at the best
only order of magnitude estimates. However, it is likely that owing to the
lack of oxygen in the atmosphere, the dissolved sulphate concentration was
low and the cations present were balanced principally by $Cl^-$ and $HCO_3^-$
(Ronov, 1971; Garrels and Perry, 1974). In addition, iron(II) was undoubtedly
an important constituent of the sea water. In an oxygenated atmosphere,
such as that of today, iron concentrations in natural waters are low owing to
the insolubility of iron(III) oxide. However, under conditions of low oxygen
concentration, iron(II) could occur in sea water at parts per million levels.
The fact that iron formations are an important rock type of the early and
middle Precambrian is probably due to a large extent to the greater mobility
of iron(II) at this time; iron(II) behaved similarly to calcium and magnesium
and was dissolved in waters and deposited as ferrous carbonate and silicate.
Later in the earth's history, as the surface environment became more oxygenated, iron would tend to travel in the suspended load of streams and behave
similarly to aluminium in the cycle.

Owing to higher levels of $p_{CO_2}$ in those early times (Holland, 1968), it is
likely that the pH of sea water was lower than at present. Some idea of the
possible range of pH can be obtained from two considerations:

(1) fossilized traces of blue-green algae are found in 3 to $3·5 \times 10^9$ year
old sedimentary rocks. The observation that these algae cannot survive in
habitats with a pH lower than 5 provides a minimum estimate for sea water
pH.

(2) there is no fossil evidence for the presence of silica secreting marine
organisms 3·5 aeons ago; thus, it is likely that sea water could have been
saturated with amorphous silica. If so, the pH would have been below about
7·5 or precipitation of sepiolite would have occurred. No deposits of this
mineral of Precambrian age have been found.

To obtain an estimate of the major cation composition of ancient sea water,
it is necessary to assume that equilibria similar to those which prevail in
present-day sea water existed. One reconstruction of ancient sea water
composition is given in Table 5.3. This water differs from modern sea water
in that it contains no dissolved oxygen, is virtually free of $SO_4^{2-}$, contains
about four times more $HCO_3^-$, has significant dissolved iron and a lower pH.
Its calcium and magnesium contents are similar to those of modern sea water;
Ronov (1971) considers that at this time the concentrations of these constituents
were either higher or lower than they are at present. There is no cogent
evidence to suggest that the sodium chloride content of ancient sea water
was like that of today's oceans. The salinity of ancient sea water could have

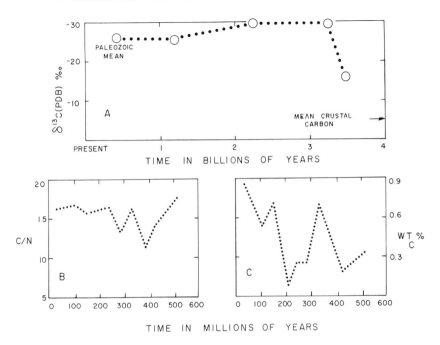

FIG. 5.19. Attributes of organic matter in sedimentary rocks as a function of age. A. Carbon isotopic composition of sedimentary organics. B. Carbon to nitrogen ratio in Phanerozoic sedimentary organics. C. Carbon content of Phanerozoic sedimentary rocks (data from Oehler et al., 1972; Trask and Patmode, 1942; Ronov, 1958).

TABLE 5.3

*Reconstruction of ancient sea water\*.* (After Garrels and Perry, 1973).

| | | | |
|---|---|---|---|
| $P_{O_2(g)}$ | $= 10^{-70.07}$ atm | $Ca^{2+}_{aq}$ | $= 10^{-1.95}$ (0.011 m) |
| $P_{CO_2(g)}$ | $= 10^{-1.20}$ atm | $Mg^{2+}_{aq}$ | $= 10^{-1.52}$ (0.030 m) |
| $P_{CH_4(g)}$ | $= 10^{-4.39}$ atm | $Fe^{2+}_{aq}$ | $= 10^{-4.13}$ (0.000074 m) |
| $P_{H_2(g)}$ | $= 10^{-0.11}$ atm | $K^+_{aq}$ | $= 10^{-2}$ (0.01 m) |
| $P_{NH_3(g)}$ | $= 10^{-6.95}$ atm | $Na^+_{aq}$ | $= 10^{-0.330}$ (0.47 m) |
| $P_{H_2S(g)}$ | $= 10^{-8.34}$ atm | $Cl^-_{aq}$ | $= 10^{-0.26}$ (0.55 m) |
| $HS^-_{aq}$ | $= 10^{-5.41}$ m | $SiO_{2aq}$ | $= 10^{-2.7}$ (0.002 m) |
| $SO^{2-}_{4aq}$ | $= 10^{-10.4}$ m | pH | $= 10^{-6.68}$ |
| $HCO^-_{3aq}$ | $= 10^{-2.02}$ (0.0094 m) | | |

\* Assumptions: Equilibrium at 25°C with $FeSiO_3$, $FeS_2$, $FeCO_3$, $CaCO_3$, $CaMg(CO_3)_2$, and amorphous $SiO_2$. Activities of $Na^+$, $K^+$, and $Cl^-$ and values of activity coefficients of ions were assumed to be the same as those in the present ocean. aq = dissolved species; g = gaseous species.

been as much as 45% greater than today's 35‰, if no NaCl was stored in evaporite deposits. Whatever the actual composition of sea water 3 to 3·5 aeons ago, it was certainly a much more reduced system than today. When did it become a "modern" sea water?

About $3·5 \times 10^9$ years ago, oxygen was stored principally in carbonate and silicate rocks and in water. Much of this oxygen would be released eventually to the earth's surface principally by photosynthetic and photodissociation reactions. The photosynthesizing prokaryotic micro-organisms at this time probably reduced $CO_2$ to carbohydrate $(CH_2O)$ in the hydrosphere by accepting electrons from reduced substances, such as iron(II) (Cloud, 1972a, b). Garrels and Perry (1974) have pointed out that during oxidation of the iron(II) carbonate in sedimentary rocks to iron(III) oxide the number of moles of $O_2$ consumed are less than those of $CO_2$ released. If the excess $CO_2$ is used in photosynthesis this will result in a net overall production of free oxygen. Initially, as oxygen was produced by photosynthesis, it was utilized in the oxidation of reduced iron minerals (e.g., iron(II) carbonate, silicate and sulphide) at the earth's surface, viz:

sulphide oxidation

$$7CO_2 + 9H_2O + 2FeS_2 \xrightarrow{\text{bacteria}} 7(CH_2O) + 4SO_4^{2-} + 4H^+ + 2Fe^{2+}$$

$$(10)$$

and

iron oxidation

$$CO_2 + 5H_2O + 4Fe^{2+} \xrightarrow{\text{bacteria}} (CH_2O) + 2Fe_2O_3 + 8H^+. \quad (11)$$

The fixation of $CO_2$ in organic matter would lower the partial pressure of $CO_2$ in the atmosphere as Precambrian time progressed. Once photo-synthesized, organic matter would undergo bacterial degradation to an unreactive residue, but owing to lack of atmospheric oxygen would not decompose completely. The degraded organic matter would be eroded, deposited and recycled in particulate form. Sulphate and $Fe_2O_3$ produced by sulphide and iron oxidation would enter the dissolved load and the suspended load of streams respectively. Sulphate would become the important dissolved sulphur species in surface waters at atmospheric oxygen partial pressures of about $10^{-68}$ atm. The presence of $FeS_2$, presumably derived from reduction of $SO_4^{2-}$, in iron formations two aeons old (James, 1954) suggests the presence of abundant sulphate in waters of iron-depositing basins. It is possible, however, that iron formations originated in semi-restricted environments in which the composition of the water differed from that of the "normal" sea water of that time (Garrels et al., 1973; Eugster and Chou, 1972). If so, the world's oceans may not have contained significant sulphate until the advent of eukaryotic photosynthetic organisms, at which time oxygen could enter the atmosphere. Licari and Cloud (1972), on the basis of stratigraphical and palaeontological criteria, consider that this event took place 1·3 to 1·6 × 10⁹ years ago.

The decrease in the organic carbon content of shales from about 2 × 10⁹ years ago to Phanerozoic time (Fig. 5.18) probably represents the change from a system in which organic matter was stored, to one in which it underwent decomposition during erosion from rocks in an oxygenated atmosphere. It is also possible that the increase in the isotopic composition of the carbon of sedimentary organic material from about $\delta^{13}C$ of $-30\%_0$ for 1·6 aeons old rocks to $-26\%_0$ for younger Precambrian rocks (Fig. 5.19A) was caused by the decreased isotopic fractionation between reduced and oxidized carbon resulting from an abrupt fall in $p_{CO_2}$ (Garrels and Perry, 1974).

The decline in sulphur content and increase in $CaSO_4$ content of sedimentary rocks during this same time (Fig. 5.18) probably arises from the transfer of sulphur from reduced reservoirs to oxidized ones, once more pointing to increased levels of atmospheric oxygen. $CaSO_4$ is easily eroded from the sedimentary mass, and consequently, its present distribution may have little to do with atmospheric $O_2$.

These conclusions are compatible with the distribution of sedimentary minerals indicative of oxidation states in the Precambrian. For example, clastic red beds representing progressive accumulation of $Fe_2O_3$ and oxygen during the Precambrian do not appear until 1·9 aeons ago, whereas detrital siderite, pyrite and uraninite are found in older deposits. These latter minerals are easily oxidized in oxygenated environments.

Once oxygen was being used in oxidation of organic matter, its net rate of accumulation would decrease and a steady state would be reached in which the mass of new organic matter deposited each year would equal the mass oxidized during weathering of rocks (Holland, 1965; Broecker, 1970; Garrels and Perry, 1974). This balance would maintain oxygen levels near the present-day values, and the continuous evolutionary change in sea water composition would cease. Apparently, this state of balance was reached in late Proterozoic time and has continued until today. The lack of progressive changes in the carbon content and the C/N ratio of Phanerozoic rocks (Fig. 5.19B, C) and the nearly constant $\delta^{13}C$ content of carbonate rocks (Fig. 5.10) from the late Proterozoic to the present support this conclusion.

## 5.5. SUMMARY AND CONCLUSIONS

An hypothesis is presented for the sedimentary lithosphere in which trends in Phanerozoic sedimentary rock types, chemistry and mineralogy were interpreted as being secondary features of the rock mass and arising from differential sedimentary cycling and long-term diagenetic processes. This hypothesis implies that the composition and rock-type ratios of sediments deposited have remained nearly the same during Phanerozoic time.

On the basis of these conclusions, a steady-state model for the atmosphere–biosphere-ocean-sediments system has been developed. This model does not require that any major evolutionary change in sea water composition took place during Phanerozoic time. However, periodic excursions from this steady state, may have developed, usually with time periods of less than tens of millions of years. On the basis of the observation that the trends in sedimentary rock compositions extend back into the late Precambrian, it has been suggested that the model may be applicable to the last 1·5 aeons of earth history.

More than $1·5 \times 10^9$ years ago, there were compositional and biological trends in sedimentary rocks which suggest that a progressive change in sea water composition occurred between 3·5 and $1·5 \times 10^9$ years ago. Emphasis has been placed on the importance of oxygen during this time as a titrating agent for the early reduced atmosphere–ocean system. It has been postulated that global temperatures gradually decreased as Precambrian time progressed. Steady state conditions were reached about $1·5 \times 10^9$ years ago and sea water composition has not evolved greatly since then.

ACKNOWLEDGEMENTS

My deepest appreciation goes to my colleague, Bob Garrels, for the years of productive association that have led to many of the ideas and concepts in this paper. Many of the diagrams in this manuscript are from our previous joint publications. I would like to thank Abraham Lerman for introducing me to the mathematics of multi-reservoir cyclic systems. I am indebted to Mary Ellen Hagenauer and Betty Faulkner for typing the manuscript, help in editing, and the bibliographic search. The Petroleum Research Fund of the American Chemical Society supported the research on which this chapter was based.

## REFERENCES

Abelson, P. H. (1966). *Proc. Nat. Acad. Sci. U.S.A.* **55**, 1365.
Bada, J. L. and Miller, S. L. (1968). *Science, N.Y.* **159**, 423.

Berkner, L. V. and Marshall, L. C. (1964). *Proc. Faraday Soc.* **37**, 122.
Berkner, L. V. and Marshall, L. C. (1965). *Proc. Nat. Acad. Sci. U.S.A.* **53**, 1169.
Berkner, L. V. and Marshall, L. C. (1966). *J. Atmos. Sci.* **23**, 133.
Berner, R. A. (1972). *In* "The Changing Chemistry of the Oceans" (D. Dyrssen and D. Jagner, eds.), pp. 347–362. Almquist and Wiksell, Stockholm.
Berner, R. A. (1973). *In* "Hydrogeochemistry", Vol. I, pp. 402–417. The Clarke Co., Washington, D.C.
Birch, F. (1965). *Bull. Geol. Soc. Amer.* **76**, 133.
Bischoff, J. L. and Ku, T. (1970). *J. Sedim. Petrol.* **40**, 960.
Bischoff, J. L. and Ku, T. (1971). *J. Sedim. Petrol.* **41**, 1008.
Bischoff, J. L. and Sayles, F. L. (1972). *J. Sedim. Petrol.* **42**, 711.
Bolin, B. (1970). *Scient. Amer.* **223**, 124.
Brinkmann, R. T. (1969). *J. Geophys. Res.* **74**, 5355.
Brock, T. D. (1967). *Science, N.Y.* **158**, 1012.
Brock, T. D., Brock, K. M., Belly, R. T. and Weiss, R. L. (1972). *Arch. Mikrobiol.* **84**, 54.
Brock, T. D. (1973). *Science, N.Y.* **179**, 480.
Broecker, W. S. (1970). *J. Geophys. Res.* **75**, 3553.
Broecker, W. S. (1971). *Quaternary Res.* **1**, 188.
Burst, J. F., Jr. (1959). *Clays Clay Miner.* **6**, 327.
Burton, J. D. and Liss, P. S. (1968). *Nature, Lond.* **220**, 905.
Calvert, S. E. (1968). *Nature, N.Y.* **219**, 910.
Cameron, E. M. and Jonasson, I. R. (1972). *Geochim. Cosmochim. Acta* **36**, 985.
Chave, K. E. (1954). *J. Geol.* **62**, 587.
Chilingar, G. V. (1956). *Bull. Amer. Ass. Petrol. Geol.* **40**, 2256.
Clarke, F. W. (1924). *Bull. U.S. Geol. Surv.* **770**, 84  p.
Cloud, P. (1972a). *Amer. J. Sci.* **272**, 537.
Cloud, P. (1972b). Abst. Soc. Econ. Geol. Symp. Precambrian Iron Formations, Duluth, Minnesota.
Conway, E. J. (1942). *Proc. R. Ir. Acad.* **48** B, no. 8.
Conway, E. J. (1943). *Proc. R. Ir. Acad.* **48** B, no. 9.
Culkin, F. (1965). *In* "Chemical Oceanography" (J. P. Riley and G. Skirrow, eds.), Vol. I, pp. 121–161. Academic Press, London.
Degens, E. T. and Epstein, S. (1962). *Bull. Amer. Ass. Petrol. Geol.* **46**, 534.
De Segonzac, D. (1965). *Bull. Serv. Carte Geol. Als. Low.* **17**, 287.
Drever, J. I. (1971a). *J. Sedim. Petrol.* **41**, 982.
Drever, J. I. (1971b). *Science, N.Y.* **172**, 1334.
Drever, J. I. (1974). *In* "The Sea" (E. D. Goldberg, ed.). Vol. 5. Interscience, New York.
Engel, A. E. J. (1963). *Science, N.Y.* **140**, 143.
Engel, A. E. J. and Engel, C. G. (1964). *In* "Advancing Frontiers in Geology and Geophysics" (A. P. Subramanian and S. Balakrishna, eds.), pp. 17–37. Osmania Univ. Press, Hyderabad, India.
Eugster, H. P. and Chou, I. (1972). Abst. Soc. Econ. Geol. Symp. Precambrian Iron Formations, Duluth, Minnesota.
Fairbridge, R. W. (1964). *In* "Problems in Palaeoclimatology" (A. E. M. Nairn, ed.). Interscience, New York.
Fanale, F. P. (1971). *Chem. Geol.* **8**, 79.
Fanning, K. A. and Pilson, M. E. Q. (1971). *Science, N.Y.* **173**, 1228.

French, B. M. (1966). *Rev. Geophys.* **4**, 223.
Galimov, E. M., Kuznetzova, N. G. and Prokhorov, U. S. (1968). *Geochemistry* **5**, 1126.
Garrels, R. M. and Mackenzie, F. T. (1969). *Science, N.Y.* **163**, 570.
Garrels, R. M. and Mackenzie, F. T. (1971). "Evolution of Sedimentary Rocks", 397 pp. W. W. Norton, New York.
Garrels, R. M. and Mackenzie, F. T. (1972). *Mar. Chem.* **1**, 27.
Garrels, R. M. and Mackenzie, F. T. (1973). *In* "Palaeooceanography" (W. Hay, ed.). SEPM Spec. Publ. (in press).
Garrels, R. M., Mackenzie, F. T. and Siever, R. (1972). *In* "Nature of the Solid Earth" (E. C. Robertson, ed.), pp. 93–121. McGraw-Hill, New York.
Garrels, R. M. and Perry, E. A. Jr. (1974) *In* "The Sea" (E. D. Goldberg, ed.), Vol. 5, Interscience, New York.
Goldberg, E. D. (1965). *In* "Chemical Oceanography" (J. P. Riley and G. Skirrow, eds.), Vol. I, pp. 163–194. Academic Press, London.
Goldich, S. S., Hedge, C. E. and Stern, T. W. (1970). *Bull. geol. Soc. Amer.* **81**, 3671.
Goldschmidt, V. M. (1933). *Fortsch. Mineral. Krist. Petrogr.* **17**, 112.
Gregor, C. B. (1968a). *Kon. Ned. Akad. Wetensch. Proc.* **71**, 22.
Gregor, C. B. (1968b). *Nature, Lond.* **219**, 360.
Grim, R. E. and Johns, W. D. (1949). *Clays Clay Miner.* **2**, 81.
Harris, R. C. (1966). *Nature, London.* **212**, 275.
Hart, R. A. (1973). "A model for chemical exchange in the basalt–seawater system of oceanic Layer II" (unpublished).
Heath, G. R. (1973). *In* "Geologic History of the Ocean" (W. W. Hay, ed.). SEPM Spec. Publ. (in press).
Heck, E. T. (1964). *W. Va. Geol. Econ. Surv. Bull.* **28**, 1.
Helgeson, H. C., Garrels, R. M. and Mackenzie, F. T. (1969). *Geochim. Cosmochim. Acta* **33**, 455.
Helgeson, H. C. and Mackenzie, F. T. (1970). *Deep-Sea Res.* **17**, 877.
Higgins, G. H, (1968). Program, 1968 Annual Meeting: Geol. Soc. Amer., Boulder, Colorado. pp. 135–136.
Holland, H. D. (1962). *In* "Petrologic Studies: A volume to honor A. F. Buddington", Geol. Soc. America, pp. 447–477.
Holland, H. D. (1964). *In* "The Origin and Evolution of Atmospheres and Oceans" (J. Brancazio and A. G. W. Cameron, eds.), pp. 86–101. John Wiley and Sons, New York.
Holland, H. D. (1965). *Proc. Nat. Acad. Sci., U.S.A.* **53**, 1173.
Holland, H. D. (1968). *In* "Origin and Distribution of the Elements" (L. H. Ahrens, ed.), pp. 949–954. Pergamon Press, New York.
Holland, H. D. (1972). *Geochim. Cosmochim. Acta* **36**, 637.
Holser, W. T. and Kaplan, I. R. (1966). *Chem. Geol.* **1**, 93.
Horn, M. K. and Adams, J. A. S. (1966). *Geochim. Cosmochim. Acta* **30**, 279.
Hurley, P. M. (1965). Thirteenth Annual Program: U.S. Atomic Energy Commision: Massachusetts Inst. Tech., no. 1381–13, 160.
James, F. J. (1954). *Econ. Geol.* **49**, 235.
Johnson, F. J. (1970). *In* "Global Effects of Environmental Pollution" (S. F. Singer, ed.), pp. 4–11. Springer, New York.
Keith, M. L. and Weber, J. N. (1964). *Geochim. Cosmochim. Acta,* **28**, 1787.
King, D. L. (1970). *J. Wat. Pollut. Control. Fed.* **42**, 2035.

Knauth, P. (1972). "Oxygen and hydrogen ratios in cherts and related rocks". PhD. Thesis, Calif. Inst. Tech.

Koczy, F. T. (1956). *Deep-Sea Res.* **3**, 279.

Kroopnick, P. (1971). "Oxygen and carbon in thè oceans and atmosphere; Stable isotopes as tracers for consumption: Production, and circulation models". Thesis, Scripps Inst. Oceanogr., La Jolla, Calif., 230 pp.

Kuenen, Ph. H. (1946). *Amer. J. Sci.* **244**, 563.

Kulp, J. L., Turekian, K. K. and Boyd, D. W. (1952). *Bull. Geol. Soc. Amer.* **63**, 701.

Lafon, G. M. and Mackenzie, F. T. (1973). *Soc. Econ. Palaeont. Miner. Spec. Publ.* (in press).

Li, Y. H. (1972). *Amer. J. Sci.* **272**, 119.

Licare, G. R. and Cloud, P. (1972). *Proc. Nat. Acad. Sci.* **69**, 2500.

Lisitizin, A. P. (1972). *Soc. Econ. Palaeont. Miner. Spec. Publ.*, no. 17, 218 pp.

Liss, P. S. and Spencer, C. P. (1970). *Geochim. Cosmochim. Acta* **34**, 1073.

Lotze, F. (1964). *In* "Problems in Palaeoclimatology" (A. E. M. Nairn, ed.), pp. pp. 491–509. Interscience, New York.

Lowenstam, H. A. (1961). *J. Geol.* **69**, 241.

Mackenzie, F. T., Garrels, R. M., Bricker, O. P. and Bickley, F. (1967). *Science* **155**, 1404.

Mackenzie, F. T. and Garrels, R. M. (1965). *Science, N.Y.* **150**, 57.

Mackenzie, F. T. and Garrels, R. M. (1966a). *Amer. J. Sci.* **264**, 407.

Mackenzie, F. T. and Garrels, R. M. (1966b). *J. Sedim. Petrol.* **36**, 1075.

McKenzie, J. A. (1972). "A mathematical model for the isotopic balance of sulphur in the oceans" (in preparation).

Mangelsdorf, P., Wilson, T. R. S. and Daniell, E. (1969). *Science, N.Y.* **165**, 171.

Manheim, F. T., Chan, K. M. and Sayles, F. L. (1970). *In* "Initial Reports of the Deep Sea Drilling Project", Vol. 5, pp. 501–511. U.S. Government Printing Office, Bashington.

Manz, R. H., Jr. (1953). *J. Geol.* **16**, 51.

Nissenbaum, A., Presley, B. J. and Kaplan, I. R. (1972). *Geochim. Cosmochim. Acta*, **36**, 1007.

Oehler, D. Z., Schopf, J. W. and Kuenvolden, K. A. (1972). *Science, N.Y.* **172**, 1246.

Perry, E. and Hower, J. (1970). *Clays Clay Miner.* **18**, 167.

Pressley, B. J. and Kaplan, I. R. (1970). *In* "Initial Reports of the Deep Sea Drilling Project", Vol. IV, pp. 415–438. U.S. Government Printing Office, Washington.

Pytkowicz, R. M. (1967). *Geochim. Cosmochim. Acta*, **31**, 63.

Pytkowicz, R. M. (1972). *In* "The Changing Chemistry of the Oceans" (D. Dyrssen and D. Jagner, eds.), pp. 147–152. Almquist and Wiksell, Stockholm.

Rees, C. E. (1970). *Earth Planet. Sci. Lett.* **7**, 366.

Ristvet, B. L., Mackenzie, F. T., Thorstenson, D. C. and Leeper, R. H. (1973). *Bull. Amer. Assoc. Petrol. Geol.* (in press).

Ronov, A. B. (1958). *Geochemistry*, no. 5, 510.

Ronov, A. B. (1959). *Geochemistry*, no. 5, 493.

Ronov, A. B. (1968). *Sedimentology*, **10**, 25.

Ronov, A. B. (1964). *Geochemistry*, no. 4, 713.

Ronov, A. B. (1971). *Ber. deutsch. Ges. geol. Wiss. A, Geol. Palaeont.* **16**, 331.

Ronov, A. B. (1972). *Sedimentology*, **19**, 157.

Ronov, A. B. and Migdisov, A. A. (1971). *Sedimentology*, **16**, 137.

Ronov, A. B., Migdisov, A. B. and Barskaya, N. V. (1969). *Sedimentology*, **13**, 179.

Rubey, W. W. (1951). *Bull. Geol. Soc. Amer.* **62**, 1111.
Rutten, M. G. (1966). *Palaeogr. Palaeoclim. Palaeocol.* **2**, 47.
Sagan, C. and Mullen, G. (1972). *Science, N.Y.* **177**, 52.
Sayles, F. L., Manheim, F. T. and Waterman, L. S. (1972). *In* "Initial Reports of the Deep Sea Drilling Project", Vol. 15. U.S. Government Printing Office, Washington (in press).
Shapiro, J. (1973). *Science, N.Y.* **179**, 382.
Siever, R. (1968a). *Sedimentology*, **11**, 5.
Siever, R. (1968b). *Earth Planet. Sci. Lett.* **5**, 106.
Sillén, L. G. (1961). *In* "Oceanography" (M. Sears, ed.), pp. 549–581. Amer. Assn. Adv. Sci., Washington.
Sillén, L. G. (1967a). *Science, N.Y.* **156**, 1189.
Sillén, L. G. (1967b). *In* "Equilibrium Concepts in Natural Water System". Adv. in Chem. Ser. 67, Amer. Chem. Soc., Washington, D.C. p.. 57–69.
Stanier, R. Y., Doudoroff, M. and Adelberg, E. A. (1963). "The Microbial World". Prentice-Hall, New York.
Stumm, W. and Morgan, J. J. (1970). "Aquatic Chemistry", 583 pp. Wiley-Interscience, New York.
Tansey, R. M. and Brock, T. D. (1972). *Proc. Nat. Acad. Sci.* **69**, 2426.
Trask, P. D. and Patnode, H. W. (1942). "Source Beds of Petroleum", 566 pp. Amer. Assn. Petrol. Geol., Tulsa.
Turekian, K. K. (1965). *In* "Chemical Oceanography" (J. P. Riley and G. Skirrow, eds.), Vol. II, pp. 81–126. Academic Press, New York.
Van Moort, J. C. (1971). *Clays Clay Miner.* **19**, 1.
Vinogradov, A. P. and Ronov, A. B. (1956a). *Geochemistry*, **2**, 123.
Vinogradov, A. P. and Ronov, A. B. (1956b). *Geochemistry*, **6**, 533.
Weaver, C. E. (1967). *Geochim. Cosmochim. Acta*, **31**, 2181.
Weaver, C. E. and Wampler, J. M. (1970). *Bull. Geol. Soc. Amer.* **81**, 3423.
Weber, J. N. (1965). *Nature, Lond.* **207**, 930.
Wollast, R. and De Broeu, F. (1971). *Geochim. Cosmochim. Acta*, **35**, 613.
Wollast, R. (1974). *In* "The Sea" (E. D. Goldberg, ed.), Vol. 5. Interscience, New York

Chapter 6

# Salinity and the Major Elements of Sea Water

T. R. S. WILSON

*Institute of Oceanographic Sciences,
Wormley, Godalming, Surrey, England*

## 6.1. INTRODUCTION

The large number of reviews on the concept of salinity which have appeared in recent years (Johnston, 1964; Cox, 1965; Horne, 1969; Johnston, 1969; Carpenter, 1972) testifies to its importance to all branches of oceanography. The purpose of this chapter is to summarize only briefly those aspects of the subject covered in the previous edition of this book (Cox, 1965), but to give a more detailed coverage of modern practice and instrumentation. The review by the late Dr. R. A. Cox is recommended to those who require a more detailed discussion of the history of the topic and of those instruments in use up to the early sixties.

The major ions of sea water have been defined (Culkin, 1965) as those which make a significant contribution to the measured salinity. In practice, this is taken to mean elements present at concentrations greater than 1 ppm in oceanic water. These are listed in Table 6.1. Silicon, although it may occur at concentrations of the order of several parts per million, is conventionally excluded because of its atypical behaviour, a consequence of its high biological activity. The major ions are commonly referred to as "conservative",

TABLE 6.1

*The major ions of sea water*

| Ion | $g kg^{-1}$ at $S = 35\%_{oo}$ | $g kg^{-1}$ chlorinity $\%_{oo}^{-1}$ |
|---|---|---|
| $Cl^-$ | 19·354 | 0·9989 |
| $SO_4^{2-}$ | 2·712 | 0·1400 |
| $Br^-$ | 0·0673 | 0·00347 |
| $F^-$ | 0·0013 | 0·000067 |
| B | 0·0045 | 0·000232 |
| $Na^+$ | 10·77 | 0·5560 |
| $Mg^{2+}$ | 1·290 | 0·0665* |
| $Ca^{2+}$ | 0·4121 | 0·02127 |
| $K^+$ | 0·399 | 0·0206 |
| $Sr^{2+}$ | 0·0079 | 0·00041 |

* Recent reported values lie between 0·06612 and 0·06692

an expression which implies that their concentrations in sea water bear a constant ratio, one to another. This constancy of composition of ocean water permits the use of the term "salinity" to describe the concentration of "sea salt" in a sample of sea water. So remarkable is this constancy that the possibility of the existence of compositional differences in sea salt from various parts of the world ocean was for many years almost completely neglected. Section 6.3 deals with the most recent findings in this field and their relation to the concept of salinity.

## 6.2. DEFINITIONS OF SALINITY

During the 19th century, chemical investigations of steadily increasing precision confirmed the apparent constancy of composition of sea salt referred to above. As a consequence of this, an International Commission set up in 1899 under Professor Martin Knudsen was able to recommend the following definition. "Salinity is to be defined to be the weight of inorganic salts in one kilogram of sea water, when all bromides and iodides are replaced by an equivalent quantity of chlorides, and all carbonates are replaced by an equivalent quantity of oxides".

The complexity of this definition reflects the difficulties which the Commission experienced in the determination of salinity. The obvious method, evaporation to dryness and weighing of the resultant salts, has been little used because of difficulties imposed by the volatility of some inorganic constituents (especially hydrogen chloride) and the tenacity with which water of crystallization is retained. As a part of the Commission's investigations, Sørensen (see Forch *et al.*, 1902) determined the salinity of nine sea water samples of various salinities, using a gravimetric method based directly on the definition given above. This was intended to allow the calculation of an empirical relationship between salinity as defined above and chlorinity (defined by the Commission as the chlorine equivalent of the total halide concentration in parts per thousand by weight, measured by titration with silver nitrate solution. The high precision Volhard method or the operationally simpler Mohr titration were used to determine this parameter).

The relationship relating salinity to chlorinity arrived at by Forch *et al.* was:

$$S\%_0 = 1 \cdot 805 \, Cl\%_0 + 0 \cdot 030$$

This was used as the working definition of salinity for some 65 years, the direct determination being abandoned for practical reasons in favour of the chlorinity determination. In effect, therefore, the official gravimetric definition was not used after 1902, all reliance being placed on the chlorinity titration and on the salinity to chlorinity relation arrived at by Knudsen using Sørensen's results. It is clear from the report of the Commission that this was deliberately intended by them for the reasons of operational convenience mentioned above.

The system worked well for many years. One drawback, however, proved to be the inaccuracy of the atomic weights available in 1900. As these were involved in the definition of chlorinity given by Forch *et al.*, small shifts in the definition would occur each time the atomic weights were revised. Because of this problem chlorinity was redefined in the following way in 1937 (Jacobsen and Knudsen, 1940):

"The number giving the chlorinity in per mille of a sea water sample is by definition identical with the number giving the mass with unit gram of Atomgewichtssilber just necessary to precipitate the halogens in 0·3285234 kg of the sea water sample.

By adopting this definition we have obtained the result that all sea water standards could disappear without causing any break in the continuity of the chlorinity determination. Even if the amount of

Atomgewichtssilber stored in Copenhagen should disappear, a break in continuity would not be involved, since a fresh supply could be made up by following the instructions given in the literature." (Atomgewichtssilber is a pure silver used in the 1938 redetermination of atomic weights. Professor Hönigschmid, who had prepared this material and performed some of the determinations used by Jacobsen and Knudsen (1940), presented a sample of 100 g of this silver to the Danish Hydrographic Laboratory as a standard for future reference.)

Until recently, the chlorinity titration remained the preferred precision method for the characterization of sea water samples. Conventionally, one of several variations of the Mohr titration was used, the silver nitrate being standardized against I.A.P.S.O. Standard Sea Water. This is now provided by the I.A.P.S.O. Standard Sea Water Service, Brook Road, Wormley, Godalming, Surrey, England, and its chlorinity standardization is traceable to the sea waters originally measured in the 1937 and 1900 definitions of chlorinity. The methods currently used for the preparation and standardization of the sea water are summarized in section 6.2.3. The water is supplied in sealed 280 ml ampoules of special resistant glass; water stored in such ampoules has been shown to retain its chlorinity unchanged over periods of years.

The chlorinity values obtained in field investigations were sometimes reported as such in the literature, and sometimes converted to salinity or density before publication by means of the tables produced by Knudsen for the 1902 Report. These tables of course included, *inter alia*, Sørensen's conclusions on the relation of chlorinity to salinity. No world-wide standard of practice for reporting results was observed. For all this earlier work it is, however, a fairly safe assumption that, unless otherwise stated, any salinity reported in an oceanographic context has been derived from a chlorinity determination by the use of Knudsen's tables.

Starting in the mid fifties this "definition" of salinity came to be questioned increasingly. The reasons for this were threefold. Firstly, the development of conductimetric salinometers meant that the chlorinity determination was no longer the only standard method for determining the salinity of sea water. Secondly, new investigations of the gravimetric salinity determination (Guntz and Kocher, 1952; Morris and Riley, 1964) threw doubt on the absolute accuracy of Sørensen's method, although not on its precision (Caritt and Carpenter, 1959). Thirdly, the small number and non-representative distribution of the sea water samples used by Sørensen inevitably cast doubt on the validity for average ocean water of the Knudsen relationship between salinity and chlorinity.

Because of these factors a new investigation of the inter-relationships between the measured parameters (chlorinity, conductivity ratio, refractive index) and the derived parameters, of which the most important are salinity and specific gravity ($\sigma_t$), was necessary. A large number of carefully stored sea water samples was assembled and analysed for chemical composition (Culkin and Cox, 1966; Morris and Riley, 1966; Riley and Tongudai, 1967) and for chlorinity and conductivity ratio (Cox et al., 1967). Investigations of refractive index (Rusby, 1967) and specific gravity (Cox et al., 1970) were also carried out on this series of samples.

The data relating conductivity ratio to chlorinity were considered by an International Joint Panel consisting of representatives of several international oceanographic organizations. The meetings of this panel were sponsored by UNESCO, which published reports on the discussions (UNESCO, 1962, 1965, 1966) and also the tables which embody the practical salinity definition in terms of chlorinity. For historical reasons, connected with the inclusion of six Baltic samples among the nine sea waters used by Sørensen, the equation found by him does not make salinity quite proportional to chlorinity (i.e. $S = 0.03\%_0$ when $Cl = 0\%_0$). It was therefore decided (UNESCO, 1962) to replace Sørensen's relation by the truly proportional expression.

$$S\%_0 = 1.80655\ Cl\%_0$$

It should be noted that near normal ocean levels of salinity ($35\%_0$) this relationship is operationally identical to that of Sørensen. At $32\%_0$ and at $38\%_0$ the difference is $0.0026\%_0$ (Lyman, 1969). At low salinities the difference is greater ($0.025\%_0$ at $6\%_0$), but since salinity usually varies rapidly in situations where such low values occur this was not regarded as a significant objection to the change. The Panel therefore recommended that in future chlorinity determined by silver nitrate titration should be reported as chlorinity, and that the above equation should be used if it was felt necessary to derive from chlorinity an "estimate" of salinity (Charnock and Crease in UNESCO, 1965).

A decision of the Joint Panel also removed the chlorinity titration from the position which it had held for 65 years as the preferred method for the characterization of the salt content of sea water. This was of course a reflection of the increasing importance of the conductimetric "salinometer". Consideration was given to the possibility of recommending that the conductivity ratio (section 6.2.2) measured by these instruments be reported, rather than the salinity derived from it. It was, however, decided that salinity should be retained for historical reasons and because of its (qualitative) simplicity. It was felt that for "most oceanographers, ... biologists and geologists ... a salinity of $33.04\%_0$ will mean something; a conductivity ratio of $0.950$ will not".

From this decision it followed that a method must be agreed for obtaining values of salinity from the measured conductivity ratio. The experimental data obtained by Cox *et al.* (1967) was used to calculate a polynomial relating the salinity (estimated from chlorinity by the relationship previously agreed) to the measured conductivity ratio at 15°C ($R_{15}$). It was found necessary to discard all results from samples deeper than 200 m as deeper waters are slightly richer in calcium (section 6.3.1.4), and thus have a slightly higher conductivity ratio relative to their chlorinity. Since most deep samples had a salinity of about 34·8‰, a discontinuity occurred in the curve relating conductivity to chlorinity when these results were not excluded.

The polynomial thus obtained was slightly adjusted by the addition of a small constant term (0·00018‰) to make the conductivity ratio of water of 35‰ and 15°C exactly equal to unity. The resultant equation (Table 6.3) was then recommended as the new "definition of salinity", intended to replace the gravimetric definition of Forch *et al.* (1902). Because some salinometers do not incorporate a 15°C thermostat bath, a second polynomial was calculated to permit the conversion of conductivity ratios measured at other temperatures ($R_t$) to the corresponding values of $R_{15}$. This second equation was derived by measuring samples of known $R_{15}$ at eight temperatures using an inductive salinometer. This range of samples was made up by mixing English Channel water with Red Sea water or distilled water in various proportions; since Red Sea water is slightly depleted in calcium relative to open ocean water (section 6.3), this may have introduced a small error in the temperature coefficient at high salinities, but this is unlikely to be large enough to be a practical problem, since such high salinities are encountered rarely, if at all, outside the Red Sea.

These recommendations of the Joint Panel were summarized in a note in each of the major oceanographic journals (see e.g. Wooster *et al.*, 1969). It is unfortunate that both this note and the tables produced from the recommended polynomial (UNESCO, 1966) refer to the latter as a new "definition of salinity" (Lyman, 1969; Tsurikova and Tsurikov, 1971; Carpenter, 1972). The polynomial merely expresses salinity in terms of the ratio of its conductivity to that of a notional water of 35‰ salinity. Since this latter cannot itself be defined by conductivity ratio measurement, it must be referred to a chlorinity titration. The new polynomial is, therefore, a reformulation in terms of conductivity ratio of the Jacobsen and Knudsen (1940) definition of chlorinity. The ultimate standard, reproducible by workers at any place at any future time, was and remains a certain weight of pure silver, of the type presented by Professor Hönigschmid to the Copenhagen Laboratory in 1938.

It appears that the words "definition of salinity in terms of conductivity" were habitually used by members of the Joint Panel at a time when it was intended to relate the results to an absolute conductivity value for water of 35‰ salinity, directly derived from primary standards of length and electric current (UNESCO, 1965). Later, however, considerable practical difficulties were experienced in making this determination to the required accuracy (1 part in $10^5$). It was therefore decided to publish the polynomial, and tables derived from it, in order to provide a working arrangement agreed throughout the oceanographic community and compatible with any eventual definition of salinity in terms of absolute conductivity. The effort to establish accurate absolute conductivity values still continues (F. Culkin, personal communication); at present, the results of Reeburgh (1965) and Thomas et al. (1934), which are referred to simple salt solutions of known conductivity, are the best available. These results are precise to about one part in $10^4$, although for reasons given by Reeburgh the absolute accuracy may be poorer than this. When the means are available, it is intended that the absolute conductivity of each batch of I.A.P.S.O. Standard Water should be determined. This invaluable international working standard solution will then be characterized both in terms of the amp and metre and in terms of Atomgewichtssilber. This will remove the possibility of small errors in the standard arising from changes in the chlorinity to conductivity relationship caused by variations in composition from one batch to another. It is important to note that variation in the composition or pH of ocean samples may introduce small but significant uncertainties in the conductivity to density relationship (section 6.2.2), and that if density values are required to the highest precision this source of error must be controlled. This may be accomplished either by direct density determination (section 6.2.6) or by ensuring empirically by analysis that errors caused by composition variation are not significant in the system under consideration.

6.2.1. SALINITY MEASUREMENT

Salinometers designed for the examination of deep sea samples must be very accurate. Because of the comparatively small range of salinity found in the oceans and the strong influence of this parameter on the specific gravity and other important physical properties of sea water, the salinity must be known to an accuracy of one part in ten thousand (i.e. 0·003‰ in an overall salinity of about 35‰). Although for some purposes an even higher precision in specific gravity might be desired, it is doubtful if any improvement in the accuracy of salinity determination would be justified. This is because uncertainties in the relationship between salinity and specific gravity begin to become important at about the 0·01‰ level of accuracy. These uncertainties arise from pH differences and from variations in the major ion ratios (section 6.3).

The optimal instrumental characteristics for studies of the salinity of estuarine waters are very different from those required for oceanic samples. As Mangelsdorf (1967) has pointed out, in large areas of Chesapeake Bay a salinity precision of $\pm 0.003\%_0$ would be wasted unless the position were specified to 40 ft, the time to 90 s and the depth to 0.25 in! In this rapidly fluctuating situation the best strategy is to obtain many measurements quickly, if necessary at some sacrifice of precision. It is thus most important to consider the instruments available and to select the one which best suits the problem in hand.

In this section the characteristics and principles of operation of various salinity instruments in current general use are described. More complete information on the performance of commercially available instruments can be found in the excellent series of Instrument Fact Sheets, produced by and available from the National Oceanographic Instrumentation Centre, Washington 20310.

### 6.2.2. CONDUCTIVITY MEASUREMENTS

As mentioned in the previous section, salinity measurement by conductivity has almost completely superseded the chemical determination of chlorinity. This is because of the increased speed, reliability and simplicity of modern salinometers and the fact that they can be used at sea by relatively unskilled personnel. Many conductivity instruments are available; some are designed for use in the laboratory and others are intended primarily for field applications. A number of specialized designs have been developed for specific research topics, such as ocean microstructure studies (Gregg and Cox, 1971).

All laboratory salinometers consist of a cell which contains the sample, an a.c. transformer bridge, a temperature compensation system and some form of null indicator.

#### 6.2.2.1. Thermostat laboratory salinometers

Conductivity is a function of temperature as well as salinity. At $35\%_0$ the effect of a temperature variation of $0.001°C$ on conductivity is equivalent to a salinity change of $0.001\%_0$. The first precision laboratory salinometers, which appeared in the mid-fifties, used thermostat baths to hold the sample temperature constant. Since it is not easy to regulate such baths to $0.001°C$, an arrangement was adopted which allowed the sample cell resistance to be compared directly with that of a similar cell filled with standard sea water immersed in the same bath. The temperature coefficients of the sample and standard are in practice sufficiently well matched to permit the use of a relatively coarse bath temperature control ($\pm 0.1$°C).

Typically, the cell used in these instruments consists of a narrow vertical U-tube of glass with platinized platinum electrodes at each end. These cells must be filled carefully to avoid trapping air bubbles. Similarly, a low bath temperature (15°C) is normally used to minimize bubble formation through outgassing. The purpose of the electrodes is to connect the bridge electronics to the solution phase. In the former, electrical energy is transferred by electron movement in metal wires: in the latter the carriers are solvated ions. The electrodes thus act as electron/ion transducers. For precise measurement of conductivity an electrode should be reversible (Ives and Jantz, 1961), stable and immune from poisoning. The platinum electrode is not ideal in these respects. It is particularly susceptible to fouling by organic material; this necessitates frequent re-platinization of the electrode surface.

The sample must be allowed to come into thermal equilibrium before measurement; to save time a number of cells (6 to 11) are usually provided. Initially, the cells are all filled with I.A.P.S.O. standard and adjustments are made to compensate for the small differences in cell constant between cells. The test cells are then drained, rinsed and each is loaded with a different sample. After equilibration, the conductivity of the water in each cell is compared with that of I.A.P.S.O. Standard Sea Water contained in the standard cell. This comparison is made by switching each cell in turn into an a.c. bridge circuit of the transformer ratio type, which is then balanced. The use of the transformer ratio circuit reduces problems associated with small leakage currents and eliminates the high stability standard resistors required by earlier designs. Balance is indicated by a null meter or a magic eye. A more detailed description of the bridge designs used in these salinometers is given by Cox (1965).

It will be appreciated that these instruments are too large to be portable. However, they are fitted as standard equipment on many research vessels, even though more modern designs are superior in regard to size, weight and ruggedness. The results given by these thermostat salinometers are at least as accurate and reliable (one standard deviation of 0.003‰) as those obtainable from any alternative design. A precision of $\pm 0.001$‰, or better, has been claimed for a recent thermostat salinometer (NOIC, 1971a).

### 6.2.2.2. Inductive laboratory salinometers

Because of problems associated with the stability of the metal electrodes, most recent precision salinometers make use of inductive cells (Fig. 6.1). These devices consist of a transmitting and a receiving coil, side by side on the same axis within a toroidal sensing head. Shielding between the coils reduces the receiver coil pick up to very low levels when the sensing head is in air. When the head is immersed in a conductive medium, such as sea water, a conductivity loop through the centre tube of the toroid links both coils. The current induced in this loop by the transmitter (10 kHz), in turn induces a current in the receiver coil; the magnitude of the received signal is proportional to the conductivity of the sea water surrounding the head.

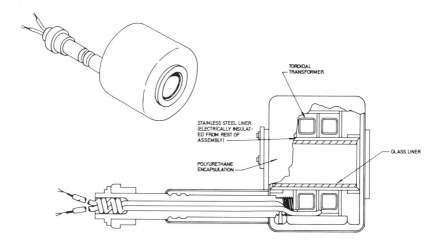

FIG. 6.1. Inductive salinity-sensing head designed for *in situ* measurement. (Redrawn from original supplied by The Plessey Company).

The first application of this principle to salinity measurement was for the *in situ* examination of coastal waters (Esterson and Pritchard, 1955); the immunity of the cell to "poisoning" by organic films and pollutant compounds gave it a special advantage in this application. It was clear, however, that the device had considerable promise as a high precision salinity sensor, and development followed rapidly with the production of both *in situ* and laboratory instruments.

A diagram of the type of conductivity cell used in recent inductive laboratory salinometers is shown in Fig. 6.2. None of these instruments are thermostatically controlled, instead temperature compensation is employed. The sensing toroid is contained with a chamber of acrylic plastic, together with the platinum resistance thermometer used for temperature measurement and as part of the compensation circuit. In use, sea water is drawn into the cell from the bottom by means of suction from an air pump. After rinsing, the cell is filled with a standard sea water which is stirred rapidly. The temperature is measured by switching the platinum resistance thermometer into a resistive bridge network. The temperature compensation network is then set to the appropriate value for this temperature (derived from a table). This network will then compensate for temperature variations of up to $\pm 3°C$ from the set value. The conductivity ratio* transformer bridge is set to the known conductivity ratio of the standard, and the indicator

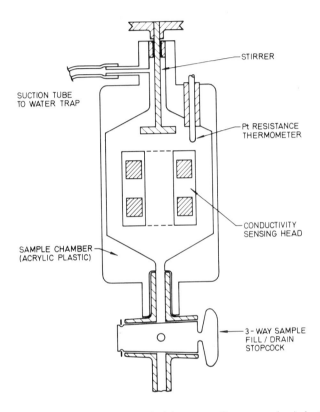

FIG. 6.2. Section through sample chamber of a laboratory salinometer using inductive sensing.

meter is nulled by adjustment of the standardization transformer. The adjustment of this transformer and of the temperature compensation switches is not changed for the rest of the sample series. The null reading of the conductivity ratio transformer dials for each sample is then found and noted, together with the sample temperature. A substandard of known salinity is

* The conductivity ratio of a sea water is defined as the ratio of its conductivity to that of water of exactly 35‰ salinity under the same conditions of pressure and temperature.

run at intervals of about 30 min so that the conductivity ratio readings may be corrected for drift. This procedure will give an accuracy adequate for most purposes; if the highest accuracy is required the procedure described by Van Landingham (1967) should be adopted. The calculation of salinity from the salinity ratio thus obtained is covered in section 6.2.3.

Instruments of this type are convenient and portable. All conform essentially to the original design described by Brown and Hamon (1961). This is still manufactured in Australia. More recent American versions are designed to higher standards of impact and vibration resistance, and as a result they are much more expensive. However, since an instrument in regular field use will inevitably be subject to rough handling and vibration, the greater expense is probably justified.

### 6.2.2.3. In situ *salinometers*

Accurate conductivity measurement *in situ* presents difficulties additional to those experienced in a laboratory environment. Conductivity is a function not only of salinity but also of temperature and pressure, and these two latter parameters must be monitored for compensation purposes. Since they measure salinity (or conductivity), temperature and depth *in situ* salinometers are usually described as STD and CTD probes. The former have an internal compensation circuit to produce a value for salinity directly from the three measured parameters. When the time constants of the sensors involved differ (Goulet and Culverhouse, 1972), this compensation circuit produces charac-teristic transient spikes on the data record if strong temperature gradients are present. In this situation it is better to employ a CTD probe, and to apply compensation for time constant differences during the calculation of salinity.

The most common type of *in situ* instrument requires a suspension cable which incorporates one or more conductor wires within the load-bearing outer sheath. At the upper end, on board ship, a means must be provided for terminating this cable on the rotating winch drum in such a way that con-nections may be made to the power supply and data recording systems. A slip-ring system, similar to the commutator of an a.c. motor, is normally used. Problems of noise generation at the sliding contact have led to the pro-duction of alternative designs using contact by mercury baths or systems of planetary gear wheels. Whatever system is used, great care must be taken to protect it from the normal deck environment, which is decidedly hostile to such devices. A proportion of STD faults are eventually traced to noise originating in the slip ring assembly. Attempts have been made to circum-vent this problem by using self-contained analogue or digital recording systems. Self-contained instruments so far available are not quite as accurate as surface recording designs (Table 6.2) and suffer from the drawback that

*Summary of the characteristics of some commercially available in situ salinity-temperature-depth systems*

| Manufacturer | Model | Parameters measured | Claimed accuracy | Range[a] From | to | Data transmission system[b] | Display and data storage system |
|---|---|---|---|---|---|---|---|
| Geodyne Division, E. G. and G. Waltham, Massachusetts 02154 | 775–21 | C T D | 0·01 0·005 0·3% FSD | 25 −2 0 | 65 35 330 | Digital,[b] conductor cable | Display octal format Binary. output to logger |
| Guildline Instruments Inc., Smith Falls, Ontario | 8010 | S T D | 0·04 0·02 0·2% FSD | 28 −2 | 40 30 | d.c. analogue voltage Multi-conductor cable | Analogue chart, logger output |
| Kieler Howaldtswerk, Kiel | Bathysonde[c] | C T | 0·02 0·02 | 20 −3 | 70 34 | FM telemetry single conductor cable | Analogue chart, logger output |
| Plessey Environmental Systems (formerly Bissett Berman Corp.), San Diego, California 92123 | 9030 | S T D | 0·05 0·05 0·25% FSD | 30 −2 0 to 0 | 40 35 500 6000 | None | No display. Records digital data on magnetic tape within probe |
|  | 9040 | S T D C | 0·02 0·02 0·25% FSD 0·03 | 30 −2 0 10 | 40 36 6000 60 | FM telemetry single conductor cable | Analogue chart, logger output |
|  | 9060 | S T D | 0·1 0·1 0·25% FSD | 30 −2 0 to 0 | 40 35 600 6000 | None | Analogue chart within probe |

[a] Range units are salinity (S) part per thousand; conductivity (C) millimhos cm$^{-1}$; temperature (T) degrees centigrade; depth (D) metres.
[b] Accuracy data supplied by manufacturers, except for Bathysonde (Seidler, 1963; Bonnet, 1971) and Geodyne 775–21 (NOIC, 1971b). Values are one standard deviation. Units as for range except where indicated.
[c] The Bathysonde described here is no longer in production, having been replaced by a more advanced design.

successful operation cannot be verified without recovering and opening the instrument package. However, such instruments are well suited to nearshore survey work from small vessels, where convenience and speed of operation are more useful than high accuracy, and a simple winch is all that is likely to be available.

An indication of the types of STD units available is given in Table 6.2. The list is not exhaustive since this is a time of rapid development and new instruments, and manufacturers, appear frequently. However, the table is intended to illustrate the present position with regard to availability and performance of equipment. It must be emphasized that most of the figures are taken from the manufacturers' literature, and probably represent the behaviour of the instrument under good conditions. In particular, to obtain these accuracies the calibration of the instruments must be continually monitored by independent means, as sudden jumps in the salinity calibration are not uncommon. These are probably caused by mechanical shock to the sensor head or to any dry solder joints which may exist in the probe electronics. They are normally of the order of one tenth of a part per thousand in salinity, and so may pass unnoticed unless a close watch is kept. It is usual to take at least one bottle sample with each STD cast, the bottle being attached to, or just above, the probe unit and triggered by messenger after equilibration of its reversing thermometers. A more convenient and less time-consuming calibration procedure involves the use of a multiple "rosette" sampler clamped to the probe unit. This device is triggered electrically from the surface when a sample is desired, and several such samples may be taken in succession on a cast. However, if a temperature calibration is desired it is still necessary to allow time for the reversing thermometers to reach equilibrium with the ambient temperature before triggering.

It will be appreciated therefore that present STD units, although very useful, are by no means ideal. Besides the problems of "spiking" and calibration jump already mentioned, the available units are still not as reliable as might be desired, and systems tend to be too complicated and expensive. Calibration and data reduction methods can also introduce error (section 6.2.7). At present, improved instruments are under development which it is hoped will approach the accuracy obtainable using the Nansen bottle and reversing thermometer, the classical tools of the oceanographer (Sverdrup et al., 1940).

Because STD systems produce large volumes of data, it is appropriate to mention the problems of handling these data. An analogue record is produced by most instruments as the measurements are made. This is essential to enable the operator to check that all systems are working. The paper charts produced are, however, an inconvenient way of storing data, and errors may be introduced when the chart readings are transcribed by hand. To avoid this, it is usual to convert the signals obtained from the underwater unit to digital form for storage on magnetic or punched paper tape; some workers have reported success in recording the FM signals directly using a good quality audio tape recorder and a reference frequency source to allow compensation for speed variation on playback (Morrison,

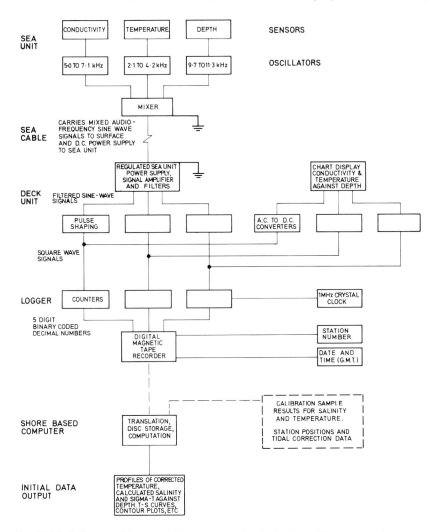

FIG. 6.3. Block diagram of data acquisition system used at the Institute of Oceanographic Sciences, Bidston.

1973). The main value of this latter system is as a back up, since the data must be digitized eventually for computation. There is usually no real advantage in postponing this step.

High precision (about 1 part in $10^5$) is necessary to avoid degradation of the data by the logging system. High sampling rates are also often required. The type of solution adopted by most workers is illustrated in Fig. 6.3. The counters are arranged to count the number of pulses of a high stability "clock" frequency for a period defined by a certain number of cycles of the appropriate FM signal. The period chosen depends on the precision required and the frequency of the FM signal, but is usually in the range 100 to 1000 Hz. If range expansion is required, this is obtained by converting the digital signal back to analogue form and displaying on additional analogue recorders. Once in digital form the signal is available for the calculation of specific gravity, $\sigma_t$ and, in the case of CTD systems, salinity. The procedures used for these calculations are mentioned in section 6.2.3.

### 6.2.3. CHLORINITY DETERMINATION

Chemical determination of halide content by titration with silver nitrate solution was for many years the usual routine method for determining the "sea salt" content of sea water samples. The results were expressed as chlorinity (section 6.2) and related to specific gravity and other important physical and chemical (section 6.3) variables by empirically derived relationships.

The most important modern application of the silver nitrate titration is in the preparation of Standard Sea Water at the Charlottenlund laboratory* of the International Association for Physical Sciences of the Ocean. A bulk sample of Atlantic water is collected near N 60°30·0′, W 22°20′ and returned to the laboratory for processing. After filtration through 0·22 μm membrane filters, the salinity of the water is adjusted by the addition of distilled water until it is close to 35‰. It is packaged in 280 ml ampoules; the final standardization is then performed on representative ampoules selected from the batch.

The method used (F. Hermann, personal communication) employs a weight burette technique for the measurement of both sample and silver nitrate. The concentration of silver nitrate solution is selected so that it is equivalent to an equal volume of the sea water. The weight burettes (Fig. 6.4) which are designed so that stopcock grease is not required, are then used to weigh out known amounts of sample and titrant, such that a very small excess of chloride remains, into a 300 ml conical flask of dark coloured glass. The burettes are rinsed into the flask with 10 ml of distilled water. After mixing for 30 min by magnetic stirrer, the flasks are left in the dark overnight. The

* This work will be transferred to Wormley, England from 1975 (see p. 368).

titration is then completed using weak silver nitrate solution, the end point being detected by the differential potentiometric method of Hermann (1951). The silver nitrate is standardized by comparison with previous batches of standard sea water, traceable to Knudsen's standards described in section 6.2.1 (Forch *et al.*, 1902; Jacobsen and Knudsen, 1940). The method forms a comparison system of exceptional precision. It has been shown to maintain its reproducibility when carried out by different workers in their own laboratories; Hermann and Culkin (personal communication) report a precision of $3.4 \times 10^{-4}$ parts per thousand in salinity for duplicate determinations carried out at Wormley and Charlottenlund, with no significant difference between the results of the two laboratories for any of the 7 different batches of standard sea water investigated.

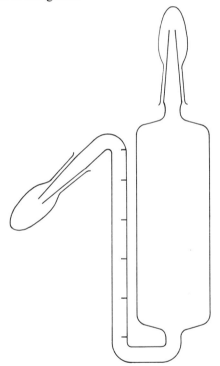

FIG. 6.4. Weight burette designed by Hermann for use in the standardization of I.A.P.S.O. Standard Sea Water.

A more conventional but less accurate technique developed by Knudsen from the method of Mohr is still of value for the measurement of chlorinity, since the capital cost of the equipment required is much less than that of more modern physical methods and it may be used at sea (for reviews of the methods available for the determination of chlorinity see Riley (1965a) and Johnston (1969)). A 15 ml Knudsen pipette is used to measure the sea water sample into the titration vessel. This pipette differs from the standard type in that, after filling, the volume of sample is defined by rotation of a 3-way stopcock fitted at the upper end. The device is thus simple to use at sea, although care must be taken not to allow stopcock grease to contaminate the inside surfaces of the pipette.

The titrant used in the Mohr method is silver nitrate

$$Cl^- + Ag^+ = AgCl_{(s)}.$$

The other halides present are similarly precipitated. Potassium chromate is added as an indicator so that, when the halides have been titrated to a low level at the end point, silver chromate is precipitated.

$$2\,Ag^+ + CrO_4^{2-} = \underset{red}{Ag_2CrO_{4(s)}}.$$

The colour change of the precipitate to brick red is not readily visible to those who are red-green colour blind. Other indicators which have been recommended include fluorescein (Miyake, 1939) and phenosafranine (Cunningham and Duedall, 1970).

The titrant is normally added from a Knudsen bulb burette. Again, this is designed so that it can be used for routine analysis at sea; the burette is filled by gravity from a large storage container, and the zero mark is defined by a three-way stopcock at the top, as in the Knudsen pipette. Since most open-ocean samples lie in a relatively small chlorinity range, the burette is designed so that much of its capacity is in the bulb. This allows the scale to be graduated in increments of 0·02 ml to improve precision. Chlorinities between 16‰ and 21‰ can be determined using silver nitrate solution of 36·75 g l$^{-1}$; outside this range the titrant strength must be adjusted to that the titre falls on scale.

Before titration commences, five drops of 10% potassium chromate solution are added to the sample. The silver nitrate is then added as a fine stream from the burette, with strong magnetic stirring to break up the silver chloride flocs. The endpoint is indicated by the precipitate turning pale brick-red for more than 30 s after the addition of the last drop increment of silver nitrate. The titrant is standardized by the titration of I.A.P.S.O. Standard Sea Water in place of the sample. This method is capable of a precision of ±0·01 Cl‰ in expert hands; since this corresponds to a burette reading precision of ±0·02 ml it will be appreciated that great care and frequent scrupulous cleaning of the burette and pipette are needed to attain this precision. If the samples are weighed into the titration beaker, precision may be improved still further; this expedient is limited to work in shore laboratories, since weighing is of course not practicable at sea.

A modification of the Knudsen titration has been suggested by Grasshoff and Wenck (1972). This uses a Metrohm incremental piston burette in place of the Knudsen burette; the absence of greased stopcocks from the system is stated to improve markedly the convenience and accuracy of shipboard analysis.

Potentiometric end point determination has been utilized by several workers to enhance the precision of the silver nitrate titration. As noted above, the method of Hermann (1951) is applied to the calibration of Standard Sea Water. Reeburgh and Carpenter (1964) used a differential electrochemical system for end point detection. The potential between two identical silver–silver chloride electrodes immersed in the titration vessel is measured. The system is brought near to the end point by titration of a weighed sample with silver nitrate using fluorescein indicator in the conventional manner. A small volume of solution about one of the electrodes is then prevented from mixing with the rest of the solution, although electrical contact is maintained. The addition of a further very small increment of titrant then causes a rise in the potential difference between the electrodes, since they are now part of a chloride concentration cell. The isolated solution is mixed into the main volume, a further small volume is isolated and the measurement is repeated with another small increment of titrant. The procedure is continued until the end point is passed, and the potential measured for each increment is plotted against volume added. The maximum potential difference change (about 15 mV) corresponds with the true end point. Reeburgh and Carpenter quote a precision of 1 part in 10 000 for their method. At ocean-water chlorinities, this corresponds to $0 \cdot 002\%_0$ in chlorinity, or about $0 \cdot 0035\%_0$ in salinity.

One of the major drawbacks of manual titration methods lies in the time taken per sample and the operator skill required. The method quoted above requires 15 min per sample, almost an order of magnitude more than the inductive salinometer. A semi-automatic method of chlorinity titration has been described (Jagner and Åren, 1970) which reduces the time per sample to 5 min while retaining high precision ($\pm 0 \cdot 004\%_0$ in chlorinity). This method uses a silver sensing electrode and a calomel reference electrode connected to the titration vessel by a salt bridge. The silver nitrate is delivered in small aliquots at 20 s intervals and the potential difference between the indicator electrodes at each addition is recorded in digital form on paper tape. A computer is used to calculate the end point by the method of Gran (1952). It seems probable that further development of this approach could reduce still further the time taken per sample; however the method is not suitable for use on board ship, since weighing of the sample is required.

6.2.4. OPTICAL METHODS

The salinity of a sea water sample may be estimated by comparison of its refractive index with that of a standard of known salinity. Since refractive index is more closely related to density than any of the other parameters used for salinity estimation (Mangelsdorf, 1967), there are theoretical grounds for believing that optical salinometers may eventually replace conductimetric devices as the preferred method for salinity estimation. Present refractometers of the Abbe and Pulfricht designs are however not sufficiently accurate for deep ocean studies, the limit of precision of these instruments being equivalent to a salinity uncertainty of about $0.05‰$ (Cox, 1965). Interferometers, in which comparison is made directly between sample and standard, are more suited to oceanographic studies, being theoretically capable of a resolution equivalent to $0.001‰$ or better. Because the temperature coefficient of refractive index varies only slightly with salinity it is not essential to control the temperature of the sea water cell; however, temperature gradients within the apparatus, and especially between the sample and standard, must be minimized by careful instrument design. Rusby (1967) has given an interesting account of the difficulties which must be overcome to obtain results of high accuracy. His results may be represented by the polynomial:

$$S = 35.00 + 5.3302 \times 10^3 \, \Delta n + 2.274 \times 10^5 \, \Delta n^2 + 3.9 \times 10^6 \, \Delta n^3$$

$$+ 10.59 \, \Delta n \, (t - 20) + 2.5 \times 10^2 \, \Delta n^2 \, (t - 20)$$

where $S$ is the salinity ($‰$), $t$ the temperature in deg C and $\Delta n$ the refractive index of the sample at $546.227$ nm minus that of a standard sea water of salinity $35‰$. (It should be noted that in the original paper, and in the UNESCO tables (second issue), the second term of this polynomial contains a typographical error.) The standard deviation of a single experimental point from this best-fit line is $0.0055‰$. Rusby attributes the scatter mainly to rinsing errors. Because of the physical design of the cells imposed by optical considerations, it was found difficult to eliminate cross-contamination between successive samples. This is also seen as the biggest potential difficulty in adapting this method for routine use in high precision salinity estimation, although Russian workers have used an interference refractometer (ITR-2) routinely at rather more modest precision levels (standard deviation $0.01‰$) (Vel'mozhnaya, 1960). No instrument has yet been designed to measure refractive index in situ. The results of Stanley (1971) show that, to be useful to oceanography, temperature and pressure must also be monitored, as these latter factors significantly influence the relationship between the refractive index and the specific gravity of sea water.

For very approximate work, a useful pocket instrument known as the temperature compensated Goldberg refractometer (American Optical, Keene, New Hampshire) has recently become available (Behrens, 1965, Manheim, 1968). The device is housed in a tube about 20 cm long. In use, a drop of the sample solution is placed on the face of a liquid-filled prism at one end of the tube, and a transparent cover hinged down to hold the drop in place. A scale is then viewed through an eyepiece at the other end of the tube. The boundary between the illuminated and dark sections of the scale gives the reading. Earlier models have a scale designed for use with sugar solutions; more recently a version has become available directly calibrated for use with salt solutions. In either case, a calibration graph must be constructed for maximum accuracy; the instrument is then capable of a precision of better than $\pm 0.5\%_0$ in the range between $0.5\%_0$ and $220\%_0$. The device is suitable for biological field work and has also been used for determination of the salinity of interstitial waters of sediments when only very small samples are available, as in the JOIDES deep drilling project (Manheim and Sayles, 1969). It should be noted that the temperature compensation built into this instrument is less effective at low temperatures (Behrens, 1965). If field measurements at temperatures below 15°C are contemplated, it may be more accurate to use the less expensive uncompensated Goldberg refractometer in conjunction with calibration graphs to correct for the temperature effect.

### 6.2.5. ELECTROCHEMICAL METHODS

Ion selective electrodes, developed within the last seven years, have aroused much interest in the field of analytical chemistry. In general, they are not well suited to direct measurement in sea water because of the complexity of the solution and the high precision normally required for meaningful measurements. (See Vol. 2, Chapter 20.) However, they have been found to be very useful in certain specialized applications (Garrels and Thompson, 1962). Mangelsdorf (1967) has discussed the general characteristics of these devices with special regard to their possible application to salinity measurement in estuaries. Brewer and Spencer (1970) used a chloride selective electrode to monitor chlorinity in their in situ device designed to measure deep sea anomalies in the fluoride to chlorinity ratio. They used a cell of the form

$$\text{Ag}|\text{AgCl} \quad \begin{array}{c} \text{Standard sea} \\ \text{water} \end{array} \quad \overset{\text{F}^-}{\underset{\phantom{.}}{\Big|}} \quad \begin{array}{c} \text{Sample} \\ \text{sea water} \end{array} \quad \overset{\text{Cl}^-}{\underset{\phantom{.}}{\Big|}} \quad \begin{array}{c} \text{Standard} \\ \text{sea water} \end{array} \quad \text{AgCl}|\text{Ag}$$

where fluoride and chloride specific membranes are represented by the

o

vertical dashed lines. It will be seen that, as long as the cell is isothermal and the $a_{F^-} : a_{Cl^-}$ ratio in the sample equals that of the standard, a zero output will result. If the ratio $(Q)$ in the sample chamber changes, a voltage will appear across the cells. This is caused by concentration-driven diffusion of either or both of the ions of interest through their appropriate membranes, such that

$$\Delta V = \Delta V_{F^-} + \Delta V_{Cl^-} \simeq 2 \cdot 303 \frac{RT}{F} \log \frac{Q}{Q_{\text{stand.}}}$$

where $R$ is the gas constant, $T$ the absolute temperature and $F$ the Faraday constant. Although the chloride electrode senses chloride ion activity rather than chlorinity, in the present case the gain in operational simplicity more than justifies the error thus introduced.

A similar design of cell has been in use for some years for the purpose of direct electrochemical salinity measurements (Koske, 1964; Gieskes, 1967). Devices using this type of cell are known as membrane salinometers and their theory has been discussed by Gieskes (1967). The membranes used are anion and cation selective, respectively.

|  | | A$^-$ | | C$^+$ | | |
|---|---|---|---|---|---|---|
| Ag\|AgCl | Standard sea water salinity $S_2$ | | Sample sea water salinity $S_1$ | | Standard sea water salinity $S_2$ | AgCl\|Ag |

The output voltage from the above cell is given by

$$\Delta V = C \frac{2RT}{F} \ln \frac{S_1}{S_2}$$

The empirical constant $C$ is determined by such factors as the permselectivity of the membranes, the use of salinity rather than ion activity, and the fact that sea water is a mixture of monovalent and divalent species. For a typical sensor of this type, $C$ has a value between 0·7 and 0·8, so that the sensitivity is about 90 mV output change for a factor of ten change in salinity. The membrane sensor has a relatively low impedance (about 5 to 10 k$\Omega$) so that the cell output may be monitored down to submillivolt levels with fairly simple equipment.

Under completely isothermal laboratory conditions the membrane sensor has been routinely used to detect salinity changes of less than 0·0001‰ corresponding to an output change of 0·1 μV (Mangelsdorf and Wilson, 1971). However, the accuracy of instruments without temperature control is about 1% of the salinity reading; the precision of replicates run as a batch is about 0·1%. Since the cell has a logarithmic output characteristic, absolute accuracy is higher at low salinities. This, and the simplicity and low cost of the device, make it suitable for use in routine estuarine studies. A practical design for a portable instrument has been described (Wilson, 1971). With a little modification, the basic sensor may be used to measure the salinity of sediment interstitial water *in situ* (Sanders *et al.*, 1965) or to obtain water column salinity profiles from a submersible instrument package (Gieskes, 1968).

### 6.2.6. DIRECT SPECIFIC GRAVITY DETERMINATION

Although sigma-$t$ is usually referred to as a measure of the "density" of sea water it is in fact defined (Knudsen, 1901) by

$$\sigma_t = (S_t - 1)\,1000$$

where $S_t$ is the specific gravity of sea water at $t°C$, relative to distilled water at $4°C$. Since the density of distilled water at $4°C$ is about 3 parts in $10^5$ below unity, the numerical value of the absolute density of a sea water sample is not quite equal to its specific gravity (Cox *et al.*, 1968). In addition, density has units of g cm$^{-3}$ whereas specific gravity is a dimensionless ratio.

Cox *et al.* (1968) have pointed out that the distilled water used by Forch *et al.* (1902) in their specific gravity determinations was of unknown isotopic composition. It was not appreciated at that time that the process of partial distillation may significantly alter the density of water by isotopic fractionation. Consequently, note was not made of the distillation procedure used, and it is not now possible to reproduce the standard used by Forch *et al.* As noted in section 2.1, the growing use of conductimetric salinometers has created a need for precise data on the relationship between relative conductivity and specific gravity. Cox *et al.* (1970) have described the apparatus which was designed for this investigation, the experimental procedures used and the results obtained. Their method depended on measuring the upthrust on a silica sinker suspended in the sample. Considerable care in design and experimental skill were necessary to obtain accurate measurements ($\pm 0.008$ in sigma-$t$). In general, the results were found to agree well with those of Forch *et al.*, whose data are, on average, systematically lower by $0.006$ in sigma-$t$ over the salinity range of normal sea water. Below $15\%_0$ salinity the deviation became larger. The reason for this is uncertain, although differences in chemical composition between the low salinity samples used by the two sets of investigators may have contributed.

The high precision method used by Cox *et al.*, is not suitable for routine work because of its complexity, and because a weighing under isothermal conditions is necessary. Hence until recently, routine direct determinations of specific gravity were not possible, except for approximate measurements using hydro-meters. The development of a new density comparison instrument (Kratky *et al.*, 1969) has changed this situation. This device uses a thermostatically controlled chamber containing a mechanical oscillator assembly. The vibra-ting element is a 2 mm diam. glass tube clamped at two nodes of oscillation. The volume of solution under test is contained within the tube. The period (T) of the system is given by

$$T = 2\pi \sqrt{\frac{Vd + M}{C}}$$

where $d$ is the density of the sample and $V$ its volume, $M$ is the mass of the vibrating system without sample and $C$ is a constant for the system. Thus, a sample may be compared with a known standard by measuring the frequencies of the oscillator loaded with each in turn

$$d_1 - d_2 = K(T_1^2 - T_2^2)$$

The apparatus constant $K$ must be found by measurement of two known standards at each working temperature.

Stability of temperature control is essential if the full accuracy of the instrument is to be realized. In addition to the temperature effect on the specific gravity of the sample, a variation of $0.01°C$ introduces an instru-mental error equivalent to $1.5 \times 10^{-6}$ in specific gravity. This is comparable with the overall precision of the instrument.

A rather confused situation has occurred over the reporting of results from this instrument. Initial results for low salinity Baltic samples (Kremling, 1971) were calculated as densities. The comparison, made in Kremling's paper, with specific gravities from Knudsen's tables is thus invalid. In addition, the tables were entered with a chlorinity value obtained from salinity (measured with a conductimetric salinometer) by means of the Knudsen relation $Cl\%_0 = (S\%_0 - 0.03)/1.805$. Crease (1971) has pointed out that such salinity values should have been converted to chlorinity using the relationship $Cl\%_0 = (S\%_0)/1.80655$ (Section 6.2.1). At low salinities this relationship estimates a higher chlorinity for a given salinity than does the older Knudsen formula. (Apparently because of a misprint, Crease (1971) states the converse of this, but the correct sense is implicit in his calculation.) Therefore, before being compared to Knudsen's tables, the results of Kremling (1971) should be corrected to specific gravity, and the additional correction proposed by Crease must also be applied. When this is done, Kremling's results become consistent with those of Cox *et al.* Both then suggest that Knudsen's tables are low for low salinities (about $0.025$ in sigma-$t$ at $5\%_0$). This agreement was later confirmed by actual determinations on the same sample by both sets of workers (Kremling, 1972).

The density comparison instrument has also been used to determine relationships between specific gravity and salinity and temperature over the salinity range 9‰ to 39‰ (Kremling, 1972). The results obtained agreed with those of Cox *et al.* (1970) in showing that Knudsen's tables are slightly low in specific gravity in this range also, the average difference of 0·013 in sigma-*t* being very slightly larger than that found by Cox *et al.*

The direct density comparison instrument used by Kremling could be applied to routine hydrographic survey work, although it appears to be rather slower in operation than the modern salinometer. It would be necessary to measure also *in situ* temperature of the sample in order to calculate its *in situ* specific gravity. It is doubtful if any advantage would be gained over the conductimetric method, unless sample composition differences sufficient to introduce excessive error in salinity-derived specific gravity values were expected. Attempts have been made (Kuenzler, 1968) to apply a similar principle to an *in situ* instrument, but effective compensation for the effects of temperature on the system (the factor K mentioned above) has apparently proved difficult. There is no doubt that such an instrument would find useful application, particularly in coastal studies and in the Baltic, where the results of Kremling (1971) have shown that the normal conductivity to sigma-*t* relationship may not hold for a small percentage of samples.

6.2.7. THE RELATIONSHIP BETWEEN SALINITY AND MEASURED PARAMETERS

These relationships, which are the basic tools of all practical oceanographers, are summarized in Table 6.3. The equations represent the experimental results of the workers mentioned and are valid for at least the oceanic range ($0° < t°C < 25°$; $30 < S‰ < 40$), except where otherwise indicated; most have become available only within the last five years, and replace older less accurate data. The original papers quoted should be consulted for detailed information on the range of validity of each polynomial and the accuracy with which it fits the experimental data. Certain of these papers (Knudsen, 1901; UNESCO, 1968) also contain extensive tables based on the relationships quoted; these allow the derived parameters to be calculated when a computer is not available.

A situation of some complexity has arisen in respect of the calculation of salinity from the measurements of *in situ* CTD probes. The conductivity value given by these instruments may be represented by $C_{s,t,p}$ where $S$, $t$ and $p$ are the *in situ* values of salinity, temperature and pressure respectively. It is important to realize that, although $C_{s,t,p}$ may be assigned units of

TABLE 6.3

*Relations between salinity ($S‰$) and various measured parameters*

| Relation | Source |
|---|---|
| 1. Chlorinity ($g\,kg^{-1}$)    See text for definition<br>$S = 1{\cdot}80655\,Cl$ | Wooster *et al.* (1969) |
| $S = 1{\cdot}805\,Cl + 0{\cdot}03$   (now obsolete)<br>Chlorosity at $t°C$ ($g\,dm^{-3}$) $= Cl \times s_t$ (see 4 below) | Knudsen (1901) |
| 2. Conductivity, relative to sea water of 35‰ salinity<br>(i) $10° < t°C < 30°$. $p = 0db$<br>$S = -0{\cdot}08996 + 28{\cdot}2970\,R_{15} + 12{\cdot}80832\,R_{15}^2 - 10{\cdot}67869\,R_{15}^3 + 5{\cdot}98624\,R_{15}^4 - 1{\cdot}32311\,R_{15}^5$<br>where $R_{15} = R_t + R_t(R_t - 1)(t - 15)[96{\cdot}7 - 72R_t + 37{\cdot}3R_t^2 - (0{\cdot}63 + 0{\cdot}21R_t^2)(t - 15)]\,10^{-5}$ | Cox *et al.* (1967)<br>and UNESCO (1968) |
| (ii) $-2° < t°C < 20°$. $p = 0db$<br>$S = -0{\cdot}5933 + 32{\cdot}4822R_0 + 3{\cdot}1106R^2 + 0{\cdot}004\sin\left(\dfrac{(R_0 - 0{\cdot}64)2\pi}{0{\cdot}59}\right)$<br>where $R_0 = R_t - \left[6 + 380\sin\left(\dfrac{(R_t + 0{\cdot}04)\pi}{1{\cdot}03}\right) + 15\sin\left(\dfrac{(R + 0{\cdot}04)3\pi}{1{\cdot}63}\right)\right][0{\cdot}0777t - 0{\cdot}000454t^2 - 0{\cdot}000018t^3]\,10^{-5}$ | Perkin and Walker (1972) |

(iii) correction from ambient pressure $p < 10^4\ db$ to $p = 0\,db$

$R_t = [g(t)f(p) + h(p)j(t)][1 + l(t)m(S)]\,10^{-2} + 1$

where $g(t) = 1.5192 - 4.5302 \times 10^{-2}\,t + 8.3089 \times 10^{-4}t^2 - 7.900 \times 10^{-6}\,t^3$

$\qquad f(p) = 1.04200 \times 10^3\,p - 3.3913 \times 10^{-8}\,p^2 + 3.300 \times 10^{-13}\,p^3$

$\qquad h(p) = 4 \times 10^{-4} + 2.577 \times 10^{-5}\,p - 2.492 \times 10^{-9}\,p^2$

$\qquad j(t) = 1.000 - 1.535 \times 10^{-1}\,t + 8.276 \times 10^{-3}\,t^2 - 1.657 \times 10^{-4}\,t^3$

$\qquad l(t) = 6.950 \times 10^{-3} - 7.6 \times 10^{-5}\,T \qquad M(S) = 35 - S$

Bradshaw and Schleicher (1965)

3. Refractive index ($n$)   $\Delta n = n_s - n_{35}$ (at 546.227 nm)

$S = 35.000 + 5.3302 \times 10^3 \Delta n + 2.274 \times 10^5 \Delta n^2 + 3.9 \times 10^6 \Delta n^3$
$\qquad + 10.59 \Delta n(t - 20) + 2.5 \times 10^2 \Delta n^2(t - 20)$

Rusby (1967) and UNESCO (1968)

4. Specific gravity ($s_t$) $9 < S < 41\%_0$ $\sigma_t = (s_t - 1)10^3$

(i) $\sigma_t = 8.009691 \times 10^2 + 5.88194 \times 10^{-2}\,t + 0.7901864\,S$
$\qquad - 8.114654 \times 10^{-3}\,t^2 - 3.253104 \times 10^{-3}\,St + 1.31708 \times 10^{-4}\,S^2$
$\qquad + 4.76004 \times 10^{-5}\,t^3 + 3.892875 \times 10^{-5}\,St^2$
$\qquad + 2.879715 \times 10^{-6}\,S^2t - 6.118315 \times 10^{-8}\,S^3$

Cox et al. (1970)

(ii) $\sigma_t = (\sigma_0 + 0.1324)[1 - A_t + B_t(\sigma_0 - 0.1324)] + \Sigma_t$

where $\sigma_0 = -6.9 \times 10^{-2} + 1.4708\ Cl - 1.570 \times 10^{-3}\ Cl^2 + 3.98 \times 10^{-5}Cl$
$\qquad A_t = 4.7867 \times 10^{-3}\,t - 9.8185 \times 10^{-5}\,t^2 + 1.0843 \times 10^{-6}\,t^3$
$\qquad B_t = 1.803 \times 10^{-5}\,t - 8.146 \times 10^{-7}\,t^2 + 1.667 \times 10^{-8}\,t^3$
$\qquad \Sigma_t = -(t - 3.98)^2\,(t + 283)\,[503.57(t + 67.26)]^{-1}$

Knudsen (1901)

$m\Omega^{-1}$ cm$^{-1}$, it is not in fact an absolute conductivity value. Before commencing a calculation, it is most important to ascertain the conditions of salinity, temperature and pressure under which the unit was calibrated, and also the value of absolute conductivity which was assumed for these conditions in setting up the instrument. Since this value is not always explicitly stated, confusion on this point can lead to significant error. For example, the values effectively assigned by various workers to $C_{35, 15, 0}$ have been shown by Walker and Chapman (1973) to vary from 42·929 $m\Omega^{-1}$ cm$^{-1}$ to 42·698 $m\Omega^{-1}$ cm$^{-1}$, a spread of half a percent, or about ten times the stated accuracy of the CTD systems employed. At least one manufacturer routinely calibrates instruments on the arbitrary assumption that $C_{35, 20, 0} = 50$, a value which is over 4% higher than that experimentally determined. Thus, it is most important that quoted *in situ* conductivity results should include information on the conditions used to calibrate the instrument, and on the absolute conductivity values used for these calculations.

Further possible error may rise from the equation used to calculate salinity from the conductivity-temperature-depth data. Although only three of these equations have appeared in the open literature (Accerboni and Mosetti, 1967; Fedorov, 1971; Perkin and Walker, 1972), many others have appeared in unpublished technical reports and similar publications of limited circulation (e.g. Ribe and Howe, 1967; Rhode, 1968; Zaburdaev et al., 1969). Greenberg (1972) and Walker and Chapman (1973) have recently compared a number of these equations. All of the eight studied derived originally from the experimental data of Thomas et al. (1934), Brown and Allentoft (1966) or Cox et al. (1967), either singly or in combination. The data of Thomas et al. has been criticized (e.g. by Cox, 1963) while that of Cox et al. (1967) does not extend below 10°C, although work is in hand to extend the data to 0°C (F. Culkin, personal communication). The most reliable relationship available for the computation of pressure corrections is that of Bradshaw and Schleicher (1965); for shallow work a linear relation based on their data is adequate. The intercomparison showed clearly that most of the equations agreed satisfactorily ($\pm 0.02\%$) over salinity and temperature ranges encountered in the open ocean. The differences increased to several times this value at higher temperatures, salinities and pressures, and at very low temperatures. Certain of the relationships differed consistently from the mean result for all, and thus are probably not reliable. Those with a stated limited range of validity showed a marked deterioration in accuracy outside this range.

This situation is obviously not satisfactory in the long term. Errors may be introduced by non-standardized calibration procedures and by differences in calculation, as well as by instrumental inaccuracy. Therefore, salinity results obtained from present CTD and STD probes must be regarded as most unreliable by normal oceanographic standards, unless the instrument is calibrated from measurements on a sample taken during the cast. The latter procedure allows a correction to be applied to bring the results from the *in situ* probe into agreement with the International Oceanographic Tables. This empirical correction compensates for the three sources of error mentioned above. It will not be possible to discard this precaution until instrumental stability is greatly improved (section 6.2.2), until reliable absolute conductivity values are available and until the validity of the relationship used to calculate the International Tables is extended below 10°C. Until this Utopian situation is attained, it would be helpful if there could be a somewhat greater degree of agreement on the empirical relationships to be used in data reduction.

## 6.3. MAJOR ELEMENTS

The major elements of sea water are listed in Table 6.1 together with their concentrations at a salinity of 35‰. This section reviews the findings of recent studies of sea water composition; although reference is made to the details of analytical techniques when necessary, the reader requiring a more complete description and discussion of the analytical techniques employed should refer to Vol 2, Chapters 19 and 20.

The concept of "conservatism" or "constancy of composition", by which the elemental composition of sea salt is regarded as constant throughout the world ocean, is an approximation which must be assumed to be valid if salinity is to be a useful parameter. If it were not operationally valid, the relation between salinity and specific gravity at constant temperature would be indeterminate, and direct determination of the specific gravity of all ocean samples would be necessary. The concept is associated with the name of Dittmar, presumably because his analyses of 77 samples collected by Buchanan on the Challenger Expedition laid the foundations for the work of Forch *et al.*, discussed in section 6.2 above. Reference to Dittmar's report (1884) shows that he summarized his results in the following way: "... the percentages of the several components are subject to only slight variations. Apart from the one success with the lime, I was not able to trace back the fluctuations to natural causes." Because of this finding he suggested that further studies, with improved analytical precision, would be useful.

It was to be another 80 years before the suggestion was carried into practice. This is a tribute to the technical difficulties of improving on Dittmar's careful work, and perhaps also to the hypnotic effect of the twin concepts of salinity and constant composition. It required the advent of the conducti-

metric salinometer with its improved precision in routine use, and the
attendant re-evaluation of the meaning of salinity described in section 6.2, to
prompt a further study of these ideas. This is not to say that analyses of major
elements were not performed; Culkin (1965) has summarized a fairly extensive body of data on individual elements. However, systematic studies on
the validity of these basic concepts, consisting of a complete analysis of many
sea water samples, have been carried out only within the last decade (Culkin
and Cox, 1966; Morris and Riley, 1966; Riley and Tongudai, 1967). The
results of these studies and other recent work are reviewed in the following
sections.

### 6.3.1. MAJOR CATIONS

#### 6.3.1.1. Sodium

The results of Culkin and Cox (1966) agree with the consensus of previous
results (Culkin, 1965) that the mean ratio of sodium (g kg$^{-1}$) to chlorinity
(parts per thousand, as defined in section 6.2) is 0·5555. Determinations of
analytical error by analysis of replicates gave a value of 0·0007 for one
standard deviation. Sodium was calculated by difference after the determination of calcium, magnesium, potassium and total cations. Riley and Tongudai
(1967) used a technique in which total alkali metals were determined gravimetrically as sulphates and the potassium (determined gravimetrically using
tetraphenyl boron) was subtracted to give a value for sodium. The value
found by this means was slightly higher at 0·5567; the precision of measurement was similar. Both sets of workers observed no significant variation with
depth or geographical position. This latter finding is supported by the results
of Mangelsdorf and Wilson (1972, and additional unpublished results);
over 900 samples, mainly from the Atlantic and Pacific oceans (all depths)
were compared to Sargasso Sea surface water using the method of difference
chromatography (see Vol. 2, Chapter 19). No variations in the sodium difference signal greater than the analytical noise level (ca. ±0·02% for this
method) were observed. It does not therefore seem that variations in this
cation are likely to contribute significant error to hydrographic measurements.

#### 6.3.1.2. Potassium

The gravimetric determination of potassium as the tetraphenylborate was
used by both Culkin and Cox and by Riley and Tongudai. Both sets of
workers found a potassium to chlorinity ratio of 0·0206 with standard
deviations of replicates of ±1% and ±0·8% respectively. No variation with
depth or position was found. Although the difference chromatographic

technique affords a rather higher precision for potassium ($\pm 0.12\%$) no significant regional anomalies were detected during analysis of the suite of 900 samples mentioned above. However, when the technique was optimized for potassium variations and repeated analyses carried out, reproducible signals corresponding to potassium variations were observed for a Sargasso Sea water column; the total spread of these apparent variations was always less than $0.1\%$ at this station. Such minute variations, which may arise through biological transport processes, are totally insignificant in hydrographic terms.

Fluxes through the sediment–water interface may be of great importance to the geochemical balance of a particular chemical species. Mangelsdorf et al. (1969) showed that the published values for the potassium content of interstitial water implied a supply of potassium to the world ocean, which, even on conservative estimates, would be four times greater than the supply due to river inflow. No known removal mechanism was capable of dealing with inputs of this magnitude, although geological evidence suggests that the potassium content of the oceans is not increasing markedly with time (Holland, 1972; see also Chapter 5). This conflict was resolved by the observation that spurious enrichments could occur as a result of unavoidable temperature rises during the sampling process, and evidence of such effects was later also provided by Bischoff et al. (1970). Improved methods of sampling (Wilson et al., 1972) have been used, in conjunction with difference chromatography, to produced more reliable gradient values for North Atlantic sediments (Sayles et al., 1973). The results obtained showed average depletions of potassium concentration relative to the other cations of about $1\%$ per 100 cm. This implies a diffusive transport of potassium from the sea to the sediments which is of the order required to balance an appreciable fraction of the worldwide river input. Further study will be needed to delineate the precise mechanism of this geochemical process, to define its geographical extent, and to study the sediment-water interactions of other major ions (see Chapter 33).

The sea is obviously very well mixed indeed with respect to the major alkali metals. This finding supports the deduction from mass balance considerations that the residence time of these cations is very considerably in excess of the mixing time of the oceans (Chapter 5).

### 6.3.1.3. Magnesium

A certain amount of uncertainty exists about the concentration of magnesium in sea water. The magnesium/chlorinity ratios of Culkin and Cox (0·06692 ± 0·00004) do not quite agree with those of Riley and Tongudai

(0·0667 ± 0·00007) or Tsunogai *et al.* (1968) (0·06676 ± 0·00006). Recently
Carpenter and Manella (1973) have suggested that the titration endpoint
used in the method of Culkin and Cox introduces a systematic overestimate
of about 1% and that matrix differences between standards and samples
introduce an error of 0·57% in the work of Riley and Tongudai. Correction
for these effects reduces the results of both sets of workers to 0·06629, close
to Carpenter and Manella's own results of 0·006626. Jagner (1967) reports a
still lower value of (0·06612 ± 0·00005). Another factor contributing to the
spread of literature values, which is large compared to the precision of the
methods used, is the difficulty of preparing accurate standard solutions of
magnesium. Jagner and Åren (1971) report a variation of 0·6% in results
obtained for total alkaline earth ions in sea water when magnesium standards
prepared from turnings, ribbons and ingots were compared. Since their
analytical method has a quoted precision about sixty times better than this,
virtually all of this variation must be due to uncertainties in the magnesium
concentrations of the standards employed. This probably arises primarily
from the presence of various amounts of oxide on the magnesium metal
samples used. The uncertainty in sea water composition so produced is
equivalent to $10^{19}$ g of magnesium in the world ocean, and is just sufficient
to be significant in studies on the colligative properties of sea water (Carpenter,
1972).

As might be expected, the degree of natural variation of the magnesium to
chlorinity ratio is much more closely defined than its absolute value. Culkin
and Cox (1966) and Riley and Tongudai (1967) both observed a slightly
larger variation (ca. ±0·1%) than would have been expected from the
precision of the analytical method used (ca. ±0·05%, found by repeated
analysis of the same sample). It was not possible to correlate the variations
observed with either depth or position, although variability is less in deeper
water. Similar results have been observed using difference chromatography.
When this latter method of analysis is used a precision of ±0·06% can be
obtained for the magnesium to total cation ratio. Results for 880 samples
from the Atlantic and Pacific oceans show a standard deviation about the
mean of rather less than ±0·09%. There is some tendency for extreme values
to occur at adjacent sampling depths, which tends to support the significance
of the observations, since the samples were deliberately analysed in random
order at each station. The small size of the variations and their distribution
pattern are in agreement with the results of Culkin and Cox (1966) and Riley
and Tongudai (1967). Forty samples from the Red Sea appeared to show
a mean magnesium content about 0·1% higher than that of the Atlantic
Atlantic and Pacific samples analysed, but no other systematic geographical
change in the Mg : Cl ratio was noted in this study. Lyakhin (1971) has reported

an apparent reduction of about 1 % close to the east coast of South America. Billings *et al.* (1969) reported a 10% variation in the magnesium to chlorinity ratio with time at a station near Bermuda. However, this result is in conflict with a large mass of experimental evidence in addition to that cited above; for instance variations of this nature would be expected to produce scatter in the relationship between salinity and density much greater than has been in fact observed. Analyses of sea water samples taken in the same area by Manella (Carpenter, 1971) agree with similar studies by the author who found no variation larger than 0·1 %. It seems likely that the scatter in the results found by Billings *et al.* (1969) arise from unknown analytical error rather than from variations in the actual sea water composition. The atomic absorption method used by these workers is not well suited to the high precision analysis of the major cations of sea water, as matrix interferences occur (Carr, 1970), and variations in the flame characteristics are difficult to control. Other recent reports of relatively large variations in major ion ratios have also originated from atomic absorption analyses (Angino *et al.*, 1966; Billings and Harriss, 1965). For the reasons noted, major cation values for sea water obtained using this technique should be treated with the deepest reserve until confirmed by analysis of the same samples by more reliable methods.

### 6.3.1.4. *Calcium*

This element has a relatively high variability due to its involvement with the biosphere and with the carbonate system in the oceans. Several calcium-containing carbonate solid phases are close to equilibrium with ocean water. The mineral forms which appear to be most important in the carbonate system are calcite and aragonite, polymorphic forms of $CaCO_3$. Dolomite $(CaMgCO_3)$ does not seem to participate in this system in spite of the fact that thermodynamic data predict that solutions of oceanic composition are supersaturated with respect to this mineral (Cloud, 1965). It is likely that calcium, unlike all other major ions with the possible exception of strontium, is regulated by the solution and precipitation of a solid phase (Goldberg, 1965); the behaviour of this element is therefore of considerable interest.

The average value of the calcium to chlorinity ratio is a matter of close agreement between various groups of workers. Culkin and Cox (1966) and Riley and Tongudai (1967) obtained values of 0·02126 and 0·02128 respectively, with a standard deviation of about 0·25%. They were able to demonstrate statistically that in deeper waters the calcium concentration is higher than that in surface waters by about 0·5%, but because their samples were relatively few and widely separated they were not able to show significant geographical variation. Tsunogai *et al.* (1968), by concentrating on a small

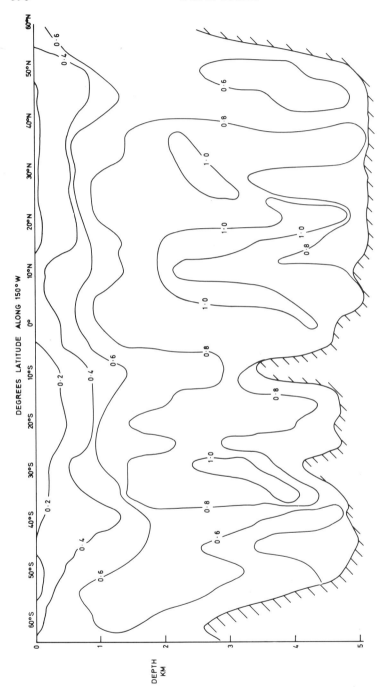

FIG. 6.5. Profile of $Ca^{2+}$ variations along longitude $170°W$, expressed as percentage differences from Sargasso Sea water.

geographical area in the Sea of Japan, were able to present for the first time information on the variation of calcium concentration between adjacent water masses and with depth in a number of water columns. The average value for the calcium to chlorinity ratio found by Tsunogai et al. was 0·02131, with a standard deviation of replicates of 0·1% and a difference between surface and deep samples of about 0·5%.

Using the difference chromatographic procedure it is possible to detect variations in the ratio of calcium to total cations of about 0·15%. Of the suite of over 900 samples collected by Professor Wangersky and co-workers during the Hudston-70 cruise, some 345 were taken in the Pacific along meridian 150°W. The percentage enrichments, relative to Sargasso sea surface water, found in these samples are shown in Fig. 6.5. These results are in broad agreement with those of previous studies and with the calculation of Li et al. (1969) which showed an enrichment of calcium in deep water relative to that in surface waters of 0·8%. Tsunogai et al. (1973) have reported calcium results for a section along 170°W which are in very reasonable agreement with Fig. 6.5. They have also compared these results to calcium values calculated from published alkalinity data. According to Wattenberg's relationship (Morcos, 1970), at constant salinity, $\Delta$ Carbonate Alkalinity $= 2\Delta[Ca^{2+}]$. The assumption is that the alkalinity variation is caused by dissolution of calcium carbonate (see Chapter 9). Tsunogai et al. were able to confirm the findings of Moberg and Revelle (1937) that Wattenberg's relationship is broadly true for the Pacific Ocean. Observed small deviations from the behaviour predicted by this relationship may be due to a contribution of calcium from sources other than carbonates, such as detrital silicate minerals. Studies made on the Bahama Banks, where calcium carbonate precipitation from sea water occurs, have shown that the calcium anomaly calculated from the Wattenberg relationship agrees satisfactorily with that found experimentally (Traganza and Szabo, 1967). The accuracy of calculation was limited to $\pm 1\%$ by the accuracy of the available alkalinity data.

Analysis of samples from the Red Sea by difference chromatography shows that waters leaving this area are depleted by 0·3% in calcium relative to inflowing waters. This value is in good agreement with that calculated by consideration of the rate of deposition of carbonates in the Red Sea (J. Milliman, personal communication) and the published estimates of water exchange with the Indian Ocean at Bab el Mandeb (see Section 15.5).

### 6.3.1.5. Strontium

The behaviour of strontium in sea water has received considerable attention in recent years because of the usefulness of the isotope strontium-90 in fallout and oceanic mixing studies. Unfortunately, there has been a wide measure of disagreement between various workers, both as to the exact strontium to chlorinity ratio and as to the natural variation in this parameter over the world ocean. Bowen (1970) has critically reviewed the literature, and has concluded that the methods used by Chow and Thompson (1955), Fabricand et al. (1962, 1967) and Culkin and Cox (1966) were probably among the most reliable. The two latter sets of workers both analysed Copenhagen Standard Sea Water (batch P33) and obtained results in good agreement. This type of intercomparison has been a regrettably infrequent feature of strontium analysis, and indeed of major ion analysis in general.

The investigators mentioned above concur in finding a ratio of strontium ($g kg^{-1}$) to chlorinity (‰) of $0.00041 \pm 0.00001$, without significant natural variation in the samples studied. This also agrees with the results of Riley and Tongudai (1967); Aleksandruk and Stepanov (1968) and Nagaya et al. (1971). Those workers who have found a significant range of variation include Andersen and Hume (1968a, b) (6%), Angino et al. (1966) (20%) and Billings et al. (1969) (12%). Consistent variation with depth or position was not found by these workers and their absolute values for the Sr:Cl ratio tend to be lower than average. Attempts to explain these findings by postulating that a variable small fraction of the strontium occurs as colloidal-sized particles (Angino et al., 1966) have been rejected by Bowen (1970) on the grounds that this would require up to $1.6$ mg $l^{-1}$ of particulate strontium. Since estimates of *total* particulate material lie below this range (Jacobs and Ewing, 1969), this hypothesis would require a very high percentage of strontium in this material, which is not consistent with observation. Even for plankton samples abnormally rich in Acantharia species, which have skeletons of celestite (strontium sulphate), strontium was less than one per cent of the dry matter weight. In view of the lack of consistency with depth of the reported variations, the lack of a geochemical mechanism adequate to explain these variations and the fact that several investigations of high precision failed to detect significant variations between samples, it seems safe to conclude that no significant variation in this ratio is proved by the above results. The variations which have been reported are probably due to uncompensated matrix interference; Brass and Turekian (1972) have reported considerable depression of the strontium atomic absorption signal by silicon, and it is not impossible that other interferences of this type have also contributed to variations in the reported value for the Sr:Cl ratio.

The most recent work on this element has been performed by isotope dilution techniques using a mass spectrometer (Bernat et al., 1972; Brass and Turekian, 1972). This method is variously abbreviated as I.D.M.S. or M.S.I.D. The measurements are more precise than the results discussed above (standard deviation $\pm 0.2\%$ or better), although the absolute accuracy so far obtained is only of the order of $\pm 1\%$ because of absolute calibration problems. The few profiles which have been published using this technique show small but apparently significant differences between the mean Sr:Cl ratio of waters from different regions, and suggest that the surface layers may possibly be lower in this ratio by up to 3%. Below the thermocline, variation with depth at any given station is usually within $\pm 1\%$ of the mean. These limited results, therefore, show that although measurable variation in the Sr:Cl ratio probably does occur, its magnitude is very much smaller than some workers had previously reported.

It is fortunate that the advent of this improved technique for strontium analysis should have coincided with the planning of the GEOSECS series of cruises. The inter-comparison and calibration of methods attendant upon this large co-operative effort has already resulted in improved information and techniques becoming available. When these techniques are applied to the large suite of samples obtained during the GEOSECS cruises the state of our knowledge regarding the variation with depth and position of strontium and several other chemical species will be much improved.

### 6.3.2. MAJOR ANIONS

#### 6.3.2.1. Chloride

The salt content of sea waters was for many years measured by determination of the chlorinity, which was originally defined as the total halide content expressed as chloride (section 6.2). Innumerable determinations of this parameter have been made. The true chloride concentration cannot be obtained from these results by simple subtraction of the chlorine equivalent of the other halides present from the chlorinity. For example, water of 35‰ salinity has a chlorinity of 19·374‰; the concentration of chloride is 19·354 $g\,kg^{-1}$ and the other halides converted to chloride add another 0·0320 $g\,kg^{-1}$ chlorine equivalent. The difference of 0·012 g of chloride per kilogram of sea water between the chlorinity and the total halide expressed as chloride arises as a consequence of Knudsen's redefinition of chlorinity (Jacobsen and Knudsen, 1940) in terms of Atomgewichtssilber (section 6.2.1), and reflects the changes in accepted atomic weights which have occurred since the first standard sea waters were set up by Forch et al. (1902). For oceanic waters the ratio of chloride to chlorinity is 0·99896, and the ratio of total halide (expressed as chloride) to chlorinity is 1·0006.

#### 6.3.2.2 Sulphate

The sulphate ($g\,kg^{-1}$) to chlorinity ratio in all ocean waters appears to be very close to 0·1400 (Bather and Riley, 1954; Morris and Riley, 1966). The method almost universally used for the determination of sulphate in sea water involves the precipitation and gravimetric determination of barium sulphate. The method is very reproducible, but probably owes its accuracy to a fortunate compensation of the errors caused by the solubility of barium sulphate and those caused by co-precipitation of calcium as calcium sulphate (Bather and Riley, 1954). No alternative method has yet proved satisfactory however, and indeed a semimicro version of the procedure of Bather and Riley has recently been adopted for studies of sulphate concentration in pore waters collected during the JOIDES deep sea drilling project (Manheim and Sayles, 1969).

Sulphate is enriched in sea ice relative to the water from which the ice is derived (Lewis and Thompson, 1950), and measurable enrichments and depletions due to this effect have been reported to exist in the North Pacific (Fukai and Shiokawa, 1955). However, the property of the sulphate ion which most influences its geochemical behaviour is its ability to act as a source of oxygen for microorganisms in anoxic environments. In general, only a super-ficial layer of marine sediment contains oxygen; below this layer the micro-biological oxidation of organic material necessitates a corresponding reduction of sulphate to sulphide. The resultant concentration gradient of sulphate within the sediment results in a loss of sulphate from the sea water to the sediment. This loss is balanced by the high input of sulphate, relative to chloride, in river waters. Because these processes are not localized geographic-ally but occur over the whole world, and because the residence time of sul-phate is much greater than the mixing time, geographical variations in the sulphate to chlorinity ratio are very small indeed (range less than $\pm 0.4\%$). Such variations are too small to affect the salinity to density relationship significantly and thus are insignificant from the standpoint of hydrographic measurements.

The deep waters of certain basins (e.g. Black Sea, Cariaco Trench, parts of the Baltic and some Fjords) may become depleted in oxygen as a result of restricted vertical water movement (see Chapter 15). In such places sulphate reducing bacteria occur in the water column, and lower sulphate to chlorinity ratios may be encountered as a result. Skopinsev *et al.* (1958) have reported depletions in this ratio of $4\%$ in samples from the central Black Sea. In most instances of this type the depletion is smaller; this fact, and the restricted exchange of depleted water with the open ocean, combine to ensure that the effect of this depletion is of only local importance.

### 6.3.2.3. Bromide

Recent studies of this element in ocean water (Culkin, 1965; Morris and Riley, 1966) are in agreement in finding that the bromide (g kg$^{-1}$) to chlorinity ratio is remarkably constant with depth and position at 0·003473 (0·0673 g kg$^{-1}$ at S = 35‰). The variation range (2%) found by the latter workers for 211 samples representative of all the major seas and oceans corresponds to 13 times the coefficient of variation for replicate analyses of the same sample. It is possible therefore that some of the variability observed reflects natural variations; however no significant difference in the ratio could be observed for samples from different oceans or with depth or latitude.

Low salinity Baltic sea waters are relatively depleted in bromine, presumably because of low bromine levels in the river waters of this region. Depletions of 6% have been observed in samples of chlorinity 2‰ (Morris and Riley, 1966).

### 6.3.2.4. Fluoride

Because of reports of anomalous fluoride: chlorinity values in some Atlantic deep waters a number of studies have been made on this element in recent years. The original reports (Greenhalgh and Riley, 1963; Riley, 1965) indicated that the mean F:Cl ratio for ocean waters was 6·7 × 10$^{-5}$ (about 1·3 mgF$^{-}$ kg$^{-1}$ of sea water) but values up to 30% in excess of this were encountered at certain stations south of Greenland and in the N.E. Mediterranean, at depths generally in excess of 2000 m. Other stations in the same general area, some quite close to the anomalous sites, showed no enrichment. Brewer et al. (1970) carried out analyses on more than 200 samples from 33 North Atlantic stations. Most of these samples showed normal fluoride levels (F:Cl = 6·9 × 10$^{-5}$, close to the values quoted above) but high values were encountered below 1000 m at a station over the Nares abyssal plain, N. of Puerto Rico, and below 5000 m at a station off the continental shelf of Newfoundland.

All these investigations utilized the lanthanum–alizarin complexone colorimetric method of Greenhalgh and Riley (1961). However Brewer et al. also measured the relative activity of F$^{-}$ in their samples using a fluoride ion specific electrode. This method showed a constant activity of fluoride ion in all samples, including those which returned anomalous colorimetric results. Further analysis of the anomalous samples by the photon activation method of Wilkniss and Linnenbom (1968) showed rather poor agreement with the colorimetric results, but did imply an anomalously high fluoride concentration.

Bewers (1971) reported uniform and normal fluoride:chlorinity ratios from nine North Atlantic stations. Kester (1971) reported slightly elevated values off the continental shelf S.W. of Newfoundland, with F:Cl ratio values generally falling on passing along a transect to the Mid-Atlantic ridge. Both these investigators used the colorimetric method. Warner (1971), using a standard addition method based on the fluoride-selective electrode, found no anomalous results in 224 samples from the North Atlantic. Caribbean and Pacific. No increases in soluble fluoride on storage over ten weeks at low pH were observed.

There is as yet no unequivocal explanation for this collection of results. Since anomalous results have only been reliably observed by the colorimetric method and do not appear when the same samples are measured by the electrode method, it is possible that some interference, as yet undiscovered, is to blame. It is, however, almost as difficult to envisage an inferfering species with the observed distribution as it is to rationalize the observations in terms of true fluoride enrichments; in addition, this explanation must ignore as unreliable the photon activation results of Brewer *et al.* (1970).

If the measurements are accepted at face value it becomes necessary to postulate that the extra fluoride is in some way inaccessible to the lanthanum fluoride electrode at pH 5·0, although available to the colometric technique at lower pH. Brewer *et al.* rejected the possibility that the extra fluoride was complexed, on the grounds that (i) suitable complexing agents are not thought to occur at sufficiently high levels in sea water, and (ii) such a mechanism ought also to provide some electrode values *lower* than normal (when the complexing agent is present in excess) a circumstance which has not yet been observed. They favoured the idea that the extra fluoride was incorporated in particulate material of colloidal size (less than 0·45 μm, since their samples were filtered through a 0·45 μm membrane filter prior to analysis). This colloidally bound fluoride would enter the water column by resuspension of bottom sediments and become available for analysis as soluble fluoride only on treatment with the complexone reagent at low pH.

No data exist on the fluoride content of colloidal suspensions in sea water, but this explanation would require levels of up to 0·3 mg colloidal fluoride per litre, which seems rather high. Also, unless the excess fluoride is present *only* as particles of less than 0·45 μm, analysis of unfiltered samples should reveal some enrichments even greater than those observed for filtered samples. This has not been found. As already noted, prolonged storage at low pH did not result in higher fluoride results by the electrode method (Warner, 1971). Bewers *et al.* (1973) have examined water samples taken in 1971 from the region south of Greenland previously reported to exhibit anomalous values. No such values were detected either by the Greenhalgh and Riley method or by the electrometric method of Warner. This apparent rapid change in fluoride levels cannot be explained in terms of volcanic injection or of mineral dissolution, since the quantities of fluoride involved are far too great. The situation therefore remains unsatisfactory, since no mechanism has yet been proposed which is consistent with the reported observations.

### 6.3.2.5 Boron

The investigations cited by Culkin (1965) agreed in finding an average value for the boron content of sea water of 35‰ salinity of about 4·5 mg kg$^{-1}$ with a spread of about $\pm 0 \cdot 23$ mg kg$^{-1}$. The more recent determinations by Gassaway (1967) and Belyayev and Ovsyangy (1969) also agree with these results. Although an inverse relationship between boron and dissolved oxygen has been reported to exist at one station (Noakes and Hood, 1961), insufficient information is available to extend this finding to wider areas or to provide a clear picture of the geographical variation of this element. The situation is complicated by reported interactions of boron with carbohydrates (Noakes and Hood, 1961; Gast and Thompson, 1958). However, more recent work (Williams and Strack, 1966) appears to rule out this possibility on the grounds that insufficient suitable organic material is available in normal sea water.

The control mechanism for boron levels in sea water has been studied by Harriss (1969), who considered that the major part of the river-dissolved input load of $4 \times 10^{11}$ g of boron per year (Livingstone, 1963) is removed by adsorption on detrital clays on passing to the higher ionic strength of sea water (evidence for this process has been found recently by Liss and Pointon, 1973). Smaller fractions, of the order of one tenth of this quantity, may be removed from sea water during the formation of authigenic silicates and by deposition in siliceous oozes. Boron contained in organic material was assumed to be returned to the water column by decay processes before being incorporated in the sediments. In atmospheric aerosols boron is known to be enriched relative to other sea water salts, but loss to the atmosphere does not represent a removal pathway from sea water since, at steady state, atmospheric boron is eventually returned to the sea by precipitation and run off (Brujewicz and Korzh, 1971).

Element balances of this type help to highlight any gaps or gross uncertainties which may exist in our knowledge of the geochemistry of the element concerned. At the present state of our understanding, element balances always involve quite sweeping assumptions. These assumptions are an indication of the nature and degree of our lack of understanding, and hence a pointer to the proper design of future studies.

## 6.4. THE STABLE ISOTOPES OF WATER AND THE MAJOR ANIONS

Most major ions, and also the oxygen and hydrogen of sea water, possess several non-radioactive isotopes. Although, the abundance ratios of the isotopes of a given element are fairly constant, small but significant changes in relative abundance may be produced as the element passes through the various physical, chemical and biological stages of its geochemical cycle. This occurs because the atomic weight differences modify the physical and chemical properties of the isotope slightly; this effect is more pronounced for the lighter elements, since the relative mass change between isotopes of such elements is greater. The small variations in isotopic composition which occur in various parts of the geochemical system may be measured with considerable precision on quite small samples by means of mass spectrometry.

### 6.4.1. STABLE ISOTOPES OF SEA WATER

In addition to the most abundant atomic components $^1H$ and $^{16}O$, natural water contains the heavy isotopes deuterium (D, which is hydrogen of mass 2), $^{17}O$ and $^{18}O$. Since the proportions of these isotopes are low (D:H $= 1.56 \times 10^{-4}$; $^{17}O:^{16}O = 0.0004$; $^{18}O:^{16}O = 0.002$), only small quantities of the heavy molecules HDO (2000 ppm.) and $H_2^{18}O$ (320 ppm) exist at equilibrium, and water molecules containing more than one anomalously heavy atom are vanishingly rare.

Because HDO and $H_2^{18}O$ have a lower vapour pressure than normal water, they tend to be enriched in the liquid phase during the processes of evaporation and precipitation. Consequently, surface sea water from equatorial areas of high evaporation is isotopically "heavier" than the deeper water below, which is derived from surface waters at higher latitudes where precipitation of meteoric water predominates. The effect is stronger for HDO than for $H_2^{18}O$. Craig (1961a) has demonstrated that $\delta D = 8\delta^{18}O + 10$ for meteoric waters, where both isotopes are expressed as per mille enrichments relative to "standard mean ocean water" (SMOW).*

Friedman et al. (1964) showed that sea-ice contained about 20‰ more deuterium than the sea water in the same area. Craig and Gordon (1965)

---

* Because absolute measurement of heavy isotope concentrations is not practicable, all results are expressed relative to SMOW (Craig, 1961b); this is itself defined by reference to the National Bureau of Standards isotopic reference standard No. 1 (a large volume sample of distilled water) by the relationships

$$D/H \, (SMOW) = 1.050 \, D/H \, (NBS \, 1)$$
$$^{18}O/^{16}O \, (SMOW) = 1.008 = {}^{18}O/^{16}O \, (NBS \, 1)$$

The "δ value" is given by

$$\delta = \left( \frac{\text{isotope ratio of sample}}{\text{isotope ratio of SMOW}} - 1 \right) \times 1000$$

have shown that neither this effect nor the run off of "light" melt waters from the land are sufficient to account for the observed relatively low deuterium and $^{18}O$ content of high latitude surface sea waters. These workers postulated that the important processes are those of evaporation and precipitation over the sea itself. During ice age maxima, removal of meteoric water from the hydrosphere to form continental glaciers might have increased the average $\delta^{18}O$ content of the sea to about $+0.9\%_0$, relative to SMOW (Craig, 1965). At times of minimum glaciation this value would fall to about $-0.6\%_0$ compared to the present day value of $-0.08\%_0$. These calculations are important since they relate to one of the most useful applications of stable isotope measurements, the establishment of a palaeotemperature scale. This technique was suggested by Urey (1947) and depends on the relationship between the oxygen isotope composition of fossil carbonates and the temperature at which they were formed (Epstein et al., 1953). This relationship must be established by study of modern carbonates. As might be expected, the method has its limitations, and some care is necessary in the interpretation of results; it is however valuable as one of the few ways in which the variation of climate over geological time can be investigated.

Deuterium and $^{18}O$ variations have also been used to study the circulation of the deep oceans (Craig and Gordon, 1966). This work necessitates high precision in measurements as the variations in $\delta$ values between deep water masses are very small. For example, $\delta^{18}O$ values vary from $0.12\%_0$ (North Atlantic deep water) to $-0.45\%_0$ (Antarctic bottom water), relative to SMOW. The results indicate that the deep waters of the Pacific and Indian oceans are formed mainly by the mixing of Antarctic bottom water with that from the Atlantic in the Weddell Sea area, although a third component may be present also. These workers, and Horibe and Ogura (1968), have demonstrated the usefulness of $\delta^{18}O$-salinity diagrams for distinguishing deep water masses. Horibe and Ogura (1968) have also shown that a linear relationship between $\delta^{18}O$ and salinity found by them, and previously also reported by Epstein and Mayeda (1953) and Redfield and Friedman (1965), could be the result of mixing of isotopically light fresh water with the oceanic surface layers. Ehhalt (1969) has demonstrated that a similar effect exists in the Baltic Sea.

### 6.4.2. OTHER STABLE ISOTOPE STUDIES

Because of differences in the metabolic behaviour of isotopes of different atomic mass, isotope fractionation effects occur during the metabolic processes of photosynthesis and respiration. Thus, it has been known for some time that dissolved oxygen at the oxygen minimum layer is enriched in $^{18}O$ (Rakestraw et al., 1951; Dole et al., 1954), although later work has shown that the degree of enrichment is not as great as had been thought (Kroopnick et al., 1972). Kroopnick et al. also studied the $\delta^{13}C$ values of dissolved carbon dioxide and were able to show that the observed oxygen consumption in deep water was consistent with the $^{13}C$ enrichment found. It was also possible to estimate that about 25% of the observed increase in total $CO_2$ was supplied by dissolution of solid carbonate material, in good agreement with previous estimates based on alkalinity and total carbon dioxide measurements (Li et al., 1969; see also Chapter 9).

Carbon and oxygen isotope measurements have also been used to establish the sequence of events which led to the production of the observed characteristic sediments of the Black Sea (Deuser, 1971). The results are consistent with the hypothesis that this basin contained fresh water until about 9000 years B.P. After this, the post glacial rise of sea levels permitted sea water to flow into the basin through the Bosporous, and the resulting salinity increase caused an increase in the $^{18}O$ content of the deposited carbonates. At about this time the deeper layers of water first became anoxic and the deposition of organic-rich sediments began. The $\delta^{13}C$ fluctuations in this material suggest that the rate of rise of salinity may have been rather unsteady. A more steady period, with salinity values close to those of the present day, seems to have been reached about 3000 years BP, although the anoxic layer may have increased in thickness since that time.

In principle, several other major elements possess a range of stable isotopes which might yield useful data, but little work has as yet appeared on this subject. The apparent variation of $^{87}Sr:^{86}Sr$ over geological time has been investigated by several groups of workers (e.g., Peterman et al., 1970; Armstrong, 1971; Brass, 1973), but an unequivocal explanation for the observed results has yet to emerge. Rees (1970) has presented an improved model for the sulphur isotope balance of the oceans, and Schwarcz et al. (1969) have proposed that the enrichment of the oceans in boron-11 may arise by an isotope effect during absorption of boron from sea water onto clay sediments. This latter explanation appears to be consistent with the conclusions of Harriss (1969) (Section 6.3.2.5) that adsorption on clays constitutes an important removal pathway for this element.

REFERENCES

Accerboni, E. and Mosetti, F. (1967). Boll. Geofis. Teor. Appl. 9, 87.
Aleksandruk, V. M. and Stepanov, A. V. (1968). Okeanologiia Akad. Nauk. SSSR, 8, 746.
Andersen, N. R. and Hume, D. N. (1968a). Anal. Chim. Acta, 40, 207.

Andersen, N. R. and Hume, D. N. (1968b). *In* "Trace Inorganics in Water" (R. F. Gould, ed.) p. 269. Am. Chem. Soc., Adv. in Chem. Series No. 73. Washington, D.C.

Angino, E. E., Billings, G. K. and Andersen, N. R. (1966). *Chem. Geol.* **1**, 145.

Armstrong, R. L. (1971). *Nature Phys. Sci.* **230**, 132.

Bather, J. M. and Riley, J. P. (1954). *J. Cons. Int. Explor. Mer.* **20**, 145.

Behrens, E. W. (1965). *J. Mar. Res.* **23**, 165.

Belyayev, L. I. and Ovsyangy, Ye. I. (1969). *Oceanology*, **8**, 734.

Bernat, M., Church, T. and Allegre, C. J. (1972). *Earth Planet. Sci. Lett.* **16**, 75.

Bewers, J. M. (1971). *Deep-Sea Res.* **18**, 237.

Bewers, J. M., Miller, G. R., Kester, D. R. and Warner, T. B. (1973). *Nature Phys. Sci.* **242**, 142.

Billings, G. K., Bricker, O. P., Mackenzie, F. T. and Brooks, A. L. (1969). *Earth Planet. Sci. Lett.* **6**, 231.

Billings, G. K. and Harriss, R. C. (1965). *Texas J. Sci.* **17**, 129.

Bischoff, J. L., Greer, R. E. and Luistro, A. O. *Science, N.Y.* **167**, 3922.

Bonnet, J.-F. (1971). *Int. Hydro. Rev.* **48**, 155.

Bowen, V. T. (1970). *In* "Reference Methods for Marine Radioactivity Studies". pp. 93–127. IAEA, Vienna.

Bradshaw, A. and Schleicher, K. E. (1965). *Deep-Sea Res.* **12**, 151.

Brass, G. W. (1973). *Trans. Amer. Geophys. Union*, **54**, 488.

Brass, G. and Turikian, K. K. (1972). *Earth Planet. Sci. Lett.* **16**, 117.

Brewer, P. G. and Spencer, D. W, (1970). Unpublished Manuscript, Woods Hole Oceanogr. Inst. Ref. No. 70–62, 9 pp.

Brewer, P. G., Spencer, D. W. and Wilkniss, R. E. (1970). *Deep-Sea Res.* **17**, 1.

Brown, N. L. and Allentoft, B. (1966). Final report to U.S. Office of Naval Research, contract NOnr 4290 (00), M.J.O. No. 2003, Bissett-Berman.

Brown, N. L. and Hamon, B. V. (1961). *Deep-Sea Res.* **8**, 65.

Brujewicz, S. W. and Korzh, V. D. (1971). *Okeanológiia Akad. Nauk, SSSR*, **11**, 414.

Carpenter, J. H. (1972). *Nat. Bur. Stand. Spec. Publ.* **351**, 393.

Carpenter, J. H. and Manella, M. E. (1973). *J. Geophys. Res.*, **78**, 3621.

Carr, B. A. (1970). *Limnol. Oceanogr*, **15**, 318.

Carritt, D. E. and Carpenter, J. H. (1959). *Nat. Acad. Sci.–Nat. Res. Counc. Publ.* **600**, 67.

Chow, T. J. and Thompson, T. G. (1955). *Analyt. Chem.* **27**, 18.

Cloud, P. E. (1965). *In* "Chemical Oceanography" (J. P. Riley and G. Skirrow, eds.) Vol. 2, pp. 127–158. Academic Press, London.

Cox, R. A. (1963). *In* "Progress in Oceanography" (M. Sears, ed.) Vol. 1, p. 243. Pergamon Press, Oxford.

Cox, R. A. (1965). *In* "Chemical Oceanography" (J. P. Riley and G. Skirrow, eds.) Vol. 1 pp. 73–120. Academic Press, London.

Cox, R. A., Culkin, F. and Riley, J. P. (1967). *Deep-Sea Res.* **14**, 203.

Cox, R. A., McCartney, M. J. and Culkin, F. (1968). *Deep-Sea Res.* **15**, 319.

Cox, R. A., McCartney, M. J. and Culkin, F. (1970). *Deep-Sea Res.* **17**, 679.

Craig, H. (1961a). *Science, N.Y.* **133**, 1702.

Craig, H. (1961b). *Science, N.Y.* **133**, 1833.

Craig, H. (1965). *In* "*Proc. Conf. Stable Isotopes Oceanogr. Studies Paleotemperatures.*"

Craig, H. and Gordon, L. T. (1965). *In* "Symposium on Marine Geochemistry", 373 pp. Univ. of Rhode Island, Occ. Publ. 3–1965.

Craig, H. and Gordon, L. I. (1966). *Trans. Amer. Geophys. Union*, **47**, 112.

Crease, J. (1971). *Nature, Lond.* **233**, 329.

Culkin, F. (1965). *In* "Chemical Oceanography" (J. P. Riley and G. Skirrow, eds.) Vol. 1, pp. 121–161. Academic Press, London.

Culkin, F. and Cox, R. A. (1966). *Deep-Sea Res.* **13**, 789.

Cunningham, C. C. and Duedall, I. W. (1970). *J. Cons. Perm. int. Explor. Mer.* **33**, 292.

Deuser, W. G. (1972). *J. Geophys. Res.* **77**, 1071.

Dittmar, W. (1884). *In* "The Voyage of H.M.S. *Challenger*" (J. Murray, ed.) Vol. 2, 51 pp. H.M.S.O., London.

Dole, M., Lane, G. A., Rudd, D. P. and Zaukelies, D. A. (1954). *Geochim. Cosmochim. Acta*, **6**, 65.

Ehhalt, D. H. (1969). *Tellus*, **21**, 429.

Epstein, S., Buchsbaum, R., Lowenstam, H. A. and Urey, H. C. (1953). *Bull. Geol. Soc. Amer.* **64**, 1315.

Epstein, S., and Mayeda, T. (1973). *Geochim. Cosmochim. Acta*, **4**, 213.

Esterson, G. and Pritchard, D. W. (1955). *Proc. 1st Conf. Coastal Eng., Inst. Counc. Wave Res.* The Engineering Foundation pp. 260–71.

Fabricand, B. P., Sawyer, R. R., Ungar, S. G. and Adler, S. (1962). *Geochim. Cosmochim. Acta*, **26**, 1023.

Fabricand, B. P., Imbimbo, E. S. and Brey, M. E. (1967). *Deep-Sea Res.*, **14**, 785.

Fedorov, K. N. (1971). *Okeanológiia Akad. Nauk SSSR*, **11**, 622.

Forch, C., Knudsen, M. and Sorensen, S. P. L. (1902). *Kgl. Dan. Vidensk. Selsk. Raekke, naturvidensk, og methem.* Afd XII, I., Skrifter, 6, 151 pp.

Friedman, I, Redfield, A. C., Schoen, B. and Harris, J. (1964). *Rev. Geophys.* **2**, 177.

Fukai, R. and Shiokawa, F. (1955). *Bull. Chem. Soc. Jap.* **38**, 636.

Garrels, R. M. and Thompson, M. E. (1962). *Amer. J. Sci.* **206**, 57.

Gassaway, J. D. (1967). *Int. J. Oceanol. Limnol.* **1**, 85.

Gast, J. A. and Thompson, T. G. (1958). *Analyt. Chem.* **30**, 1549.

Gieskes, J. M. T. M. (1967). *Kiel. Meeresforsch.* **23**, 75.

Gieskes, J. M. T. M. (1968). *Tech. Rep., Inst. für Meereskunde, Universität Kiel, Abt. Meereschemie*, 17pp.

Goldberg, E. D. (1965). *In* "Chemical Oceanography" (J. P. Riley and G. Skirrow, eds.) Vol. 1, pp. 163–196. Academic Press, London.

Goulet, J. R. and Culverhouse, B. J. (1972). *J. Geophys. Res.* **77**, 4588.

Gran, G. (1952). *Analyst*, **77**, 661.

Grasshoff, K. and Wenk, A. (1972). *J. Cons. int. Explor. Mer.* **34**, 522.

Greenberg, D. A. (1972). Can. Ocean Data Centre, M.S.B., Dept. of Environment, Canada, 25 pp.

Gregg, M. C. and Cox, C. S. (1971). *Deep-Sea Res.* **18**, 925.

Greenhalge, R. and Riley, J. P. (1961). *Analyt. Chim. Acta*, **25**, 179.

Greenhalge, R. and Riley, J. P. (1963). *Nature, Lond.* **197**, 371.

Guntz, A. A. and Kocher, J. (1952). *C.R. Acad. Sci. Paris*, **234**, 2300.

Harriss, R. C. (1969). *Nature, Lond.* **223**, 290.

Hermann, F. (1951). *J. Cons. Perm. Int. Explor. Mer.* **17**, 223.

Holland, H. D. (1972). *Geochim. Cosmochim. Acta*, **36**, 637.

Horibe, Y. and Ogura, N. (1968). *J. Geophys. Res.* **73**, 1239.

Horne, R. A. (1969). "Marine Chemistry" 568 pp. Wiley-Interscience, New York.

Ives, D. J. G. and Jantz, G. J. (1961). *In* "Reference Electrodes" (D. J. G. Ives and G. J. Jantz, eds.) pp. 1–70. Academic Press, New York and London.

Jacobs, M. B. and Ewing, M. (1969). *Science, N.Y.* **163**, 380.

Jacobsen, J. P. and Knudsen, M. (1940). *Pub. Sci. Ass. d'Oceanographie Physique,* No. 7, 38 pp.

Jagner, D. (1967). "Report on the Chemistry of Sea Water" III. Unpublished manuscript, Göteborg, 6 pp.

Jagner, D. and Åren, K. (1970). *Analyt. Chim. Acta,* **52**, 491.

Jagner, D. and Åren, K. (1971). *Analyt. Chim. Acta,* **57**, 185.

Johnston, R. (1964). *Oceanogr. Mar. Biol. Ann. Rev.* **2**, 97.

Johnston, R. (1969). *Oceanogr. Mar. Biol. Ann. Rev.* **7**, 31.

Kester, D. R. (1971). *Deep-Sea Res.* **18**, 1123.

Knudsen, M. (1901). Hydrographic Tables, Tutein og Koch, Copenhagen, 63 pp.

Koske, P. H. (1964). *Kiel. Meersforsch.* **20**, 138.

Kratky, O., Leopold, H. and Stablinger, H. (1969). *Z. Angew. Phys.* **27**, 273.

Kremling, K. (1971). *Nature, Lond.* **229**, 109.

Kremling, K. (1972). *Deep-Sea Res.* **19**, 377.

Kroopnick, P., Weiss, R. F., and Craig, H. (1972). *Earth Planet. Sci. Lett.* **16**, 103.

Kuenzler, H. W. (1968). Unpublished report, Rep 68–3, Fluid Dynamics Laboratory, M.I.T.

Lewis, G. J. and Thompson, T. G. (1950). *J. Mar. Res.* **9**, 211.

Li, Y-H., Takahashi, T. and Broeker, W. S. (1969). *J. Geophys. Res.* **74**, 5507.

Liss, P. S. and Pointin, M. J. (1973). *Geochim. Cosmochim. Acta,* **37**, 1493.

Livingstone, D. A. (1963). *U.S. Geol. Survey Prof. Paper* 440 G.

Lyakhin, Yu. I. (1971). *Okeanologiia Akad. Nauk SSSR,* **11**, 635.

Lyman, J. (1969). *Limnol. Oceanogr.* **14**, 928.

Mangelsdorf, P. C. jr. (1967). *In* "Estuaries" (G. H. Lauff, ed.), Publ. 83. A.A.A.S., Washington, D.C.

Mangelsdorf, P. C. and Wilson, T. R. S. (1971). *J. Phys. Chem.* **75**, 1418.

Mangelsdorf, P. C. and Wilson, T. R. S. (1972). *Trans. Amer. Geophys. Union,* **53**, 402.

Mangelsdorf, P. C., Wilson, T. R. S. and Daniell, E. (1969). *Science, N.Y.* **165**, 171.

Manheim, F. T. (1968). *J. Sediment. Petrology,* **38**, 668.

Manheim, F. T. and Sayles, F. L. (1969). *In* "International Report, Deep-Sea Drilling Project" (M. Ewing *et al.,* eds.), Vol. 1, 403.

Miyake, Y. (1939). *Bull. Chem. Soc. Jap.* **14**, 29.

Moberg, E. G. and Revelle, R. R. D. (1937). *Int. Assn. Phys. Oceanogr. Procès-verb.* **2**, 153.

Morcos, S. A. (1970). *J. Cons. Perm. Int. Explor. Mer.* **33**, 126.

Morris, A. W. and Riley, J. P. (1964). *Deep-Sea Res.* **11**, 899.

Morris, A. W. and Riley, J. P. (1966). *Deep-Sea Res.* **13**, 699.

Morrison, G. K. (1973). *Deep-Sea Res.,* **20**, 665.

Nagaya, Y., Nakamura, K. and Saiki, M. (1971). *J. Oceanogr. Soc. Jap.* **27**, 20.

N.O.I.C. (1971a). National Oceanographic Instrumentation Center, Resources and Facilities. U.S. Govt. Printing Office, publ. 428–723–1971. Washington, D.C.

N.O.I.C. (1971b). National Oceanographic Instrumentation Centre, Instrument Fact Sheet I.F.S. 71012 7 pp. Washington, D.C.

Noakes, J. E. and Hood, D. W. (1961). *Deep-Sea Res.* **8**, 121.

Perkin, R. G. and Walker, E. R. (1972). *J. Geophys. Res.* **77**, 6618.

Peterman, Z. E., Hedge, C. E. and Tourtelot, H. A. (1970). *Geochim. Cosmochim. Acta,* **34**, 105.

Rakestraw, N. W., Rudd, D. P. and Dole, M. (1951). *J. Amer. Chem. Soc.* **73**, 2976.

Redfield, A. C. and Friedman, I. (1965). *In* "Proc. Symp. Marine Geochemistry, University of Rhode Island" (D. R. Schink and J. T. Corless, eds.) pp. 149–169. Univ. of Rhode Island.

Reeburgh, W. S. (1965). *J. Mar. Res.,* **23**, 187.

Reeburgh, W. S. and Carpenter, J. H. (1964). *Limnol. Oceanogr.* **9**, 589.

Rees, C. E. (1970). *Earth Planet. Sci. Lett.* **7**, 366.

Rhode, J. (1968). Dissertation, University of Kiel.

Ribe, R. L. and Howe, J. G. (1967). Unpublished, Report, Oceanographic Instrumentation Center Naval Oceanographic Office, Washington, D.C.

Riley, J. P. (1965a). *In* "Chemical Oceanography" (J. P. Riley and G. Skirrow, eds.) Vol. 2, pp. 295–411. Academic Press, London.

Riley, J. P. (1965b). *Deep-Sea Res.* **12**, 214.

Riley, J. P. and Tonguadi, M. (1967). *Chem. Geol.* **2**, 263.

Rusby, J. S. M. (1967). *Deep-Sea Res.* **14**, 427.

Sanders, H. L., Mangelsdorf, P. C. and Hampson, G. R. (1965). *Limnol. Oceanogr.* **10**, R216.

Sayles, F. L., Wilson, T. R. S., Hume, D. N., and Mangelsdorf, P. C. (1973). *Science, N.Y.* **181**, 154.

Schwarcz, H. P., Agyei, E. K. and McMullen, C. C. (1969). *Earth Planet. Sci. Lett.* **6**, 1.

Seidler, G. (1963). *Deep-Sea Res.* **10**, 269.

Skopintsev, B. A., Gubin, F. A., Vorobeva, R. V. and Verschinina, O. A. (1958). *C.R. Acad. Nauk., SSSR,* **119**, 121.

Stanley, E. M. (1971). *Deep-Sea Res.* **18**, 833.

Sverdrup, H. U., Johnson, M. W. and Fleming, R. H. (1940). "The Oceans", 1087 pp. Prentice-Hall, New Jersey.

Thomas, B. D., Thomson, T. G. and Utterback, C. L. (1934). *J. Cons. Int. Explor. Mer.* **9**, 28.

Traganza, E. D. and Szabo, B. J. (1967). *Limnol. Oceanogr.* **12**, 281.

Tsunogai, S., Nishimura, M. and Nakanya, S. (1968). *Talanta,* **15**, 385.

Tsunogai, S., Yamahata, H., Kudo, S. and Saito, O. (1973). *Deep-Sea Res.* **20**, 717.

Tsurikova, A. P. and Tsurikov, V. L. (1971). *Oceanologiia Akad.Nauk, SSSR,* **11**, 339.

U.N.E.S.C.O. (1962). *Tech. Papers in Mar. Sci.* No. 1, 29 pp.

U.N.E.S.C.O. (1965). *Tech. Papers in Mar. Sci.* No. 4, 29 pp.

U.N.E.S.C.O. (1966). International Oceanographic Tables, National Institute of Oceanography of Great Britain, and U.N.E.S.C.O. 118 pp.

U.N.E.S.C.O. (1968). International Oceanographic Tables, National Institute of Oceanography of Great Britain, and U.N.E.S.C.O. 128 pp.

Urey, H. C. (1947). *J. Chem. Soc.* 562.

Van Landingham, J. W. (1967). *J. Ocean. Technol.* **2**, 53.

Vel'mozhnaya, Yu. A. (1960). *Tr. Morsk. Gidrofiz. Inst. Akad. Nauk Ukr. SSR,* **22**, 26. (U.S. Hydrographic Office Translation TRANS. 158 (1962)).

Walker, E. R. and Chapman, K. D. (1973). Pacific Marine Science Report 73–5. 52 pp. Unpublished Manuscript, Pacific Marine Sciences Directorate, Victoria, B.C.

Warner, T. B. (1971). *Deep-Sea Res.* **18**, 1225.

Wilkniss, P. E. and Linnenbom, V. J. (1968). *Limnol. Oceanogr.* **13**, 530.

Williams, P. M. and Strack, P. M. (1966). *Limnol. Oceanogr.* **11**, 401.

Wilson, T. R. S. (1971). *Limnol. Oceanogr.* **16**, 581.

Wilson, T. R. S., Sayles, F. L., Mangelsdorf, P. C. (1972). *Trans. Amer. Geophys. Union*, **53**, 293.

Wooster, W. S., Lee, A. J., and Dietrich, G. (1969). *Limnol. Oceanogr.* **14**, 437.

Zaburdaev, V. I., Dobruskina, I. M., Lyashenko, E. F. and Poplavskaya, M. G. (1969). Oceanic Hydrological Investigation Report, No. 1. Akad. Sci., Ukr. SSR, Sevastopol, 5 pp.

Chapter 7

# Minor Elements in Sea Water

## PETER G. BREWER[*]

*Woods Hole Oceanographic Institution,*
*Woods Hole, Massachusetts, U.S.A.*

* Contribution No. 3203 from the Woods Hole Oceanographic Institution.

415

## 7.1. INTRODUCTION

This chapter is concerned with the minor, or trace, elements in sea water. These elements are conventionally considered to be those which exist at concentrations of 1 mg kg$^{-1}$ or less. The eleven major ions of sea water are discussed elsewhere in this volume, as are the dissolved gases, the nutrient elements C, N, O, P, Si and radioactive isotopes although these are undeniably trace species. Some 61 elements remain, ranging from lithium (Z = 3) to bismuth (Z = 83). These exhibit an enormous range of chemical reactions and enter widely into the biochemical and geochemical cycles of the ocean. Chemical speciation and the analytical chemistry of sea water will not be discussed in detail as these topics are considered in other chapters. However, it will be necessary to refer occasionally to these aspects for the sake of clarity.

The format followed here will be to discuss the abundances of the various elements in sea water, their modes of introduction to, and removal from the sea and their distribution within the oceanic water column. In a few instances it will be possible to discuss the rates at which these processes occur and to correlate these rates with specific chemical pathways. There is a voluminous literature on these subjects, much of it contradictory. However, remarkably little of it has been devoted to precise quantitative aspects of the distribution of these elements in the sea.

## 7.2. ABUNDANCES AND RESIDENCE TIMES

The abundances of the elements in sea water are listed in Table 7.1 together with estimates of their residence times. Richards (1956) has criticized the use of such tables on the grounds that they "tacitly imply that the reported concentrations are somehow representative of the whole ocean" and that such average values may be meaningless because of the large variations both geographically and with depth, and the analytical and sampling errors prevalent in the data at that time. Indeed, he concluded that up to 1956 there had been no valid analysis of deep water samples for the following elements (and certainly no detailed profiles permitting the estimation of oceanic fluxes): Sb, Ba, Cd, Cs, Ce, Cr, Co, Ga, Ge, La, Pb, Hg, Mo, Ni, Sc, Se, Ag, Tl, Th, W, Sn, V, Y and Zn. Since that time our knowledge has improved, although even now there are no reliable oceanic profiles for many of these elements. The concentrations of these elements are unlikely to vary by as much as an order of magnitude except under extreme conditions, and this means that very precise analytical methods are required to investigate their distributions, oxidation states and chemical speciations.

TABLE 7.1

*The abundances of the chemical elements in sea water and estimates of their residence times*

| Element | Chemical species | Total concentration (molar) | Total concentration ($\mu g\ l^{-1}$) | Residence time (years) | | |
|---|---|---|---|---|---|---|
| | | | | Barth | Goldberg and Arrhenius | Most recent estimate* |
| H | $H_2O$ | 55 | $1 \cdot 1 \times 10^8$ | — | — | — |
| He | He(gas) | $1 \cdot 7 \times 10^{-9}$ | $6 \cdot 8 \times 10^{-3}$ | — | — | — |
| Li | $Li^+$ | $2 \cdot 6 \times 10^{-5}$ | 180 | $1 \cdot 2 \times 10^7$ | $1 \cdot 9 \times 10^7$ | $2 \cdot 3 \times 10^6$ |
| Be | $BeOH^+$ | $6 \cdot 3 \times 10^{-10}$ | $5 \cdot 6 \times 10^{-3}$ | — | — | — |
| B | $B(OH)_3, B(OH)_4^-$ | $4 \cdot 1 \times 10^{-4}$ | 4440 | — | — | $1 \cdot 3 \times 10^7$ |
| C | $HCO_3^-, CO_3^{2-} . CO_2$ | $2 \cdot 3 \times 10^{-3}$ | $2 \cdot 8 \times 10^4$ | — | — | — |
| N | $N_2, NO_3^-, NO_2^-, NH_4^+$ | $1 \cdot 07 \times 10^{-2}$ | $1 \cdot 5 \times 10^5$ | — | — | — |
| O | $H_2O, O_2$ | 55 | $8 \cdot 8 \times 10^8$ | — | — | — |
| F | $F^-, MgF^+$ | $6 \cdot 8 \times 10^{-5}$ | $1 \cdot 3 \times 10^3$ | — | — | $5 \cdot 2 \times 10^5$ |
| Ne | Ne (gas) | $7 \times 10^{-9}$ | $1 \cdot 2 \times 10^{-1}$ | — | — | — |
| Na | $Na^+$ | $4 \cdot 68 \times 10^{-1}$ | $10 \cdot 77 \times 10^6$ | $2 \cdot 1 \times 10^8$ | $2 \cdot 6 \times 10^8$ | $6 \cdot 8 \times 10^7$ |
| Mg | $Mg^{2+}$ | $5 \cdot 32 \times 10^{-2}$ | $12 \cdot 9 \times 10^5$ | $2 \cdot 2 \times 10^7$ | $4 \cdot 5 \times 10^7$ | $1 \cdot 2 \times 10^7$ |
| Al | $Al(OH)_4^-$ | $7 \cdot 4 \times 10^{-8}$ | 2 | $3 \cdot 1 \times 10^3$ | $1 \times 10^2$ | $1 \cdot 0 \times 10^2$ |
| Si | $Si(OH)_4$ | $7 \cdot 1 \times 10^{-5}$ | $2 \times 10^6$ | $3 \cdot 5 \times 10^4$ | $1 \times 10^4$ | $1 \cdot 8 \times 10^4$ |
| P | $HPO_4^{2-}, PO_4^{3-}, H_2PO_4^-$ | $2 \times 10^{-6}$ | 60 | — | — | $1 \cdot 8 \times 10^5$ |
| S | $SO_4^{2-}, NaSO_4^-$ | $2 \cdot 82 \times 10^{-2}$ | $9 \cdot 05 \times 10^5$ | — | — | — |
| Cl | $Cl^-$ | $5 \cdot 46 \times 10^{-1}$ | $18 \cdot 8 \times 10^6$ | — | $1 \times 10^8$ | $1 \times 10^8$ |
| Ar | Ar (gas) | $1 \cdot 1 \times 10^{-7}$ | $4 \cdot 3$ | — | — | — |
| K | $K^+$ | $1 \cdot 02 \times 10^{-2}$ | $3 \cdot 8 \times 10^5$ | $1 \cdot 1 \times 10^7$ | $1 \cdot 1 \times 10^7$ | $7 \times 10^6$ |
| Ca | $Ca^{2+}$ | $1 \cdot 02 \times 10^{-2}$ | $4 \cdot 12 \times 10^5$ | $1 \times 10^6$ | $8 \times 10^6$ | $1 \times 10^6$ |
| Sc | $Sc(OH)_3^0$ | $1 \cdot 3 \times 10^{-11}$ | $6 \times 10^{-4}$ | — | — | $4 \times 10^4$ |
| Ti | $Ti(OH)_4^0$ | $2 \times 10^{-8}$ | 1 | — | — | $1 \cdot 3 \times 10^4$ |
| V | $H_2VO_4^-, HVO_4^{2-}$ | $5 \times 10^{-8}$ | $2 \cdot 5$ | — | — | $8 \times 10^4$ |
| Cr | $Cr(OH)_3, CrO_4^{2-}$ | $5 \cdot 7 \times 10^{-9}$ | $0 \cdot 3$ | — | — | $6 \times 10^3$ |
| Mn | $Mn^{2+}, MnCl^+$ | $3 \cdot 6 \times 10^{-9}$ | $0 \cdot 2$ | — | — | $1 \times 10^4$ |

P

Table 7.1—continued

| Element | Chemical species | Total concentration (molar) | $\mu g\,l^{-1}$ | Residence time (years) | | |
|---|---|---|---|---|---|---|
| | | | | Barth | Goldberg and Arrhenius | Most recent estimate* |
| Fe | $Fe(OH)_2^+$, $Fe(OH)_4^-$ | $3.5 \times 10^{-8}$ | 2 | — | — | $2 \times 10^2$ |
| Co | $Co^{2+}$ | $8 \times 10^{-10}$ | 0·05 | — | — | $3 \times 10^4$ |
| Ni | $Ni^{2+}$ | $2.8 \times 10^{-8}$ | 1·7 | $1.5 \times 10^4$ | $1.8 \times 10^4$ | $9 \times 10^4$ |
| Cu | $CuCO_3^0$, $CuOH^+$ | $8 \times 10^{-9}$ | 0·5 | $4.3 \times 10^4$ | $5 \times 10^4$ | $2 \times 10^4$ |
| Zn | $ZnOH^+$, $Zn^{2+}$, $ZnCO_3^0$ | $7.6 \times 10^{-8}$ | 4·9 | | | $2 \times 10^4$ |
| Ga | $Ga(OH)_4^-$ | $4.3 \times 10^{-10}$ | 0·03 | | | $1 \times 10^4$ |
| Ge | $Ge(OH)_4$ | $6.9 \times 10^{-10}$ | 0·05 | | | — |
| As | $HAsO_4^{2-}$, $H_2AsO_4^-$ | $5 \times 10^{-8}$ | 3·7 | | | $5 \times 10^4$ |
| Se | $SeO_3^{2-}$ | $2.5 \times 10^{-9}$ | 0·2 | | | $2 \times 10^4$ |
| Br | $Br^-$ | $8.4 \times 10^{-4}$ | $6.7 \times 10^4$ | | | $1 \times 10^8$ |
| Kr | Kr (g) | $2.4 \times 10^{-9}$ | 0·2 | | | — |
| Rb | $Rb^+$ | $1.4 \times 10^{-6}$ | 120 | $6.1 \times 10^6$ | $2.7 \times 10^5$ | $4 \times 10^6$ |
| Sr | $Sr^{2+}$ | $9.1 \times 10^{-5}$ | $8 \times 10^4$ | $1 \times 10^7$ | $1.9 \times 10^7$ | $4 \times 10^6$ |
| Y | $Y(OH)_3^0$ | $1.5 \times 10^{-11}$ | $1.3 \times 10^{-3}$ | | | — |
| Zr | $Zr(OH)_4^0$ | $3.3 \times 10^{-10}$ | $3 \times 10^{-2}$ | | | — |
| Nb | | $1 \times 10^{-10}$ | $1 \times 10^{-2}$ | | | — |
| Mo | $MoO_4^{2-}$ | $1 \times 10^{-7}$ | 10 | $2.1 \times 10^6$ | $5 \times 10^5$ | $2 \times 10^5$ |
| Tc | | | | | | — |
| Ru | | | | | | |
| Rh | | | | | | |
| Pd | | | | | | |
| Ag | $AgCl_2^-$ | $4 \times 10^{-10}$ | 0·04 | $2.5 \times 10^5$ | $2.1 \times 10^6$ | $4 \times 10^4$ |
| Cd | $CdCl_2^0$ | $1 \times 10^{-9}$ | 0·1 | | | — |
| In | $In(OH)_2^+$ | $0.8 \times 10^{-12}$ | $1 \times 10^{-4}$ | | | — |
| Sn | $SnO(OH)_3^-$ | $8.4 \times 10^{-11}$ | $1 \times 10^{-2}$ | | | — |
| Sb | $Sb(OH)_6^-$ | $2 \times 10^{-9}$ | 0·24 | | | $7 \times 10^3$ |
| Te | $HTeO_3^-$ | | | | | — |

| Element | Species | | | | | |
|---|---|---|---|---|---|---|
| I | $IO_3^-$, $I^-$ | $5 \times 10^{-7}$ | 60 | — | — | $4 \times 10^5$ |
| Xe | Xe (gas) | $3{\cdot}8 \times 10^{-10}$ | $5 \times 10^{-2}$ | — | — | — |
| Cs | $Cs^+$ | $3 \times 10^{-9}$ | $0{\cdot}4$ | — | — | $6 \times 10^5$ |
| Ba | $Ba^{2+}$ | $1{\cdot}5 \times 10^{-7}$ | 2 | $5 \times 10^4$ | $8{\cdot}4 \times 10^4$ | $4 \times 10^4$ |
| La | $La(OH)_3^0$ | $2 \times 10^{-11}$ | $3 \times 10^{-3}$ | — | — | $6 \times 10^2$ |
| Ce | $Ce(OH)_3^0$ | $1 \times 10^{-10}$ | $1 \times 10^{-3}$ | — | — | — |
| Pr | $Pr(OH)_3^0$ | $4 \times 10^{-12}$ | $6 \times 10^{-4}$ | — | — | — |
| Nd | $Nd(OH)_3^0$ | $1{\cdot}9 \times 10^{-11}$ | $3 \times 10^{-3}$ | — | — | — |
| Pm | $Pm(OH)_3^0$ | — | — | — | — | — |
| Sm | $Sm(OH)_3^0$ | $3 \times 10^{-12}$ | $0{\cdot}5 \times 10^{-4}$ | — | — | — |
| Eu | $Eu(OH)_3^0$ | $9 \times 10^{-13}$ | $0{\cdot}1 \times 10^{-4}$ | — | — | — |
| Gd | $Gd(OH)_3^0$ | $4 \times 10^{-12}$ | $7 \times 10^{-4}$ | — | — | — |
| Tb | $Tb(OH)_3^0$ | $9 \times 10^{-13}$ | $1 \times 10^{-4}$ | — | — | — |
| Dy | $Dy(OH)_3^0$ | $6 \times 10^{-12}$ | $9 \times 10^{-4}$ | — | — | — |
| Ho | $Ho(OH)_3^0$ | $1 \times 10^{-12}$ | $2 \times 10^{-4}$ | — | — | — |
| Er | $Er(OH)_3^0$ | $4 \times 10^{-12}$ | $8 \times 10^{-4}$ | — | — | — |
| Tm | $Tm(OH)_3^0$ | $8 \times 10^{-13}$ | $2 \times 10^{-4}$ | — | — | — |
| Yb | $Yb(OH)_3^0$ | $5 \times 10^{-12}$ | $8 \times 10^{-4}$ | — | — | — |
| Lu | $Lu(OH)_3^0$ | $9 \times 10^{-13}$ | $2 \times 10^{-4}$ | — | — | — |
| Hf | | $4 \times 10^{-11}$ | $7 \times 10^{-3}$ | — | — | — |
| Ta | | $1 \times 10^{-11}$ | $2 \times 10^{-3}$ | — | — | — |
| W | $WO_4^{2-}$ | $5 \times 10^{-10}$ | $0{\cdot}1$ | $5{\cdot}6 \times 10^{-2}$ | $2 \times 10^3$ | $1{\cdot}2 \times 10^5$ |
| Re | $ReO_4^-$ | $2 \times 10^{-11}$ | $4 \times 10^{-3}$ | — | — | — |
| Os | | — | — | — | — | — |
| Ir | | — | — | — | — | — |
| Pt | | — | — | — | — | — |
| Au | $AuCl_2^-$ | $2 \times 10^{-11}$ | $4 \times 10^{-3}$ | — | — | $2 \times 10^5$ |
| Hg | $HgCl_4^{2-}$, $HgCl_2^0$ | $1{\cdot}5 \times 10^{-10}$ | $3 \times 10^{-2}$ | — | — | $8 \times 10^4$ |
| Tl | $Tl^+$ | $5 \times 10^{-11}$ | $1 \times 10^{-2}$ | — | — | — |
| Pb | $PbCO_3^0$, $Pb(CO_3)_2^{2-}$ | $2 \times 10^{-10}$ | $3 \times 10^{-2}$ | — | — | $4 \times 10^2$ |
| Bi | $BiO^+$, $Bi(OH)_2^+$ | $1 \times 10^{-10}$ | $2 \times 10^{-2}$ | — | — | — |
| Po | $PoO_3^{2-}$ - $PoO(OH)_2^0$ (?) | | | | | |

Table 7.1—*continued*

| Element | Chemical species | Total concentration (molar) | $\mu g\,l^{-1}$ | Barth | Residence time (years) Goldberg and Arrhenius | Most recent estimate* |
|---|---|---|---|---|---|---|
| At | | | | | | |
| Rn | Rn (gas) | $2 \cdot 7 \times 10^{-21}$ | $6 \times 10^{-13}$ | — | — | — |
| Fr | | | | | | |
| Ra | $Ra^{2+}$ | $3 \times 10^{-16}$ | $7 \times 10^{-8}$ | — | — | — |
| Ac | | | | | — | — |
| Th | $Th(OH)_4^0$ | $4 \times 10^{-11}$ | $1 \times 10^{-2}$ | — | — | $2 \times 10^2$ |
| Pa | | $2 \times 10^{-16}$ | $5 \times 10^{-8}$ | — | — | — |
| U | $UO_2(CO_3)_2^{4-}$ | $1 \cdot 4 \times 10^{-8}$ | $3 \cdot 2$ | — | — | $3 \times 10^6$ |

* Data for the most recent estimate taken principally from Goldberg *et al.* (1971).

The concept of the residence time of an element in sea water was introduced by Barth (1952), and is given, in years, as

$$\tau = \frac{A}{dA/dt} \qquad (7.1)$$

where $A$ is the total amount of the element dissolved in the ocean, and $dA/dt$ is the amount introduced to, or removed from, the ocean each year assuming a steady state. Barth used data on river input and could calculate only a few values because of the lack of accurate input data. Goldberg and Arrhenius (1958) carried out a major re-investigation of the problem using instead, estimates of the rates at which elements are removed from sea water calculated from their rates of incorporation into pelagic sediments. Their conclusions were in remarkable agreement with those of Barth, in spite of the many assumptions involved, and subsequent investigations have not altered the broad picture. Figure 7.1 shows a plot of the logarithm of the residence time against atomic number, and in spite of the large gaps in the data the expected periodicity can be observed. The alkali and alkaline earth metals (excepting Be with its marked covalent characteristics) have long residence times which decrease with increasing atomic number following the decrease of the effective hydrated ionic radii and hydration energies. The elements Al, Ti, Cr, Fe and Th, which readily form insoluble hydroxides, have short residence times, and within each transition series there is a tendency for residence time to increase with increasing atomic number.

Quoted residence times should be viewed with caution, since they are based on the assumption that complete mixing of the input takes place in times that are short compared to the residence times. There is general agreement that deep water circulation of the ocean takes place on a time scale of ca. $1 \times 10^3$ years, and the residence times derived for Al ($10^2$ years), Fe ($2 \times 10^2$ years), La ($6 \times 10^2$ years) and Th ($2 \times 10^2$ years) are significantly shorter than this. More importantly, the concept implies that material is introduced from rivers directly into the deep ocean, and thus no account is taken of estuarine removal processes which are very important for some elements. Goldberg et al. (1971) have derived residence times for elements in surface waters, $\tau s_x$, where

$$\tau s_x = \frac{C_x(\text{surface})}{C_x(\text{deep})} \tau s_{Cl^-} \qquad (7.2)$$

where $C_x(\text{surface})$ and $C_x(\text{deep})$ are the concentrations of element $x$ in surface and deep waters respectively, and $\tau s_{Cl^-}$ represents the residence time of chloride in surface waters in which it is assumed to be a conservative property i.e. that it is transported only by mixing. Since the residence time of water in the surface mixed layer is ca. 10–20 years, then silicon and phosphorus,

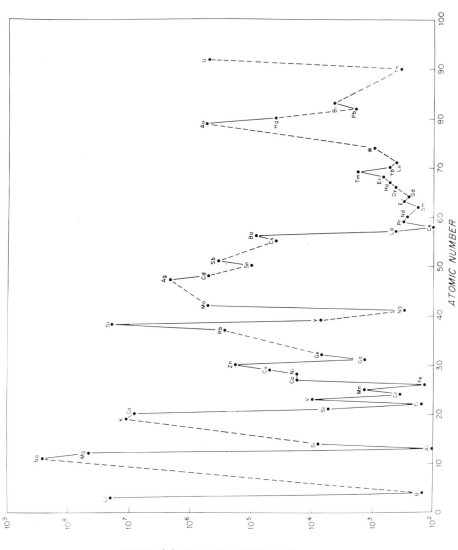

Fig. 7.1. The residence time of several elements in sea water plotted against atomic number.

which exhibit variations of more than an order of magnitude with depth, will have residence times in surface water of less than 2 years, and barium and radium would be removed in 5 to 10 years.

## 7.3. INPUT AND REMOVAL

Detailed information on the transport of elements to, and removal from, the ocean is given elsewhere in this book. However, for convenience a brief summary is given below.

### 7.3.1. INPUT

An excellent review of the known mass fluxes of weathered material has been given by Garrels and Mackenzie (1971). They have estimated that $2.5 \times 10^{16}$ g of material are added to the ocean each year, and that 90% of this load ($2.25 \times 10^{16}$ g yr$^{-1}$) is transported by rivers, $0.42 \times 10^{16}$ g yr$^{-1}$ in the dissolved state and $1.83 \times 10^{16}$ g yr$^{-1}$ as solids. Other fluxes are $0.2 \times 10^{16}$ g yr$^{-1}$ transported by ice, of which 90% is from the Antarctic continent; the atmospheric dust flux is only some $0.0006 \times 10^{16}$ g yr$^{-1}$. It should be noted that the particulate load of rivers has a distribution that is markedly skewed both geographically and with time; the quantity of material carried during brief periods of flood far exceeds the normal transport, and the rivers of South East Asia (e.g. the Mekong) supply some 80% of the global total, but contribute only 38% of the dissolved solids (Garrels and Mackenzie, 1971).

*River input.* Data on the chemical composition of river waters has been compiled by Livingstone (1963) and by Durum and Haffty (1963). Figure 7.2. which has been redrawn from that by Durum and Haffty shows the ranges of concentrations found for 17 minor elements in the world's largest rivers, together with the median values for North America. The data include both the dissolved and particulate load, and it should be noted that only 5 elements (Al, Fe, Mn, Ba, Si) have concentrations that range much above 100 μg l$^{-1}$, the remainder having median values of 10 μg l$^{-1}$ or less.

Although the chemical processes that occur at the river/ocean boundary are poorly understood they are of great importance if a proper interpretation is to be given to the residence times quoted in Table 7.1. In river plumes there is a transition from a medium of low to one of high ionic strength, and generally from low to high pH. In addition, there is ample opportunity for adsorption–desorption and precipitation-dissolution reactions to occur. Liss and Spencer (1970) have studied the removal of silica in a Welsh estuary by examining the deviation from linearity of silica-salinity diagrams, and a similar study has been carried out by Wollast and de Broeu (1971) for the

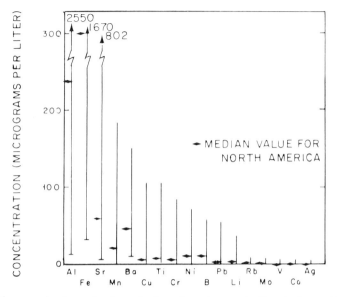

Fig. 7.2. The range in observed values of minor element concentrations in the world's large rivers (Durum and Haffty, 1963). Permission of Pergamon Press.

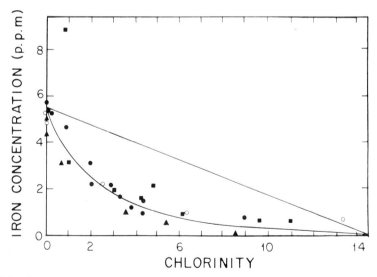

Fig. 7.3. Dissolved iron vs. chlorinity of samples from the Mullica River and Great Bay, New Jersey (Coonley et al., 1971). Permission of Elsevier.

Scheldt estuary. Coonley *et al.* (1971) have determined dissolved iron in a New Jersey river and observed a marked depletion at higher salinities, as is shown in Fig. 7.3; this depletion is attributable to the oxidation of Fe(II) to Fe(III) and the subsequent hydrolysis, flocculation and settling.

Not all estuarine processes lead to a net removal of chemical elements from the incoming river water. Kharkar *et al.* (1968) have conducted an elegant study of the stream supply of dissolved Ag, Mo, Sb, Se, Cr, Co, Rb and Cs to the oceans. Samples from the Mississippi and some lesser United States rivers, together with samples from the Rhone and Amazon, were filtered and after freeze-drying the minor element concentrations were determined by neutron activation analysis. A summary of their data is shown in Table 7.2.

TABLE 7.2

*The minor element composition of river waters (Kharkar et al., 1968)*

| Element | Average concentration $\mu g\, l^{-1}$ | Range $\mu g\, l^{-1}$ |
|---|---|---|
| Ag | 0·30 | 0·10–0·55 |
| Sb | 1·1 | 0·27–4·90 |
| Cr | 1·4 | 0·10–3·08 |
| Co | 0·19 | 0·037–0·44 |
| Rb | 1·1 | 0·56–1·61 |
| Cs | 0·02 | 0·011–0·043 |
| Se | 0·20 | 0·114–0·348 |
| Mo | 1·8 | 0·44–5·67 |

In addition, they investigated estuarine adsorption and desorption processes by carrying out laboratory studies with montmorillonite, illite, kaolinite, ferric oxide, manganese dioxide, hydrated ferric oxide and peat. These solid phases were equilibrated with radioactive tracers ($^{60}$Co, $^{110m}$Ag, $^{75}$Se, $^{51}$Cr, $^{99}$Mo) in distilled water at stable element concentrations approximating to those of stream water. The sample was then filtered and placed in natural sea water, and again allowed to reach equilibrium. The results showed net desorption of Co, Ag and Se from the clays and the manganese dioxide because of the displacement of the surface-held metals by the calcium and magnesium ions present in sea water. On the basis of their data they calculated the supply of "soluble" Co, Ag and Se to the oceans (Table 7.3) and proposed modified oceanic residence times. These conclusions have not yet been verified by experiments in the field.

*Other inputs.* Inputs of minor elements other than those from rivers are difficult to assess, but they may be important locally, as, for example, off the African and Australian coasts where desert dust storms transport large

TABLE 7.3
*Dissolved, desorbable and total supply of Co, Ag and Se to the oceans*
*(taken from Kharkar et al., 1968)*

| Element | Average in stream $\mu g \, l^{-1}$ | Absorbed in stream and desorbed in sea water $\mu g \, l^{-1}$ | Total "soluble" load $\mu g \, l^{-1}$ |
|---|---|---|---|
| Co | 0·15 | 0·39 | 0·54 |
| Ag | 0·31 | 0·02 | 0·33 |
| Se | 0·20 | 0·04 | 0·24 |

quantities of material, or for specific elements volatilized by the industrial activities of man.

Hoffman *et al.* (1969) have examined the distribution of V, Cu and Al in the lower atmosphere between California and Hawaii, and reported average values as follows: Cu, 13·7 ng m$^{-3}$; V, 0·17 ng m$^{-3}$; Al, 18 ng m$^{-3}$ (for samples from Hawaii). On the basis of the ratios of the concentrations of these elements to that of sodium they postulated that the particulate matter had a non-marine source and suggested that the Al and V were derived from continental weathering. The source of the high Cu concentrations was unknown,

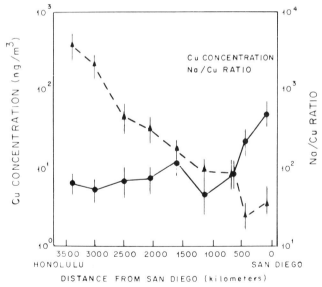

FIG. 7.4. Variation in the copper concentration (continuous line) and the sodium/copper ratio (broken line) for atmospheric particulate matter with distance from San Diego to Honolulu (from Hoffman *et al.*, 1969). Permission of the American Chemical Society.

although the variation of the Na/Cu ratio with distance from land (Fig. 7.4) pointed to the probability of an industrial origin. Lead aerosols in the marine atmosphere have been investigated by Chow *et al.* (1969) who reported data from the Pacific in the range $0.3–1.5$ ng m$^{-3}$.

Although vulcanism may have been a significant source of trace metals in the past it does not seem to be so at present. Horn and Adams (1966) have re-examined the classical geochemical balance given by Goldschmidt (1933) in which the amount of primary igneous rock which has been weathered is compared to the composition of the sediments and sea water. They concluded that 10 elements (Cl, S, Mn, Br, B, Pb, As, Mo, I, Se) could not be brought into balance; of these, the volatile elements may well have a volcanic origin. Boström and Peterson (1966) and Boström *et al.* (1969) have reported large areas of alumina-poor and ferro-manganese-rich sediments derived from hydrothermal precipitates on the East Pacific Rise.

### 7.3.2. REMOVAL

For an element in abundant supply, if no other removal process is operative, the concentration in the sea should be controlled by the solubility of its least soluble compound. However, as Goldschmidt (1937) noted, most metals are markedly undersaturated with respect to their least soluble compounds, and the supply of metals during geological time has been more than sufficient to attain saturation. Clearly, some other removal mechanisms must be operating, and Goldschmidt suggested that adsorption on ferric oxide precipitates was a likely process.

Krauskopf (1956), in a classic, but now dated, paper, undertook a major examination of the factors controlling the concentration of 13 elements (Zn, Cu, Pb, Bi, Cd, Ni, Co, Hg, Ag, Cr, Mo, W, V) in sea water. He computed solubility products for their likely most insoluble compounds and suggested that only for Sr and Ba might saturation be a control. In anoxic environments the precipitation of metallic sulphides possibly controls Cu, Zn, Hg, Ag, Cd, Bi and Pb. However, since oceanic anoxic environments are rare, it is unlikely that this mechanism exerts a significant overall control in the ocean. Adsorption onto ferric oxide, manganese dioxide, apatite, montmorillonite, dried plankton and peat moss was tested, and found to be a possible control for Zn, Cu, Pb, Bi, Cd, Hg, Ag and Mo. Finally, an assessment of biological removal mechanisms was made from the known concentration factors of marine organisms, from which it was concluded that V, Ni and possibly Co, W and Mo, were controlled by these means. Krauskopf's work may be criticized on many grounds—the lack of careful pH and temperature controls, the excessive ratios of adsorbent to solution and the poor analytical techniques that were available at that time. Nonetheless, it remains a coherent attack on an important problem.

Removal of trace elements from sea water by inorganic processes takes place by adsorption and precipitation. Precipitation of an element through the exceeding of a solubility product occurs but rarely in the oceans, although it may be significant in estuaries. Adsorption, or scavenging, by suspended sediment and sinking detrital particles is the principal inorganic removal mechanism that has been invoked. The nature of this process makes it difficult to evaluate its quantitative oceanic significance, since chemical analyses of suspended particles do not distinguish between adsorbed and intrinsic components, and laboratory studies of uptake onto pure solid phases or natural sediments can approximate only roughly to the natural conditions. Most experimental work has been carried out with radionuclides, and equilibrium between the isotope and the stable element is not always achieved.

Adsorption is attributable to both weak electrostatic attraction, dependent on the surface charge of the solid phase at the appropriate pH, and to stronger specific interactions involving some degree of chemical bonding (see Chapter 4). Thus, the selectivity sequence will vary with the solid phase; the probable sequence for decreasing uptake by deep-sea sediments as deduced from laboratory radio-isotope experiments is $^{147}$Pm $> ^{106}$Ru $> ^{54}$Mn $> ^{95}$Zr/ Nb $> ^{59}$Fe $> ^{65}$Zn $> ^{86}$Rb $> ^{137}$Cs, U, Pu $> ^{90}$Sr $> ^{45}$Ca (Duursma, 1969). The corresponding sequence for uptake by ferro-manganese nodules based on direct chemical analyses appears to be Co $>$ Ni $>$ Cu $>$ Zn $>$ Ba $>$ Sr $>$ Ca $>$ Mg. For both solid phases the greater selectivity for transition metals than for alkali metals is apparent. The most direct evidence for adsorption as a removal mechanism comes from comparisons of laboratory studies of the uptake of radionuclides with field observations of the fate of radioactive waste (Duursma and Gross, 1971). It is possible to define a distribution coefficient for an adsorptive process as the ratio of the amount of adsorbed radionuclide present in unit volume of dry sediment to that in unit volume of sea water at equilibrium. Distribution coefficients for several isotopes for various size fractions of sediment are shown in Table 7.4 and the data for the activity found in natural sediments is given for comparison. The correlations strongly suggest that removal occurs by an adsorptive process.

Removal of trace elements by plankton can take place through assimilation by organisms, with subsequent transport as faecal material, or by the body of the organism on sinking. Adsorption onto the organic detritus present in sea water can also occur. Calculations based on simple uptake show that the fact that an element is concentrated in various planktonic species does not necessarily lead to large variations in the dissolved element concentration.

*Distribution of sorbed radionuclides and total radioactivity in different size fractions in three types of sediment\* by permission of the National Academy of Sciences.*

| Sediments Origin | Size fractions μm | % wt | Distribution coefficients for each size fraction (value × 10²) | | | | | | | | | |
|---|---|---|---|---|---|---|---|---|---|---|---|---|
| | | | $^{90}$Sr | $^{137}$Cs | $^{106}$Ru | $^{59}$Fe | $^{65}$Zn | $^{60}$Co | $^{147}$Pm | $^{54}$Mn | $^{65}$Zr/Nb | $^{144}$Ce |
| Dutch Wadden Sea* | >64 | 51·1±4·9 | 0·0 | 0·0 | 0·0 | 0·0 | 0·0 | 5·8 | 0·0 | 0·0 | 0·3 | 0·4 |
| | 32–64 | 21·7±4·5 | 0·0 | 1·6 | 3·1 | 76·0 | 0·0 | 3·7 | 0·0 | 1·4 | 42·0 | 3·3 |
| | 16–32 | 9·5±3·2 | 0·0 | 5·4 | 0·0 | 53·0 | 260·0 | 65·0 | 0·0 | 35·0 | 66·0 | 120·0 |
| | 8–16 | 5·6±2·0 | 0·0 | 12·0 | 7·3 | 370·0 | 490·0 | 59·0 | 0·0 | 8·0 | 1040·0 | 950·0 |
| | 4–8 | 8·4±1·9 | 0·0 | 16·0 | 5·2 | 510·0 | 380·0 | 430·0 | 320·0 | 480·0 | 1220·0 | 540·0 |
| | <4 | 3·7±0·8 | 26·0 | 6·2 | 4·7 | 540·0 | 112·0 | 220·0 | 280·0 | 97·0 | 670·0 | 124·0 |
| Mediterranean off Monaco* | >64 | 0 | — | — | — | — | 5·2† | — | — | — | — | — |
| | 32–64 | 3·0±1·3 | 15·0 | 2·8 | 41·0 | 15·0 | 68·0 | 380·0 | 24·0 | 19·0 | 117·0 | 32·0 |
| | 16–32 | 15·2±4·2 | 2·5 | 1·3 | 11·0 | 101·0 | 140·0 | 540·0 | 0·8 | 41·0 | 290·0 | 73·0 |
| | 8–16 | 36·5±8·6 | 7·6 | 1·4 | 22·0 | 118·0 | 150·0 | 730·0 | 4·5 | 61·0 | 150·0 | 82·0 |
| | 4–8 | 39·1±7·5 | 9·3 | 2·3 | 34·0 | 183·0 | 140·0 | 820·0 | 101·0 | 76·0 | 310·0 | 147·0 |
| | <4 | 6·2±1·9 | 5·9 | 0·5 | 7·8 | 63·0 | 97·0 | 140·0 | 130·0 | 23·0 | 160·0 | 41·0 |

| Sediments Origin | Size fractions μm | % wt | Radioactivity μCi/g dry weight |
|---|---|---|---|
| Irish Sea Pipeline Outlet, Sellafield‡ | 100–200 | 26·0 | $2·0 \times 10^{-4}$ |
| | 50–100 | 27·4 | $3·0 \times 10^{-4}$ |
| | 20–50 | 22·6 | $9·0 \times 10^{-4}$ |
| | 10–20 | 14·0 | $2·5 \times 10^{-3}$ |
| | 4–10 | 6·1 | $4·6 \times 10^{-3}$ |
| | <4 | 4·0 | $6·2 \times 10^{-3}$ |

Specific radioactivity $= 1·3 \pm 0·4 \times 10^{-6}\ \mu\text{Ci/cm}^2$

\* The Dutch Wadden Sea and Mediterranean sediments were suspended in radionuclide-enriched seawater for one month (30 g sediment in 20 litres of seawater); size fractions were separated by sedimentation techniques (Duursma and Eisma, unpublished).

† For $^{65}$Zn, another Mediterranean sediment was used with 3·3‰ >64 μm; 7·6‰ 32–64 μm; 49·2‰ 16–32 μm; 24·9‰ 8–16 μm; 12·7‰ 4–8 μm; and 2·3‰ <4 μm.

‡ The Irish Sea sediments were exposed in the field to radionuclides discharged near Sellafield (Jones, 1960).

Spencer and Brewer (1969) have investigated the distribution of Cu, Zn and Ni in the Sargasso Sea and the Gulf of Maine. They calculated that in the Gulf of Maine phosphorus uptake by phytoplankton would be $4.34$ g m$^{-2}$ for the season from March to October. The Cu:P ratio in phytoplankton is approximately $6.5 \times 10^{-3}$, indicating that Cu uptake by phytoplankton should be 28 mg m$^{-2}$ over the same period. Assuming a dissolved Cu concentration of 3 µg Cu l$^{-1}$ then a 50 m deep water column contains 150 mg Cu m$^{-2}$. If the extreme (and unverifiable) assumption is made that copper is lost solely to the sediments, and if it is further assumed that none is regenerated in the water column or added by mixing, then the surface waters would be depleted of copper by 18% at the end of the season. In practice no such depletion has been observed.

Concentration factors for marine plankton may be defined as the ratio of the concentration of an element in an organism to that concentration directly available from the organism's environment. For phytoplankton this refers to the accumulation of trace elements directly from sea water. The accumulation of trace elements by marine organisms is discussed briefly in Section 7.9 (see also Table 7.5). For a more detailed treatment of the subject

TABLE 7.5

*Elements accumulated by phytoplankton to levels at least $10^3$ times those in sea water (Lowman et al., 1971).*

| Element | Concentration factor | Element | Concentration factor |
|---------|---------------------|---------|---------------------|
| Al | $1 \times 10^5$ | Pb | $4 \times 10^4$ |
| As | — | Mn | $4 \times 10^3$ |
| Be | $1 \times 10^3$ | Ni | $5 \times 10^3$ |
| C | $4 \times 10^3$ | Nb | $1 \times 10^3$ |
| Cd | — | P | $3 \times 10^4$ |
| Ce | $9 \times 10^4$ | Pu | $2.6 \times 10^3$ |
| Cr | $2 \times 10^3$ | Sc | $2 \times 10^3$ |
| Co | $1 \times 10^3$ | Ag | $2 \times 10^4$ |
| Cu | $3 \times 10^4$ | Zn | $2 \times 10^4$ |
| I | — | Zr | $6 \times 10^4$ |
| Fe | $4 \times 10^4$ | | |

the reader should refer to Lowman *et al.* (1971). The biologically driven flux of an element through the ocean is determined not only by the concentration factor in phytoplankton, but by grazing and excretion of the phytoplankton crop by zooplankton. The vertically migrating zooplankton add a further complication. Figure 7.5, based on the work of Kuenzler (1965), shows the factors involved in the vertical biological transport of iron. If these calcula-

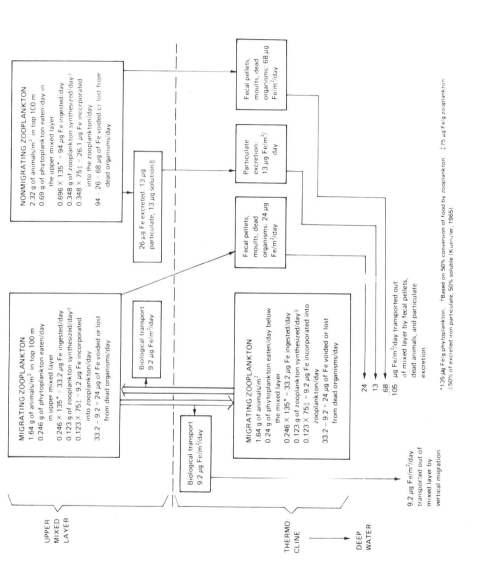

FIG. 7.5. Block diagram of the vertical transport of iron by marine organisms in the Northeastern Pacific, based on Kuenzler (1965). Reproduced by permission of the National Academy of Sciences.

tions are correct, then *ca.* 90% of the flux of iron through the thermocline is controlled by the flux of faecal pellets, dead organisms and moulted exoskeletons of zooplankton species.

There have been few detailed examinations of these processes. Boothe and Knauer (1972) have determined the concentrations of As, Cd, Co, Cr, Cu, Fe, Mn, Pb and Zn in a brown alga and in the faecal material of a crab feeding on it. With the exception of Cr and Cd, which showed lower concentrations in the faecal material than in the alga, all other elements showed enrichment factors of 2·3–14 in the faeces. This enrichment of trace elements in the faeces lends support to the suggestion by Polikarpov (1966) that primary consumers accumulate only a small portion of their radionuclide (and hence, minor element) content from food, and that adsorption from the surrounding water is the principal mechanism by which minor elements accumulate in body tissues. Polikarpov's theory is by no means proven, and Lowman *et al.* (1971) have pointed out that the data to support this theory comes from elements for which low concentration factors exist in the food organism. Martin (1970) has examined the transport of trace metals by moulted copepod exoskeletons. He compared analyses of 10 surface zooplankton samples with data from 12 samples collected at depths of 100 m or more and found that average values for Cu, Fe, Mn, Ni, Pb, Sr and Zn were higher in the deep samples. It was suggested that the greater availability of food in the surface layers resulted in more rapid moulting of the zooplankton, and thus less time was available for adsorption of metals from the water by the exoskeleton. A summary of the data is given in Table 7.6. Direct analyses of the moults,

TABLE 7.6

*Variations in the elemental composition of zooplankton from surface and deep (>99 m) waters (Martin, 1970).*

| Element | Concentration ($\mu g\ g^{-1}$ ash) Surface | Deep |
|---|---|---|
| Pb | 117 | 183 |
| Zn | 657 | 1809 |
| Fe | 2900 | 4200 |
| Cd | 16 | 15 |
| Co | 44 | 37 |
| Cu | 115 | 132 |
| Ni | 100 | 150 |
| Mn | <70 | 88 |
| Sr | 890 | 1140 |
| Ca | 103900 | 105000 |
| Mg | 45700 | 46200 |

sampled in the water column, are not available. The moults sink rapidly and the rare large particles are not caught with conventional sampling techniques.

Turekian *et al.* (1973) have attempted to assess the rate of removal of trace elements by pteropod tests. Pteropod tests are aragonitic and because they are more soluble than the calcitic tests of foraminifera and coccoliths, they dissolve at depths greater than *ca.* 3500 m, releasing their associated metals upon dissolution. Living pteropods were sampled, ashed and analyzed for Fe, Ce, La, Sm, Eu, Th, Sc, Cr, Co, Sb and Se. A comparison of the minor element analyses of pteropods with those of non-calcareous plankton showed a marked similarity. The elemental correlations for both sets of analyses showed that the association with iron was of major importance. It was suggested that iron hydroxides associated with the tests scavenged dissolved elements from sea water. The low Ce:La, Co:Cr and Th:La ratios of the tests as compared to those of detrital clay material were characteristic of this authigenic component. The rate of supply of $CaCO_3$ to the sediments is *ca.* 2 g $CaCO_3$ cm$^{-2}$ (1000 yr)$^{-1}$, carrying with it 600 μg Fe cm$^{-2}$ (1000 yr)$^{-1}$ and the associated trace elements. The $CaCO_3$ phase alone appeared to be pure, and to transport comparatively minor quantities. The absolute flux of minor elements was not estimated, although the addition of iron, as a solid phase component, to the deep water by this process is *ca.* 2 μg l$^{-1}$ (1000 yr)$^{-1}$.

## 7.4. SAMPLING AND STORAGE OF SEA WATER SAMPLES

The analytical chemistry of sea water is discussed in Chapter 19; however, a brief discussion of sampling and storage is necessary here in order to provide a basis for the discussion of data for the various elements available in the literature.

### 7.4.1. SAMPLING

The vast majority of oceanic water samples have been collected with the brass Nansen bottle, usually lined with an inert substance such as Teflon. These bottles are not ideal, and imperfections in the inert inner coating may lead to reaction of the brass with sea water as in (7.3).

$$2 (Cu, Zn) + O_2 + 4 HCO_3^- \rightarrow 2 (Cu, Zn)^{2+} + 4 CO_3^{2-} + 2 H_2O \quad (7.3)$$
brass

This effect is illustrated by data given by Park (1968) which show (Table 7.7) that a flaw in the coating of the bottle at 7100 m, resulted in a depletion of *ca.* 0·16 ml $O_2$ l$^{-1}$ and an increase in the alkalinity of 0·04 meq l$^{-1}$.

TABLE 7.7

*Effect of defective coating of water bottle on composition of sample (taken from Park, 1968). Deep sea temperature and chemical concentrations at hydrographic station HA-56 of R/V Yaquina Cruise YALOC-66, at 50° 28′N, 176° 14′W. (Nansen bottles were coated with Teflon, but the coating of the bottle at 7100 m was faulty. Compare the chemical concentrations at 7100 m with that of other depths).*

| Depth (m) | Temperature (°C) | Salinity (‰) | Oxygen (ml l⁻¹) | pH | Alkalinity (meq l⁻¹) |
|---|---|---|---|---|---|
| 6900 | 1·81 | 34·693 | 3·65 | 7·87 | 2·58 |
| 7000 | 1·84 | 34·692 | 3·67 | 7·86 | 2·57 |
| 7100 | 1·84 | 34·685 | 3·52 | 7·90 | 2·61 |
| 7200 | 1·86 | 34·693 | 3·69 | 7·85 | 2·58 |

Unfortunately copper and zinc were not also determined in these samples. Depletion in oxygen sampled in Nansen bottles has been observed many times, notably by Spencer (1972) who found a difference of 0·07 ml $O_2$ $l^{-1}$ between samples taken from Nansen bottles with supposedly good coatings and non-metallic Niskin samplers. Both Bowen (1968) and Spencer and Brewer (1970) have reviewed the problems connected with the selection of materials for the fabrication of sampling devices.

A basic requirement of a sampling device is that it should deliver from the appropriate depth a representative sample that has neither mixed with the surrounding sea water after the sampler has closed, nor reacted with the walls of the sampler. Metallic devices are clearly unsuitable for the determination of trace metals, but may be used for the collection of samples for gas analysis (other than oxygen), or for analysis for other non-metallic species. A reversing bottle of the Nansen type, fabricated of polycarbonate, has been made by the Hydrobios Co., Kiel, Germany, and the British National Institute of Oceanography produces a polypropylene bottle fitted with neoprene end-caps. Neoprene should be avoided for trace metal work because of the large quantities of zinc which may be leached from it by sea water. The sampling device currently in general use in the United States is the P.V.C. Niskin bottle, available from General Oceanics Inc., Miami, Florida, in sizes up to 30 l. This sampler consists of a P.V.C. cylindrical barrel fitted with lucite end caps and closed with an internal spring, usually Teflon coated. It would be desirable to eliminate the internal spring and it seems probable that this will be done in future designs.

Robertson (1968a) has assessed the probable contribution of many materials to the trace element levels found in sea water samples. He determined the concentrations of Sc, Cr, Fe, Co, Cu, Zn, Ag, Sb, Cs and Hf not only in sea water itself, but also in commonly used constructional materials, such as Teflon, Plexiglass, P.V.C., surgical rubber, neoprene, nylon, and polyethylene. In addition, a variety of chemical reagents, sealants, and papers were also examined. He concluded that rubber materials contain unsatisfactory levels of zinc, and that nylon is unsatisfactory because of its high level of cobalt (0·14%). He found P.V.C. to contain significant levels of zinc, iron, antimony and copper, although leaching of these elements into a sea water sample could not be detected. Teflon and Plexiglass were found to be remarkably free from minor element contamination. Containers of borosilicate glass or quartz, other than "Spectrosil", were found to have undesirably high levels of antimony, iron and zinc. Polyethylene bottles were judged to be the most desirable containers for the storage of sea water samples.

### 7.4.2. STORAGE

Many workers have cited the fact that their data were lower than previously reported values as evidence that their samples were not contaminated. However, it does not necessarily mean that their samples were unaltered since adsorption of minor elements on to the walls of the container may have significantly altered their concentrations in the sample.

Robertson (1968b) examined, by means of radioactive tracers, the adsorption of several trace elements (Sc, Fe, Zn, Co, Sr, Rb, Ag, In, Sb, Cs, U) onto various container surfaces. His data are shown diagrammatically in Fig. 7.6. The experiments were made in Pyrex glass bottles at pH 8·0 and in polyethylene bottles at pH 8·0 and 1·5. The data clearly show serious losses of In, Sc, Fe, Ag, U and Co on storage of unacidified samples. Samples stored in the acidified condition in polyethylene gave the minimum loss, although adsorption of Sc and U was still significant.

Confirmation of these data has been obtained by direct analysis of oceanic samples from a test station in the Northeast Pacific by Spencer et al. (1970). At this station a profile of 16 water samples was collected. These were split into filtered and unfiltered groups and stored in polyethylene bottles in both the acidified (HCl, pH 1·8) and unacidified states. The samples were analysed for Sr, Cs, Rb, Co, Ni, Sc, Cu, Sb, Zn, Fe and U both initially and after up to 100 days. The data for the inter-comparison of the analyses of acidified and unacidified samples are given in Table 7.8. Because of the low concentration of particulate matter in ocean water, none of the elements, with the exception of iron, showed any consistent difference between filtered and unfiltered samples. It should be noted that this would not be true for estuarine and coastal waters in which the particulate load is much higher. The data in Table 7.8 indicate that zinc, cobalt and iron were significantly higher in acidified samples, but it was not possible to establish whether there was a significant difference for copper and scandium. It should be clear from the foregoing that storage in polyethylene in the acidified condition represents the preferred procedure for many trace elements.

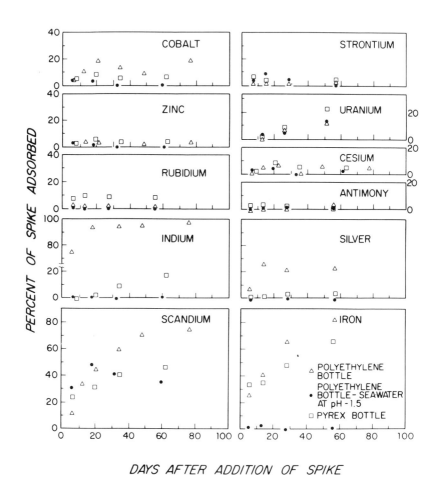

DAYS AFTER ADDITION OF SPIKE

FIG. 7.6. The adsorption of radiotracers from sea water on to various container surfaces (Robertson, 1968b). Permission of Elsevier.

TABLE 7.8

*Intercomparison of sea water analyses for various elements carried out by four laboratories\* in order to assess the difference between storage of samples in the acidified, and in the unacidified condition.*

| Element | Laboratory | Mean acidified $\mu$g kg$^{-1}$ | Mean unacidified, $\mu$g kg$^{-1}$ | % diff. | Probability (null hypothesis) | % Diff. Robertson (1968b) |
|---------|------------|---------|-----------|------|-------------|----------|
| Sb | Battelle | 0·217 | 0·212 | 2 | >0·10 | 2 |
| Co | Battelle | 0·029 | 0·021 | 30 | <0·0001 | 20 |
| Sc | Battelle | 0·00056 | 0·0005 | 14 | >0·10 | 30 |
| Cu | Woods Hole | 1·37 | 1·12 | 18 | 0·088 | |
| Ni | Woods Hole | 2·29 | 2·39 | −4 | >0·10 | |
| Fe | Woods Hole | 5·57 | 1·50 | 73 | <0·0001 | 80 |
| Zn | Woods Hole | 3·78 | 1·73 | 54 | <0·0001 | 5 |
| Zn | Battelle | 2·78 | 1·67 | 40 | <0·0001 | 5 |
| U | Yale | 3·18 | 3·14 | 2 | >0·10 | 0 |
| U | Battelle | 3·40 | 3·29 | 3 | 0·02 | 0 |
| Sr | Battelle | 8·02 | 8·02 | | >0·10 | |
| Sr | Yale | 7·89 | 7·96 | −1 | >0·10 | |
| Sr | Woods Hole | 8·11 | 8·03 | 1 | >0·10 | |
| Cs | Battelle | 0·29 | 0·30 | −2 | >0·10 | 2 |
| Cs | Yale | 0·33 | 0·33 | | >0·10 | 2 |
| Cs | Scripps | | 0·28 | | | |

\* From Spencer *et al.* (1970). Participating laboratories were Battelle Northwest, Yale University, Woods Hole Oceanographic Institution and Scripps Institution of Oceanography. The probability (null hypothesis) that the means of the two sets of analyses are the same is given and the percentage difference between the two sets of data is compared to the difference found by Robertson (1968b) from radio tracer experiments. By permission of the American Geophysical Union.

7.4.3. ANALYSES

There have been few attempts to assess the reliability of sea water trace element analyses. Cooper (1958) pointed out many of the problems of storage and contamination, and suggested an international programme for the exchange of standard samples in order to compare data from different laboratories.

Brewer and Spencer (1970) have carried out a trace element intercalibration study in which 26 participating laboratories analysed two sea water samples both of which were acidified and stored in polyethylene bottles. Data for 15 elements (Sb, Cd, Cs, Co, F, Cu, Fe, Pb, Mn, Ni, Ag, Sr, Rb, U, Zn) were reported by two or more laboratories thus permitting comparison. Each sample was drawn from a large (200 l) polyethylene drum into numbered 2 l polyethylene bottles which were sent to participating laboratories in a random sequence. Inspection of the data revealed no trend with time of analysis or with bottle number, and it appears that the sample was homogeneous and stable with respect to storage. The results however were disturbing: only 5 elements were determined (within the entire laboratory set) with a coefficient of variation of less than 10%, and none of these 5 were transition metals (Table 7.9). No one method of analysis was found to be

TABLE 7.9

*Coefficients of variation of the analyses of several trace elements in sea water as determined by a number of laboratories (Brewer and Spencer, 1970).*

| Element | Coefficient of variation (%) | Abundance ($\mu g\ kg^{-1}$) |
|---------|------------------------------|------------------------------|
| Sr | 2·5 | 8100 |
| F | 2·8 | 1350 |
| Rb | 5·4 | 121 |
| Cr | 5·5 | 0·3 |
| U | 5·6 | 3·3 |
| Sb | 10·2 | 0·4 |
| Zn | 18·4 | 5 |
| Cu | 20·4 | 3 |
| Co | 22·0 | 0·1 |
| Mn | 24·5 | 1·5 |
| Fe | 25·8 | 14 |
| Pb | 29·2 | 5 |
| Ni | 33·5 | 2 |

superior, but two interesting points emerged; firstly, the methods based on solvent extraction for Co, Fe and Zn gave higher results than "total" methods based on the neutron activation analysis of dried sea salts, thus negating the suggestion that the solvent extraction techniques might measure only part of the element present, and secondly, the analyses for iron indicated that calibration of the iron determination was a greater source of error than the analysis itself in any one laboratory, i.e. the determinations were precise but not accurate.

If this study really reflects the "state of the art" then it is evident that care must be exercised when comparison is made between data from different laboratories, using samples obtained from different parts of the ocean. In particular, it is imprudent to draw conclusions from weak regional trends that may be apparent in data from disparate sources. The perils of doing this are clearly demonstrated by Fig. 7.7 which shows the various analyses for zinc (plotted against bottle sequence number) obtained during the intercalibration study. The data range from 0·02 to 21·2 $\mu g\ kg^{-1}$.

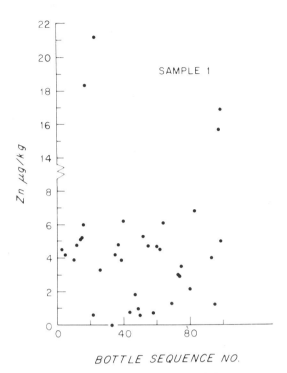

FIG. 7.7. Reported analyses of zinc, in sub-samples of a single homogeneous sea water sample, from laboratories participating in a trace element intercalibration study (Brewer and Spencer, 1970).

## 7.5. TREATMENT OF OCEANIC CHEMICAL DATA

Once information on a minor element has been compiled with respect to its distribution either with depth or geographic location, it is necessary to treat the data in such a way as to reveal the factors controlling the distribution. This is a fundamental problem of oceanography. The general equation governing the concentration, $c$, of a chemical species at any point in the ocean is given by

$$\frac{\partial c}{\partial t} = K_x \frac{\partial^2 c}{\partial x^2} + K_y \frac{\partial^2 c}{\partial y^2} + K_z \frac{\partial^2 c}{\partial z^2} - \frac{u \partial c}{\partial x} - \frac{v \partial c}{\partial y} - \frac{w \partial c}{\partial z} + J = 0 \quad (7.4)$$

for an ocean at a steady state with respect to time. $u$, $v$ and $w$ are the three dimensional components of the advective velocity in the directions $x$, $y$ and $z$; $K_x$, $K_y$ and $K_z$ are the corresponding eddy diffusion coefficients and $J$ is the net local rate of *in situ* accumulation (i.e. production rate – consumption rate) of the chemical species. For a conservative tracer $J = 0$, and for a chemically conservative radioactive tracer $J$ is equal to $-\lambda t$ where $\lambda$ is the radioactive decay constant. This equation could, in principle, be solved by means of six chemical tracers each of which had independent source and sink functions, if the condition that the ocean is in a steady state is satisfied (see also Section 1.7). In practice, certain simplifications have to be made by means of finite difference methods, or by the restriction of continuous advection–diffusion models to one or two dimensions.

### 7.5.1. BOX MODELS

Box, or discontinuous models have been used by Riley (1951), Keeling and Bolin (1967, 1968) and Holland (1971) among others, to describe the distribution of chemical variables within the ocean. The model proposed by Riley (1951) described the distribution of oxygen, phosphate and nitrate in the Atlantic ocean; 76 areal squares were selected such that the centre of each was 1000 km from the centre of the next, and to each square was assigned advective properties (upwelling, downwelling, horizontal transport) consistent with the physical oceanographic information. Eddy diffusion coefficients were also assigned on the basis of the observed salinity distribution, and up to seven depth layers were considered. This model was used in conjunction with the observed phosphate, nitrate and oxygen distributions and estimates of the regional rates of uptake and release resulting from productivity and decomposition at depth. It was then assumed that stability was attained with respect to transfer between boxes and the final computed distributions were compared with field data. The aim when using such models is not to reproduce a "picture" of the ocean, but to examine the sensitivity of the system to perturbation of its controls (advection, diffusion and productivity), and to clarify the effects of the various processes.

A model of this kind could be applied to any trace species about which a certain amount of regional information is available if realistic assumptions can be made regarding the biological and geochemical controls on its distribution. For the elements with which this chapter is concerned, only the stochastic model of the distribution of nickel in the Atlantic Ocean will be examined (Spencer, 1969).

Spencer pointed out that although many trace elements are known to be concentrated by marine organisms by factors of up to $10^5$ over their dissolved levels, direct evidence for the non-conservative distribution of these elements in sea water is scarce. The question arises as to whether the low rates of metal uptake by plankton, which are probably accompanied by relatively high rates of regeneration and excretion in the upper layers, are sufficient to cause observable regional differences in dissolved trace metal concentrations. Nickel was selected as a test element for the model since existing data (Schutz and Turekian, 1965a) revealed a significant difference in the dissolved nickel concentrations of North Atlantic surface water ($2{\cdot}0$ μg Ni kg$^{-1}$) and deep water ($3{\cdot}0$ μg Ni kg$^{-1}$). A fractile diagram illustrating this difference is shown in Fig. 7.8. The model used to examine this distribution was based on a first order absorbing Markov chain, consisting of a matrix of transition probabilities describing the transfer of an atom of nickel between certain defined states in the ocean. These states were assumed to be the dissolved and particulate forms of the element within each box. Transfer of the nickel

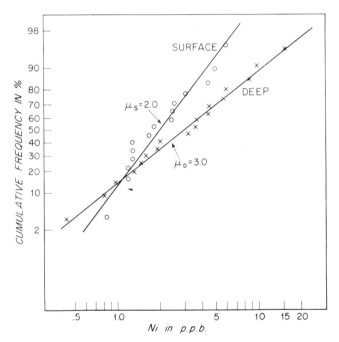

Fig. 7.8. Fractile diagram of dissolved nickel in North Atlantic surface and deep waters, based on the data of Schutz and Turekian (1965a). From Spencer (1969) by permission of the author.

between states and boxes was assumed to be governed by first order conditional probabilities, and each state was transient. In addition to these states, absorbing states (e.g. the ocean bottom) were included as nickel traps. The system is illustrated schematically in Fig. 7.9. Turbulent mixing was not considered, and estimates of vertical advection were those given by Riley

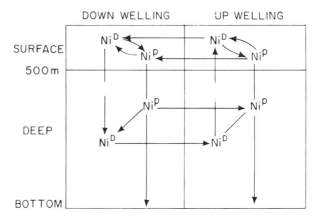

FIG. 7.9. Schematic representation of the stochastic model for nickel distribution in the Atlantic Ocean, from Spencer (1969). $Ni^D$ and $Ni^P$ refer respectively to dissolved and particulate nickel. Permission of the author.

(1951). Estimates of horizontal advection were selected from a variety of sources. It was assumed that phytoplankton take up nickel (i.e. transform it into a particulate state) in direct proportion to the nickel concentration in sea water, and that the Ni/C ratio of phytoplankton is $1 \times 10^{-4}$. Open ocean productivity has been estimated as $50–100 \text{ g Cm}^{-2} \text{ yr}^{-1}$ and thus, in these areas, the probability of a nickel atom transferring from the dissolved to the particulate state is 0·005 to 0·01, a value compatible with the lack of observed biologically-caused seasonal variations in the trace element content of surface waters. Deep-water re-solution of the particulate nickel was assumed initially to be 85 %, and the remaining 15 % was considered to be trapped in the sediment. It was assumed that the sources of nickel were the Amazon, the Congo and the rivers of North America.

Once the states and transfer probabilities had been assigned the computer model was cycled until it reached stability. Tests of the sensitivity of the model to changes in the variables were interesting; for instance doubling the rates of horizontal advection in the surface North Atlantic produced changes in the predicted nickel concentration of less than 1 %. The most sensitive feature of the model was the estimate of the rate of regeneration of nickel from particles in the surface layers. With regeneration rates of up to 60 %, differences in nickel concentration of a factor of two could be generated. However, at 80 % surface regeneration, regional variations were obscured. It is clear that the mere fact that an element is concentrated by phytoplankton does not ensure its non-conservative distribution in sea water; in fact, this may depend upon the efficiency of its removal from the surface layers by organisms higher in the food chain or on its incorporation into the more resistant tissues of the primary concentrator.

## 7.5.2. CONTINUOUS MODELS

Continuous models have recently received significant attention and, in particular, the one-dimensional (vertical) advection-diffusion model is widely used. The model was first described by Wyrtki (1962), and developed by Munk (1966) and Craig (1969). It is applicable when the horizontal advective velocities, or the horizontal concentration gradients, are zero. This condition implies that the model is applicable only over linear regions of the potential temperature-salinity diagram. In this case Equation (7.4) reduces to

$$K_z \frac{\partial^2 c}{\partial z^2} - \frac{w \partial c}{\partial z} + J = \frac{\partial c}{\partial t} = 0 \qquad (7.5)$$

for a steady state ocean. If one assumes that the ratio $(z^*)$ of the vertical eddy diffusion coefficient $(K_z)$ to the vertical advective velocity $(w)$ is constant over the mixing interval $(z_m)$ being studied, then for a stable conservative tracer $(J = 0)$ the solution of this equation is

$$(c - c_o) = (c_m - c_o)\frac{(e^{z/z^*} - 1)}{(e^{z_m/z^*} - 1)} \qquad (7.6)$$

where $c_o$ is the concentration at the lower boundary, $z_o$, and $c$ is the concentration at depth $z$, $z$ being positive upward. Defining $f(z)$ as

$$f(z) = \frac{(e^{z/z^*} - 1)}{(e^{z_m/z^*} - 1)} \qquad (7.7)$$

then for a non-conservative tracer with an *in situ* production or consumption rate invariant with depth, the solution of (7.5) is given by

$$(c - c_o) = (c_m - c_o)f(z) + J/w(z - z_m)f(z) \qquad (7.8)$$

For a non-conservative tracer with $J$ varying exponentially with depth (as is frequently found to be the case for the distribution of dissolved oxygen) then

$$J = J_0 e^{-\mu z} \qquad (7.9)$$

and the solution is

$$(c - c_o) = (c_m - c_o)f(z) + \frac{J_0/w}{\mu(1 + \mu z^*)}[(1 - e^{-\mu z}) - (1 - e^{-\mu z_m})f(z)] \qquad (7.10)$$

The model has been applied to describe the vertical profiles of $O_2$, $CO_2$ and $^{14}C$ (Craig 1969) and the minor elements Ba (Bacon and Edmond 1972) and Mn (Spencer and Brewer 1971). As an example it will be used to describe the vertical profile of copper in the Northeast Pacific Ocean reported by Spencer et al. (1970). The hydrographic conditions at this station have been described by Craig and Weiss (1970). The station is at 28° 29′N, 121° 38′W in 4200 m of water, the salinity minimum is at 150 m, and below this the salinity increases continuousiy to the bottom. The potential temperature–salinity diagram is essentially linear between boundary values at 860 m and 4000 m, and between these depths the salinity and potential temperature ($\theta$) data have been fitted with the mixing parameter $z^*$ (equation (7.6)). The least squares fit gave $z^*(S\text{‰}) = 0.856 \pm 0.05$ km and $z^*(\theta) = 0.875 \pm 0.025$ km, and a mean value of $z^* = 0.866$ km was selected as characterizing this mixing interval. If $w$ is taken as approximately 5 m yr$^{-1}$ then $K = 1.4$ cm$^2$ s$^{-1}$; both figures are in reasonable agreement with present thinking. The data are shown in Fig. 7.10. The vertical profile of dissolved copper at this station is shown in

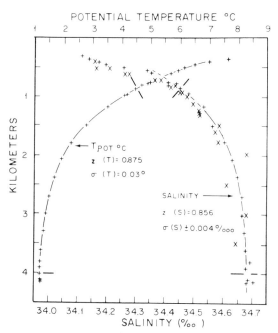

FIG. 7.10. Salinity and potential temperature vs. depth at 28°29′N, 121°38′W. The depth interval 860–4000 m is a linear region of the potential temperature–salinity diagram and the data from this interval has been fitted with the mixing parameter $z^*$ (Craig and Weiss, 1970). Permission of the American Geophysical Union.

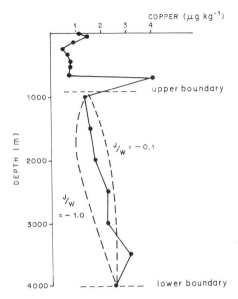

FIG. 7.11. The vertical distribution of dissolved copper in the Pacific Ocean (28°29′N, 121°38′W) showing *in situ* consumption of Cu in the mixing interval from 860–4000 m by means of a vertical advection–diffusion model. A best fit to the data is given by a consumption parameter of $J/w = -0\cdot4\,\mu\mathrm{g\,kg^{-1}\,km^{-1}}$.

Fig. 7.11. It can be seen that the shape of the profile is generally concave between the two boundary points defined by breaks in the potential temperature–salinity diagram, and that knowing $z^*$ it is necessary to invoke an *in situ* consumption term in order to fit the data to equation (7.5). The scatter in the data is such that no great accuracy can be claimed; however, a consumption parameter $(J/w)$ of between $-1$ and $-0\cdot1\,\mu\mathrm{g\,kg^{-1}\,km^{-1}}$ is indicated, and a best fit is obtained with $J/w = -0\cdot4\,\mu\mathrm{g\,kg^{-1}\,km^{-1}}$. If $w = 5\,\mathrm{m\,yr^{-1}}$ then the rate of *in situ* consumption of copper in the deep ocean is $2\times10^{-3}$ $\mu\mathrm{g\,kg^{-1}\,yr^{-1}}$. If the data and the assumption are correct, then this would indicate a deep water residence time of copper of *ca.* 1000 years. The functional form of the *in situ* consumption term is unknown; a common but unproven explanation is that it takes account of scavenging by settling particles.

## 7.6. THE CHEMICAL ELEMENTS IN SEA WATER

In this section a survey is given of the known distribution of the chemical elements in sea water, taking the elements in order of their groups in the periodic table.

7.6.1. GROUP I ELEMENTS Li, Na, K, Rb, Cs

Of the five alkali metals, Na ($10\cdot76$ g kg$^{-1}$ at $35\cdot0\%_0$ salinity) and K ($0\cdot387$ g kg$^{-1}$ at $35\cdot0\%_0$ salinity) are major ions of sea water. All the elements have long residence times and are characterized by a lack of chemical reactivity in solution and by their existence almost entirely as solvated (monovalent) cations. The bulk of the evidence suggests that these elements are conservative in sea water; that is that they are covariant with salinity over the normal oceanographic range.

Chow and Goldberg (1962) determined Li in 14 samples of Pacific water by means of mass spectrometric isotope dilution analysis and obtained a mean value of $173 \pm 2\cdot3$ µg kg$^{-1}$ (normalized to salinity $= 35\cdot0\%_0$). Earlier work on single samples by Marchand (1855), Thomas and Thompson (1933), Goldschmidt et al. (1933), Strock (1936), Bardet et al. (1937) and Ishibashi and Kurata (1939) yielded values ranging from 72–200 µg kg$^{-1}$. Riley and Tongudai (1964) have reported analyses of 27 samples from all oceans and find $183 \pm 3$ µg kg$^{-1}$, again normalized to $35\cdot0\%_0$ salinity. Two samples from the Black Sea showed a greater range, having a Li/Cl$\%_0$ ratio of $9\cdot69 \times 10^{-6}$ at 5 m depth and $9\cdot02 \times 10^{-6}$ at 100 m depth. The mean Li/Cl$\%_0$ ratio of all of the samples was $9\cdot39 \pm 0\cdot17 \times 10^{-6}$.

Rubidium and caesium have been determined by many workers and have been found to exhibit conservative behaviour. The factors affecting the oceanic distribution of Cs are of importance in determining the fate of the fission produced nuclide $^{137}$Cs. The earliest reliable analyses were those of Smales and Salmon (1955) who reported 120 µg Rb l$^{-1}$ and $0\cdot5$ µg Cs l$^{-1}$ in North Atlantic water. Bolter et al. (1964) analyzed a large number of oceanic samples and found average concentrations of 125 µg Rb l$^{-1}$ and $0\cdot30$ µg Cs l$^{-1}$, normalized to $35\cdot00\%_0$ salinity. Riley and Tongudai (1966) found $119 \pm 4$ µg Rb l$^{-1}$ and $0\cdot55 \pm 0\cdot06$ µg Cs l$^{-1}$, again normalized to $35\cdot00\%_0$. No trend with depth was found for three bulked samples of surface, intermediate and deep North Atlantic water, and no seasonal changes were detected in Irish Sea samples. Folsom (Spencer et al., 1970) has determined Cs in North East Pacific waters by flame photometry and reports $0\cdot280 \pm 0\cdot003$ µg Cs l$^{-1}$. The oceanic data for Cs thus fall into two groups, ca. $0\cdot5$ µg Cs l$^{-1}$ and ca. $0\cdot3$ µg Cs l$^{-1}$. It is unlikely that this is due to real variability, and analytical discrepancies probably account for the range.

7.6.2. GROUP II ELEMENTS Be, Mg, Ca, Sr, Ba, Ra

Beryllium, because of its largely covalent behaviour, differs considerably from the rest of the Group II elements. However, it does form a cation, $[Be(H_2O)_2]^{2+}$ (Cotton and Wilkinson, 1972), and may form polynuclear hydrolysis products such as $Be_3(OH)_3^{3+}$ (Stumm and Morgan, 1970). Beryllium has been detected in sea water (Merrill et al., 1960; Ishibashi et al., 1956) in the range 0·001 to 0·03 $\mu g\,l^{-1}$ however; no information exists on its distribution either geographically or with depth. $^7Be$ ($t_{\frac{1}{2}} = 53$ days) is generated in the atmosphere by the cosmic ray flux and is rapidly removed from its stratospheric reservoir to the ocean in particulate form. The half-life of the isotope is too short to permit its use for tracing horizontal motions. However, it is a useful tracer of vertical diffusion within the thermocline (Silker et al., 1968; Silker, 1972). Most studies based on its use have assumed conservative behaviour of the nuclide, this being justified by its observed distribution and the fact that only 5–10 % of the radioactivity is separable from the sea water by filtration. $^{10}Be$ ($t_{\frac{1}{2}} = 2·5 \times 10^6$ years) is also produced in the atmosphere and appears to be transferred rapidly to the sediments (Burton, 1965). The non-radiogenic removal rate is probably such that the particulate settling effects are observable only on a time scale between the mean lives of these two nuclides.

Magnesium (1·294 $g\,kg^{-1}$ at 35·0‰ salinity) and calcium (0·413 $g\,kg^{-1}$ at 35·0‰ salinity) are major ions of sea water, and as such bear a nearly constant ratio to chlorinity. Because of the downward flux of calcareous tests, deep waters tends to be enriched in calcium (ca. 1·5 %) with respect to surface waters (see Chapter 6). Strontium (0·008 $g\,kg^{-1}$ at 35·0‰ salinity) is also considered to be a major ion of sea water and a good deal of effort has been devoted to the question of whether its oceanic concentration is significantly perturbed by biological effects (see Chapter 6). The best estimate would seem to be that the variation is ca. 1·5 % (Brass and Turekian, 1972).

Barium has recently been the subject of detailed investigation, and a great deal of information is now being compiled on its distribution. The interest in the element stems in large part from the suggestion by Koczy (1958) that barium may be used as a chemical analogue of radium. $^{226}Ra$ ($t_{\frac{1}{2}} = 1622$ years) and $^{228}Ra$ ($t_{\frac{1}{2}} = 6·7$ years) are both important oceanic tracers. However, their usefulness is limited by the fact that no stable isotope of radium exists; this makes it difficult to resolve radiogenic effects from those of biological and chemical origin. Koczy (1958) found that surface waters are significantly depleted in $^{226}Ra$. He concluded that it is scavenged from the surface and transported to deep water which is thereby enriched with it. If barium behaves identically to (i.e. as a stable isotope of) radium, then these non-radiogenic processes may be evaluated.

Barium has been determined in sea water by many workers, notably Chow and Goldberg (1960), Turekian and Johnson (1966), Wolgemuth and Broecker (1970) and Bacon and Edmond (1972). The earlier data have largely been superseded, and it is now recognized that barium ranges from ca.

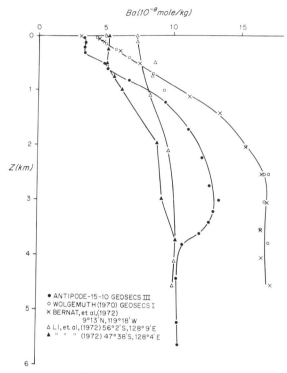

FIG. 7.12. Barium profiles from the Pacific and Indian–Antarctic Oceans (Bacon and Edmond, 1972). Permission of the North-Holland Publishing Company.

4–20 $\mu g \, kg^{-1}$ in oceanic profiles. Figure 7.12 shows barium profiles from the Pacific and Indian-Antarctic Oceans. Bacon and Edmond (1972) have shown that a plot of barium, silicate, or specific alkalinity against salinity, at the GEOSECS III test station in the southwest Pacific, is linear. They attribute this to simple mixing, and a one dimensional vertical advection–diffusion model suggested that over the depth range (1000–3400 m) for which the model was applicable, there was no *in situ* production or consumption of barium, silicate or alkalinity. Edmond (1970) has shown that $^{226}$Ra data for the Antarctic Circumpolar Current are strongly correlated with silicate, and since neither barium no radium are known to have any specific biochemical functionality, it appears that both elements are incorporated into the siliceous skeletal phases of marine organisms. The correlation of barium with radium now suggests that it may be difficult to distinguish between the non-radiogenic effects of the radium distribution and those arising from radioactive decay. $^{226}$Ra–Ba plots are found to be essentially linear, as are barium–silicate plots, thus indicating that the deep water loss of radium by decay is obscured by the *in situ* liberation from sinking siliceous skeletal material. Bacon and Edmond have suggested that the "primary" flux of radium, resulting from the decay of sedimentary thorium, is $1\cdot7 \times 10^{-14}$ g $^{226}$Ra cm$^{-2}$ year$^{-1}$. If the Ra:Si ratio in siliceous tests is $10^{-11}$ then the "cyclic" flux of radium is $1\cdot2 \times 10^{-14}$ g $^{226}$Ra cm$^{-2}$ year$^{-1}$, and the relative size of the two fluxes indicates the importance of the biologically derived vertical transport of the element in the ocean. The correlation of Ba with $^{228}$Ra has not yet been examined.

### 7.6.3. GROUP III ELEMENTS B, Al, Ga, In, Tl

All the Group III elements, except boron with its markedly covalent characteristics, have extraordinarily short residence times in sea water. This is no doubt a consequence of their rapid hydrolysis and of the low solubility of their hydrolysis products.

Boron is a major element of sea water ($0\cdot004$ g kg$^{-1}$ at $35\cdot0\%_0$ salinity) and has been shown to be conservative (Greenhalgh, unpublished data) (see also Chapter 6).

Aluminium is of great interest because of its importance in mineral formation. Its solution chemistry is dominated by its hydroxy species (e.g. $Al(OH)_4^-$), and generally, analytical data do not distinguish between dissolved, colloidal, and even finely divided particulate material. Simmons *et al.* (1953) determined aluminium in ten coastal surface samples by a fluorimetric technique and found a range of 0–7 µg Al l$^{-1}$. Sackett and Arrhenius (1962) used a modification of this method for an extensive survey of samples from the Weddell Sea and from the Californian coastal waters. They reported average values of $1\cdot0$ µg Al l$^{-1}$ for dissolved Al and $1\cdot5$ µg Al l$^{-1}$ for the particulate forms of the element. The wide range of particulate concentrations (0–120 µg Al l$^{-1}$) in near-shore waters led them to suggest that particulate Al could be used as a tracer for the mixing of river and sea water. Joyner (1964) examined particulate Al and Fe associated with run-off plumes of the Columbia River and found that particulate Al was inversely covariant with salinity. He concluded that for the Columbia River "the dilution of sea water by coastal runoff was resolved with much greater sensitivity by means of aluminium analyses than with salinity". These notions are, as the authors well realized, only first order approximations. No element in its particulate form, with its flotation or its gravitational settling, can be used as a conservative property, especially one with such complex and unstable hydrolysis products as aluminium. More desirably, the deviation from a linear trend with salinity may be used to give important information on removal rates and authigenic mineral formation.

Q

Feely *et al.* (1971) have examined the distribution of particulate Al in the Gulf of Mexico. They analyzed 72 samples from 8 vertical profiles and obtained a mean concentration of $2 \cdot 0 \ \mu g \ Al \ l^{-1}$. The ratio of aluminosilicate material to total suspended matter was found to be $0 \cdot 14$ at the surface, $0 \cdot 08$ at mid-depths and up to $0 \cdot 30$ close to the bottom because of the resuspension of sediment. A minimum residence time of 3 years was suggested for aluminosilicate material injected at the surface, indicating the rapid removal of the products of river input.

The remaining Group III elements Ga, In, Tl have received little attention, possibly because their low abundance and rapid removal from sea water have not stimulated much oceanographic interest. Culkin and Riley (1958) have determined gallium in sea water and reported an average value of $0 \cdot 030 \pm 0 \cdot 007 \ \mu g \ l^{-1}$ for six surface samples. Two deeper North Atlantic samples gave $0 \cdot 037 \ \mu g \ l^{-1}$ (250 m) and $0 \cdot 023 \ \mu g \ l^{-1}$ (3750 m), not significantly different from the surface concentrations. Indium has been determined in sea water by Matthews and Riley (1970a) who reported $1 \cdot 20 \pm 0 \cdot 06 \ ng \ In \ l^{-1}$ for 5 replicate analyses of coastal water from North Wales. Open ocean concentrations were found to be a factor of ten lower than this; for a North Atlantic station they reported the following data: 2 m depth, $0 \cdot 31 \ ng \ l^{-1}$; 500 m, $0 \cdot 12 \ ng \ l^{-1}$; 1000 m, $0 \cdot 10 \ ng \ l^{-1}$; 2000 m, $0 \cdot 11 \ ng \ l^{-1}$. The trend towards surface enrichment of In is difficult to explain, but the data are convincingly lower than those for inshore waters.

Matthews and Riley (1970b) have also determined thallium in sea water, and again coastal waters appear to be enriched in this element compared with oceanic waters. A sample from the Irish Sea was found to have $19 \ ng \ Tl \ l^{-1}$, whereas samples from a station in the Bay of Biscay had $10 \cdot 1 \pm 0 \cdot 6 \ ng \ Tl \ l^{-1}$. One sample from a depth of 500 m had an anomalous concentration of $16 \cdot 6 \ ng \ Tl \ l^{-1}$.

### 7.6.4. GROUP IV ELEMENTS C, Si, Ge, Sn, Pb

The oceanic chemistries of carbon and silicon, which are of major importance is discussed elsewhere (see Chapters 9, 11 and 12). Little attention has been paid to the remaining elements in this group.

In its chemical properties germanium somewhat resembles silicon. The Ge:Si ratio in sea water is greater than that in crustal rocks. Burton and Riley (1958) have reported analyses for Ge in sea water on 8 surface samples which were found to have $0 \cdot 06 \pm 0 \cdot 01 \ \mu g \ l^{-1}$. Two deeper samples from the North Atlantic were found to have $0 \cdot 06 \ \mu g \ l^{-1}$ (240–270 m) and $0 \cdot 07 \ \mu g \ l^{-1}$ (3660–3900 m) indicating the constancy of the element concentration with depth.

Little information exists on the distribution of tin in the marine environment. Shimizu and Ogata (1963) reported $1 \cdot 8 \ \mu g \ l^{-1}$ for Japanese coastal waters, and Hamaguchi *et al.* (1964) found 0·8 µg l$^{-1}$ for a sample from 500 m in the Northwestern Pacific. Recent work by Smith and Burton (1972) records significantly lower values; their data, which are shown in Table 7.10, clearly indicate the decrease in dissolved tin on going from a coastal to an oceanic environment. The evidence suggests that the concentration varies considerably in coastal regions in which rapid removal from the sea water occurs, probably by adsorption on to clays, a concept consistent with the low residence time of tin in the ocean.

TABLE 7.10

*Distribution of tin in sea water (Smith and Burton, 1972)*

| Sample | Salinity ‰ | Tin (µg kg$^{-1}$) |
|---|---|---|
| Southampton Water | 34·1 | 0·040 ± 0·007 |
| English Channel | 35·6 | 0·033 ± 0·001 |
| (49°28′N 4°40′W) | | |
| Gulf of Naples | 37·3 | 0·022 ± 0·002 |
| N.E. Atlantic | | |
| (33°35′N 13°39′W) | 36·8 | 0·010 ± 0·003 |
| (22°38′N 20°01′W) | 36·8 | 0·008 ± 0·003 |

The distribution of stable lead in natural waters has received little attention in recent years despite concern over the problem of heavy metal contamination and the effects it may have on the biological populations of the sea. Few marine laboratories have developed reliable techniques for low level lead determinations, and many of the published data for lead in sea water and marine organisms are probably overestimates.

From the limited data available from isotope dilution measurements, it is apparent that natural concentrations of lead in the deep oceans are quite low, ranging from 0·02–0·04 µg Pb l$^{-1}$ (Tatsumoto and Patterson, 1963a, b). However, the input of stable lead to the marine environment as a result of the industrial activities of man has steadily increased and is now estimated to exceed that due to natural weathering processes. As a consequence, lead is one of the few elements for which discernible perturbations in natural concentration levels in open ocean waters can be attributed to anthropogenic sources. Lead concentrations which are ten to one hundred times higher than natural levels are frequently observed in nearshore waters adjacent to heavily industrialized areas (Preston *et al.*, 1972; Piotrowicz *et al.*, 1972), and careful work by Chow and Patterson (1966) on open ocean samples has revealed significantly elevated lead concentrations in surface waters. These authors have pointed out that the depth distribution of oceanic lead bears a close resemblance to that of nuclear debris washed from the troposphere. In contrast to depth profiles of the chemically similar elements barium and radium which show marked depletions in surface waters, lead concentrations are high in the mixed layer and decrease rapidly with depth to uniformly low levels. The modern origin of the lead found in surface waters is supported by measurements of lead isotope ratios.

Evidence exists to suggest that lead introduced to the surface layer of the ocean is rapidly transported to the deep water and ultimately to the sediments. Chow and Patterson (1962) have suggested that the patterns of lead isotope variations in the authigenic component of pelagic sediments reflect the age of major drainage basins on the continents and that the passage times for lead in the oceans must be relatively short in order to produce the observed inhomogeneities. These authors have suggested biological transport mechanisms and, using a steady state model of the mixed layer, have estimated that the surface layer residence time for lead is seven years. This is in reasonable agreement with similar estimates of surface residence times for lead-210, which range from 2–5 years. On the basis of the high settling velocities calculated for lead-210 in surface waters of the Western Pacific, Tsunogai (1973) has proposed that vertical transport of lead occurs via incorporation into a manganese oxide phase. Recently, Craig et al. (1973) have presented data which suggest that scavenging of lead-210 by particulate matter occurs in the deep ocean. However, the use of radio-lead measurements to elucidate transport and removal mechanisms is complicated by the lack of information on the speciation of lead in sea water.

The input of lead to the ocean needs special consideration in view of the large scale recent mobilization of lead to the atmosphere by man. From the average rate of chemical deposition of lead in sediments during the Pleistocene, Chow and Patterson (1962) have estimated the pre-industrial input of dissolved lead to have been approximately $1 \cdot 1 \times 10^{10} \, g \, yr^{-1}$ in the northern hemisphere. Although the input from river runoff has increased by a factor of three because of chemical leaching resulting from agricultural practice, it is small relative to the estimated total artificial input of $3 \cdot 9 \times 10^{11} \, g \, yr^{-1}$ which includes inputs of $1 \cdot 4 \times 10^{11} \, g \, yr^{-1}$ from the introduction of industrial chemicals to rivers and $2 \cdot 5 \times 10^{11} \, g \, yr^{-1}$ from washout of lead aerosols produced by the combustion of lead alkyls.

Although the major fraction of the lead introduced to the atmosphere as aerosols is redeposited near the source (Chow et al., 1970), it is clear that the smaller size fractions of airborne lead particulates are transported over long distances and thus constitute a major flux of lead to the open ocean (National Academy of Sciences, 1971). A high open ocean atmospheric lead burden has been indicated by Hoffman (1971), who reported mean lead concentration in marine aerosols over the mid-Pacific which are one hundred times higher than those expected from crustal abundances. Hoffman and Duce* have found that the average atmospheric lead content over the open North Atlantic is $10 \, ng \, m^{-3}$, and Piotrowicz et al. (1972) have estimated an average dry deposition rate of $10^{-14} \, g \, Pb \, cm^{-2} \, s^{-1}$ on the sea surface.

### 7.6.5. GROUP V ELEMENTS N, P, As, Sb, Bi

Because of their roles as micronutrient elements, nitrogen and phosphorus will be discussed in Chapter 11. Molecular nitrogen is treated in Chapter 8.

Arsenic possesses some chemical similarities to phosphorus and has recently received attention because of its possible role in metabolic processes as a phosphate analogue, and its interference in conventional phosphate analysis through the formation of a molybdenum blue colour. Antimony and bismuth show more marked cationic behaviour than does arsenic, but should be considered as metalloids rather than as true metals.

Early work on arsenic in sea water by Gautier (1903) and Rakestraw and Lutz (1933) yielded values up to $75 \, \mu g \, As \, l^{-1}$; however, this data is now considered to be inaccurate. More recent work has been carried out by Armstrong and Harvey (1951), Smales and Pate (1952), Young et al. (1959), Sugawara et al. (1962), Portmann and Riley (1964) and Johnson and Pilson (1972). All recent work indicates a total concentration of $0.8-8 \, \mu g \, As \, l^{-1}$. At equilibrium in oxygenated sea water, arsenic should exist as $HAsO_4^{2-}$, but in reducing environments it will be present as species of $As^{3+}$. Sugawara et al. (1962) and Kanamori (1962) have shown that $HAsO_4^{2-}$ is, indeed, the dominant chemical species by a factor of 3–4, but that traces of $HAsO_2^-$ are also present. These probably originate by biological reduction of $HAsO_4^{2-}$; the low concentration is probably the result of the slow kinetics of the oxidation of the reduced species.

Kanamori (1962) carried out analyses for total arsenic on 160 samples from the Pacific and showed that samples from below a depth of 100 m contained 1.6 times the concentration of As present in surface samples, principally because of variation in $HAsO_4^{2-}$. Little variation with depth of $HAsO_2^-$ (ca. 12% of the total arsenic) was noted.

Johnson and Pilson (1972) have described a detailed investigation of 178 samples from 25 stations in the western North Atlantic. The method used was a colorimetric technique in which phosphate + arsenate was determined, and phosphate alone was determined separately after reduction of $HAsO_4^{2-}$ to $HAsO_2^-$ which will not form a molybdenum blue. Their data showed a deep water enrichment of $HAsO_4^{2-}$ by a factor of 1.5, presumably a result of biological stripping from surface water and regeneration at depth. The average surface concentration was $2.1 \, \mu g \, As \, l^{-1}$ and the average deep water

* Cited in Piotrowicz et al. (1972).

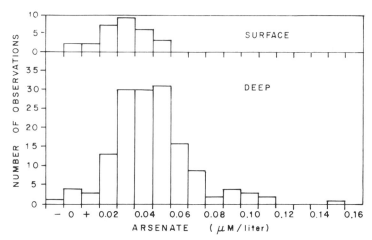

FIG. 7.13. Histogram of dissolved arsenate data for sea water from the North Atlantic and Caribbean (Johnson and Pilson, 1972). Permission of the Sears Foundation.

concentration was $3 \cdot 3$ $\mu g \, As \, l^{-1}$. The data are shown as a histogram in Fig. 7.13. The technique is subject to the criticism that it is insensitive in the presence of large quantities of phosphate. However, it has the advantage that analyses may be performed at sea. These conclusions are in reasonable agreement with those of other workers and provide evidence of the systematic variability of As in sea water.

Antimony occurs in sea water at a level of $0 \cdot 1 – 0 \cdot 4$ $\mu g \, Sb \, l^{-1}$, and the bulk of the evidence shows it to be constant both geographically and with depth. Schutz and Turekian (1965a) determined Sb in 75 oceanic samples by neutron activation analysis and reported an average value of $0 \cdot 33$ $\mu g \, Sb \, l^{-1}$. However, higher values were found in the Gulf of Mexico, the South West Atlantic and the Indian Ocean. They noted that samples high in antimony were also high in cobalt, although the correlation between the two elements was not strong.

The investigation by Schutz and Turekian (1965a) was part of a major attack on the trace element problem, and the paper is of considerable importance. However, the data obtained is severely limited by the sampling techniques. Some samples were obtained with an "epoxy-lined barrel", some with an unlined galvanized barrel and some surface samples with the pump used for cooling the hydro-winch. The samples were filtered through $0 \cdot 45$ $\mu m$ pore size Millipore filters, pipetted into 125 ml Pyrex bottles fitted with glass stoppers, and stored at their natural pH values for periods of up to two years. No adsorption of Co, Ag, Cs and Sc was detected (cf. Robertson, 1968b), but the sampling system does leave much to be desired.

Portmann and Riley (1966) developed a technique for the determination of Sb, based on co-precipitation with manganese dioxide followed by colorimetric determination with rhodamine B, and applied this method to 7 Irish Sea samples. Their results ranged from $0\cdot13$–$0\cdot40$ μg Sb $l^{-1}$. The most recent data is that of Robertson (Spencer et al., 1970; Brewer et al., 1972) which was obtained by neutron activation analysis of dried sea salt. Profiles from the Eastern Pacific and the Sargasso Sea both gave a mean value of $0\cdot21$ μg $l^{-1}$ with no significant difference between surface and depth.

Bismuth is known to occur in sea water at a level of $0\cdot01$–$0\cdot04$ μg Bi $l^{-1}$, but information on its oceanic distribution is non-existent. Brooks (1960) analysed a single 250 l sample of South Atlantic surface water by ion-exchange concentration of the bismuth chloro-anion at low pH and obtained a value of $0\cdot02$ μg Bi $l^{-1}$. The sample concentration process took 100 days. Portmann and Riley (1966) analysed five 10 l samples and found $0\cdot024$–$0\cdot042$ μg Bi $l^{-1}$ for near shore samples. Two North Atlantic samples contained $0\cdot033$ μg Bi $l^{-1}$ (surface) and $0\cdot015$ μg Bi $l^{-1}$ (2000 m). The lower value at depth, if significant, implies a removal of Bi from deep ocean water.

### 7.6.6. GROUP VI ELEMENTS S, Se, Te, Po

Sulphur, as the sulphate ion ($2\cdot712$ g kg$^{-1}$ at $35\cdot00$ S‰) is a major component of sea water and as such is discussed in Chapter 6.

Selenium is present in sea water at a concentration of ca. $0\cdot1$ μg Se kg$^{-1}$. Its low residence time ($2 \times 10^4$ years) reflects its active and rapid removal from sea water. The higher concentrations of Se found in near-shore and shelf sediments (ca. $0\cdot55$ μg g$^{-1}$) as opposed to deep sea clays (ca. $0\cdot15$ μg g$^{-1}$) give further support to this view. Early work on Se in sea water (Goldschmidt and Strock, 1935; Ishibashi et al., 1953) gave unrealistically high values in the range 4–6 μg Se $l^{-1}$. Schutz and Turekian (1965) found an average concentration of $0\cdot09$ μg Se $l^{-1}$ in their survey of 23 oceanic samples, with a narrow range of abundance from $0\cdot052$ μg Se $l^{-1}$ in the Antarctic to $0\cdot11$ μg Se $l^{-1}$ in the Caribbean. Chau and Riley (1965) found $0\cdot34$–$0\cdot50$ μg Se $l^{-1}$ in Irish Sea water.

Little is known of Se in marine organisms, and although it is toxic to many animals, it has beneficial effects at low concentration, being associated with vitamin E (Horwitt, 1965). Theory suggests that in oxygenated sea water Se exists as $SeO_3^{2-}$; this is strongly adsorbed by iron hydroxides (Geering et al., 1968). Under reducing conditions, elemental Se is the stable form and relatively high concentrations of Se occur in euxinic sediments. No experimental work on the oxidation state of Se in sea water has been carried out.

The concentration of tellurium in sea water is unknown. All isotopes of polonium are radioactive and their distribution is discussed in Chapter 18;

Po is strongly concentrated by marine organisms, enrichment factors of $2 \times 10^6$ being recorded (Folsom *et al.*, 1972).

7.6.7. GROUP VII ELEMENTS F, Cl, Br, I, At.

The halogen group of elements is of fundamental oceanographic interest. The ocean is the world's major reservoir of chlorine ($19 \cdot 35 \mathrm{~g} \, \mathrm{Cl}^- \, \mathrm{kg}^{-1}$ at $35 \cdot 0\%_0$ salinity) and historically the determination of chloride in sea water has laid the foundations of oceanography as a means for the determination of salinity. Bromide ($0 \cdot 067 \mathrm{~g} \, \mathrm{kg}^{-1}$ at $35 \cdot 0\%_0$ salinity) is a major ion of sea water and is extracted commercially; fluoride ($1 \cdot 3 \times 10^{-3} \mathrm{~g} \, \mathrm{kg}^{-1}$ at $35 \cdot 0\%_0$ salinity) lies on the borderline between the major and minor constituents (see Chapter 6). It is currently a matter of some debate as to whether fluoride is a conservative component of sea water or not (Greenhalgh and Riley, 1963; Riley, 1965; Brewer *et al.*, 1970). There is no isotope of astatine sufficiently long-lived for it to be detected in the ocean.

Iodine is thus the only halogen present in true trace quantities in the ocean (*ca.* $0 \cdot 5 \times 10^{-6}$ M). Pfaff (1825) first detected iodine in sea water, but it was not until 1916 that Winkler obtained a value for the total iodine concentration ($38 \, \mu\mathrm{g} \, \mathrm{l}^{-1}$) which is acceptable by modern standards. He suggested that both iodate ($\mathrm{IO}_3^-$) and iodide ($\mathrm{I}^-$) are present in sea water, and Shaw and Cooper (1957) postulated that HIO is the dominant oxidized species, a proposal contested by Sugawara (1957). Sugawara and Terada (1958) experimentally verified that both iodate and iodide are present in sea water and showed that HIO is an unstable species. These results, confirmed by Johannesson (1958), were accepted by Shaw and Cooper (1958). The presence of even trace quantities of $\mathrm{I}^-$ is evidence of significant redox disequilibrium in the marine iodine system.

Sillén (1961) showed that for the reaction:

$$\mathrm{IO}_3^- + 6\mathrm{H}^+ + 6e^- = \mathrm{I}^- + 3\mathrm{H}_2\mathrm{O} \quad [\mathrm{pK}_a = 110 \cdot 1]$$

in oxygenated sea water of pE $12 \cdot 5$ and pH $8 \cdot 1$ the equilibrium ratio $[\mathrm{IO}_3^-]/[\mathrm{I}^-]$ should be $10^{13 \cdot 5}$. Redox disequilibrium is often brought about by biological activity through photosynthesis, and the presence of detectable iodide in sea water is evidence for significant involvement of iodine in the oceanic biological cycle. Recent work by Tsunogai and co-workers (Tsunogai and Henmi, 1971; Tsunogai, 1971) and by Wong and Brewer (1974) has indicated that iodate is the principal form of iodine in sea water, varying from $0 \cdot 2$ to $0 \cdot 5 \, \mu\mathrm{M}$ and that iodide ranges from $0 \cdot 2 \, \mu\mathrm{M}$ to below the detection limit (*ca.* $0 \cdot 1 \, \mu\mathrm{M}$). Wong and Brewer (1974) determined iodate at three stations in the South Atlantic by direct iodometric titration of sea water. The vertical profiles of iodate and those of phosphate and nitrate at 36° 0·0′S, 45° 0·0′W in the Argentine Basin are shown in Fig. 7.14. The correlation between the iodate and the nutrient data is good ($IO_3^-$ : $NO_3^-$ correlation coefficient 0·95) thus lending support to the suggestion that iodine is transported from surface to deep water by sinking organic matter. A stoichiometric

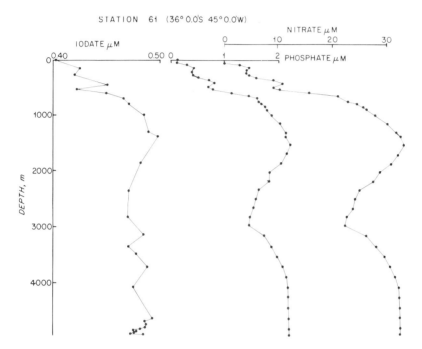

FIG. 7.14. Vertical profiles of iodate, phosphate and nitrate at a station in the Argentine Basin (Wong and Brewer, 1974). Permission of the Sears Foundation.

model may be constructed from this data giving a mole ratio of $NO_3^-$ : $PO_4^{3-}$ : $IO_3^-$ = 357:23:1. If iodate is being regenerated in deep water in stoichiometric proportion to the nutrients, then from the known rates of deep water oxygen consumption (ca. $1 \times 10^{-3}$ ml $O_2$ kg$^{-1}$ yr$^{-1}$) it is possible to calculate a rate of deep water iodate production (ca. $2·5 \times 10^{-5}$ μmol $IO_3^-$ kg$^{-1}$ yr$^{-1}$). This data indicates that only 10% of the deep water iodate is of cyclic (i.e. oxidative) origin and that 90% is conservative.

The stoichiometric oxidation model of iodate probably does not completely explain the behaviour of iodine in the ocean. Data given by Wong and Brewer (1974) for the Angola Basin (2° 0·0′S, 4° 33·0′W) reveal a different pattern. The surface iodate concentrations are only 0·27 μM, and the correlation with the nutrients is poor. The low surface concentrations are probably caused by the reduction of iodate to iodide, but the poor correlation with the nutrients is difficult to explain. It seems likely that the iodine system is only loosely coupled to the nutrient distribution and that in the water at the depth of intense oxygen minimum ($<1 \, ml \, O_2 \, l^{-1}$) at this station the regenerative reactions are separated. In the Argentine Basin where distributions are controlled by horizontal advection rather than by *in situ* production, the differing modes of production are obscured.

The oceanic precursor of $IO_3^-$ is probably $I^-$, and determinations of $I^-$ in sea water are of considerable interest. $I^-$ is released from organic matter in the deep water and is then oxidized to iodate. Thus, the oceanic profile of iodide should be a "mirror image" of that for iodate; that this is approximately the case is shown by iodate and iodide data from the Pacific (Tsunogai, 1971) in Fig. 7.15. If no process other than oxidation is operating, then the oceanic consumption rate of iodide should match the oceanic production rate of iodate, thus providing an elegant test of oceanic models. Moreover, if the rate-limiting step is the kinetics of the oxidation of the iodide ion, then these rates can be tested by laboratory experiment and extrapolated to match the

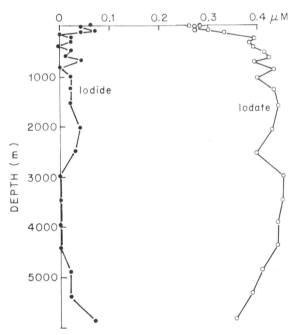

FIG. 7.15. The vertical distribution of dissolved iodide and iodate in the North Pacific (12°01′N, 158°02′E), from Tsunogai (1971). Permission of Pergamon Press.

low oceanic rates. However, it is unlikely that the situation is so simple. Price and Calvert (1973) have reported extraordinarily high concentrations of iodide (up to 1990 µg g$^{-1}$) in near shore marine sediments, and they found a marked correlation of iodide with organic carbon. The high ratios of I:C relative to those found in marine organisms led them to postulate scavenging of iodide from sea water by organic debris. If this process is occurring it would certainly lead to differences in the iodide and iodate consumption and production rates.

Tsunogai (1971) has suggested that the high iodide concentrations in sediment interstitial waters (ca. 3·5 µM) relative to oceanic bottom waters (ca. 0·04 µM) may lead to a significant diffusive flux of iodide out of the sediments. The flux proposed by Tsunogai is 4·4 × 10$^{-4}$ M m$^{-2}$ yr$^{-1}$. This is extraordinarily large, being 440 times the input from rivers and 88 times the input from rain. Direct evidence of a detectable iodide flux from the sediments into the water column is scarce, and it is difficult to separate the contribution derived from the proposed vertical diffusive flux from that resulting from advective transported iodine-rich bottom water. Calculations show that the residence time of iodide in sea water is ca. 1 × 10$^3$ years and that the "half-life" with respect to oxidation is several tens of years.

Liss et al, (1973) consider that the iodide–iodate couple acts as a pE indicator for sea water. The values of pE obtained from this calculation (ca. 10·5) contrasts with the generally accepted belief that the pE of sea water is maintained at ca. 12·5 by the oxygen–water couple. It does not appear likely that the iodine system may be used in this way since the evidence presented here points strongly to biologically promoted disequilibrium in the iodine system. The data of Liss et al. do give some evidence of a diffusive iodide flux out of the sediments in the San Pedro and Santa Barbara basins off Southern California where advective processes are restricted.

### 7.6.8. Zn, Cd, Hg

Zn, Cd and Hg are not transition elements in the fullest sense since their $d$ electrons are too strongly bound to be removed in chemical reactions. They comprise a subgroup of considerable interest through their marked interaction with the marine biosphere, yet remarkably little is known of their distribution in sea water.

The concentration of zinc in sea water ranges from 1–50 µg l$^{-1}$, the mean value being close to 3 µg l$^{-1}$. Despite many measurements, little is yet known about the factors controlling the distribution of zinc. Spencer and Brewer (1971) have investigated the distribution of Zn, Cu and Ni in the Sargasso Sea and the Gulf of Maine. Samples were filtered, and stored in polyethylene bottles at their natural pH; this may have led to some loss by adsorption;

however, even if there is a systematic error there still remains a general trend. No seasonal concentration variation could be detected for these elements. In fact, the concentrations in summer in the Gulf of Maine were slightly higher than in winter because of the influence of terrestrial run off and the reduced mixing in the stratified water column. A distinct difference in Zn, Cu and Ni was found between slope water to the north and west of the Gulf Stream (Zn, $10 \mu g \, l^{-1}$; Cu, $2 \cdot 5 \mu g \, l^{-1}$; Ni, $1 \cdot 6 \mu g \, l^{-1}$) and the Sargasso Sea water (Zn, $2 \cdot 6 \mu g \, l^{-1}$; Cu, $0 \cdot 6 \mu g \, l^{-1}$; Ni, $0 \cdot 8 \mu g \, l^{-1}$). An eddy which had spun off the Gulf Stream and into the Sargasso Sea and which contained a core of slope water was found to maintain these trace elements differences over a period of several months. Zirino and Healy (1971) have made anodic stripping voltammetric analyses for Zn in northeast Pacific waters. They concluded that surface waters (above 125 m depth) had significantly lower Zn concentration than did deeper waters, and they attributed this to biological removal. Their data are shown in Table 7.11. Zn values obtained after

TABLE 7.11

*Zinc concentrations in the North-East Tropical Pacific Ocean (Zirino and Healy, 1971).*

| Depth (m) | No. of samples | Mean concentration $(\mu g \, l^{-1})$ |
|-----------|----------------|-------------------------------|
| 0         | 28             | 1·6                           |
| 1–40      | 52             | 1·2                           |
| 41–125    | 44             | 1·2                           |
| 126–1000  | 105            | 2·2                           |
| 0–1000    | 229            | 1·7                           |

filtration were *ca.* 45% lower, and a substantial fraction of this particulate Zn was released to solution at pH 5·6, suggesting that it is weakly adsorbed. Topping (1969) has measured Zn in Northwestern Indian Ocean waters; his data from 59 samples range from $48 \cdot 4 – 3 \cdot 9 \mu g \, Zn \, l^{-1}$, indicating the possibility that the Zn concentrations in Indian Ocean waters are higher than those in the Atlantic or Pacific Oceans. The sampling bottle which he used had neoprene end-caps which were covered with polyethylene sheet to prevent leakage of Zn into the sample. Nevertheless, Zn contamination is a possible explanation of the generally high values. Riley and Taylor (1972) have determined Zn in 101 samples from the Northeast Atlantic. Their data range from $22 \cdot 0 – 2 \cdot 5 \mu g \, Zn \, l^{-1}$ and show a trend towards surface enrichment. From the foregoing it is apparent that, although considerable variability exists in the distribution of Zn in sea water, few coherent explanations have been offered to describe this variability.

The speciation of Zn in sea water is markedly pH dependent (see Section 3.4.7); at pH 8·1 $Zn(OH)_2^0$ is the predominant species, whereas $Zn^{2+}$ becomes important below pH 7·8 (Zirino and Yamamoto, 1972). This change in charge should lead to a difference in adsorptive behaviour and thus affect the mode of removal of Zn from sea water.

The marine distribution of mercury is of considerable topical interest because of concern over mercury poisoning and the accumulation of methyl mercury in fish. The concentration of mercury in ocean water is close to 0·02 µg Hg $l^{-1}$. Hosohara (1961) has reported concentrations as high as 0·27 µg$l^{-1}$ in Pacific deep water, although more recent data indicate that this may be anomalous. Leatherland et al. (1971) found 0·013–0·018 µg $l^{-1}$ in four unfiltered Northeast Atlantic surface samples and three out of five deep samples were similar in concentration, the remaining two being markedly lower. These data are similar to those obtained by Weiss et al. (1972) who reported 0·012–0·027 µg $l^{-1}$ for a station in the Eastern Pacific. Close to shore (60 km) concentrations as high as 0·173 µg $l^{-1}$ were found, these possibly being associated with higher particulate concentrations.

Mercury is strongly adsorbed onto particulate matter and sediments. Living phytoplankton do not take up Hg in large quantities. However, adsorption onto dead cells is rapid, and it seems likely that oceanic concentrations are buffered by scavenging in this manner, and that much of the Hg brought into the coastal zone by run-off is immobilized by adsorption. Burton and Leatherland (1971) found 0.02 µg Hg $l^{-1}$ in samples from the English Channel, whereas Smith et al. (1971) found 0·045–2·85 µg Hg $l^{-1}$ in the lower Thames River, of which > 80% was associated with particulate matter.

The flux of mercury from fresh water sediments is dominated by the microbial transformation of inorganic Hg to methyl mercury ($CH_3 \cdot Hg^+$) and dimethyl mercury ($CH_3 \cdot Hg \cdot CH_3$). Jenson and Jernelov (1969) and Wood et al. (1968) have studied the methylation process and have shown that methyl mercury is taken up rapidly by living organisms; the volatility of the dimethyl form results in its transfer to the atmosphere.

The industrial use of mercury has resulted in the mobilization of large quantities of the element. Weiss et al. (1971) have examined the Hg content of Greenland snow and report concentrations of 0·035–0·075 µg Hg $l^{-1}$ for the period from 800 B.C. to 1952 and 0·087–0·23 µg Hg$l^{-1}$ from 1952 to 1965. The accumulated industrial usage of mercury is now close to $5 \times 10^5$ tons (Gavis and Ferguson, 1972), and the rate of release to the environment is currently an order of magnitude greater than the flux due to natural weathering.

Cadmium has also received significant attention as a pollutant element through its uptake by shellfish. The Japanese "itai-itai" disease was found to be caused by excessive human intake of cadmium derived from Cd polluted water and fish from areas adjacent to a Cd processing factory. In spite of the considerable interest in Cd in marine organisms, very little is known of its distribution in sea water. Abdullah *et al.* (1972) found up to $4.2$ µg Cd $1^{-1}$ in the River Avon estuary, with a marked depletion to $<0.2$ µg Cd $1^{-1}$ in Irish Sea Waters. Rapid removal of Cd to the sediments is indicated. The industrial output of Cd is approximately $10^4$ tons per year, and although much of this may be transported atmospherically, no increase in Cd in oceanic surface waters has been detected. Riley and Taylor (1972) reported $0.07–0.71$ µg Cd $1^{-1}$ in North Atlantic waters.

### 7.6.9. THE TRANSITION ELEMENTS

#### 7.6.9.1. *Sc, Ti, V, Cr, Mn, Fe, Co, Ni, Cu*

The transition elements have played a central role in chemistry, and the application of crystal and ligand field theory has contributed much to the understanding of the way in which these elements are bound in complexes. A great deal of attention has been focused on transition elements in sea water; even so, the data on their oceanic distributions are still often inaccurate and capable of no simple explanation.

Scandium has been determined by Schutz and Turekian (1965a) in open ocean samples as $<0.004$ µg Sc $1^{-1}$. More recent data obtained by Robertson (Spencer *et al.*, 1970; Brewer *et al.*, 1972) have shown that the oceanic range of dissolved Sc is from $0.0002–0.001$ µg Sc $1^{-1}$. The element is similar in its chemical characteristics to the lanthanides and, geochemically, should be associated with silicates. It is slightly enriched in deep waters; at a station in the Sargasso Sea the mean Sc concentration of all samples above 2000 m was $0.00064$ µg Sc $1^{-1}$. and below 2000 m was $0.00092$ µg Sc $1^{-1}$; there is $<1\%$ probability that these means are the same.

Titanium should exist only at low concentrations in sea water because it is rapidly hydrolyzed. There are no exact data although Griel and Robinson (1952) report up to $1.9$ µg Ti $1^{-1}$.

Vanadium is present in sea water in the range $0.2–4$ µg V $1^{-1}$. The element is concentrated in marine organisms (Ascidians and Tunicates), but evidence of its systematic non-conservative behaviour is lacking. Sugawara and co-workers (Sugawara *et al.*, 1953; Naito and Sugawara, 1957; Sugawara *et al.*, 1963) report $1.7–2.8$ µg V $1^{-1}$ for Pacific waters. Chan and Riley (1966) found $1.8$ µg V $1^{-1}$ in an Irish Sea sample; Morris (1968) found $0.86 \pm 0.14$ µg V $1^{-1}$ in coastal water off North Wales. The only detailed vertical oceanic profiles of V are those given by Riley and Taylor (1972) who found essentially random variations between 0–4 µg V l$^{-1}$ in the tropical northeast Atlantic Ocean.

The distribution and ultimate fate of chromium discharged by rivers to the oceans is poorly known, but is thought to be critically determined by the differences in chemical behaviours of the trivalent and hexavalent species. Most of the chromium available for weathering in crustal rocks occurs as Cr(III)-substituted spinels, silicates, and aluminosilicates (Thayer, 1956). Oxidation to Cr(VI) during weathering may occur under conditions of high pH; however, the predominant oxidation state of stable chromium in river waters is uncertain.

The transport and removal of $^{51}$Cr introduced as $Cr_2O_7^{2-}$ to the Columbia River by the Hanford reactor complex has been studied in detail by a number of investigators. Nelson *et al.* (1966) have found that most of the $^{51}$Cr remains in solution in anionic form, presumably dissociating to $CrO_4^{2-}$ at the natural pH of the river. Cutshall (1967) concluded from uptake studies of $^{51}$Cr(VI) and $^{51}$Cr(III) on Columbia River sediments that reduction of $^{51}$Cr(VI) to a trivalent species would result in its rapid removal to particulate matter and sediments. Curl *et al.* (1965) have demonstrated that roughly 20% of the $^{51}$Cr from Columbia River water is retained by membrane filters (pore size 0·65 µm) and that the percentage of $^{51}$Cr attached to particles gradually increases downstream, indicating that $^{51}$Cr must be scavenged by hydro-colloids (as $CrO_4^{2-}$) or by adsorption (as Cr(III) species). Rapid sorption of Cr(III) onto glass beads and plankton from sea water has been demonstrated by Curl *et al.* (1965). However, Cr(VI) is not appreciably adsorbed by common components of particulate matter in sea water (Krauskopf, 1956). Johnson (1965) found that particulate-bound $^{51}$Cr is not reversibly desorbed in sea water.

Observations of appreciable concentrations of soluble $^{51}$Cr in sea water samples collected several hundred kilometres from the mouth of the Columbia River (Cutshall *et al.*, 1966), along with indications of the highly adsorbable nature of Cr(III) species have thus been taken as evidence that $CrO_4^{2-}$ is the predominant species in sea water. However, generalization of the behaviour of this artificially introduced nuclide is unwarranted in view of the uncertainty of the form of naturally weathered chromium and the tendency of Cr(III) to undergo extensive hydrolysis reactions to form kinetically non-labile hydroxy complexes which may persist in solution.

Existing data on the concentration of chromium in sea water are sparse and limited primarily to inshore surface waters (Table 7.12). In general, chromium concentrations fall within the range of 0·1–0·5 µg Cr l$^{-1}$, although a few investigators have reported higher values in the range of 1–2 µg Cr l$^{-1}$ (Black and Mitchell, 1952; Chau *et al.*, 1968; and Schutz and Turekian,

TABLE 7.12
Data on the concentration of chromium in sea water

| Reference | Samples | Concentration $(\mu g \, l^{-1})$ |
|---|---|---|
| Black and Mitchell (1952) | 3, British Coastal | 1–2·5 |
| Ishibashi and Shigematsu (1950) | 3, Japanese Coastal | 0·04–0·07 |
| Loveridge et al. (1960) | 6, Coastal | 0·13–0·25 |
| Chuecas and Riley (1966) | 2, Irish Sea | 0·46 |
| Morris (1968) | 1, British Coastal | 0·29 |
| Chau et al. (1968) | 1, Hong Kong | 1·59 |
| Fukai and Vas (1967) | 5, Mediterranean | 0·25–0·96 |
| Elderfield (1970) | 10, British Coastal | 0·31–0·65 |

1965a). These measurements were made without regard to the problem of oxidation state. Several authors have considered the thermodynamic aspects of chromium speciation in sea water (Elderfield, 1970; Fukai and Huynh-Ngoc, 1968; and Cutshall, 1967), and concluded that the principal species should be $Cr(OH)_2^+$ and $CrO_4^{2-}$, with the latter clearly predominant under oxic conditions:

$$Cr(H_2O)_4(OH)_2^+ \rightleftharpoons CrO_4^{2-} + 6H^+ + 2H_2O + 3e^-; \log K = +65 \quad (7.11)$$

Attempts to support these conclusions with analytical data have produced contradictions in the literature, even though in some cases, similar or identical analytical methods were used (Table 7.12). Using an iron hydroxide coprecipitation technique, Fukai (1967) and Fukai and Vas (1967), found significant amounts of both chromium (III) and (VI) in waters of the Ligurian Sea. The Cr(VI) species was the generally predominant form, although a range of 20–95% of the total chromium content was reported. However, Elderfield (1970) used the same technique to determine Cr(III) and Cr(VI) in coastal waters off Wales and found that Cr(III) constituted over 90% of the total dissolved chromium in all but two of the samples examined. Kuwamoto and Murai (1969) employed a bismuth sub-nitrate coprecipitation technique and found dissolved chromium to be approximately equally distributed between the two oxidation states in surface Pacific waters between 170°W and 155°E. Stanford (1971) detected significant amounts of Cr(VI) using a solvent extraction–neutron activation technique, but was unable to demonstrate the existence of Cr(III) in waters off the Oregon coast. Recently, Frew (unpublished work) has measured significant concentration of Cr(III) at all depths up to 5,000 m in two Sargasso Sea profiles using a gas chromatographic technique.

Although much of the reported oxidation state variation of chromium is undoubtedly a consequence of limitations of the various analytical methods used, it seems clear that redox equilibrium is not attained for the Cr(III)–Cr(VI) couple in sea water, and that *in situ* oxidation-reduction processes may be important controlling factors during the partitioning of chromium between soluble and particulate phases and, consequently, its removal from the water column.

The marine chemistry of manganese is complicated and is dominated by redox changes, Mn(II) being unstable in normal oxygenated sea water with respect to oxidation to Mn(IV) with the resultant formation of solid $MnO_2$. The existence of manganese nodules on the sea floor is attributable to this oxidation process. The solution chemistry of Mn(II) in anoxic basins is of importance (see pp. 472–5), but few accurate determinations of dissolved Mn have been made in normal oceanic waters. Roña *et al.* (1962) have determined dissolved Mn in Gulf of Mexico and Pacific Ocean waters, and have reported data varying from $0.2$–$8.6$ µg Mn $l^{-1}$ with no apparent systematic trend. They attempted to separate ionic and non-ionic (i.e. "organically bound") fractions by a dialysis technique, but contamination problems were such that in some samples the "total" Mn was less than the "complexed" fraction. Morris (1968) found $1.86 \pm 0.10$ µg Mn $l^{-1}$ in Welsh coastal waters, Topping (1969) found $0.80$–$1.30$ µg Mn $l^{-1}$ in the north east Indian Ocean and Riley and Taylor (1972) reported $1.0$–$5.0$ dg Mn $l^{-1}$ in the north east tropical Atlantic. More recently, Bender (unpublished work) has found significantly lower values (*ca.* $0.10$–$0.30$ dg Mn $l^{-1}$) in the north west Atlantic.

The high concentration of dissolved Mn in the interstitial waters of reducing sediments leads to an upward flux of Mn(II) from the sediments; this manganese is oxidized to $MnO_2$ at and above levels in the sedimentary columns at which oxidizing conditions prevail. The kinetics of the oxidation of Mn(II) have been studied by Hem (1963) and Morgan (1964) and the process has been found to be slow below pH $8.5$. Morgan has suggested the following reaction scheme:

$$Mn(II) + O_2 \rightarrow MnO_2 \text{ (s)} \qquad (7.12)$$

$$Mn(II) + MnO_2 \text{ (s)} \rightleftharpoons Mn(II) \cdot MnO_2 \text{ (s)} \qquad (7.13)$$

$$Mn(II) \cdot MnO_2 \text{(s)} + O_2 \rightarrow 2\,MnO_2 \text{ (s)} \qquad (7.14)$$

The reaction is autocatalytic with respect to $MnO_2$, is first order with respect to Mn(II) and $P_{O_2}$, and is dependent on $[OH^-]^2$. In recent work (Elert and Brewer, unpublished) the apparent rate expression

$$-\frac{d\,[Mn(II)]}{dt} = K\,[Mn(II)]\,[MnO_2]\,[OH^-]^2\,[O_2], \qquad (7.15)$$

has been derived from an elementary kinetic scheme.

Expressing $[O_2]$ in the conventional oceanographic units of ml $l^{-1}$ and the Mn(II) removal rate in moles $l^{-1}$ min $^{-1}$ they found

$$K = 1.2 \times 10^{11} \, l^3 \, mol^{-3} \, min^{-1} \quad (25°C)$$

This constant is applicable only to oxidation in the presence of pure $MnO_2$. The presence of other solid phases, such as sedimentary material or ferric hydroxides should be expected to affect the reaction markedly.

The chemistry of iron is also controlled by redox processes and considerable controversy exists over the speciation of the ferric complexes found in natural waters. Determinations of Fe in sea water are numerous (Table 7.13); the

TABLE 7.13
*Some recent determinations of iron in sea water.*

| Authors | Data ($\mu g$ Fe $l^{-1}$) | |
|---|---|---|
| Menzel and Spaeth (1962) | 0·1–6·2 | Sargasso Sea |
| Topping (1969) | 0·1–61·8 | Indian Ocean |
| Spencer *et al.* (1970) | 2·7–9·8 | North East Pacific |
| Brewer *et al.* (1972) | 0·9–19·8 | Sargasso Sea |
| Riley and Taylor (1972) | 4·1–15·3 | North East Atlantic |

data range from 0 to $> 10$ $\mu g$ Fe $l^{-1}$ depending on the proximity to shore and to a source of input. The kinetics of the oxidation of Fe(II) in sea water are rapid above pH 6 (Stumm and Lee, 1961). The reaction is first order with respect to $[Fe(II)]$ and $[O_2]$, and second order with respect to $[OH^-]$. Thus, reduced iron introduced into sea water will be rapidly removed as $Fe(OH)_3$ or FeOOH.

Spencer *et al.* (1970) have reported a maximum of dissolved iron between 1000 and 2000 m depth in the north east Pacific which appears to be correlated with the excess $^3He$ injected into the water column from the crest of the East Pacific Rise. Similarly, Brewer *et al.* (1972) have found an iron maximum in the Sargasso Sea close to a depth of 3200 m, again correlated with excess $^3He$ and associated with water originating from the Norwegian Sea via flow over the Iceland–Scotland ridge. Further (unpublished) work has shown that the most significant accumulation of iron in this water mass occurs in the Gibbs Fracture Zone of the mid-Atlantic Ridge. The formation of oxidized iron species in mid-water appears to result in locally anomalous profiles, and the data are by no means conclusive. If these data are correct, then there are a series of deep water inputs of iron from active ridge sites, these inputs being rapidly removed (on a time scale of tens of years) from sea water by oxidative processes.

Menzel and Spaeth (1962) have determined particulate and dissolved iron over a one year period in the Sargasso Sea off Bermuda in order to examine correlations between seasonal variations in primary productivity and the chemistry of iron. It had previously been demonstrated (Menzel and Ryther, 1961) that the availability of iron might be a limiting factor in the growth of phytoplankton in the tropical Atlantic. Unfortunately, the results of such experiments *in vitro* cannot be readily applied to direct observations of the ocean. No correlation between iron concentration and primary production was found, and the ratio of particulate to soluble iron was roughly 1:1 at all seasons.

Cobalt is of oceanographic interest both in its biological context of association with vitamin $B_{12}$, and through its accumulation in manganese nodules. Unequivocal evidence of its systematic non-conservative behaviour in sea water is lacking. Schutz and Turekian (1965) found an average value of $0.27$ µg Co $l^{-1}$ for oceanic samples with low values ($0.078$ µg Co $l^{-1}$) from the Gulf of Mexico, higher values ($0.15$ µg Co $l^{-1}$) in the western Atlantic and up to $0.6$ µg Co $l^{-1}$ in the Indian Ocean. Both silver and cobalt showed a trend towards increasing concentration with depth in areas of high productivity. In further work Schutz and Turekian (1965b) have examined the distribution of Co, Ni and Ag in sea water around Pacific Antarctica. Concentrations in the range $0.01$ to $0.34$ µg Co $l^{-1}$ were found, the values increasing northwards. Calculations were made on Co and Ag fluxes based on removal by manganese nodules and diatoms, and input through glacial weathering and volcanic activity. This model is reasonable, and certainly represents a systematic attempt to sample and interpret data on a rational geochemical basis. More recent work by Robertson (1970) reports much lower values for Co in sea water, and the trends reported by Schutz and Turekian could not be recognized. A comparison of the two sets of data are given in Table 7.14. Robertson concluded that the observed variations in Co distribution are erratic and cannot be fully explained by reasonable estimates of biological uptake.

Nickel occurs in sea water in concentrations close to $2$ µg Ni $l^{-1}$. The most sophisticated examination of the distribution of Ni is the stochastic model developed by Spencer (1969) (page 440) based on the data of Schutz and Turekian (1965a). Other information is minimal; Riley and Taylor (1972) reported $1.1$–$4.0$ µg Ni $l^{-1}$ in the North East Atlantic with no discernible trend with depth. Spencer and Brewer (1969), in their study of a Gulf Stream eddy found significant differences between slope water ($1.6$ µg Ni $l^{-1}$) and Sargasso Sea water ($0.8$ µg Ni $l^{-1}$). The slope water trapped in the eddy centre maintained its Ni concentration from June to October, indicating the lack of very short term controls on its distribution.

Copper is present in sea water at a level of *ca.* $1.0$ µg Cu $l^{-1}$. In high concentrations it is toxic to marine organisms, and has widespread use in anti-

TABLE 7.14

Regional average concentrations of cobalt in oceanic waters, from Robertson (1970) and Schutz and Turekian (1965a) (ng l$^{-1}$)

| Region | Robertson | Region | Schutz and Turekian |
|---|---|---|---|
| Northcentral Pacific | 9–7 | Central Pacific | 750 |
| Bering Sea | 30 | | |
| Tropical Northeast Pacific | 20 | East Pacific | 180 |
| Northeast Pacific– | | | |
| Oregon Coastal | 46 | | |
| Tropical Northwest Atlantic | 29 | Northwest Atlantic | 210 |
| Tropical Northeast Atlantic | 21 | Northeast Atlantic | 130 |
| Irish Sea | 41 | Southwest Atlantic | 220 |
| North Sea | 37 | Southeast Atlantic | 250 |
| English Channel | 40 | | |
| Mediterranean Sea | 20 | | |
| Monaco Coastal | 41 | | |
| | | Caribbean Sea | 78 |
| | | Gulf of Mexico | 840 |
| | | Labrador Sea | 160 |
| | | Indian Ocean | 1400 |
| | | Antarctic | 31 |

fouling paints. At normal oceanic concentrations no such toxicity is of course observed, and Cu may be an essential trace element for growth. Topping (1969) and Riley and Taylor (1972) report data on the distribution of Cu in oceanic samples, but no significant trends were found in the data. In studies of metals in a Gulf Stream eddy Spencer and Brewer (1969) found Cu to co-vary with Ni and Zn (slope water, 2·4 μg Cu l$^{-1}$; Sargasso Sea water, 0·6 μg Cu l$^{-1}$). Alexander and Corcoran (1967) have determined "ionic, particulate and total soluble copper" in waters off Florida; inshore waters had twice the total Cu concentration (20 μg Cu l$^{-1}$) of off-shore waters (10 μg l$^{-1}$).

This is in general agreement with the results of analyses of inshore vs. off-shore waters observed by Spencer and Brewer (1969) further to the north, but the absolute amounts are far higher and could not be explained by dilution of the northern waters by simple mixing. The separation between "ionic" and "total" copper was accomplished by solvent extraction of the "ionic" species, and extraction after digestion with perchloric acid for the "total" copper. The average Cu concentrations of a test sample found in this way were: "ionic" 1·13 μg Cu l$^{-1}$ and "total" 6·10 μg Cu l$^{-1}$. These large differences would need further investigation; no indication was given of the specific organic ligands or functional groups responsible for the binding.

7.6.9.2. *Y, Zr, Nb, Mo, Tc, Ru, Rh, Pd, Ag*

Little information has been obtained on the elements of the second and third transition series. No data exist on the distributions of Y, Zr, Nb, Ru, Rh or Pd in sea water; no stable isotope of Tc exists.

Molybdenum occurs in sea water as molybdate ion, $MoO_4^{2-}$, at a level of *ca.* 10 μg Mo l$^{-1}$. Early work reported data ranging from 0·4–16 μg Mo l$^{-1}$ and has been reviewed by Chan and Riley (1966). Most work has been carried out on single samples, and only recently have more widespread investigations taken place. Head and Burton (1970) found 9·4–12·4 μg Mo l$^{-1}$ in estuarine and coastal waters, the distribution being essentially conservative. Riley and Taylor (1972) report 2·1–18.8 μg Mo l$^{-1}$ in north east Atlantic samples with no discernible trend. Bertine (1972) has studied the deposition of Mo in anoxic waters; she suggested that under anoxic conditions $MoO_2^+$ is the stable form of the element, and with increasing sulphide concentration thiooxymolybdates should occur (eg. $Mo_2O_6S^{2-}$, $MoO_2S_2^-$). Laboratory experiments showed that co-precipitation of Mo with iron sulphide was the primary removal mechanism on a short time scale; reduction to Mo(V) and subsequent adsorption onto solid phases took place more slowly. Direct precipitation was not observed.

Silver has been determined in sea water, but the true concentration cannot yet be ascertained. Early work by Schutz and Turekian (1965a, b) recorded 0·1–0·7 μg Ag l$^{-1}$ with significant variability between oceans, and low values for the Antarctic regions. Robertson (1971) has recently reported much lower values (average 0·008 μg Ag l$^{-1}$) for Pacific waters, thus casting doubt on earlier work. The difference is not trivial in view of the toxic effect of Ag on marine organisms, and the increasing use of silver iodide as a cloud-seeding agent; the lower concentrations in sea water could well be modified locally by this mechanism.

### 7.6.9.3. The Lanthanides

The distribution of the lanthanide elements in sea water has been investigated by Goldberg *et al.* (1963). Balashov and Kitrov, (1961), and Høgdahl *et al.* (1968). Since lanthanide fractionation occurs in crustal weathering it seems likely that there are regional differences in the supply of lanthanides to the ocean. However, analyses of marine sediments reveal a fairly uniform pattern (Haskin and Gehl, 1962), and Høgdahl *et al.* suggested that the concentrations in sea water should reflect this difference.

The earlier data published by Balashov and Kitrov (1961) for Indian Ocean water were obtained by colorimetric analysis and are probably too high (La = 0·3 μg l$^{-1}$). The analyses by Goldberg *et al.* (1963) on two samples from off the coast of Southern Californian yielded much lower values (La

$= 3 \times 10^{-3}\,\mu g\,l^{-1}$) with a distinct vertical trend; the sample from a depth of 4000 m having a lanthanide concentration four times that of a sample from 100 m. The samples were not filtered, and the determinations were carried out by neutron activation analysis. The data given by Høgdahl et al. (1968) are the most complete, and the mean values found by them for North Atlantic Deep Water are given in Table 7.15. The abundance relative to lanthanides in chondrites is shown in Figure 7.16 for a single sample, and it is apparent

TABLE 7.15

*Mean values of the absolute abundances of the lanthanide elements in North Atlantic Deep Water.*

| Element | Abundance $(ng\,l^{-1})$ | Element | Abundance $(ng\,l^{-1})$ |
|---|---|---|---|
| La | 3·4 | Er | 0·87 |
| Nd | 2·8 | Lu | 0·15 |
| Gd | 0·70 | Pr | 0·64 |
| Ho | 0·22 | Eu | 0·13 |
| Yb | 0·82 | Dy | 0·91 |
| Ce | 1·2 | Tm | 0·17 |
| Sm | 0·45 | Y | 1·33 |
| Tb | 0·14 | | |

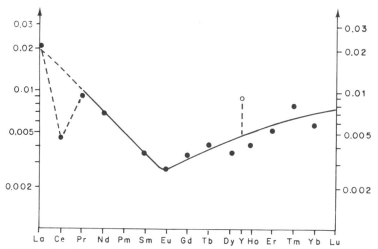

FIG. 7.16. Lanthanides $(\mu g\,l^{-1})$ in sea water (07°30′N, 45°30′W) from 4339 m depth relative to ppm lanthanides in chondrites (Høgdahl et al., 1968). Permission of the American Chemical Society.

that significant depletion of Ce has occurred. Since Ce is enriched in manganese nodules it seems probable that scavenging of Ce(III) from sea water by $MnO_2$ has occurred, and Goldberg (1961) has suggested that in this process oxidation takes place and Ce is incorporated into the $MnO_2$ lattice as $Ce^{4+}$. In addition to this single element depletion, the lanthanide pattern shows a minimum at Eu, again probably associated with the redox chemistry of Eu and the preferential uptake of the heavier elements by manganese nodules and phosphorites.

### 7.6.9.4. *Hf, Ta, W, Re, Os, Ir, Pt, Au*

Very little can be said about the distribution of the elements of this transition series in sea water, and accurate information for the elements other than gold is almost totally lacking.

The concentration of tungsten in Irish Sea surface water has been found to be $0.116\,\mu g\,W\,l^{-1}$. (Chan and Riley, 1967). Rhenium was measured in four Pacific Ocean surface samples by Scadden (1969) who found $0.0084$ $\mu g\,l^{-1}$ with a coefficient of variation of $17\%$. Olafsson and Riley (1972) have recently analysed six samples of North Atlantic water and found $0.0027$–$0.0058$ $\mu g\,Re\,l^{-1}$.

Analyses of gold in sea water are of historical interest, principally through the efforts of Haber (1928), who carried out extensive investigations for the German government after World War I in the hope that economic recovery of gold from sea water might be possible. The element is present at approximately $0.005\,\mu g\,Au\,l^{-1}$, and no systematic regional trends have been observed The data from various investigators are shown in Table 7.16.

TABLE 7.16

*Determinations of gold in sea water.*

| Concentration ($\mu g\,l^{-1}$) | Reference |
| --- | --- |
| 0.004 | Haber (1928) |
| 0.1–0.2 | Caldwell (1938) |
| 0.02–2.0 | Stark (1943) |
| 0.002–44 | Putnam (1953) |
| 0.01–0.5 | Hummel (1957) |
| 0.009 | Brooks (1960) |
| 0.068 ± 0.003 | Weiss and Lai (1963) |
| 0.004–0.027 | Schutz and Turekian (1965a) |

## 7.7. Anoxic Basins

Anoxic conditions prevail in only an insignificant proportion of the world ocean, but are common in enclosed basins (e.g. fjords, lakes). The Black Sea is anoxic below depths of $ca.\,200$ m, and truly marine anoxic basins are the Cariaco Trench, and the Gotland Deep of the Baltic. The biochemical and physical factors leading to the creation of anoxic conditions are discussed in Chapter 16. Briefly, the phenomenon occurs when the rate of consumption of deep water oxygen by micro-organisms exceeds the rate of supply of oxygen by downward mixing of aerated waters. In the absence of oxygen, bacteria turn to other terminal electron acceptors in order to oxidize organic matter and first nitrate and nitrite and then sulphate are reduced. The reduction of sulphate to sulphide results in the lowering of the redox potential of the water and many minor elements are thereby reduced to lower oxidation states; it may be calculated that Fe, Mn, Se, As, Cr, I, Mo, V and U would change their oxidation states on undergoing the transition from the normal environment (oxic) to a reducing (anoxic or euxinic) one.

The redox potential, here expressed as pE, of oxygenated waters is controlled by the reaction

$$\tfrac{1}{4}O_2(g) + H^+ + e^- = \tfrac{1}{2}H_2O \qquad (pE^\circ = 20\cdot75) \qquad (7.16)$$

and for the pH and oxygen levels found in normal aerated sea water pE $= 12\cdot5$. Under anoxic conditions in the presence of $H_2S$ the pE is controlled by the sulphide–sulphate couple as (7.17)

$$\tfrac{1}{8}SO_4^{2-} + \tfrac{9}{8}H^+ + e^- = \tfrac{1}{8}HS^- + \tfrac{1}{2}H_2O \qquad (pE^\circ = -4\cdot13) \qquad (7.17)$$

and the calculated pE for the natural water is close to $-4\cdot2$. Spencer and Brewer (1971) have calculated the pE values at various depths in the Black Sea and their data are shown in Table 7.17. These reactions define the equilibrium boundary conditions of the system, and other redox couples will be driven by these reactions to equilibrium with the sea water at rates dependent upon the kinetic restraints of the individual reactions. An outstanding example of this process may be seen in the behaviour of manganese in anoxic basins (Spencer and Brewer, 1971; Spencer $et\ al.$, 1972). In oxygenated waters, Mn(IV) predominates as $MnO_2$ (s) and reported concentrations of Mn(II) are very low ($ca.$ $0\cdot04$–$1\cdot0\ \mu g\ kg^{-1}$) in agreement with (7.18).

$$MnO_2(s) + 4H^+ + 2e^- = Mn^{2+} + 2H_2O \qquad (pE^\circ = 40\cdot84) \quad (7.18)$$

In the presence of sulphide Mn(II) is formed, and the equilibrium solubility of manganese is thereby greatly increased. Figure 7.17 shows a vertical profile of dissolved manganese in the Black Sea. Analyses from several stations

TABLE 7.17

Calculated values of the pE of Black Sea waters from various depths
(Spencer and Brewer, 1971)

| Depth | T °C | S‰ | pH | $O_2$ (ml l$^{-1}$) | $p_{O_2}$ (atm.) | pE |
|-------|------|-----|-----|------|------|-----|
| Station 1431, $\frac{1}{4}O_2(g) + H^+ + e^- = \frac{1}{2}H_2O$, $pE^\circ = 20\cdot75$ | | | | | | |
| 0 | 4·71 | 17·596 | 8·30 | 8·52 | 0·20 | 12·30 |
| 25 | 0·29 | 18·230 | 8·26 | 7·77 | 0·182 | 12·33 |
| 48 | 6·45 | 18·337 | 8·26 | 7·52 | 0·177 | 12·33 |
| 69 | 6·57 | 18·275 | 8·27 | 7·68 | 0·185 | 12·33 |
| 89 | 6·76 | 18·321 | 8·26 | 7·58 | 0·183 | 12·33 |
| 122 | 7·70 | 18·726 | 8·18 | 6·46 | 0·160 | 12·40 |
| 146 | 8·39 | 19·976 | 7·75 | 0·93 | 0·023 | 12·62 |
| 163 | 8·49 | 20·518 | 7·70 | 0·42 | 0·0106 | 12·58 |
| Station 1448, $\frac{1}{8}SO_4^{2-} + \frac{9}{8}H^+ + e^- = \frac{1}{8}HS^- + \frac{1}{2}H_2O$, $pE^\circ = -4\cdot13$ | | | | $H_2S$, $\mu M$ | | |
| 144 | 8·61 | 21·209 | 7·79 | 13·6 | | $-4\cdot13$ |
| 500 | 8·88 | 22·086 | 7·69 | 197·1 | | $-4\cdot26$ |
| 1100 | 8·96 | 22·339 | 7·60 | 364·5 | | $-4\cdot20$ |
| 1500 | 9·03 | 22·359 | 7·58 | 440·9 | | $-4\cdot18$ |

have been combined in a single profile by normalizing the depths to a reference density surface, in this case indicated at $\sigma_\theta = 16\cdot41$ which represents the boundary between oxygenated and sulphide-containing waters. A marked increase in dissolved manganese can be seen, reaching a maximum level of 450 μg kg$^{-1}$ some 40–50 m below the interface and decreasing to 250 μg kg$^{-1}$ in the deep water. It seems likely that in this particular instance the upper limit on solubility is controlled by the formation of $MnCO_3$. Assuming an activity coefficient of 0·3 for $Mn^{2+}$, then deep water $Mn^{2+} = 10^{-5\cdot8}$ M; the calculated ion association product for $MnCO_3$ is then $10^{-10\cdot6}$ which is close to the solubility product of rhodochrosite ($10^{-10\cdot8}$). Although the relative abundance of manganese in the two zones is controlled by solubility, the shape of the profile is controlled by dynamic and kinetic effects. Deep water manganese is transported upwards by mixing, and precipitates as $MnO_2$ at the interface. These particles then sink and are redissolved in the anoxic zone. Spencer and Brewer (1971) have described this phenomenon by means of a one-dimensional vertical advection-diffusion model. The mixing parameter, $z^*$, over the linear $\theta - S‰$ region through the interface, was found to be 0·09 km. Assuming a deep water input of 190 km$^3$ yr$^{-1}$ into the Black Sea from the Bosphorus, and that this input

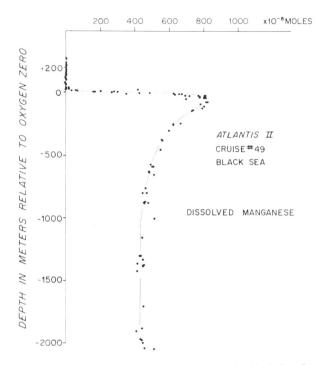

FIG. 7.17. The vertical distribution of dissolved Mn(II) in the Black Sea. Permission of the American Geophysical Union.

advects uniformly upwards to the surface mixed layer, they calculated a vertical advective velocity of $0.5 \text{ m yr}^{-1}$, and a vertical eddy diffusion coefficient of $44 \text{ m}^2 \text{ yr}^{-1}$ ($0.014 \text{ cm}^2 \text{ sec}^{-1}$). The manganese profile over the same depth region was used in conjunction with the solution of the equation for a stable non-conservative tracer allowing for an exponentially decreasing *in situ* rate of production. The derived production parameter was

$$J = 4.5 \times 10^{-4} \, e^{0.038z} \, \mu\text{g kg}^{-1} \, \text{yr}^{-1} \qquad (7.19)$$

The median value was $J = 17 \, \mu\text{g kg}^{-1} \, \text{yr}^{-1}$ at a depth of 2 metres above the oxygen-sulphide boundary. From this data it is possible to calculate the upward flux of dissolved manganese, i.e.,

$$\text{Flux Mn}_{\text{dissolved}} = K \frac{dc}{dz} + wc = 675 \text{ mg m}^{-2} \text{ yr}^{-1} \qquad (7.20)$$

and also the manganese supply necessary to maintain the profile in a steady state by dissolution of the particulate flux, *viz*:

$$\text{Flux Mn}_{\text{particulate}} = \int_{2000}^{295} 4 \cdot 5 \times 10^{-4} \, e^{0 \cdot 038z} \, dz = 875 \, \text{mg m}^{-2} \, \text{yr}^{-1} \quad (7.21)$$

The difference between the two values corresponds to the annual input of particulate manganese to the Black Sea from rivers and from organic detritus. Brewer and Spencer (1974) have shown that this can be contributed by 22% of the river input, the remaining 78% of the detrital load being deposited on shelf areas and close to the river mouth. The deep water acts as a trap for manganese, by means of the precipitation and dissolution cycle,

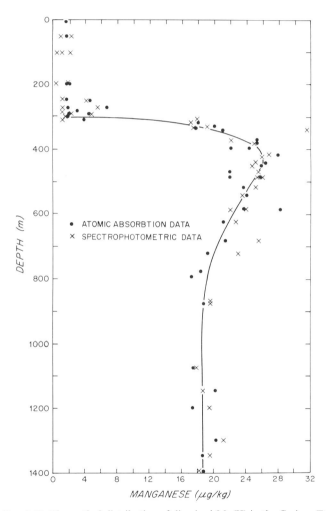

FIG. 7.18. The vertical distribution of dissolved Mn(II) in the Cariaco Trench.

leading to an increase of dissolved manganese in the deep water of $ca.$ $0.1 \, \mu g \, kg^{-1} \, yr^{-1}$ until saturation with the least soluble compound is reached. These calculations provide the most detailed treatment so far given of trace metal chemistry under anoxic conditions. The conclusions are quite general; Brewer (unpublished work) has described the chemistry of manganese in the Cariaco Trench in similar fashion. A vertical profile of Mn(II) in the Cariaco Trench is shown in Fig. 7.18, and the similarity to the Black Sea profile is striking, although the maximum concentration is only $ca. 23 \, \mu g \, kg^{-1}$. Advection-diffusion model calculations show that for an advective velocity of $5 \, m \, yr^{-1}$ (indicating a deep water residence time of 200 years) the annual input of particulate manganese required to maintain the profile is close to $270 \, mg \, m^{-2} \, yr^{-1}$. The similarity of this estimate to that for the Black Sea reflects the similar concentrations of particulate manganese found in surface waters of the two basins.

Elements other than manganese have not been described in such detail. Vertical profiles of Fe, Cu, Zn, Ni and Co in the Black Sea are shown in Figs. 7.19–7.23 in which they are plotted relative to a constant density level

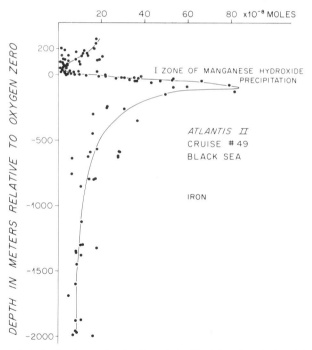

FIG. 7.19. The vertical distribution of dissolved iron in the Black Sea. Permission of the American Geophysical Union.

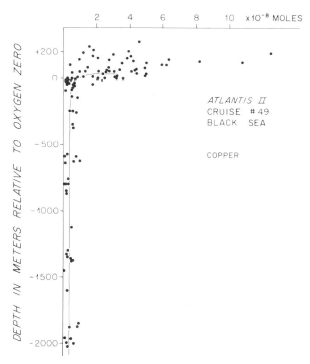

FIG. 7.20. The vertical distribution of dissolved copper in the Black Sea. Permission of the American Geophysical Union.

in the same way as was used for manganese (see above). The sharp increase of iron observed at the oxygen zero boundary clearly illustrates the effects of reduction of Fe(III) to Fe(II). However, a simple model, as used for manganese cannot be used to treat the profile. Possibly this is because of the precipitation of a sulphide phase as the ion association product calculated $(10^{-6.8})$ is close to the solubility product of FeS $(10^{-6.3})$. Cu and Zn would be expected to precipitate as sulphides in the deep water, and this can clearly be seen in Figs. 7.20 and 7.21. However, the amount of dissolved metal in the water is clearly well in excess of that predicted from equilibrium with CuS and ZnS from their solubility products $(10^{-25.3}$ and $10^{-14.7}$ respectively). This observation is supported by the conclusions of Schwarzenbach and Widmer (1963) who found that compounds with very small solubility products tend to form stable metal-sulphide complexes in solution. Although the analytical errors associated with the determinations of Co and Ni are large, the available evidence suggests that these elements show no marked variation on going from anoxic to an oxygenated environment.

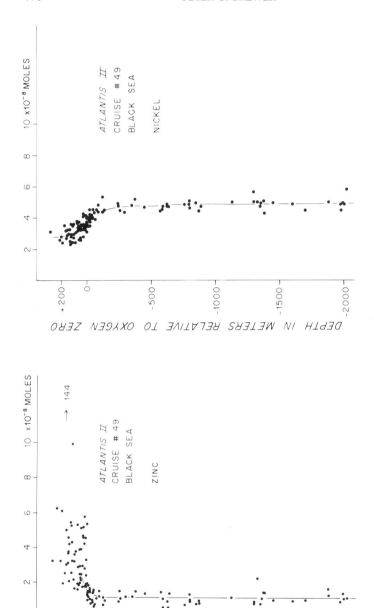

FIG. 7.21. The vertical distribution of dissolved Zn in the Black Sea. Permission of the American Geophysical Union.

FIG. 7.22. The vertical distribution of dissolved nickel in the Black Sea. Permission of the American Geophysical Union.

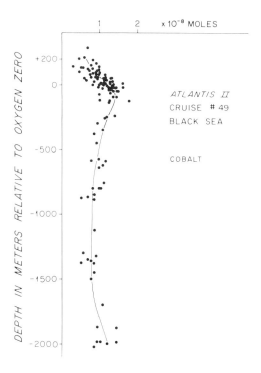

FIG. 7.23. The vertical distribution of dissolved cobalt in the Black Sea. Permission of the American Geophysical Union.

The distribution of trace elements in the particulate form (i.e. trapped by a 0·45 μm pore size filter) in the Black Sea has been investigated by Spencer et al. (1972). The experimental procedure was to filter 6–10 l of sea water through a pre-weighed filter which was later dried, pressed into the form of a pellet and analyzed by instrumental neutron activation analysis. The elements detected were Mn, Fe, Co, Hg, Sc, Zn, La and Sb and their distribution with respect to depth gave evidence of the presence of both detrital particles and the products of redox reactions. The vertical profile of particulate manganese is shown in Fig. 7.24 and the peak resulting from precipitation of $MnO_2$ (60 μg Mn $kg^{-1}$ of sea water) is clearly seen. The vertical profile of particulate Zn also exhibits a maximum some 30 m below the oxygen-sulphide boundary, probably because of the precipitation of ZnS. Predictably, the distributions of Sc and La were unaffected by redox processes, and could be shown to be the result of detrital input.

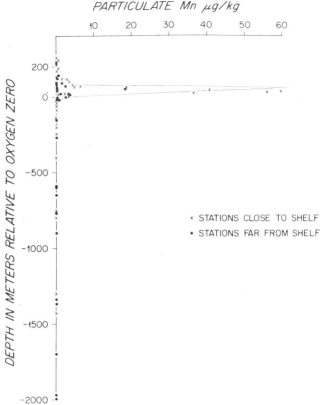

FIG. 7.24. The vertical distribution of particulate manganese in the Black Sea (by permission of the American Geophysical Union).

## 7.8. RED SEA BRINES

Probably the most extraordinary example of anomalies in the minor element composition of sea water is to be found in the hot brines of the Red Sea. These brines were first located near 21°17′N, 38°02′E (Swallow and Crease, 1965), at 2000 m depth, in a series of depressions of the central rift of the Red Sea. Their chemical composition has been investigated by Brewer *et al.* (1965) and Miller *et al.* (1966), and is discussed in detail in the text edited by Degens and Ross (1969).

The primary brine source is located in the Atlantis(II) Deep, and in February 1971 the deepest brine had a temperature of 59·2°C and was increasing in temperature by 0.63°C per year. The ionic composition of the brine is markedly different to that of sea water, the salinity of the deepest brine layer being *ca.*

257‰, Magnesium and sulphate are depleted and calcium is enriched relative to sea water. The minor element composition of the brines which is dominated by iron and manganese is given in Table 7.18. The hottest brine contains approximately 80 mg kg$^{-1}$ of both iron and manganese, presumably

TABLE 7.18

The major and minor element composition of the Red Sea deep brines (g kg$^{-1}$).

| Element | Atlantis (II) Deep hot brine | Atlantis (II) Deep intermediate brine | Discovery Deep brine |
|---|---|---|---|
| Cl | 156·03 | 80·06 | 155·3 |
| Br | 0·128 | 0·101 | 0·119 |
| SO$_4$ | 0·840 | 2·26 | 0·695 |
| Si | 0·0276 | — | 0·0035 |
| Na | 92·60 | 46·90 | 93·05 |
| Ca | 5·15 | 2·47 | 5·12 |
| K | 1·87 | 1·07 | 2·14 |
| Mg | 0·764 | — | 0·810 |
| Sr | $4·8 \times 10^{-2}$ | $2·7 \times 10^{-2}$ | 0·046 |
| Fe | $8·1 \times 10^{-2}$ | $2 \times 10^{-4}$ | $2·7 \times 10^{-4}$ |
| Mn | $8·2 \times 10^{-2}$ | $8·2 \times 10^{-2}$ | $5·46 \times 10^{-2}$ |
| Zn | $5·4 \times 10^{-3}$ | $1·52 \times 10^{-4}$ | $7·7 \times 10^{-4}$ |
| Cu | $2·6 \times 10^{-4}$ | $1·72 \times 10^{-5}$ | $7·5 \times 10^{-5}$ |
| Co | $1·6 \times 10^{-4}$ | $8·0 \times 10^{-7}$ | $1·29 \times 10^{-4}$ |
| Ni | — | $1·2 \times 10^{-6}$ | $3·42 \times 10^{-4}$ |
| Pb | $6·3 \times 10^{-4}$ | $8·8 \times 10^{-6}$ | $1·65 \times 10^{-4}$ |

leached from sedimentary rocks by geothermal waters. Sulphides have been detected in the metalliferous sediments associated with these brines, but the brine pool itself contains no oxygen or $H_2S$. It is possible that the pE of the system is controlled by Fe and Mn. Immediately above the hottest brine layer is a layer of intermediate salinity (ca. 130‰) and temperature (49·7°, February 1971), and above this layer is a mixing zone (ca. 30 m) over which it merges into the Red Sea Bottom Water (RSBW).

The brines constitute a highly stratified system in which redox transformations and precipitation phenomena may be observed. At the 59°/49° interface iron is precipitated and manganese is retained in solution; much of the Zn, Cu, Co and Pb appear to be removed, probably by adsorption on the solid ferruginous phase. No change in dissolved manganese concentration can be detected here, in spite of the change in salinity. At the 49°/RSBW interface most of the manganese is precipitated. However, because of the slow kinetics of the oxidation of Mn(II), the removal is incomplete and some Mn(II) leaks out into the Red Sea Bottom Water (Brewer, unpublished work). The separation of Fe and Mn at the two interfaces is clearly the result of the lower oxidation potential of Fe(II). The precipitates formed in this manner contribute to the metalliferous sediments underlying the brine area, and indicate an ore body in the process of active formation.

## 7.9. Interaction of Minor Elements with Marine Organisms

All marine organisms have in their body tissues and skeletons detectable amounts of the more common minor elements of sea water. The concentrations of these elements in the organism are frequently enhanced several orders of magnitude over those in the surrounding sea water. The mode of accumulation of these metals, and their effect on the organisms, will be discussed briefly here. In view of the large literature on the subject, the choice of the topics discussed must necessarily be selective; a comprehensive survey would fill a volume in itself.

### 7.9.1. CONCENTRATIONS IN MARINE ORGANISMS

The classic work of Vinogradov (1953) covers information on the chemical composition of marine organisms up to 1944. More recent work has focussed on the simultaneous determination of a multiplicity of elements in samples of natural populations. Martin and Knauer (1973) have reported on the elemental composition of phytoplankton from Monterey Bay, California. They point out that collection of such material in sufficient quantities for analysis is often difficult, and that the sample is often contaminated by suspended sediment particles or chips of rust and paint from sampling gear and ships. Most samples contain some zooplankton, thus making it difficult to examine elemental levels between members of the food web. The rates of uptake of minor elements will depend upon the amounts of metal available, population dynamics and age, and physical variables such as temperature, salinity and light intensity. The samples examined by Martin and Knauer (1973) were collected with a 76 μm mesh net (and thus micro-planktonic organisms are omitted) and the organic fractions of the phytoplankton were analyzed separately from the siliceous frustules. The data shown in Table 7.19 are classified on a chemical rather than a taxonomic basis into samples containing no significant Ti, and samples with Ti and Sr concentrators. The lack of taxonomic data on the samples makes the information less useful than it would be with a complete characterization. However, the amounts reported are comparable to those in other more limited and specific studies.

In contrast to these studies of plankton from ocean water containing "normal" levels of trace metals, phytoplankton grown in culture media are exposed to much higher metal concentrations. Riley and Roth (1971)

TABLE 7.19
The elemental composition of phytoplankton (Martin and Knauer, 1973).

| Element | *Organic fraction ($\mu$g g$^{-1}$ dry wt) | | | Siliceous frustules ($\mu$g g$^{-1}$ of dry silica) | |
|---|---|---|---|---|---|
| | I | II | III | I | II |
| Na*† | 138 | 106 | 89 | — | — |
| K*† | 13 | 13 | 11 | — | — |
| Mg*† | 16 | 15 | 11 | — | — |
| Ca*† | 6·5 | 5·3 | 6·5 | — | — |
| Sr | 147 | 119 | 697 | — | — |
| Ba | 33 | 19 | 287 | — | — |
| Al | 110 | 444 | 38 | 620 | 2550 |
| Fe | 224 | 1510 | 231 | 220 | 560 |
| Mn | 6·1 | 13·3 | 7·7 | ND | 4·3 |
| Ti | ND | 27 | ND | 115 | 400 |
| Cr | ND | 3·9 | — | — | — |
| Cu | 3·2 | 7·4 | 11 | 5·6 | 9 |
| Ni | 1·9 | 7·8 | 2·3 | — | — |
| Zn | 19 | 122 | 24 | 5 | 10 |
| Ag | 0·2 | 0·6 | ND | — | — |
| Cd | 1·5 | 1·6 | 3·9 | — | — |
| Pb | ND | 7·2 | 9·2 | — | — |
| Hg | 0·19 | 0·16 | 0·16 | — | — |
| SiO$_2$ (%) | 10 | 15 | 11 | — | — |

ND = Not detected            Group I:    No detectable Ti
* Median values only given    Group II:   Detectable Ti
† Expressed as mg g$^{-1}$     Group III: Samples high in Sr

have analysed fifteen species of phytoplankton grown in culture media containing orders of magnitude more Fe, Mn, Ag, Cu, Co, Pb, Sn, Cr, V and Zn than does natural sea water. The metals were buffered by the presence of chelating agents. They reported that the trace metal content of the algae increased with increasing concentrations of metals in the medium, but that there were no trends which could be correlated with the taxonomy of the organism. The conclusion that phytoplankton trace element levels reflect the background metal level of the medium is by no means clear cut. A comparison of the composition of the culture with that of a natural phytoplankton population showed that the natural sample had much higher Al, Ti, Sr and Pb concentrations, perhaps indicating the importance of micro-particles of clay material in the natural sample. No toxicity of metals in the culture medium to the organisms was observed.

Although some generalizations can be made regarding the elemental composition of phytoplankton, the composition of zooplankton is more com-

TABLE 7.20

The elemental composition of zooplankton (Nicholls et al., 1959).

| Sample | Ash % dry | Element (μg g⁻¹ in ash) | | | | | | | | | | | | | | | | |
|---|---|---|---|---|---|---|---|---|---|---|---|---|---|---|---|---|---|---|
| | | B | Mo | V | Ni | Co | Ti | Cr | Pb | Sn | Cu | Hg | Cd | Ag | Mn | Fe | Zn | Al |
| Coelenterata | | | | | | | | | | | | | | | | | | |
|   *Cyanea capillata* | 63·0 | 100 | 3 | 5 | <1 | 3 | 3 | <1 | 6 | 4 | 13 | | | | | | | |
| Ctenophora | | | | | | | | | | | | | | | | | | |
|   *Beröe cucumis* | 70·3 | 115 | 3 | 8 | <1 | <1 | <1 | <1 | 6 | 7 | 700 | | | | | | | |
| Mollusca, Pteropoda | | | | | | | | | | | | | | | | | | |
|   *Limacina retroversa* | 64·2 | 50 | 8 | 85 | 2 | 20 | 5 | <1 | 200 | <1 | 30 | | | | | | | |
|   *Clione limacina* | 33·3 | 90 | <1 | 16 | 4 | 30 | <1 | | 65 | 20 | 70 | | | | | | | |
| Mollusca, Cephalopoda | | | | | | | | | | | | | | | | | | |
|   *Ommastrephes illicebrosa* | 7·7 | 420 | 4 | 4 | <1 | 3 | 6 | <1 | 5 | 3 | 2700 | | | | | | | |
| Arthropoda, Copepoda | | | | | | | | | | | | | | | | | | |
|   *Centropages typicus & hamatus* | 22·8 | 160 | 10 | 16 | 55 | 4 | 3 | 260 | 1300 | 50 | 600 | | | | | | | |
|   *Calanus finmarchicus* | 17·6 | 760 | 12 | 21 | 165 | 3 | 10 | <1 | 575 | <1 | 1350 | | | | | | | |
|   *Calanus finmarchicus* with ~50% neritic phytoplankton | 36·3 | 325 | 36 | 54 | 70 | 80 | 70 | | 75 | 90 | 300 | | | | | | | |
| Arthropoda, Euphausiacea | | | | | | | | | | | | | | | | | | |
|   *Euphausia krohnii* | 18·6 | 440 | 17 | 45 | 6 | 5 | 9 | <1 | 20 | <1 | 600 | | | | | | | |
| Chaetognatha | | | | | | | | | | | | | | | | | | |
|   *Sagitta elegans* | 21·6 | 130 | 3 | 13 | 480 | 110 | 3 | 1 | 300 | 20 | 1100 | | | | | | | |
| Tunicata | | | | | | | | | | | | | | | | | | |
|   *Salpa fusiformis* | 77·1 | 50 | 1 | 7 | 60 | 1 | 1 | 1 | 3 | 8 | 500 | | | | | | | |
| *Mixed zooplankton (Median concentrations from Martin & Knauer 1973) | | | | | 8·4 | | | | 2·1 | | 11·5 | 0·11 | 2·3 | 0·26 | 4·3 | 100 | 180 | 15 |

* Data given in μg g⁻¹ dry weight of sample. The samples were collected on a transect between Hawaii and Monterey, and consisted of "predominantly copepods".

plex. Inter-species differences are greater, and in any one organism the composition will depend upon the surrounding water, the principal food source and the age and recent history of the specimen. Nicholls *et al.* (1959) have reported analyses of ten species of zooplankton for B, Mo, V, Ni, Co, Ti, Cr, Pb, Sn, Cu (Table 7.20). No general oceanic conclusions can be drawn from such a limited survey. However, Ni accumulation appeared to be typical of copepods, and, on the basis of fresh weight accumulations, Ni and Co accumulation in *Sagitta* and *Limacina* was striking. Martin and Knauer (1973) have analyzed mixed zooplankton populations from the Pacific and their data are also included in Table 7.20.

## 7.9.2. EFFECTS ON PRODUCTIVITY

It has long been realized that, although the primary controls on photosynthetic production of organic matter in sea water are the availability of nutrient anions and light, other, more complex, factors also play an important role. This distinction between waters which are "good" or "bad" in their ability to support phytoplankton growth has often been attributed to the effects of trace metal species, and the presence or absence of organic chelators, which might alter the activity of the free metal ion. The subject has recently been reviewed by Siegel (1971).

An early paper by Spencer (1957) dealt with the effects of Cu on *Phaeodactylum tricornutum* in the presence of (ethylenedinitrilo) tetraacetic acid. Spencer pointed out that Cu is toxic to this organism in concentrations in excess of $2 \times 10^{-6}$ M, but that when buffered by the chelator, concentrations of $\Sigma$Cu up to $5 \times 10^{-3}$ M could be tolerated. In metal-free media the organism was found to grow poorly, and Spencer suggested that a low finite level of the free metal ion, buffered by the large number of competing equilibria, was necessary for growth. Johnston (1963, 1964) has discussed in detail the effects of adding artificial chelators to natural sea water, and found that this could improve "bad" waters. He concluded that "the supply of chelating substances is frequently the most crucial aspect of phytoplankton nutrition in sea water". These concepts are illustrated by observations made by Barber and Ryther (1969) in their study of the factors affecting primary production in the Cromwell Current upwelling. They found that the upwelling brought to the surface nutrient rich water that was poor in its ability to support phytoplankton growth. Surface water one degree north or south of the upwelling site, i.e. "aged" upwelled water, was not deficient in this respect. The addition of nutrients, trace metals, vitamins or amino acids did not improve the quality of the "bad" water. However, the addition of (ethylenedinitrilo) tetraacetic acid or an extract of an undefined zooplankton led to an increase in growth. They concluded that a probable explanation of the regional productivity differences was that organisms must condition the newly upwelled water by the addition of natural products having a chelating function before optimum growth can occur. The evidence is not substantial, and it is uncertain whether trace metals do play any part in this effect.

More recent work has extended these observations. Barber *et al.* (1971) have investigated the productivity of waters at 15° S off the coast of Peru during a period of active upwelling. Two stations were found at which the specific growth rate was unusually low and a systematic set of enrichment experiments was carried out at them. The addition of a trace metal mixture (Cu, Zn, Co, Mn, Mo) together with Fe and (ethylene %-dinitrilo) tetraacetic acid markedly improved the water; stripping natural organic compounds out by passing the water through a charcoal column caused a tenfold decrease in specific growth rates, but the original growth rate could be restored by the addition of the chelator. Addition of deferriferrioxamine B (so as to give a highly stable iron(III) complex) caused a reduction in specific growth rates, but addition of excess iron could restore this deficiency. Menzel and Ryther (1961) have shown that the availability of iron may be the initial limiting factor for growth of phytoplankton in the Sargasso Sea.

In summary, it is now well established that low density inoculum populations of phytoplankton grow poorly in a nutrient-rich deep ocean water medium (Barber, 1973). This phenomenon is best explained on the hypothesis that the metabolic products generated by a larger population can alter the unfavourable medium. These products appear to be functional analogues of well known chelators. In an artificial growth medium these chelators may be obtained from the undefined "soil extract" (as in Erdschreiber medium), but can be replaced to a large extent with synthetic ligands (Provasoli *et al.*, 1957).

### 7.9.3. CALCULATION AND OBSERVATIONS OF METAL SPECIATION

Although the circumstantial evidence that metal chelates play a significant role in the ocean is quite strong, direct evidence for this is less so. No single chelator has yet been identified as being specifically responsible for the observed effects; in fact, a multiplicity of complex and labile molecules are probably present.

Two investigative techniques have been used: computational and analytical. The attempts to use equilibrium calculations on a multi-metal/ligand system have been reviewed in Chapter 3. The work is sophisticated and the most complete treatment so far (Morel and Morgan, 1972) has considered a system of 20 metals and 31 ligands, resulting in 788 soluble species and 83 possible solids. The model showed that at the normal oceanic concentrations of the various ligands, inorganic associations would be completely predominant over organic complexes. In a natural system the equilibrium model may not be applicable as the specific organic ligands naturally present are not known.

However, calculations of this type show that caution is necessary in making interpretations of experiments in which high concentrations of a single ligand are added to a complex sea water system.

The analytical evidence is based upon either difference measurements of metals in the sample both before and after treatment with a powerful oxidizing agent, or on the response of a mercury coated graphite electrode in anodic stripping voltammetry.

Buglio et al. (1961) have reported that a significant fraction of the Cu in sea water is in a dissolved but undialysable form. Slowey et al. (1967) found that up to 56% of the Cu in a coastal sample was extractable into a non-polar solvent. Increased $Cu^{2+}$ after oxidation of organic matter with perchloric acid or U.V. radiation has been reported by Corcoran and Alexander (1964), Slowey and Hood (1966), Williams (1969) and Foster and Morris (1971). The data given by Foster and Morris (1971) are the most consistent; these workers found between 8 and 40% of the total Cu to be associated with organic matter in coastal waters, but there was no direct correspondence between it and the total amount of dissolved organic matter. The general validity of these results seems proven, but the trace element intercalibration study conducted by Brewer and Spencer (1970) revealed a wide discrepancy in the ability of many laboratories to distinguish small differences in the trace element composition of sea water samples. Experimental artefacts cannot yet be ruled out.

Anodic stripping voltammetry (Matson et al., 1965) has been used for the determination in sea water of a number of elements, such as Cu, Zn, Pb, and Cd (Fitzgerald, 1970) and Bi and Sb (Gilbert and Hume, 1973) and holds promise for the simple and rapid determination of these elements in small (ca. 50 ml) samples. Matson (1968) has suggested that this technique may be used as a tool for the investigation of the chemical speciation of an ion in solution, and in particular that of organic complexes. On the basis or absence of double stripping peaks, peak potential shifts and peak shape changes, he postulated that distinctions could be made between free metal ions, labile organic complexes and non-labile complexes. As an example of this behaviour, he reported a double stripping peak for a solution of $2 \times 10^{-7}$ M Cu in $5 \times 10^{-5}$ M (ethylenedinitrilo) tetraacetic acid. A subsequent re-investigation by Hume and Carter (1972) showed that the double peaks were irreproducible and were probably related to adsorption on to the electrode surface. They concluded that it was unlikely that quantitative information could be obtained about the nature of complex metal ion species in natural water samples using anodic stripping voltammetry with a mercury coated graphite electrode. The technique does have considerable use both in direct analysis of ionic species and in the investigation of the formation of complexes with simple inorganic ligands (e.g., carbonate and sulphate complexes) (Stumm and Bilinski, 1972).

7.9.4. EFFECTS OF HEAVY METALS ON HIGHER MARINE ORGANISMS

Studies of the effect of heavy metals on marine organisms have been carried out since the last century, interest being aroused by such obvious effects as oysters turning green as a result of excessive accumulation of Cu (Boyce and Herdman, 1898). Each organism will exhibit a complex series of metabolic responses to exposure to differing metal concentrations, and no simple conclusions can be reached. The subject has been discussed in the excellent review by Bryan (1971), and an extensive bibliography has been compiled by Eisler (1973).

### 7.9.4.1. *Uptake, storage and excretion*

Marine organisms absorb, ingest, excrete, store and regulate their body burden of metals. Absorption from solution through the gills, or over the body surface, is a major means of uptake. Bryan (1971) has reported that although the Zn level in the blood of the lobster *Homarus vulgaris* is approximately $7 \,\mu g \, ml^{-1}$ ($10^3$ times that in sea water), active transport from sea water into the blood takes place through the gills. The concentration gradient is maintained by the bonding of blood Zn to proteins such as haemocyanin. The absorption is probably a two stage process, and *Homarus vulgaris*, placed in zinc-rich sea water will absorb, store and excrete elevated levels of the element. In contrast to this, *Homarus vulgaris* does not appear to take up appreciable amounts of Mn (Bryan and Ward, 1965). Lobsters placed in normal sea water containing $2 \,\mu g \, Mn \, l^{-1}$ absorbed ca. $0.4 \,\mu g \, Mn \, day^{-1}$ from sea water, and about $24.6 \,\mu g \, day^{-1}$ from food. Starved lobsters placed in sea water containing $10^2$ and $10^3 \,\mu g \, Mn \, l^{-1}$ did not take up greater quantities of Mn.

Absorption from the stomach fluid and digestive tract has been shown to be important for Zn uptake by the oyster *Ostrea edulis* (Preston and Jeffries, 1969) from ingested particles. Zn is adsorbed rapidly from the stomach fluid of the lobster *Homarus vulgaris* (Bryan, 1964) and accumulation of Zn increases the levels in the urine, excretory organs, hepatopancreas and gills. The removal of Zn stored in the hepatopancreas probably takes place via the blood and excretory organs rather than the gut. In contrast, the injection of $10 \,\mu g$ of Mn into the stomach resulted in only a small temporary increase in the concentration in the muscle.

Of the three principal metal removal mechanisms, i.e. via gut, urine and diffusion through the body surface, the urine and rectal fluids appear to be the major pathway. Crustaceans excrete Zn, Cu, Co, Mn and Hg in the urine. The gut is an important pathway for Zn removal in freshwater crayfish, and for Cu in cyprid larva of the barnacle (Bernard and Lane, 1961). Temporary removal of metals can take place by storage in a particular organ, such as the hepatopancreas, with subsequent gradual removal. Some metals remain bound. The mussel *Mytilus edulis* absorbs iron directly from sea water until the total iron content is *ca.* 1% above which there is no further uptake. The iron is not eliminated on transfer of the animal to clean water (Hobden, 1969).

It is important to consider whether organisms are potentially capable of regulating their body metal burden. Bryan (1966) reports that the crab (*Carcinus maenus*) can regulate its body level of Zn. After exposure to sea water containing 500 times the normal concentration of Zn, the concentrations of Zn in the blood and the whole animal increased by only factors of 2, and 4 respectively. In contrast, the oyster (*Crassostrea virginica*) absorbs Pb from the surrounding sea water in almost exact proportion to the ambient levels (Pringle *et al.*, 1968) and can lose this Pb later on exposure to normal sea water.

### 7.9.4.2. *Toxicity*

Because of the widespread use of Cu and Hg in marine anti-fouling paints, some aspects of the toxicity of heavy metals to marine organisms have been studied in detail. The degree of toxicity will vary with the metal and the organism, but as a crude generalization the order of toxicity is Hg > Ag > Cu > Cd > Zn > Pb > Cr > Ni > Co. Portmann (1968) has conducted static 48-hour toxicity studies on the pink shrimp (*Pandalus montagui*), brown shrimp (*Crangon crangon*), shore crab (*Carcinus maenus*) and cockle (*Cardium edule*). In these studies animals were acclimatized at 15°C for three days, and then placed in aquaria containing varying concentrations of heavy metals (Hg, Cu, Zn and Ni). The data for the approximate 48 hour LD-50 concentration are given in Table 7.21. Such compilations are, of course, over-simplifications. Portmann (1968) was able to show that the effect of temperature on Cu and Hg toxicity to the brown shrimp and cockle was very pronounced. When the temperature was reduced to 5°C the brown shrimp tolerance increased by a factor of 5, and in the cockle the tolerance increased by a factor of $\sim 100$. Changes in salinity are of similar importance.

The lethal dose of a metal is not the only statistic of concern. Sub-lethal effects of exposure to heavy metals may be marked. The oyster (*Crassostrea virginica*) has been shown to turn visibly green after long exposure to sea water containing 25 μg Cu $l^{-1}$ (Shuster and Pringle, 1969). Zn at a concentration of 60 μg Zn $l^{-1}$ has been shown to cause 25% abnormalities in the larvae of the sea urchin *Lytechinus pictus*. Inhibitory effects on growth of all marine organisms may be caused by metal concentrations only slightly above their normal environmental levels.

Mandelli (1969) has examined the inhibitory effects of Cu on several species of dinoflagellates and diatoms in culture. *Coccochloris elabans* would

TABLE 7.21

*Approximate static 48-hour 15°C LD -50 value for toxicity of heavy metals in sea water to various marine organisms (Portmann, 1968).*

|  | Metal concentration (mg $l^{-1}$) | | | |
| --- | --- | --- | --- | --- |
| Organism | Hg | Cu | Zn | Ni |
| Pink Shrimp | 0·1 | 0·2 | 10·0 | 200·0 |
| (*Pandalus montagui*) |  |  |  |  |
| Brown Shrimp | 6·0 | 30·0 | 100·0 | 150·0 |
| (*Crangon crangon*) |  |  |  |  |
| Shore Crab | 1·0 | 100·0 | 12 | 300·0 |
| (*Carcinus maenus*) |  |  |  |  |
| Cockle | 10·0 | 1·0 | 200 | 500·0 |
| (*Cardium edule*) |  |  |  |  |

not grow in the presence of 30 µg Cu $l^{-1}$, *Skeletonema costatum* was inhibited at 50 µg Cu $l^{-1}$. Dilling *et al.* (1926) found that Pb in sea water does not retard metamorphosis of plaice embryos, but does retard growth of the fish. Growth of the pluteal larvae of the sea urchin *Paracentrotus lividus* is inhibited at a Cu concentration of only 10 − 20 µg $l^{-1}$, and a Cu concentration of 50 µg $l^{-1}$ is lethal.

The mode of action of heavy metals on biological systems is thought to be principally through enzyme systems, although extraordinary concentrations may result in direct tissue damage.

ACKNOWLEDGEMENTS

I wish to thank D. W. Spencer, N. M. Frew, J. W. Murray and G. Wong for much helpful advice and discussion. This work was supported by N.S.F. Grant GA-13574 and by the U.S. Atomic Energy Commission Contract AT (11-1)-3566.

REFERENCES

Abdullah, M. I., Royle, L. G. and Morris, A. W. (1972). *Nature, Lond.,* **235**, 158.
Alexander, J. E. and Corcoran, E. F. (1967). *Limnol. Oceanogr.* **12**, 236.
Armstrong, F. A. J. and Harvey, H. W. (1951). *J. Mar. Biol. Ass. U.K.* **29**, 145.
Bacon, M. P. and Edmond, J. M. (1972). *Earth Planet Sci. Letters,* **16**, 66.
Balashov, Yu. and Kitrov, L. M. (1961). *Geochemistry (U.S.S.R.),* **9**, 877.
Barber, R. T. (1973). *In* "Trace Metals and Metals Organic Interactions in Natural Waters" (P. C. Singer, ed.). Ann Arbor Science Publishers, Ann Arbor, 380 pp.
Barber, R. T. and Ryther, J. H. (1969). *J. Exp. Mar. Biol. Ecol.* 3, 191.
Barber, R. T., Dugdale, R. C., MacIsaac, J. J. and Smith, R. L. (1971). *Inv. Pesq.* **35**, 171.
Bardet, J., Tchakirien, A. and Lagrange, R. (1937). *C.R. Acad. Sci., Paris,* **204**, 443.
Barth, T. W. (1952). "Theoretical Petrology". John Wiley, New York.
Bernard, F. J. and Lane, C. E. (1961). *Biol. Bull. Mar. Biol. Lab., Woods Hole,* **121**, 438.

Bernat, M., Church, T. and Allègrè, C. J. (1972). *Earth Planet. Sci. Letters*, **16**, 75.
Bertine, K. K. (1972). *Mar. Chem.* **1**, 43
Black, W. A. P. and Mitchell, R. L. (1952). *J. Mar. Biol. Ass. U.K.* **30**, 575.
Bolter, E., Turekian, K. K. and Schutz, D. F. (1964). *Geochim. Cosmochim. Acta*, **28**, 1459.
Boothe, P. N. and Knauer, G. A. (1972). *Limnol. Oceanogr.* **17**, 270.
Boström, K. and Peterson, M. N. A. (1966). *Econ. Geol.* **61**, 1258.
Boström, K., Peterson, M. N. A., Joensuu, O. and Fisher, D. E. (1969), *J. Geophys. Res.* **74**, 3261.
Bougis, P. (1965). *C.r. hebd. Seanc. Acad. Sci., Paris*, **260**, 2929.
Bowen, V. T. (1968). *In* "Reference Methods for Marine Radioactivity Studies." (Y. Nishiwaki and R. Fukai eds.), International Atomic Energy Agency, Vienna.
Boyce, R. and Herdman, W. A. (1898). *Proc. Roy. Soc. Lond.* **62**, 30.
Brass, G. W. and Turekian, K. K. (1972). *Earth Planet. Sci. Letters*, **16**, 117.
Brewer, P. G., Riley, J. P. and Culkin, F. (1965). *Deep-Sea Res.* **12**, 497.
Brewer, P. G., Spencer, D. W. and Wilkniss, P. E. (1970). *Deep-Sea Res.* **17**, 1.
Brewer, P. G. and Spencer, D. W. (1970). *Woods Hole Oceanogr. Inst. Tech. Rept.* **70–62**, 74 pp.
Brewer, P. G., Spencer, D. W. and Robertson, D. E. (1972). *Earth Planet. Sci. Letters* **16**, 111.
Brewer, P. G. and Spencer, D. W. (1974). *In* "The Black Sea: Its Geology, Chemistry and Biology". *Amer. Ass. Petrol. Geol. Mem.* Vol. 20.
Brewer, P. G. and Spencer, D. W. (1974). *Mem. Am. Ass. Petrol. Geol.* (in press).
Brooks, A. R. (1960). *Analyst.* **85**, 745.
Bryan, G. W. (1964). *J. Mar. Biol. Ass. U.K.* **44**, 549.
Bryan, G. W. (1966). *In* "Symposium on Radioecological Concentration Processes", Pergamon Press, pp 1005–1016.
Bryan, G. W. (1971). *Proc. Roy. Soc. Lond.* **177B**, 389.
Bryan, G. W. and Ward, E. (1965). *J. Mar. Biol. Ass. U.K.* **45**, 65.
Buglio, B., Rona, E. and Hood, D. W. (1961). *U.S. Atomic Energy Commission Rept. Ref.* **61–19A**, 52 pp.
Burton, J. D. and Riley, J. P. (1958). *Nature, Lond.* **181**, 179.
Burton, J. D. (1965). *In* "Chemical Oceanography", (J. P. Riley and G. Skirrow, eds.), Vol. II, Academic Press, London and New York.
Burton, J. D. and Leatherland, T. M. (1971). *Nature, Lond.* **231**, 440.
Caldwell, W. E. (1938). *J. Chem. Educ.* **15**, 507.
Chan, K. M. and Riley, J. P. (1967). *Analyt. Chim. Acta*, **39**, 103.
Chan, K. M. and Riley, J. P. (1966). *Analyt. Chim. Acta*, **35**, 365.
Chau, Y. K. and Riley, J. P. (1965). *Analyt. Chim. Acta*, **33**, 36.
Chau, Y. K., Sim, S. S. and Wong, Y. H. (1968). *Analyt. Chim. Acta*, **43**, 13.
Chow, T. J. and Goldberg, E. D. (1960). *Geochim. Cosmochim. Acta*, **20**, 192.
Chow, T. J. and Goldberg, E. D. (1962). *J. Mar. Res.* **20**, 163.
Chow, T. J. and Patterson, C. C. (1962). *Geochim. Cosmochim. Acta*, **26**, 263.
Chow, T. J. and Patterson, C. C. (1966). *Earth Planet. Sci. Letters*, **1**, 397.
Chow, T. J., Earl, J. L. and Bennett, C. F. (1969). *Envir. Sci. Tech.* **3**, 737.
Chow, T. J., Motto, R. H. and Chilko, D. M. (1970). *Envir. Sci. Tech.* **4**, 318.
Chuecas, L. and Riley, J. P. (1966). *Analyt. Chim. Acta*, **34**, 337.
Coonley, L. S., Baker, E. B. and Holland, H. D. (1971). *Chem. Geol.* **7**, 51.
Cooper, L. H. N. (1958). *J. Mar. Res.* **17**, 128.

Corcoran, E. E. and Alexander, J. E. (1964). *Bull. Mar. Sci. Gulf Caribb.* **14**, 594.
Cotton, F. A. and Wilkinson, G. (1972). "Advanced Inorganic Chemistry", 3rd ed. Wiley Interscience, New York.
Craig, H. (1969). *J. Geophys. Res.* **74**, 5491.
Craig, H. and Weiss, R. F. (1970). *J. Geophys. Res.* **75**, 7641.
Craig, H., Krishnaswami, S. and Somayajulu (1973). *Earth Planet. Sci. Lett.* **17**, 295
Culkin, F. and Riley, J. P. (1958). *Nature, Lond.* **181**, 179.
Curl, H. Jr., Cutshall, N. and Osterberg, C. (1965). *Nature, Lond.* **205**, 275.
Cutshall, N. H. (1967). Ph.D. thesis, Oregon State University, Corvallis.
Cutshall, N., Johnson, V. and Osterberg, C. (1966). *Science,* **152**, 202.
Degens, E. T. and Ross, D. A. (eds) (1969). "Hot Brines and Recent Heavy Metal Deposits in the Red Sea." Springer-Verlag, New York.
Dilling, W. S., Healy, C. W. and Smith, W. C. (1926). *Annals Appl. Biol.* **13**, 168.
Durum, W. H. and Haffty, J. (1963). *Geochim. Cosmochim. Acta,* **27**, 1.
Duursma, E. K. (1969). *In* "Annual Report of the International Laboratory of Marine Radioactivity", Tech. Rep. Ser. No. 98. International Atomic Energy Agency Laboratory, Monaco.
Duursma, E. K. and Gross, M. G. (1971). *In* "Radioactivity in the Marine Environment", National Academy of Sciences, Washington, D.C., p. 147.
Edmond, J. M. (1970). *J. Geophys. Res.* **75**, 5286.
Eisler, R. (1973). *U.S. Environmental Protection Agency Report,* No. **EPA-R3-007**, February, 1973.
Elderfield, H. (1970). *Earth Planet. Sci. Letters,* **9**, 10.
Feely, R. A., Sackett, W. M. and Harris, J. E. (1971). *J. Geophys. Res.* **76**, 5893.
Fitzgerald, W. F. (1970). Ph.D. Thesis, Mass. Inst. Technology, 130 pp.
Folsom, T. R., Wong, K. M. and Hodge, V. F. (1972). "Symp. Natural Radiation Environment II". Houston, Texas.
Foster, P. and Morris, A. W. (1971). *Deep-Sea Res.* **18**, 231.
Fukai, R. (1967). *Nature, Lond.* **213**, 901.
Fukai, R. and Huynh-Ngoc (1968). International Atomic Energy Agency, "Radioactivity in the Sea", Publication No. 22, p. 1.
Fukai, R. and Vas, D. (1967). *J. Oceanog. Soc. Jap.* **23**, 298.
Garrels, R. M. and Mackenzie, F. T. (1971). "Evolution of Sedimentary Rocks". Norton, New York. 397 pp.
Gautier, A. (1903). *C.R. Acad. Sci., Paris,* **137**, 232.
Gavis, J. and Ferguson, J. F. (1972). *Water Res.* **6**, 989.
Geering, H. R., Cary, E. E., Jones, L. H. P. and Allaway, W. H. (1968). *Soil. Sci. Soc. Amer. Proc.* **32**, 35.
Gilbert, T. R. and Hume, D. N. (1973). *Analyt. Chim. Acta,* **65**, 451.
Goldberg, E. D. (1961). *In* "Oceanography" (M. Sears, ed.), Am. Ass. Advance. Sci., Publ 67, Washington, D.C.
Goldberg, E. D. and Arrhenius, G. O. S. (1958). *Geochim. Cosmochim. Acta,* **13**, 153.
Goldberg, E. D., Broecker, W. S., Gross, M. G. and Turekian, K. K. (1971). *In* "Radioactivity in the Marine Environment", Natl. Acad. Sciences, Washington.
Goldberg, E. D., Koide, M., Schmitt, R. A. and Smith, R. H. (1963). *J. Geophys. Res.* **68**, 4209.
Goldschmidt, V. M. (1933). *Fortsch. Min. Krist. Petrog.* **17**, 112.
Goldschmidt, V. M. (1937). *J. Chem. Soc.* **655**.
Goldschmidt, V. M. and Strock, L. W. (1935). *Nachr. Ges. Wiss. Gottingen, Math. Phys. NF, Kl.* **IV, 1**, 123.

Goldschmidt, V. M., Berman, H., Hauptman, H. and Peters, C. (1933). *Nach. Ges. Wiss. Gottingen, Math, Phys. Kl.* **IV (N.S.)**, **2**, 235.
Greenhalgh, R. and Riley, J. P. (1963). *Nature, Lond.* **197**, 371.
Griel, J. V. and Robinson, R. J. (1952). *J. Mar. Res.* **11**, 172.
Haber, F. (1928). *Z. Ges. Erdk. Berl.* **3**, 3.
Hamaguchi, H., Kuroda, R., Onuma, N., Kawabuchi, K., Mitsubayashi, T. and Hosohara, K. (1964). *Geochim. Cosmochim. Acta*, **28**, 1039.
Haskin, L. A. and Gehl, M. A. (1962). *J. Geophys. Res.* **67**, 2537.
Head, P. C. and Burton, J. D. (1970). *J. Mar. Biol. Ass. U.K.* **50**, 439.
Hem, J. D. (1963). U.S. Geological Survey, Water Supply Paper 1667-A, 52.
Hobden, D. J. (1969). *J. Mar. Biol. Ass. U.K.* **49**, 661.
Hoffman, G. L. (1971). Ph.D. thesis, Univ. Hawaii, Honolulu (unpubl.).
Hoffman, G. L., Duce, R. A. and Zoller, W. H. (1969). *Envir. Sci. Tech.* **3**, 1207.
Holland, W. R. (1971). *Tellus*, **23**, 371.
Horn, M. K, and Adams, J. A. S. (1966). *Geochim. Cosmochim. Acta*, **30**, 279.
Horwitt, M. H. (1965). *Fed. Proc.* **24**, 68.
Hosohara, K. (1961). *J. Chem. Soc. Japan, Pure Chem. Sect.* **82**, 1107.
Høgdahl, O. T., Melsom, S. and Bowen, V. T. (1968). *Adv. Chem. Ser.* **73**, 308.
Hume, D. N. and Carter, J. N. (1972). *Chem. Analityczna*, **17**, 747.
Hummel, R. W. (1957). *Analyst.* **82**, 483.
Ishibashi, M. and Kurata, K. (1939). *J. Chem. Soc. Japan*, **60**, 1109.
Ishibashi, M., Shigematsu, T. and Nakagawa, Y. (1953). *Rec. Oceanogr. Wks. Jap. N.S.* **1**, (ii), 44.
Ishibashi, M., Shigematsu, T. and Nakagawa, Y. (1956). *Bull. Inst. Chem. Res., Kyoto Univ.* **34**, 210.
Jensen, S. and Jernelov, A. (1969). *Nature, Lond.* **223**, 753.
Johnannesson, J. K. (1958). *Nature, Lond.* **182**, 251.
Johnson, D. L. and Pilson, M. E. Q. (1972). *J. Mar. Res.* **30**, 140.
Johnson, V. (1965). *In* "Ecological Studies of Radioactivity in the Columbia River Estuary and Adjacent Pacific Ocean", (C. Osterberg, ed.), Oregon State Univ. Prog. Rep. 65–14, Corvallis.
Johnston, R. (1963). *J. Mar. Biol. Ass. UK* **43**, 427.
Johnston, R. (1964). *J. Mar. Biol. Ass. U.K.* **44**, 87.
Jones, R. F. (1960). *Limnol. Oceanogr.* **5**, 312.
Joyner, T. (1964). *J. Mar. Res.* **22**, 259.
Kanamori, S. (1962). Ph.D. thesis, Nagoya Univ., 113 pp.
Keeling, C. D. and Bolin, B. (1967). *Tellus*, **19**, 566.
Keeling, C. D. and Bolin, B. (1968). *Tellus*, **20**, 17.
Kharkar, D. P., Turekian, K. K. and Bertine, K. K. (1968). *Geochim. Cosmochim. Acta*, **32**, 285.
Koczy, F. F. (1958). *In* "Proc. 2nd Intern. Conf. Peaceful Uses At. Energy", Geneva, **18**, 357.
Krauskopf, K. B. (1956). *Geochim. Cosmochim. Acta*, **9**, 1.
Kuenzler, E. J. (1965). U.S.A.E.C. document NYO-3145-1.
Kuwamoto, T. and Murai, S. (1969). *In* "Preliminary Report of the Hakuho Maru Cruise KH-684" (Y. Horibe, ed.), p. 72.
Leatherland, T. M., Burton, J. D., McCartney, M. J. and Culkin, F. (1971). *Nature, Lond.* **232**, 112.
Li, Y.-H., Ku, T.-L., Mathieu, G. G. and Wolgemuth, K. (1972). *Earth Planet. Sci. Letters* (in press).

Liss, P. S., Herring, J. R. and Goldberg, E. D. (1973). *Nature, Lond.* **242**, 108.
Liss, P. S. and Spencer. C. P. (1970). *Geochim. Cosmochim. Acta.* **34**, 1073.
Livingstone, D. A. (1963). *U.S. Geol. Surv. Profess. Paper,* **440-G**.
Loveridge, B. A., Milner, G. W. C., Barnett, G. A., Thomas, A. N. and Henry, W. M. (1960). *U.K. At. Energy Authority Res. Grp Rept AERE,* **R3323**, 6 pp.
Lowman, F. G., Stevenson, R. A., McClin Escalera, R., Lugo Ufret, S. (1967). *In* "Radiological concentration processes", (B. Aberg and F. P. Hungate, eds.). Pergamon Press, New York, p. 735.
Lowman, F. G., Rice, T. R. and Richards, F. A. (1971). *In* "Radioactivity in the Marine Environment", National Academy of Sciences. Washington, p. 161.
Mandelli, E. F. (1969). *Mar. Sci.* **14**, 47.
Marchand, E. (1855). *Mem. Acad. Med. (Paris),* **19**, 121.
Martin, J. H. (1970). *Limnol. Oceanogr.* **15**, 756.
Martin, J. H. and Knauer, G. A. (1973). *Geochim. Cosmochim. Acta,* **37**, 1639.
Matson, W. R., Roe, D. K. and Carritt, D. E. (1965). *Analyt. Chem.* **37**, 1594.
Matson, W. R. (1968). Ph.D. Thesis, Mass. Inst. Technology, 150 pp.
Matthews, A. D. and Riley, J. P. (1970). a, *Analyt. Chim. Acta,* **51**, 287.
Matthews, A. D. and Riley, J..P. (1970. b, *Chem. Geol.* **6**, 149.
Menzel, D. W. and Ryther, J. H. (1961). *Deep-Sea Res.* **7**, 276.
Menzel, D. W. and Spaeth, J. P. (1962). *Linmol. Oceanogr.* **7**, 155.
Merrill, J. R., Lyden, E. F. X., Honda, M. and Arnold, J. (1960). *Geochim. Cosmochim. Acta,* **18**, 108.
Miller, A. R., Densmore, C. D., Degens, E. T., Hathaway, J. C., Manheim, F. G., McFarlin, P. F., Pocklington, R. and Jokela, A. (1966). *Geochim. Cosmochim. Acta,* **30**, 341.
Mokyesvskaya, V. V. (1961). *Dokl. Acad. Nauk., S.S.S.R.,* Ser. A, **137**, 1445.
Morel, F. and Morgan, J. (1972). *Environ. Sci. Technol.* **6**, 58.
Morgan, J. J. (1964). Ph.D. thesis, Harvard University, Cambridge, Mass.
Morris, A. W. (1968). *Analyt. Chim. Acta,* **42**, 397.
Munk, W. (1966). *Deep-Sea Res.* **13**, 707.
Naito, H. and Sugawara, K. (1957). *Bull Chem. Soc. Japan,* **30**, 799.
National Academy of Sciences (1971). *Marine Environmental Quality,* Ocean Sci. Comm., Nat. Acad. Sci.-Nat. Res. Council, 107 pp.
Nelson, J. L. *et al.* (1966). *In* "Disposal of Radioactive Wastes into the Seas, Oceans and Surface Waters". International Atomic Energy Agency, Vienna, p. 139.
Nicholls, G. D., Curl, H. C. and Bowen, V. T. (1959). *Limnol. Oceanogr.* **4**, 472.
Nozaki, Y. and Tsunogai, S. (1973). *Earth Planet. Sci. Lett.* **20**, 88.
Olafsson, J. and Riley, J. P. (1972). *Chem. Geol.* **10**, 227.
Park, K. (1968). *Deep-Sea Res.* **15**, 721.
Pfaff, C. (1825). *Jb. Chem. Phys.* **15**, 378.
Piotrowicz, S. R., Ray, B. J., Hoffman, G. L. and Duce, R. A. (1972). *J. Geophys. Res.* **77**, 5243.
Polikarpov, G. G. (1966). "Radioecology of Aquatic Organisms". Reinhold, New York. 314 pp.
Portmann, J. E. (1968). *Meeresuntersuchungen,* **17**, 247.
Portmann, J. E. and Riley, J. P. (1964). *Analyt. Chim. Acta,* **31**, 509.
Portmann, J. E. and Riley, J. P. (1966a). *Analyt. Chim. Acta,* **35**, 35.
Portmann, J. E. and Riley, J. P. (1966b). *Analyt. Chim. Acta,* **34**, 201.
Preston, A. and Jeffries, D. F. (1969). *In* "Environmental Contamination by Radioactive Materials", International Atomic Energy Agency, Vienna, p. 183.
Preston, A., Jeffries, D. F., Dutton, J. W. R., Harvey, B. R. and Steele, A. K. (1972). *Environ. Pollution,* **3**, 69.

Price, N. B. and Calvert, S. B. (1973). *Geochim. Cosmochim. Acta*, (in press).
Pringle, B. H., Hissong, D. E. Katz, E. L. and Mulawka, S. T. (1968). *J. Sanit. Engng Dir. Am. Soc. Cir. Engrs*, **94**, 455.
Provasoli, L., McLaughlin, J. J. A. and Droop, M. R. (1957). *Archiv. für Mikrobiol.* **25**, 392.
Putnam, G. L. (1953). *J. Chem. Ed.* **30**, 259.
Rakestraw, N. W. and Lutz, F. B. (1933). *Biol. Bull. Woods Hole,* **65**, 397.
Richards, F. A. (1956). *Geochim. Cosmochim. Acta,* **10**, 241.
Riley, G. A. (1951). *Bull. Bingham Oceanogr. Coll.* **13**, 707–730.
Riley, J. P. and Tongudai, M. (1964). *Deep-Sea Res.* **11**, 563.
Riley, J. P. (1965). *Deep-Sea Res.* **12**, 219.
Riley, J. P. and Roth, I. (1971). *J. Mar. Biol. Ass. U.K.* **50**, 721.
Riley, J. P. and Tongudai, M. (1966). *Chem. Geol.* **1**, 291.
Riley, J. P. and Taylor, D. (1972). *Deep-Sea Res.* **19**, 307.
Robertson, D. E. (1968a). *Analyt. Chem.* **40**, 1067.
Robertson, D. E. (1968b). *Analyt. Chim. Acta,* **42**, 533.
Robertson, D. E. (1970). *Geochim. Cosmochim. Acta,* **34**, 553.
Robertson, D. E. (1971). *In* "Pacific Northwest Laboratory Annual Report for 1970". Battelle Memorial Institute, Richland, Washington.
Rona, E., Hood, D. W., Muse, L. and Buglio, B. (1962). *Limnol. Oceanogr.* **7**, 201.
Rozhanskaya, L. I. (1963). *Tr. Sevastopol. Biol. Sta.* **16**, 467.
Sackett, W. and Arrhenius, G. (1962). *Geochim. Cosmochim. Acta,* **26**, 955.
Scadden, E. M. (1969). *Geochim. Cosmochim. Acta,* **33**, 633.
Schutz, D. F. and Turekian, K. K. (1965a). *Geochim. Cosmochim. Acta,* **29**, 259.
Schutz, D. F. and Turekian, K. K. (1956b). *J. Geophys. Res.* **70**, 5519.
Schwarzenbach, G. and Widmer, M. (1963). *Helv. Chim. Acta,* **46**, 2613.
Shaw, T. I. and Cooper, L. H. N. (1957). *Nature, Lond.* **180**, 250.
Shaw, T. I. and Cooper, L. H. N. (1958). *Nature, Lond.* **182**, 251.
Shimizu, K. and Ogata, N. (1963). *Bunseki Kagaku,* **12**, 526.
Shuster, C. N. and Pringle, B. H. (1969). *Proc. natl. Shellfish Ass.* **59**, 91.
Siegel, A. (1971). *In* "Organic Compounds in Aquatic Environments" (S. J. Faust and J. V. Hunter, eds.), Marcel Dekker, New York, 638 pp.
Silker. W. B. (1972). *Earth Planet. Sci. Letters,* **16**, 131.
Silker. W. B., Robertson, D. E., Rieck, H. G. Perkins, R. W. and Prospero, J. (1968). *Science,* **161**, 879.
Sillén, L. G. (1961). *In* "Oceanography" (M. Sears, ed.), Amcr. Ass. Advance, Sci., Publ. 67, Washington, D.C.
Simmons, L. H., Monaghan, P. H. and Taggart, M. S. (1953). *Analyt. Chem.* **25**, 989.
Skopintsev, B. A. and Popova, T. P. (1963). *Sov. Oceanogr.* **3**, 15.
Slowey, J. F. and Hood, D. W. (1966). *U.S. Atomic Energy Commission Rept No.,* AT(40-1)-2799. 163 pp.
Slowey, J. F., Hood, D. W. and Jeffrey, L. M. (1967). *Nature, Lond.* **214**, 377.
Smales, A. A. and Pate, B. D. (1952). *Analyst.* **77**, 188.
Smales, A. A. and Salmon, L. (1955). *Analyst.* **80**, 37.
Smith, J. D. and Burton, J. D. (1972). *Geochim. Cosmochim. Acta,* **36**, 621.
Smith, J. D., Nicholson and Moore, P. J. (1971). *Nature, Lond.* **232**, 393.
Spencer, C. P. (1957). *J. Gen. Microbiol.* **16**, 282.
Spencer, D. W. and Brewer, P. G. (1969). *Geochim. Cosmochim. Acta,* **33**, 325.
Spencer, D. W. (1969). *U.S. Atomic Energy Commission, Report* **No. NYO-1918-170.**
Spencer, D. W. (1972). *Earth Planet. Sci. Letters,* **16**, 91.
Spencer, D. W. and Brewer, P. G. (1970). *Crit. Rev. Solid State Sci.* **1**, 409.
Spencer, D. W. and Brewer, P. G. (1971). *J. Geophys. Res.* **76**, 5877.

Spencer, D. W., Robertson, D. E., Turekian, K. K. and Folsom, T. R. (1970). *J. Geophys. Res.* **75**, 7688.

Spencer, D. W., Brewer, P. G. and Sachs, P. L. (1972). *Geochim. Cosmochim. Acta*, **36**, 71.

Stanford, H. M. (1971). M.Sc. Thesis, Oregon State University, Corvallis.

Stark, W. (1943). *Helv. Chim. Acta*, **26**, 424.

Strock, L. W. (1936). *Nach. Ges. Wiss. Göttingen. Math. Phys. Kl.* IV (N.S.) **1**, 171.

Stumm, W. and Lee, G. F. (1961). *Ind. Eng. Chem.* **53**, 143.

Stumm, W. and Morgan, J. J. (1970). *Aquatic Chemistry*. 583 pp. Wiley Interscience, New York.

Stumm, W. and Bilinski, H. (1972). In "Advances in Water Pollution Research, Sixth International Conference". Pergamon Press, 1973, pp. 39–49.

Sugawara, K. (1957). *In* "Proc. Regional Symp. Phys. Oceanogr." Tokyo.

Sugawara, K. and Terada, K. (1958). *Nature*, **182**, 250.

Sugawara, K., Tanaka, M. and Naito, H. (1953). *Bull Chem. Soc. Japan*, **26**, 417.

Sugawara, K., Terada, K., Kanamori, S., Kanamori, N. and Okabe, S. (1962). *J. Earth Sci., Nagoya Univ.* **10**, 34.

Sugawara, K., Terada, K., Kanamori, S. and Okobe, S. (1963). *J. Earth Sci., Nagoya Univ.* **10**, 34.

Swallow, J. C. and Crease, J. (1965). *Nature, Lond.* **205**, 165.

Tatsumoto, M. and Patterson, C. (1963a). *In* "Earth Science and Meteoritics", (J. Geiss and E. Goldberg, eds.), North Holland Publishing Co., Amsterdam, p. 74.

Tatsumoto, M. and Patterson, C. (1963b). *Nature, Lond.* **199**, 350.

Thayer, T. P. (1956). *In* "Chromium", (M. J. Udy, ed.), Vol. 1, Reinhold, New York.

Thomas, B. D. and Thompson, T. G. (1933). *Science*, **77**, 547.

Topping, G. (1969). *J. Mar. Res.* **27**, 318.

Tsunogai, S. (1971). *Deep-Sea Res.* **18**, 913.

Tsunogai, S. and Henmi, T. (1971). *J. Oceanogr. Soc. Japan*, **27**, 67.

Tsunogai, S. and Nozacki, Y. (1971). *Geochem. J.* **5**, 165.

Turekian, K. K. and Johnson, D. G. (1966). *Geochim. Cosmochim. Acta*, **30**, 1153.

Turekian, K. K., Katz, A. and Chan, L. (1973). *Limnol. Oceanogr.* **18**, 240.

Vinogradov, A. P. (1953). "The Elementary Composition of Marine Organisms", Sears Foundation Marine Research Memoir 2, New Haven, Connecticut.

Weiss, H. V. and Lai, M. G. (1963). *Analyt. Chim. Acta*, **28**, 242.

Weiss, H. V., Koide, M. and Goldberg, E. D. (1971). *Science*, **174**, 692.

Weiss, H. V., Yamamoto, S., Crozier, T. E. and Mathewson, J. H. (1972). *Envir. Sci. Tech.* **6**, 644.

Williams, P. M. (1969). *Limnol. Oceanogr.* **14**, 156.

Winkler, L. H. (1916). *Z. Angew. Chim.* **29**, 205.

Wolgemuth, K. and Broecker, W. S. (1970). *Earth Planet. Sci. Letters*, **8**, 372.

Wollast, R. and deBroeu, F. (1971). *Geochim. Cosmochim. Acta*, **35**, 613.

Wong, G. T. F. and Brewer, P. G. (1974). *J. mar. Res.* **32**, 25.

Wood, J. M., Kennedy, F. S. and Rosen, C. G. (1968). *Nature, Lond.* **220**, 173.

Wyrtki, K. (1962). *Deep-Sea Res.* **9**, 11.

Young, E. C., Smith, D. G. and Langille, W. M. (1959). *J. Fish. Res. Bd., Can.* **16**, 7.

Zirino, A. and Healy, M. L. (1971). *Limnol. Oceanogr.* **16**, 773.

Zirino, A. and Yamamoto, S. (1972). *Limnol. Oceanogr.* **17**, 661.

Chapter 8

# Dissolved Gases Other Than CO$_2$

## DANA R. KESTER

*Graduate School of Oceanography, University of Rhode Island*
*Kingston, Rhode Island 02881, U.S.A.*

## 8.1. INTRODUCTION

Studies of dissolved gases have played a major role in understanding physical and chemical processes in the ocean. Much of this work has been reviewed by Richards (1957, 1965). Historically, oxygen has been the most widely studied gas in the ocean. However, before attempts are made to separate the effects of physical and biological processes in producing oceanic dissolved oxygen variations, it is useful to examine the occurrence and behaviour of "unreactive" gases such as nitrogen, argon, and the other noble gases. The distributions of noble gases in the ocean have given an insight into physical processes at the air–sea interface as well as the radiogenic input of helium via the seafloor. The occurrence in the ocean of trace gases such as methane and carbon monoxide has been important in helping to evaluate the global cycle of these gases. The role of $CO_2$ is considered separately in Chapter 9 because of its importance in the pH buffer system of sea water. The atmosphere is the major source of gases to the ocean; hence, consideration must be given to the abundance of gases in the atmosphere and to their transfer to the ocean.

## 8.2. BASIC CONCEPTS

### 8.2.1. COMPOSITION OF GASES IN THE ATMOSPHERE

The atmosphere is a homogeneous mixture of the major gases ($N_2$, $O_2$, and Ar) and of the unreactive minor ones (Ne, He, Kr, and Xe). Water vapour is the most variable component of the atmosphere. Unstable minor gases (e.g. CO, $NO_2$ and $CH_4$) which are produced by biological processes and by man's activity will vary with proximity to their sources and sinks.

The concept of partial pressure is a useful means of representing the composition of a gaseous mixture such as the atmosphere. In the early nineteenth century Dalton showed that the total pressure, $p_t$, exerted by a mixture of gases contained in a volume is the sum of the partial pressures of the constituent gases (Castellan, 1971). For the atmosphere Dalton's Law of partial pressures leads to the expression

$$p_t = p_{N_2} + p_{O_2} + p_{Ar} + p_{H_2O} \qquad (8.1)$$

If ideal behaviour is assumed the partial pressure of each gas is

$$p_G = \frac{n_G RT}{V}, \qquad (8.2)$$

where $n_G$ is the number of moles of the gas, $G$, contained in a volume $V$, $R$ is the gas constant (0·082053 l atm mol$^{-1}$ K$^{-1}$) and $T$ is the absolute temperature.

The amount of each gas in dry air can be expressed as its mole fraction, $f_G = n_G/\Sigma n_G$, as in Table 8.1. The plus and minus values in column 2 reflect the constancy and precision to which this composition is known. It is evident from equation (8.2) that the relative proportions of these gases are numerically the same on a mole fraction, a partial pressure, or a volumetric basis, within the limits of ideality. The non-ideal behaviour of gases may be approximately described by the Van der Waals equation of state

$$\left( p_G + \frac{n_G^2 a}{V^2} \right)(V - n_G b) = n_G RT \tag{8.3}$$

where $a$ is a coefficient related to intermolecular attraction and $b$ accounts for the finite volume and compressibility of the molecules. The molar volume of atmospheric gases at 0°C and 1 atm (standard temperature and pressure, STP) departs from ideality (22.414 l mol$^{-1}$) by $+0.1\%$ for He and $-0.6\%$ for Xe (Table 8.1). Thus, ideality may be assumed in all but the most accurate work. Equations of state which are more exact than equation (8.3) have been developed (Beattie, 1961). However, under conditions close to STP, the Van der Waals equation is adequate to at least 0·1%.

TABLE 8.1

*Abundance and properties of atmospheric gases.*

| Gas | Mole fraction in dry air* ($f_G$) | Van der Waals' coefficients† | | Molar volume at STP‡ (l mol$^{-1}$) |
|-----|-----|-----|-----|-----|
| | | $a$ | $b$ | |
| N$_2$ | 0·78080 ± 0·00004 | 1·390 | 0·03913 | 22·391 |
| O$_2$ | 0·20952 ± 0·00002 | 1·360 | 0·03183 | 22·385 |
| Ar | (9·34 ± 0·01) × 10$^{-3}$ | 1·345 | 0·03219 | 22·386 |
| CO$_2$ | (3·3 ± 0·1) × 10$^{-4}$ | 3·592 | 0·04267 | 22·296 |
| Ne | (1·818 ± 0·004) × 10$^{-5}$ | 0·2107 | 0·01709 | 22·421 |
| He | (5·24 ± 0·004) × 10$^{-6}$ | 0·03412 | 0·02370 | 22·436 |
| CH$_4$ | 2 × 10$^{-6}$ | 2·253 | 0·04278 | 22·356 |
| Kr | (1·14 ± 0·01) × 10$^{-6}$ | 2·318 | 0·03978 | 22·350 |
| CO | (0·1–0·2) × 10$^{-6}$ | 1·485 | 0·03985 | 22·387 |
| N$_2$O | 5 × 10$^{-7}$ | 3·782 | 0·04415 | 22·288 |
| Xe | (8·7 ± 0·1) × 10$^{-8}$ | 4·194 | 0·05105 | 22·277 |

* From Glueckauf (1951) and Valley (1965).
† From Weast (1969); units of $a$ are atm l$^2$ mol$^{-2}$ and $b$ are l mol$^{-1}$.
‡ Based on equation (8.3).

With the exception of $CH_4$, CO, and $N_2O$, the available data indicate that the mole fractions in Table 8.1 do not vary geographically or with altitude up to at least 95 km (Glueckauf 1951; Cook, 1961). The water vapour content of the atmosphere, however, does vary significantly. These variations are generally expressed in terms of a relative humidity and a temperature. For example, at 80% relative humidity at 15°C, the partial pressure of water is $0.80 \, p_0 = 10.2$ mm of Hg, where $p_0$ is the vapour pressure of water at the particular temperature (Weast, 1969). For a barometric pressure of 758.0 mm of Hg, water comprises about 1.3% of the total gases—an abundance greater than that of argon. The partial pressure of the other gases may be found by reducing the barometric pressure to that of "dry air", 747.8 mm of Hg, and using the mole fractions in Table 8.1. This procedure is summarized by the equation

$$p_G = \left[ p_t - \frac{h}{100} p_0 \right] f_G, \qquad (8.4)$$

where $h$ is the percentage relative humidity.

TABLE 8.2
*Isotopic abundances of atmospheric gases (after Cook, 1961).*

| Element | Mass number | Mol % | Element | Mass number | Mol % |
|---------|-------------|-------|---------|-------------|-------|
| H (in $H_2O$) | 1 | 99.98 | Kr | 78 | 0.354 |
| H (in $H_2O$) | 2 | 0.02 | Kr | 80 | 2.27 |
| He | 3 | $1.1 \times 10^{-4}$ | Kr | 82 | 11.56 |
| He | 4 | 100.0 | Kr | 83 | 11.55 |
| C (in $CO_2$) | 12 | 98.9 | Kr | 84 | 56.90 |
| C (in $CO_2$) | 13 | 1.1 | Kr | 86 | 17.37 |
| C (in $CO_2$) | 14 | $9.5 \times 10^{-13}$ | | | |
| N | 14 | 99.62 | Xe | 124 | 0.096 |
| N | 15 | 0.38 | Xe | 126 | 0.090 |
| O | 16 | 99.757 | Xe | 128 | 1.919 |
| O | 17 | 0.039 | Xe | 129 | 26.44 |
| O | 18 | 0.204 | Xe | 130 | 4.08 |
| Ne | 20 | 90.92 | Xe | 131 | 21.18 |
| Ne | 21 | 0.257 | Xe | 132 | 26.89 |
| Ne | 22 | 8.82 | Xe | 134 | 10.44 |
| Ar | 36 | 0.337 | Xe | 136 | 8.87 |
| Ar | 38 | 0.063 | | | |
| Ar | 40 | 99.600 | | | |

The stable isotopic composition of atmospheric gases is summarized in Table 8.2. The values in Tables 8.1 and 8.2 along with a knowledge of the total amount of the atmosphere ($5.0 \times 10^{21}$ g or $1.71 \times 10^{20}$ mol; Cook, 1961) enable estimates of the total atmospheric abundance of a gas, or its isotope, to be made.

### 8.2.2. DISSOLUTION OF GASES IN SEA WATER

The concept of partial pressure may also be applied to gas molecules dissolved in an aqueous solution. Henry's Law provides a relationship between the partial pressure* of a gas in solution, $P_G$, and its concentration, $c_G$:

$$P_G = K_G c_G \tag{8.5}$$

where $K_G$ is the Henry's Law constant. The magnitude of this constant will depend on the particular gas, the solution, the temperature and the total pressure being considered, as well as on the units selected for the partial pressure (atm, mm of Hg, etc.) and the "concentration" (molar, molal, mole fraction, etc.). Gaseous equilibrium is attained between the atmosphere and sea water when the partial pressure of the gas is the same in the two phases (i.e. $P_G = p_G$). Combination of Henry's Law with the condition of equilibrium gives for the solubility of a gas in sea water ($c_G^*$).

$$c_G^* = \frac{1}{K_G} p_G \tag{8.6}$$

## 8.3. SOLUBILITY OF GASES IN SEA WATER

### 8.3.1. SOLUBILITY PARAMETERS

Equation (8.6) permits gas solubility to be expressed in a number of ways. The Henry's Law constant based on a mass–mass concentration scale such as molality or mole fraction is perhaps the most fundamental expression of gas solubility in that its magnitude does not reflect the composition of a specific gas phase, nor the temperature and pressure dependences of the volume of the solution. Alternatively, the Bunsen coefficient, $\beta_G$, is frequently used. This is the volume of the gas at STP (e.g. in $cm^3$) which can be absorbed by a unit volume of *solution* (e.g. also in $cm^3$) when the partial pressure of the gas is

---

* Partial pressures will be used in place of the thermodynamically correct fugacities, because the differences between these quantities under normal conditions are only about $0.03\%$ (Klots and Benson, 1963). Henry's Law is traditionally expressed in terms of a mole fraction concentration scale; however, the proportionality between partial pressure and concentration is not restricted to the mole fraction scale, so that the concentration scale may be selected for maximum convenience.

one atmosphere. It should be noted that this definition requires that the volume of the solution after the dissolution of the gas be converted to its value prior to dissolution. This can be accomplished by the use of the appropriate partial molal volumes (Weiss, 1970). The Bunsen coefficient is conceptually similar (except for the units used) to both $c_G^*$ (when $p_G = 1$) and the inverse of the Henry's Law constant.

For practical oceanographic purposes the most convenient solubility expression for atmospherically homogeneous gases is the "air solubility", the value of $c_G^*$ when $P_G$ is taken to be the partial pressure in air with a relative humidity of $100\%$ at a total pressure of 760 mm Hg. It is possible to correct the partial pressure of a gas for conditions of total pressures and relative humidities other than the standard values by

$$p'_G = p_G \frac{p_t - p_s(h/100)}{760 - p_s}, \tag{8.7}$$

where $p'_G$ is the corrected partial pressure and $p_s$ is the vapour pressure of water at the temperature and salinity under consideration. The appropriate barometric pressure and relative humidity under natural conditions are not easy to obtain. Atkinson (1973) observed a five hour time lag in the response of dissolved nitrogen gas in the upper 12 m of the mixed layer during a drop in atmospheric pressure under natural conditions. Hence the instantaneous pressure may not be appropriate. The relative humidity near to the air–water interface is probably greater than that of the bulk air, although its actual value is uncertain because of disequilibrium between the interface and the bulk air or bulk sea water. Once a parcel of water is out of contact with the atmosphere, it is difficult to ascertain the conditions under which it was equilibrated. For general purposes it is, therefore, reasonable to accept the standard partial pressure obtained from equation (8.4) when $p_t = 760$ mm Hg and $h = 100\%$, using the potential temperature of the sample for the evaluation of $p_0$.

In the past, some solubility data for gases in sea water have been referred to pressures based on a dry atmosphere. This choice is probably not realistic for the air–sea interface, and if one desires to eliminate the effect of the water vapour pressure variations from gas solubilities, either the Henry's Law constant or a Bunsen coefficient would be preferable to a "dry air solubility concentration". Because of these different conventions, it is important to specify the exact meaning of a particular solubility.

The commonly used units for $c_G^*$ have been $cm^3$ (at STP) $l^{-1}$ (formerly ml $l^{-1}$). It should be noted that in this mode of expression no allowance is made for the temperature and pressure dependence of the volume of sea water. This shortcoming was of little account when analytical precision and accuracy were low. However, in recent years the quality of analytical methods has improved considerably. Thus, for dissolved oxygen, the Winkler method has been refined to provide precisions, and presumably accuracies, of about 0·1 % under laboratory conditions (Carpenter, 1965a; Carritt and Carpenter, 1966 see Chapter 19). These improved methods have also been employed under field conditions in the hope of obtaining similarly reliable results. To take full advantage of this precision it is desirable to express the amount of sea water in terms of weight units, since these are pressure and temperature independent (e.g. as $cm^3$ of gas at STP per kg of sea water) (Craig and Weiss, 1968; Weiss, 1970). It has also been recommended that the volumetric unit of gas at STP be replaced by the μg atomic weight unit for consistency with other chemical parameters, such as the nutrients, used in oceanography (Helland–Hansen et al., 1948; Carpenter, 1966). In this chapter the solubility of gases in sea water will be expressed by their equilibrium concentration, $c_G^*$, in moles of gas $kg^{-1}$ of sea water, based on air at 760 mm Hg total pressure with a 100 % relative humidity. This air solubility expressed in $\mu mol\,kg^{-1}$ or $nmol\,kg^{-1}$ has the advantages considered above, and can be interpreted by those with chemically based backgrounds who are unfamiliar with oceanographic literature. The molar volume of gases at STP (Table 8.1) may be used to convert volumetric to mole units.

Conventional methods for the determination of dissolved oxygen in sea water under field conditions do not provide an exact assessment of the quantity of sea water in which the oxygen has been determined. The water is generally contained in either a modified Erlenmeyer flask (Green and Carritt, 1966), or a B.O.D. bottle, of known volume, but the temperature of that water when its oxygen is removed by the manganous hydroxide precipitate is not known. This temperature will be somewhere between the *in situ* and the ambient, a range which can produce an uncertainty of about 0·5 % in the oxygen concentration depending on the particular field conditions. The reporting of concentrations on a per kilogram basis correctly places the burden of quantifying this uncertainty on the analyst who is in a better position to assess it than are those who may subsequently utilize the results.

8.3.2. SOLUBILITY OF ATMOSPHERIC GASES IN SEA WATER

Considerable effort was expended in the 1960's to obtain accurate values for the solubility of gases in sea water. These data are of special importance because departures of ocean water from equilibrium with atmospheric gases provides information on many types of processes. For example, supersaturation by non-reactive gases such as nitrogen or argon may be related to the role which entrapped air bubbles play in gas exchange. Supersaturation by oxygen in waters equilibrated with non-reactive gases may reflect photo-synthetic production of oxygen. Considerations such as these require reliable solubility data.

A sensitive gasometric technique for solubility measurements in sea water was developed by Murray *et al.* (1969). They determined the solubility of nitrogen over the range of 0·6 to 31°C and from 0 to about 36‰ salinity. There was excellent agreement between their values and those of Douglas (1964, 1965).

The solubility of oxygen in sea water has been an issue of considerable controversy (Richards, 1965). Studies in recent years have largely removed the uncertainties in $c_{O_2}^*$ which are produced by systematic errors resulting from the use of analytical methods having good precision but poor accuracy. Carpenter (1965a, b), Green (1965), and Carritt and Carpenter (1966) system-atically examined the sources of errors in the Winkler technique, and they developed modifications which improve the accuracy of oxygen measure-ments. The solubility of oxygen in sea water was redetermined by Green (1965), Carpenter (1966), and Murray and Riley (1969). Murray and Riley (1969) found good agreement between oxygen solubilities based on chemical Winkler measurements and on the gasometric method of Murray *et al.* (1969). The results of Carpenter (1966) and of Murray and Riley (1969) were in good agreement, whereas the solubility relationship used by Green (Green and Carritt, 1967) yielded values which were about 1% higher than the other two sets of data at low temperature. However, most of Green's experimental data were within 0·3% of the results of both Carpenter and Murray and Riley. Most of the discrepancies in recent tabulations of oxygen solubilities have resulted from the selection of interpolation functions to describe the data.

There have been three recent investigations of the solubility of argon in sea water (Douglas, 1964, 1965; Murray and Riley, 1970; Weiss, 1971a). The results by Douglas and by Murray and Riley were in good agreement in pure water and in sea water below 15°C. Significant deviations (up to 1·2%) between these two sets of data occurred, however, at temperatures greater than 20°C. The measurements by Weiss at a few selected intermediate sali-nities confirmed the values which he extrapolated from the results of Douglas which only covered the salinity range 28–38‰. The solubilities of helium and neon in sea water were measured by Weiss (1971b) using the micro-gasometric method.

The solubility data for nitrogen, oxygen, and argon have been summarized by Weiss (1970), who selected those data which he judged to be most reliable. He developed a series of equations for representing the temperature and salinity dependences of these gas solubilities in sea water. The functional relationships used by Weiss were expressed in terms of the Bunsen coefficient; they conformed to the empirical Setchénow equation for the salinity dependence at constant temperature:

$$\ln (\beta)_T = b_1 + b_2 S\%_o \tag{8.8}$$

where $b_1$ and $b_2$ are constants for the particular temperature. The temperature dependence at constant salinity was based on the integrated form of the Van't Hoff equation:

$$\ln(\beta)_S = a_1 + a_2/T + a_3 \ln T \tag{8.9}$$

where $T$ is absolute temperature and $a_1, a_2$ and $a_3$ are constants for a particular salinity. These functional forms were combined, $b_2$ was expressed as a quadratic function of temperature, and temperature was numerically scaled by a factor of 100 to produce the relationship

$$\ln \beta = A_1 + A_2 (100/T) + A_3 \ln (T/100)$$
$$+ S\%_o [B_1 + B_2 (T/100) + B_3 (T/100)^2] \tag{8.10}$$

A similar expression was also obtained for the air solubility in units of $cm^3 \, l^{-1}$ (and of $cm^3 \, kg^{-1}$) by incorporating an additional temperature term so as to allow for the temperature dependences of the vapour pressure and the density of sea water, thus

$$\ln c^* = A_1 + A_2(100/T) + A_3 \ln (T/100) + A_4(T/100)$$
$$+ S\%_o [B_1 + B_2 (T/100) + B_3 (T/100)^2] \tag{8.11}$$

The numerical values for the constants in these equations depend on the gas considered and the desired expression of solubility (e.g. Bunsen coefficient, air solubility in $cm^3 \, l^{-1}$, $cm^3 \, kg^{-1}$, or $mol \, kg^{-1}$).

Table 8.3 summarizes the solubility constants used by Weiss to represent the temperature and salinity dependence of $c_G^*$ according to equation (8.11). The unit of $c_G^*$ to be preferred for future use in marine chemistry is $\mu mol$ (or $nmol$) $kg^{-1}$; however, it is likely that the commonly used unit $cm^3 \, l^{-1}$ will persist, pending revision of major reference sources and for this reason the constants are quoted in both sets of units. The equations developed by Weiss are useful in that they provide a convenient and systematic means of calculating a gas solubility in sea water. Table 8.4 gives brief compilation of gas solubilities at $35\%_o$ salinity based on equation (8.11). A more extensive tabulation is given in the Appendix to this Chapter (Tables A8.1–A8.5).

It is important to know whether the use of an equation such as (8.11) smooths out some of the actual variability of gas solubilities with temperature and salinity. For dissolved oxygen, values calculated from the equation given by Weiss frequently depart from the measured values by as much as $0.03 \, cm^3 \, l^{-1}$. Carpenter (1966) estimated the accuracy of his determina-

TABLE 8.3

Solubility of gases in sea water with the constants for equation (8.11) to yield $c_G^*$ in $\mu mol\ kg^{-1}$ (values in parentheses yield $cm^3\ l^{-1}$) relative to air at 760 mm Hg total pressure at 100% relative humidity.

| Gas | Source of experimental data | $A_1$ | $A_2$ | $A_3$ | $A_4$ | $B_1$ | $B_2$ | $B_3$ |
|---|---|---|---|---|---|---|---|---|
| $N_2$ | Douglas (1964, 1965) | −173·2221 | 254·6078 | 146·3611 | −22·0933 | −0·054052 | 0·027266 | −0·0038430 |
|  | Murray et al. (1969) | (−172·4965) | (248·4262) | (143·0738) | (−21·7120) | (−0·049781) | (0·025018) | (−0·0034861) |
| $O_2$ | Carpenter (1966) | −173·9894 | 255·5907 | 146·4813 | −22·2040 | −0·037362 | 0·016504 | −0·0020564 |
|  | Murray and Riley (1969) | (−173·4292) | (249·6339) | (143·3483) | (−21·8492) | (−0·033096) | (0·014259) | (−0·0017000) |
| Ar | Douglas (1964, 1965) | −174·3732 | 251·8139 | 145·2337 | −22·2046 | −0·038729 | 0·017171 | −0·0021281 |
|  | Weiss (1971a) | (−173·5146) | (245·4510) | (141·8222) | (−21·8020) | (−0·034474) | (0·014934) | (0·0017729) |
| Ne | Weiss (1971b) | −166·8040 | 225·1946 | 140·8863 | −22·6290 | −0·127113 | 0·079277 | −0·0129095 |
|  |  | (−160·2630) | (211·0969) | (132·1657) | (−21·3165) | (−0·122883) | (0·077055) | (−0·0125568) |
| He | Weiss (1971b) | −163·4207 | 216·3442 | 139·2032 | −22·6202 | −0·44781 | 0·023541 | −0·0034266 |
|  |  | (−152·9405) | (196·8840) | (126·8015) | (−20·6767) | (−0·040543) | (0·021315) | (0·0030732) |

TABLE 8.4

*Solubilities of $N_2$, $O_2$, Ar, Ne and He in sea water at 35‰ salinity based on equation (8.11) and Table 8.3 (see the Appendix for a more complete tabulation).*

| | $\mu mol\,kg^{-1}$ | | | $nmol\,kg^{-1}$ | |
|---|---|---|---|---|---|
| $t\,^\circ C$ | $c^*_{N_2}$ | $c^*_{O_2}$ | $c^*_{Ar}$ | $c^*_{Ne}$ | $c^*_{He}$ |
| 0 | 616·4 | 349·5 | 16·98 | 7·88 | 1·77 |
| 5 | 549·6 | 308·1 | 15·01 | 7·55 | 1·73 |
| 10 | 495·6 | 274·8 | 13·42 | 7·26 | 1·70 |
| 15 | 451·3 | 247·7 | 12·11 | 7·00 | 1·68 |
| 20 | 414·4 | 225·2 | 11·03 | 6·77 | 1·66 |
| 25 | 383·4 | 206·3 | 10·11 | 6·56 | 1·65 |
| 30 | 356·8 | 190·3 | 9·33 | 6·36 | 1·64 |

tions to be about $0.006\,cm^3\,l^{-1}$ and Murray and Riley (1969) estimated theirs to be about $0.01\,cm^3\,l^{-1}$. In Fig. 8.1 two sets of measured solubilities for oxygen at approximately 20° and 5°C are compared. This figure also illustrates the applicability of the Setchénow relationship. At 20°C no common systematic departures from Weiss's equation are shown by the two sets of solubility data. However, at 5°C the Weiss equation predictions differ from the observed values of both sets of data by about $+0.03\,cm^3\,l^{-1}$ at low salinity and $-0.03\,cm^3\,l^{-1}$ at 20‰ salinity. The precision of Carpenter's replicate measurements is remarkably good and within his estimated limits of $\pm0.1\%$ accuracy. There are, however, some systematic differences between the results of the two sets of analyses which exceed their estimated accuracies.

A similar comparison of the variations of the measured solubilities with the Weiss equation predictions at various temperatures (Fig. 8.2) for 36.80‰ salinity shows that Weiss's equation may be low by about $0.01\,cm^3\,l^{-1}$ at 5°, but it agrees with the observed values at 35°C. At 27·42‰ salinity there is a wide divergence between the two sets of solubility data below 8°C, but Weiss's equation provides a reasonable fit for temperatures greater than 8°C.

For most oceanographic applications Weiss's equation for oxygen provides a useful and adequate fit to the solubility data. The existing solubility data are not sufficiently detailed nor consistent between analysts to establish significant departures of less than 0·5% from the functional form selected by Weiss.

The solubilities of krypton and xenon in sea water are less well known than are those of the gases discussed thus far. Benson (1965) gave values for $c^*_{Kr}$ and $c^*_{Xe}$ based on smoothed data from König (1963). Wood and Caputi (1966) reported values of the Henry's Law constant for these gases in distilled

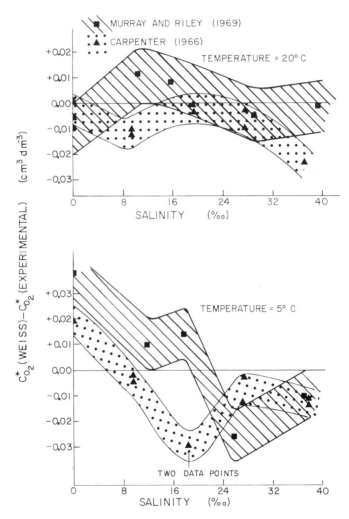

FIG. 8.1. Comparison of measured oxygen solubilities with those calculated from equation (8.11) as a function of salinity at two temperatures. The width of the textured bands represents the estimated accuracies of the experimental data.

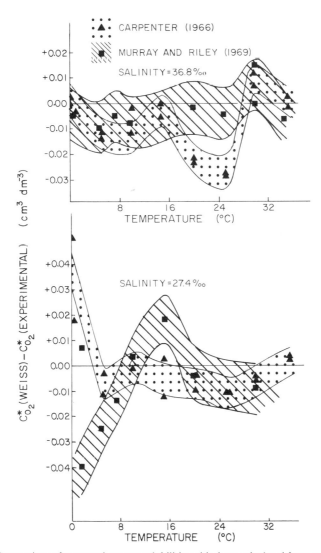

FIG. 8.2. Comparison of measured oxygen solubilities with those calculated from equation (8.11) as a function of temperature at two salinities. The width of the textured bands represents the estimated accuracies of the experimental data.

water and in sea water (salinity $= 34.6‰$) at about 1, 26, and 48°C. Pro visional values of $c^*_{Kr}$ and $c^*_{Xe}$ were calculated from their results at various temperatures and salinities assuming that ln $c^*_G$ varied linearly with salinity and quadratically with the scaled temperature, $T/100$ (Table 8.5). These values are 15–30% higher than those based on König's data. Wood and Caputi

TABLE 8.5

*Solubilities of krypton and xenon in sea water relative to air at 760 mm Hg total pressure and 100% relative humidity based on the data of Wood and Caputi (1966).*

$c^*_{Kr}$ (nmol kg$^{-1}$)

| $t°C$ | 0 | 10 | 20 | 30 | Salinity (‰) 33 | 35 | 37 | 40 |
|---|---|---|---|---|---|---|---|---|
| 0 | 5·8 | 5·2 | 4·8 | 4·4 | 4·2 | 4·1 | 4·1 | 4·0 |
| 5 | 4·9 | 4·5 | 4·1 | 3·7 | 3·6 | 3·6 | 3·5 | 3·4 |
| 10 | 4·2 | 3·8 | 3·5 | 3·2 | 3·1 | 3·1 | 3·0 | 3·0 |
| 15 | 3·6 | 3·3 | 3·1 | 2·8 | 2·8 | 2·7 | 2·7 | 2·6 |
| 20 | 3·2 | 2·9 | 2·7 | 2·5 | 2·5 | 2·4 | 2·4 | 2·3 |
| 25 | 2·8 | 2·6 | 2·4 | 2·3 | 2·2 | 2·2 | 2·1 | 2·1 |
| 30 | 2·5 | 2·3 | 2·2 | 2·0 | 2·0 | 2·0 | 1·9 | 1·9 |

$c^*_{Xe}$ (nmol kg$^{-1}$)

| $t°C$ | 0 | 10 | 20 | 30 | Salinity (‰) 33 | 35 | 37 | 40 |
|---|---|---|---|---|---|---|---|---|
| 0 | 0·90 | 0·83 | 0·76 | 0·69 | 0·67 | 0·66 | 0·65 | 0·63 |
| 5 | 0·74 | 0·68 | 0·63 | 0·58 | 0·57 | 0·56 | 0·55 | 0·53 |
| 10 | 0·61 | 0·56 | 0·52 | 0·48 | 0·47 | 0·46 | 0·45 | 0·44 |
| 15 | 0·51 | 0·47 | 0·44 | 0·40 | 0·39 | 0·39 | 0·38 | 0·37 |
| 20 | 0·44 | 0·40 | 0·37 | 0·35 | 0·34 | 0·33 | 0·33 | 0·32 |
| 25 | 0·37 | 0·35 | 0·32 | 0·30 | 0·29 | 0·29 | 0·28 | 0·28 |
| 30 | 0·32 | 0·30 | 0·28 | 0·26 | 0·26 | 0·25 | 0·25 | 0·25 |

(1966) attributed their higher results to more complete equilibration in their measurements than was achieved by other workers. In a subsequent section it will be shown that, in order for noble gas data to be useful for examining oceanic processes, the solubilities and the oceanic concentrations must be known to within 1%. More extensive high accuracy measurements of the solubilities of Kr and Xe are required.

## 8.4. AIR–SEA GAS EXCHANGE

### 8.4.1. MODELS FOR GAS EXCHANGE

A simple conceptual model of air–sea gas exchange is useful in identifying some of the factors affecting this process. It comprises three regions (see Fig. 8.3 and Chapter 9)—a turbulent atmospheric phase in which the partial

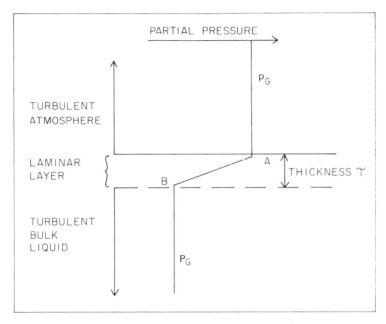

FIG. 8.3. Schematic illustration of the laminar layer model for gas exchange.

pressure, $p_G$, is uniform, a turbulent bulk liquid region with a uniform $P_G$, and a laminar layer separating these two turbulent regions (Whitman, 1923). The laminar layer is a region in which any motion of the liquid is parallel to the air–liquid interface. According to this model the rate determining step for gas exchange is the molecular diffusion of the gas through the laminar layer. Liquid in the laminar layer is undoubtedly exchanging with the turbulent region, but on the time scale required for molecular diffusion, the layer may be regarded as being permanent with an average thickness, $\tau$. The partial pressure of the gas at the upper boundary of the laminar layer, point $A$, may be taken to be $p_G$. At the lower boundary, point $B$, it is $P_G$. Assuming a linear gradient of partial pressure within the laminar layer, Fick's first law of diffusion combined with Henry's Law relating gas concentration and partial pressure predicts that the rate of transfer of the gas into the liquid will be

$$\frac{dG}{dt} = \frac{AD_G(p_G - P_G)}{K_G\tau} \tag{8.12}$$

where $A$ is the interfacial area, $D_G$ is the molecular diffusion coefficient of the gas, and $K_G$ is its Henry's Law constant. This simple model indicates that the rate of gas exchange depends on the difference in partial pressure between the atmosphere and the turbulent liquid and on the thickness of the laminar layer.

A more general relationship than equation (8.12) for the rate of gas exchange is

$$\frac{dG}{dt} = \frac{Ae_G(p_G - P_G)}{K_G} \tag{8.13}$$

where $e_G$ is an empirical parameter called the exit coefficient or exchange velocity. The laminar layer model (equation (8.12)) interprets $e_G$ as the ratio of $D_G$ to $\tau$. Other models are available, however, which assign a different significance to $e_G$. Higbie (1935) considered a case in which the rate determining factor was the time, $\theta$, for which a volume element of liquid persists at the air–liquid interface. For this model,

$$e_G = 2\sqrt{\frac{D_G}{\pi\theta}} \tag{8.14}$$

The existence of a laminar layer at the air–sea interface under some conditions is supported by observations of thermal structure in the upper few centimetres, Ewing and McAlister (1960) observed a boundary layer less than 1 mm thick which was 0·6°C cooler than the underlying bulk region due to evaporative heat loss.

Schink et al. (1968) have suggested an alternative to the laminar layer model in which the vertical eddy diffusion coefficient varies from the molecular value at the air–sea boundary to a wind related value in the mixed layer. This approach could provide a more realistic description of gas exchange than does the boundary layer model over a range of sea states, but it has not yet been formally developed.

8.4.2. EXPERIMENTAL STUDIES OF GAS EXCHANGE

Downing and Truesdale (1955) have examined the rate of oxygen exchange between air and water. They assessed the effects of wind, waves, temperature, and surface contaminants on this rate. Their results showed that the exchange velocity remained nearly constant for wind speeds between 0–3 m s$^{-1}$

(measured at 5 cm above the liquid surface), and then began to increase rapidly for wind speeds between 3–13 m s$^{-1}$ (Fig. 8.4). They noted that at 3 m s$^{-1}$ the roughness of the water surface increased with the formation of small wavelets and ripples. They suggested that the air flow over the liquid

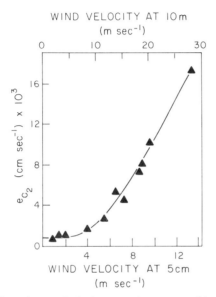

FIG. 8.4. Variation of the exchange velocity for oxygen in sea water with wind speed in the presence of mechanically produced ripples 2 cm high at a frequency of 204 ripples per minute (after Downing and Truesdale, 1955).

may change from laminar to turbulent at this speed. It should be noted that their wind speeds measured at a height of 5 cm were about one-half the magnitude of the corresponding wind speed at the standard meteorological height of 10 m.

Downing and Truesdale also observed increases in the exchange velocity as mechanically generated waves varied from 3–11 cm in height, or as the period decreased from 1·7 to 0·8 sec. They found a linear increase in $e_{O_2}$ with temperature; the values at 25°C were about twice those at 5°C for sea water. The effect of oil films at the air–water interface depended on the air flow conditions. In the absence of wind, oil films between 10$^{-6}$ and 10$^{-4}$ cm thick had little effect on the rate of gas exchange; thicknesses greater than 10$^{-4}$ cm, however, produced proportional decreases in the exchange velocity. In the presence of wind the rate of exchange varied for all film thicknesses between 10$^{-6}$ and 10$^{-2}$ cm. They concluded that it was unlikely that oil pollution in natural waters would produce films greater than 10$^{-4}$ cm, and hence that

S

such pollution was not a significant factor in the rate of gas exchange. Downing and Truesdale recognized that their laboratory study probably did not quantitatively simulate conditions in the natural environment. Nevertheless, their study provides insight into the sensitivity of gas exchange to various parameters.

Kanwisher (1963) duplicated some of Downing and Truesdale's experiments and found similar results for the effect of wind speed and waves on the rate of exchange. His experiments indicated thicknesses of the laminar layer generally in the range of 0·002–0·020 cm. The smaller thicknesses occurred with increased turbulence (stirring) of the liquid and with increased wind speed over the liquid surface. He estimated that the flux was roughly proportional to the square of the wind speed.

### 8.4.3. FIELD STUDIES OF GAS EXCHANGE

Redfield (1948) used oxygen data obtained in the Gulf of Maine over a 15 month period to evaluate atmospheric exchange under natural conditions. He calculated an oxygen balance between successive sets of observations using changes in the phosphate concentration to determine the net biological production or consumption of oxygen. He found that during the spring, oxygen was transferred from the sea to the atmosphere, primarily as a result of photosynthetic production. In the summer, warming of the waters produced further evolution of oxygen to the atmosphere. Oxygen was regained by the water during the winter; about 40% of the amount exchanged was required for the oxidation of organic matter, and 60% was absorbed by the cooling of the waters.

Redfield characterized the exchange rate by an exchange coefficient, $E$, which in terms of equations (8.12) and (8.13) is

$$E_{O_2} = \frac{D_{O_2}}{K_{O_2}\tau} = \frac{e_{O_2}}{K_{O_2}} \tag{8.15}$$

His values for $E_{O_2}$ ranged from about $3 \times 10^6$ cm$^3$ month$^{-1}$ m$^{-2}$ atm$^{-1}$ for the exchange from the ocean to the atmosphere during the spring and summer to $13 \times 10^6$ cm$^3$ month$^{-1}$ m$^{-2}$ atm$^{-1}$ during the winter transfer from the atmosphere to the ocean.

Pytkowicz (1964) determined the exchange rates for oxygen off the Oregon coast. Between the winter and summer periods he observed departures in the oxygen–phosphate correlation which were produced by a loss of oxygen from the ocean to the atmosphere. His analysis of the data permitted not only the evaluation of $E_{O_2} = 3 \times 10^6$ cm$^3$ month$^{-1}$ m$^{-2}$ atm$^{-1}$, in good agreement with Redfield's value for the summer in the Gulf of Maine, but it also yielded vertical eddy diffusion coefficients ranging from 2·3 g cm$^{-1}$ s$^{-1}$ at 10 m to 0·1 g cm$^{-1}$ s$^{-1}$ at 30 m.

A third approach to the evaluation of gas exchange rates under field conditions based on radon-222 has been developed by Broecker (1965). Radium-226 present in sea water decays to produce $^{222}$Rn which has a half-life of 3·85 days. Surface waters become deficient in $^{222}$Rn because of its loss to the atmosphere, and the rate of this exchange along with that of vertical mixing in surface waters can be obtained from the vertical profiles of $^{222}$Rn and from $^{226}$Ra concentrations. Broecker (1966) has applied this approach to the northwest Pacific Ocean, and has derived a laminar layer thickness of 0·003 cm and an eddy diffusion coefficient of 50 g cm$^{-1}$ s$^{-1}$ for the upper 50 m. Other measurements of radon in surface waters yielded exchange velocities (also referred to as "piston velocities") of $(7 \pm 3) \times 10^{-3}$ cm s$^{-1}$ in the northeast Pacific (Broecker and Kaufman, 1970) and of $2 \times 10^{-3}$ cm s$^{-1}$ northeast of Barbados (Broecker and Peng, 1971). These values correspond to laminar layer thicknesses of $0·002 \pm 0·001$ cm and 0·007 cm respectively. Broecker and Peng (1971), using the gas exchange rate based on $^{222}$Rn, have calculated that the rate of exchange of oxygen is $2 \times 10^6$ cm$^3$ month$^{-1}$ m$^{-2}$ atm$^{-1}$, a value which is consistent with the results of Redfield (1948) and of Pytkowicz (1964). Future studies of $^{222}$Rn will undoubtedly yield valuable information on the rates of oceanic gas exchange with the atmosphere.

The exchange velocities for oxygen in sea water obtained in field and laboratory studies show remarkable consistency (Table 8.6). The rate of exchange may vary by a factor of six in response to sea state and seasonal changes.

TABLE 8.6

*Summary of oxygen exchange velocities determined by various methods.*

| Investigator | Method | $e_{O_2}$ cm s$^{-1}$ |
|---|---|---|
| Downing and Truesdale (1955) | Laboratory study | $4–12 \times 10^{-3}$ |
| Kanwisher (1963) | Laboratory study | $2–14 \times 10^{-3}$ |
| Redfield (1948) | Seasonal budget | $4–17 \times 10^{-3}$ |
| Pytkowicz (1964) | $O_2$–$PO_4$ relationships | $4 \times 10^{-3}$ |
| Broecker and Peng (1971) | Radon gradients | $3 \times 10^{-3}$ |

### 8.4.4. ROLE OF AIR BUBBLES IN GAS EXCHANGE

A quantitative consideration of the mechanism by which gases are exchanged across the sea surface will require an assessment of the role played by air bubbles. Some fundamental observations in this regard were made by Wyman *et al.* (1952) who examined the effect of hydrostatic pressure on the rate of solution and on compositional changes of air bubbles. Their results were consistent with relationships based on molecular diffusion. The radii of dissolving bubbles decreased nearly linearly with time, and the rate of decrease varied with depth from 5 to 50 m. The air bubbles became progressively enriched with $N_2$ because, even though the molecular diffusion constants for $N_2$ and $O_2$ are nearly equal, the smaller Henry's Law constant for $O_2$ compared with that for $N_2$ leads to a more rapid rate of solution for $O_2$.

Liebermann (1957) compared the rates of solution of stationary bubbles and freely ascending bubbles. Free bubbles exhibited a "diffusion coefficient" which was twice as large as that of stationary bubbles. This difference probably represents differences in the interfacial boundary between the bubble and the bulk liquid under stationary and dynamic conditions.

Kanwisher (1963) estimated that air bubbles were submerged as much as 20 m below the sea surface during a storm in the North Atlantic Ocean. At this depth the partial pressure of gases in an air bubble will be three times that at the air–sea interface. Kanwisher applied the laminar layer diffusion model to the region surrounding an air bubble 0·05 cm in diameter. He concluded from a consideration of measurements of the flux and the difference in partial pressure that the thickness of the laminar layer around the bubble was about 0·0010–0·0015 cm. It is reasonable to expect that the turbulent motion near a bubble rising through the liquid will reduce the thickness of the laminar layer relative to that of a free interface. Gas exchange between a bubble and sea water is more rapid than is that across the sea surface, because the laminar layer is thinner and the partial pressure is increased by the depth of submergence.

An acoustic technique was used by Medwin (1970) to examine bubble size populations to depths of about 15 m. For sea state 1 he concluded that two populations of bubbles were present. One size range was generally less than 40 μm in radius and these were thought to be associated with atmospheric particles deposited in the sea. The other was approximately 100 μm in radius and may represent bubbles formed at the air–sea interface by collapsing waves.

The observed bubble populations may be combined with a knowledge of the rate of bubble dissolution to assess their significance in air–sea gas exchange. Atkinson (1973) showed that the flux of dissolved gases from air bubbles was comparable in magnitude to that which would enter the sea surface as a result of a barometric change of about 2 % in 12 hours (i.e. equivalent to that occurring during the passage of a typical storm). Furthermore, the rate of exchange due to air bubbles can be several orders of magnitude greater than the seasonal fluxes observed by Redfield (1948) and Pytkowicz (1964), hence bubbles are likely to be significant in gas exchange. It is likely that air bubbles are more effective in transferring gases from the atmosphere to the ocean than vice versa because of the hydrostatic pressure when they are submerged. Additional work is required to obtain a more comprehensive understanding of the mechanism by which gases are exchanged between the ocean and the atmosphere.

## 8.5. Non-Reactive Gases in the Ocean

### 8.5.1. PROCESSES AFFECTING NON-REACTIVE GASES

The noble gases and N$_2$ are generally regarded as non-reactive in oceanic waters. Some qualification of this assumption may be necessary for N$_2$ and He, but it provides a useful starting point for considering these gases in the ocean. The oceanic distribution of non-reactive gases is determined by physical processes and by the effects of temperature and salinity on their solubility. In contrast, the distribution of oxygen, is influenced by biological as well as physical processes. There is considerable interest in using the distributions of the unreactive gases to assist in obtaining a quantitative separation of the effects of physical and biological processes on the distribution of oxygen. Early work on this problem was carried out by Buch (1929), Rakestraw and Emmel (1938a), Hamm and Thompson (1941), and Sugawara and Tochikubo (1955). The advent of analytical techniques such as gas chromatography and mass spectrometry, which can be more specific and sensitive than gasometric methods, is leading to further advances in this area of investigation.

The most convenient way of representing variations in non-reactive gases is to quote their solubility under conditions of a conventionally defined standard atmosphere (760 mm of Hg total pressure and 100% relative humidity). Observed gas concentrations are usually related to their solubilities by either the degree of saturation, $\sigma_G$, or the saturation anomaly, $\Delta_G$; both are expressed as percentages, viz.

$$\sigma_G = \frac{c_G}{c_G^*} \times 100 \qquad \text{and} \qquad \Delta_G = \frac{c_G - c_G^*}{c_G^*} \times 100 \qquad (8.16)$$

The air solubility $c_G^*$ has been previously defined and $c_G$ is the observed gas concentration. It should be noted that for oceanographic purposes $c_G^*$ should be evaluated at the potential, rather than in situ, temperature in order to compare values from various depths. The use of $\sigma_G$ or $\Delta_G$ permits an examination of the distribution of gases independent of the effects which temperature and salinity produce on their concentrations at saturation.

There are a variety of physical processes which can cause the concn. of non-reactive gases to depart from their solubility as defined by $c_G^*$.

1. Departures from standard atmospheric pressure during equilibration.
2. Partial dissolution of air bubbles.
3. Air injection (i.e. the complete dissolution of air bubbles).
4. Differential heat and gas exchange.
5. Mixing of waters of different temperatures.
6. Radiogenic or primordial addition of He.

In principle, the addition of rain water (saturated with atmospheric gases) or ice and glacial melt water (devoid of atmospheric gases) could affect $\sigma_G$ and $\Delta_G$, but such effects are likely to be obscured by other processes at the air–sea interface except under extensive ice cover. The radiogenic production of $^{40}$Ar from $^{40}$K in sea water is not significant; Revelle and Suess (1962) estimated that it would take $10^{11}$ years to generate the amount of $^{40}$Ar in the ocean from the $^{40}$K in sea water. The effects of many of these physical processes on dissolved gases in the ocean have been considered by Carritt (1954), Benson (1965), Bieri et al. (1966), Craig et al. (1967), Bieri (1971) and Craig and Weiss (1971).

It is evident from equation (8.4) that departures from the standard atm. pressure as a result of barometric pressure or relative humidity changes affect all gases by an equal percentage. If sea water is equilibrated with air at 30°C, 80% relative humidity, and 760 mm Hg total pressure, $\Delta_G$ will be $+0.9\%$ for all atmospheric gases; this effect decreases at lower temperatures.

When an air bubble is submerged in the ocean, it will tend to dissolve because of the increase of hydrostatic pressure. If the bubble only partially dissolves before escaping back to the atmosphere it will increase $\sigma_G$ and $\Delta_G$ by an equal percentage for all gases. This process is equivalent to an increase in total pressure. The criterion for partial dissolution is that the composition of gases in the bubble remains nearly unchanged by the dissolution. Under these conditions, bubbles at a depth of 1 m can produce values of $\Delta_G = +10\%$ for all atmospheric gases. If the air bubble dissolves completely (air injection), each dissolved gas is affected differently, as will be shown below.

Many of the physical processes affecting $\Delta_G$ are related to the solubilities of different gases and to their variations with temperature. Figure 8.5 summarizes some of the solubility characteristics of gases in sea water. From the magnitudes of these solubilities we see that the mole proportion of each gas in sea water at equilibrium with air, $f_{SW}$, is considerably different from that of air (Table 8.7). This difference in composition produces large variations in $\Delta_G$ for different gases when air is injected into sea water.

Another factor illustrated in Fig. 8.5 is the variation in $c_G^*$ with temperature.

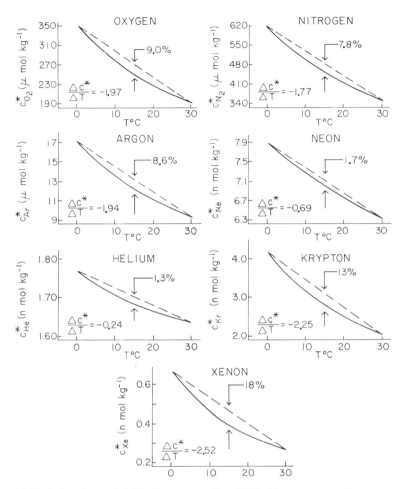

FIG. 8.5. Variation of gas solubilities in sea water ($S = 35‰$) with temperature. The percentage deviation from linearity at 15°C shows the value of $\Delta_G$ produced by mixing equal portions of sea water equilibrated at 0° and 30°C. The $\Delta c^*/\Delta T$ is the slope of $c_G^*$ versus $T$ at 15°C expressed as a percentage change in $c_G^*$ per degree C. This quantity is numerically equivalent to $d \ln c_G^*/dT$, and it represents the change in $\Delta_G$ produced by a 1°C decrease in temperature without gas exchange.

If sufficient heat is withdrawn from equilibrated sea water at 15°C to produce a 1°C decrease in temperature without an exchange of dissolved gases (as in radiative heat loss) $\Delta_G$ will range from $-0.24\%$ for He to $-2.5\%$ for Xe. A 1°C increase arising from radiative or geothermal heating will produce effects of similar magnitude but of opposite sign. If waters are heated too rapidly for the maintenance of gas exchange equilibrium and then removed

DANA R. KESTER

TABLE 8.7

Comparison of the mole proportion ($f$) of gases dissolved in saturated sea water and in dry air, and the saturation anomaly, $\Delta_{air}$, resulting from the injection of 1 cm³ of air at STP per kg of sea water. (All sea water values are based on 15°C and 35‰ salinity.)

| Gas | N₂ | O₂ | Ar | CO₂ | Ne | He | Kr | Xe |
|---|---|---|---|---|---|---|---|---|
| $f_{air}$ | 0·780 | 0·209 | 0·009 | 0·0003 | $18·2 \times 10^{-6}$ | $5·2 \times 10^{-6}$ | $1·1 \times 10^{-6}$ | $0·09 \times 10^{-6}$ |
| $f_{sw}$ | 0·626 | 0·343 | 0·016 | 0·014 | $9·7 \times 10^{-6}$ | $2·3 \times 10^{-6}$ | $3·8 \times 10^{-6}$ | $0·54 \times 10^{-6}$ |
| $\Delta_{air}$ | +7·7% | +3·8% | +3·5% | +0·1% | +11·6% | +13·8% | +1·8% | +1·0% |

from contact with the atmosphere (as may occur at high latitudes or in regions of downwelling) their saturation anomalies will follow a pattern similar to that arising from radiative heat transfer, but the magnitudes will be moderated by the partial exchange with the atmosphere. A second effect caused by temperature variations occurs during mixing. If equilibrated sea waters (35‰ salinity) at 0° and 30°C are mixed in equal proportions, the nonlinearity of $c_G^*$ with temperature will produce saturation anomalies ranging from 1·3% for He to 18% for Xe.

The differences in sensitivity of different gases to various processes enables methods to be devised for using several gases simultaneously to evaluate the magnitudes of these processes. For example, the anomalies for Xe and Kr are not sensitive to air injection, but they are sensitive to the mixing of waters of different temperatures, or to differential heat and to gas exchange. In contrast, the helium anomaly is not sensitive to the latter two processes, but it is a good indicator of air injection. Because departures from the standard atmospheric pressure and partial dissolution of air bubbles affect all gases equally, these two processes cannot be distinguished from each other by an examination of saturation anomalies. It should be noted that for all the processes considered, the anomalies for argon and oxygen are affected nearly equally; hence argon is the best non-reactive gas for differentiating between the biological and the physical processes affecting oxygen.

### 8.5.2. OCCURRENCE OF DISSOLVED NITROGEN GAS IN THE OCEAN

Rakestraw and Emmel (1938a) determined the nitrogen content at six locations in the northwest Atlantic Ocean at depths ranging from 0–4500 m. They found values of $\Delta_{N_2}$ (evaluated from their own solubility data) between $-5\%$ and $+8\%$. There were neither distinct trends with depth nor differences associated with different water masses.

Hamm and Thompson (1941) examined the degree of saturation in the northeast Pacific Ocean of the sum of all gases other than $O_2$ and $CO_2$; this sum is primarily $N_2$ with a small contribution from Ar. For depths between 100 and 2000 m they found values of $\Delta_{(N_2 + Ar)}$ from $-3\%$ to $+2\%$ (in terms of the solubility data obtained by Fox, 1907). However, they observed that between 0 and 100 m systematically lower values for $\Delta_{(N_2 + Ar)}$ of $-9\%$ to $-3\%$ occurred. These low values were attributed to the stripping of $N_2$ and Ar from sea water by photosynthetically produced bubbles of oxygen. The low $\Delta_{(N_2 + Ar)}$ values were sometimes associated with $\Delta_{O_2}$ of as much as $+8\%$ at depths of 20–30 m. The formation of $O_2$ bubbles at these depths would require $\Delta_{O_2} \approx +200\%$ to $300\%$; hence their explanation implies that the stripping should occur in the upper 1 m and that the ($N_2$ + Ar) deficient water should be transported down to 20–30 m—this seems unlikely. The influence of oxygen bubbles on gas exchange requires more critical consideration than has been given heretofore.

Biochemical processes are known by which $N_2$ may be converted to organically bound nitrogen and, ultimately, nitrate (Fogg and Stewart, 1965), and $N_2$ may also be produced during the oxidation of organic matter under anoxic conditions (Fry, 1955; Chapter 11). Thus, it may not be strictly correct to regard $N_2$ as a non-reactive gas in the marine environment. Benson and Parker (1961) examined this question by measuring the ratio of $N_2$ to Ar mass spectrometrically for various locations in the North and South Atlantic Ocean. They were able to calculate the percentage change in $N_2$ which occurred between the time of isolation of the sample from the atmosphere by oceanic transport and the time of sampling. They found that 38 of 67 samples (57%) from various locations and depths were within 0·5% of their initial $N_2$ concentration, and all but two values were within 2%. Thus, their observations indicate that for oceanic waters $N_2$ may be considered to be non-reactive to within about 1%.

The $N_2/Ar$ ratio was used by Richards and Benson (1961) to examine $N_2$ production in two anoxic regions—the Cariaco Trench and the Dramsfjord. They found systematically higher values of $N_2$ relative to Ar in the sulphide-bearing waters of these regions. Their results indicated values of $\Delta_{N_2}$ of $+2·3\%$ for the Cariaco Trench and $+3·6\%$ for the Dramsfjord.

Linnenbom et al. (1966) carried out extensive measurements of $N_2$ in ocean water during the development of gas chromatographic methods for sea water analysis. Their values, summarized in Table 8.8, show an essentially normal distribution of variations from saturation. They noted that the spread of their observations, as measured by the standard deviation, was two to three times their analytical precision which may imply variability of $\Delta_{N_2}$ due to oceanic processes. However, they did not note any systematic variations which they could correlate with other oceanographic variables.

TABLE 8.8

*Saturation anomalies for nitrogen obtained by Linnenbom et al. (1966) relative to the solubility values of Rakestraw and Emmel (1938b).*

| Region | $\Delta_{N_2}$ | |
|---|---|---|
| | Mean | % Std. dev. |
| Greenland Sea | $-0·6\%$ | $3·1\%$ |
| Andaman Sea | $+0·1\%$ | $3·0\%$ |
| North of Puerto Rico | $-0·6\%$ | $1·8\%$ |

During the HUDSON 70 expedition Atkinson (1972) measured N$_2$ concn. in the South Atlantic and examined variations among the principal water masses (Table 8.9). The $\Delta_{N_2}$ values for surface waters were highly variable, although their average was close to that for equilibrium with the standard atmospheric pressure. Antarctic Intermediate Water was distinctly less

TABLE 8.9

*Saturation anomalies from data by Atkinson (1972) for N$_2$ in the major water masses of the South Atlantic Ocean relative to the solubility relationship given by Weiss (1970).*

| Water mass | $\Delta_{N_2}$ | |
| | Mean | % Std. dev. |
| --- | --- | --- |
| All surface waters ($z < 10$ m) | −0·17% | 3·4% |
| South Atlantic central water | −0·48% | 1·3% |
| Antarctic intermediate water | −1·42% | 3·0% |
| North Atlantic deep water | +0·32% | 1·9% |
| Antarctic bottom water | −0·04% | 2·2% |

saturated with N$_2$ than were other subsurface waters. Atkinson attributed these lower values to the lower barometric pressure in the vicinity of the Antarctic Convergence during the winter.

Craig *et al.* (1967) determined nitrogen and other gases in waters from the South Pacific Ocean. They found a mean value of $\Delta_{N_2} = +1\cdot9\%$ with a standard deviation of 1·1% and a range of −0·3 to +4·1%. Their saturation anomaly estimates were based on the solubility data given by Douglas (1965); these are about 1·3% lower than those obtained by Rakestraw and Emmel (1938b).

Differences in the solubility data employed and in the method of reporting the results make a detailed comparison of the various studies of N$_2$ difficult at the 1% level. It is clear that ocean waters are generally within one or two percent of saturation for N$_2$ and that anoxic environments may contain 2–4% more N$_2$ than is required for saturation. Craig *et al.* (1967) noted that the spread of $\Delta_{N_2}$ has generally decreased as the analytical precision for N$_2$ has been improved.

### 8.5.3. OCCURRENCE OF NOBLE GASES IN THE OCEAN

During the 1960's several programmes were initiated to study the abundances of noble gases in sea water (König *et al.*, 1964; Mazor *et al.*, 1964; Bieri *et al.*, 1964, 1966; Craig *et al.*, 1967). The early data produced conflicting results for the degree of saturation of noble gases in ocean waters, and uncertainties in the basic solubility data further obscured quantitative interpretation. Nevertheless, refinements of sampling and analytical techniques in recent years have provided consistent and useful information on the occurrence of these gases in the ocean.

Some of these results are summarized in Table 8.10 which lists, by region, the means and standard deviations of the saturation anomalies for Ne, Ar, and He. These parameters were calculated from the data reported by Craig *et al.* (1967) and by Bieri *et al.* (1968) using the solubility relationships given by Weiss (1970). Oceanic waters are typically within 5% of saturation relative to the standard atmospheric pressure for these non-reactive gases.

The simultaneous use of noble gas saturation anomalies may be illustrated by considering the quantitative approach presented by Craig and Weiss (1971) to separate the effects of three physical processes—*effective* pressure variations, differential heat and gas exchange and air injection—from three noble gas anomalies. The saturation anomaly was assumed to be composed of three components each expressed as a percentage of $c_G^*$

$$\Delta_G = \delta_P + \delta_T + \delta_a \tag{8.17}$$

The anomaly caused by differences in the effective and standard atmospheric pressures may be written

$$\delta_P = \left( \frac{p_{\text{atm}} - (h/100)\, p_0}{760 - p_0} - 1 \right) 100 + \frac{z_e}{10} \tag{8.18}$$

where $z_e$ is the depth in cm at which the water is equilibrated with air bubbles and the other symbols have the meanings previously defined. The possible effects of relative humidities different from 100% have been included in this discussion for completeness even though their significance has not been established (see section 8.3.1). Since $\delta_P$ does not depend on a specific property of the dissolved gas, it will not be possible to determine $p_{\text{atm}}$, $h$, or $z_e$ separately. The anomaly arising from a change in temperature, $\Delta T$, (without gas exchange) is given approximately by

$$\delta_T = -100 \frac{d \ln c_G^*}{dT} \Delta T \tag{8.19}$$

The $\Delta T$ is the observed potential temperature minus the temperature at the time of equilibration. This term does not take into account anomalies caused by the mixing of waters of different temperatures; it applies to changes in temperature such as those caused by radiative heating or cooling and geothermal heating. The anomaly which results from air injection is

$$\delta_a = \frac{(f_{\text{air}})_G}{c_G^*} \frac{100}{22\,390} a \tag{8.20}$$

TABLE 8.10

Average saturation anomalies for Ne, Ar, and He expressed as percentages. The ± values indicate one standard deviation; the number of data points is given in parentheses under each value.

| Region | Surface waters (mixed layer) | | | Deep waters (greater than 1000 m)[‡] | | |
|---|---|---|---|---|---|---|
| | $\Delta_{Ne}$ | $\Delta_{Ar}$ | $\Delta_{He}$ | $\Delta_{Ne}$ | $\Delta_{Ar}$ | $\Delta_{He}$ |
| Atlantic[*] (0–20° N) | +4·5 ± 1·1 (9) | +2·0 ± 2·2 (9) | +4·6 ± 1·4 (9) | +5·2 ± 2·7 (13) | +1·8 ± 1·8 (13) | +6·0 ± 1·4 (13) |
| Drake passage[*] | +4·4 ± 1·0 (5) | +2·6 ± 1·0 (5) | +5·7 ± 2·0 (5) | +4·6 ± 2·3 (6) | −0·3 ± 3·3 (6) | +5·5 ± 3·1 (6) |
| South Pacific[†] (25° S–60° S) | — | −0·5 ± 1·2 (5) | — | — | −1·5 ± 1·3 (5) | — |
| Equatorial Pacific[†] (10° N–13° S) | +4·6 ± 2·1 (2) | −3·0 ± 0·3 (2) | — | +6·2 ± 1·4 (7) | −2·5 ± 2·0 (8) | — |
| Equatorial Pacific[*] (10° N–13° S) | +2·4 ± 1·6 (9) | +1·2 ± 2·1 (9) | +2·9 ± 1·0 (9) | +3·3 ± 1·3 (20) | −2·0 ± 1·6 (20) | +7·1 ± 1·7 (20) |
| North Pacific[*] (25° N–35° N) | +1·8 ± 0·8 (8) | −0·3 ± 1·0 (8) | +3·2 ± 0·6 (8) | +1·9 ± 1·8 (14) | −3·9 ± 1·6 (12) | +6·3 ± 1·4 (14) |

[*] Data from Bieri et al. (1968).
[†] Data from Craig et al. (1967).
[‡] For the Drake Passage, greater than 100 m.

where $a$ is the volume of air (cm³) at STP injected per kg of sea water, and the 22 390 is the molar volume of air in cm³ at STP. Thus,

$$\Delta_G = \delta_P - 100 \frac{d \ln c_G^*}{dT} \Delta T + \frac{(f_{air})_G}{223 \cdot 9 c_G^*} a \qquad (8.21)$$

The application of equation (8.21) may be illustrated by considering the anomalies for the Atlantic mixed layer (Table 8.10). Typical values for the temperature and salinity are 26°C and 36‰ respectively. The coefficients for the $\Delta T$ and $a$ terms of equation (8.21) may be obtained under these conditions for each gas from the solubility data (equation (8.11) or Tables in the Appendix) to obtain

$$\Delta_{Ne} = +4 \cdot 5 = \delta_P + 0 \cdot 629 \Delta T + 12 \cdot 47a,$$

$$\Delta_{Ar} = +2 \cdot 0 = \delta_P + 1 \cdot 62 \Delta T + 4 \cdot 22a$$

and

$$\Delta_{He} = +4 \cdot 6 = \delta_P + 0 \cdot 180 \Delta T + 14 \cdot 16a.$$

The solution of these equations is $\delta_P = -2 \cdot 8\%$, $\Delta T = +1 \cdot 7°C$, and $a = 0 \cdot 50$ cm³ air at STP kg⁻¹.

A positive value of $\delta_P$ can be attributed to one or more of three factors. A barometric pressure of 768 mm Hg, equilibration with air bubbles at a depth of 10 cm, or a relative humidity of 65% at 25°C will cause $\delta_P = +1\%$. However, only a value of $p_{atm} < 760$ mm Hg can produce a negative $\delta_P$ under natural conditions. Craig and Weiss (1971) pointed out that unreasonably large negative values of $\delta_P$ may indicate the analytical loss of a portion of the gas phase from a sample. The use of these simultaneous equations to assess physical processes requires accurate data, and the validity of any deductions hinges on the selection of the process components for $\Delta_G$. This analysis of saturation anomalies treats the aqueous solution as a closed system and does not include the effects of mixing.

Studies of the noble gas distributions have shown air injection to be an important factor in contributing dissolved gases to ocean water. Thus, Craig and Weiss (1971) and Bieri (1971) found that deep Pacific and Atlantic waters had been injected with 0·5 to 1·0 cm³ of air at STP per kg of sea water.

In addition to the effects which have already been considered, helium may be introduced into deep waters of the ocean from the underlying seafloor. Revelle and Suess (1962) suggested that He produced from the radioactive decay of uranium and thorium in oceanic sediments and rocks may diffuse upward to the sea water. Suess and Wänke (1965) subsequently calculated that deep North Pacific waters may contain up to 18% more He than do deep Indian Ocean waters; however, the data available at that time indicated only a 6% increase between the deep Indian and North Pacific Oceans. Helium introduced to the ocean through the seafloor has been termed "excess helium", ie the amount by which it exceeds that introduced by processes affecting the other noble gases.

Craig and Weiss (1971) analysed the saturation anomalies for He, Ne, and Ar, and they concluded that all of the anomalies in the deep North Atlantic could be attributed to air injection. They considered $\Delta_{He}$ in the Pacific to be about $+7\%$ of which about $+3\%$ was excess helium and $+4\%$ was due to air injection. Bieri (1971) also considered the relative effects of air injection and helium addition from the seafloor. His analysis of data from the eastern equatorial and southeast Pacific indicated $\Delta_{He} = +11\cdot8\%$ of which $+8\cdot5\%$ was due to air injection and $+3\cdot3\%$ was due to excess helium.

Studies of the non-reactive gases in the ocean have provided valuable information on the significance of various processes. It is likely that results obtained during the GEOSECS program will provide a more detailed and definitive view of these processes than is presently available. The existing data indicate that ocean waters are within $4\%$ of saturation for $N_2$ and Ar, whereas, as a result of air injection, Ne and He show slightly larger departures from saturation. An additional several percent supersaturation of helium occurs in the deep Pacific Waters as a result of radiogenic production.

### 8.5.4. HELIUM-3 IN THE OCEAN

Another possible way in which the amount of radiogenic helium-4 in the ocean can be evaluated is via the $^3$He/$^4$He ratio. In air $^3$He/$^4$He $= 1\cdot34 \times 10^{-6}$ (Cockett and Smith, 1973; Clarke et al., 1969). In principle, it should be possible to determine the radiogenic $^4$He in sea water from a decrease in the $^3$He/$^4$He ratio. This approach has several advantages over the method which makes use of $^4$He saturation anomalies alone because factors such as air injection and solubility uncertainties will nearly cancel out. Measurements have shown that, contrary to what might have been expected, deep ocean waters have higher $^3$He/$^4$He ratios than does the atmosphere. These results are normally expressed as isotope anomalies in the following manner:

$$\delta(^3He) = \frac{(^3He/^4He)_{sw} - (^3He/^4He)_{air}}{(^3He/^4He)_{air}} \times 100\%$$

The values of $\delta(^3He)$ obtained by Clarke et al. (1969, 1970) and by Jenkins et al. (1972) are illustrated in Fig. 8.6. The proportion of $^3$He in deep waters increases in the progression from North to South Atlantic and South to North Pacific. Thus, there is evidently a source of $^3$He to deep ocean waters which increases the $^3$He/$^4$He ratio in spite of the radiogenic alpha particle production of excess helium-4.

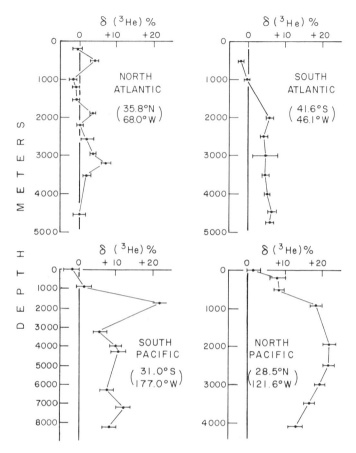

FIG. 8.6. Helium-3 isotope anomalies in four major ocean basins obtained by Clarke *et al.* (1969, 1970) and by Jenkins *et al.* (1972).

There are a number of possible sources of this $^3$He which do not have a corresponding $^4$He complement. One which may occur in the lithosphere is a $^6$Li$(n, \alpha)^3H$ reaction brought about by neutrons associated with cosmic rays. The tritium (half-life 12·5 years) will decay to $^3$He. This source is not believed to be of primary significance (Johnson and Axford, 1969). Tritium is also produced in the atmosphere by a $^{14}$N$(n, {}^3H)^{12}C$ reaction and this will also lead to the formation of $^3$He. A third possible source is primordial material (i.e. that which has not been subjected to lithospheric chemical fractionation) which contains a greater proportion of $^3$He than do components of the earth's crust. For example, helium constitutes 23 % of the mass of the universe, and in meteorites in which He is 1000 times more concentrated than in crustal material, the $^3He/^4He$ ratio is as high as 0·2 (Cockett and Smith, 1973; Aldrich and Nier, 1948). The addition at the earth's surface of primordial material either from within the earth or from extra-terrestrial sources could be a source of $^3He$. In addition, it is thought that high latitude auroral precipitation of solar plasma is a significant source of $^3He$ to the atmosphere (Johnson and Axford, 1969).

Clarke *et al.* (1969, 1970) and Jenkins *et al.* (1972) have suggested that high oceanic values of $^3He/^4He$ result from both the contribution of $^3He$ from tritium decay in the upper layers and the submarine injection of primordial $^3He$; the latter is probably responsible for the deep maxima which are evident especially in the Pacific Ocean. They believe that the relative enrichment of $^3He$ at 2000 m may be caused by injection at the crest of the mid-ocean ridge system. The contribution to the deep water of $^3He$ from tritium decay in water masses formed at high latitudes has been considered by Fairhall (1969) and by Craig and Clarke (1970); however, it is likely that this source is insufficient to explain the observed values of $\delta(^3He)$ of up to 20%.

It is likely that more information on the oceanic distribution of helium isotopes will be acquired in the near future. These results will provide a more definite indication of the major factors which control helium geochemistry. The present state of our knowledge of the processes involved in the occurrence of helium in deep Pacific waters is summarized by Table 8.11.

TABLE 8.11

*Portions of helium in deep Pacific waters contributed by various sources.*

| Process | $^4He$ nmol kg$^{-1}$ | $^3He$ fmol kg$^{-1}$* | Percentage relative to solubility | |
| | | | $^4He$ | $^3He$ |
| --- | --- | --- | --- | --- |
| Atmospheric solubility | 1·75 | 2·32 | 100 | 100 |
| Air injection | 0·07 | 0·09 | 4 | 4 |
| Radiogenic alpha production | 0·05 | — | 3 | — |
| Primordial | — | 0·60 | — | 26 |

* femto moles = $10^{-15}$ moles.

The results given in this table were calculated for the conditions which are representative of the Pacific $^3He/^4He$ maximum: $\delta(^3He) = 20\%$, $\Delta_{He} = 7\%$, 2°C, and 34·8‰ salinity. In accordance with the previous analysis of saturation anomalies the $\Delta_{He}$ has been partitioned into 4% arising from air injection and 3% originating from radiogenic production. The saturation anomaly for total helium in sea water, $\Delta_{He}$, stems entirely from $^4He$ and it is not enhanced significantly by the primordial $^3$He. The total saturation anomaly of $^3$He in this calculation relative to its air solubility is 30%.

One question which has not been resolved is whether any of the excess $^4$He presently attributed to radiogenic production could be primordial $^4$He accompanying the primordial $^3$He. In order to answer this question it is necessary to know the $^3$He/$^4$He ratio in the primordial (e.g. mantle) source. If this ratio is similar to that in meteorites, the primordial source will not be significant for $^4$He.

In summary, the studies of non-reactive gases in the ocean have provided valuable information on various processes. It is likely that it will be possible to draw even more detailed and definitive conclusions about these processes when the results of the GEOSECS Program become available. The existing data indicate that ocean waters are within 4% of saturation for $N_2$ and Ar, whereas Ne and He show slightly larger departures from saturation because of air injection. An additional several percent supersaturation of helium occurs in deep Pacific waters; this arises from radiogenic production. The enrichment of these waters with respect to $^3$He is thought to be mainly of primordial origin.

## 8.6. Dissolved Oxygen in the Ocean

### 8.6.1. distribution of dissolved oxygen

The vertical distribution of dissolved oxygen may be illustrated by profiles and vertical sections selected from three of the major ocean basins (Fig. 8.7). The oxygen concentration at the upper boundary of the ocean is established primarily by exchange with the atmosphere, and thus it is determined largely by the solubility of oxygen in sea water. Two remarkable features of the oxygen distribution are the presence of an oxygen minimum region and the relatively high oxygen concentrations in deep waters. These features of the oxygen distribution are the result of a balance between biological oxygen consumption and physical transport of oceanic waters. Knowledge of these processes may be gained through a study of the oxygen distribution.

Several aspects of oceanic circulation are reflected in the oxygen distribution. Antarctic Intermediate Water is evident as an intrusion of oxygen-rich water in the Atlantic and Pacific Oceans from the surface at 50° S to about 800 m at 20° S. North Atlantic Deep Water (NADW) appears as a mass of oxygen-rich water extending from 0–2000 m at 60°N to an oxygen concentration maximum at about 3000 m in the south equatorial Atlantic. The contributions of NADW and Antarctic Bottom Water flowing northward in the deep Pacific result in a progressive decrease in the oxygen content from south to

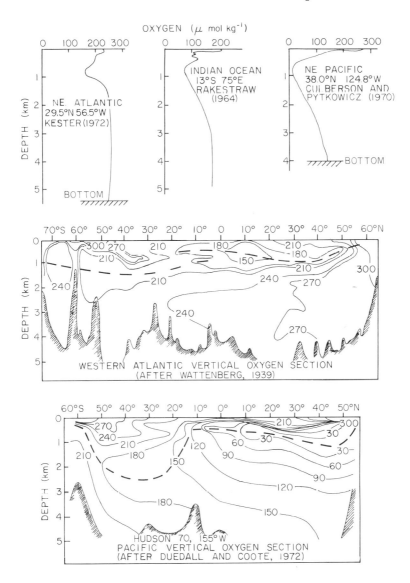

FIG. 8.7. Vertical distribution of oxygen in the ocean. Upper: profiles from three ocean basins; Centre: north–south vertical section through the western Atlantic; Lower: north–south vertical section through the central Pacific. The dashed line represents the depth of the oxygen minimum surface.

north. The intermediate depth waters of the North Pacific are substantially more deficient in oxygen than are their counterparts in the North Atlantic.

Horizontal variability in the oxygen minimum layer is illustrated in Fig. 8.8. In the Pacific, the intense minimum off Central and South America is advected westward by the North and South Equatorial Currents, and a similar horizontal gradient occurs off Africa.

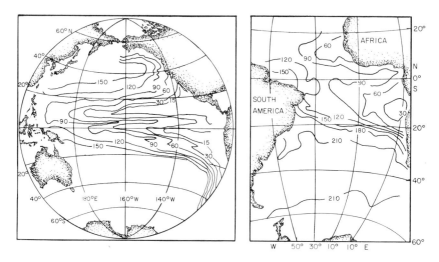

FIG. 8.8. Horizontal variation in oxygen concentration at a depth near the oxygen minimum. Left: Oxygen (in µmol kg$^{-1}$) in the Pacific on the 125 cl ton$^{-1}$ thermosteric anomaly surface which has a depth of approximately 300–500 m in most regions (after Reid, 1965). Right: Oxygen (in µmol kg$^{-1}$) in the South Atlantic at a depth of 400 m (after Wattenberg, 1939).

### 8.6.2. PROCESSES RESPONSIBLE FOR THE OXYGEN DISTRIBUTION

The distribution of oxygen within the ocean is the net result of biological consumption and physical replenishment by advection and mixing. The development of a quantitative interpretation of these processes is aided by examining the oxygen which has been *utilized* rather than the amount which *remains*. In this way it is possible to eliminate differences in the oxygen distribution which arise from the conditions when different water masses were previously in contact with the atmosphere. The ultimate approach for accounting for these different initial conditions is to assess the effects of various physical processes by the simultaneous determination of non-reactive gases along with oxygen. The contribution of processes such as air injection, thermal effects, and atmospheric pressure fluctuations to the oxygen content of oceanic waters may be determined as described in Section 8.5.3.

A less comprehensive, but sufficiently accurate, alternative is to use a single non-reactive gas as a reference standard for the estimation of the initial oxygen content. It is evident from Fig. 8.5 and Table 8.7 that Ar is the best reference gas for oxygen because of their similar solubility behaviour. These two gases will be affected nearly equally by physical processes which produce deviations from solubility equilibrium. Accordingly, the initial oxygen concentration, $c^i_{O_2}$, of a portion of water may be evaluated from its temperature, salinity and observed argon concentration, $c_{Ar}$ from the expression (Benson, 1965)

$$c^i_{O_2} = \frac{c^*_{O_2}}{c^*_{Ar}} \times c_{Ar}. \qquad (8.22)$$

The oxygen consumed since the waters were last in contact with the atmosphere is given by

$$c^i_{O_2} - c_{O_2} = \frac{c^*_{O_2}}{c^*_{Ar}} \times c_{Ar} - c_{O_2} \qquad (8.23)$$

where $c_{O_2}$ is the observed oxygen concentration. Refinements in recent years of gas chromatographic determinations of argon may lead to extensive measurements which will permit the application of equation (8.23) to the oceanic oxygen distribution.

In the absence of non-reactive gas measurements, Redfield (1942) employed the parameter which he designated "apparent oxygen utilization" (AOU), viz.,

$$AOU = c^*_{O_2} - c_{O_2} \qquad (8.24)$$

The qualification of this quantity as "apparent" provides recognition that exact equilibrium relative to a standard atmospheric pressure does not always occur at the sea surface, and that the non-linearity of $c^*_{O_2}$ with temperature can introduce a departure of AOU from the actual consumption. Surface water oxygen data indicate that in most regions of the ocean, $c^i_{O_2}$ is within a few percent of $c^*_{O_2}$ at times when deep and intermediate waters are likely to form (Richards and Corwin, 1956; Pytkowicz, 1968; Kester and Pytkowicz, 1968). However, at upwelling locations, surface oxygen concentrations may be up to 20% less than saturation (Broenkow, 1965; Kester and Pytkowicz, 1968). The mixing of representative water types produces a relatively small bias in AOU values (Pytkowicz and Kester, 1966; Pytkowicz, 1968). The approximation involved in using AOU as an estimate of oxygen consumption can be obtained from information on oceanic variations of argon (cf. equations 8.23 and 8.24). Argon in deep waters is generally within 4% of saturation, hence $c^*_{O_2}$ will differ from $c^i_{O_2}$ by a similar amount.

The primary aim of oxygen utilization measurements is the determination of the locations in the ocean at which such utilization occurs. Redfield (1942) examined the AOU and phosphate distributions along a north–south transect through the Atlantic, and he found that much of the oxygen consumption in Antarctic Intermediate Water occurred near its source at depths less than a few hundred metres. Only between the latitudes of 20° S to 25° N did he find it necessary to consider additional consumption at depths of 600–800 m as the water was transported northward. He concluded that this additional consumption resulted from the sinking of decomposing organisms from the overlying surface waters.

Linear correlations between salinity and oxygen content occur for intermediate waters in the eastern South Pacific and South Atlantic regions (Menzel and Ryther, 1968; Menzel, 1970). These observations indicate that the decrease in oxygen as the intermediate waters proceed from their source results primarily from mixing rather than from *in situ* oxidation of organic material. Even though most of the oxygen utilization occurs in the upper few hundred metres, care should be used in drawing the conclusion that further consumption at greater depths does not affect the oxygen distribution. The regions of the ocean which thus far show a quasi-conservative behaviour of oxygen are those of relatively dynamic intermediate waters and of intense horizontal gradients produced by the adjacent upwelling regions (Fig. 8.8).

A quantitative approach to the resolution of the effects of physical and biological processes in the vertical distribution of oxygen was presented by Wyrtki (1962) in an analysis of the oxygen minimum region. He developed a model in which the steady state distribution was maintained by vertical advection, vertical eddy diffusion, and a rate of *in situ* oxygen consumption, $R$, which decreased exponentially with depth *viz*,

$$A_z \frac{\partial^2 c_{O_2}}{\partial z^2} - w \frac{\partial c_{O_2}}{\partial z} = R = R_0\, e^{-\alpha z} \tag{8.25}$$

In this expression the vertical eddy diffusion coefficient, $A_z$, is assumed to be independent of the depth, $z$; The vertical advection velocity is denoted by $w$, $R_0$ is the rate of oxygen utilization at the upper boundary of the model, and $\alpha$ accounts for the rate at which $R$ decreases exponentially with $z$. This model requires that horizontal processes have no effect on the oxygen distribution within the depth range considered. The model assumes that the oxygen concentrations at the upper and lower boundaries are maintained by processes such as atmospheric exchange and horizontal transport. Wyrtki applied this model to an oxygen profile in the Indian Ocean (15° S, 95° E) at depths between 600 and 3000 m. He evaluated the ratio of $w/A_z$ from the vertical temperature profile; the value of $\alpha$ was based on the results of Riley

(1951) for the Atlantic Ocean, and the ratio $R_0/A_z$ was adjusted to provide the best fit with the observed profile (Fig. 8.9).

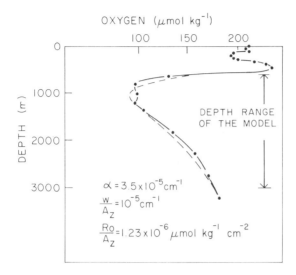

FIG. 8.9. Application of the vertical advection-eddy diffusion model to an oxygen profile in the Indian Ocean (after Wyrtki, 1962).

The ability of this simple model to fit the vertical oxygen distribution provides insight into the effects of various processes. It is interesting to examine the effects on the oxygen profile caused by variations in the parameters $A_z$, $w$, $R_0$ and $\alpha$. Wyrtki considered the problem of oxygen consumption occurring in one body of water which is then transported horizontally and mixed into a second body where no further consumption occurs. He found that the horizontal attenuation of the oxygen minimum under these conditions was much more rapid than is observed in most regions of the ocean. Thus, it appears that *in situ* oxygen consumption is required to account for the world-wide occurrence of the oxygen minimum region; however, in certain areas where the oxygen minimum is relatively intense and near the surface (e.g., off the west coasts of Africa and South America) horizontal mixing may predominate over *in situ* utilization.

The vertical advection-eddy diffusion model has been developed further and applied to various chemical distributions by Craig (1969, 1971) who demonstrated that the dramatic change in depth of the oxygen minimum between 60°S and 5°S in the Pacific Ocean (Fig. 8.7) is consistent with a vertical model which includes *in situ* consumption. Culberson (1972) applied the vertical model to an oxygen profile in the northeast Pacific Ocean. He considered several alternatives for the variation in the rate of oxygen consumption with depth. An exponentially decreasing $R$ with depth produced a better fit to the profiles than did a constant $R$ or one which varied linearly with depth. Culberson compared the value for $R$ derived from the model with estimates based on enzyme analysis of the respiratory electron transport system in plankton made by Packard (1969) (Fig. 8.10). The vertical model

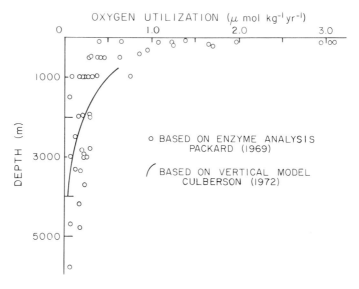

FIG. 8.10. Comparison of the rate of oxygen utilization as a function of depth in the North Pacific Ocean derived from a vertical advection-eddy diffusion model and estimated from enzyme measurements (after Culberson, 1972).

yielded values of $R$ which were consistent in magnitude and depth dependence with those obtained by enzyme analysis.

It should be realized that the vertical advection–eddy diffusion model is an oversimplification of oceanic processes. Although there may be some places in the ocean where this is a reasonable approximation, there are undoubtedly instances where it fits observed distributions even though its basic assumptions and the values of the parameters which must be assumed do not correspond with physical reality. For example, if there are horizontal contributions to the distribution, or if a parameter such as $A_z$ or $w$ varies vertically, these effects will be absorbed in the $R_0$ or $\alpha$ terms. Nevertheless, the model provides a useful means of inter-relating the distributions of temperature, salinity, oxygen and other biologically active constituents.

The ultimate quantitative model of oceanic distributions must embody a three dimensional description of the processes involved. The classical study by Riley (1951) of the oxygen, phosphate and nitrate distributions in the Atlantic Ocean is a major step toward such a description. He expressed the distribution of a variable in terms of vertical and horizontal velocities, a lateral and vertical eddy diffusion coefficient, and a production or consumption term for the non-conservative variables such as oxygen. Thus, for oxygen the general equation (Sverdrup et al., 1942) became

$$\frac{\partial c_{O_2}}{\partial t} = \frac{\partial}{\partial x}\left\{\frac{A_1}{\rho}\frac{\partial c_{O_2}}{\partial x}\right\} + \frac{\partial}{\partial y}\left\{\frac{A_1}{\rho}\frac{\partial c_{O_2}}{\partial y}\right\} + \frac{\partial}{\partial z}\left\{\frac{A_z}{\rho}\frac{\partial c_{O_2}}{\partial z}\right\}$$

$$- \left\{u\frac{\partial c_{O_2}}{\partial x} + v\frac{\partial c_{O_2}}{\partial y} + w\frac{\partial c_{O_2}}{\partial z}\right\} + R_{O_2}, \qquad (8.26)$$

where $t$ is time; $A_1$ and $A_z$ are respectively lateral and vertical eddy diffusion coefficients; $\rho$ is density; $u$, $v$, and $w$ are the $x$, $y$, and $z$ components of velocity respectively; and $R_{O_2}$ is the rate of oxygen production (a negative value corresponds to utilization). The Atlantic Ocean was divided into 76 areas and a method of finite differences was used in the application of the distribution equation. The velocity components were obtained by geostrophic calculation, the principle of continuity, and a knowledge of water masses. The vertical eddy diffusion coefficient was assumed to be a function of vertical shear and stability. The lateral eddy diffusion coefficient was evaluated from the salinity distribution. When these parameters were used for the oxygen distribution, Riley was able to determine the rate of oxygen consumption. Some of his results are summarized in Table 8.12. One conclusion

TABLE 8.12

*Parameters for the distribution of variables in the Atlantic Ocean (From Riley, 1951).*

| $\sigma_T$ Surface | Average depth (m) | $R_{O_2}$ ml l$^{-1}$yr$^{-1}$ | $A_1 \times 10^{-6}$ g cm$^{-1}$ s$^{-1}$ | $A_z$ g cm$^{-1}$s$^{-1}$ |
|---|---|---|---|---|
| 26·5 | 200 | −0·21 | 57 | 0·30 |
| 26·7 | 280 | −0·08 | 18 | 0·16 |
| 26·9 | 370 | −0·050 | 14 | 0·12 |
| 27·1 | 510 | −0·055 | 10 | 0.38 |
| 27·3 | 700 | −0·035 | 2·5 | 0·31 |
| 27·5 | 1000 | −0·013 | 2·4 | 0·13 |
| 27·7 | 1250 | −0·005 | 8 | 2·6 |
| — | 1500 | −0·0016 | 0·22 | 0·43 |
| — | 2000 | −0·0013 | 0·23 | 0·45 |
| — | 3000 | −0·00013 | 0·81 | 0·35 |
| — | 4000 | −0·00013 | 2·8 | 0·54 |

drawn from Riley's analysis was that, of the total organic matter produced by phytoplankton in the surface layers, $90\%$ is utilized in the upper 200 m; the remainder sinks to deeper waters where it leads to oxygen consumption by respiration and bacterial decomposition. It was also concluded that the oxygen minimum lies between the region of the maximum rate of consumption and the region of the minimal rate of replenishment.

### 8.6.3. RELATIONSHIP OF OXYGEN TO OTHER CHEMICAL PARAMETERS

A significant aspect of the oceanic oxygen distribution is the correlation it shows when expressed as AOU with several other chemical parameters such as the phosphate, nitrate, and carbon dioxide concentrations and the pH. These correlations lead to a stoichiometric model for photosynthesis and respiration in the ocean which brings together many of the variations of biologically active constituents. This model may be represented as:

$$\underbrace{(CH_2O)_{106}\,(NH_3)_{16}\,H_3PO_4}_{\text{organic matter}} + 138O_2 \underset{\text{photosynthesis}}{\overset{\text{oxidation}}{\rightleftharpoons}}$$

$$106CO_2 + 16HNO_3 + H_3PO_4 + 122H_2O, \qquad (8.27)$$

in which the formula for organic matter represents the statistical empirical average composition. Organic carbon is represented by carbohydrate units; organically bound nitrogen is expressed as ammonia, although in reality it exists as proteins, amino acids, urea, etc. and organically bound phosphorus is represented by phosphoric acid. These representations are selected only to provide mass and redox balances for the primary organically active elements. The significance of this model is that it provides a quantitative basis for the examination of the effects of variations in parameters such as oxygen, nitrate, and phosphate.

The basis for this stoichiometric model was the work of Redfield (1934, 1958) and Redfield *et al.* (1963) in which the correlations of various biological elements were examined. They found that the observed changes of oxygen, phosphate, nitrate, and $CO_2$ concn. for many areas of the ocean were close to those predicted from the stoichiometric coefficients based on a statistical composition for plankton. Departures in the observed oxygen–phosphate relationship from the predictions of equation (8.27) have been related to the preformed phosphate concentration, i.e. the amount of phosphate which existed in the water when it was previously in contact with the atmosphere. Application of this conservative oceanic parameter has been useful in distinguishing the effects of physical and biological processes in the distribution of oxygen (Redfield, 1942; Sugiura, 1964, 1965; Pytkowicz and Kester, 1966; Pytkowicz, 1968). The correlation between AOU and $CO_2$ concn. is more complex because of processes other than organic decomposition affecting the $CO_2$ content of deep waters. Nevertheless, Culberson and Pytkowicz (1970) found agreement between the stoichiometric model and the observed AOU–$CO_2$ variations after allowing for differences in alkalinity and in initial $CO_2$ concentrations. Oceanic variations in pH and AOU have also been shown to be closely linked to each other (Park, 1968; Culberson, 1972).

### 8.6.4. CONTINUOUS *in situ* OXYGEN MEASUREMENTS

The development of continuous recording *in situ* instruments and probes for the measurement of salinity, temperature, depth, and oxygen (STD-$O_2$) provides a new dimension in the resolution and study of oceanic distributions. It is becoming possible to resolve chemical features, the existence of which had hardly been appreciated, or which previously could only be inferred statistically (Cooper, 1967). Lambert *et al.* (1973) reported some of the initial observations of continuous oxygen profiles in several oceanographic regions. In one study of this type the oxygen distribution between 0 and 1500 m was mapped in a 40 $km^2$ region of the mid-Atlantic (22° N, 35° W) (Kester *et al.*, 1973). A body of water was observed in the main thermocline which contained 10% less oxygen than did adjacent waters at the same depth. This oxygen minimum feature was distinct from the main minimum; it was approximately 80 m thick and extended 15 km in the east–west direction and more than 37 km in the north–south direction. This parcel of water shifted westward by 4 km in a 27 hour period. The reduced oxygen content of this body of water probably reflects either a greater rate of oxygen utilization than that in the surrounding water or the intrusion of water from a different source.

An example of the type of information provided by the *in situ* sensors is illustrated in Fig. 8.11 for a location south of Iceland (60° N, 22° W). Oxygen profiles in this region were typified by a series of extrema above the oxygen minimum and relatively smooth, nearly exponential, curves below the minimum. This pattern suggested that, above the minimum, horizontal processes play a significant role in the oxygen distribution, whereas below the minimum, vertical mixing predominates. Three profiles obtained over a two hour period (Fig. 8.11) showed a 40 m vertical shift in the temperature and oxygen features as well as some progressive changes in the shapes of the profiles. These observations indicate the complex temporal and spatial variability of oceanic distributions.

Continuous oxygen profiles reveal the extreme gradients which can occur

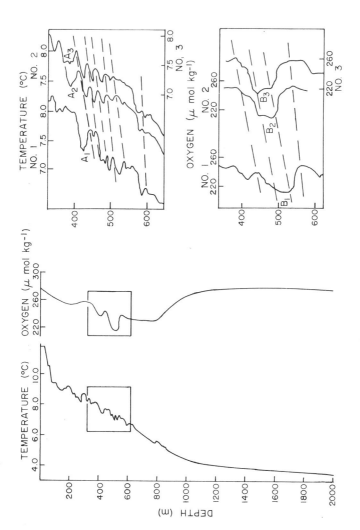

Fig. 8.11. Vertical temperature and oxygen profiles at a location 200 miles south of Iceland obtained by continuous *in situ* measurements. The section between 350 and 600 m is shown with an expanded scale on the right for three successive casts over a two hour period. The horizontal shift in the temperature and oxygen scales for these three casts is proportional to their separation in time. The dashed lines and the designators A and B trace the change in depth and shape of corresponding features on the three profiles (after Lambert *et al.*, 1973).

between adjacent water masses. Vertical changes of up to 18 $\mu$mol kg$^{-1}$ per 10 m have been observed at depths of 600 m; such gradients can persist only in the absence of vertical mixing. The continuous profiles also provide an extreme test for modelling chemical distributions; a vertical mixing model which can fit a continuous profile has much less freedom than one which fits eight or twelve points in the mixing range.

## 8.7. MINOR REACTIVE GASES IN THE OCEAN

### 8.7.1. NITROUS OXIDE

Most studies of nitrous oxide ($N_2O$) in the marine environment have been made with the aim of examining its distribution in relation to large scale mixing processes. The main source of atmospheric $N_2O$ is considered to be microbial activity in soil; unlike nitric oxide (NO) and nitrogen dioxide ($NO_2$) it is not believed to have an anthropogenic source (Schütz et al., 1970). The principal mechanism for the removal of $N_2O$ from the earth's surface is photochemical dissociation in the stratosphere (Bates and Hays, 1967). The gas is chemically unreactive in the lower atmosphere, and it would be expected that its distribution in the ocean would reflect the hemispherical distribution of land masses.

Craig and Gordon (1963) were the first investigators to examine the occurrence of $N_2O$ in the ocean. They found $N_2O$ concentrations ranging from 7 to 12 nmol kg$^{-1}$ in three aggregate water samples from the South Pacific Ocean. Junge et al. (1971a) obtained an extensive set of $N_2O$ measurements for the North Atlantic Ocean at 30° W from the equator to 60° N. Sea water at depths of 500–1000 m contained about 11 nmol kg$^{-1}$ which was nearly twice the amount which would be expected on the basis of equilibrium being established with the atmospheric partial pressure of $N_2O$. Argon data (Hahn, 1972) indicate that these values cannot be explained by physical processes such as air injection or thermal effects. Samples collected from depths of 1000–3000 m were generally 50–70% saturated relative to atmospheric $N_2O$. Surface waters of the North Atlantic are thus a source of $N_2O$ to the atmosphere. The mechanism for $N_2O$ production in surface ocean water has not yet been established, although biochemical processes in the nitrogen cycle are a likely possibility. The processes which deplete $N_2O$ from deep and intermediate ocean water are also unknown. Junge and Hahn (1971) have examined vertical profiles of $N_2O$ in the water column southeast of Iceland; these showed smaller vertical gradients than those observed at lower latitudes.

The distribution of $N_2O$ in the atmosphere has not been completely characterized. Between 1966 and 1969 a 20% increase of $N_2O$ was observed by Schütz et al. (1970) in Mainz, Germany. This was interpreted as reflecting a possible global, or at least a hemispheric, increase. Junge et al. (1971a) observed low $N_2O$ partial pressures in the marine atmosphere associated with air masses from arid west Africa and they suggested that $N_2O$ may be removed from the atmosphere in the Sahara Desert. Measurements by Schütz et al. (1970) of atmospheric $N_2O$ near Johannesburg, South Africa (26° S) indicated that the $N_2O$ partial pressure in the southern hemisphere is higher than that inferred by Craig and Gordon (1963). Considerably more information will be required before the geochemistry of $N_2O$ is understood.

8.7.2. GASEOUS HYDROCARBONS

Of the low molecular weight hydrocarbons occurring in the marine environment most attention has been focused on methane ($CH_4$); some attention has also been given to the low molecular weight paraffins and olefins (e.g. ethane ($CH_3$—$CH_3$), ethylene ($CH_2$=$CH_2$), propane ($CH_3$—$CH_2$—$CH_3$), propylene ($CH_3$—$CH$=$CH_2$), and iso-butane ($[CH_3]_2$—$CH$—$CH_3$)). The partial pressure of methane in the atmosphere is relatively constant, although in restricted localities near to sources higher concentrations may be found. This gas is formed primarily through biochemical processes and is destroyed photochemically. Both the estimated fluxes and carbon-14 measurements suggest that the residence time of methane in the troposphere is about 100 years (Junge, 1963). However, Weinstock and Niki (1972) have recently postulated that $CH_4$ is oxidized by hydroxyl radicals in the troposphere at a rate which implies a residence time of about 1·5 years. The wide divergence between these two estimates makes it difficult to give a figure for the residence time in the lower atmosphere.

Surface ocean waters of tropical northern latitudes contain a relatively constant amount, 1·8 nmol kg$^{-1}$, of $CH_4$ (Swinnerton et al., 1969; Lamontagne et al., 1971). The corresponding marine air contains 1·4 ppm of methane. Thus, methane concentration in these surface waters is close to equilibrium with the atmospheric partial pressure. Little is known about the vertical distribution of methane in the ocean. However, Swinnerton and Linnenbom (1967) have found a methane maximum at depths of 30–100 m in the Gulf of Mexico.

Methane concentrations much higher than those corresponding with atmospheric equilibrium occur in coastal waters subject to pollution and in anoxic marine environments. Swinnerton et al. (1969) found $CH_4$ concentrations as high as 90–360 nmol kg$^{-1}$ in the Potomac River; these high concentrations were attributed to pollution. The air in this region contained about twice as much methane as does the marine atmosphere. Atkinson and Richards (1967) examined the methane concentration in several anoxic basins: Lake Nitinat, the Cariaco Trench, and the Black Sea. The concentrations found ranged between 5 and 40 $\mu mol\,kg^{-1}$ and generally increased with depth in a manner similar to that of hydrogen sulphide. This methane was attributed to the anaerobic fermentation of carbohydrates, but it was not determined whether this process occurred in the anoxic water column or in the underlying sediments.

Initial reports indicate that $C_2$—$C_4$ hydrocarbons are present in marine waters; higher concentrations occur in anoxic waters than in open ocean surface waters (Swinnerton and Linnenbom, 1967; Linnenbom and Swinnerton, 1968). Wilson et al. (1970) found that ethylene and propylene can be produced by the effect of light on natural organic substances in sterile sea water. There is, however, insufficient information on the gaseous hydrocarbons other than methane to be certain which processes determine their distribution in the marine environment.

### 8.7.3. CARBON MONOXIDE

The atmospheric and marine chemistry of carbon monoxide have received considerable attention in recent years because of their obvious importance in connection with man's activities. Man's production of CO during the late 1960's has been estimated to be about $2\cdot6 \times 10^{14}\,g\,yr^{-1}$ (Robinson and Robbins, 1969). Thus, the total amount of CO present in the atmosphere could be produced anthropogenically in 2–3 years. Measurements of CO partial pressures in the troposphere (Junge et al., 1971b) indicate higher values in the northern hemisphere $(0\cdot1–0\cdot2 \times 10^{-6}\,atm)$ than in the southern hemisphere $(0\cdot04–0\cdot12 \times 10^{-6}\,atm)$, a feature consistent with a possible anthropogenic influence. No substantial evidence has been obtained which indicates a global increase in CO in the lower atmosphere. By implication therefore, a sink must exist to balance the anthropogenic input.

It might be assumed that the ocean would absorb some of the CO produced by man. Robinson and Robbins (1968) measured the CO in marine air between the United States and New Zealand, and they found the highest CO concentrations above the central portion of the North and South Pacific They also observed occasional diurnal increases in atmospheric CO at about the time of sunset. They concluded that ocean surface waters may be a source of CO to the atmosphere. Linnenbom and Swinnerton (1968) measured CO in marine air and in surface water of the northwest Atlantic Ocean. They found that the CO concentration in these waters exceeded the equilibrium value set by the prevailing atmospheric partial pressure, and thus that the ocean represents a natural source of CO. Diurnal increases in CO concn. of 5–9 fold have been observed in the surface waters of the northern tropical Atlantic and Pacific Oceans, with max. concn. occurring in the afternoon (Swinnerton et al., 1970a; Lamontagne et al., 1971; Swinnerton and Lamontagne, 1974). Carbon monoxide production appears to be most significant in the upper 40 m of the water column (Swinnerton et al., 1970b). Carbon monoxide may be produced photochemically from dissolved organic carbon in sea water (Wilson et al., 1970). However, biochemical production is believed to be more significant, and the production is highly sensitive to light intensity, being less during cloudy and overcast conditions than in direct sunlight (Lamontagne et al., 1971). Surface waters are thus a source of CO rather than a sink relative to the atmosphere.

Evaluation of the flux of CO between the ocean and the atmosphere requires a knowledge of the CO solubility in sea water. Selected values of the Henry's Law constant $K_{CO}$ based on the microgasometric measurements made by Douglas (1967) are given in Table 8.13. The equilibrium concn. of

TABLE 8.13

*Henry's Law constant for carbon monoxide in seawater based on the solubility data of Douglas (1967).*

$K_{CO}$ [atm kg-sea water(mol-CO)$^{-1}$]

| | Salinity (‰) | | | |
|---|---|---|---|---|
| $T°C$ | 28 | 31 | 34 | 37 |
| 0 | 762 | 780 | 799 | 819 |
| 5 | 857 | 875 | 895 | 916 |
| 10 | 947 | 966 | 987 | 1009 |
| 15 | 1033 | 1054 | 1076 | 1100 |
| 20 | 1118 | 1141 | 1166 | 1192 |
| 25 | 1199 | 1225 | 1252 | 1280 |
| 30 | 1274 | 1298 | 1324 | 1350 |

CO in 37‰ salinity sea water at 25°C exposed to air with a $0.15 \times 10^{-6}$ atm partial pressure of CO is:

$$c_{CO}^* = \frac{1}{1280} 0.15 \times 10^{-6} = 0.117 \text{ nmol kg}^{-1}$$

The CO concentrations in tropical Pacific surface waters lie in the range 0.9 to 3.6 nmol kg$^{-1}$ (Lamontagne et al., 1971)—about 8–30 fold supersaturation.

The question of a global sink for CO remains. Oceanically derived CO must augment the anthropogenic source of the gas. One way of analysing the characteristics of the sink being sought is to consider the residence time of CO in the troposphere implied by the various fluxes. In the stratosphere carbon monoxide is oxidized to $CO_2$ by hydroxyl radicals. If the stratosphere is indeed the sink for the CO in the troposphere, a residence time of the order of decades is implied for CO in the lower atmosphere because of the low rate of mixing between these two regions. Combining an estimate of the oceanic and anthropogenic fluxes yields a total mean residence time of 1·3 yr (Junge et al., 1971b). Junge et al. (1971b) have proposed a model for atmospheric CO distribution which required some unspecified tropospheric sink and which yielded a residence time of about a year. Weinstock (1969) used the ratio of $^{14}CO/^{12}CO$ in the troposphere to obtain a residence time of 0·1 yr for CO in the lower atmosphere. The range of these residence times indicates the unsatisfactory nature of these estimates of the global CO cycle.

Weinstock and Niki (1972) resolved the discrepancies in the preceding analysis by considering a CO cycle in which there are tropospheric, oceanic and anthropogenic sources of CO and a tropospheric sink. They deduced that the natural tropospheric source has a magnitude of $5 \times 10^{15} \text{ g CO yr}^{-1}$ which is 20 times greater than the anthropogenic source and 60 times greater than the oceanic source, but which is consistent with a radiocarbon residence time of 0·1 yr. They showed that the tropospheric source of CO could be the reaction of methane with hydroxyl radicals. This process also represents a sink for methane which must be taken into account when the global balance of $CH_4$ is considered; the consumption of $CH_4$ in the formation of CO by the mechanism proposed by Weinstock and Niki is comparable to the estimated terrigenous production of methane. They also used hydroxyl radicals to explain the tropospheric sink of CO:

$$CO + OH\cdot \rightarrow CO_2 + H\cdot$$

Their model requires a tropospheric concentration of hydroxyl radicals which is consistent with photochemical models. A direct determination of the hydroxyl radical concentration in the troposphere would provide a key test of their model.

Present knowledge of the CO cycle is summarized in Fig. 8.12. Inman et al. (1971) estimated that the consumption of CO by soil bacteria in the United States alone is several times larger than the anthropogenic production. If these estimates and those of Weinstock and Niki for the hydroxyl reactions are correct, the anthropogenic and oceanic sources will not perturb the global balance significantly, and the apparently higher CO values in the northern

T

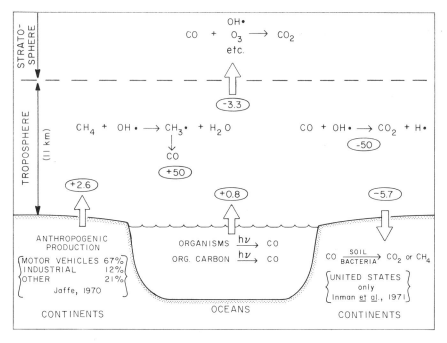

FIG. 8.12. Major processes in the global cycle of carbon monoxide. The numbers in ellipses are the annual flux of CO associated with each process in units of $10^{14}$ g y$^{-1}$. Positive values represent sources to the troposphere and negative values are sinks.

compared with the southern hemisphere cannot be attributed to man-made sources.

## 8.8. AREAS OF FUTURE WORK

Substantial progress has been made during the past ten years in our knowledge of dissolved gases in the ocean. Work is in progress which will lead to further advances in our understanding. The GEOSECS program will provide the most comprehensive and consistent set of data for the oceanic distribution of dissolved gases as well as other chemical constituents. The quantitative interpretation of chemical distributions in terms of physical, biological, and geochemical processes will be improved. Additional information will be gained through the use of continuous *in situ* sensors.

The solubilities of krypton and xenon must be determined more definitively before these gases can be used to examine processes affecting nonreactive gases. The solubility data for the minor atmospheric gases are based on measurements in which their partial pressures were near the total pressure of 1 atm. Bieri and Koide (1972) have suggested that at the low partial pressures of these gases in air there may be different molecular interactions which alter their solubilities from those observed when the nearly pure gases are equilibrated with sea water; evidence for this phenomenon has apparently been observed for CO (Meadows and Spedding, 1974). Clarification of this possibility must be obtained, because an error of a few percent in the solubilities of neon and helium could substantially alter the estimated magnitude of air injection for some water masses.

It is likely that further work on the mechanism and rate of gas exchange between the ocean and the atmosphere will lead to a better understanding of oceanic processes. Radon should continue to be very useful in establishing the rates of exchange of waters and gases near the air–sea interface.

The role of the ocean in the global cycle of gases with relatively short atmospheric residence times will be an important consideration in evaluating the impact of man's activity on environmental conditions.

## ACKNOWLEDGEMENTS

The author would like to express his appreciation to Michael E. Q. Pilson and Robert A. Duce for many stimulating discussions, to R. F. Weiss and J. P. Riley for their suggestions on the manuscript, to Marilyn A. Maley and James F. Holzgraf for their assistance in preparing the manuscript, and to R. M. Pytkowicz for suggesting this review be prepared. Portions of this work were supported by Office of Naval Research Contract No. N00014-68-A-0215-0003 and National Science Foundation Grant GX-33777.

TABLE A8.1

Solubility of nitrogen in sea water ($\mu$ mol $kg^{-1}$) based on equation (8.11).

| T°C | Salinity (‰) | | | | | | | | | | | | |
|---|---|---|---|---|---|---|---|---|---|---|---|---|---|
| | 0 | 4 | 8 | 12 | 16 | 20 | 24 | 28 | 31 | 33 | 35 | 37 | 39 |
| −1 | 844·8 | 817·2 | 790·5 | 764·6 | 739·6 | 715·5 | 692·1 | 669·4 | 653·0 | 642·2 | 631·6 | 621·2 | 611·0 |
| 0 | 822·7 | 796·0 | 770·1 | 745·1 | 721·0 | 697·6 | 674·9 | 653·0 | 637·1 | 626·6 | 616·4 | 606·3 | 596·4 |
| 1 | 801·5 | 775·7 | 750·7 | 726·5 | 703·1 | 680·5 | 658·6 | 637·3 | 621·9 | 611·8 | 601·9 | 592·1 | 582·5 |
| 2 | 781·2 | 756·3 | 732·1 | 708·7 | 686·0 | 664·1 | 642·8 | 622·3 | 607·3 | 597·5 | 587·9 | 578·4 | 569·1 |
| 3 | 761·9 | 737·7 | 714·3 | 691·6 | 669·7 | 648·4 | 627·9 | 607·9 | 593·4 | 583·9 | 574·6 | 565·4 | 556·3 |
| 4 | 743·5 | 720·1 | 697·4 | 675·4 | 654·2 | 633·5 | 613·6 | 594·3 | 580·2 | 571·0 | 561·9 | 553·0 | 544·2 |
| 5 | 725·8 | 703·1 | 681·1 | 659·8 | 639·2 | 619·2 | 599·8 | 581·0 | 567·4 | 558·4 | 549·6 | 541·0 | 532·4 |
| 6 | 708·9 | 686·9 | 665·6 | 644·9 | 624·9 | 605·5 | 586·7 | 568·5 | 555·2 | 546·5 | 537·9 | 529·5 | 521·2 |
| 7 | 692·6 | 671·3 | 650·6 | 630·5 | 611·1 | 592·3 | 574·0 | 556·3 | 543·4 | 535·0 | 526·7 | 518·5 | 510·4 |
| 8 | 677·2 | 656·5 | 636·4 | 616·9 | 598·0 | 579·7 | 562·0 | 544·8 | 532·2 | 524·0 | 515·9 | 508·0 | 500·1 |
| 9 | 662·3 | 642·2 | 622·7 | 603·7 | 585·4 | 567·6 | 550·4 | 533·6 | 521·4 | 513·4 | 505·6 | 497·8 | 490·2 |
| 10 | 648·0 | 628·4 | 609·5 | 591·1 | 573·2 | 555·9 | 539·2 | 522·9 | 511·0 | 503·3 | 495·6 | 488·1 | 480·7 |
| 11 | 634·3 | 615·3 | 596·9 | 579·0 | 561·6 | 544·8 | 528·5 | 512·7 | 501·1 | 493·5 | 486·1 | 478·7 | 471·5 |
| 12 | 621·1 | 602·6 | 584·7 | 567·3 | 550·4 | 534·0 | 518·1 | 502·7 | 491·5 | 484·1 | 476·8 | 469·7 | 462·6 |
| 13 | 608·4 | 590·5 | 573·0 | 556·1 | 539·6 | 523·7 | 508·2 | 493·2 | 482·3 | 475·1 | 468·0 | 461·0 | 454·2 |
| 14 | 596·3 | 578·8 | 561·8 | 545·3 | 529·3 | 513·8 | 498·7 | 484·1 | 473·4 | 466·4 | 459·5 | 452·7 | 446·0 |
| 15 | 584·6 | 567·5 | 551·0 | 534·9 | 519·3 | 504·2 | 489·5 | 475·3 | 464·8 | 458·0 | 451·3 | 444·7 | 438·1 |
| 16 | 573·3 | 556·7 | 540·6 | 525·0 | 509·8 | 495·0 | 480·7 | 466·8 | 456·6 | 450·0 | 443·4 | 436·9 | 430·6 |
| 18 | 552·0 | 536·2 | 520·9 | 506·0 | 491·6 | 477·5 | 463·9 | 450·7 | 441·0 | 434·6 | 428·4 | 422·2 | 416·1 |
| 20 | 532·2 | 517·2 | 502·6 | 488·4 | 474·7 | 461·3 | 448·3 | 435·7 | 426·4 | 420·4 | 414·4 | 408·5 | 402·7 |
| 22 | 513·8 | 499·5 | 485·6 | 472·1 | 459·0 | 446·2 | 433·8 | 421·7 | 412·9 | 407·1 | 401·4 | 395·8 | 390·2 |
| 24 | 496·6 | 483·0 | 469·7 | 456·8 | 444·2 | 432·0 | 420·2 | 408·6 | 400·2 | 394·6 | 389·2 | 383·8 | 378·5 |
| 26 | 480·5 | 467·5 | 454·8 | 442·5 | 430·5 | 418·8 | 407·4 | 396·4 | 388·3 | 383·0 | 377·7 | 372·6 | 367·5 |
| 28 | 465·4 | 453·0 | 440·8 | 429·0 | 417·5 | 406·3 | 395·4 | 384·8 | 377·1 | 372·0 | 366·9 | 362·0 | 357·1 |
| 30 | 451·3 | 439·4 | 427·7 | 416·4 | 405·4 | 394·6 | 384·2 | 374·0 | 366·5 | 361·6 | 356·8 | 352·1 | 347·4 |
| 32 | 438·0 | 426·6 | 415·4 | 404·5 | 393·9 | 383·6 | 373·6 | 363·8 | 356·6 | 351·9 | 347·3 | 342·7 | 338·2 |

TABLE A8.2

Solubility of oxygen in sea water ($\mu\ mol\ kg^{-1}$) based on equation (8.11).

| T°C | Salinity (‰) | | | | | | | | | | | | |
|---|---|---|---|---|---|---|---|---|---|---|---|---|---|
| | 0 | 4 | 8 | 12 | 16 | 20 | 24 | 28 | 31 | 33 | 35 | 37 | 39 |
| −1 | 469·7 | 455·5 | 441·7 | 428·3 | 415·4 | 402·8 | 390·6 | 378·8 | 370·2 | 364·6 | 359·0 | 353·5 | 348·2 |
| 0 | 456·4 | 442·7 | 429·4 | 416·5 | 404·0 | 391·9 | 380·1 | 368·7 | 360·4 | 354·9 | 349·5 | 344·2 | 339·0 |
| 1 | 443·8 | 430·6 | 417·7 | 405·3 | 393·2 | 381·5 | 370·1 | 359·0 | 351·0 | 345·7 | 340·5 | 335·4 | 330·3 |
| 2 | 431·7 | 418·9 | 406·5 | 394·5 | 382·8 | 371·5 | 360·5 | 349·8 | 342·0 | 336·9 | 331·8 | 326·9 | 322·0 |
| 3 | 420·2 | 407·9 | 395·9 | 384·2 | 372·9 | 361·9 | 351·3 | 340·9 | 333·4 | 328·5 | 323·6 | 318·8 | 314·1 |
| 4 | 409·3 | 397·3 | 385·7 | 374·4 | 363·5 | 352·9 | 342·6 | 332·5 | 325·2 | 320·4 | 315·7 | 311·1 | 306·5 |
| 5 | 398·8 | 387·2 | 375·9 | 365·0 | 354·4 | 344·1 | 334·1 | 324·4 | 317·4 | 312·7 | 308·1 | 303·6 | 299·2 |
| 6 | 388·7 | 377·5 | 366·6 | 356·0 | 345·8 | 335·8 | 326·1 | 316·7 | 309·8 | 305·3 | 300·9 | 296·5 | 292·2 |
| 7 | 379·1 | 368·2 | 357·7 | 347·4 | 337·5 | 327·8 | 318·4 | 309·3 | 302·6 | 298·3 | 294·0 | 289·7 | 285·5 |
| 8 | 369·9 | 359·4 | 349·1 | 339·2 | 329·6 | 320·2 | 311·1 | 302·2 | 295·8 | 291·5 | 287·3 | 283·2 | 279·2 |
| 9 | 361·1 | 350·9 | 341·0 | 331·3 | 322·0 | 312·9 | 304·0 | 295·4 | 289·2 | 285·0 | 281·0 | 277·0 | 273·0 |
| 10 | 352·6 | 342·7 | 333·1 | 323·7 | 314·6 | 305·8 | 297·2 | 288·9 | 282·8 | 278·8 | 274·8 | 271·0 | 267·1 |
| 11 | 344·5 | 334·9 | 325·5 | 316·5 | 307·6 | 299·1 | 290·7 | 282·6 | 276·7 | 272·8 | 269·0 | 265·2 | 261·5 |
| 12 | 336·7 | 327·3 | 318·3 | 309·5 | 300·9 | 292·6 | 284·5 | 276·6 | 270·8 | 267·0 | 263·3 | 259·6 | 256·0 |
| 13 | 329·2 | 320·1 | 311·3 | 302·8 | 294·4 | 286·3 | 278·5 | 270·8 | 265·2 | 261·5 | 257·9 | 254·3 | 250·8 |
| 14 | 322·0 | 313·2 | 304·7 | 296·3 | 288·2 | 280·4 | 272·7 | 265·3 | 259·8 | 256·2 | 252·7 | 249·2 | 245·8 |
| 15 | 315·1 | 306·5 | 298·2 | 290·1 | 282·3 | 274·6 | 267·1 | 259·9 | 254·6 | 251·1 | 247·7 | 244·3 | 240·9 |
| 16 | 308·5 | 300·1 | 292·0 | 284·2 | 276·5 | 269·1 | 261·8 | 254·7 | 249·6 | 246·2 | 242·8 | 239·5 | 236·3 |
| 18 | 295·9 | 288·0 | 280·3 | 272·9 | 265·6 | 258·5 | 251·7 | 245·0 | 240·1 | 236·9 | 233·7 | 230·6 | 227·5 |
| 20 | 284·2 | 276·7 | 269·5 | 262·4 | 255·5 | 248·8 | 242·3 | 235·9 | 231·2 | 228·2 | 225·2 | 222·2 | 219·3 |
| 22 | 273·4 | 266·3 | 259·4 | 252·7 | 246·1 | 239·7 | 233·5 | 227·5 | 223·0 | 220·1 | 217·3 | 214·4 | 211·6 |
| 24 | 263·3 | 256·5 | 250·0 | 243·6 | 237·3 | 231·3 | 225·4 | 219·6 | 215·4 | 212·6 | 209·9 | 207·2 | 204·5 |
| 26 | 253·8 | 247·4 | 241·2 | 235·1 | 229·2 | 223·4 | 217·7 | 212·2 | 208·2 | 205·6 | 202·9 | 200·4 | 197·8 |
| 28 | 245·0 | 238·9 | 232·9 | 227·1 | 221·5 | 215·9 | 210·6 | 205·3 | 201·5 | 198·9 | 196·4 | 194·0 | 191·5 |
| 30 | 236·7 | 230·9 | 225·2 | 219·7 | 214·2 | 209·0 | 203·8 | 198·8 | 195·1 | 192·7 | 190·3 | 188·0 | 185·7 |
| 32 | 228·9 | 223·4 | 217·9 | 212·6 | 207·5 | 202·4 | 197·5 | 192·7 | 189·2 | 186·9 | 184·6 | 182·3 | 180·1 |

TABLE A8.3

Solubility of argon in sea water ($\mu$ mol kg$^{-1}$) based on equation (8.11)

| $T°C$ | \multicolumn{13}{c}{Salinity (‰)} | | | | | | | | | | | | |
|---|---|---|---|---|---|---|---|---|---|---|---|---|---|
| | 0 | 4 | 8 | 12 | 16 | 20 | 24 | 28 | 31 | 33 | 35 | 37 | 39 |
| −1 | 22·87 | 22·17 | 21·49 | 20·84 | 20·20 | 19·58 | 18·98 | 18·40 | 17·98 | 17·70 | 17·43 | 17·16 | 16·90 |
| 0 | 22·24 | 21·56 | 20·91 | 20·27 | 19·66 | 19·06 | 18·48 | 17·92 | 17·51 | 17·24 | 16·98 | 16·72 | 16·46 |
| 1 | 21·63 | 20·98 | 20·35 | 19·73 | 19·14 | 18·56 | 18·00 | 17·46 | 17·07 | 16·81 | 16·55 | 16·30 | 16·05 |
| 2 | 21·05 | 20·42 | 19·81 | 19·22 | 18·64 | 18·09 | 17·54 | 17·02 | 16·64 | 16·39 | 16·14 | 15·90 | 15·66 |
| 3 | 20·50 | 19·89 | 19·30 | 18·73 | 18·17 | 17·63 | 17·11 | 16·60 | 16·23 | 15·98 | 15·75 | 15·51 | 15·28 |
| 4 | 19·97 | 19·39 | 18·81 | 18·26 | 17·72 | 17·20 | 16·69 | 16·20 | 15·84 | 15·60 | 15·37 | 15·14 | 14·92 |
| 5 | 19·47 | 18·90 | 18·34 | 17·81 | 17·29 | 16·78 | 16·29 | 15·81 | 15·46 | 15·23 | 15·01 | 14·79 | 14·57 |
| 6 | 18·99 | 18·43 | 17·90 | 17·38 | 16·87 | 16·38 | 15·90 | 15·44 | 15·10 | 14·88 | 14·66 | 14·45 | 14·24 |
| 7 | 18·52 | 17·99 | 17·47 | 16·96 | 16·47 | 16·00 | 15·53 | 15·09 | 14·76 | 14·54 | 14·33 | 14·12 | 13·92 |
| 8 | 18·08 | 17·56 | 17·06 | 16·57 | 16·09 | 15·63 | 15·18 | 14·75 | 14·43 | 14·22 | 14·02 | 13·81 | 13·61 |
| 9 | 17·65 | 17·15 | 16·66 | 16·19 | 15·73 | 15·28 | 14·84 | 14·42 | 14·11 | 13·91 | 13·71 | 13·51 | 13·32 |
| 10 | 17·24 | 16·76 | 16·28 | 15·82 | 15·38 | 14·94 | 14·52 | 14·11 | 13·81 | 13·61 | 13·42 | 13·23 | 13·04 |
| 11 | 16·85 | 16·38 | 15·92 | 15·47 | 15·04 | 14·62 | 14·21 | 13·81 | 13·51 | 13·32 | 13·14 | 12·95 | 12·77 |
| 12 | 16·47 | 16·02 | 15·57 | 15·13 | 14·71 | 14·30 | 13·90 | 13·52 | 13·23 | 13·05 | 12·86 | 12·68 | 12·50 |
| 13 | 16·11 | 15·67 | 15·23 | 14·81 | 14·40 | 14·00 | 13·61 | 13·24 | 12·96 | 12·78 | 12·60 | 12·43 | 12·25 |
| 14 | 15·77 | 15·33 | 14·91 | 14·50 | 14·10 | 13·71 | 13·34 | 12·97 | 12·70 | 12·53 | 12·35 | 12·18 | 12·01 |
| 15 | 15·43 | 15·01 | 14·60 | 14·20 | 13·81 | 13·44 | 13·07 | 12·71 | 12·45 | 12·28 | 12·11 | 11·94 | 11·78 |
| 16 | 15·11 | 14·70 | 14·30 | 13·91 | 13·53 | 13·17 | 12·81 | 12·46 | 12·21 | 12·04 | 11·88 | 11·72 | 11·56 |
| 18 | 14·50 | 14·11 | 13·73 | 13·36 | 13·01 | 12·66 | 12·32 | 11·99 | 11·75 | 11·59 | 11·44 | 11·28 | 11·13 |
| 20 | 13·93 | 13·56 | 13·20 | 12·86 | 12·52 | 12·19 | 11·87 | 11·55 | 11·32 | 11·17 | 11·03 | 10·88 | 10·73 |
| 22 | 13·40 | 13·05 | 12·71 | 12·38 | 12·06 | 11·75 | 11·44 | 11·14 | 10·93 | 10·78 | 10·64 | 10·50 | 10·37 |
| 24 | 12·91 | 12·58 | 12·25 | 11·94 | 11·63 | 11·34 | 11·04 | 10·76 | 10·55 | 10·42 | 10·28 | 10·15 | 10·02 |
| 26 | 12·45 | 12·13 | 11·83 | 11·53 | 11·23 | 10·95 | 10·67 | 10·40 | 10·21 | 10·08 | 9·95 | 9·82 | 9·70 |
| 28 | 12·01 | 11·71 | 11·42 | 11·14 | 10·86 | 10·59 | 10·32 | 10·07 | 9·88 | 9·75 | 9·63 | 9·51 | 9·39 |
| 30 | 11·61 | 11·32 | 11·04 | 10·77 | 10·51 | 10·25 | 10·00 | 9·75 | 9·57 | 9·45 | 9·33 | 9·22 | 9·10 |
| 32 | 11·23 | 10·95 | 10·69 | 10·43 | 10·17 | 9·93 | 9·69 | 9·45 | 9·28 | 9·16 | 9·05 | 8·94 | 8·83 |

TABLE A8.4

Solubility of neon in sea water (nmol kg$^{-1}$) based on equation (8.11)

| T°C | Salinity (‰) | | | | | | | | | | | | |
|---|---|---|---|---|---|---|---|---|---|---|---|---|---|
|  | 0 | 4 | 8 | 12 | 16 | 20 | 24 | 28 | 31 | 33 | 35 | 37 | 39 |
| −1 | 10·16 | 9·88 | 9·61 | 9·34 | 9·08 | 8·84 | 8·59 | 8·36 | 8·18 | 8·07 | 7·96 | 7·85 | 7·74 |
| 0 | 10·03 | 9·76 | 9·49 | 9·24 | 8·98 | 8·74 | 8·50 | 8·27 | 8·10 | 7·99 | 7·88 | 7·78 | 7·67 |
| 1 | 9·91 | 9·64 | 9·39 | 9·13 | 8·89 | 8·65 | 8·42 | 8·19 | 8·03 | 7·92 | 7·81 | 7·71 | 7·60 |
| 2 | 9·79 | 9·53 | 9·28 | 9·04 | 8·80 | 8·56 | 8·34 | 8·11 | 7·95 | 7·85 | 7·74 | 7·64 | 7·54 |
| 3 | 9·68 | 9·43 | 9·18 | 8·94 | 8·71 | 8·48 | 8·26 | 8·04 | 7·88 | 7·78 | 7·68 | 7·57 | 7·47 |
| 4 | 9·57 | 9·33 | 9·08 | 8·85 | 8·62 | 8·40 | 8·18 | 7·97 | 7·81 | 7·71 | 7·61 | 7·51 | 7·41 |
| 5 | 9·47 | 9·23 | 8·99 | 8·76 | 8·54 | 8·32 | 8·10 | 7·90 | 7·74 | 7·64 | 7·55 | 7·45 | 7·35 |
| 6 | 9·37 | 9·13 | 8·90 | 8·67 | 8·46 | 8·24 | 8·03 | 7·83 | 7·68 | 7·58 | 7·49 | 7·39 | 7·30 |
| 7 | 9·27 | 9·04 | 8·81 | 8·59 | 8·38 | 8·17 | 7·96 | 7·76 | 7·62 | 7·52 | 7·43 | 7·33 | 7·24 |
| 8 | 9·18 | 8·95 | 8·73 | 8·51 | 8·30 | 8·10 | 7·90 | 7·70 | 7·56 | 7·46 | 7·37 | 7·28 | 7·19 |
| 9 | 9·09 | 8·86 | 8·65 | 8·43 | 8·23 | 8·03 | 7·83 | 7·64 | 7·50 | 7·40 | 7·31 | 7·22 | 7·13 |
| 10 | 9·00 | 8·78 | 8·57 | 8·36 | 8·16 | 7·96 | 7·76 | 7·58 | 7·44 | 7·35 | 7·26 | 7·17 | 7·08 |
| 11 | 8·91 | 8·70 | 8·49 | 8·29 | 8·09 | 7·89 | 7·70 | 7·52 | 7·38 | 7·29 | 7·20 | 7·12 | 7·03 |
| 12 | 8·83 | 8·62 | 8·41 | 8·21 | 8·02 | 7·83 | 7·64 | 7·46 | 7·33 | 7·24 | 7·15 | 7·07 | 6·98 |
| 13 | 8·75 | 8·54 | 8·34 | 8·14 | 7·95 | 7·77 | 7·58 | 7·40 | 7·27 | 7·19 | 7·10 | 7·02 | 6·93 |
| 14 | 8·67 | 8·47 | 8·27 | 8·08 | 7·89 | 7·71 | 7·53 | 7·35 | 7·22 | 7·14 | 7·05 | 6·97 | 6·89 |
| 15 | 8·60 | 8·40 | 8·20 | 8·01 | 7·83 | 7·65 | 7·47 | 7·30 | 7·17 | 7·08 | 7·00 | 6·92 | 6·84 |
| 16 | 8·53 | 8·33 | 8·14 | 7·95 | 7·77 | 7·59 | 7·42 | 7·24 | 7·12 | 7·04 | 6·96 | 6·88 | 6·80 |
| 18 | 8·39 | 8·20 | 8·01 | 7·83 | 7·65 | 7·48 | 7·31 | 7·14 | 7·02 | 6·94 | 6·86 | 6·78 | 6·71 |
| 20 | 8·26 | 8·07 | 7·89 | 7·71 | 7·54 | 7·37 | 7·21 | 7·05 | 6·93 | 6·85 | 6·77 | 6·70 | 6·62 |
| 22 | 8·13 | 7·95 | 7·78 | 7·60 | 7·44 | 7·27 | 7·11 | 6·95 | 6·84 | 6·76 | 6·69 | 6·61 | 6·54 |
| 24 | 8·01 | 7·84 | 7·67 | 7·50 | 7·33 | 7·17 | 7·02 | 6·86 | 6·75 | 6·68 | 6·60 | 6·53 | 6·46 |
| 26 | 7·90 | 7·73 | 7·56 | 7·40 | 7·24 | 7·08 | 6·93 | 6·78 | 6·66 | 6·59 | 6·52 | 6·45 | 6·38 |
| 28 | 7·79 | 7·62 | 7·46 | 7·30 | 7·14 | 6·99 | 6·84 | 6·69 | 6·58 | 6·51 | 6·44 | 6·37 | 6·30 |
| 30 | 7·69 | 7·52 | 7·36 | 7·20 | 7·05 | 6·90 | 6·75 | 6·60 | 6·50 | 6·43 | 6·36 | 6·29 | 6·22 |
| 32 | 7·59 | 7·43 | 7·27 | 7·11 | 6·96 | 6·81 | 6·66 | 6·52 | 6·42 | 6·35 | 6·28 | 6·21 | 6·15 |

TABLE A8.5

Solubility of helium in sea water (nmol kg$^{-1}$) based on equation (8.11).

| T°C | Salinity (‰) | | | | | | | | | | | | |
|---|---|---|---|---|---|---|---|---|---|---|---|---|---|
| | 0 | 4 | 8 | 12 | 16 | 20 | 24 | 28 | 31 | 33 | 35 | 37 | 39 |
| −1 | 2·20 | 2·15 | 2·10 | 2·04 | 2·00 | 1·95 | 1·90 | 1·86 | 1·82 | 1·80 | 1·78 | 1·76 | 1·74 |
| 0 | 2·18 | 2·13 | 2·08 | 2·03 | 1·98 | 1·94 | 1·89 | 1·84 | 1·81 | 1·79 | 1·77 | 1·75 | 1·73 |
| 1 | 2·17 | 2·12 | 2·07 | 2·02 | 1·97 | 1·93 | 1·88 | 1·84 | 1·80 | 1·78 | 1·76 | 1·74 | 1·72 |
| 2 | 2·16 | 2·11 | 2·06 | 2·01 | 1·96 | 1·92 | 1·87 | 1·83 | 1·79 | 1·77 | 1·75 | 1·73 | 1·71 |
| 3 | 2·14 | 2·10 | 2·05 | 2·00 | 1·95 | 1·91 | 1·86 | 1·82 | 1·79 | 1·77 | 1·74 | 1·72 | 1·70 |
| 4 | 2·13 | 2·08 | 2·04 | 1·99 | 1·94 | 1·90 | 1·85 | 1·81 | 1·78 | 1·76 | 1·74 | 1·72 | 1·70 |
| 5 | 2·12 | 2·07 | 2·02 | 1·98 | 1·93 | 1·89 | 1·84 | 1·80 | 1·77 | 1·75 | 1·73 | 1·71 | 1·69 |
| 6 | 2·11 | 2·06 | 2·02 | 1·97 | 1·92 | 1·88 | 1·84 | 1·80 | 1·76 | 1·74 | 1·72 | 1·70 | 1·68 |
| 7 | 2·10 | 2·05 | 2·01 | 1·96 | 1·92 | 1·87 | 1·83 | 1·79 | 1·76 | 1·74 | 1·72 | 1·70 | 1·68 |
| 8 | 2·09 | 2·04 | 2·00 | 1·95 | 1·91 | 1·86 | 1·82 | 1·78 | 1·75 | 1·73 | 1·71 | 1·69 | 1·67 |
| 9 | 2·08 | 2·03 | 1·99 | 1·94 | 1·90 | 1·86 | 1·82 | 1·78 | 1·75 | 1·73 | 1·71 | 1·69 | 1·67 |
| 10 | 2·07 | 2·02 | 1·98 | 1·94 | 1·89 | 1·85 | 1·81 | 1·77 | 1·74 | 1·72 | 1·70 | 1·68 | 1·66 |
| 11 | 2·06 | 2·02 | 1·97 | 1·93 | 1·89 | 1·84 | 1·80 | 1·76 | 1·74 | 1·72 | 1·70 | 1·68 | 1·66 |
| 12 | 2·05 | 2·01 | 1·96 | 1·92 | 1·88 | 1·84 | 1·80 | 1·76 | 1·73 | 1·71 | 1·69 | 1·67 | 1·66 |
| 13 | 2·04 | 2·00 | 1·96 | 1·91 | 1·87 | 1·83 | 1·79 | 1·75 | 1·72 | 1·71 | 1·69 | 1·67 | 1·65 |
| 14 | 2·04 | 1·99 | 1·95 | 1·91 | 1·87 | 1·83 | 1·79 | 1·75 | 1·72 | 1·70 | 1·68 | 1·67 | 1·65 |
| 15 | 2·03 | 1·99 | 1·94 | 1·90 | 1·86 | 1·82 | 1·78 | 1·74 | 1·71 | 1·70 | 1·68 | 1·66 | 1·64 |
| 16 | 2·02 | 1·98 | 1·94 | 1·90 | 1·86 | 1·82 | 1·78 | 1·74 | 1·71 | 1·69 | 1·68 | 1·66 | 1·64 |
| 18 | 2·01 | 1·97 | 1·92 | 1·88 | 1·84 | 1·81 | 1·77 | 1·73 | 1·70 | 1·69 | 1·67 | 1·65 | 1·63 |
| 20 | 2·00 | 1·95 | 1·91 | 1·87 | 1·84 | 1·80 | 1·76 | 1·72 | 1·70 | 1·68 | 1·66 | 1·64 | 1·63 |
| 22 | 1·98 | 1·94 | 1·90 | 1·86 | 1·83 | 1·79 | 1·75 | 1·72 | 1·69 | 1·67 | 1·66 | 1·64 | 1·62 |
| 24 | 1·97 | 1·93 | 1·89 | 1·86 | 1·82 | 1·78 | 1·75 | 1·71 | 1·68 | 1·67 | 1·65 | 1·63 | 1·62 |
| 26 | 1·96 | 1·92 | 1·88 | 1·85 | 1·81 | 1·77 | 1·74 | 1·70 | 1·68 | 1·66 | 1·64 | 1·63 | 1·61 |
| 28 | 1·95 | 1·91 | 1·88 | 1·84 | 1·80 | 1·77 | 1·73 | 1·70 | 1·67 | 1·66 | 1·64 | 1·62 | 1·61 |
| 30 | 1·94 | 1·90 | 1·87 | 1·83 | 1·80 | 1·76 | 1·73 | 1·69 | 1·67 | 1·65 | 1·64 | 1·62 | 1·60 |
| 32 | 1·93 | 1·90 | 1·86 | 1·82 | 1·79 | 1·75 | 1·72 | 1·69 | 1·66 | 1·65 | 1·63 | 1·62 | 1·60 |

REFERENCES

Aldrich, L. T. and Nier, A. O. (1948). *Phys. Rev.* **74**, 1590.

Atkinson, L. P. (1972). "Air bubbles in an oceanic mixed layer: Effect on gas concentrations and air–sea gas exchange". Ph.D. Thesis, Dalhousie University, Halifax, Nova Scotia, 117 pp.

Atkinson, L. P. (1973). *J. Geophys. Res.* **78**, 962.

Atkinson, L. P. and Richards, F. A. (1967). *Deep-Sea Res.* **14**, 673.

Bates, D. R. and Hays, P. B. (1967). *Planet. Space Sci.* **15**, 189.

Beattie, J. A. (1961). *In* "Argon, Helium and the Rare Gases" (G. A. Cook, ed.), Vol. I, pp. 251–312. Interscience Publishers, New York.

Benson, B. B. (1965). *In* "Occasional Publication No. 3" (D. R. Schink and J. T. Corless, eds.), pp. 91–107. Univ. Rhode Island, Kingston.

Benson, B. B. and Parker, P. D. M. (1961). *Deep-Sea Res.* **7**, 237.

Bieri, R., Koide, M. and Goldberg, E. D. (1964). *Science*, **146**, 1035.

Bieri, R. H., Koide, M. and Goldberg, E. D. (1966). *J. Geophys. Res.* **71**, 5243.

Bieri, R., Koide, M. and Goldberg, E. D. (1968). *Earth Planet. Sci. Letts*, **4**, 329.

Bieri, R. H. (1971). *Earth Planet. Sci. Lett.* **10**, 329.

Bieri, R. H. and Koide, M. (1972). *J. Geophys. Res.* **77**, 1667.

Broecker, W. S. (1965). *In* "Symposium on Diffusion in Oceans and Fresh Waters" (T. Ichiye, ed.), pp. 116–145. Lamont Geological Observatory, Palisades, N.Y.

Broecker, W. S. (1966). *In* "Abstracts of Papers, Second International Oceanographic Congress", pp. 62–63. U.S.S.R. Acad. Sci., Moscow.

Broecker, W. S. and Kaufman, A. (1970). *J. Geophys. Res.* **75**, 7679.

Broecker, W. S. and Peng, T.-H. (1971). *Earth Planet. Sci. Lett.* **11**, 99.

Broenkow, W. W. (1965). *Limnol. Oceanogr.* **10**, 40.

Buch, K. (1929). *J. Cons. int. Explor. Mer.* **4**, 162.

Carpenter, J. H. (1965a). *Limnol. Oceanogr.* **10**, 135.

Carpenter, J. H. (1965b). *Limnol. Oceanogr.* **10**, 141.

Carpenter, J. H. (1966). *Limnol. Oceanogr.* **11**, 264.

Carritt, D. E. (1954). *Deep-Sea Res.* **2**, 59.

Carritt, D. E. and Carpenter, J. H. (1966). *J. Mar. Res.* **24**, 286.

Castellan, G. W. (1971). "Physical Chemistry", 2nd edition, 866 pp. Addison-Wesley, Reading, Mass.

Clarke, W. B., Beg, M. A. and Craig, H. (1969). *Earth Planet. Sci. Lett.* **6**, 213.

Clarke, W. B., Beg, M. A. and Craig, H. (1970). *J. Geophys. Res.* **75**, 7676.

Cockett, A. H. and Smith, K. C. (1973). *In* "Comprehensive Inorganic Chemistry" (A. F. Trotman-Dickenson, ed.), Vol. 1, pp. 139–211. Pergamon Press, Oxford.

Cook, G. A. (1961). *In* "Argon, Helium and the Rare Gases" (G. A. Cook, ed.) Vol. I, pp. 35–64, Interscience Publishers, New York.

Cooper, L. H. N. (1967). *Sci. Prog., Oxf.* **55**, 73.

Craig, H., Weiss, R. F. and Clarke, W. B. (1967). *J. Geophys. Res.* **72**, 6165.

Craig, H. (1969). *J. Geophys. Res.* **74**, 5491.

Craig, H. (1971). *J. Geophys. Res.* **76**, 5078.

Craig, H. and Clarke, W. B. (1970). *Earth Planet. Sci. Lett.* **9**, 45.

Craig, H. and Gordon, L. I. (1963). *Geochim. Cosmochim. Acta*, **27**, 949.

Craig, H. and Weiss, R. F. (1968). *Earth Planet. Sci. Lett.* **5**, 175.

Craig, H. and Weiss, R. F. (1971). *Earth Planet. Sci. Lett.* **10**, 289.

Culberson, C. H. (1972). "Processes Affecting the Oceanic Distribution of Carbon Dioxide". Ph.D. Thesis, Oregon State University, Corvallis, 178 pp.

Culberson, C. and Pytkowicz, R. M. (1970). *J. Oceanogr. Soc. Japan*, **26**, 95.

Douglas, E. (1964). *J. Phys. Chem.* **68**, 169.

Douglas, E. (1965). *J. Phys. Chem.* **69**, 2608.

Douglas, E. (1967). *J. Phys. Chem.* **71**, 1931.

Downing, A. L. and Truesdale, G. A. (1955). *J. Appl. Chem.* **5**, 570.

Duedall, I. W. and Coote, A. R. (1972). *J. Geophys. Res.* **77**, 2201.

Ewing, G. and McAlister, E. D. (1960). *Science*, **131**, 1374.

Fairhall, A. W. (1969). *Earth Planet. Sci. Lett.* **7**, 249.

Fogg, G. E. and Stewart, W. D. P. (1965). *Sci. Prog. Oxf.* **53**, 191.

Fox, C. J. J. (1907). *Publ. Circon. Con. Perm. Intern. Exp. Mer,* No. 41.

Fry, B. A. (1955). "The Nitrogen Metabolism of Micro-Organisms", 166 pp. Methuen, London.

Glueckauf, E. (1951). *In* "Compendium of Meteorology" (T. F. Malone, ed.), pp. 3–10. American Meteorological Society, Boston.

Green, E. J. (1965). "A Redetermination of the Solubility of Oxygen in Sea Water and Some Thermodynamic Implications of the Solubility Relations." Ph.D. Thesis, Mass. Inst. Tech., Cambridge, 137 pp.

Green, E. J. and Carritt, D. E. (1966). *Analyst*, **91**, 207.

Green, E. J. and Carritt, D. E. (1967). *J. Mar. Res.* **25**, 140.

Hahn, J. (1972). *In* "The Changing Chemistry of the Oceans" (D. Dyrssen and D. Jagner, eds.), pp. 53–69. John Wiley, New York.

Hamm, R. E. and Thompson, T. G. (1941). *J. Mar. Res.* **4**, 11.

Helland-Hansen, B., Jacobsen, J. P. and Thompson, T. G. (1948). *Publ. Sci. Assoc. Oceanogr. Phys.* No. 9, 28 pp.

Higbie, R. (1935). *Amer. Inst. Chem. Eng. Trans.* **31**, 365.

Inman, R. E., Ingersoll, R. B. and Levy, E. A. (1971). *Science*, **172**, 1229.

Jaffee, L. S. (1970). *In* "Global Effects of Environmental Pollution" (S. F. Singer, ed.), pp. 34–49. Springer-Verlag, New York.

Jenkins, W. J., Beg, M. A., Clarke, W. B., Wangersky, P. J. and Craig, H. (1972). *Earth Planet Sci. Lett.* **16**, 122.

Johnson, H. E. and Axford, W. I. (1969). *J. Geophys. Res.* **74**, 2433.

Junge, C. E. (1963). "Air Chemistry and Radioactivity", 382 pp. Academic Press, New York.

Junge, C., Bockholt, B., Schütz, K. and Beck, R. (1971a). *In* "Meteor Forschungsergebnisse Reihe B: Meteorologie und Aeronomie", No. 6, pp. 1–11.

Junge, C., Seiler, W. and Warneck, P. (1971b). *J. Geophys. Res.* **76**, 2866.

Junge, C. and Hahn, J. (1971). *J. Geophys. Res.* **76**, 8143.

Kanwisher, J. (1963). *Deep-Sea Res.* **10**, 195.

Kester, D. R. (1972). "Physical and Chemical Data Report, Cruise 114, R/V TRIDENT." Univ. Rhode Island, Kingston.

Kester, D. R. and Pytkowicz, R. M. (1968). *J. Geophys. Res.* **73**, 5421.

Kester, D. R., Crocker, K. T. and Miller, G. R. (1973). *Deep-Sea Res.* **20**, 409.

Klots, C. E. and Benson, B. B. (1963). *J. Mar. Res.* **21**, 48.

König, H. (1963). *Z. Natur.* **18a**, 363.

König, H., Wänke, H., Bien, G. S., Rakestraw, N. W. and Suess, H. E. (1964). *Deep Sea Res.* **11**, 243.

Lambert, R. B., Kester, D. R., Pilson, M. E. Q. and Kenyon, K. E. (1973). *J. Geophys. Res.* **78**, 1479.

Lamontagne, R. A., Swinnerton, J. W. and Linnenbom, V. J. (1971). *J. Geophys. Res.* **76**, 5117.

Lieberman, L. (1957). *J. Appl. Phys.* **28**, 205.

Linnenbom, V. J., Swinnerton, J. W. and Cheek, C. H. (1966). "Statistical evaluation of gas chromatography for the determination of dissolved gases in sea water", U.S. Naval Research Laboratory Report 6344, Washington, D.C., 16 pp.

Linnenbom, V. J. and Swinnerton, J. W. (1968). *In* "Symposium on Organic Matter in Natural Waters" (D. W. Hood, ed.), pp. 455–467. Univ. of Alaska, College, Alaska.

Mazor, E., Wasserburg, G. J. and Craig, H. (1964). *Deep-Sea Res.* **11**, 929.

Meadows, R. W. and Spedding, D. J. (1974). *Tellus*, **26**, 143.

Medwin, H. (1970). *J. Geophys. Res.* **75**, 599.

Menzel, D. W. (1970). *Deep-Sea Res.* **17**, 751.

Menzel, D. W. and Ryther, J. H. (1968). *Deep-Sea Res.* **15**, 327.

Murray, C. N. and Riley, J. P. (1969). *Deep-Sea Res.* **16**, 311.

Murray, C. N. and Riley, J. P. (1970). *Deep-Sea Res.* **17**, 203.

Murray, C. N., Riley, J. P. and Wilson, T. R. S. (1969). *Deep-Sea Res.* **16**, 297.

Packard, T. T. (1969). "The estimation of the oxygen utilization rate in seawater from the activity of the respiratory electron transport system in plankton". Ph.D. Thesis, University of Washington, Seattle, 115 pp.

Park, P. K. (1968). *J. Oceanol. Soc. Korea*, 3, 1.

Pytkowicz, R. M. (1964). *Deep-Sea Res.* **11**, 381.

Pytkowicz, R. M. (1968). *J. Oceanogr. Soc. Japan* **24**, 21.

Pytkowicz, R. M. and Kester, D. R. (1966). *Deep-Sea Res.* **13**, 373.

Rakestraw, N. W. (1964). *In* "Recent Researches in the Fields of Hydrosphere, Atmosphere and Nuclear Geochemistry, Ken Sugawara Festival Volume" (Y. Miyake and T. Koyama, eds.), pp. 243–255. Maruzen Co., Tokyo.

Rakestraw, N. W. and Emmel, E. V. M. (1938a). *J. Mar. Res.* **1**, 207.

Rakestraw, N. W. and Emmel, E. V. M. (1938b). *J. Phys. Chem.* **42**, 1211.

Redfield, A. C. (1934). *In* "James Johnstone Memorial Volume" (R. J. Daniel, ed.), pp. 176–192. University Press, Liverpool.

Redfield, A. C. (1942). *Pap. phys. Oceanogr.* **No. 9**, 22 pp.

Redfield, A. C. (1948). *J. Mar. Res.* **7**, 347.

Redfield, A. C. (1958). *Amer. Scient.* **46**, 205.

Redfield, A. C., Ketchum, B. H. and Richards, F. A. (1963). *In* "The Sea: Ideas and Observations on Progress in the Study of the Seas" (M. N. Hill, ed.), Vol. II, pp. 26–77. Interscience Publishers, New York.

Reid, J. L. (1965). "The Johns Hopkins Oceanographic Studies", No. 2, 85 pp. Johns Hopkins Press, Baltimore.

Revelle, R. and Suess, H. E. (1962). *In* "The Sea: Ideas and Observations on Progress in the Study of the Seas" (M. N. Hill, ed.), Vol. I, pp. 313–321. Interscience Publishers, New York.

Richards, F. A. (1957). *In* "Treatise on Marine Ecology and Paleoecology" (J. W. Hedgpeth, ed.), Vol. II, pp. 185–238. Memoir 67, Geological Society of America, New York.

Richards, F. A. (1965). *In* "Chemical Oceanography" (J. P. Riley and G. Skirrow, eds.), Vol. I, pp. 197–225. Academic Press, London.

Richards, F. A. and Benson, B. B. (1961). *Deep-Sea Res.* **7**, 254.

Richards, F. A. and Corwin, N. (1956). *Limnol. Oceanogr.* **1**, 263.

Riley, G. A. (1951). *Bull. Bingham Oceanogr. Coll.* **13**, 1.

Robinson, E. and Robbins, R. C. (1968). *Antarct. J. U.S.* **3**, 194.

Robinson, E. and Robbins, R. C. (1969). "Sources, Abundance and Fate of Gaseous Atmospheric Pollutants; Supplement", Stanford Research Institute Project PR-6755, Stanford Univ., California.

Schink, D. R., Guinasso, N. L. and Charnell, R. L. (1968). *Amer. Geophys. Un., Trans.* **49**, 205.

Schütz, K., Junge, C., Beck, R. and Albrecht, B. (1970). *J. Geophys. Res.* **75**, 2230.

Suess, H. E. and Wänke, H. (1965). In "Progress in Oceanography" (M. Sears, ed.), Vol. 3, pp. 347–353, Pergamon Press, New York.

Sugawara, K. and Tochikubo, I. (1955). *J. Earth Sci., Nagoya Univ.* **3**, 77.

Sugiura, Y. (1964). In "Recent Researches in the Fields of Hydrosphere, Atmosphere and Nuclear Geochemistry, Ken Sugawara Festival Volume" (Y. Miyake and T. Koyama, eds.), pp. 49–63. Maruzen, Co. Tokyo.

Sugiura, Y. (1965). *Pap. Meteorol. Geophys.* **15**, 208.

Sverdrup, H. U., Johnson, M. W. and Fleming, R. H. (1942). "The Oceans", 1087 pp. Prentice-Hall, Englewood Cliffs, New Jersey.

Swinnerton, J. W., and Lamontagne, R. A. (1974). *Tellus*, **26**, 136.

Swinnerton, J. W., Linnenbom, V. J. and Cheek, C. H. (1969). *Envir. Sci. Tech.* **3**, 836.

Swinnerton, J. W., Linnenbom, V. J. and Lamontagne, R. A. (1970a). *Science*, **167**, 984.

Swinnerton, J. W., Linnenbom, V. J. and Lamontagne, R. A. (1970b). *N.Y. Acad. Sci., Annals*, **174**, 96.

Swinnerton, J. W. and Linnenbom, V. J. (1967). *Science*, **156**, 1119.

Valley, S. L. (1965). "Handbook of Geophysics and Space Environments." McGraw-Hill, New York, 683 pp.

Wattenberg, H. (1939). "Atlas zu: Die Verteilung des Sauerstoffs im Atlantischen Ozean". *Wiss. Ergebn. dtsch. atlant. Exped. Meteor* 1925–27, **9**, 81 pp. Verlag Walter de Gruyter, Berlin.

Weast, R. C. (1969). "Handbook of Chemistry and Physics", 50th edition, 2356 pp. Chemical Rubber Co., Cleveland. Ohio.

Weinstock, B. (1969). *Science*, **166**, 224.

Weinstock, B. and Niki, H. (1972). *Science*, **176**, 290.

Weiss, R. F. (1970). *Deep-Sea Res.* **17**, 721.

Weiss, R. F. (1971a). *Deep-Sea Res.* **18**, 225.

Weiss, R. F. (1971b). *J. Chem. Eng. Data*, **16**, 235.

Whitman, W. G. (1923). *Chem. Metall. Eng.* **29**, 146.

Wilson, D. F., Swinnerton, J. W. and Lamontagne, R. A. (1970). *Science*, **168**, 1577.

Wood, D. and Caputi, R. (1966). "Technical Report No. 988", 14 pp. U.S. Naval Radiological Defense Laboratory, San Francisco.

Wyman, J., Scholander, P. F., Edwards, G. A. and Irving, L. (1952). *J. Mar. Res.* **11**, 47.

Wyrtki, K. (1962). *Deep-Sea Res.* **9**, 11.

# Appendix

Tables of physical and chemical constants relevant to marine chemistry

TABLE 1

TABLE 1

*Some physical properties of pure water (after Dorsey, 1940)*

| | |
|---|---|
| Molecular weight | 18·0153 |
| Heat of formation | 285·89 kJmol$^{-1}$ (at 25°C and 1 atm) |
| Ionic dissociation constant | 10$^{-4}$ $M^{-1}$ (at 25°C and 1 atm) |
| Heat of ionization | 55·71 kJmol$^{-1}$ (at 25°C and 1 atm) |
| Viscosity | 8·949 mP (at 25°C and 1 atm) |
| Velocity of sound | 1496·3 ms$^{-1}$ (at 25°C and 1 atm) |
| Density | 0·9979751 g cm$^{-3}$ (at 25°C and 1 atm) |
| Freezing point | 0°C (at 1 atm) |
| Boiling point | 100°C (at 1 atm) |
| Isothermal compressibility | 45·6 × 10$^{-6}$ atm$^{-1}$ (at 25°C over the range 1–10 atm) |
| Specific heat at constant volume | 4·1786 int.J (g°C)$^{-1}$ (at 25°C and 1 atm) |
| Thermal conductivity | 0·00598 W cm$^{-1}$ °C$^{-1}$ (at 20°C and 1 atm) |
| Temperature of maximum density | 3·98°C (at 1 atm) |
| Dielectric constant | 81·0 (at 1 atm, 17°C, and 60 MHz) |
| Electrical conductivity | Less than 10$^{-8}$ Ω$^{-1}$ cm$^{-1}$ (at 25°C and 1 atm) |

TABLE 2

Concentrations of the major ions in sea water of various salinities $(g\ kg^{-1})$*

| Salinity (‰) | Na$^+$ | Mg$^{2+}$ | Ca$^{2+}$ | K$^+$ | Sr$^{2+}$ | B | Cl$^-$ | SO$_4^{2-}$ | Br$^-$ | F$^-$ | HCO$_3^-$ |
|---|---|---|---|---|---|---|---|---|---|---|---|
| 5 | 1·539 | 0·185 | 0·058 | 0·057 | 0·001 | 0·001 | 2·763 | 0·387 | 0·010 | 0·0002 | 0·020 |
| 10 | 3·078 | 0·370 | 0·118 | 0·114 | 0·002 | 0·001 | 5·527 | 0·775 | 0·019 | 0·0004 | 0·041 |
| 15 | 4·617 | 0·555 | 0·177 | 0·171 | 0·003 | 0·002 | 8·290 | 1·162 | 0·029 | 0·0005 | 0·061 |
| 20 | 6·156 | 0·739 | 0·235 | 0·228 | 0·005 | 0·003 | 11·054 | 1·550 | 0·038 | 0·0007 | 0·081 |
| 25 | 7·695 | 0·924 | 0·294 | 0·285 | 0·006 | 0·003 | 13·817 | 1·937 | 0·048 | 0·0009 | 0·101 |
| 30 | 9·234 | 1·109 | 0·353 | 0·342 | 0·007 | 0·004 | 16·581 | 2·325 | 0·058 | 0·0011 | 0·122 |
| 31 | 9·542 | 1·146 | 0·365 | 0·353 | 0·007 | 0·004 | 17·133 | 2·402 | 0·059 | 0·0011 | 0·126 |
| 32 | 9·850 | 1·183 | 0·377 | 0·365 | 0·007 | 0·004 | 17·685 | 2·480 | 0·062 | 0·0012 | 0·130 |
| 33 | 10·157 | 1·220 | 0·388 | 0·376 | 0·008 | 0·004 | 18·239 | 2·557 | 0·063 | 0·0012 | 0·134 |
| 34 | 10·465 | 1·257 | 0·400 | 0·388 | 0·008 | 0·004 | 18·791 | 2·635 | 0·065 | 0·0012 | 0·137 |
| 35 | 10·773 | 1·294 | 0·412 | 0·399 | 0·008 | 0·004 | 19·344 | 2·712 | 0·067 | 0·0013 | 0·142 |
| 36 | 11·081 | 1·331 | 0·424 | 0·410 | 0·008 | 0·005 | 19·897 | 2·789 | 0·069 | 0·0013 | 0·146 |
| 37 | 11·389 | 1·368 | 0·435 | 0·422 | 0·008 | 0·005 | 20·449 | 2·867 | 0·071 | 0·0013 | 0·150 |
| 38 | 11·696 | 1·405 | 0·447 | 0·433 | 0·009 | 0·005 | 21·002 | 2·944 | 0·073 | 0·0014 | 0·154 |
| 39 | 12·004 | 1·442 | 0·459 | 0·445 | 0·009 | 0·005 | 21·555 | 3·022 | 0·075 | 0·0014 | 0·158 |
| 40 | 12·312 | 1·479 | 0·471 | 0·456 | 0·009 | 0·005 | 22·107 | 3·099 | 0·077 | 0·0015 | 0·162 |
| 41 | 12·620 | 1·516 | 0·482 | 0·467 | 0·009 | 0·005 | 22·660 | 3·177 | 0·079 | 0·0015 | 0·166 |
| 42 | 12·928 | 1·553 | 0·494 | 0·479 | 0·009 | 0·005 | 23·213 | 3·254 | 0·081 | 0·0015 | 0·170 |

* Cations concentrations; averages of mean results of Cox and Culkin (1967) and Riley and Tongudai (1967). Sulphate and bromide concentration based on mean values from Morris and Riley (1966).

TABLE 3

*Preparation of artificial sea water* $(S = 35 \cdot 00\%_0)$

| Lyman and Fleming (1940) (g.) | | Kalle (1945) (g.) | |
|---|---|---|---|
| NaCl | 23·939 | NaCl | 28·566 |
| $MgCl_2$ | 5·079 | $MgCl_2$ | 3·887 |
| $Na_2SO_4$ | 3·994 | $MgSO_4$ | 1·787 |
| $CaCl_2$ | 1·123 | $CaSO_4$ | 1·308 |
| KCl | 0·667 | $K_2SO_4$ | 0·832 |
| $NaHCO_3$ | 0·196 | $CaCO_3$ | 0·124 |
| KBr | 0·098 | KBr | 0·103 |
| $H_3BO_3$ | 0·027 | $SrSO_4$ | 0·0288 |
| $SrCl_2$ | 0·024 | $H_3BO_3$ | 0·0282 |
| NaF | 0·003 | | |
| Water to | 1 kg | Water to | 1 kg |

Kester *et al.* (1967)

| A. Gravimetric salts | $g\,kg^{-1}$ |
|---|---|
| NaCl | 23·926 |
| $Na_2SO_4$ | 4·008 |
| KCl | 0·667 |
| $NaHCO_3$ | 0·196 |
| KBr | 0·098 |
| $H_3BO_3$ | 0·026 |
| NaF | 0·003 |

B. Volumetric salts (standardized by Mohr method)

| | Approx. molarity | Use volume equivalent to |
|---|---|---|
| $Mg_2 6H_2O$ | 1·0 M | $1·297\,g\,Mg\,kg^{-1}$ |
| $CaCl_2 2H_2O$ | 1·0 M | $0·406\,g\,Ca\,kg^{-1}$ |
| $SrCl_2 6H_2O$ | 0·1 M | $0·0133\,g\,Sr\,kg^{-1}$ |

C Water to 1 kg

*Note*: (i) Allowance must be made for water of crystallization of any of the salts used.
(ii) After aeration the pH should lie between 7·9 and 8·3.

TABLE 4

*Collected conversion factors*

| Conversion | | Factor | Reciprocal |
|---|---|---|---|
| $\mu$g $NO_3^-$ | $\longrightarrow \mu$g N | 0·2259 | 4·427 |
| $\mu$g $NO_2^-$ | $\longrightarrow \mu$g N | 0·3045 | 3·286 |
| $\mu$g $NH_3$ | $\longrightarrow \mu$g N | 0·8225 | 1·216 |
| $\mu$g $NH_4^+$ | $\longrightarrow \mu$g N | 0·7764 | 1·287 |
| $\mu$g $PO_4^{3-}$ | $\longrightarrow \mu$g P | 0·3261 | 3·066 |
| $\mu$g $P_2O_5$ | $\longrightarrow \mu$g P | 0·4364 | 2·291 |
| $\mu$g $SiO_2$ | $\longrightarrow \mu$g Si | 0·4675 | 2·139 |
| $\mu$g $SiO_4^{4-}$ | $\longrightarrow \mu$g Si | 0·3050 | 3·278 |
| $\mu$g N | $\longrightarrow \mu$g-at N | 0·07138 | 14·008 |
| $\mu$g P | $\longrightarrow \mu$g-at. P | 0·03228 | 30·975 |
| $\mu$g Si | $\longrightarrow \mu$g-at. Si | 0·03560 | 28·09 |

TABLE 5

*Table for conversion of weights of nitrogen, phosphorus and silicon expressed in terms of $\mu$g into $\mu$g-at.*

| $\mu$g N, P, or Si$^{-1}$ | $\mu$g-at N$l^{-1}$ | $\mu$g-at. P $l^{-1}$ | $\mu$g-at. Si $l^{-1}$ |
|---|---|---|---|
| 1 | 0·071 | 0·032 | 0·036 |
| 2 | 0·143 | 0·065 | 0·071 |
| 3 | 0·214 | 0·097 | 0·107 |
| 4 | 0·286 | 0·129 | 0·142 |
| 5 | 0·357 | 0·161 | 0·178 |
| 6 | 0·428 | 0·194 | 0·214 |
| 7 | 0·500 | 0·226 | 0·249 |
| 8 | 0·571 | 0·258 | 0·284 |
| 9 | 0·643 | 0·291 | 0·320 |
| 10 | 0·714 | 0·323 | 0·356 |
| 20 | 1·428 | 0·646 | 0·712 |
| 30 | 2·142 | 0·968 | 1·068 |
| 40 | 2·856 | 1·291 | 1·424 |
| 50 | 3·569 | 1·614 | 1·780 |
| 60 | 4·283 | 1·937 | 2·136 |
| 70 | 4·997 | 2·260 | 2·492 |
| 80 | 5·711 | 2·582 | 2·848 |
| 90 | 6·425 | 2·905 | 3·204 |
| 100 | 7·139 | 3·228 | 3·560 |

TABLE 6

Solubility of oxygen (C) in sea water ($cm^3\,dm^{-3}$) with respect to an atmosphere of 20·95% oxygen and 100% relative humidity at a total atmospheric pressure of 760 mm Hg. (UNESCO 1973)*

| T (°C) | Salinity (‰) | | | | | | | | | | | | | | |
|---|---|---|---|---|---|---|---|---|---|---|---|---|---|---|---|
| | 0 | 5 | 10 | 15 | 20 | 25 | 30 | 31 | 32 | 33 | 34 | 35 | 36 | 37 | 38 |
| 0 | 10·22 | 9·87 | 9·54 | 9·22 | 8·91 | 8·61 | 8·32 | 8·27 | 8·21 | 8·16 | 8·10 | 8·05 | 7·99 | 7·94 | 7·88 |
| 1 | 9·94 | 9·60 | 9·28 | 8·97 | 8·68 | 8·39 | 8·11 | 8·05 | 8·00 | 7·94 | 7·89 | 7·84 | 7·78 | 7·73 | 7·68 |
| 2 | 9·67 | 9·35 | 9·04 | 8·74 | 8·45 | 8·17 | 7·90 | 7·85 | 7·79 | 7·74 | 7·69 | 7·64 | 7·59 | 7·53 | 7·48 |
| 3 | 9·41 | 9·10 | 8·80 | 8·51 | 8·23 | 7·96 | 7·70 | 7·65 | 7·60 | 7·55 | 7·50 | 7·45 | 7·40 | 7·35 | 7·30 |
| 4 | 9·16 | 8·86 | 8·57 | 8·29 | 8·02 | 7·76 | 7·51 | 7·46 | 7·41 | 7·36 | 7·31 | 7·26 | 7·22 | 7·17 | 7·12 |
| 5 | 8·93 | 8·64 | 8·36 | 8·09 | 7·83 | 7·57 | 7·33 | 7·28 | 7·23 | 7·18 | 7·14 | 7·09 | 7·04 | 7·00 | 6·95 |
| 6 | 8·70 | 8·42 | 8·15 | 7·89 | 7·64 | 7·39 | 7·15 | 7·11 | 7·06 | 7·01 | 6·97 | 6·92 | 6·88 | 6·83 | 6·79 |
| 7 | 8·49 | 8·22 | 7·95 | 7·70 | 7·45 | 7·22 | 6·98 | 6·94 | 6·89 | 6·85 | 6·81 | 6·76 | 6·72 | 6·67 | 6·63 |
| 8 | 8·28 | 8·02 | 7·76 | 7·52 | 7·28 | 7·05 | 6·82 | 6·78 | 6·74 | 6·69 | 6·65 | 6·61 | 6·57 | 6·52 | 6·48 |
| 9 | 8·08 | 7·83 | 7·58 | 7·34 | 7·11 | 6·89 | 6·67 | 6·63 | 6·59 | 6·54 | 6·50 | 6·46 | 6·42 | 6·38 | 6·34 |
| 10 | 7·89 | 7·64 | 7·41 | 7·17 | 6·95 | 6·73 | 6·52 | 6·48 | 6·44 | 6·40 | 6·36 | 6·32 | 6·28 | 6·24 | 6·20 |
| 11 | 7·71 | 7·47 | 7·24 | 7·01 | 6·80 | 6·58 | 6·38 | 6·34 | 6·30 | 6·26 | 6·22 | 6·18 | 6·14 | 6·10 | 6·07 |
| 12 | 7·53 | 7·30 | 7·08 | 6·86 | 6·65 | 6·44 | 6·24 | 6·21 | 6·17 | 6·13 | 6·09 | 6·05 | 6·01 | 5·98 | 5·94 |
| 13 | 7·37 | 7·14 | 6·92 | 6·71 | 6·50 | 6·31 | 6·11 | 6·07 | 6·04 | 6·00 | 5·96 | 5·93 | 5·89 | 5·85 | 5·82 |
| 14 | 7·20 | 6·98 | 6·77 | 6·57 | 6·37 | 6·17 | 5·99 | 5·95 | 5·91 | 5·88 | 5·84 | 5·80 | 5·77 | 5·73 | 5·70 |
| 15 | 7·05 | 6·84 | 6·63 | 6·43 | 6·24 | 6·05 | 5·87 | 5·83 | 5·79 | 5·76 | 5·72 | 5·69 | 5·65 | 5·62 | 5·58 |
| 16 | 6·90 | 6·69 | 6·49 | 6·30 | 6·11 | 5·93 | 5·75 | 5·71 | 5·68 | 5·64 | 5·61 | 5·58 | 5·54 | 5·51 | 5·48 |
| 17 | 6·75 | 6·55 | 6·36 | 6·17 | 5·99 | 5·81 | 5·64 | 5·60 | 5·57 | 5·53 | 5·50 | 5·47 | 5·43 | 5·40 | 5·37 |
| 18 | 6·61 | 6·42 | 6·23 | 6·05 | 5·87 | 5·69 | 5·53 | 5·49 | 5·46 | 5·43 | 5·40 | 5·36 | 5·33 | 5·30 | 5·27 |
| 19 | 6·48 | 6·29 | 6·11 | 5·93 | 5·75 | 5·59 | 5·42 | 5·39 | 5·36 | 5·33 | 5·29 | 5·26 | 5·23 | 5·20 | 5·17 |
| 20 | 6·35 | 6·17 | 5·99 | 5·81 | 5·64 | 5·48 | 5·32 | 5·29 | 5·26 | 5·23 | 5·20 | 5·17 | 5·14 | 5·10 | 5·07 |
| 21 | 6·23 | 6·05 | 5·87 | 5·70 | 5·54 | 5·38 | 5·22 | 5·19 | 5·16 | 5·13 | 5·10 | 5·07 | 5·04 | 5·01 | 4·98 |
| 22 | 6·11 | 5·93 | 5·76 | 5·60 | 5·44 | 5·28 | 5·13 | 5·10 | 5·07 | 5·04 | 5·01 | 4·98 | 4·95 | 4·92 | 4·89 |
| 23 | 5·99 | 5·82 | 5·65 | 5·49 | 5·34 | 5·18 | 5·04 | 5·01 | 4·98 | 4·95 | 4·92 | 4·89 | 4·87 | 4·84 | 4·81 |

## Table 6 cont.

*Solubility of oxygen (C) in sea water ($cm^3\ dm^{-3}$) with respect to an atmosphere of 20·95% oxygen and 100% relative humidity at a total atmospheric pressure of 860 mm Hg. (UNESCO, 1973)\**

| $T$ (°C) | \multicolumn{15}{c}{Salinity (‰)} |
| | 0 | 5 | 10 | 15 | 20 | 25 | 30 | 31 | 32 | 33 | 34 | 35 | 36 | 37 | 38 |
|---|---|---|---|---|---|---|---|---|---|---|---|---|---|---|---|
| 24 | 5·88 | 5·71 | 5·55 | 5·39 | 5·24 | 5·09 | 4·95 | 4·92 | 4·89 | 4·86 | 4·84 | 4·81 | 4·78 | 4·75 | 4·73 |
| 25 | 5·77 | 5·61 | 5·45 | 5·30 | 5·15 | 5·00 | 4·86 | 4·84 | 4·81 | 4·78 | 4·75 | 4·73 | 4·70 | 4·67 | 4·65 |
| 26 | 5·66 | 5·51 | 5·35 | 5·20 | 5·06 | 4·92 | 4·78 | 4·75 | 4·73 | 4·70 | 4·67 | 4·65 | 4·62 | 4·59 | 4·57 |
| 27 | 5·56 | 5·41 | 5·26 | 5·11 | 4·97 | 4·83 | 4·70 | 4·67 | 4·65 | 4·62 | 4·60 | 4·57 | 4·54 | 4·52 | 4·49 |
| 28 | 5·46 | 5·31 | 5·17 | 5·03 | 4·89 | 4·75 | 4·62 | 4·60 | 4·57 | 4·55 | 4·52 | 4·50 | 4·47 | 4·45 | 4·42 |
| 29 | 5·37 | 5·22 | 5·08 | 4·94 | 4·81 | 4·67 | 4·55 | 4·52 | 4·50 | 4·47 | 4·45 | 4·42 | 4·40 | 4·37 | 4·35 |
| 30 | 5·28 | 5·13 | 4·99 | 4·86 | 4·73 | 4·60 | 4·47 | 4·45 | 4·43 | 4·40 | 4·38 | 4·35 | 4·33 | 4·31 | 4·28 |
| 31 | 5·19 | 5·05 | 4·91 | 4·78 | 4·65 | 4·53 | 4·40 | 4·38 | 4·36 | 4·33 | 4·31 | 4·28 | 4·26 | 4·24 | 4·22 |
| 32 | 5·10 | 4·96 | 4·83 | 4·70 | 4·58 | 4·45 | 4·33 | 4·31 | 4·29 | 4·26 | 4·24 | 4·22 | 4·20 | 4·17 | 4·15 |

\* Based on measurements by Carpenter (1966) and Murray and Riley (1969a) fitted by Weiss (1970) to the thermodynamically consistent equation:

$$\ln C = A_1 + A_2(100/T) + A_3 \ln(T/100) + A_4(T/100) + S‰[B_1 + B_2(T/100) + B_3(T/100)^2]$$

where

| $A_1$ | $A_2$ | $A_3$ | $A_4$ | $B_1$ | $B_2$ | $B_3$ |
|---|---|---|---|---|---|---|
| −173·4292 | 249·6339 | 143·3483 | −21·8492 | −0·033096 | 0·014259 | −0·0017000 |

and $T$ and $S‰$ are the absolute temperature (K) and salinity in parts per mille respectively.

TABLE 7

*Solubility of nitrogen in sea water ($cm^3 dm^{-3}$) with respect to an atmosphere of 78·084 % nitrogen and 100 % relative humidity at a total pressure of 760 mm Hg (Weiss (1970) from data by Murray and Riley (1969b)).*

| | Salinity ‰ | | | | | | | | |
|---|---|---|---|---|---|---|---|---|---|
| $T(°C)$ | 0 | 10 | 20 | 30 | 34 | 35 | 36 | 38 | 40 |
| −1 | — | — | 16·28 | 15·10 | 14·65 | 14·54 | 14·44 | 14·22 | 14·01 |
| 0 | 18·42 | 17·10 | 15·87 | 14·73 | 14·30 | 14·19 | 14·09 | 13·88 | 13·67 |
| 1 | 17·95 | 16·67 | 15·48 | 14·38 | 13·96 | 13·86 | 13·75 | 13·55 | 13·35 |
| 2 | 17·50 | 16·26 | 15·11 | 14·04 | 13·64 | 13·54 | 13·44 | 13·24 | 13·05 |
| 3 | 17·07 | 15·87 | 14·75 | 13·72 | 13·32 | 13·23 | 13·13 | 12·94 | 12·76 |
| 4 | 16·65 | 15·49 | 14·41 | 13·41 | 13·03 | 12·93 | 12·84 | 12·66 | 12·47 |
| 5 | 16·26 | 15·13 | 14·09 | 13·11 | 12·74 | 12·65 | 12·56 | 12·38 | 12·21 |
| 6 | 15·88 | 14·79 | 13·77 | 12·83 | 12·47 | 12·38 | 12·29 | 12·12 | 11·95 |
| 8 | 15·16 | 14·14 | 13·18 | 12·29 | 11·95 | 11·87 | 11·79 | 11·62 | 11·46 |
| 10 | 14·51 | 13·54 | 12·64 | 11·80 | 11·48 | 11·40 | 11·32 | 11·17 | 11·01 |
| 12 | 13·90 | 12·99 | 12·14 | 11·34 | 11·04 | 10·96 | 10·89 | 10·74 | 10·60 |
| 14 | 13·34 | 12·48 | 11·67 | 10·92 | 10·63 | 10·56 | 10·49 | 10·35 | 10·21 |
| 16 | 12·83 | 12·01 | 11·24 | 10·53 | 10·25 | 10·19 | 10·12 | 9·99 | 9·86 |
| 18 | 12·35 | 11·57 | 10·84 | 10·16 | 9·90 | 9·84 | 9·77 | 9·65 | 9·52 |
| 20 | 11·90 | 11·16 | 10·47 | 9·82 | 9·57 | 9·51 | 9·45 | 9·33 | 9·21 |
| 22 | 11·48 | 10·78 | 10·12 | 9·50 | 9·26 | 9·21 | 9·15 | 9·03 | 8·92 |
| 24 | 11·09 | 10·42 | 9·79 | 9·20 | 8·98 | 8·92 | 8·87 | 8·76 | 8·65 |
| 26 | 10·73 | 10·09 | 9·49 | 8·92 | 8·71 | 8·65 | 8·60 | 8·50 | 8·39 |
| 28 | 10·38 | 9·77 | 9·20 | 8·66 | 8·45 | 8·40 | 8·35 | 8·25 | 8·15 |
| 30 | 10·06 | 9·48 | 8·93 | 8·41 | 8·21 | 8·16 | 8·12 | 8·02 | 7·92 |
| 32 | 9·76 | 9·20 | 8·67 | 8·18 | 7·99 | 7·94 | 7·89 | 7·80 | 7·71 |
| 34 | 9·48 | 8·94 | 8·43 | 7·96 | 7·77 | 7·73 | 7·68 | 7·59 | 7·51 |
| 36 | 9·21 | 8·69 | 8·20 | 7·75 | 7·57 | 7·53 | 7·48 | 7·40 | 7·31 |
| 38 | 8·95 | 8·46 | 7·99 | 7·55 | 7·38 | 7·33 | 7·29 | 7·21 | 7·13 |
| 40 | 8·71 | 8·23 | 7·78 | 7·36 | 7·19 | 7·15 | 7·11 | 7·03 | 6·95 |

The solubility at any value of salinity and temperature in the above range can be calculated if the following constants are substituted in the equation below (Table 6).

| $A_1$ | $A_2$ | $A_3$ | $A_4$ | $B_1$ | $B_2$ | $B_3$ |
|---|---|---|---|---|---|---|
| −172·4965 | 248·4262 | 143·0738 | −21·7120 | −0·049781 | 0·025018 | −0·003486 |

## TABLE 8

*Solubility of argon in sea water ($cm^{-3} dm^{-3}$) with respect to an atmosphere of 0·934 % argon and 100 % relative humidity at a total atmosphere pressure of 760 mm Hg (Weiss (1970) from data by Douglas (1964, 1965)).*

| T (°C) | Salinity ‰ 0 | 10 | 20 | 30 | 34 | 35 | 36 | 38 | 40 |
|---|---|---|---|---|---|---|---|---|---|
| −1 | —— | —— | 0·4456 | 0·4156 | 0·4042 | 0·4014 | 0·3986 | 0·3931 | 0·3877 |
| 0 | 0·4980 | 0·4647 | 0·4337 | 0·4048 | 0·3937 | 0·3910 | 0·3883 | 0·3830 | 0·3777 |
| 1 | 0·4845 | 0·4524 | 0·4224 | 0·3944 | 0·3837 | 0·3811 | 0·3785 | 0·3733 | 0·3682 |
| 2 | 0·4715 | 0·4405 | 0·4115 | 0·3845 | 0·3741 | 0·3716 | 0·3691 | 0·3641 | 0·3592 |
| 3 | 0·4592 | 0·4292 | 0·4012 | 0·3750 | 0·3650 | 0·3625 | 0·3601 | 0·3552 | 0·3505 |
| 4 | 0·4474 | 0·4184 | 0·3912 | 0·3659 | 0·3562 | 0·3538 | 0·3515 | 0·3468 | 0·3422 |
| 5 | 0·4360 | 0·4080 | 0·3817 | 0·3572 | 0·3478 | 0·3455 | 0·3432 | 0·3387 | 0·3342 |
| 6 | 0·4252 | 0·3980 | 0·3726 | 0·3488 | 0·3397 | 0·3375 | 0·3353 | 0·3309 | 0·3265 |
| 8 | 0·4049 | 0·3794 | 0·3555 | 0·3331 | 0·3246 | 0·3225 | 0·3204 | 0·3162 | 0·3121 |
| 10 | 0·3861 | 0·3622 | 0·3397 | 0·3186 | 0·3106 | 0·3086 | 0·3066 | 0·3027 | 0·2989 |
| 12 | 0·3688 | 0·3463 | 0·3251 | 0·3053 | 0·2977 | 0·2958 | 0·2939 | 0·2902 | 0·2866 |
| 14 | 0·3528 | 0·3316 | 0·3116 | 0·2929 | 0·2857 | 0·2839 | 0·2822 | 0·2787 | 0·2752 |
| 16 | 0·3380 | 0·3180 | 0·2991 | 0·2814 | 0·2746 | 0·2729 | 0·2712 | 0·2679 | 0·2647 |
| 18 | 0·3242 | 0·3053 | 0·2875 | 0·2707 | 0·2642 | 0·2626 | 0·2610 | 0·2579 | 0·2548 |
| 20 | 0·3114 | 0·2935 | 0·2766 | 0·2607 | 0·2546 | 0·2531 | 0·2516 | 0·2486 | 0·2457 |
| 22 | 0·2995 | 0·2825 | 0·2665 | 0·2514 | 0·2455 | 0·2441 | 0·2427 | 0·2399 | 0·2371 |
| 24 | 0·2883 | 0·2722 | 0·2570 | 0·2426 | 0·2371 | 0·2357 | 0·2344 | 0·2317 | 0·2291 |
| 26 | 0·2779 | 0·2626 | 0·2481 | 0·2344 | 0·2292 | 0·2279 | 0·2266 | 0·2241 | 0·2215 |
| 28 | 0·2681 | 0·2535 | 0·2398 | 0·2268 | 0·2217 | 0·2205 | 0·2193 | 0·2169 | 0·2144 |
| 30 | 0·2588 | 0·2450 | 0·2319 | 0·2195 | 0·2147 | 0·2136 | 0·2124 | 0·2101 | 0·2078 |
| 32 | 0·2502 | 0·2370 | 0·2245 | 0·2127 | 0·2081 | 0·2070 | 0·2059 | 0·2037 | 0·2015 |
| 34 | 0·2420 | 0·2294 | 0·2175 | 0·2062 | 0·2019 | 0·2008 | 0·1997 | 0·1976 | 0·1955 |
| 36 | 0·2342 | 0·2222 | 0·2109 | 0·2001 | 0·1959 | 0·1949 | 0·1939 | 0·1919 | 0·1899 |
| 38 | 0·2269 | 0·2154 | 0·2046 | 0·1943 | 0·1903 | 0·1893 | 0·1883 | 0·1864 | 0·1845 |
| 40 | 0·2199 | 0·2090 | 0·1986 | 0·1888 | 0·1849 | 0·1840 | 0·1831 | 0·1812 | 0·1794 |

The solubility at any value of salinity and temperature in the above range can be calculated if the following constants are substituted in the equation below (Table 6).

| $A_1$ | $A_2$ | $A_3$ | $A_4$ | $B_1$ | $B_2$ | $B_3$ |
|---|---|---|---|---|---|---|
| −173·5146 | 245·4510 | 141·8222 | −21·8020 | −0·034474 | 0·014934 | −0·0017729 |

## TABLE 9

*Literature citations for solubilities of other gases in sea water*

| Gas | Reference |
|---|---|
| Carbon dioxide | Murray and Riley (1971); see also Chapter 9, Table 9. |
| Helium | Weiss (1971); see also Chapter 8, Table A8.5. |
| Neon | Weiss (1971); see also Chapter 8, Table A8.4 |
| Krypton | Wood and Caputi (1966); see also Chapter 8, Table 8.5. |
| Xenon | Wood and Caputi (1966); see also Chapter 8, Table 8.5. |
| Carbon monoxide | Douglas (1967); see also Chapter 8, Table 8.12. |
| Hydrogen | Crozier and Yamamoto (1974). |

TABLE 10

The density of artificial sea water as a function of temperature and chlorinity*
(Millero and Lepple, 1973)

| Cl (‰) | 0°C | 5°C | 10°C | 15°C | 20°C | 25°C | 30°C | 35°C | 40°C |
|---|---|---|---|---|---|---|---|---|---|
| 0 | 0·999868 | 0·999992 | 0·999728 | 0·999129 | 0·998234 | 0·997075 | 0·995678 | 0·994063 | 0·992247 |
| $3·42_6$ | 1·004944 | 1·004959 | 1·004599 | 1·003921 | 1·002962 | 1·001744 | 1·000295 | 0·998643 | 0·996783 |
| $6·05_5$ | 1·008665 | 1·008705 | 1·008292 | 1·007566 | 1·006575 | 1·005335 | 1·003868 | 1·00219C | 1·000307 |
| $8·17_4$ | 1·011851 | 1·011731 | 1·011265 | 1·010502 | 1·009472 | 1·008201 | 1·006707 | 1·005013 | 1·003113 |
| $11·69_5$ | 1·016982 | 1·016758 | 1·016208 | 1·015368 | 1·014275 | 1·012949 | 1·011407 | 1·009669 | 1·007745 |
| $13·67_3$ | 1·019835 | 1·019564 | 1·018970 | 1·018102 | 1·016986 | 1·015641 | 1·014087 | 1·012346 | 1·010406 |
| $16·33_3$ | 1·023703 | 1·023352 | 1·022695 | 1·021772 | 1·020611 | 1·019229 | 1·017642 | 1·015866 | 1·013920 |
| $19·05_6$ | 1·027648 | 1·027227 | 1·026511 | 1·025538 | 1·024335 | 1·022921 | 1·021311 | 1·019528 | 1·017564 |
| $21·53_7$ | 1·031240 | 1·030774 | 1·029989 | 1·028941 | 1·027731 | 1·026307 | 1·024658 | 1·022890 | 1·020925 |

* These densities are relative to those tabulated by Kell (1967) for pure water assuming the density of pure water is $1·000000 \text{ g ml}^{-1}$ at 3·98°C.

TABLE 11

*The expansibility of artificial sea water as a function of temperature and chlorinity,*
$\alpha \times 10^6$ *(deg.$^{-1}$) (Millero and Lepple, 1973)*

| Cl‰ | 0°C | 5°C | 10°C | 15°C | 20°C | 25°C | 30°C | 35°C | 40°C |
|---|---|---|---|---|---|---|---|---|---|
| 0·000 | −68·1 | 16·0 | 87·9 | 150·7 | 206·6 | 257·0 | 303·1 | 345·7 | 385·4 |
| 3·426 | −46·9 | 35·5 | 105·1 | 165·4 | 218·7 | 266·7 | 310·7 | 351·8 | 391·0 |
| 6·055 | −28·0 | 49·4 | 115·2 | 172·7 | 224·1 | 271·0 | 314·8 | 356·4 | 396·7 |
| 8·174 | −14·8 | 60·4 | 124·4 | 180·5 | 230·7 | 276·6 | 319·4 | 359·9 | 398·8 |
| 11·695 | 8·2 | 79·2 | 140·2 | 194·1 | 242·7 | 287·2 | 328·3 | 366·7 | 402·6 |
| 13·673 | 18·4 | 88·1 | 147·5 | 199·6 | 246·6 | 289·8 | 330·1 | 368·6 | 405·7 |
| 16·333 | 36·1 | 102·3 | 159·2 | 209·4 | 254·8 | 296·8 | 335·8 | 372·8 | 407·9 |
| 19·056 | 51·0 | 115·2 | 170·2 | 218·5 | 262·2 | 302·4 | 339·9 | 375·2 | 408·8 |
| 21·537 | 61·9 | 127·6 | 181·6 | 227·5 | 267·9 | 304·9 | 340·1 | 375·0 | 410·7 |

$\alpha = -1/d(\partial d/\partial t)$ where $d$ is the density of the sea water.

TABLE 12

*The isothermal compressibility of sea water at 1 atm as a function of salinity and
temperature (Lepple and Millero, 1971)*

| | $\beta \times 10^{-6}$ (bar$^{-1}$) | | | | | | | | |
|---|---|---|---|---|---|---|---|---|---|
| S(‰) | 0°C | 5°C | 10°C | 15°C | 20°C | 25°C | 30°C | 35°C | 40°C |
| 0·00 | 50·886 | 49·171 | 47·811 | 46·736 | 45·895 | 45·250 | 44·774 | 44·444 | 44·243 |
| 6·14 | 50·07 | 48·42 | 47·10 | 46·09 | 45·31 | 44·71 | 44·26 | 43·92 | 43·75 |
| 11·80 | 49·25 | 47·70 | 46·43 | 45·41 | 44·66 | 44·13 | 43·68 | 43·34 | 43·19 |
| 14·75 | 48·84 | 47·30 | 46·11 | 45·15 | 44·38 | 43·83 | 43·43 | 43·13 | 43·01 |
| 21·01 | 48·14 | 46·71 | 45·59 | 44·63 | 43·92 | 43·45 | 42·96 | 42·71 | 42·63 |
| 24·52 | 47·63 | 46·25 | 45·17 | 44·29 | 43·61 | 42·98 | 42·68 | 42·33 | 42·23 |
| 29·38 | 47·01 | 45·62 | 44·62 | 43·74 | 43·17 | 42·56 | 42·24 | 41·96 | 41·86 |
| 34·25 | 46·49 | 45·17 | 44·15 | 43·32 | 42·69 | 42·18 | 41·88 | 41·69 | 41·55 |
| 35·00 | 46·32 | 45·03 | 44·02 | 43·19 | 42·58 | 42·11 | 41·78 | 41·49 | 41·48 |
| 39·00 | 45·84 | 44·62 | 43·63 | 42·80 | 42·30 | 41·73 | 41·53 | 41·23 | 41·15 |

TABLE 13

Observed values for the change in the specific volume of sea water from 0° to T°C at various pressures and salinities. Unit of specific volume = $10^{-6}\ cm^3\ g^{-1}$. (Cox et al., 1970)

| P, bars absolute → | \( S = 35.00‰ \) | | | | | | \( S = 30.50‰ \) | | | | \( S = 39.50‰ \) | | | |
|---|---|---|---|---|---|---|---|---|---|---|---|---|---|---|
| | 8.3 | 201.3 | 401.2 | 601.0 | 800.9 | 1000.8 | 8.3 | 201.3 | 601.0 | 1000.9 | 8.3 | 201.3 | 601.0 | 1000.8 |
| S, ‰ → | 35.000 | 35.004 | 35.005 | 35.002 | 35.002 | 35.002 | 30.502 | 30.504 | 30.506 | 30.510 | 39.503 | 39.502 | 39.504 | 39.507 |
| pH (1 bar, 25°C) → | 7.91 | 7.95 | 7.94 | 7.94 | 8.00 | 7.96 | 8.06 | 7.98 | 8.03 | 8.00 | 8.22 | 8.18 | 8.13 | 8.16 |
| T(°C) ↓ | | | | | | | | | | | | | | |
| -2.000 | — | — | -277.1 | -356.9 | -424.3 | -480.5 | — | — | -341.9 | -472.3 | — | — | -370.6 | -489.4 |
| -1.000 | — | -97.5 | — | — | — | — | — | -86.9 | — | — | — | -107.1 | — | — |
| 0.000 | 0 | 0 | 0 | 0 | 0 | 0 | 0 | 0 | 0 | 0 | 0 | 0 | 0 | 0 |
| 2.000 | 132.2 | 224.9 | 310.0 | 383.2 | 444.9 | 497.6 | 106.9 | 204.7 | 368.6 | 489.0 | 155.8 | 245.2 | 394.4 | 504.3 |
| 4.000 | 311.2 | 489.5 | 652.0 | 791.7 | 910.6 | 1012.5 | 262.7 | 450.5 | 766.4 | 998.5 | 355.8 | 527.9 | 815.7 | 1026.8 |
| 6.000 | 535.0 | 791.0 | 1023.0 | 1225.3 | 1396.8 | 1544.7 | 464.7 | 734.7 | 1189.8 | 1523.0 | 599.1 | 846.2 | 1259.1 | 1556.1 |
| 8.000 | 801.0 | 1127.3 | 1424.4 | 1683.3 | 1902.4 | 2094.3 | 712.1 | 1055.7 | 1637.2 | 2064.2 | 883.0 | 1198.6 | 1726.2 | 2117.5 |
| 10.000 | 1107.1 | 1498.0 | 1854.4 | 2163.3 | 2427.7 | 2660.1 | 1000.9 | 1412.3 | 2106.9 | 2623.2 | 1205.8 | 1582.7 | 2216.4 | 2686.7 |
| 12.000 | 1452.7 | 1901.6 | 2312.4 | 2668.5 | 2971.6 | 3243.6 | 1330.2 | 1802.6 | 2603.7 | 3197.3 | 1566.5 | 1999.4 | 2729.5 | 3274.6 |
| 14.000 | 1836.3 | 2336.9 | 2796.2 | 3198.3 | 3535.9 | 3827.4 | 1698.8 | 2226.3 | 3123.4 | 3790.3 | 1962.6 | 2446.4 | 3264.0 | 3872.9 |
| 16.000 | 2255.5 | 2804.0 | 3306.4 | 3745.3 | 4119.2 | 4448.9 | 2104.7 | 2682.5 | 3665.3 | 4398.0 | 2394.3 | 2923.1 | 3818.4 | 4489.2 |
| 18.000 | 2709.3 | 3299.8 | 3840.9 | 4315.0 | 4721.5 | 5075.4 | 2547.1 | 3169.6 | 4228.6 | 5021.7 | 2858.7 | 3428.0 | 4394.0 | 5119.5 |
| 20.000 | 3196.3 | 3823.9 | 4400.2 | 4906.4 | 5341.0 | 5719.9 | 3022.8 | 3685.2 | 4815.6 | 5661.0 | 3355.8 | 3961.8 | 4994.4 | 5764.6 |
| 22.000 | 3717.0 | 4376.5 | 4984.1 | 5516.1 | 5975.8 | 6378.6 | 3533.0 | 4230.0 | 5421.9 | 6315.4 | 3883.4 | 4522.1 | 5610.2 | 6423.2 |
| 24.000 | 4268.2 | 4957.5 | 5591.4 | 6151.3 | 6630.1 | 7051.0 | 4075.6 | 4804.1 | 6049.9 | 6985.9 | 4442.8 | 5108.6 | 6246.4 | 7100.4 |
| 26.000 | 4850.1 | 5564.4 | 6223.6 | 6803.5 | 7303.4 | 7738.2 | 4649.9 | 5403.9 | 6698.9 | 7673.4 | 5031.4 | 5721.8 | 6901.7 | 7789.7 |
| 28.000 | 5461.6 | 6197.0 | 6877.1 | 7472.2 | 7990.8 | 8439.8 | 5255.2 | 6032.2 | 7367.9 | 8374.8 | 5648.6 | 6358.8 | 7576.4 | 8493.4 |
| 30.000 | 6102.8 | 6855.3 | 7554.3 | 8165.9 | 8693.8 | 9159.8 | 5889.4 | 6687.1 | 8056.6 | 9091.8 | 6294.8 | 7021.5 | 8269.4 | 9211.1 |

TABLE 14

*Specific gravity and percentage volume reduction of sea water\* under pressure*
*(amended from Cox, 1965)*

| Pressure (db) | Specific gravity | % decrease in volume |
|---|---|---|
| 0 | 1·02813 | 0·000 |
| 100 | 1·02860 | 0·046 |
| 200 | 1·02908 | 0·093 |
| 500 | 1·03050 | 0·231 |
| 1,000 | 1·03285 | 0·460 |
| 2,000 | 1·03747 | 0·909 |
| 3,000 | 1·04199 | 1·349 |
| 4,000 | 1·04640 | 1·778 |
| 5,000 | 1·05071 | 2·197 |
| 6,000 | 1·05494 | 2·609 |
| 7,000 | 1·05908 | 3·011 |
| 8,000 | 1·06314 | 3·406 |
| 9,000 | 1·06713 | 3·794 |
| 10,000 | 1·07104 | 4·175 |

\* Salinity, 35·00‰; Temperature 0°C.

TABLE 15

*Percentage reduction in volume of sea water under a pressure of 1,000 db at various temperatures and salinities. (After Cox, 1965).*

| $S‰$ | Temperature (°C) | | | |
|---|---|---|---|---|
| | 0 | 10 | 20 | 30 |
| 0 | 0·500 | 0·470 | 0·451 | 0·440 |
| 10 | 0·486 | 0·459 | 0·442 | 0·432 |
| 20 | 0·474 | 0·448 | 0·432 | 0·423 |
| 30 | 0·462 | 0·438 | 0·424 | 0·415 |
| 35 | 0·457 | 0·433 | 0·419 | 0·411 |
| 40 | 0·450 | 0·428 | 0·415 | 0·407 |

TABLE 16

*Thermal expansion of sea water under pressure $(10^{-6} cm^3 (\,^{\circ}C)^{-1})$. (Bradshaw and Schleicher, 1970)*

| Pressure (bars) | Temperature (°C) | | | |
|---|---|---|---|---|
| | 0 | 10 | 20 | 30 |
| | | $S = 30.50\%_0$ | | |
| 1 | 39 | 155 | 246 | 324 |
| 500 | 158 | 229 | 290 | 346 |
| 1000 | 240 | 284 | 323 | 362 |
| | | $S = 35.00\%_0$ | | |
| 1 | 52 | 162 | 251 | 327 |
| 500 | 166 | 234 | 293 | 347 |
| 1000 | 244 | 286 | 325 | 363 |
| | | $S = 39.50\%_0$ | | |
| 1 | 65 | 170 | 256 | 329 |
| 500 | 174 | 239 | 296 | 348 |
| 1000 | 248 | 289 | 326 | 363 |

TABLE 17a*

*Velocity of sound in sea water†*

| Pressure (db) | Temperature (°C) | | | | | | |
|---|---|---|---|---|---|---|---|
| | 0 | 5 | 10 | 15 | 20 | 25 | 30 |
| 0 | 1449·3 | 1471·0 | 1490·4 | 1507·4 | 1522·1 | 1534·8 | 1545·8 |
| 1000 | 1465·8 | 1487·4 | 1506·7 | 1523·7 | 1538·5 | 1551·3 | 1562·5 |
| 2000 | 1482·4 | 1504·0 | 1523·2 | 1540·2 | 1555·0 | 1567·9 | 1579·2 |
| 3000 | 1499·4 | 1520·7 | 1538·6 | 1555·6 | | | |
| 4000 | 1516·5 | 1537·7 | 1555·2 | 1572·2 | | | |
| 5000 | 1533·9 | 1554·8 | 1571·9 | 1588·9 | | | |
| 6000 | 1551·5 | 1572·1 | | | | | |
| 7000 | 1569·3 | | | | | | |
| 8000 | 1587·3 | | | | | | |
| 9000 | 1605·4 | | | | | | |
| 10000 | 1623·5 | | | | | | |

* Reproduced by permission of U.S. Navy Oceanographic Office.
† Velocities in $m\,s^{-1}$; pressures in decibars above atmosphere. Salinity 35‰. For other salinities see Table 17b.
For detailed tables of the velocity of sound in sea water, see U.S. Naval Oceanographic Office (1962) and Bark *et al.* (1964).

TABLE 17b*

*Effect of salinity on sound velocity†*

| $S‰$ | Temperature (°C) | | | | | | |
|------|------|------|------|------|------|------|------|
|      | 0    | 5    | 10   | 15   | 20   | 25   | 30   |
| 30   | −7·0 | −6·7 | −6·5 | −6·2 | −5·9 | −5·6 | −5·3 |
| 32   | −4·2 | −4·0 | −3·9 | −3·7 | −3·5 | −3·4 | −3·2 |
| 33   | −2·8 | −2·7 | −2·6 | −2·5 | −2·4 | −2·2 | −2·1 |
| 34   | −1·4 | −1·3 | −1·3 | −1·2 | −1·2 | −1·1 | −1·1 |
| 35   | 0    | 0    | 0    | 0    | 0    | 0    | 0    |
| 36   | 1·4  | 1·3  | 1·3  | 1·2  | 1·2  | 1·1  | 1·1  |
| 37   | 2·8  | 2·7  | 2·6  | 2·5  | 2·4  | 2·3  | 2·1  |
| 38   | 4·2  | 4·1  | 3·9  | 3·7  | 3·6  | 3·4  | 3·2  |
| 40   | 7·0  | 6·8  | 6·5  | 6·2  | 6·0  | 5·7  | 5·3  |

* Reproduced by permission of U.S. Navy Oceanographic Office.

Corrections to be applied to the values in Table 17a for salinities other than 35‰·

TABLE 18

Specific heat of sea water at constant pressure ($J\,g^{-1}\,{}^\circ C^{-1}$) at various salinities and temperatures (Millero et al., 1973).

| Salinity, ‰ | 0°C | 5°C | 10°C | 15°C | 20°C | 25°C | 30°C | 35°C | 40°C |
|---|---|---|---|---|---|---|---|---|---|
| 0 | 4·2174 | 4·2019 | 4·1919 | 4·1855 | 4·1816 | 4·1793 | 4·1782 | 4·1779 | 4·1783 |
| 5 | 4·1812 | 4·1679 | 4·1599 | 4·1553 | 4·1526 | 4·1513 | 4·1510 | 4·1511 | 4·1515 |
| 10 | 4·1466 | 4·1354 | 4·1292 | 4·1263 | 4·1247 | 4·1242 | 4·1248 | 4·1252 | 4·1256 |
| 15 | 4·1130 | 4·1038 | 4·0994 | 4·0982 | 4·0975 | 4·0977 | 4·0992 | 4·0999 | 4·1003 |
| 20 | 4·0804 | 4·0730 | 4·0702 | 4·0706 | 4·0709 | 4·0717 | 4·0740 | 4·0751 | 4·0754 |
| 25 | 4·0484 | 4·0428 | 4·0417 | 4·0437 | 4·0448 | 4·0462 | 4·0494 | 4·0508 | 4·0509 |
| 30 | 4·0172 | 4·0132 | 4·0136 | 4·0172 | 4·0190 | 4·0210 | 4·0251 | 4·0268 | 4·0268 |
| 35 | 3·9865 | 3·9842 | 3·9861 | 3·9912 | 3·9937 | 3·9962 | 4·0011 | 4·0031 | 4·0030 |
| 40 | 3·9564 | 3·9556 | 3·9590 | 3·9655 | 3·9688 | 3·9718 | 3·9775 | 3·9797 | 3·9795 |

TABLE 19

*The relative partial equivalent heat capacity of sea salt* $(cal(eq\ deg)^{-1})$ *(Millero et al., 1973a)*

| Salinity (‰) | Temperature (°C) | | | | | | |
|---|---|---|---|---|---|---|---|
| | 0 | 5 | 10 | 15 | 20 | 25 | 30 |
| 0 | 0 | 0 | 0 | 0 | 0 | 0 | 0 |
| 5 | 2·8 | 3·1 | 3·4 | 3·7 | 4·0 | 4·3 | 4·6 |
| 10 | 4·7 | 5·0 | 5·4 | 5·7 | 5·9 | 6·3 | 6·6 |
| 15 | 6·9 | 7·1 | 7·3 | 7·5 | 7·7 | 7·9 | 8·1 |
| 20 | 9·5 | 9·4 | 9·4 | 9·4 | 9·3 | 9·3 | 9·2 |
| 25 | 12·2 | 11·9 | 11·5 | 11·2 | 10·8 | 10·5 | 10·2 |
| 30 | 15·3 | 14·6 | 13·8 | 13·1 | 12·3 | 11·6 | 10·8 |
| 35 | 18·6 | 17·4 | 16·2 | 15·0 | 13·8 | 12·6 | 11·3 |
| 40 | 22·2 | 20·5 | 18·7 | 17·0 | 15·2 | 13·4 | 11·7 |

TABLE 20

*Thermal conductivity* $(K\ in\ 10^{-5}\ W\ cm\ deg^{-1})$ *of sea water* $(S = 34·994‰)$ *as a function of temperature and pressure. (After Castelli et al., 1974)*\*

| Pressure (p) (bars) | Temperature $(t°C)$ | | | |
|---|---|---|---|---|
| | 1·82 | 10 | 20 | 30 |
| 200 | 563 | 578 | 594 | 605 |
| 400 | 570 | 585 | 601 | 613 |
| 600 | 578 | 592 | 609 | 619 |
| 800 | 585 | 599 | 615 | 627 |
| 1000 | 591 | 606 | 622 | 634 |
| 1200 | 596 | 613 | 628 | 641 |
| 1400 | 602 | 618 | 634 | 647 |

$K = 5·5286 \times 10^{-3} + 3·4025 \times 10^{-7}P + 1·8364 \times 10^{-7}t - 3·3058 \times 10^{-9}t$

\* Other data have been published by Caldwell (1974).

TABLE 21

*Freezing point of sea water* $(T_f)$ *at atmospheric pressure based on the data of Doherty and Kester* (*1974*).

| $S\%_{00}$ | $T_f$ (°C) | $S\%_{00}$ | $T_f$ (°C) | $S\%_{00}$ | $T_f$ (°C) |
|---|---|---|---|---|---|
| 5 | −0·275 | 17 | −0·918 | 29 | −1·582 |
| 6 | −0·328 | 18 | −0·973 | 30 | −1·638 |
| 7 | −0·381 | 19 | −1·028 | 31 | −1·695 |
| 8 | −0·434 | 20 | −1·082 | 32 | −1·751 |
| 9 | −0·487 | 21 | −1·137 | 33 | −1·808 |
| 10 | −0·541 | 22 | −1·192 | 34 | −1·865 |
| 11 | −0·594 | 23 | −1·248 | 35 | −1·922 |
| 12 | −0·648 | 24 | −1·303 | 36 | −1·979 |
| 13 | −0·702 | 25 | −1·359 | 37 | −2·036 |
| 14 | −0·756 | 26 | −1·414 | 38 | −2·094 |
| 15 | −0·810 | 27 | −1·470 | 39 | −2·151 |
| 16 | −0·864 | 28 | −1·526 | 40 | −2·209 |

The freezing point at *in situ* pressure is given by
$T_f(°C) = -0·0137 - 0·051990\ S\%_{00} - 0·00007225\ (S\%_{00})^2 - 0·000758z$ where $z$ is the depth in metres.

TABLE 22

*Boiling point elevation of sea water* $(S = 35·00\%)$ *at various temperatures,* (*Stoughton and Lietzke, 1967*)

| Temp. (°C) | 30 | 40 | 50 | 60 | 70 | 80 | 90 | 100 |
|---|---|---|---|---|---|---|---|---|
| Vap. press. (atm) | 0·042 | 0·073 | 0·122 | 0·197 | 0·309 | 0·469 | 0·694 | 1·003 |
| Elevation of B.P (°C) | 0·325 | 0·350 | 0·377 | 0·405 | 0·433 | 0·463 | 0·493 | 0·524 |
| Temp. (°C) | 120 | 140 | 160 | 180 | 200 | 220 | 240 | 260 |
| Vap. press. (atm) | 1·965 | 3·577 | 6·119 | 9·931 | 15·407 | 22·99 | 33·18 | 46·52 |
| Elevation of B.P (°C) | 0·590 | 0·660 | 0·735 | 0·817 | 0·906 | 1·003 | 1·111 | 1·232 |

TABLE 23

*Osmotic pressure and vapour depression of sea water at 25°C (Robinson, 1954)*

| | | | | | | Chlorinity | | | | | |
|---|---|---|---|---|---|---|---|---|---|---|---|
| | 12 | 13 | 14 | 15 | 16 | 17 | 18 | 19 | 20 | 21 |
| Osmotic pressure (atm) | 15·51 | 16·85 | 18·19 | 19·55 | 20·91 | 22·28 | 23·366 | 25·06 | 26·47 | 27·89 |
| Vap. press. lowering* $\times 10^2$ | 1·139 | 1·237 | 1·334 | 1·433 | 1·532 | 1·631 | 1·732 | 1·832 | 1·936 | 2·039 |

$(p^o - p)/p^o$ where $p$ and $p^o$ are the vapour pressures of sea water and pure water respectively ($p^o = 23\cdot75$ mm at 25°C).

TABLE 24

*Surface tension of clean sea water (in $N\,m^{-1}$) at various salinities and temperatures (from data by Krümmel (1900)\* and others (After Fleming and Revelle, 1939)*

| $S\%_{00}$ | Temperature (°C) | | | |
|---|---|---|---|---|
| | 0 | 10 | 20 | 30 |
| 0 | $75 \cdot 64 \times 10^{-3}$ | $74 \cdot 20 \times 10^{-3}$ | $72 \cdot 76 \times 10^{-3}$ | $71 \cdot 32 \times 10^{-3}$ |
| 10 | 75·86 | 74·42 | 72·98 | 71·54 |
| 20 | 76·08 | 74·64 | 73·20 | 71·76 |
| 30 | 76·30 | 74·86 | 73·42 | 71·98 |
| 35 | 76·41 | 74·97 | 73·53 | 72·09 |
| 40 | 76·52 | 75·08 | 73·64 | 72·20 |

Surface tension $(N\,m^{-1}) = 10^3\,(75 \cdot 64 - 0 \cdot 144t + 0 \cdot 0221\,S\%_{00})$

\* Measurements made on bubbles below the surface, they therefore take no account of the effects of surface contamination which may be very considerable (e.g. see Lumby and Folkard, 1956 and Vol. 2, pp. 233–4.

TABLE 25

*The viscosity of sea water ($\eta$) at various salinities and temperatures (in centipoises) computed from values for distilled water ($\eta_0$) by Korson et al. (1969) using equations developed by Millero (1974)*

| Salinity ‰ | Temperature °C | | | | | | | | | | | | | | | |
|---|---|---|---|---|---|---|---|---|---|---|---|---|---|---|---|---|
| | 0 | 2 | 4 | 6 | 8 | 10 | 12 | 14 | 16 | 18 | 20 | 22 | 24 | 26 | 28 | 30 |
| 0 | 1·7916 | 1·6739 | 1·5681 | 1·4725 | 1·3857 | 1·3069 | 1·2349 | 1·1691 | 1·1087 | 1·0532 | 1·0020 | 0·9547 | 0·9109 | 0·8703 | 0·8326 | 0·7975 |
| 5 | 1·8049 | 1·6868 | 1·5808 | 1·4849 | 1·3979 | 1·3189 | 1·2466 | 1·1807 | 1·1200 | 1·0644 | 1·0129 | 0·9655 | 0·9215 | 0·8807 | 0·8428 | 0·8076 |
| 10 | 1·8180 | 1·6995 | 1·5930 | 1·4968 | 1·4095 | 1·3302 | 1·2576 | 1·1913 | 1·1304 | 1·0745 | 1·0228 | 0·9751 | 0·9309 | 0·8900 | 0·8519 | 0·8165 |
| 15 | 1·8312 | 1·7122 | 1·6054 | 1·5087 | 1·4210 | 1·3412 | 1·2685 | 1·2018 | 1·1407 | 1·0845 | 1·0327 | 0·9847 | 0·9402 | 0·8991 | 0·8608 | 0·8252 |
| 20 | 1·8445 | 1·7251 | 1·6178 | 1·5208 | 1·4325 | 1·3525 | 1·2794 | 1·2125 | 1·1513 | 1·0945 | 1·0424 | 0·9942 | 0·9495 | 0·9082 | 0·8697 | 0·8339 |
| 25 | 1·8579 | 1·7380 | 1·6302 | 1·5327 | 1·4442 | 1·3638 | 1·2903 | 1·2231 | 1·1614 | 1·1046 | 1·0522 | 1·0036 | 0·9588 | 0·9172 | 0·8786 | 0·8426 |
| 30 | 1·8713 | 1·7509 | 1·6427 | 1·5448 | 1·4560 | 1·3751 | 1·3012 | 1·2338 | 1·1717 | 1·1146 | 1·0619 | 1·0132 | 0·9682 | 0·9263 | 0·8875 | 0·8513 |
| 32 | 1·8767 | 1·7563 | 1·6478 | 1·5497 | 1·4607 | 1·3797 | 1·3057 | 1·2379 | 1·1758 | 1·1186 | 1·0658 | 1·0171 | 0·9719 | 0·9300 | 0·8910 | 0·8547 |
| 34 | 1·8823 | 1·7643 | 1·6528 | 1·5545 | 1·4652 | 1·3843 | 1·3101 | 1·2423 | 1·1800 | 1·1227 | 1·0698 | 1·0210 | 0·9757 | 0·9336 | 0·8945 | 0·8582 |
| 36 | 1·8876 | 1·7696 | 1·6578 | 1·5594 | 1·4701 | 1·3888 | 1·3146 | 1·2465 | 1·1841 | 1·1267 | 1·0737 | 1·0248 | 0·9793 | 0·9372 | 0·8981 | 0·8617 |
| 38 | 1·8932 | 1·7752 | 1·6630 | 1·5644 | 1·4748 | 1·3934 | 1·3189 | 1·2508 | 1·1883 | 1·1308 | 1·0778 | 1·0286 | 0·9831 | 0·9409 | 0·9017 | 0·8651 |
| 40 | 1·8986 | 1·7805 | 1·6680 | 1·5692 | 1·4795 | 1·3980 | 1·3233 | 1·2551 | 1·1925 | 1·1348 | 1·0817 | 1·0325 | 0·9869 | 0·9446 | 0·9053 | 0·8686 |
| 42 | 1·9041 | 1·7861 | 1·6732 | 1·5741 | 1·4842 | 1·4026 | 1·3278 | 1·2595 | 1·1967 | 1·1389 | 1·0857 | 1·0363 | 0·9906 | 0·9483 | 0·9089 | 0·8721 |

Viscosity of pure water $\eta_t$ at temperature $t\,°C$ is given by $\log \dfrac{\eta_t}{\eta_{20}} = \dfrac{1\cdot1709(20 - t) - 0\cdot001827(t - 20)^2}{t + 89\cdot93}$ where $\eta_{20}$ is the viscosity at 20 °C.

Viscosity of sea water calculated from ratio $\dfrac{\eta}{\eta_t^0} = 1 + A\,Cl_t^{\frac{1}{2}} + B\,Cl_t$.

Where $Cl_t$ is the volume chlorinity ($Cl_t = Cl_{\text{‰}}^0 \times \text{density}$) and $A = 0\cdot000366, 0\cdot001403$ and $B = 0\cdot002756, 0\cdot003416$ at 5° and 25°C; constants at other temperatures obtained by linear interpolation or extrapolation.

According to Matthäus (1972) the change in dynamic viscosity ($\Delta\eta_p$, centipoises) produced by increase in pressure ($P$, kg cm$^{-2}$) at temperature, $T\,°C$) can be calculated from the expression

$$\Delta\eta_p = -1\cdot7913 \times 10^{-4}\,P + 9\cdot5182 \times 10^{-8}\,P^2 + P(1\cdot3550 \times 10^{-5}\,T - 2\cdot5853 \times 10^{-7}\,T^2 - P^2(6\cdot0833 \times 10^{-9}\,T - 1\cdot1652 \times 10^{-10}\,T^2)$$

* The assistance of Miss J. Wolfe with the computations is gratefully acknowledged.

TABLE 26

Relative viscosity of Standard Sea Water ($S = 35.00‰$) at various temperatures and pressures. (Stanley and Batten, 1969)

| Pressure, kg cm$^{-2}$ | $\eta_p/\eta_1$ at $-0.024°C$ | $\eta_p/\eta_1$ at $2.219°C$ | $\eta_p/\eta_1$ at $6.003°C$ | $\eta_p/\eta_1$ at $10.013°C$ | $\eta_p/\eta_1$ at $15.018°C$ | $\eta_p/\eta_1$ at $20.013°C$ | $\eta_p/\eta_1$ at $29.953°C$ |
|---|---|---|---|---|---|---|---|
| 176 | 0.9828 | 0.9852 | 0.9891 | 0.9914 | 0.9949 | 0.9977 | 0.9997 |
| 352 | 0.9709 | 0.9742 | 0.9814 | 0.9876 | 0.9926 | 0.9972 | 0.0001 |
| 527 | 0.9620 | 0.9670 | 0.9766 | 0.9843 | 0.9900 | 0.9978 | 1.0031 |
| 703 | 0.9560 | 0.9626 | 0.9735 | 0.9821 | 0.9915 | 0.9998 | 1.0071 |
| 878 | 0.9533 | 0.9598 | 0.9733 | 0.9836 | 0.9932 | 1.0040 | 1.0131 |
| 1055 | 0.9526 | 0.9600 | 0.9750 | 0.9874 | 0.9964 | 1.0070 | 1.0179 |
| 1230 | 0.9533 | 0.9637 | 0.9767 | 0.9902 | 1.0014 | 1.0107 | 1.0244 |
| 1406 | 0.9559 | 0.9673 | 0.9821 | 0.9961 | 1.0073 | 1.0166 | 1.0313 |

Where $\eta_p/\eta_{1_1}$ is the ratio of the viscosity at pressure $p$ (kg cm$^{-2}$) relative to that at 1 atm.

U

TABLE 27

*Specific conductivity of sea water\* (Weyl (1964). From data by Thomas et al.,(1934)*

| $S\%_{oo}$ | Temperature (°C) | | | | | |
|---|---|---|---|---|---|---|
| | 25 | 20 | 15 | 10 | 5 | 0 |
| 10 | 17·345 | 15·628 | 13·967 | 12·361 | 10·816 | 9·341 |
| 20 | 32·188 | 29·027 | 25·967 | 23·010 | 20·166 | 17·456 |
| 30 | 46·213 | 41·713 | 37·351 | 33·137 | 29·090 | 25·238 |
| 31 | 47·584 | 42·954 | 38·467 | 34·131 | 29·968 | 26·005 |
| 32 | 48·951 | 44·192 | 39·579 | 35·122 | 30·843 | 26·771 |
| 33 | 50·314 | 45·426 | 40·688 | 36·110 | 31·716 | 27·535 |
| 34 | 51·671 | 46·656 | 41·794 | 37·096 | 32·588 | 28·298 |
| 35 | 53·025 | 47·882 | 42·896 | 38·080 | 33·457 | 29·060 |
| 36 | 54·374 | 49·105 | 43·996 | 39·061 | 34·325 | 29·820 |
| 37 | 55·719 | 50·325 | 45·093 | 40·039 | 35·190 | 30·579 |
| 38 | 57·061 | 51·541 | 46·187 | 41·016 | 36·055 | 31·337 |
| 39 | 58·398 | 52·754 | 47·278 | 41·990 | 36·917 | 32·094 |

\* Conductivity in millimho $cm^{-1}$.

TABLE 28

*Effect of pressure on the conductivity of sea water\* (after Bradshaw and Schleicher, 1965)*

| Temp. | Pressure (db) | S‰ | | | Temp. | S‰ | | |
|---|---|---|---|---|---|---|---|---|
| | | 31 | 35 | 39 | | 31 | 35 | 39 |
| 0°C | 1,000 | 1·599 | 1·556 | 1·512 | 15°C | 1·032 | 1·008 | 0·985 |
| | 2,000 | 3·089 | 3·006 | 2·922 | | 1·996 | 1·951 | 1·906 |
| | 3,000 | 4·475 | 4·345 | 4·233 | | 2·895 | 2·830 | 2·764 |
| | 4,000 | 5·759 | 5·603 | 5·448 | | 3·731 | 3·646 | 3·562 |
| | 5,000 | 6·944 | 6·757 | 6·569 | | 4·506 | 4·403 | 4·301 |
| | 6,000 | 8·034 | 7·817 | 7·599 | | 5·221 | 5·102 | 4·984 |
| | 7,000 | 9·031 | 8·787 | 8·543 | | 5·879 | 5·745 | 5·612 |
| | 8,000 | 9·939 | 9·670 | 9·401 | | 6·481 | 6·334 | 6·187 |
| | 9,000 | 10·761 | 10·469 | 10·178 | | 7·031 | 6·871 | 6·711 |
| | 10,000 | 11·499 | 11·188 | 10·877 | | 7·529 | 7·358 | 7·187 |
| 5°C | 1,000 | 1·368 | 1·333 | 1·298 | 20°C | 0·907 | 0·888 | 0·868 |
| | 2,000 | 2·646 | 2·578 | 2·510 | | 1·755 | 1·718 | 1·680 |
| | 3,000 | 3·835 | 3·737 | 3·639 | | 2·546 | 2·492 | 2·438 |
| | 4,000 | 4·939 | 4·813 | 4·686 | | 3·282 | 3·212 | 3·142 |
| | 5,000 | 5·960 | 5·807 | 5·655 | | 3·964 | 3·879 | 3·795 |
| | 6,000 | 6·901 | 6·724 | 6·547 | | 4·594 | 4·496 | 4·399 |
| | 7,000 | 7·764 | 7·565 | 7·366 | | 5·174 | 5·064 | 4·954 |
| | 8,000 | 8·552 | 8·333 | 8·114 | | 5·706 | 5·585 | 5·464 |
| | 9,000 | 9·269 | 9·031 | 8·794 | | 6·192 | 6·060 | 5·929 |
| | 10,000 | 9·915 | 9·661 | 9·408 | | 6·633 | 6·492 | 6·351 |
| 10°C | 1,000 | 1·183 | 1·154 | 1·125 | 25°C | 0·799 | 0·783 | 0·767 |
| | 2,000 | 2·287 | 2·232 | 2·177 | | 1·547 | 1·516 | 1·485 |
| | 3,000 | 3·317 | 3·237 | 3·157 | | 2·245 | 2·200 | 2·156 |
| | 4,000 | 4·273 | 4·170 | 4·067 | | 2·895 | 2·837 | 2·780 |
| | 5,000 | 5·159 | 5·034 | 4·910 | | 3·498 | 3·429 | 3·359 |
| | 6,000 | 5·976 | 5·832 | 5·688 | | 4·056 | 3·976 | 3·896 |
| | 7,000 | 6·728 | 6·565 | 6·402 | | 4·571 | 4·481 | 4·390 |
| | 8,000 | 7·415 | 7·236 | 7·057 | | 5·045 | 4·945 | 4·845 |
| | 9,000 | 8·041 | 7·847 | 7·652 | | 5·478 | 5·369 | 5·261 |
| | 10,000 | 8·608 | 8·400 | 8·192 | | 5·872 | 5·756 | 5·640 |

\* Percentage increase compared with the conductivity at one atmosphere.

TABLE 29

*Conductivity ratio of sea water at 15°C ($R_{15}$) and 20°C ($R_{20}$) relative to sea water of salinity 35·000‰. (From data in UNESCO, 1966).*

| $S‰$ | 15°C | 20°C | $S‰$ | 15°C | 20°C |
|---|---|---|---|---|---|
| 29·50 | 0·85795 | 0·8583 | 36·00 | 1·02545 | 1·0254 |
| 30·00 | 0·87101 | 0·8714 | 36·50 | 1·03814 | 1·0380 |
| 30·50 | 0·88404 | 0·8844 | 37·00 | 1·05079 | 1·0506 |
| 31·00 | 0·89705 | 0·8973 | 37·50 | 1·06341 | 1·0632 |
| 31·50 | 0·91002 | 0·9103 | 38·00 | 1·07601 | 1·0758 |
| 32·00 | 0·92296 | 0·9232 | 38·50 | 1·08858 | 1·0883 |
| 32·50 | 0·93588 | 0·9361 | 39·00 | 1·10112 | 1·1008 |
| 33·00 | 0·94876 | 0·9489 | 39·50 | 1·11364 | 1·1133 |
| 33·50 | 0·96160 | 0·9617 | 40·00 | 1·12613 | 1·1257 |
| 34·00 | 0·97444 | 0·9745 | 40·50 | 1·13849 | 1·1381 |
| 34·50 | 0·98724 | 0·9873 | 41·00 | 1·15103 | 1·1505 |
| 35·00 | 1·00000 | 1·0000 | 41·50 | 1·16344 | 1·1629 |
| 35·50 | 1·01275 | 1·0127 | 42·00 | 1·17583 | 1·1752 |

For 15°C $S‰ = -0.08996 + 28.29720 R_{15} + 12.80832 R_{15}^2 - 10.678969 R_{15}^3 + 5.98624 R_{15}^4 - 132311 R_{15}^5$.

TABLE 30

*Correction values ( $\times 10^4$) to be applied to conductivity ratios measured at temperatures differing from 20°C to correct them to ratios at 20°C (to be used only in conjunction with 20°C ratios in Table 29). (After UNESCO, 1966)*

| Measured ratio | Temperature (°C) | | | | | | | | |
|---|---|---|---|---|---|---|---|---|---|
| | 10 | 12 | 14 | 16 | 18 | 20 | 22 | 24 | 26 |
| 0·85 | 80 | 62 | 45 | 29 | 14 | 0 | −14 | −26 | −38 |
| 0·90 | 56 | 44 | 32 | 21 | 10 | 0 | −9 | −18 | −27 |
| 0·95 | 29 | 23 | 17 | 11 | 5 | 0 | −5 | −10 | −14 |
| 1·00 | 0 | 0 | 0 | 0 | 0 | 0 | 0 | 0 | 0 |
| 1·05 | −33 | −25 | −19 | −12 | −6 | 0 | 5 | 11 | 15 |
| 1·10 | −69 | −54 | −39 | −25 | −12 | 0 | 11 | 22 | 32 |
| 1·15 | −109 | −85 | −62 | −40 | −19 | 0 | 18 | 35 | 50 |

TABLE 31

*Light absorption of typical sea waters. Extinction for 10 cm path length. (After Clarke and James, 1939)*

| Sample | Wavelength Å | | | | | | | |
|---|---|---|---|---|---|---|---|---|
| | 3600 | 4000 | 5000 | 5200 | 6000 | 7000 | 7500 | 8000 |
| Pure water | 0·001 | 0·001 | 0·002 | 0·002 | 0·010 | 0·025 | 0·115 | 0·086 |
| Artificial sea water | 0·011 | 0·003 | 0·005 | 0·007 | 0·010 | 0·025 | 0·115 | 0·086 |
| Ocean water, unfiltered | 0·012 | 0·009 | 0·007 | 0·008 | 0·011 | 0·025 | 0·115 | 0·086 |
| Continental slope waters, unfiltered | 0·052 | 0·030 | 0·011 | 0·010 | 0·012 | 0·035 | 0·130 | 0·088 |
| Continental slope waters, filtered | 0·016 | 0·010 | 0·005 | 0·005 | 0·012 | 0·030 | 0·115 | 0·086 |
| Inshore water unfiltered | 0·055 | 0·042 | 0·028 | 0·026 | 0·035 | 0·052 | 0·140 | 0·100 |
| Inshore water, filtered | 0·015 | 0·010 | 0·005 | 0·005 | 0·010 | 0·025 | 0·110 | 0·086 |

TABLE 32

*Differences between the extinctions of sea waters and pure water\*. (From data by Clarke and James, 1939)*

| Sample | Wavelength Å | | | | | | | |
|---|---|---|---|---|---|---|---|---|
| | 3600 | 4000 | 5000 | 5200 | 6000 | 7000 | 7500 | 8000 |
| Artificial sea water | 0·010 | 0·002 | 0·003 | 0·005 | nil | nil | nil | nil |
| Ocean water unfiltered | 0·011 | 0·008 | 0·005 | 0·006 | 0·001 | nil | nil | nil |
| Continental slope water, unfiltered | 0·051 | 0·029 | 0·009 | 0·008 | 0·002 | 0·010 | 0·015 | 0·002 |
| Continental slope water, filtered | 0·015 | 0·009 | 0·003 | 0·003 | 0·002 | 0·005 | nil | nil |
| Inshore water, unfiltered | 0·054 | 0·041 | 0·026 | 0·024 | 0·025 | 0·027 | 0·025 | 0·015 |
| Inshore water, filtered | 0·014 | 0·009 | 0·003 | 0·003 | nil | nil | nil | nil |

\* $E_{SW(10\ cm)} - E_{PW(10\ cm)}$

Note: The values given in Tables 31 and 32 for unfiltered inshore waters should be taken as no more than a rough indication, since actual values vary widely with time and location.

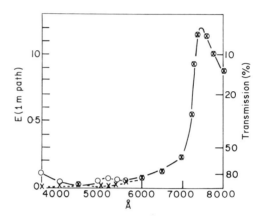

Fɪɢ. 1. Extinction (1 m path) against wavelength. Solid line—filtered ocean water. Broken line—pure water. (After Clarke and James, 1939).

TABLE 33

Refractive index differences ($\Delta n$) for sea water at a wavelength of 589.3 nm at various temperatures and salinities. ($\Delta n = (n-1.30000) \cdot 10^5$). (Matthäus, 1974)

| T[°C] | | | | | | | | | | | S [‰] | | | | | | | | | | |
| --- | --- | --- | --- | --- | --- | --- | --- | --- | --- | --- | --- | --- | --- | --- | --- | --- | --- | --- | --- | --- | --- |
| | 0 | 2 | 4 | 6 | 8 | 10 | 12 | 14 | 16 | 18 | 20 | 22 | 24 | 26 | 28 | 30 | 32 | 34 | 36 | 38 | 40 |
| 0 | 3402 | 3441 | 3481 | 3520 | 3559 | 3598 | 3637 | 3677 | 3716 | 3755 | 3794 | 3833 | 3873 | 3912 | 3951 | 3990 | 4029 | 4069 | 4108 | 4147 | 4186 |
| 1 | 3400 | 3439 | 3478 | 3517 | 3556 | 3595 | 3634 | 3674 | 3713 | 3752 | 3791 | 3830 | 3869 | 3908 | 3947 | 3986 | 4025 | 4064 | 4103 | 4142 | 4181 |
| 2 | 3398 | 3437 | 3476 | 3515 | 3553 | 3592 | 3631 | 3670 | 3709 | 3748 | 3787 | 3826 | 3865 | 3904 | 3942 | 3981 | 4020 | 4059 | 4098 | 4137 | 4176 |
| 3 | 3395 | 3434 | 3473 | 3511 | 3550 | 3589 | 3628 | 3666 | 3705 | 3744 | 3783 | 3821 | 3860 | 3899 | 3938 | 3976 | 4015 | 4054 | 4093 | 4131 | 4170 |
| 4 | 3392 | 3431 | 3469 | 3508 | 3547 | 3585 | 3624 | 3662 | 3701 | 3740 | 3778 | 3817 | 3855 | 3894 | 3933 | 3971 | 4010 | 4048 | 4087 | 4126 | 4164 |
| 5 | 3389 | 3427 | 3466 | 3504 | 3543 | 3581 | 3620 | 3658 | 3697 | 3735 | 3773 | 3812 | 3850 | 3889 | 3927 | 3966 | 4004 | 4043 | 4081 | 4120 | 4158 |
| 6 | 3385 | 3424 | 3462 | 3500 | 3538 | 3577 | 3615 | 3653 | 3692 | 3730 | 3768 | 3807 | 3845 | 3883 | 3922 | 3960 | 3998 | 4037 | 4075 | 4113 | 4152 |
| 7 | 3381 | 3419 | 3458 | 3496 | 3534 | 3572 | 3610 | 3648 | 3687 | 3725 | 3763 | 3801 | 3839 | 3878 | 3916 | 3954 | 3992 | 4030 | 4068 | 4107 | 4145 |
| 8 | 3377 | 3415 | 3453 | 3491 | 3529 | 3567 | 3605 | 3643 | 3681 | 3719 | 3757 | 3795 | 3833 | 3871 | 3909 | 3948 | 3986 | 4024 | 4062 | 4100 | 4138 |
| 9 | 3372 | 3410 | 3448 | 3486 | 3524 | 3562 | 3600 | 3638 | 3675 | 3713 | 3751 | 3789 | 3827 | 3865 | 3903 | 3941 | 3979 | 4017 | 4055 | 4093 | 4130 |
| 10 | 3367 | 3405 | 3443 | 3481 | 3518 | 3556 | 3594 | 3632 | 3669 | 3707 | 3745 | 3783 | 3821 | 3858 | 3896 | 3934 | 3972 | 4010 | 4047 | 4085 | 4123 |
| 11 | 3362 | 3399 | 3437 | 3475 | 3512 | 3550 | 3588 | 3625 | 3663 | 3701 | 3738 | 3776 | 3814 | 3851 | 3889 | 3927 | 3964 | 4002 | 4040 | 4077 | 4115 |
| 12 | 3356 | 3394 | 3431 | 3469 | 3506 | 3544 | 3581 | 3619 | 3656 | 3694 | 3732 | 3769 | 3807 | 3844 | 3882 | 3919 | 3957 | 3994 | 4032 | 4069 | 4107 |
| 13 | 3350 | 3387 | 3425 | 3462 | 3500 | 3537 | 3575 | 3612 | 3649 | 3687 | 3724 | 3762 | 3799 | 3837 | 3874 | 3911 | 3949 | 3986 | 4024 | 4061 | 4098 |
| 14 | 3344 | 3381 | 3418 | 3456 | 3493 | 3530 | 3568 | 3605 | 3642 | 3679 | 3717 | 3754 | 3791 | 3829 | 3866 | 3903 | 3941 | 3978 | 4015 | 4053 | 4090 |
| 15 | 3337 | 3374 | 3411 | 3449 | 3486 | 3523 | 3560 | 3597 | 3635 | 3672 | 3709 | 3746 | 3783 | 3821 | 3858 | 3895 | 3932 | 3969 | 4006 | 4044 | 4081 |
| 16 | 3330 | 3367 | 3404 | 3441 | 3478 | 3515 | 3552 | 3590 | 3627 | 3664 | 3701 | 3738 | 3775 | 3812 | 3849 | 3886 | 3923 | 3960 | 3997 | 4035 | 4072 |
| 17 | 3323 | 3360 | 3397 | 3434 | 3470 | 3507 | 3544 | 3581 | 3618 | 3655 | 3692 | 3729 | 3766 | 3803 | 3840 | 3877 | 3914 | 3951 | 3988 | 4025 | 4062 |
| 18 | 3315 | 3352 | 3389 | 3425 | 3462 | 3499 | 3536 | 3573 | 3610 | 3647 | 3684 | 3721 | 3757 | 3794 | 3831 | 3868 | 3905 | 3942 | 3979 | 4016 | 4052 |
| 19 | 3307 | 3344 | 3380 | 3417 | 3454 | 3491 | 3527 | 3564 | 3601 | 3638 | 3675 | 3711 | 3748 | 3785 | 3822 | 3858 | 3895 | 3932 | 3969 | 4006 | 4042 |
| 20 | 3298 | 3335 | 3372 | 3408 | 3445 | 3482 | 3518 | 3555 | 3592 | 3629 | 3665 | 3702 | 3739 | 3775 | 3812 | 3849 | 3885 | 3922 | 3959 | 3995 | 4032 |
| 21 | 3290 | 3326 | 3363 | 3399 | 3436 | 3473 | 3509 | 3546 | 3582 | 3619 | 3656 | 3692 | 3729 | 3765 | 3802 | 3838 | 3875 | 3912 | 3948 | 3985 | 4021 |
| 22 | 3281 | 3317 | 3354 | 3390 | 3427 | 3463 | 3500 | 3536 | 3573 | 3609 | 3646 | 3682 | 3719 | 3755 | 3792 | 3828 | 3865 | 3901 | 3938 | 3974 | 4011 |
| 23 | 3271 | 3308 | 3344 | 3380 | 3417 | 3453 | 3490 | 3526 | 3562 | 3599 | 3635 | 3672 | 3708 | 3745 | 3781 | 3817 | 3854 | 3890 | 3927 | 3963 | 3999 |
| 24 | 3261 | 3298 | 3334 | 3370 | 3407 | 3443 | 3479 | 3516 | 3552 | 3588 | 3625 | 3661 | 3697 | 3734 | 3770 | 3806 | 3843 | 3879 | 3915 | 3952 | 3988 |
| 25 | 3251 | 3288 | 3324 | 3360 | 3396 | 3433 | 3469 | 3505 | 3541 | 3578 | 3614 | 3650 | 3686 | 3723 | 3759 | 3795 | 3831 | 3868 | 3904 | 3940 | 3976 |
| 26 | 3241 | 3277 | 3313 | 3349 | 3386 | 3422 | 3458 | 3494 | 3530 | 3566 | 3603 | 3639 | 3675 | 3711 | 3747 | 3783 | 3820 | 3856 | 3892 | 3928 | 3964 |
| 27 | 3230 | 3266 | 3302 | 3338 | 3375 | 3411 | 3447 | 3483 | 3519 | 3555 | 3591 | 3627 | 3663 | 3699 | 3736 | 3772 | 3808 | 3844 | 3880 | 3916 | 3952 |
| 28 | 3219 | 3255 | 3291 | 3327 | 3363 | 3399 | 3435 | 3471 | 3507 | 3543 | 3579 | 3615 | 3651 | 3687 | 3723 | 3759 | 3796 | 3832 | 3868 | 3904 | 3940 |
| 29 | 3208 | 3244 | 3279 | 3315 | 3351 | 3387 | 3423 | 3459 | 3495 | 3531 | 3567 | 3603 | 3639 | 3675 | 3711 | 3747 | 3783 | 3819 | 3855 | 3891 | 3927 |
| 30 | 3196 | 3232 | 3268 | 3303 | 3339 | 3375 | 3411 | 3447 | 3483 | 3519 | 3555 | 3591 | 3627 | 3662 | 3698 | 3734 | 3770 | 3806 | 3842 | 3878 | 3914 |

TABLE 34

Refractive index differences ($\Delta n$) for sea water of salinity 35·00‰ at various temperatures and wavelengths ($\Delta n = (n - 1\cdot30000) \cdot 10^5$). (After Matthäus, 1974)

| T[°C] | Wavelength (nm) | | | | | | | | | | | | | | |
|---|---|---|---|---|---|---|---|---|---|---|---|---|---|---|---|
| | 404·7 | 435·8 | 457·9 | 467·8 | 480·0 | 488·0 | 501·7 | 508·5 | 514·5 | 546·1 | 577·0 | 579·1 | 589·3 | 632·8 | 643·8 |
| 0 | 5099 | 4840 | 4684 | 4621 | 4549 | 4504 | 4433 | 4400 | 4372 | 4240 | 4130 | 4124 | 4091 | 3961 | 3929 |
| 1 | 5094 | 4835 | 4679 | 4616 | 4544 | 4500 | 4428 | 4395 | 4367 | 4235 | 4126 | 4119 | 4086 | 3956 | 3925 |
| 2 | 5089 | 4830 | 4674 | 4611 | 4529 | 4495 | 4423 | 4390 | 4362 | 4230 | 4121 | 4114 | 4081 | 3951 | 3920 |
| 3 | 5084 | 4825 | 4669 | 4606 | 4534 | 4489 | 4418 | 4385 | 4357 | 4225 | 4115 | 4109 | 4076 | 3946 | 3914 |
| 4 | 5078 | 4819 | 4664 | 4601 | 4528 | 4484 | 4412 | 4379 | 4351 | 4219 | 4110 | 4103 | 4070 | 3941 | 3909 |
| 5 | 5072 | 4814 | 4658 | 4595 | 4522 | 4478 | 4407 | 4374 | 4345 | 4213 | 4104 | 4097 | 4065 | 3935 | 3903 |
| 6 | 5066 | 4807 | 4552 | 4589 | 4516 | 4472 | 4400 | 4367 | 4339 | 4207 | 4098 | 4091 | 4058 | 3929 | 3897 |
| 7 | 5060 | 4801 | 4645 | 4582 | 4510 | 4465 | 4394 | 4361 | 4333 | 4201 | 4091 | 4085 | 4052 | 3922 | 3890 |
| 8 | 5053 | 4794 | 4639 | 4576 | 4503 | 4459 | 4387 | 4354 | 4326 | 4194 | 4085 | 4078 | 4045 | 3916 | 3884 |
| 9 | 5046 | 4787 | 4632 | 4569 | 4496 | 4452 | 4380 | 4347 | 4319 | 4187 | 4078 | 4071 | 4038 | 3909 | 3877 |
| 10 | 5039 | 4780 | 4624 | 4561 | 4489 | 4444 | 4373 | 4340 | 4312 | 4180 | 4071 | 4064 | 4031 | 3901 | 3869 |
| 11 | 5031 | 4773 | 4617 | 4554 | 4481 | 4437 | 4366 | 4332 | 4304 | 4172 | 4063 | 4056 | 4023 | 3894 | 3862 |
| 12 | 5023 | 4765 | 4609 | 4546 | 4473 | 4429 | 4358 | 4325 | 4297 | 4164 | 4055 | 4048 | 4016 | 3886 | 3854 |
| 13 | 5015 | 4757 | 4601 | 4538 | 4465 | 4421 | 4350 | 4317 | 4288 | 4156 | 4047 | 4040 | 4008 | 3878 | 3846 |
| 14 | 5007 | 4748 | 4592 | 4529 | 4457 | 4412 | 4341 | 4308 | 4280 | 4148 | 4039 | 4032 | 3999 | 3869 | 3837 |
| 15 | 4998 | 4740 | 4584 | 4521 | 4448 | 4404 | 4333 | 4300 | 4271 | 4139 | 4030 | 4023 | 3991 | 3861 | 3829 |
| 16 | 4989 | 4731 | 4575 | 4512 | 4439 | 4395 | 4324 | 4291 | 4262 | 4130 | 4021 | 4014 | 3982 | 3852 | 3820 |
| 17 | 4980 | 4721 | 4566 | 4503 | 4430 | 4386 | 4314 | 4281 | 4253 | 4121 | 4012 | 4005 | 3972 | 3843 | 3811 |
| 18 | 4971 | 4712 | 4556 | 4493 | 4421 | 4376 | 4305 | 4272 | 4244 | 4111 | 4002 | 3995 | 3963 | 3833 | 3801 |
| 19 | 4961 | 4702 | 4546 | 4483 | 4411 | 4366 | 4295 | 4262 | 4234 | 4102 | 3993 | 3986 | 3953 | 3823 | 3791 |
| 20 | 4951 | 4692 | 4536 | 4473 | 4401 | 4356 | 4385 | 4252 | 4224 | 4092 | 3982 | 3976 | 3943 | 3813 | 3781 |

TABLE 34 cont.

| T[°C] | λ[nm] 0.4047 | 0.4358 | 0.4579 | 0.4678 | 0.4800 | 0.4880 | 0.5017 | 0.5085 | 0.5145 | 0.5461 | 0.5770 | 0.5791 | 0.5893 | 0.6328 | 0.6438 |
|---|---|---|---|---|---|---|---|---|---|---|---|---|---|---|---|
| 21 | 4940 | 4682 | 4526 | 4463 | 4390 | 4346 | 4275 | 4242 | 4214 | 4081 | 3972 | 3965 | 3933 | 3803 | 3771 |
| 22 | 4930 | 4671 | 4515 | 4452 | 4380 | 4335 | 4264 | 4231 | 4203 | 4071 | 3961 | 3955 | 3922 | 3792 | 3760 |
| 23 | 4919 | 4660 | 4504 | 4441 | 4369 | 4324 | 4253 | 4220 | 4192 | 4060 | 3951 | 3944 | 3911 | 3781 | 3749 |
| 24 | 4908 | 4649 | 4493 | 4430 | 4358 | 4313 | 4242 | 4209 | 4181 | 4048 | 3939 | 3932 | 3900 | 3770 | 3738 |
| 25 | 4896 | 4637 | 4482 | 4419 | 4346 | 4302 | 4230 | 4197 | 4169 | 4037 | 3928 | 3921 | 3888 | 3759 | 3727 |
| 26 | 4884 | 4626 | 4470 | 4407 | 4334 | 4290 | 4219 | 4186 | 4157 | 4025 | 3916 | 3909 | 3877 | 3747 | 3715 |
| 27 | 4872 | 4614 | 4458 | 4395 | 4322 | 4278 | 4207 | 4174 | 4145 | 4013 | 3904 | 3897 | 3865 | 3735 | 3703 |
| 28 | 4860 | 4601 | 4445 | 4382 | 4310 | 4265 | 4194 | 4161 | 4133 | 4001 | 3892 | 3885 | 3852 | 3722 | 3690 |
| 29 | 4847 | 4589 | 4433 | 4370 | 4297 | 4253 | 4182 | 4149 | 4120 | 3988 | 3879 | 3872 | 3840 | 3710 | 3678 |
| 30 | 4834 | 4576 | 4420 | 4357 | 4284 | 4240 | 4169 | 4136 | 4108 | 3975 | 3866 | 3859 | 3827 | 3697 | 3665 |

## TABLE 35

Absolute refractive index of sea water ($S = 35.00‰$) as a function of temperature, pressure and wavelength. (Stanley, 1971)

| Pressure | Temperature (°C) | | | | | | |
|---|---|---|---|---|---|---|---|
| | 0·03 | 5·03 | 10·03 | 15·02 | 20·00 | 24·99 | 29·98 |
| | | | 6328 Å | | | | |
| Atm. | 1·34015 | 1·33977 | 1·33935 | 1·33899 | 1·33850 | 1·33795 | 1·33737 |
| 352 kg cm$^2$ | 1·34539 | 1·34487 | 1·34431 | 1·34388 | 1·34331 | 1·34270 | 1·34207 |
| 703 kg cm$^{-2}$ | 1·35025 | 1·34962 | 1·34896 | 1·34844 | 1·34780 | 1·34713 | 1·34647 |
| 1055 kg cm$^{-2}$ | 1·35481 | 1·35403 | 1·35380 | 1·35269 | 1·35200 | 1·35129 | 1·35059 |
| 1406 kg cm$^{-2}$ | — | 1·35813 | 1·35738 | 1·35668 | 1·35592 | 1·35519 | 1·35443 |
| | | | 5017 Å | | | | |
| Atm | 1·34455 | 1·34455 | 1·34422 | 1·34379 | 1·34327 | 1·34272 | 1·34215 |
| 352 kg cm$^2$ | 1·35008 | 1·34969 | 1·34924 | 1·34873 | 1·34813 | 1·34757 | 1·34694 |
| 703 kg cm$^{-2}$ | 1·35507 | 1·35450 | 1·35394 | 1·35333 | 1·35269 | 1·35208 | 1·35137 |
| 1055 kg cm$^{-2}$ | 1·35953 | 1·35891 | 1·35834 | 1·35764 | 1·35695 | 1·35632 | 1·35561 |
| 1406 kg cm$^{-2}$ | — | 1·36314 | 1·36241 | 1·36166 | 1·36095 | 1·36019 | 1·35946 |

TABLE 36

Velocity of light ($\lambda = 589.3$ nm) in sea water at 1 atm (km s$^{-1}$) (Sager, 1974)

| $S\%_{\circ}$ | Temperature (°C) | | | | | | | | |
|---|---|---|---|---|---|---|---|---|---|
| | 0 | 5 | 10 | 15 | 20 | 25 | 30 | 35 | 40 |
| 0 | 224,732 | 224,749 | 224,785 | 224,837 | 224,904 | 224,985 | 225,080 | 225,185 | 225,305 |
| 2.5 | 224,650 | 224,668 | 224,705 | 224,759 | 224,827 | 224,909 | 225,004 | 225,110 | 225,230 |
| 5.0 | 224,567 | 224,588 | 224,626 | 224,681 | 224,749 | 224,832 | 224,928 | 225,035 | 225,156 |
| 7.5 | 224,485 | 224,507 | 224,547 | 224,603 | 224,672 | 224,756 | 224,852 | 224,960 | 225,081 |
| 10.0 | 224,402 | 224,426 | 224,468 | 224,524 | 224,595 | 224,679 | 224,776 | 224,885 | 225,006 |
| 12.5 | 224,319 | 224,346 | 224,388 | 224,446 | 224,518 | 224,603 | 224,700 | 224,810 | 224,931 |
| 15.0 | 224,236 | 224,265 | 224,309 | 224,368 | 224,441 | 224,527 | 224,625 | 224,735 | 224,857 |
| 17.5 | 224,154 | 224,185 | 224,230 | 224,290 | 224,364 | 224,450 | 224,549 | 224,660 | 224,782 |
| 20.0 | 224,072 | 224,104 | 224,151 | 224,212 | 224,287 | 224,374 | 224,473 | 224,585 | 224,707 |
| 22.5 | 223,990 | 224,024 | 224,072 | 224,134 | 224,210 | 224,297 | 224,398 | 224,510 | 224,633 |
| 25.0 | 223,907 | 223,943 | 223,994 | 224,057 | 224,133 | 224,221 | 224,322 | 224,435 | 224,559 |
| 27.5 | 223,825 | 223,863 | 223,915 | 223,979 | 224,056 | 224,145 | 224,247 | 224,360 | 224,485 |
| 30.0 | 223,743 | 223,783 | 223,836 | 223,901 | 223,979 | 224,069 | 224,171 | 224,285 | 224,411 |
| 32.5 | 223,661 | 223,703 | 223,758 | 223,823 | 223,903 | 223,993 | 224,096 | 224,211 | 224,336 |
| 35.0 | 223,579 | 223,623 | 223,679 | 223,746 | 223,826 | 223,917 | 224,020 | 224,136 | 224,262 |
| 37.5 | 223,498 | 223,543 | 223,600 | 223,669 | 223,749 | 223,841 | 223,945 | 224,061 | 224,188 |
| 40.0 | 223,416 | 223,463 | 223,521 | 223,591 | 223,673 | 223,765 | 223,870 | 223,986 | 224,114 |

## REFERENCES

Bark, L. S., Ganson, P. P. and Meister, N. A. (1964), "Tables of the Velocity of Sound in Sea Water" Pergamon, Oxford.

Bradshaw, A. and Schleicher, K. E. (1965). *Deep-Sea Res.* **12**, 151.

Bradshaw, A. and Schleicher, K. E. (1970). *Deep-Sea Res.* **16**, 691.

Caldwell, D. T. (1974). *Deep-Sea Res.* **21**, 131.

Carpenter, J. H. (1966). *Limnol. Oceanogr.* **11**, 264.

Castelli, V. J., Stanley, E. M. and Fischer, E. C. (1974). *Deep-Sea Res.* **21**, 311.

Clarke, G. L. and James, H. R. (1939). *J. Opt. Soc. Amer.* **29**, 43.

Cox, R. A. (1965). *In* "Chemical Oceanography" (J. P. Riley and G. Skirrow, eds), Vol. I. Academic Press, London.

Cox R. A. and Culkin, F. (1967). *Deep-Sea Res.* **13**, 789.

Cox, R. A., McCartney, M. J. and Culkin, F. (1970). *Deep-Sea Res.* **17**, 679.

Crozier, T. E. and Yamamoto, S. (1974). *J. Chem. Eng. Data*, **19**, 242.

Doherty, B. T. and Kester, D. R. (1974), *J. Mar. Res.* **32**, 285

Dorsey, N. E. (1940). "Properties of Ordinary Water-substance". Reinhold, New York.

Douglas, E. (1964). *J. Phys. Chem.* **68**, 169.

Douglas, E. (1965). *J. Phys. Chem.* **69**, 2608.

Douglas, E. (1967). *J. Phys. Chem.* **71**, 1931.

Fleming, R. H. and Revelle, R. R. (1939). "Recent Marine Sediments" (N. Trask ed.). Amer. Soc. Petrol. Geol., Tulsa, Oklahoma.

Kalle, K. (1945). In "Probleme der Kosmischen Physik" 2nd Edn., Vol. 23. Leipzig.

Korson, L., Drost-Hansen, W. and Millero, F. J. (1969). *J. Phys. Chem.* **73**, 34.

Kester, D. R., Duedall, I. W., Connor, D. N. and Pytkowicz, R. M. (1967). *Limnol. Oceanogr.* **12**, 176.

Krümmel, O. (1900). *Wiss. Meeresuntersuch.* **5**, 9.

Lepple, F. K. and Millero, F. J. (1971). *Deep-Sea Res.* **18**, 1233.

Lumby, J. R. and Folkard, A. R. (1956). *Bull. Inst. Océanogr. Monaco,* **1080**, 1.

Lyman, J. and Fleming, R. H. (1940). *J. Mar. Res.* **3**, 134.

Matthäus, W. (1972). *Beitr. Meeresk.* **29**, 93.

Matthäus, W. (1974). *Beitr. Meeresk.* **33**, 73.

Millero, F. J. (1974). *In* "The Sea" (E. D. Goldberg ed.), Vol. 5. Interscience, New York.

Millero, F. J. and Lepple, F. K. (1973). *Mar. Chem.* **1**, 89.

Millero, F. J., Hansen, L. D. and Hoff, E. V. (1973b). *J. Mar. Res.* **31**, 21.

Millero, F. J., Perron G. and Desnoyers, J. E. (1973). *J. Geophys. Res.* **78**, 4499.

Morris, A. W. and Riley, J. P. (1966). *Deep-Sea Res.* **13**, 689.

Murray, C. N. and Riley, J. P. (1969a). *Deep-Sea Res.* **16**, 311.

Murray, C. N. and Riley, J. P. (1969b). *Deep-Sea Res.* **16**, 297.

Murray, C. N. and Riley, J. P. (1971). *Deep-Sea Res.* **18**, 533.

Riley, J. P. and Tongudai, M. (1967). *Chem. Geol.* **2**, 263.

Robinson, R. A. (1954). *J. Mar. Biol. Ass. U.K.* **33**, 449.

Sager, G. (1974). *Beitr. Meeresk.* **33**, 68.

Stanley, E. M. (1971). *Deep-Sea Res.* **18**, 833.

Stanley, E. M. and Batten, R. C. (1969). *J. Geophys. Res.* **74**, 3415.

Stoughton, R. W. and Lietzke, M. H. (1967). *J. Chem. Eng. Data,* **12**, 101.

Thomas, B. D., Thompson, T. G. and Utterback, C. L. (1934). *J. Cons. Int. Explor.*

*Mer*, **9**, 28.

UNESCO (1966). International Oceanographic Tables, Vol. 1. National Institute of Oceanography, Wormley, Surrey, England.

UNESCO (1973). International Oceanographic Tables. Vol. 2. National Institute of Oceanography, Wormley, Surrey, England.

U.S. Naval Oceanogr. Office (1962) Tables of sound speed in sea water. Publ. SP58 U.S. Naval Oceanographic Office, Washington, U.S.A.

U.S. Navy (1961). Tables for the velocity of sound in sea water. Bureau of ships reference NObsr 81564 S-7001-0307, Washington, U.S.A.

Weiss, R. F. (1970). *Deep-Sea Res.* **17**, 721.

Weiss, R. F. (1971). *J. Chem. Eng. Data*, **16**, 235.

Weyl, P. (1964). *Limnol. Oceanogr.* **9**, 75.

Wilson, W. D. (1960). *J. Acoust. Soc. Amer.* **32**, 1357.

Wood, D. and Caputi, R. (1966). "Technical Report No. 988" U.S. Naval Radiological Defense Laboratory, San Francisco.

# Subject Index

*(Numbers in bold type indicate the page on which a subject is treated most fully.)*